KA 22.50

Organometallic Compounds of Ruthenium and Osmium

Organometallic Compounds of

Ruthenium and Osmium

Edited by

G. R. Knox
University of Strathclyde

LONDON NEW YORK
CHAPMAN AND HALL

First published 1985 by Chapman and Hall Ltd
11 New Fetter Lane, London EC4P 4EE
733 Third Avenue, New York NY 10017

Phototypeset in the United States of America by
Mack Printing Company, Easton, Pennsylvania 18042
Printed in Great Britain by J.W. Arrowsmith Ltd, Bristol

ISBN 0 412 26850 7

Library of Congress Cataloging in
Publication Data

Main entry under title:

Organometallic compounds of ruthenium
and osmium.
(Chapman and Hall chemistry
sourcebooks)
Includes index.
1. Organoruthenium compounds —
Handbooks, manuals, etc.
2. Organoosmium compounds —
Handbooks, manuals, etc.
I. Knox, G.R., 1934- .
QD412.R9'74 1985 547'.05632 85-410

ISBN 0-412-26850-7 (U.S.)

British Library Cataloguing in
Publication Data

Organometallic compounds of ruthenium
and osmium. — (Chapman and Hall
chemistry sourcebooks)
1. Organometallic compounds
2. Transition metal compounds
I. Knox, G.R.
547'.056 QD411

ISBN 0-412-26850-7

Contents

Preface	*page* vii
Introduction	ix
Os Osmium	1
Ru Ruthenium	71
Name Index	211
Molecular Formula Index	231
CAS Registry Number Index	265

This Sourcebook has been reproduced in part from the Dictionary of Organometallic Compounds. If you find it useful and you would like full coverage, then why not buy the Main Work for your library

DICTIONARY OF ORGANOMETALLIC COMPOUNDS

The recent growth in organometallic chemistry has led to an explosion in the number of organometallic compounds discovered or synthesized.
How can scientists working in this area locate reliable data on all really important organometallics?
The answer lies in the new Dictionary of Organometallic Compounds, published in November 1984.

- provides within 15,000 Entries an extensive yet selective compilation of the most important organometallic compounds
- for each compound, details of structure, physical and chemical properties, reactions and references are provided in clearly set out and easily scanned Entries
- offers access to compounds by structure, molecular formula, chemical name, or CAS Registry Number
- the structure index contains reduced images of all the structure diagrams in the section — a unique feature which allows users to 'browse', and to appreciate quickly the full range of types of compounds which have been synthesized
- provides a continually up-to-date information system with Annual Supplements
- entries compiled by subject experts under supervision of a prestigious international advisory board
- a time-saving, cost-saving and reliable resource essential to all users of organometallic compounds

From a review by F.A. Cotton, Texas A & M University
'This extraordinarily useful compendium is one that no chemical enterprise having the slightest involvement in organometallic chemistry would want to be (or would be wise to be) without.'

Publication date: November 1984 ISBN: 0 412 24710 0
286 x 213mm 3,000 pages in three volumes

For further information please write to The Promotion Department, Chapman and Hall, 11 New Fetter Lane, London EC4P 4EE.

CHAPMAN AND HALL
11 New Fetter Lane, London EC4P 4EE
733 Third Avenue, New York NY 10017

Preface

This is one of the first volumes to be published in the series of *Chapman and Hall Chemistry Sourcebooks* which provides carefully tailored information to workers in specialized areas of chemistry. The information contained in this book is derived from the *Dictionary of Organometallic Compounds*, published in November 1984.

Organoruthenium and -osmium compounds constitute an important group of organometallics. They display a fascinating array of structural types, including a variety of cluster compounds, and they have uses as catalysts and reagents. Organoruthenium compounds in particular are widely used in catalytic processes. The varied properties of these compounds will ensure that this particular compendium will reach a wide readership.

The databank on the properties of organometallic compounds, which is represented in its current form by the *Dictionary of Organometallic Compounds* and its subset publications such as this volume, will be kept continuously up-to-date. Supplements to the main *Dictionary* will appear annually and revised editions of this *Sourcebook* will be published from time to time as demands permits.

G.R. Knox

Introduction

1. Using the Sourcebook

The *Sourcebook* is divided into element sections: within each section the arrangement of entries is in order of molecular formula according to the Hill convention (i.e. C, then H, then other elements in alphabetical sequence of element symbol; where no carbon is present, the elements including H are ordered strictly alphabetically).

Every entry is numbered to assist ready location and the entry number consists of a metal element symbol followed by a five-digit number.

Indexes

There are three printed indexes: a name index which lists every compound name or synonym in alphabetical order; a molecular formula index which lists all molecular formulae, including those of derivatives, in Hill convention order; and a CAS registry number index listing all CAS numbers included in the *Sourcebook* in serial order. All indexes refer to the entry number. In the name index an entry number which follows immediately upon an index term means that the term itself is used as the entry name but an entry number which is preceded by the word 'see' means that the term is a synonym to an entry name. In all three indexes an entry number which is preceded by the word 'in' refers the reader to a specified stereoisomer or derivative which is to be found embedded within the particular entry.

In addition to the three printed indexes, each element section is preceded by a graphical structure index allowing the rapid visual location of compounds of interest. The structure index reproduces all structure diagrams present in that element section in reduced size and printed in entry number order.

The following paragraphs summarize important considerations in compiling the information in this *Sourcebook*. For more detailed information, see the Introduction to the *Dictionary of Organometallic Compounds* from which this *Sourcebook* derives.

2. Compound Selection

In compiling this *Sourcebook* the aim has been to include from the primary literature up to mid 1983:

(1) Compounds representative of all important structural types (typically, the parent member of each series, where known, together with a selection of its homologues).

(2) Any compound with an established use, such as in catalysis, as a synthetic reagent or starting material.

(3) Other compounds of particular chemical, structural, biological or historical interest, especially those thought to exhibit unusual bonding characteristics.

Some compounds which are not considered sufficiently important to justify separate entries of their own have been included as derivatives in the entries of other compounds. These may include for example:

(1) Organic derivatives in the classical sense.

(2) Donor-acceptor complexes.

(3) The various salts of an anion or cation. In nearly every case, the entry for an ionic substance refers to the naked anion or cation, and the molecular formula, molecular weight and CAS registry number given for the main entry are those of the ion, in agreement with current CAS practice. Salts of the ion with various counterions are then treated as derivatives and the molecular formulae of all of these are given.

(4) Oligomeric compounds. Where a compound is known in several states of molecular aggregation these are all included in the one entry, which usually refers to the monomer. Compounds which are known only in dimeric form are entered as such, but the hypothetical monomers are included as derivatives to ensure that the names and molecular formulae of the monomeric forms occur in the indexes.

All names and molecular formulae recorded for derivatives occur in the Name and Molecular Formula Indexes respectively.

3. Chemical Names and Synonyms

The naming of organometallic compounds is frequently problematic and so in selecting the range of alternative names to present for each compound or derivative, editorial policy has been to report names which are found in the literature, including *Chemical Abstracts*, and not to attempt to impose a system of nomenclature. The editorial generation of new names has therefore been kept to a minimum required by consistency. Most names given in the *Sourcebook* are those given in the original paper(s) and in *Chemical Abstracts*.

Names corresponding to those used by CAS during

the 8th, 9th, and 10th collective index periods (1967-71, 1972-6 and 1977-81 respectively) are labelled with the suffixes 8CI, 9CI and 10CI respectively.

4. Toxicity and Hazard Information

Toxicity and hazard information is highlighted by the sign ▷ which also appears in the indexes.

All organometallic compounds should be treated as if they have dangerous properties.

The information contained in the *Sourcebook* has been compiled from sources believed to be reliable. No warranty, guarantee or representation is made by the Publisher as to the correctness or sufficiency of any information herein, and the Publisher assumes no responsibility in connection therewith.

The specific information in this publication on the hazardous and toxic properties of certain compounds is intended to alert the reader to possible dangers associated with the use of those compounds. The absence of such information should not, however, be taken as an indication of safety in use or misuse.

5. Bibliographic References

The selection of references is made with the aim of facilitating entry into the literature for the user who wishes to locate more detailed information about a particular compound. Reference contents are frequently indicated using mnemonic suffixes. In general recent references are preferred to older ones, and the number of references quoted does not necessarily indicate the relative importance of a compound.

Journal abbreviations generally follow the practice of *Chemical Abstracts Service Source Index* (CASSI). In patent references, no distinction is made between patent applications and granted patents.

6. Sources of Further Information

The following books and review series provide more information about various aspects of organometallic chemistry. Lists of reviews specific to organic compounds of particular metals may be found in the introductory sections of the metals concerned.

General

Comprehensive Organometallic Chemistry, Wilkinson, G. *et al.* Eds, Pergamon, Oxford, 1982. This book represents the most complete and up to date review of the whole subject. In addition to sections for each element there are chapters on the use of organometallics in organic synthesis and catalysis.

Comprehensive Inorganic Chemistry, Trotman-Dickenson, A.F. *et al.* Eds, Pergamon, Oxford, 1973. Contains information about organometallics as well as discussions of oxidation states, coordination chemistry and analysis of the metals.

Gmelins Handbuch der Anorganischen Chemie, 8th Edn, Springer-Verlag, Berlin. Some volumes of Gmelin covering organometallic compounds have been updated relatively recently and can therefore be consulted for comprehensive data on some types of organometallics. Some Gmelin element sections, however, are many years out of date.

Houben-Weyls Methoden der Organischen Chemie, 4th Edn, Band XIII, *Metallorganische Verbindungen*, Thieme-Verlag, Stuttgart.

The Chemistry of the Carbon-Metal Bond, Hartley, F.R. and Patai, S. Eds, Wiley, New York, 1982-. Contains sections on the synthesis, analysis and thermochemistry of various classes of organometallic compounds.

Transition-Metal Complexes of Phosphorus, Arsenic and Antimony Ligands, McAuliffe, C.A. Ed., Macmillan, London, 1973.

Methods of Elemento-Organic Chemistry, Kocheshkov, K.A. Ed., North Holland, Amsterdam, 1967.

MTP International Review of Science: Inorganic Chemistry, Series 2, Emeléus, H.J. Ed., Butterworths, London; University Park Press, Baltimore, 1974-5.

Advances in Organometallic Chemistry, Academic Press, 1964-.

Annual Surveys of Organometallic Chemistry, Elsevier, 1964-7.

Organometallic Chemistry Reviews, Elsevier, 1966-7.

Organometallic Chemistry Reviews, Section A: Subject Reviews 1968-72.

Organometallic Chemistry Reviews, Section B: Annual Surveys 1968-74.

Journal of Organometallic Chemistry: This incorporates reviews and surveys after the discontinuation of the two series of *Organometallic Chemistry Reviews*.

Organometallic Chemistry, 1972-, (Specialist Periodical Reports), RSC.

Coordination Chemistry Reviews, Elsevier, 1966-.

Progress in Inorganic Chemistry, Interscience, 1959-.

Advances in Inorganic Chemistry and Radiochemistry, Academic Press, 1959-.

Analysis

Scott's Standard Methods of Chemical Analysis, Furman, N.H. Ed., 6th Edn, Van Nostrand, New York, 1962.

Crompton, T.R., *Chemical Analysis of Organometallic Compounds*, Academic Press, London, 1973.

Spectroscopy

Nuclear Magnetic Resonance Spectroscopy of Nuclei Other than the Proton, Axenrod, T. and Webb, G.A. Eds, Wiley, London, 1974.

NMR and the Periodic Table, Harris, R.K. and Mann, B.E. Eds, Academic Press, London, 1978.

^{13}C NMR Data for Organometallic Compounds, Mann, B.E. and Taylor, B.F. Eds, Academic Press, London, 1981.

Spectroscopic Properties of Inorganic and Organometallic Compounds, 1968-, (Specialist Periodical Reports), RSC.

Handling

Shriver, D.F., *The Manipulation of Air-Sensitive Compounds*, McGraw-Hill, 1969.

Organometallic Syntheses, Academic Press, New York, 1965, Vol. 1.

Os Osmium

G. R. Knox

Osmium (Fr., Ger.), Osmio (Sp., Ital.), Осмий (Osmii) (Russ.), オスミウム (Japan.)

Atomic Number. 76

Atomic Weight. 190.2

Electronic Configuration. [Xe] $5d^6\,6s^2$

Oxidation States. Usually −2, −1, 0, +1 or +2 in organometallics.

Coordination Number. Commonly 5 and 6 in organometallics.

Colour. Osmocene is almost colourless and its derivatives are usually very pale yellow. There is a general trend for comparable derivatives of osmium to exhibit a paler colour than their ruthenium (and iron) analogues; many simple diamagnetic osmium carbonyl complexes are colourless, cream or pale yellow.

Availability. Anhydrous $OsCl_3$ and OsO_4 are the commonest organometallic precursors; osmocene and $Os_3(CO)_{12}$ are commercially available. Belonging to the precious metals group, the cost of osmium derivatives is high.

Handling. Chromatographic techniques are widely used for separating and purifying the coloured derivatives of osmium.

Toxicity. Osmium compounds, especially OsO_4, are highly toxic.

Isotopic Abundance. ^{186}Os, 1.59%; ^{187}Os, 1.64%; ^{188}Os, 13.3%; ^{189}Os, 16.1%; ^{190}Os, 26.4%; ^{192}Os, 41.0%.

Spectroscopy. ^{187}Os has $I = \frac{1}{2}$ but has a very low relative receptivity and ^{189}Os has $I = \frac{3}{2}$. Very few studies have been carried out using osmium as an nmr probe.

Analysis. See under Ruthenium.

References. The chemistry of osmium is generally considered in relation to that of ruthenium; for reviews see under Ruthenium.

$Br_5OsCO^{\ominus\ominus}$

Os-00001

$Br_5Os(CO)^{\ominus\ominus\ominus}$

Os-00002

$Cl_5Os(CO)^{\ominus\ominus}$

Os-00003

$Cl_5Os(CO)^{\ominus\ominus\ominus}$

Os-00004

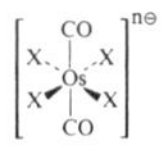

X = Br, n = 1

Os-00005

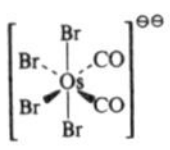

Os-00006

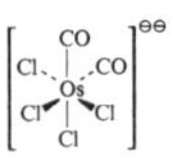

Os-00007

$[I_2Os(CO)_2]_n$

Os-00008

As Os-00005 with
X = I, n = 1

Os-00009

As Os-00005 with
X = I, n = 2

Os-00010

$Os(CO)_2(NO)_2$

Os-00011

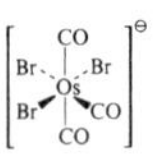

Os-00012

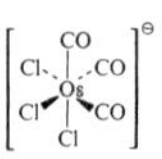

fac-form

Os-00013

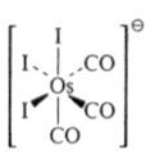

fac-form

Os-00014

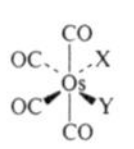

cis-form

X = Y = Br

Os-00015

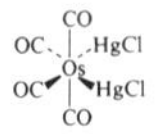

Os-00016

As Os-00015 with
X = Y = Cl

Os-00017

$HOs(CO)_4^{\ominus}$

Os-00018

As Os-00015 with
X = Y = H

Os-00019

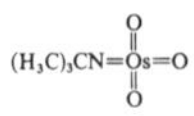

Os-00020

As Os-00015 with
X = Y = I

Os-00021

$Os(CO)_4^{\ominus\ominus}$

Os-00022

$HOs(CO)_5^{\oplus}$

Os-00023

As Os-00015 with
X = Me, Y = Br

Os-00024

As Os-00015 with
X = Me, Y = I

Os-00025

As Os-00015 with
X = Me, Y = H

Os-00026

$Os(CO)_5$

Os-00027

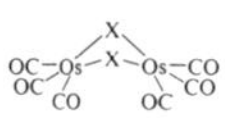

X = Br

Os-00028

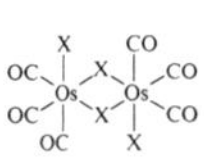

X = Br

Os-00029

As Os-00028 with
X = Cl

Os-00030

As Os-00029 with
X = Cl

Os-00031

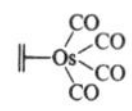

Os-00032

As Os-00015 with
X = Et, Y = Br

Os-00033

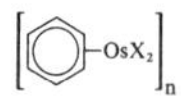

X = Br

Os-00034

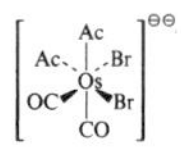

Os-00035

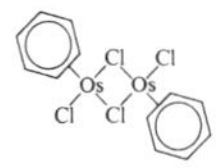

Dimer

Os-00036

As Os-00034 with
X = I

Os-00037

As Os-00015 with
X = Y = Me

Os-00038

As Os-00028 with
X = I

Os-00039

As Os-00029 with
X = I

Os-00040

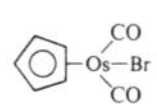

Os-00041

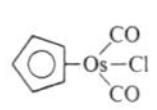

Os-00042

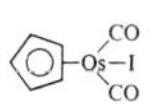

Os-00043

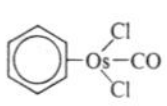

Os-00044

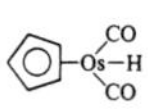

Os-00045

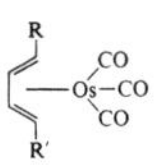

R = R′ = H

Os-00046

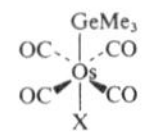

X = Br

Os-00047

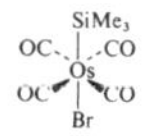

Os-00048

As Os-00047 with
X = I

Os-00049

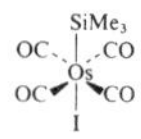

Os-00050

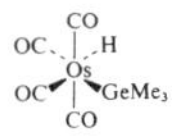

Os-00051

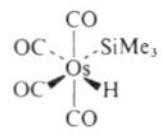

Os-00052

$BrOs(CO)_4—Os(CO)_4Br$ (*diax*)

Os-00053

$ClOs(CO)_4—Os(CO)_4Cl$ (*diax*)

Os-00054

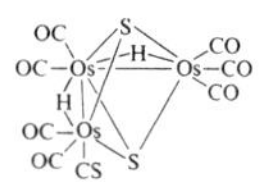

Os-00055

$(OC)_4OsH—OsH(CO)_4$

Os-00056

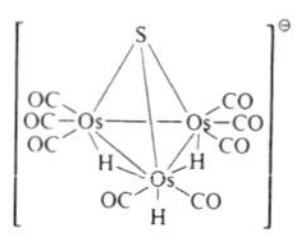

Os-00057

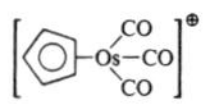

Os-00058

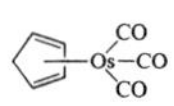

Os-00059

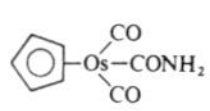

Os-00060

As Os-00046 with
$R = CH_3$, R′ = H

Os-00061

As Os-00015 with
X = Y = Et

Os-00062

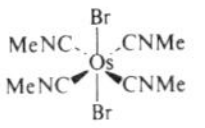

Os-00063

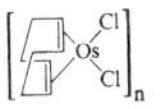

Os-00064

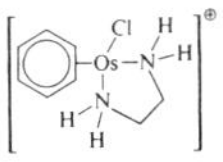

Os-00065

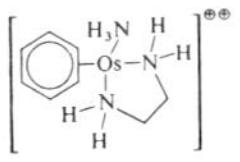

Os-00066

Os-00067

$IOs(CO)_4$—$Os(CO)_4I$ (*diax*)

Os-00068

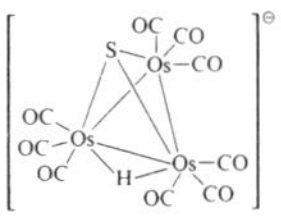

Os-00069

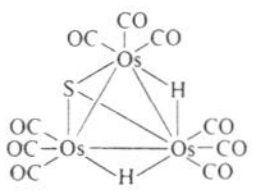

Os-00070

$BrOs(CO)_4$—$Os(CO)_4Me$

Os-00071

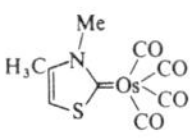

Os-00072

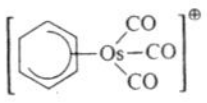

Os-00073

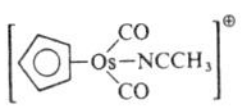

Os-00074

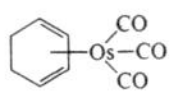

Os-00075

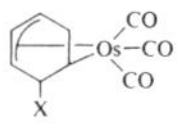

X = H

Os-00076

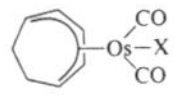

X = Br

Os-00077

As Os-00077 with
X = I

Os-00078

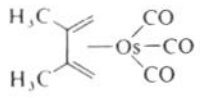

Os-00079

As Os-00046 with
R = R′ = CH_3

Os-00080

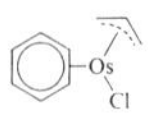

Os-00081

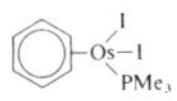

Os-00082

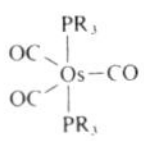

R = OMe

Os-00083

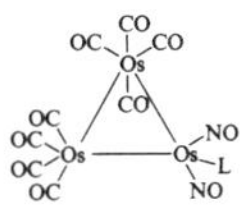

L = CO

Os-00084

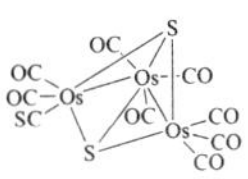

Os-00085

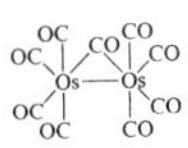

Os-00086

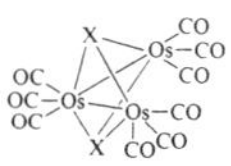

X = S

Os-00087

As Os-00087 with
X = Se

Os-00088

As Os-00087 with
X = Te

Os-00089

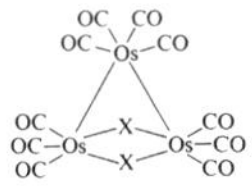

X = Br

Os-00090

As Os-00090 with
X = Cl

Os-00091

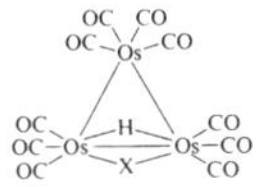

X = Br

Os-00092

As Os-00092 with
X = Cl

Os-00093

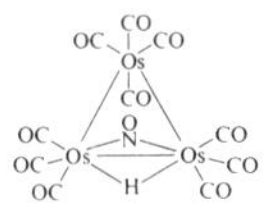

Os-00094

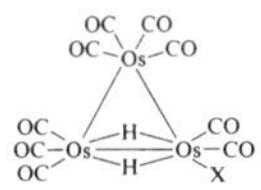

X = CO

Os-00095

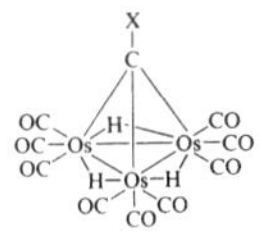

X = Br

Os-00096

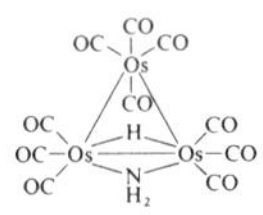

Os-00097

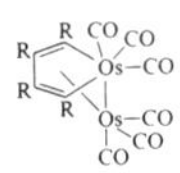

R = H

Os-00098

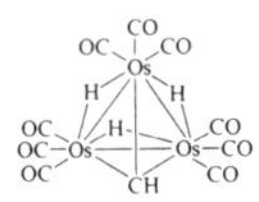

Os-00099

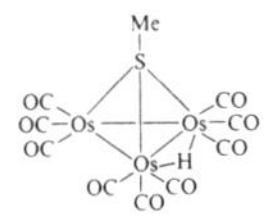

Os-00100

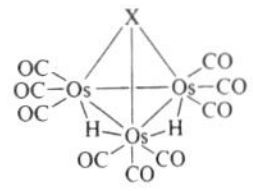

X = NMe

Os-00101

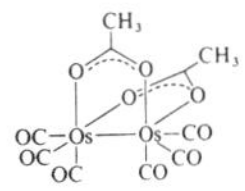

Os-00102

As Os-00076 with
X = CN

Os-00103

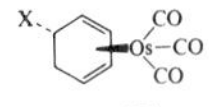

exo-form

X = CN

Os-00104

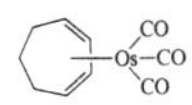

Os-00105

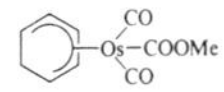

Os-00106

As Os-00104 with
X = OMe

Os-00107

As Os-00015 with
X = Y = $-COCH_2CH_3$

Os-00108

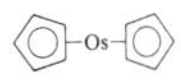

Os-00109

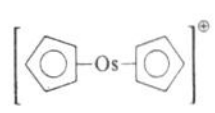

Os-00110

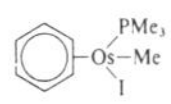

Os-00111

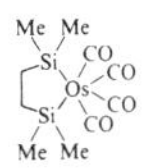

Os-00112

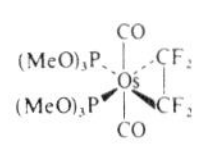

Os-00113

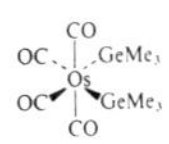

Os-00114

$(OC)_4Os(PbMe_3)_2$

Os-00115

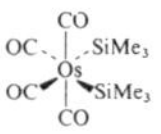

Os-00116

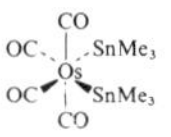

Os-00117

As Os 00090 with
X = I

Os-00118

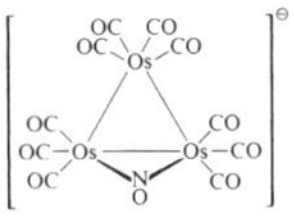

Os-00119

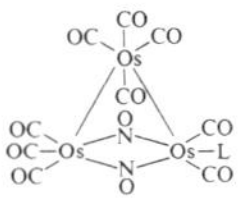

L = CO

Os-00120

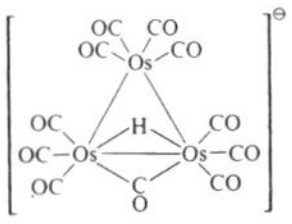

Os-00121

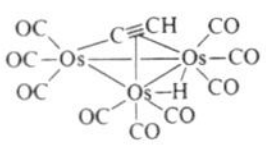

Os-00122

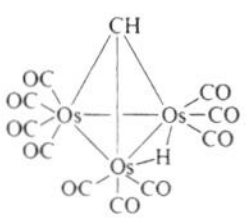

Os-00123

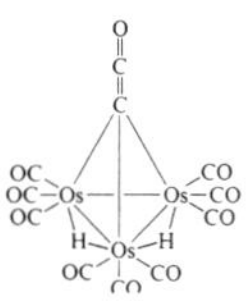

Os-00124

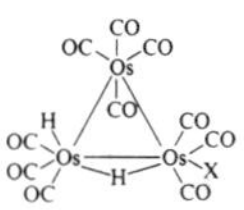

X = CO

Os-00125

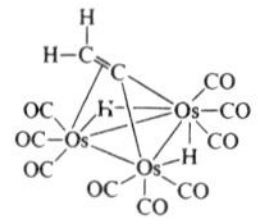

Os-00126

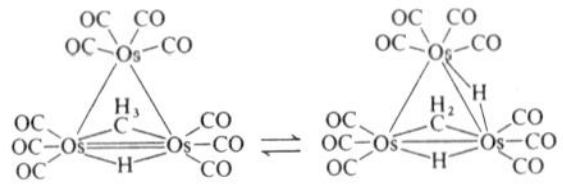

Os-00127

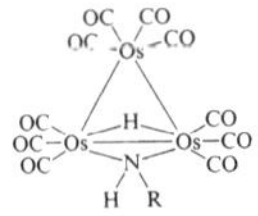

R = Me

Os-00128

As Os-00096 with
X = CH_3

Os-00129

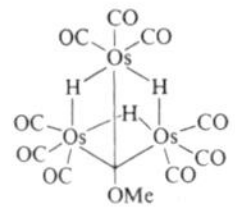

Os-00130

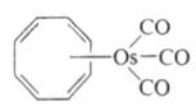

Os-00131

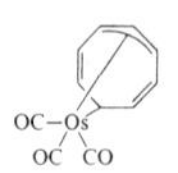

Os-00132

As Os-00084 with
L = $P(OMe)_3$

Os-00133

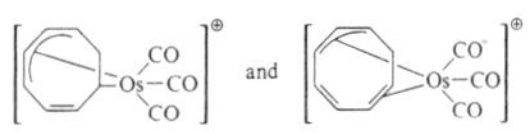

Os-00134

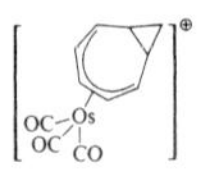

Os-00135

Os-00136

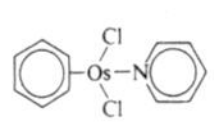

Os-00137

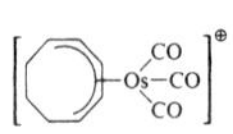

Os-00138

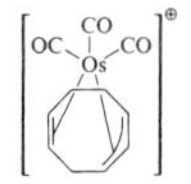

Os-00139

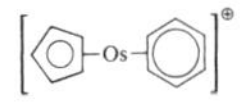

Os-00140

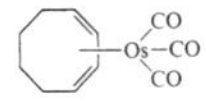

Os-00141

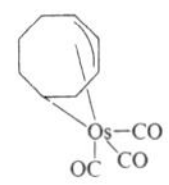

Os-00142

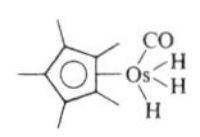

Os-00143

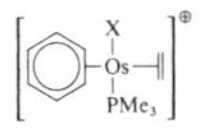

X = I

Os-00144

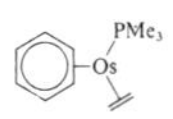

Os-00145

As Os-00144 with
X = H

Os-00146

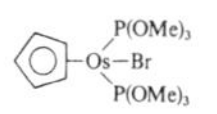

Os-00147

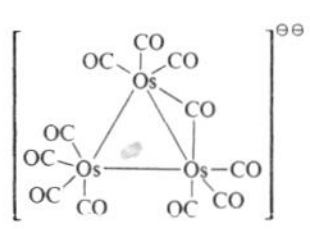

Os-00148

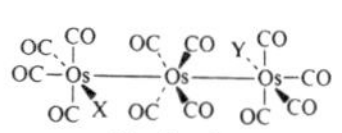

X = Y = Br

Os-00149

As Os-00149 with
X = Y = Cl

Os-00150

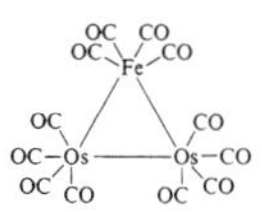

Os-00151

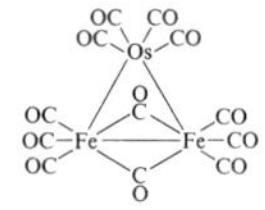

Os-00152

$MnOs_3H(CO)_{12}$

Os-00153

$RhOs_3H(CO)_{12}$

Os-00154

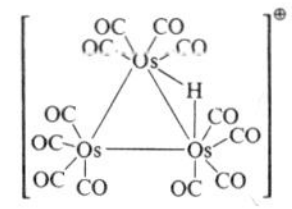

Os-00155

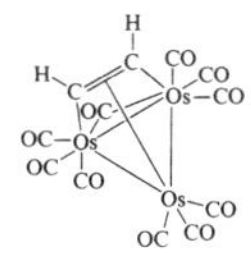

Os-00156

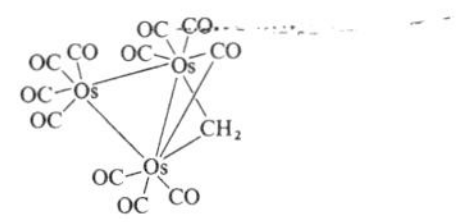

Os-00157

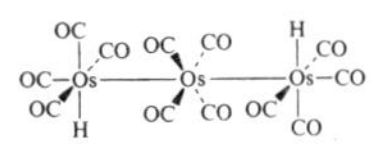

Os-00158

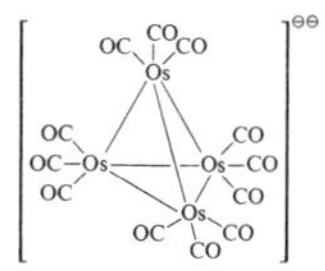

2H atoms not shown

Os-00159

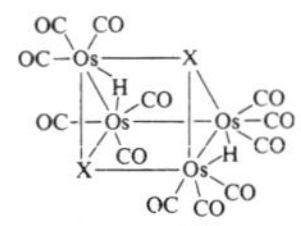

X = S

Os-00160

As Os-00160 with
X = Se

Os-00161

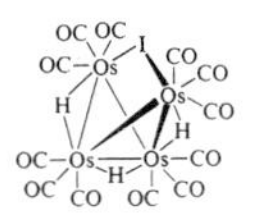

Os-00162

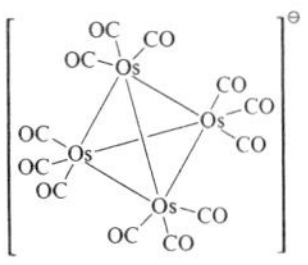

Os-00163

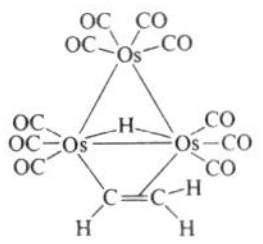

Os-00164

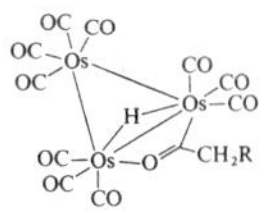

R = H

Os-00165

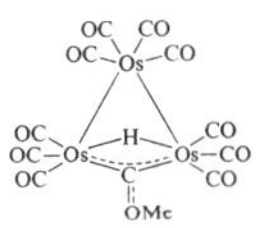

Os-00166

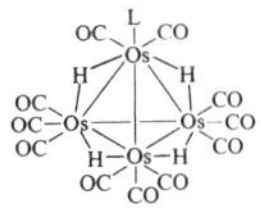

L = CO

Os-00167

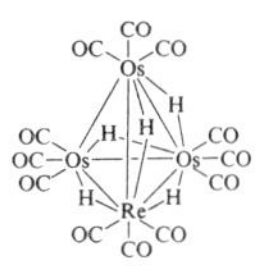

Os-00168

As Os-00092 with
X = SEt

Os-00169

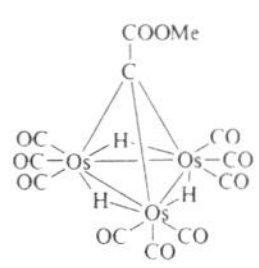

Os-00170

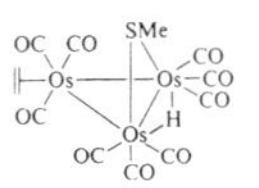

Os-00171

As Os-00120 with
L = $P(OMe)_3$

Os-00172

As Os-00120 with
L = Me_3N

Os-00173

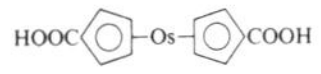

Os-00174

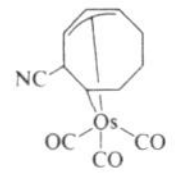

Os-00175

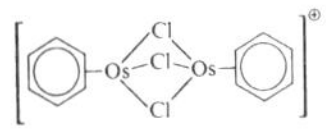

Os-00176

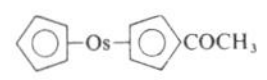

Os-00177

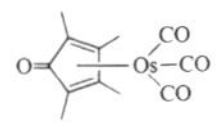

Os-00178

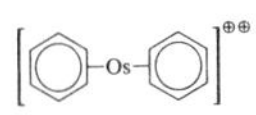

Os-00179

Os-00180

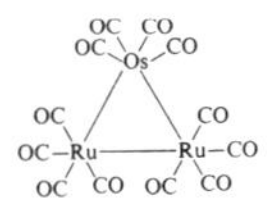

Presumed structure

Os-00181

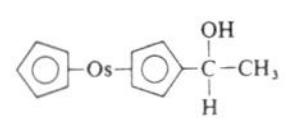

Os-00182

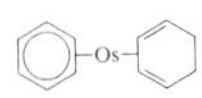

Os-00183

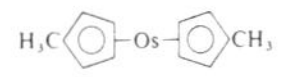

Os-00184

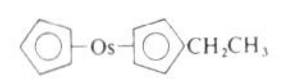

Os-00185

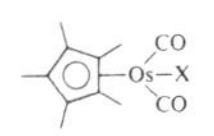

X = Br

Os-00186

As Os-00186 with
X = I

Os-00187

As Os-00186 with
X = H

Os-00188

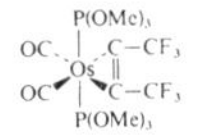

Os-00189

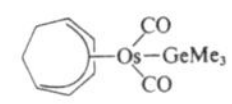

Os-00190

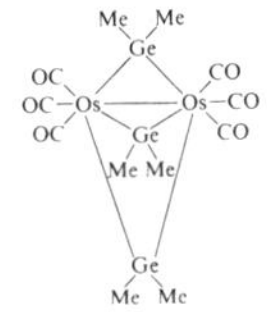

Os-00191

$Os(CNMe)_6^{\oplus\oplus}$

Os-00192

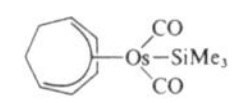

Os-00193

As Os-00144 with
X = Me

Os-00194

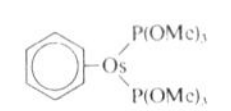

Os-00195

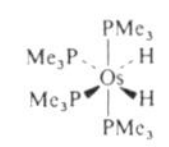

Os-00196

As Os-00149 with
X = Y = I

Os-00197

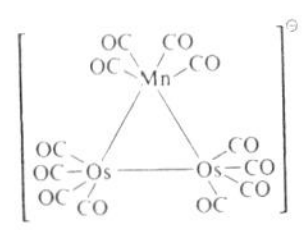

Os-00198

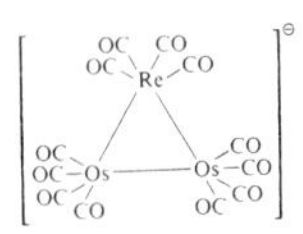

Os-00199

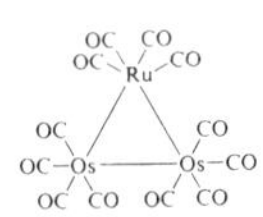

Os-00200

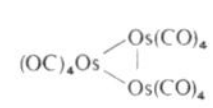

Os-00201

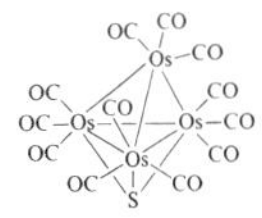

Os-00202

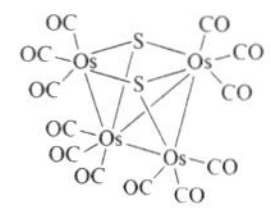

Os-00203

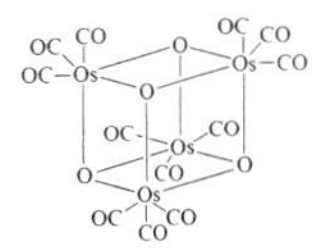

Os-00204

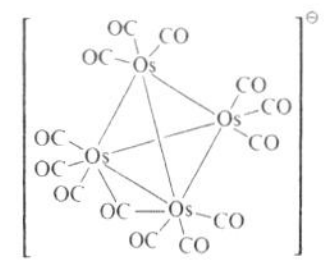

Os-00205

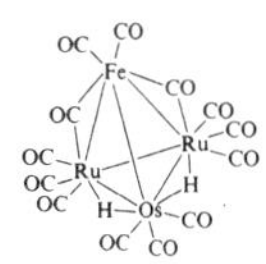

Os-00206

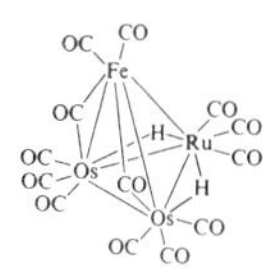

Os-00207

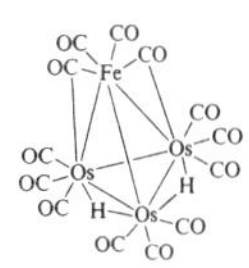

Os-00208

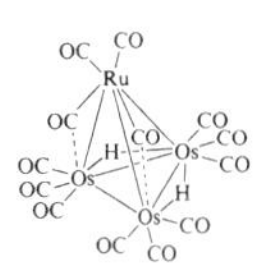

Os-00209

$H_2Os_4(CO)_{13}$

Os-00210

As Os-00149 with
X = Br, Y = Me

Os-00211

$MnOs_3H_3(CO)_{13}$

Os-00212

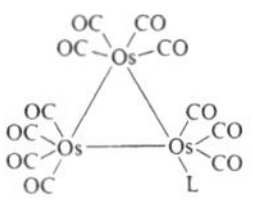
L = MeCN

Os-00213

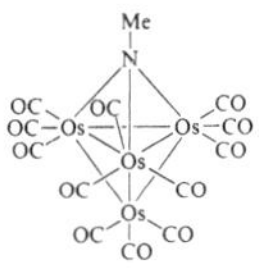

Os-00214

$ReOs_3H_3(CO)_{13}$

Os-00215

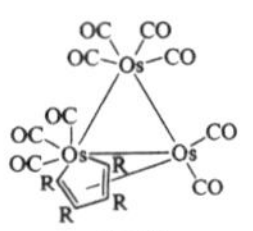
R = H

Os-00216

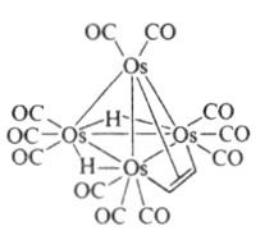

Os-00217

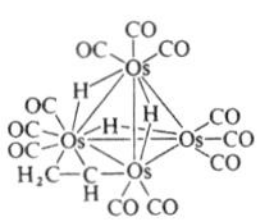

Os-00218

$[H_4Os_4(CO)_{11}(COOMe)]^{\ominus}$

Os-00219

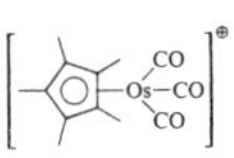

Os-00220

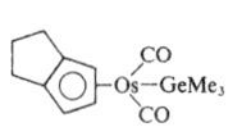

Os-00221

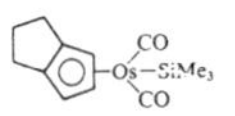

Os-00222

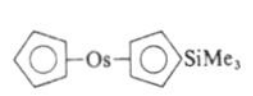

Os-00223

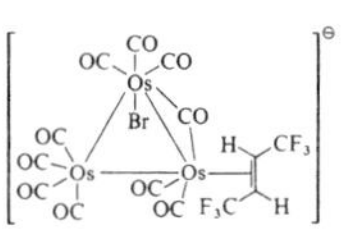

Os-00224

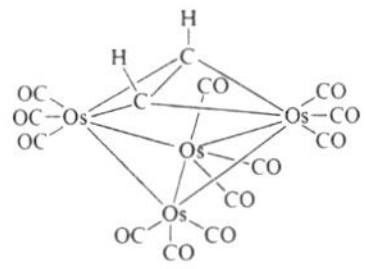

Os-00225

$Os_3(CO)_{10}(NCCH_3)_2$

Os-00226

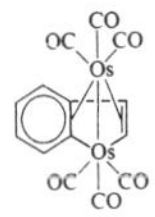

Os-00227

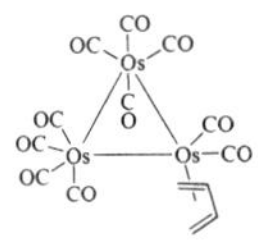

Os-00228

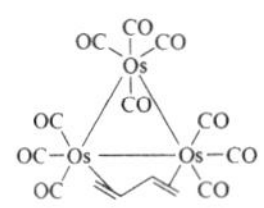

Os-00229

As Os-00149 with
X = Y = Me

Os-00230

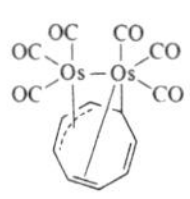

Os-00231

As Os-00213 with
L = $P(OMe)_3$

Os-00232

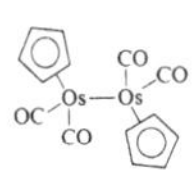

Os-00233

As Os-00095 with
X = $(H_3C)_3CNC-$

Os-00234

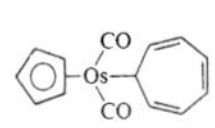

Os-00235

As Os-00098 with
R = CH_3

Os-00236

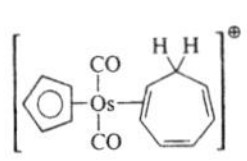

Os-00237

As Os-00167 with
L = $P(OMe)_3$

Os-00238

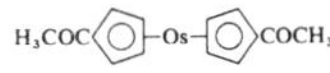

Os-00239

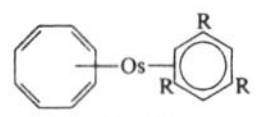
R = H

Os-00240

Os-00241

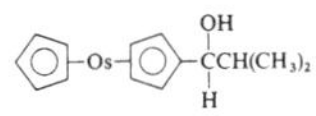

Os-00242

$[Me_3SiOs(CO)_4]_2$

Os-00243

Os-00244

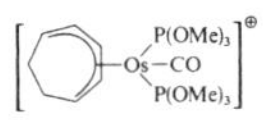

Os-00245

$(OC)_5Mn-Os(CO)_4-Mn(CO)_5$

Os-00246

$(OC)_5Re-Os(CO)_4-Re(CO)_5$

Os-00247

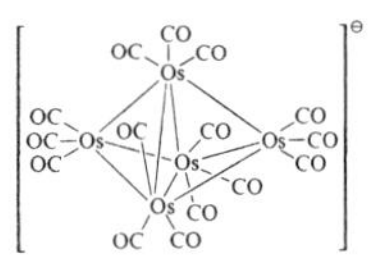
H not shown

Os-00248

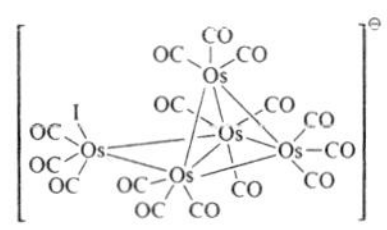

Os-00249

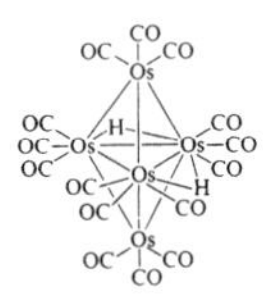

Os-00250

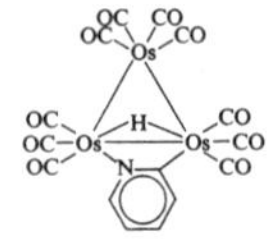

Os-00251

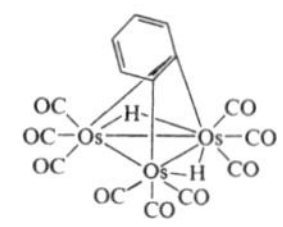

Os-00252

As Os-00101 with
X = NPh

Os-00253

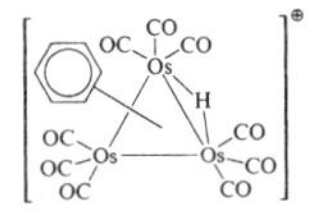
Probable structure

Os-00254

As Os-00101 with
X = PPh

Os-00255

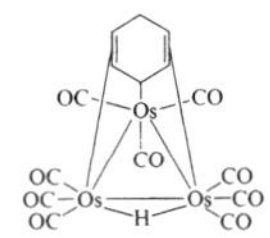

Os-00256

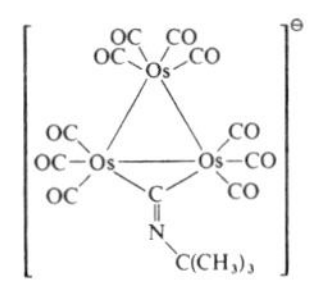
H not shown

Os-00257

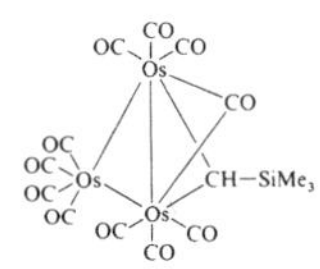

Os-00258

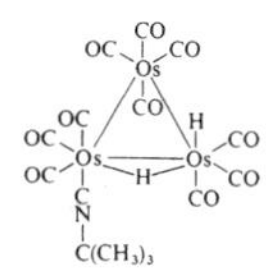

Os-00259

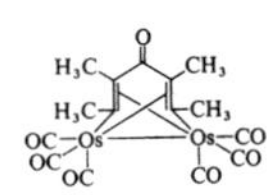

Os-00260

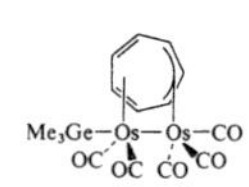

Os-00261

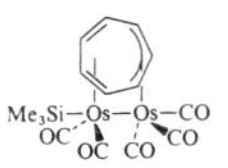

Os-00262

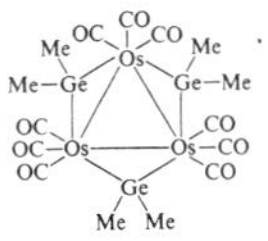

Os-00263

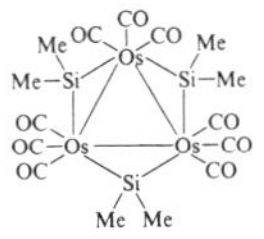

Os-00264

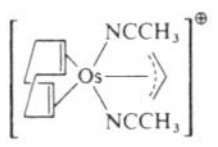

Os-00265

$BrOs[P(OMe)_3]_5^{\oplus}$

Os-00266

$ClOs[P(OMe)_3]_5^{\oplus}$

Os-00267

$Os[P(OMe)_3]_5$

Os-00268

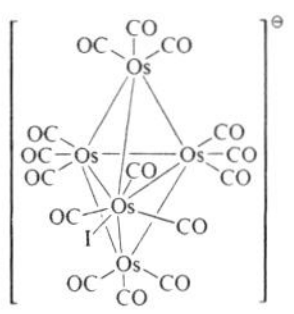

Os-00269

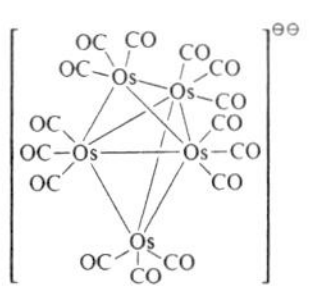

Os-00270

$MnOs_3H(CO)_{16}$

Os-00271

$ReOs_3H(CO)_{16}$

Os-00272

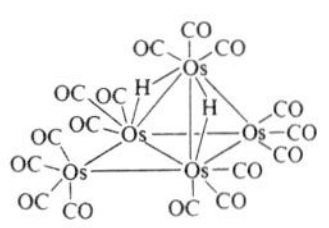

Os-00273

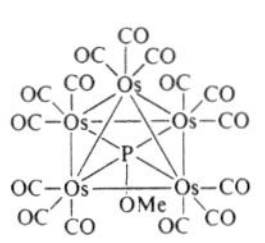

Os-00274

As Os-00092 with
X = SPh

Os-00275

As Os-00128 with
R = Ph

Os-00276

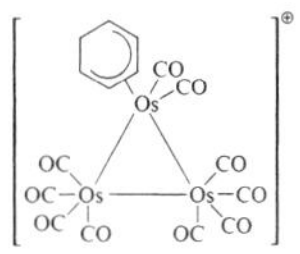

Os-00277

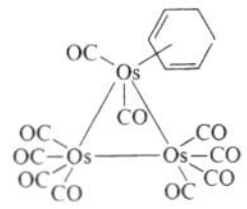

Os-00278

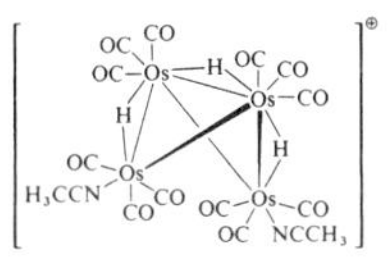

Os-00279

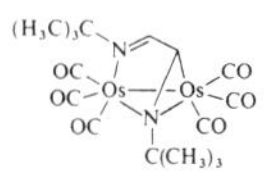

Os-00280

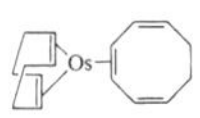

Os-00281

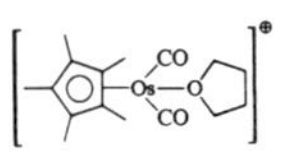

Os-00282

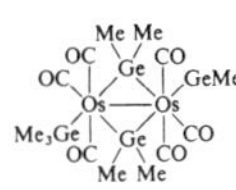

Os-00283

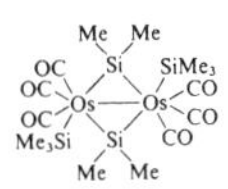

Os-00284

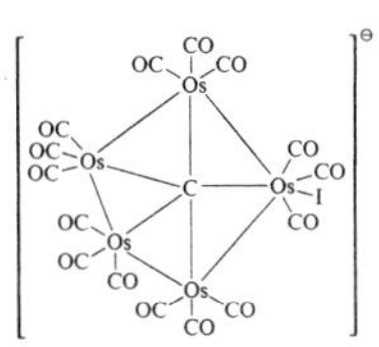

Os-00285

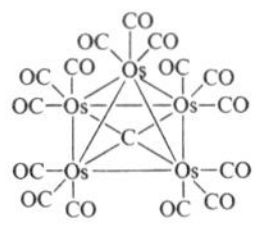

Os-00286

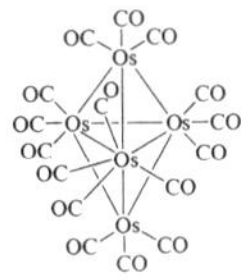

Os-00287

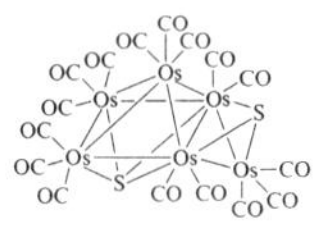

Os-00288

$Os_6O_6(CO)_{16}$

Os-00289

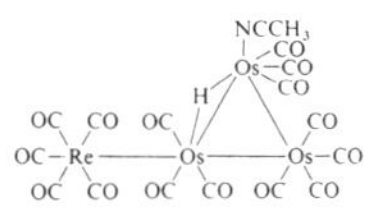

Os-00290

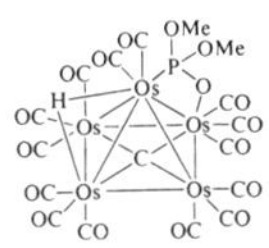

Os-00291

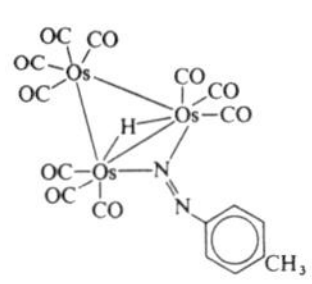

Os-00292

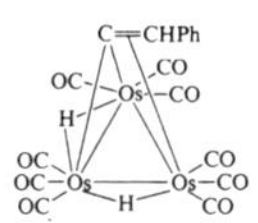

Os-00293

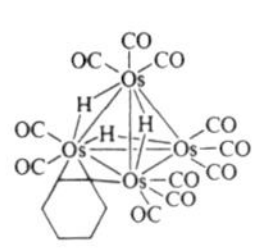

Os-00294

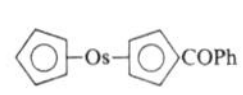

Os-00295

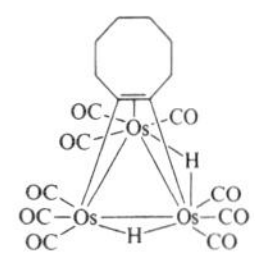

Os-00296

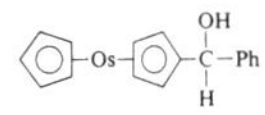

Os-00297

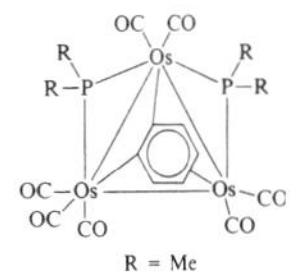

Os-00298

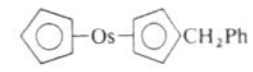

Os-00299

As Os-00240 with
R = CH_3

Os-00300

$HOs_6(CO)_{18}^{\ominus}$

Os-00301

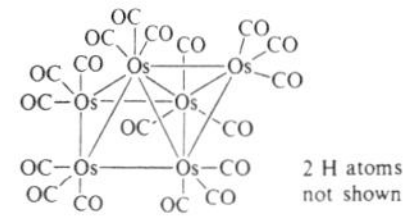

Os-00302

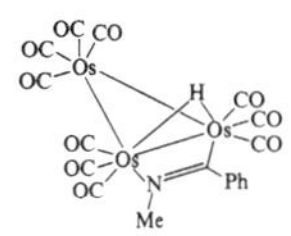

Os-00303

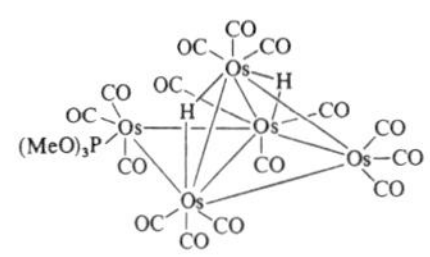

Os-00304

$Ph_3POsCl_5^{\ominus}$

Os-00305

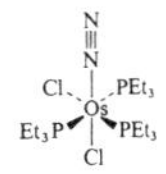

Os-00306

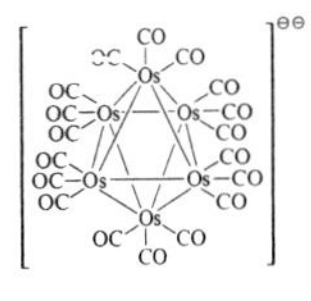

Os-00307

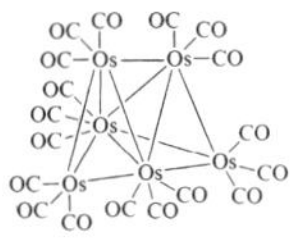

Os-00308

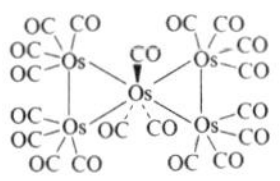

Os-00309

$H_2Os_7(C)(CO)_{19}$

Os-00310

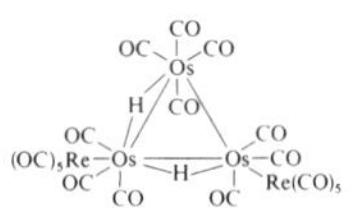

Os-00311

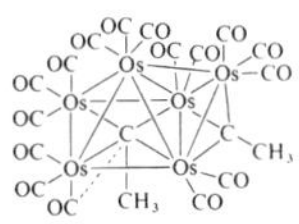

Os-00312

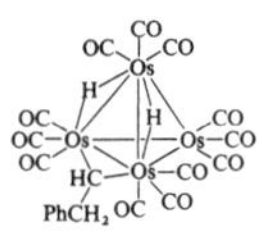

Os-00313

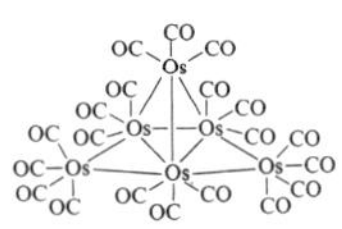

Suggested structure

Os-00314

$Os_7(CO)_{20}{}^{\ominus\ominus}$

Os-00315

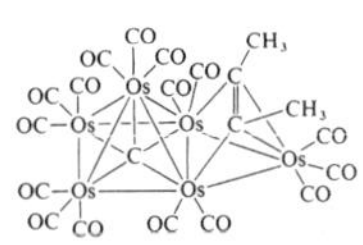

Os-00316

$Ph_3POsBr_2(CO)_3$

Os-00317

$Ph_3POsCl_2(CO)_3$

Os-00318

$Ph_3POsI_2(CO)_3$

Os-00319

$Ph_3POsClH(CO)_3$

Os-00320

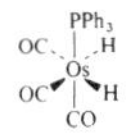

Os-00321

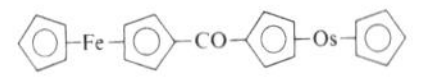

Os-00322

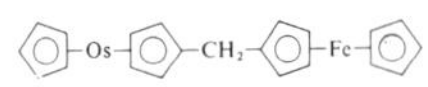

Os-00323

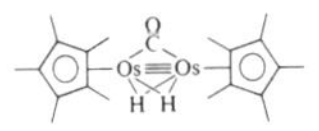

Os-00324

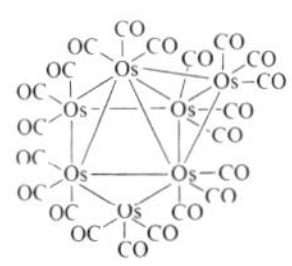

Os-00325

$Ph_3POs(CO)_4$

Os-00326

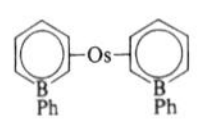

Os-00327

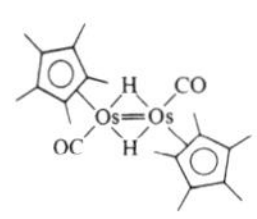

Os-00328

$Os_8(C)(CO)_{21}$

Os-00329

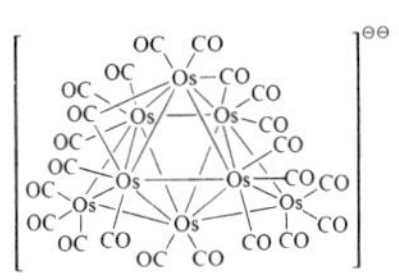

Os-00330

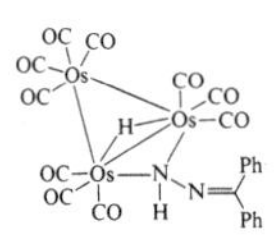

Os-00331

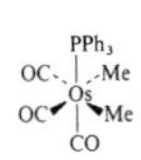

Os-00332

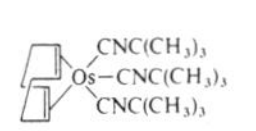

Os-00333

$Os_8(CO)_{23}$

Os-00334

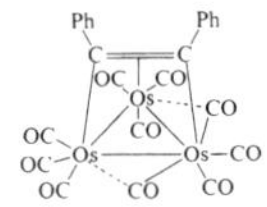

Os-00335

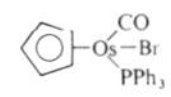

Os-00336

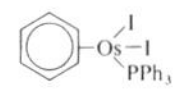

Os-00337

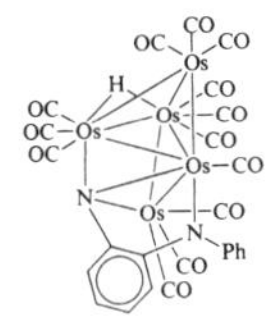

Os-00338

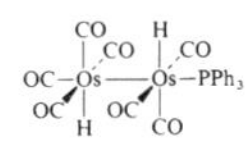

Os-00339

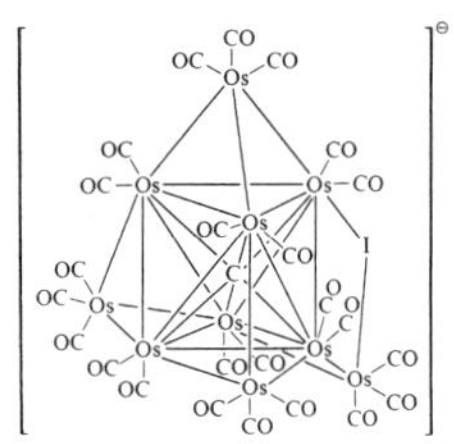

Os-00340

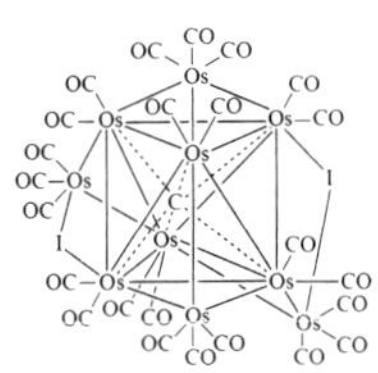

Os-00341

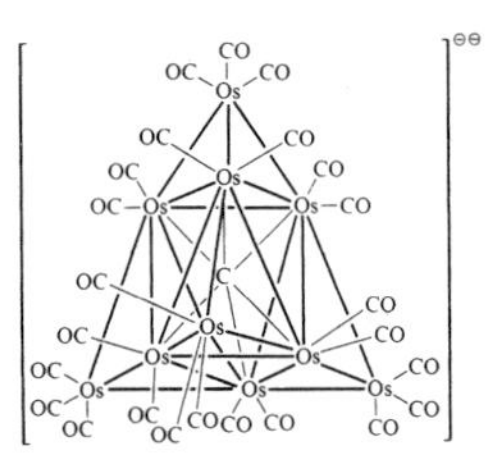

Os-00342

As Os-00095 with
X = PPh_3

Os-00343

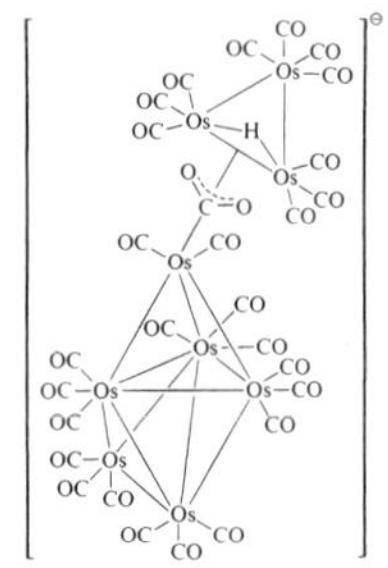

Os-00344

As Os-00125 with
X = PPh_3

Os-00345

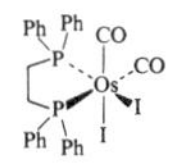

Os-00346

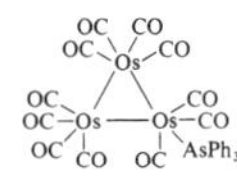

Os-00347

As Os-00213 with
L = PPh_3

Os-00348

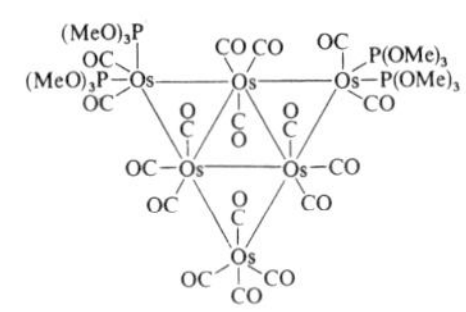

Os-00349

$BrOs[P(OEt)_3]_5{}^{\oplus}$

Os-00350

$ClOs[P(OEt)_3]_5{}^{\oplus}$

Os-00351

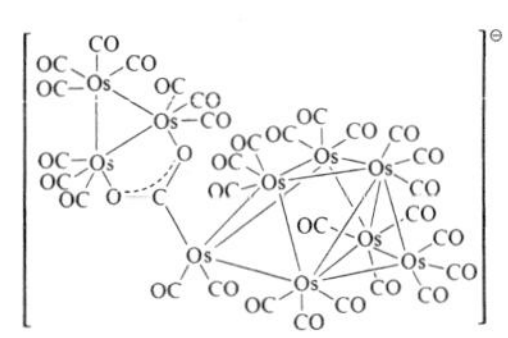

Os-00352

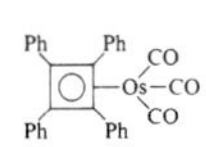

Os-00353

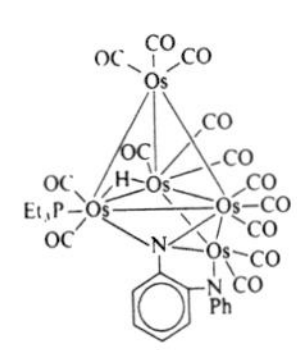

Os-00354

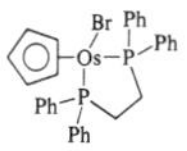

Os-00355

As Os-00098 with
R = Ph

Os-00356

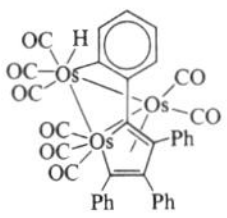

Os-00357

$(Ph_3As)_2OsBr_3(\equiv N)$

Os-00358

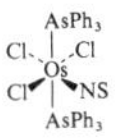

Os-00359

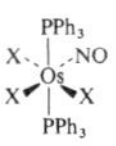

X = Br

Os-00360

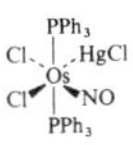

Os-00361

As Os-00360 with
X = Cl

Os-00362

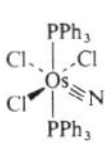

Os-00363

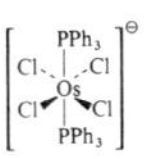

Os-00364

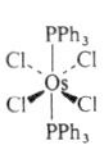

Os-00365

$(Ph_3P)_2Os(NO)_2\ (C_{2v})$

Os-00366

$Os[P(OEt)_3]_6^{\oplus\oplus}$

Os-00367

As Os-00216 with
R = Ph

Os-00368

As Os-00298 with
R = Ph

Os-00369

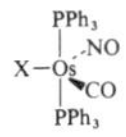

X = Cl

Os-00370

As Os-00370 with
X = H

Os-00371

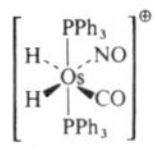

Os-00372

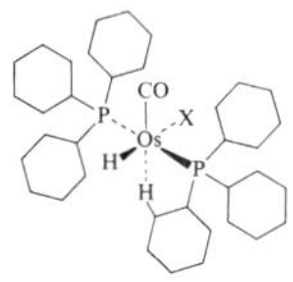

X = Br

Os-00373

As Os-00373 with
X = Cl

Os-00374

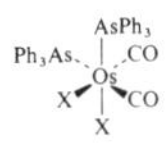

X = Br

Os-00375

As Os-00375 with
X = Cl

Os-00376

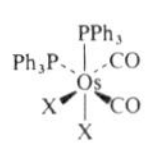

X = Br

Os-00377

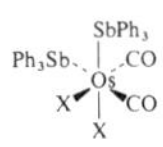

X = Br

Os-00378

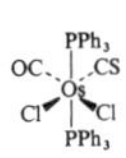

Os-00379

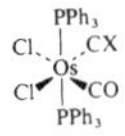

X = Se

Os-00380

As Os-00380 with
X = Te

Os-00381

As Os-00377 with
X = Cl

Os-00382

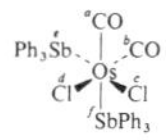

ab-dicarbonyl-*cd*-dichloro-*ef*-distibine-*form*

Os-00383

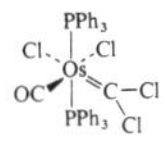

Os-00384

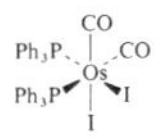

Os-00385

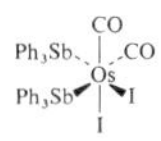

Os-00386

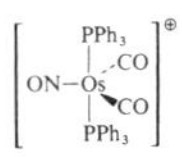

Os-00387

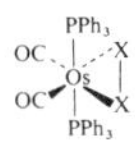

X = S

Os-00388

As Os-00388 with
X = Se

Os-00389

$Os(O_2)(CO)_2(PPh_3)_2$

Os-00390

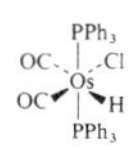

Os-00391

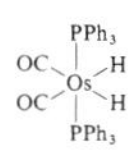

Os-00392

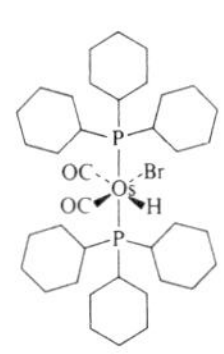

Os-00393

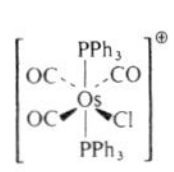

Os-00394

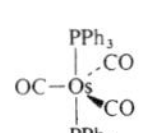

Os-00395

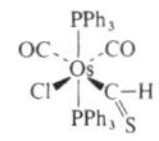

Os-00396

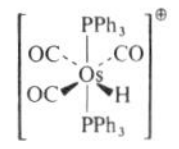

Os-00397

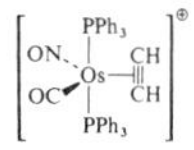

Os-00398

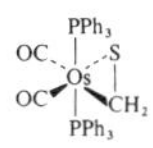

Os-00399

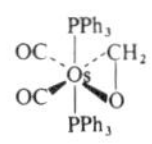

Os-00400

As Os-00370 with
X = –COOMe

Os-00401

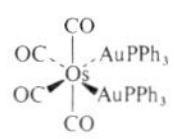

Os-00402

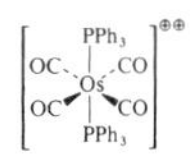

Os-00403

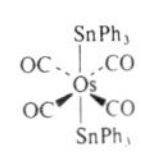

Os-00404

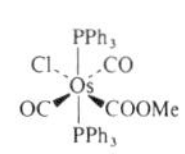

Os-00405

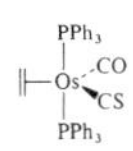

Os-00406

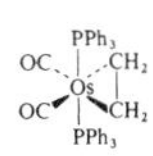

Os-00407

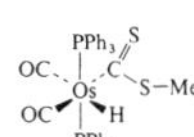

Os-00408

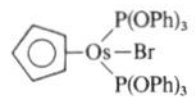

Os-00409

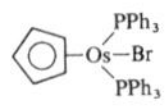

Os-00410

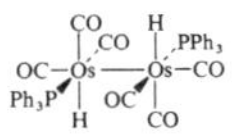

Os-00411

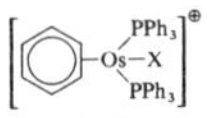

X = I

Os-00412

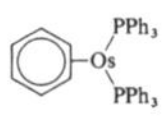

Os-00413

As Os-00412 with
X = H

Os-00414

$(Ph_3P)_2OsEt_2(CO)_2$

Os-00415

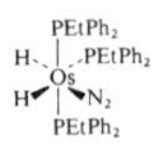

Os-00416

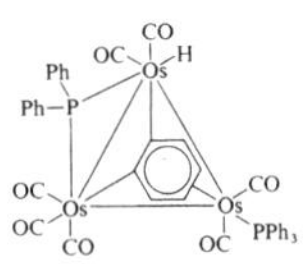

Os-00417

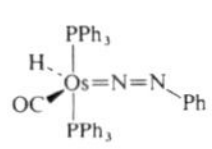

Os-00418

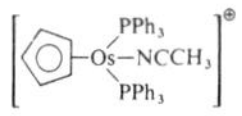

Os-00419

As Os-00412 with
X = Me

Os-00420

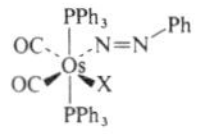

X = Br

Os-00421

As Os-00421 with
X = Cl

Os-00422

As Os-00421 with
X = I

Os-00423

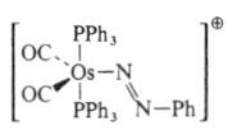

Os-00424

As Os-00421 with
X = H

Os-00425

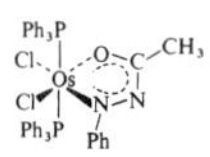

Os-00426

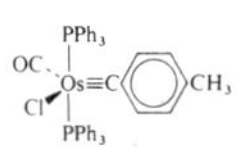

Os-00427

$Os_3(CO)_{10}(PPh_3)_2$

Os-00428

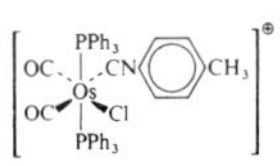

Os-00429

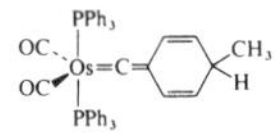

Os-00430

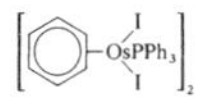

Os-00431

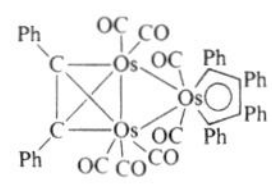

Os-00432

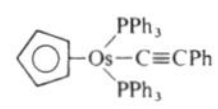

Os-00433

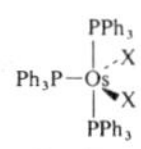

X = Br

Os-00434

As Os-00434 with
X = Cl

Os-00435

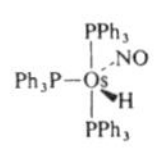

Os-00436

$H_4Os(PPh_3)_3$

Os-00437

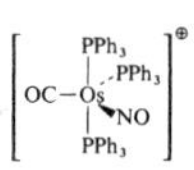

Os-00438

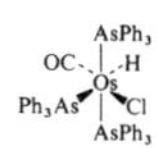

Os-00439

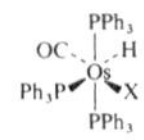

X = Br

Os-00440

As Os-00440 with
X = Cl

Os-00441

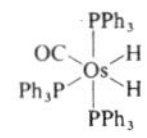

Os-00442

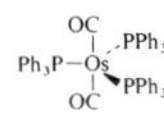

Os-00443

$Os_3(CO)_9(PPh_3)_3$

Os-00444

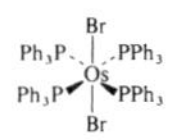

Os-00445

$HOs(PF_3)_4^{\ominus}$

Os-00446

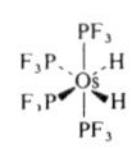

Os-00447

$Os(PF_3)_4^{\ominus\ominus}$

Os-00448

$Os(PF_3)_5$

Os-00449

$CBr_5OOs^{\ominus\ominus}$ **Os-00001**

Pentabromocarbonylosmate(2−)

$Br_5OsCO^{\ominus\ominus}$

M 617.730 (ion)
Paramagnetic.
Di-K salt: [64279-31-8].
CBr_5K_2OOs M 695.927
Dark-brown solid.
Bis(tetramethylammonium) salt: [64279-36-3].
$C_9H_{24}Br_5N_2OOs$ M 766.021
Dark-brown cryst. (EtOH).

Johannsen, F.H. *et al*, *J. Organomet. Chem.*, 1977, **135**, 221 (*synth, ir, raman, uv*)

$CBr_5OOs^{\ominus\ominus\ominus}$ **Os-00002**

Pentabromocarbonylosmate(3−)
[64279-35-2]

$Br_5Os(CO)^{\ominus\ominus\ominus}$

M 617.730 (ion)
Tri-Cs salt:
CBr_5Cs_3OOs M 1016.447
Dark-red cryst. (H_2O).

Cleare, M.J. *et al*, *J. Chem. Soc. (A)*, 1969, 372 (*synth, ir, raman*)
Johannsen, F.H. *et al*, *Z. Anorg. Allg. Chem.*, 1977, **436**, 143 (*synth*)

$CCl_5OOs^{\ominus\ominus}$ **Os-00003**

Carbonylpentachloroosmate(2−), 9CI
Pentachlorocarbonylosmate(2−)

$Cl_5Os(CO)^{\ominus\ominus}$

M 395.475 (ion)
Di-Cs salt: [56027-95-3].
CCl_5Cs_2OOs M 661.286
Orange solid.

Cleare, M.J. *et al*, *J. Chem. Soc. (A)*, 1970, 2788 (*synth, ir*)
Cleare, M.J. *et al*, *Spectrochim. Acta, Part A*, 1971, **28**, 2019 (*ir*)

$CCl_5OOs^{\ominus\ominus\ominus}$ **Os-00004**

Carbonylpentachloroosmate(3−), 8CI

$Cl_5Os(CO)^{\ominus\ominus\ominus}$

M 395.475 (ion)
Tri-Cs salt: [17568-67-1].
CCl_5Cs_3OOs M 794.192
Orange cryst.

Cleare, M.J. *et al*, *J. Chem. Soc. (A)*, 1969, 372 (*synth, ir*)

$C_2Br_4O_2Os^{\ominus}$ **Os-00005**

Tetrabromodicarbonylosmate(1−), 9CI

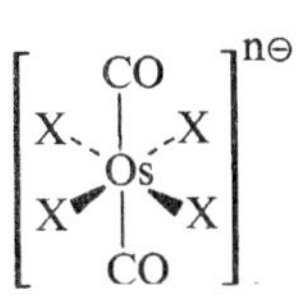

X = Br, n = 1

M 565.837 (ion)

trans-form
Tetrabutylammonium salt: [55399-52-5].
$C_{18}H_{36}Br_4NO_2Os$ M 808.304
Dark-red needles (CH_2Cl_2). Sol. Me_2CO, $CHCl_3$, CH_2Cl_2, mod. sol. EtOH.

Preetz, W. *et al*, *J. Organomet. Chem.*, 1975, **86**, 397 (*synth, ir, uv*)

$C_2Br_4O_2Os^{\ominus\ominus}$ **Os-00006**

Tetrabromodicarbonylosmate(2−)

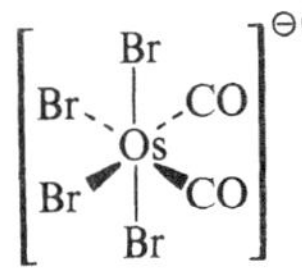

M 565.837 (ion)
Di-Cs salt:
$C_2Br_4Cs_2O_2OS$ M 689.507
Yellow cryst.
Bis(tetraethylammonium) salt: [58428-54-9].
$C_{18}H_{40}Br_4N_2O_2Os$ M 826.342
No props. reported.

Cleare, M.J. *et al*, *J. Chem. Soc. (A)*, 1969, 372 (*synth, ir, raman*)
Frickenschmidt, H.C. *et al*, *J. Organomet. Chem.*, 1978, **152**, 111 (*synth*)

$C_2Cl_4O_2Os^{\ominus\ominus}$ **Os-00007**

Dicarbonyltetrachloroosmate(2−), 9CI

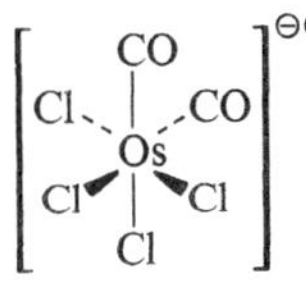

M 388.033 (ion)

cis-form
Di-Cs salt: [18400-02-7].
$C_2Cl_4Cs_2O_2Os$ M 653.844
Pale-yellow cryst.
Bis(tetraethylammonium) salt: [58428-52-7].
$C_{18}H_{40}Cl_4N_2O_2Os$ M 648.538
Light-yellow cryst.

Cleare, M.J. *et al*, *J. Chem. Soc. (A)*, 1969, 372 (*synth, ir*)
Johanssen, F.H. *et al*, *J. Organomet. Chem.*, 1975, **102**, 527 (*synth*)

$C_2I_2O_2Os$ **Os-00008**

Dicarbonyldiiodoosmium
Osmium dicarbonyl diiodide

$[I_2Os(CO)_2]_n$

M 500.030
Polymeric. Yellow cryst. ir ν_{CO} 2119, 2045, 1988 cm^{-1} (nujol).

Hales, L.A.W. *et al*, *J. Chem. Soc. (A)*, 1967, 1932.

$C_2I_4O_2Os^{\ominus}$ **Os-00009**

Dicarbonyltetraiodoosmate(1−), 9CI
As Tetrabromodicarbonylosmate(1−), Os-00005 with

X = I, n = 1

M 753.839 (ion)

***trans*-form**

Tetrabutylammonium salt: [55399-54-7].
$C_{18}H_{36}I_4NO_2S$ M 838.166
Deep-green platelets (CH_2Cl_2). Sol. Me_2CO, CH_2Cl_2, $CHCl_3$, DMF, mod. sol. EtOH.

Preetz, W. *et al*, *J. Organomet. Chem.*, 1975, **86**, 397 (*synth, ir, uv*)

$C_2I_4O_2Os^{\ominus\ominus}$ Os-00010

Dicarbonyltetraiodoosmate(2–), 9CI

As Tetrabromodicarbonylosmate(1–), Os-00005 with

X = I, n = 2

M 753.839 (ion)

***trans*-form**

Di-Cs salt: [55399-57-0].
$C_2Cs_2I_4O_2Os$ M 1019.650
Yellow-green. Insol. org. solvs., mod. sol. H_2O.

Bis(tetrabutylammonium)salt:
$C_{34}H_{72}I_4N_2O_2Os$ M 1238.773
Yellow-green needles. Sol. Me_2CO, $CHCl_3$, CH_2Cl_2, spar. sol. EtOH.

Preetz, W. *et al*, *J. Organomet. Chem.*, 1975, **86**, 397 (*synth, ir, uv*)

$C_2N_2O_4Os$ Os-00011

Dicarbonyldinitrosylosmium, 9CI

[42718-59-2]

$Os(CO)_2(NO)_2$

M 306.233

Characterised spectroscopically.

Crichton, O. *et al*, *J. Chem. Soc., Dalton Trans.*, 1973, 1321.

$C_3Br_3O_3Os^{\ominus}$ Os-00012

Tribromotricarbonylosmate(1–), 9CI

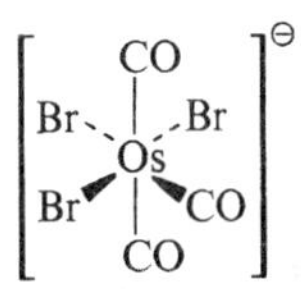

M 513.943 (ion)

Cs salt: [18400-28-7].
$C_3Br_3CsO_3Os$ M 646.849
White cryst. (EtOH).

***fac*-form**

Tetrabutylammonium salt: [58800-94-5].
$C_{19}H_{36}Br_3NO_3Os$ M 756.410
White.

***mer*-form**

Tetrabutylammonium salt: [58776-23-1]. Bright-yellow.

Cleare, M.J. *et al*, *J. Chem. Soc.* (*A*), 1969, 372 (*synth, ir, raman*)
Preetz, W. *et al*, *J. Organomet. Chem.*, 1976, **104**, 79 (*synth, ir, raman*)

$C_3Cl_3O_3Os^{\ominus}$ Os-00013

Tricarbonyltrichloroosmate(1–), 9CI

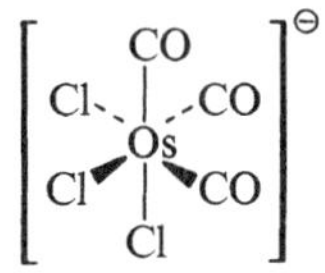

fac-form

M 380.590 (ion)

***fac*-form**

Tetrabutylammonium salt: [58800-92-3].
$C_{19}H_{36}Cl_3NO_3Os$ M 623.057
White cryst. (EtOH).

***mer*-form**

Tetrabutylammonium salt: [58755-32-1]. Yellow.

Cleare, M.J. *et al*, *J. Chem. Soc.* (*A*), 1969, 372 (*synth, ir, raman*)
Preetz, W. *et al*, *J. Organomet. Chem.*, 1976, **104**, 79 (*synth, ir, raman*)

$C_3I_3O_3Os^{\ominus}$ Os-00014

Tricarbonyltriiodoosmate(1–), 9CI

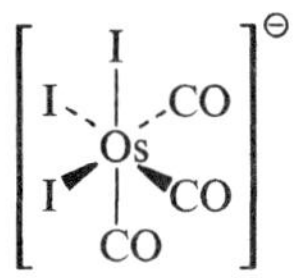

fac-form

M 654.945 (ion)

***fac*-form**

Tetrabutylammonium salt: [55399-59-2].
$C_{19}H_{36}I_3NO_3Os$ M 897.412
Bright-yellow cryst. (EtOH).

***mer*-form**

Tetrabutylammonium salt: [58776-25-3]. Yellow.

Cleare, M.J. *et al*, *J. Chem. Soc.* (*A*), 1969, 372 (*synth, ir, raman*)
Preetz, W. *et al*, *J. Organomet. Chem.*, 1975, **86**, 397; 1976, **104**, 79 (*synth, ir, raman*)

$C_4Br_2O_4Os$ Os-00015

Dibromotetracarbonylosmium, 9CI, 8CI

[14878-21-8]

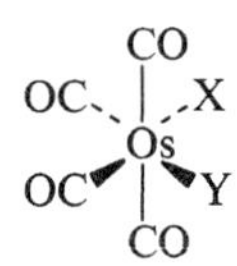

cis-form

X = Y = Br

M 462.050

***cis*-form** [18953-16-7]

Yellow cryst. Mp 317° dec.

Hales, L.A.W. *et al*, *J. Chem. Soc.* (*A*), 1967, 1389, 1932 (*synth, ir*)
L'Eplattenier, F. *et al*, *Inorg. Chem.*, 1967, **6**, 2092 (*synth, ir*)

$C_4Cl_2Hg_2O_4Os$ Os-00016

Tetracarbonylbis(chloromercurio)osmium, 8CI

[21710-55-4]

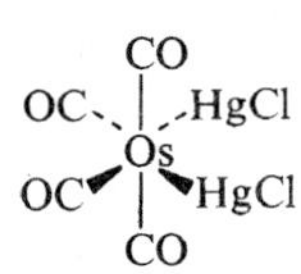

M 774.328

White solid. Mp ca. 235° dec.

Bradford, C.W. *et al*, *J. Chem. Soc. (A)*, 1968, 2456 (*synth, ir, raman*)

$C_4Cl_2O_4Os$ Os-00017

Tetracarbonyldichloroosmium, 9CI

Dichlorotetracarbonylosmium

[22873-63-8]

As Dibromotetracarbonylosmium, Os-00015 with

X = Y = Cl

M 373.148

trans-form [42230-39-7]

Colourless cryst. (CH_2Cl_2). Mp 300°.

L'Eplattenier, F. *et al*, *Inorg. Chem.*, 1967, **6**, 2092 (*synth, ir*)
Collman, J.P. *et al*, *Inorg. Chem.*, 1974, **13**, 1 (*synth, ir*)
Tyler, D.R. *et al*, *J. Am. Chem. Soc.*, 1980, **102**, 3022 (*synth*)

$C_4HO_4Os^{\ominus}$ Os-00018

Tetracarbonylhydroosmate(1–)

$HOs(CO)_4^{\ominus}$

M 303.250 (ion)

Identity not certain, prepns. by different routes have different props.

Na salt: [22533-60-4].

C_4HNaO_4Os M 326.239

Strong nucleophile. Cream air-stable solid. Sol. H_2O.

L'Eplattenier, F., *Inorg. Chem.*, 1969, **8**, 965 (*synth, ir*)
George, R.D. *et al*, *J. Chem. Soc., Dalton Trans.*, 1973, 972 (*synth, ir*)

$C_4H_2O_4Os$ Os-00019

Tetracarbonyldihydroosmium, 9CI, 8CI

[22372-70-9]

As Dibromotetracarbonylosmium, Os-00015 with

X = Y = H

M 304.257

Colourless air-stable liq. Mp –40° to –38°. pK_a 12.8 (MeOH, 25°).

Muetterties, E.L., *Transition Metal Hydrides*, Marcel Dekker, New York, 1971.
Kaesz, H.D. *et al*, *Chem. Rev.*, 1972, **72**, 231; *Prog. Inorg. Chem.*, 1979, **25**, 145 (*rev*)
Collman, J.P. *et al*, *Inorg. Chem.*, 1974, **13**, 1 (*synth, ir*)
Moss, J.R. *et al*, *Inorg. Chem.*, 1977, **16**, 75 (*synth*)
Vancea, L. *et al*, *J. Organomet. Chem.*, 1977, **134**, 219 (*cmr*)
Walker, H.W. *et al*, *J. Am. Chem. Soc.*, 1979, **101**, 7428.
Clark, W.J. *et al*, *Inorg. Chem.*, 1982, **21**, 3955 (*synth, ir, pmr*)
Nicholls, J.N. *et al*, *J. Chem. Soc., Dalton Trans.*, 1982, 1395 (*synth, ir*)

Having problems with locating a compound? Have you checked the indexes?

$C_4H_9NO_3Os$ Os-00020

Trioxo(*tert*-butylnitrido)osmium

[2-Methyl-2-propanaminato(2-)]trioxoosmium, 9CI. *Trioxo(*tert*-butylimido)osmium.* N-tert-*Butyl osmiamate*

[50381-48-1]

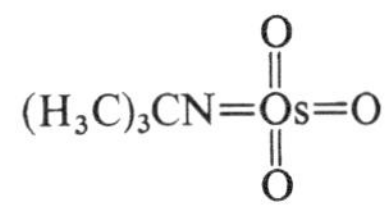

M 309.320

Reacts with a variety of alkenes to give, after reductive cleavage of the intermed. osmate esters, vicinal amino alcohols. Bright-yellow needles (pentane at –10°). Mp 112° dec. Sublimes.

Milas, N.A. *et al*, *J. Am. Chem. Soc.*, 1959, **81**, 6089 (*synth*)
Clifford, A.F. *et al*, *Inorg. Synth.*, 1960, **6**, 204 (*synth*)
Miskowski, V. *et al*, *Mol. Phys.*, 1974, **28**, 747 (*uv*)
Chong, A.O. *et al*, *J. Am. Chem. Soc.*, 1977, **99**, 3420 (*synth, use*)
Patrick, D.W. *et al*, *J. Org. Chem.*, 1978, **43**, 2628 (*synth, use*)
Hentges, S.G. *et al*, *J. Org. Chem.*, 1980, **45**, 2257 (*use*)
Fieser, M. *et al*, *Reagents for Organic Synthesis*, Wiley, 1967-83, **6**, 641 (*use*)

$C_4I_2O_4Os$ Os-00021

Tetracarbonyldiiodoosmium

[17685-65-3]

As Dibromotetracarbonylosmium, Os-00015 with

X = Y = I

M 556.051

cis-form [17632-05-2]

Yellow cryst.

trans-form [25340-26-5]

Yellow cryst.

L'Eplattenier, F. *et al*, *Inorg. Chem.*, 1967, **6**, 2092 (*synth, ir*)
Pankowski, M. *et al*, *J. Organomet. Chem.*, 1969, **19**, 393 (*synth, ir*)
Vancea, L. *et al*, *J. Organomet. Chem.*, 1977, **134**, 219 (*synth, cmr*)

$C_4O_4Os^{\ominus\ominus}$ Os-00022

Tetracarbonylosmate(2–), 9CI

$Os(CO)_4^{\ominus\ominus}$

M 302.242 (ion)

Usually generated and used *in situ*.

Di-Na salt: [27857-55-2].

$C_4H_4Na_2Os$ M 288.255

Air-sensitive cream cryst.

Collman, J.P. *et al*, *Inorg. Chem.*, 1974, **13**, 1 (*synth*)
Clark, W.J. *et al*, *Inorg. Chem.*, 1982, **21**, 3955 (*synth*)

$C_5HO_5Os^{\oplus}$ Os-00023

Pentacarbonylhydroosmium(1+), 8CI

$HOs(CO)_5^{\oplus}$

M 331.260 (ion)

Hexafluorophosphate: [30306-42-4].

$C_5HF_6O_5OsP$ M 476.224

White solid.

Lewis, J. *et al*, *J. Chem. Soc. (A)*, 1970, 2967 (*synth, ir, pmr*)

$C_5H_3BrO_4Os$ Os-00024

Bromotetracarbonylmethylosmium, 9CI, 8CI

As Dibromotetracarbonylosmium, Os-00015 with

X = Me, Y = Br

M 397.180

cis-form [27659-99-0]

White needles (pentane).

L'Eplattenier, F. *et al*, *Helv. Chim. Acta*, 1970, **53**, 1091 (*synth, ir, pmr*)
Kelland, J.W. *et al*, *J. Organomet. Chem.*, 1978, **149**, 185 (*synth*)

$C_5H_3IO_4Os$ Os-00025

Tetracarbonyliodomethylosmium, 9CI, 8CI

As Dibromotetracarbonylosmium, Os-00015 with

X = Me, Y = I

M 444.181

cis-form [22605-82-9]

Colourless cryst. (heptane). Mp 105-6° dec.

L'Eplattenier, F. *et al*, *Inorg. Chem.*, 1969, **8**, 965 (*synth, ir, pmr*)
Kelland, J.W. *et al*, *J. Organomet. Chem.*, 1978, **149**, 185 (*synth*)

$C_5H_4O_4Os$ Os-00026

Tetracarbonylhydromethylosmium, 9CI

As Dibromotetracarbonylosmium, Os-00015 with

X = Me, Y = H

M 318.284

cis-form [22639-03-8]

Colourless. Mp <25°.

Evans, J. *et al*, *J. Am. Chem. Soc.*, 1976, **98**, 4000 (*synth*)
Clark, W.J. *et al*, *Inorg. Chem.*, 1982, **21**, 3955 (*synth, ir, pmr*)

C_5O_5Os Os-00027

Pentacarbonylosmium

Osmium carbonyl, 9CI. *Osmium pentacarbonyl*

[16406-49-8]

$Os(CO)_5$

M 330.252

Colorless to pale-yellow cryst. melting to a mobile yellow liq. Mp 2-2.5°.

Rushman, P. *et al*, *Organometallics*, 1983, **2**, 693 (*synth, ir, cmr*)

$C_6Br_2O_6Os_2$ Os-00028

Di-μ-bromohexacarbonyldiosmium

[62793-34-4]

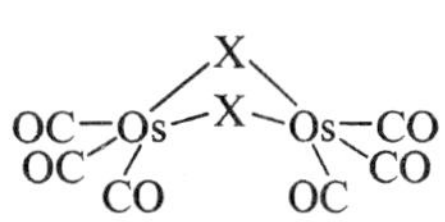

X = Br

M 708.270

Yellow prisms. Mp 102-5°. $Bp_{0.01}$ 60° subl.

Moss, J.R. *et al*, *J. Chem. Soc., Dalton Trans.*, 1977, 89 (*synth, ir, ms*)

$C_6Br_4O_6Os_2$ Os-00029

Di-μ-bromodibromohexacarbonyldiosmium

Tetrabromohexacarbonyldiosmium

[17731-88-3]

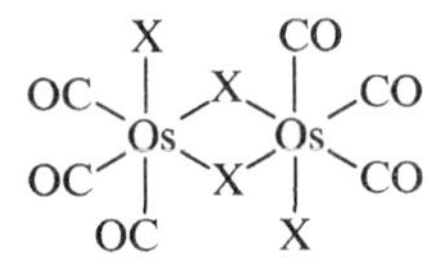

X = Br

M 868.078

White cryst. (CH_2Cl_2). Mp 315°. Bp 190° subl. *in vacuo*.

Monomer: Dibromotricarbonylosmium.

$C_3Br_2O_3Os$ M 434.039

Unknown.

Hales, L.A.W. *et al*, *J. Chem. Soc. (A)*, 1967, 1932 (*synth, ir*)
Choi, H.W. *et al*, *Inorg. Chem.*, 1981, **20**, 2664 (*synth, ms, ir*)

$C_6Cl_2O_6Os_2$ Os-00030

Hexacarbonyldi-μ-chlorodiosmium, 10CI

[62976-02-7]

As Di-μ-bromohexacarbonyldiosmium, Os-00028 with

X = Cl

M 619.368

Yellow prisms. Mp 82° dec. $Bp_{0.01}$ 60° subl.

Moss, J.R. *et al*, *J. Chem. Soc., Dalton Trans.*, 1977, 89 (*synth, ir, ms*)

$C_6Cl_4O_6Os_2$ Os-00031

Hexacarbonyldi-μ-chlorodichlorodiosmium, 9CI

Hexacarbonyltetrachlorodiosmium

[17633-40-8]

As Di-μ-bromodibromohexacarbonyldiosmium, Os-00029 with

X = Cl

M 690.274

Two isomers known which are prob. structural isomers, not dimorphs.

α-form

Colourless cryst. (C_6H_6 or CCl_4). Mp 266°. In $CHCl_3$ soln. isomerises to β-form.

β-form

Colourless cryst. ($CHCl_3$). Mp 285°.

Monomer: Tricarbonydichloroosmium.

$C_3Cl_2O_3Os$ M 345.137

Unknown.

Hales, L.A.W. *et al*, *J. Chem. Soc. (A)*, 1967, 1932 (*synth, ir*)
Bruce, M.I. *et al*, *J. Chem. Soc. (A)*, 1969, 987 (*synth, ir*)
Cleare, M.J. *et al*, *J. Chem. Soc. (A)*, 1969, 372 (*synth, ir, raman*)

$C_6H_4O_4Os$ Os-00032

Tetracarbonyl(η^2-ethene)osmium, 10CI

Tetracarbonyl(ethylene)osmium

[68569-22-2]

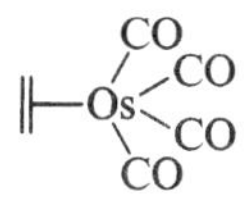

M 330.295

No phys. props. reported.

Norton, J.R. *et al*, *Adv. Chem. Ser.*, 1978, **167**, 170.

$C_6H_5BrO_4Os$ Os-00033

Bromotetracarbonylethylosmium, 8CI

[27882-35-5]

As Dibromotetracarbonylosmium, Os-00015 with

X = Et, Y = Br

M 411.207

No phys. props. reported.

L'Eplattenier, F. *et al*, *Helv. Chim. Acta*, 1970, **53**, 1091 (*synth, ir*)

$C_6H_6Br_2Os$ Os-00034

η^6-Benzenedibromoosmium

π-Benzeneosmium dibromide

[75701-03-0]

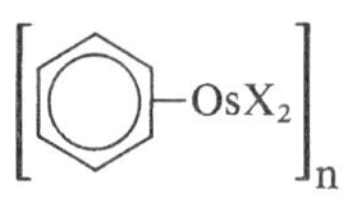

X = Br

M 428.121

Dimeric or polymeric. Yellow.

Hung, Y. *et al*, *Inorg. Chem.*, 1981, **20**, 457 (*synth*)

$C_6H_6Br_2OsO_4^{\ominus\ominus}$ Os-00035

Diacetyldibromodicarbonylosmate(2−), 8CI

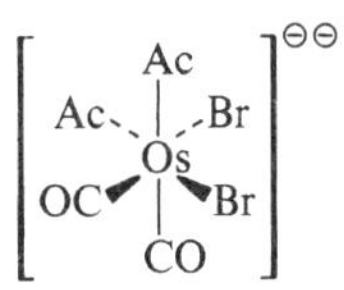

M 492.119 (ion)

Bis(bromomagnesium) salt: [23254-68-4].

$C_6H_6Br_4Mg_2OsO_4$ M 700.537

Colourless cryst. + 5THF (THF/heptane).

L'Eplattenier, F., *Inorg. Chem.*, 1969, **8**, 965 (*synth, ir, pmr*)

$C_6H_6Cl_2Os$ Os-00036

η^6-Benzenedichloroosmium

π-Benzeneosmium dichloride

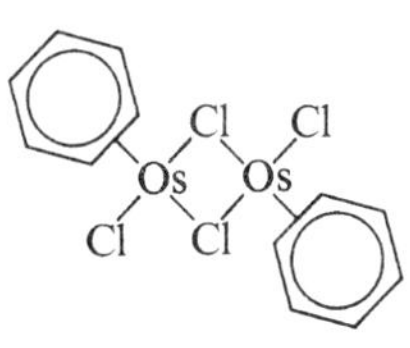

Dimer

M 339.219

Dimeric or polymeric.

Dimer: [69462-16-4]. *Bis(η^6-benzene)di-μ-chlorodichlorodiosmium,* 10CI. *Bis(η^6-benzene)tetrachlorodiosmium.*

$C_{12}H_{12}Cl_4Os_2$ M 678.439

Yellow cryst. solid (MeCN/Et$_2$O). Mp 164-6°, 206-9° dec.

MeCN complex: [72765-28-7]. *(Acetonitrile)(η^6-benzene)dichloroosmium.*

$C_8H_9Cl_2NO_5$ M 270.069

Yellow cryst. (MeCN/Et$_2$O). Reported in one ref. to be formed under identical conditions to the uncomplexed dimer.

Winkhaus, G. *et al*, *Z. Naturforsch., B*, 1966, **21**, 1109 (*synth, ir, pmr*)
Rubezhov, A.Z. *et al*, *Izv. Akad. Nauk SSSR, Ser. Khim.*, 1974, 1902 (*synth, pmr, ir*)
Bennett, M.A. *et al*, *Inorg. Chem.*, 1980, **19**, 1014 (*synth*)
Arthur, T. *et al*, *J. Organomet. Chem.*, 1981, **208**, 369 (*synth*)
Hung, Y. *et al*, *Inorg. Chem.*, 1981, **20**, 457 (*synth, pmr, uv, ir*)

$C_6H_6I_2Os$ Os-00037

(η^6-Benzene)diiodoosmium

(π-Benzene)osmium diiodide

[75701-04-1]

As η^6-Benzenedibromoosmium, Os-00034 with

X = I

M 522.122

Dimeric or polymeric. Red substance. Sol. DMF, DMSO.

Winkhaus, G. *et al*, *Z. Naturforsch., B*, 1966, **21**, 1109 (*synth, pmr, ir*)
Hung, Y. *et al*, *Inorg. Chem.*, 1981, **20**, 457 (*synth*)

$C_6H_6O_4Os$ Os-00038

Tetracarbonyldimethylosmium, 9CI, 8CI

[59599-35-8]

As Dibromotetracarbonylosmium, Os-00015 with

X = Y = Me

M 332.311

***cis*-form** [22639-01-6]

Colourless prisms. Mp 65-6°.

Vancea, L. *et al*, *J. Organomet. Chem.*, 1977, **134**, 219 (*cmr*)
Jordan, R.F. *et al*, *J. Am. Chem. Soc.*, 1979, **101**, 4853 (*cmr*)
Clark, W.J. *et al*, *Inorg. Chem.*, 1982, **21**, 3955 (*synth, ir, pmr*)

$C_6I_2O_6Os_2$ Os-00039

Hexacarbonyldi-μ-iododiosmium

[22391-77-1]

As Di-μ-bromohexacarbonyldiosmium, Os-00028 with

X = I

M 802.271

Characterised spectroscopically.

Bruce, M.I. *et al*, *J. Chem. Soc. (A)*, 1969, 987 (*synth, ir, ms*)

$C_6I_4O_6Os_2$ Os-00040

Hexacarbonyldi-μ-iododiiododiosmium

[17632-08-5]

As Di-μ-bromodibromohexacarbonyldiosmium, Os-00029 with

X = I

M 1056.080
Yellow cryst. (C_6H_6).

Hales, L.A.W. *et al*, *J. Chem. Soc.* (*A*), 1967, 1932 (*synth*, *uv*)

$C_7H_5BrO_2Os$ Os-00041

Bromodicarbonyl(η^5-2,4-cyclopentadien-1-yl)osmium

η^5-Cyclopentadienylbromodicarbonylosmium. Cyclopentadienyldicarbonylosmium bromide

[81554-88-3]

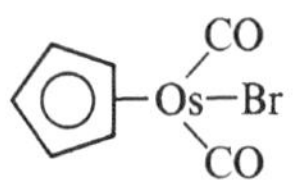

M 391.219
Pale-yellow cryst. (hexane/CH_2Cl_2). Mp 120-2°.

Fischer, E.O. *et al*, *Z. Naturforsch*, *B*, 1962, **17**, 274 (*synth*)
Hoyano, J.K. *et al*, *Inorg. Chem.*, 1982, **21**, 3095 (*synth*, *ir*, *pmr*, *ms*)

$C_7H_5ClO_2Os$ Os-00042

Dicarbonylchloro(η^5-2,4-cyclopentadien-1-yl)osmium, 10CI

Cyclopentadienyldicarbonylchloroosmium. η^5-Cyclopentadienyldicarbonylosmium chloride

[72768-82-2]

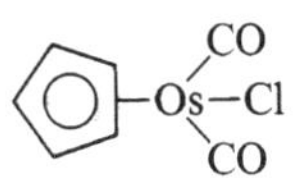

M 346.768

Behrens, H. *et al*, *Z. Naturforsch.*, *B*, 1979, **34**, 1477 (*synth*, *ir*, *pmr*)

$C_7H_5IO_2Os$ Os-00043

Dicarbonyl(η^5-2,4-cyclopentadien-1-yl)iodoosmium

Cyclopentadienylosmiumdicarbonyl iodide. (η^5-Cyclopentadienyl)dicarbonyliodoosmium

[81554-97-4]

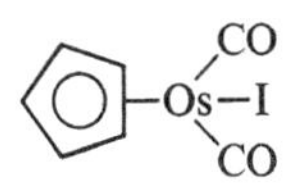

M 438.220
Pale-yellow cryst. (CH_2Cl_2/heptane). Mp 141-2°.

Hoyano, J.K. *et al*, *Inorg. Chem.*, 1982, **21**, 3095 (*synth*, *ir*, *pmr*, *ms*)

$C_7H_6Cl_2OOs$ Os-00044

(η^6-Benzene)carbonyldichloroosmium, 10CI

[75700-97-9]

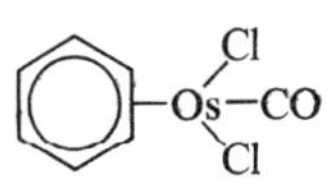

M 367.230
Orange needles (MeOH aq.).

Hung, Y. *et al*, *Inorg. Chem.*, 1981, **20**, 457 (*synth*)

$C_7H_6O_2Os$ Os-00045

Dicarbonyl(η^5-2,4-cyclopentadien-1-yl)hydroosmium, 9CI

Dicarbonyl-π-cyclopentadienylhydridoosmium. Hydrodicarbonyl-π-cyclopentadienylosmium. Cyclopentadienyldicarbonylosmium hydride

[42442-16-0]

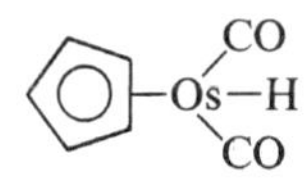

M 312.323
Pale-yellow or colourless thermally stable oil. Mp ca. 15°. $Bp_{0.01}$ 25°.

Humphries, A.P. *et al*, *J. Chem. Soc.*, *Dalton Trans.*, 1975, 1710 (*synth*, *ir*, *pmr*)
Hoyano, J.K. *et al*, *Inorg. Chem.*, 1982, **21**, 3095 (*synth*, *pmr*, *ir*, *ms*)

$C_7H_6O_3Os$ Os-00046

(η^4-1,3-Butadiene)tricarbonylosmium

[75756-64-8]

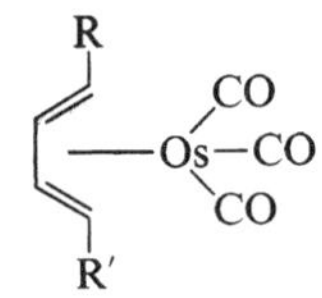

R = R′ = H

M 328.323
Colourless solid. Mp 22°.

Zobl-Ruh, S. *et al*, *Helv. Chim. Acta*, 1980, **63**, 773 (*synth*, *pmr*, *cmr*, *ir*, *ms*)

$C_7H_9BrGeO_4Os$ Os-00047

Bromotetracarbonyl(trimethylgermyl)osmium, 8CI

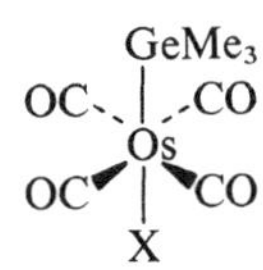

X = Br

M 499.840

***trans*-form** [30846-59-4]
Pale-yellow cryst. Mp 89-91°.

Stone, F.G.A. *et al*, *J. Chem. Soc.* (*A*), 1971, 458 (*synth*, *ir*, *pmr*)

$C_7H_9BrO_4OsSi$ Os-00048

Bromotetracarbonyl(trimethylsilyl)osmium, 8CI

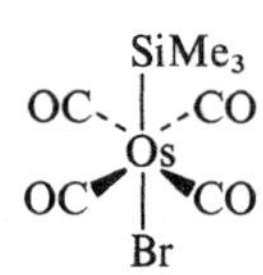

M 455.335

***trans*-form** [30848-52-3]
Pale-yellow cryst. Mp 100-1°.

Stone, F.G.A. *et al*, *J. Chem. Soc.* (*A*), 1971, 458 (*synth*, *ir*, *pmr*)

$C_7H_9GeIO_4Os$ Os-00049

Tetracarbonyliodo(trimethylgermyl)osmium, 8CI

As Bromotetracarbonyl(trimethylgermyl)osmium, Os-00047 with

X = I

M 546.840

trans-form [30846-60-7]

Yellow cryst.

Stone, F.G.A. *et al*, *J. Chem. Soc.* (*A*), 1971, 458 (*synth*, *ir*)

$C_7H_9IO_4OsSi$ Os-00050

Tetracarbonyliodo(trimethylsilyl)osmium, 8CI

[30848-53-4]

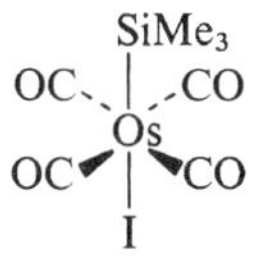

M 502.336

Stable pale-yellow cryst. Mp 94-6°. Dimerises with loss of CO on boiling in hexane.

Ash, M.J. *et al*, *J. Chem. Soc.* (*A*), 1971, 458 (*synth*, *ir*, *pmr*)

$C_7H_{10}GeO_4Os$ Os-00051

Tetracarbonylhydro(trimethylgermyl)osmium, 8CI

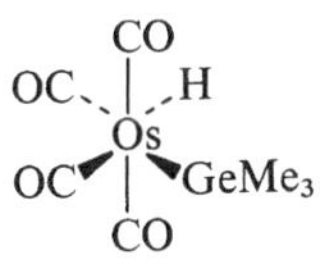

M 420.944

cis-form [34307-28-3]

Colourless liq. $Bp_{0.01}$ ca 20°.

Knox, S.A.R. *et al*, *J. Chem. Soc.* (*A*), 1971, 2874 (*synth*, *ir*, *pmr*)

$C_7H_{10}O_4OsSi$ Os-00052

Tetracarbonylhydro(trimethylsilyl)osmium, 8CI

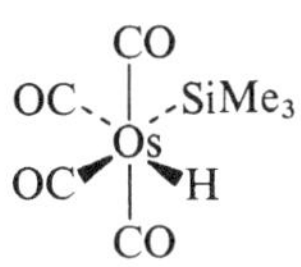

M 376.439

cis-form [25930-76-1]

Colourless liq. Mp −55°. $Bp_{0.01}$ 20°.

Knox, S.A.R. *et al*, *J. Chem. Soc.* (*A*), 1970, 3147 (*synth*, *ir*)

$C_8Br_2O_8Os_2$ Os-00053

Dibromooctacarbonyldiosmium, 8CI

[25685-07-8]

$BrOs(CO)_4$—$Os(CO)_4Br$ (*diax*)

M 764.291

Pale-yellow prisms (heptane). Mp 121-6°.

Moss, J.R. *et al*, *J. Chem. Soc.*, *Dalton Trans.*, 1979, 89 (*synth*, *ir*, *ms*)

$C_8Cl_2O_8Os_2$ Os-00054

Octacarbonyldichlorodiosmium, 8CI

[25685-06-7]

$ClOs(CO)_4$—$Os(CO)_4Cl$ (*diax*)

M 675.389

V. pale-yellow prisms (heptane). Mp 118° dec.

Moss, J.R. *et al*, *Inorg. Chem.*, 1977, **16**, 75.

Moss, J.R. *et al*, *J. Chem. Soc.*, *Dalton Trans.*, 1977, 89 (*synth*, *ir*, *ms*)

$C_8H_2O_7Os_3S_3$ Os-00055

(Carbonothioyl)heptacarbonyldi-μ-hydrodi-μ_3-thioxotriosmium, 10CI

[73555-60-9]

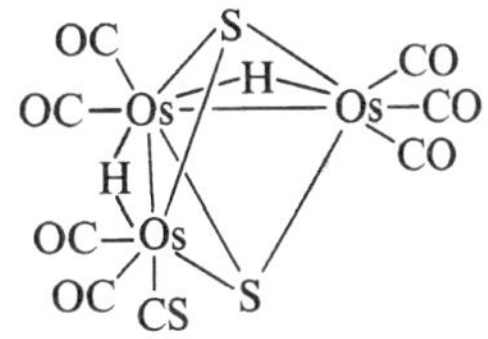

M 876.880

Amber cryst. (CH_2Cl_2/hexane).

Broadhurst, P.V. *et al*, *J. Organomet. Chem.*, 1980, **187**, 141 (*synth*, *pmr*, *struct*)

$C_8H_2O_8Os_2$ Os-00056

Octacarbonyldihydrodiosmium, 8CI

[25685-05-6]

$(OC)_4OsH$—$OsH(CO)_4$

M 606.499

Colourless oil. $Bp_{0.01}$ 40°.

Moss, J.R. *et al*, *J. Chem. Soc.*, *Chem. Commun.*, 1969, 800; *Inorg. Chem.*, 1977, **16**, 75 (*synth*, *ir*, *pmr*, *ms*)

$C_8H_3O_8Os_3S^{\ominus}$ Os-00057

Octacarbonyldi-μ-hydrohydro-μ_3-thioxotriosmate(1−)

Octacarbonyltrihydro-μ_3-thioxotrisomate(1−)

[74325-98-7]

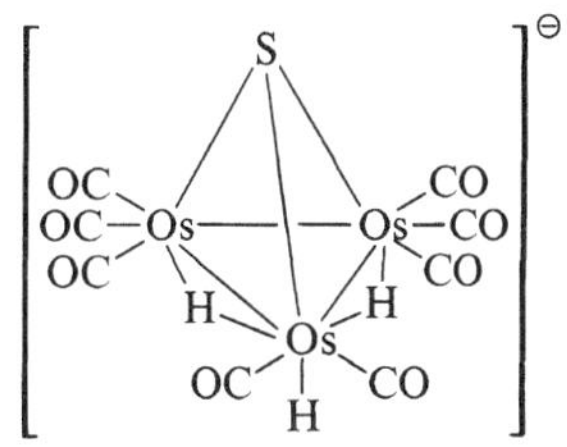

M 829.767 (ion)

No phys. props. reported.

Johnson, B.F.G. *et al*, *J. Organomet. Chem.*, 1980, **190**, C91 (*synth*, *pmr*)

$C_8H_5O_3Os^{\oplus}$ Os-00058

Tricarbonyl(η^5-2,4-cyclopentadien-1-yl)osmium(1+)

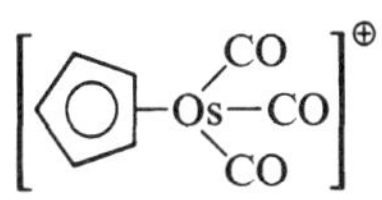

M 339.326 (ion)

Tetrafluoroborate: [81554-85-0].

$C_8H_5BF_4O_3Os$ M 426.129

Colourless cryst. (Me_2CO/Et_2O). Dec. >315°.

Behrens, H. *et al*, *Z. Naturforsch.*, *B*, 1979, **34**, 1477 (*synth*, *ir*, *pmr*)

Hoyano, J.K. *et al*, *Inorg. Chem.*, 1982, **21**, 3095 (*synth*, *ir*, *pmr*)

$C_8H_6O_3Os$ Os-00059

Tricarbonyl[(1,2,3,4-η)-1,3-cyclopentadiene]osmium, 9CI
[57512-16-0]

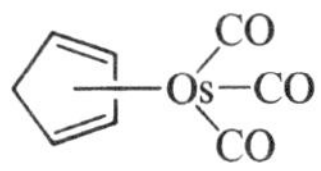

M 340.334
Thermally unstable yellow oil, dec. by air. $Bp_{0.01}$ 25°.

Humphries, A.P. *et al*, *J. Chem. Soc., Dalton Trans.*, 1975, 1710 (*synth, ir, ms*)

$C_8H_7NO_3Os$ Os-00060

(Aminocarbonyl)dicarbonyl(η⁵-2,4-cyclopentadien-1-yl)osmium, 10CI
[72763-49-6]

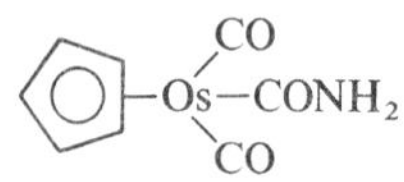

M 355.348
Colourless solid. Mp 120°.

Behrens, H. *et al*, *Z. Naturforsch., B*, 1979, **34**, 1477 (*synth, pmr, ms, ir*)

$C_8H_8O_3Os$ Os-00061

Tricarbonyl[(1,2,3,4-η)-1,3-pentadiene]osmium, 10CI
As (η⁴-1,3-Butadiene)tricarbonylosmium, Os-00046 with

R = CH_3, R′ = H

M 342.349
***syn*-form** [75756-66-0]
Colourless.

Zobl-Ruh, S. *et al*, *Helv. Chim. Acta*, 1980, **63**, 773 (*synth, cmr, ms*)

$C_8H_{10}O_4Os$ Os-00062

Tetracarbonyldiethylosmium, 8CI
As Dibromotetracarbonylosmium, Os-00015 with

X = Y = Et

M 360.365
***cis*-form** [27882-33-3]
Colourless liq.

Clark, W.J. *et al*, *Inorg. Chem.*, 1982, **21**, 3955 (*synth, ir, pmr*)

$C_8H_{12}Br_2N_4Os$ Os-00063

Dibromotetrakis(isocyanomethane)osmium, 9CI
Dibromotetrakis(methyl isocyanide)osmium

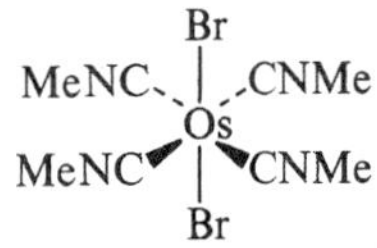

M 514.218
***trans*-form** [42608-08-2]
Orange prisms (EtOH). Mp >300°.

Chatt, J. *et al*, *J. Chem. Soc., Dalton Trans.*, 1973, 1433 (*synth, spectra*)

$C_8H_{12}Cl_2Os$ Os-00064

Dichloro[(1,2,5,6-η)-1,5-cyclooctadiene]osmium, 10CI
[39395-16-9]

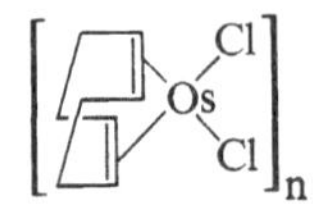

M 369.289
Polymeric. Yellow-brown solid.

Winkhaus, G. *et al*, *Z. Naturforsch., B*, 1966, **21**, 1109 (*synth*)
Schrock, R.R. *et al*, *J. Chem. Soc., Dalton Trans.*, 1974, 951 (*synth*)

$C_8H_{14}ClN_2Os^{\oplus}$ Os-00065

(η⁶-Benzene)chloro(1,2-ethanediamine-*N,N′*)osmium(1+), 10CI

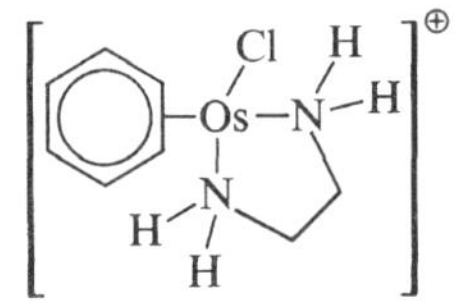

M 363.865 (ion)
Hexafluorophosphate: [75700-75-3].
$C_8H_{14}ClF_6N_2OsP$ M 508.829
Yellow cryst.

Hung, Y. *et al*, *Inorg. Chem.*, 1981, **20**, 457 (*synth, uv*)

$C_8H_{17}N_3Os^{\oplus\oplus}$ Os-00066

Ammine(η⁶-benzene)(1,2-ethanediamine-*N,N′*)osmium(2+), 10CI

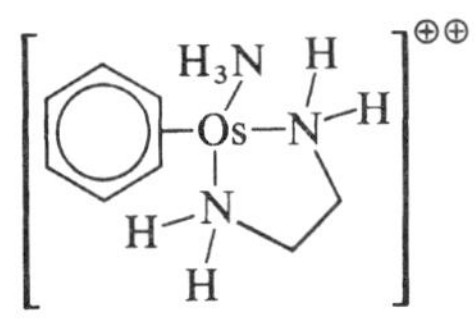

M 345.442 (ion)
Bis(hexafluorophosphate): [75700-79-7].
$C_8H_{17}F_{12}N_3OsP_2$ M 635.371
No phys. props. reported.

Hung, Y. *et al*, *Inorg. Chem.*, 1981, **20**, 457 (*synth, uv*)

$C_8H_{18}N_2O_2Os$ Os-00067

Bis[2-methyl-2-propanaminoato(2−)]dioxoosmium, 9CI
Dioxobis(tert-butylimido)osmium
[63174-13-0]

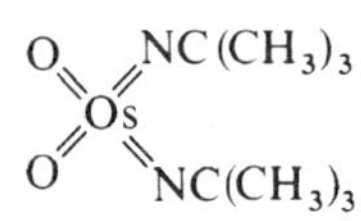

M 364.442
Reagent for the stereospecific vicinal diamination of mono- and *trans*-disubstitued alkenes. Yellow cryst. Mp 119-20°.

Chong, A.O. *et al*, *J. Am. Chem. Soc.*, 1977, **99**, 3420 (*synth, use*)
Nugent, W.A. *et al*, *J. Am. Chem. Soc.*, 1979, **101**, 7265 (*cryst struct*)
Fieser, M. *et al*, *Reagents for Organic Synthesis*, Wiley, 1967-83, **8**, 206 (*use*)

$C_8I_2O_8Os_2$ **Os-00068**

Octacarbonyldiiododiosmium

[22587-71-9]

$IOs(CO)_4$—$Os(CO)_4I$ (*diax*)

M 858.292

Characterised spectroscopically.

Bruce, M.I. *et al*, *J. Chem. Soc.* (*A*), 1969, 987 (*synth, ir*)

$C_9HO_9Os_3S^{\ominus}$ **Os-00069**

Nonacarbonyl-μ-hydro-μ_3-thioxotriosmate(1–)

[68796-19-0]

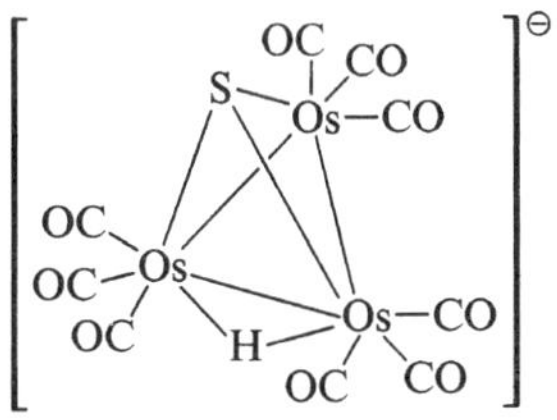

M 855.762 (ion)

Bis(triphenylphosphine)iminium salt: [69085-99-0].

$C_{45}H_{31}NO_9Os_3P_2S$ M 1394.349

Yellow.

Johnson, B.F.G. *et al*, *Acta Crystallogr., Sect. B*, 1978, **34**, 3767 (*cryst struct*)

Johnson, B.F.G. *et al*, *J. Organomet. Chem.*, 1981, **213**, 249 (*synth, ir, pmr*)

$C_9H_2O_9Os_3S$ **Os-00070**

Nonacarbonyldi-μ-hydro-μ_3-thioxotriosmium

*1,1,1,2,2,2,3,3,3-Nonacarbonyl-1,2;1,3-di-μ-hydrido-μ-3-sulfido-*triangulo-*triosmium*

[38979-82-7]

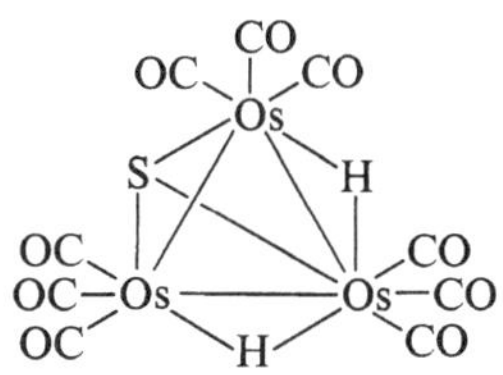

M 856.769

Cryst. (CH_2Cl_2).

Forster, A. *et al*, *J. Organomet. Chem.*, 1976, **104**, 225 (*pmr, cmr*)

Johnson, B.F.G. *et al*, *J. Chem. Soc., Dalton Trans.*, 1979, 616 (*struct*)

$C_9H_3BrO_8Os_2$ **Os-00071**

Bromooctacarbonylmethyldiosmium, 10CI

[66705-59-7]

$BrOs(CO)_4$—$Os(CO)_4Me$

M 699.422

Pale-yellow air-stable cryst. Corresponding chloro and iodo compds. also known.

Kelland, J.W. *et al*, *J. Organomet. Chem.*, 1978, **149**, 185 (*synth, ir, pmr*)

$C_9H_7NO_4OsS$ **Os-00072**

Tetracarbonyl[3,4-dimethyl-2(3*H*)-thiazolylidene]osmium, 9CI

[57512-74-0]

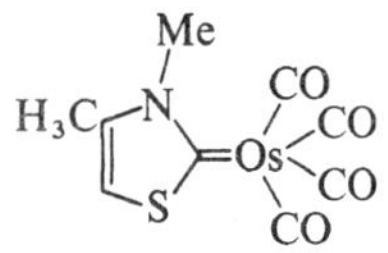

M 415.419

Yellow cryst. (CH_2Cl_2/hexane). Mp 97-100°.

Green, M. *et al*, *J. Chem. Soc., Dalton Trans.*, 1975, 939 (*synth, ir, pmr*)

$C_9H_7O_3Os^{\oplus}$ **Os-00073**

Tricarbonyl[(1,2,3,4,5-η)-2,4-cyclohexadien-1-yl]osmium(1+), 10CI

[46134-89-8]

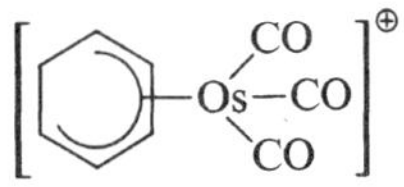

M 353.353 (ion)

Tetrafluoroborate: [33678-05-6].

$C_9H_7BF_4O_3Os$ M 440.156

White solid.

Kane-Maguire, L.A.P., *J. Chem. Soc.* (*A*), 1971, 1602 (*synth, ir*)

Bryan, E.G. *et al*, *J. Organomet. Chem.*, 1977, **129**, C19 (*synth*)

Burrows, A.L. *et al*, *J. Organomet. Chem.*, 1980, **194**, C11 (*props*)

$C_9H_8NO_2Os^{\oplus}$ **Os-00074**

(Acetonitrile)dicarbonyl(η^5-2,4-cyclopentadien-1-yl)osmium(1+)

Dicarbonyl(cyanomethane)(cyclopentadienyl)osmium(1+)

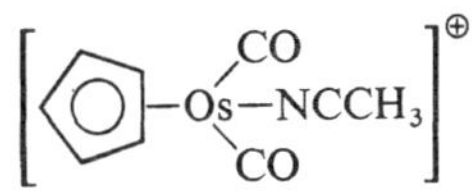

M 352.368 (ion)

Tetrafluoroborate: [81555-04-6].

$C_9H_8BF_4NO_2Os$ M 439.171

Off-white cryst. (Me_2CO/Et_2O). Dec. at 125-6°.

Hoyano, J.K. *et al*, *Inorg. Chem.*, 1982, **21**, 3095 (*synth, ir, pmr*)

$C_9H_8O_3Os$ **Os-00075**

Tricarbonyl[(1,2,3,4-η)-1,3-cyclohexadiene]osmium, 10CI

[63166-12-1]

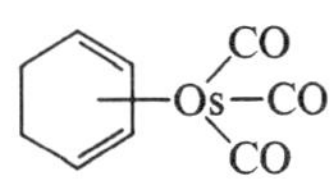

M 354.360

Characterised spectroscopically.

Burrows, A.L. *et al*, *J. Organomet. Chem.*, 1980, **194**, C11 (*synth, pmr, ir*)

$C_9H_8O_3Os$ Os-00076

Tricarbonyl[(1,3,4,5-η)-4-cyclohexene-1,3-diyl]osmium, 10CI

[75535-78-3]

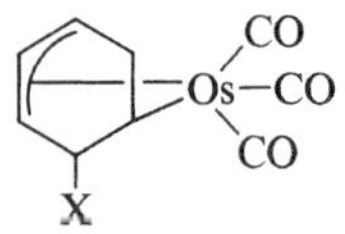

X = H

M 354.360

Characterised spectroscopically. Not separated from Tricarbonyl[(1,2,3,4-η)-1,3-cyclohexadiene]osmium, Os-00075, isomerises to it at 80°.

Burrows, A.L. *et al*, *J. Organomet. Chem.*, 1980, **194**, C11 (*synth*, *pmr*, *ir*)

$C_9H_9BrO_2Os$ Os-00077

Bromodicarbonyl[(1,2,3,4,5-η)-2,4-cycloheptadien-1-yl]osmium, 9CI

[59423-95-9]

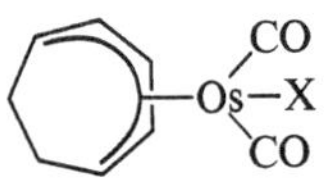

X = Br

M 419.273

Pale-yellow cryst. (CH_2Cl_2/hexane). Mp 114°.

Knox, S.A.R. *et al*, *J. Chem. Soc., Dalton Trans.*, 1976, 552 (*synth*, *ir*, *pmr*)

$C_9H_9IO_2Os$ Os-00078

Dicarbonyl[(1,2,3,4,5-η)-2,4-cycloheptadien-1-yl]iodoosmium, 9CI

[59423-90-4]

As Bromodicarbonyl[(1,2,3,4,5-η)-2,4-cycloheptadien-1-yl]osmium, Os-00077 with

X = I

M 466.273

Yellow cryst. (CH_2Cl_2/hexane). Mp 122-4°.

Knox, S.A.R. *et al*, *J. Chem. Soc., Dalton Trans.*, 1976, 552 (*synth*, *ir*, *pmr*)

$C_9H_{10}O_3Os$ Os-00079

Tricarbonyl[(1,2,3,4-η)-2,3-dimethyl-1,3-butadiene]osmium, 10CI

[72707-23-4]

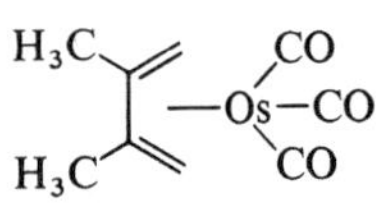

M 356.376

Colourless compd.

Ferrari, R.P. *et al*, *J. Organomet. Chem.*, 1979, **182**, 245 (*synth*, *ir*, *ms*)

$C_9H_{10}O_3Os$ Os-00080

Tricarbonyl[(2,3,4,5-η)-2,4-hexadiene]osmium, 10CI

[72707-18-7]

As (η^4-1,3-Butadiene)tricarbonylosmium, Os-00046 with

R = R′ = CH_3

M 356.376

Colourless compd.

Ferrari, R.P. *et al*, *J. Organomet. Chem.*, 1979, **182**, 245 (*synth*, *ir*, *ms*)

$C_9H_{11}ClOs$ Os-00081

(η^6-Benzene)chloro(η^3-2-propenyl)osmium, 9CI

(η^3-Allyl)(η^6-benzene)chloroosmium

[53886-39-8]

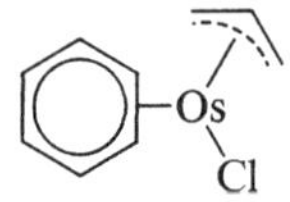

M 344.839

Yellow-green cryst. (C_6H_6/heptane). Mp 188-94° dec.

Rubezhov, A.Z. *et al*, *Izv. Akad. Nauk SSSR, Ser. Khim.*, 1974, 1902 (*synth*, *ir*, *pmr*)

Nesmeyanov, A.N. *et al*, *J. Organomet. Chem.*, 1978, **164**, 259 (*synth*, *ir*, *pmr*)

$C_9H_{15}I_2OsP$ Os-00082

(η^6-Benzene)diiodo(trimethylphosphine)osmium

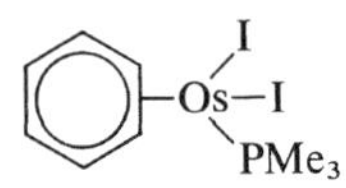

M 598.200

No phys. props. reported.

Werner, H. *et al*, *J. Organomet. Chem.*, 1980, **194**, C7 (*synth*)

$C_9H_{18}O_9OsP_2$ Os-00083

Tricarbonylbis(trimethyl phosphite-*P*)osmium

Tricarbonylbis(phosphorous acid)osmium hexamethyl ester, 8CI

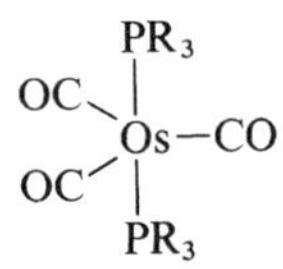

R = OMe

M 522.383

***trans*-form** [22377-52-2]

Colourless cryst. (hexane). Mp 60°.

Bruce, M.I. *et al*, *J. Chem. Soc. (A)*, 1969, 987 (*synth*, *ir*, *pmr*)

$C_9N_2O_{11}Os_3$ Os-00084

Nonacarbonyldinitrosyltriosmium, 10CI

[64973-41-7]

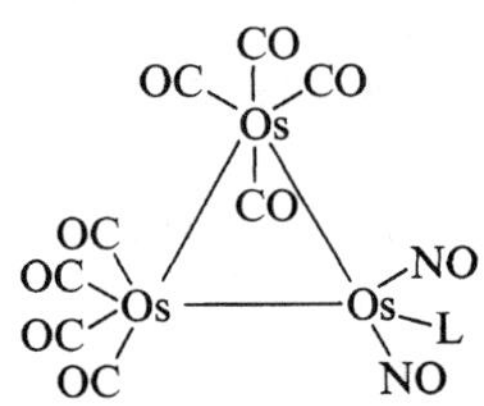

L = CO

M 882.706

Dark-red cryst. (pentane).

Bhaduri, S. *et al*, *J. Chem. Soc., Dalton Trans.*, 1979, 557 (*synth*, *ir*, *props*, *cmr*, *ms*)

$C_9O_8Os_3S_3$ Os-00085

(Carbonothioyl)octacarbonyldi-μ_3-thioxotriosmium, 10CI

[76000-21-0]

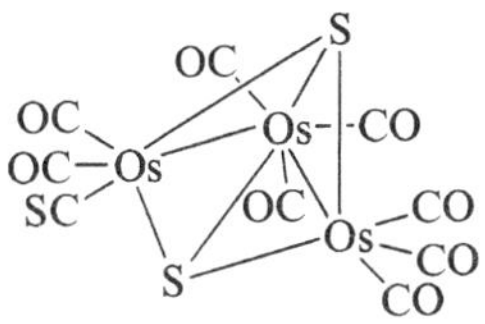

M 902.874

Amber cryst. (hexane). Mp 127-8°.

Broadhurst, P.V. *et al*, *J. Organomet. Chem.*, 1980, **194**, C35 (*synth, ir, ms, struct*)

$C_9O_9Os_2$ Os-00086

μ-Carbonyloctacarbonyldiosmium, 8CI

Nonacarbonyldiosmium. Diosmium nonacarbonyl. Diosmium enneacarbonyl

[28411-13-4]

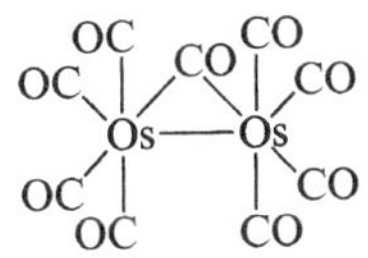

M 632.494

Thermally-sensitive and photosensitive orange-yellow cryst. Sol. hydrocarbons, unlike $Fe_2(CO)_9$. Mp 64-7° dec.

Moss, J.R. *et al*, *J. Chem. Soc., Dalton Trans.*, 1977, 95 (*synth, ir, ms*)

$C_9O_9Os_3S_2$ Os-00087

Nonacarbonyldi-μ_3-thioxotriosmium, 10CI

[72282-40-7]

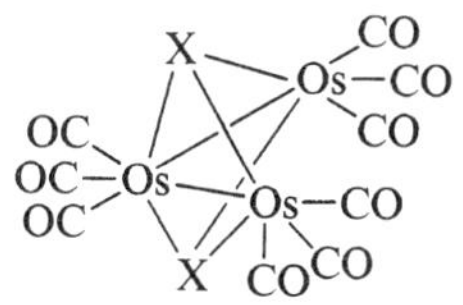

X = S

M 886.814

Yellow cryst. (hexane).

Johnson, B.F.G. *et al*, *J. Chem. Soc., Chem. Commun.*, 1979, 719 (*synth, ir*)

Adams, R.D. *et al*, *J. Am. Chem. Soc.*, 1982, **104**, 4115 (*synth*)

Adams, R.D. *et al*, *Organometallics*, 1983, **2**, 144 (*struct*)

$C_9O_9Os_3Se_2$ Os-00088

Nonacarbonyldi-μ_3-selenoxotriosmium, 10CI

[72282-41-8]

As Nonacarbonyldi-μ_3-thioxotriosmium, Os-00087 with

X = Se

M 980.614

Johnson, B.F.G. *et al*, *J. Chem. Soc., Chem. Commun.*, 1979, 719 (*synth, ir*)

Johnson, B.F.G. *et al*, *Acta Crystallogr., Sect. B*, 1981, **37**, 1731 (*struct*)

$C_9O_9Os_3Te_2$ Os-00089

Nonacarbonyldi-μ_3-telluroxotriosmium, 10CI

[72282-42-9]

As Nonacarbonyldi-μ_3-thioxotriosmium, Os-00087 with

X = Te

M 1077.894

Johnson, B.F.G. *et al*, *J. Chem. Soc., Chem. Commun.*, 1979, 719 (*synth, ir*)

$C_{10}Br_2O_{10}Os_3$ Os-00090

Di-μ-bromodecacarbonyltriosmium, 8CI

[28109-19-5]

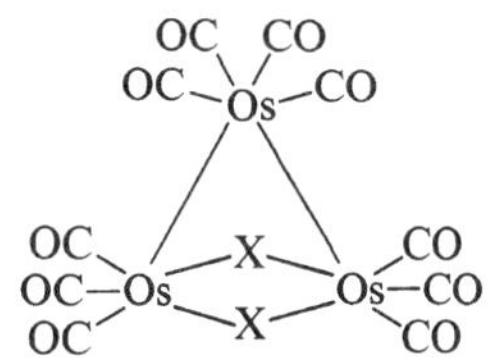

X = Br

M 1010.512

Yellow cryst.

Deeming, A.J. *et al*, *J. Chem. Soc. (A)*, 1970, 897 (*synth, ir*)

Bryan, E.G. *et al*, *J. Chem. Soc., Chem. Commun.*, 1977, 329 (*cmr*)

Vaglio, G.A., *J. Organomet. Chem.*, 1979, **169**, 83 (*ms*)

Aleksandrov, G.G. *et al*, *Koord. Khim.*, 1980, **6**, 626 (*struct*)

$C_{10}Cl_2O_{10}Os_3$ Os-00091

Decacarbonyldi-μ-chlorotriosmium, 10CI

[64387-08-2]

As Di-μ-bromodecacarbonyltriosmium, Os-00090 with

X = Cl

M 921.610

Yellow plates. $Bp_{0.15}$ 80° subl.

Deeming, A.J. *et al*, *J. Chem. Soc. (A)*, 1970, 897 (*synth, ir*)

Bryan, E.G. *et al*, *J. Chem. Soc., Chem. Commun.*, 1977, 329 (*cmr*)

$C_{10}HBrO_{10}Os_3$ Os-00092

μ-Bromodecacarbonyl-μ-hydrotriosmium

[61199-98-2]

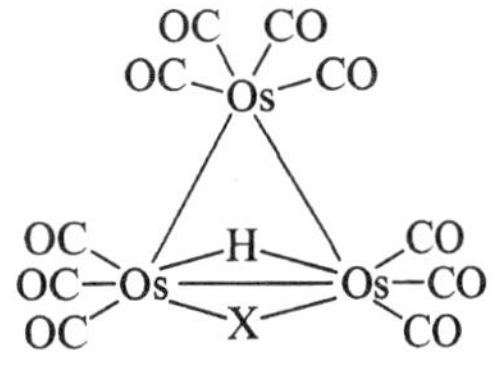

X = Br

M 931.616

Yellow cryst. (hexane).

Bryan, E.G. *et al*, *J. Chem. Soc., Dalton Trans.*, 1977, 1328 (*synth*)

Bryan, E.G. *et al*, *J. Chem. Soc., Dalton Trans.*, 1978, 196 (*cmr*)

Churchill, M.R. *et al*, *Inorg. Chem.*, 1979, **18**, 3261 (*struct*)

$C_{10}HClO_{10}Os_3$ Os-00093

Decacarbonyl-μ-chloro-μ-hydrotriosmium, 10CI

[12557-93-6]

As μ-Bromodecacarbonyl-μ-hydrotriosmium, Os-00092 with

X = Cl

M 887.165

Yellow cryst. (hexane).

Bryan, E.G. *et al*, *J. Chem. Soc., Dalton Trans.*, 1977, 1328 (*synth*)
Bryan, E.G. *et al*, *J. Chem. Soc., Dalton Trans.*, 1978, 196 (*cmr*)
Churchill, M.R. *et al*, *Inorg. Chem.*, 1979, **18**, 1926 (*struct*)

$C_{10}HNO_{11}Os_3$ Os-00094

Decacarbonyl-μ-hydro-μ-nitrosyltriosmium, 10CI

[73238-03-6]

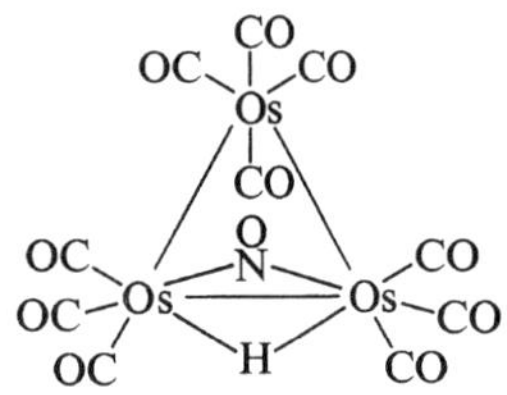

M 881.718

Characterised spectroscopically.

Johnson, B.F.G. *et al*, *J. Chem. Soc., Dalton Trans.*, 1980, 99 (*synth, ir*)
Stevens, R.E. *et al*, *J. Am. Chem. Soc.*, 1981, **103**, 4981 (*synth*)

$C_{10}H_2O_{10}Os_3$ Os-00095

Decacarbonyldi-μ-hydrotriosmium, 9CI

1,1,1,2,2,2,3,3,3-Decacarbonyl-2,3-dihydrotriosmium

[41766-80-7]

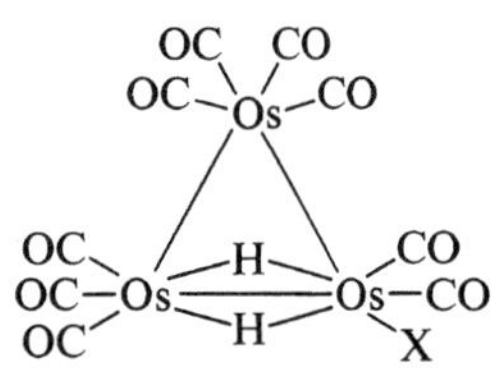

X = CO

M 852.720

Dark-red or intense purple-black cryst. (pentane).

Johnson, B.F.G. *et al*, *J. Chem. Soc. (A)*, 1968, 2859 (*synth, ir, pmr, ms*)
Knox, S.A.R. *et al*, *J. Am. Chem. Soc.*, 1975, **97**, 709.
Orpen, A.G. *et al*, *J. Chem. Soc., Chem. Commun.*, 1978, 723 (*struct*)
Broach, R.W. *et al*, *Inorg. Chem.*, 1979, **18**, 314 (*struct*)
Johnson, B.F.G., *Transition Metal Clusters*, Wiley, N.Y., 1980 (*rev*)
Aime, S. *et al*, *J. Organomet. Chem.*, 1981, **213**, 207 (*cmr*)
Johnson, B.F.G. *et al*, *Adv. Inorg. Chem. Radiochem.*, 1981, **24**, 225 (*rev*)

$C_{10}H_3BrO_9Os_3$ Os-00096

[μ_3-(Bromomethylidyne)]nonacarbonyltri-μ-hydrotriosmium, 10CI

[73746-96-0]

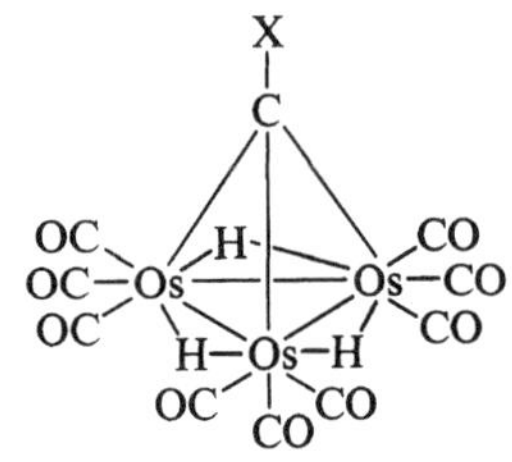

X = Br

M 917.632

Pale-yellow cryst. (CH_2Cl_2/hexane).

Keister, J.B. *et al*, *Inorg. Chem.*, 1980, **19**, 2304 (*synth, ir, pmr, ms*)

$C_{10}H_3NO_{10}Os_3$ Os-00097

μ-Amidodecacarbonyl-μ-hydrotriosmium, 10CI

[63373-89-7]

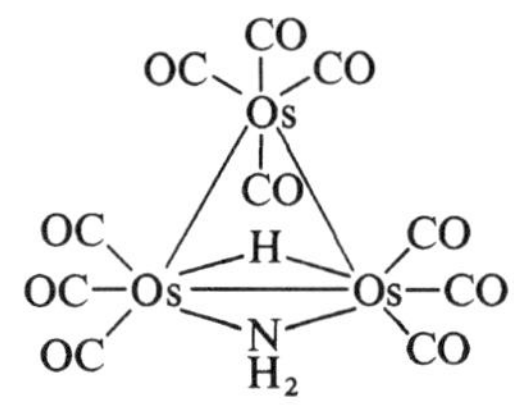

M 867.734

Yellow cryst. (CH_2Cl_2/pentane).

Bryan, E.G. *et al*, *J. Chem. Soc., Dalton Trans.*, 1977, 1328 (*synth, ir, pmr*)
Süss-Fink, G., *Z. Naturforsch., B*, 1980, **35**, 454 (*synth, ir, pmr*)

$C_{10}H_4O_6Os_2$ Os-00098

[μ-[(1,2,3,4-η:1,4-η)-1,3-Butadiene-1,4-diyl]]hexacarbonyldiosmium, 9CI

Hexacarbonyl(osmacyclopentadiene)osmium

[59464-98-1]

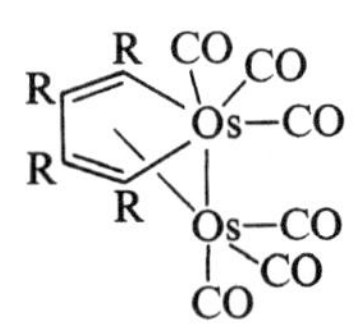

R = H

M 600.538

Colourless cryst.

Ferrari, R.P. *et al*, *Gazz. Chim. Ital.*, 1975, **105**, 939 (*synth, ir, ms*)

$C_{10}H_4O_9Os_3$ Os-00099

Nonacarbonyltri-μ-hydro-μ_3-methylidynetriosmium, 10CI

[64012-09-5]

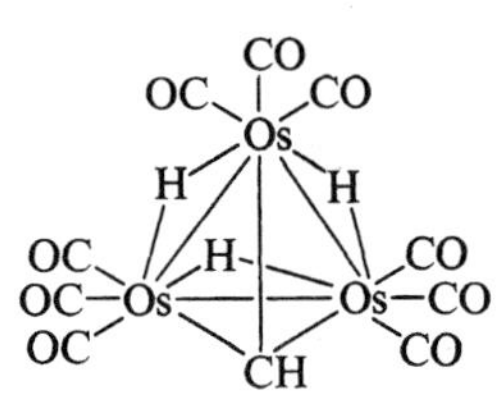

M 838.736

Almost colourless cryst. (pentane).

Calvert, R.B. *et al*, *J. Am. Chem. Soc.*, 1977, **99**, 5225 (*synth, pmr, cmr, ms*)
Azam, K.A. *et al*, *J. Chem. Soc., Dalton Trans.*, 1981, 91 (*synth, ir, pmr*)

$C_{10}H_4O_9Os_3S$ Os-00100

Nonacarbonyl-μ-hydro[μ₃-(methanethiolato)]triosmium, 10CI

[68796-12-3]

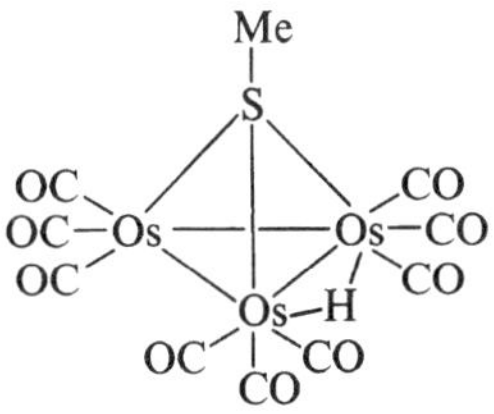

M 870.796
Yellow solid.

Johnson, B.F.G. *et al*, *J. Organomet. Chem.*, 1981, **213**, 249 (*synth*)

$C_{10}H_5NO_9Os_3$ Os-00101

Nonacarbonyldi-μ-hydro[μ_3-[methanaminato(2−)]]triosmium, 10CI

[76581-49-2]

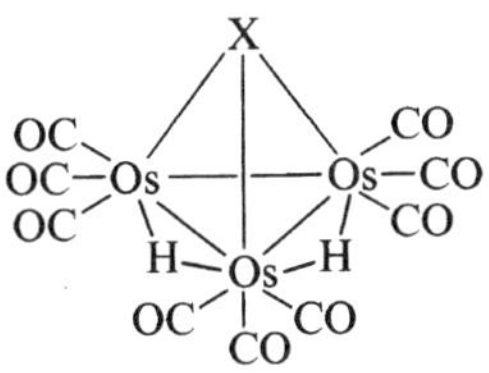

X = NMe

M 853.751
No phys. props. reported. Dec. at 198° to Dodecacarbonyl[μ_3-[methanaminato(2−)]]tetraosmium, Os-00214 .

Lin, Y.C. *et al*, *J. Am. Chem. Soc.*, 1981, **103**, 1216 (*synth*)
Lin, Y.C. *et al*, *J. Organomet. Chem.*, 1981, **213**, C41 (*struct, pmr, ir*)

$C_{10}H_6O_{10}Os_2$ Os-00102

Bis[μ-(acetato-*O*:*O*′)]hexacarbonyldiosmium, 9CI

[35163-44-1]

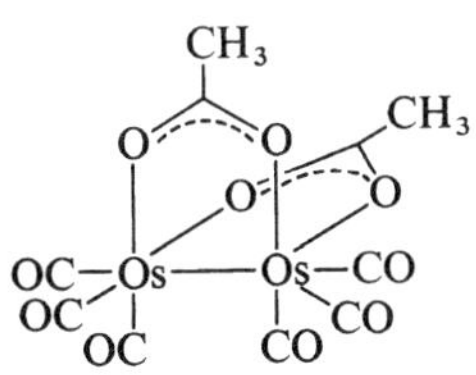

M 666.551
White cryst. (C_6H_6/pet. ether).

Crooks, G.R. *et al*, *J. Chem. Soc. (A)*, 1969, 2761 (*synth, ir, pmr*)
Bullitt, J.G. *et al*, *Inorg. Chim. Acta*, 1971, **5**, 406 (*struct*)

$C_{10}H_7NO_3Os$ Os-00103

Tricarbonyl[(1,3,4,5-η)-2-cyano-4-cyclohexene-1,3-diyl]osmium, 10CI

[75535-80-7]

As Tricarbonyl[(1,3,4,5-η)-4-cyclohexene-1,3-diyl]osmium, Os-00076 with

X = CN

M 379.370
No phys. props. reported.

Burrows, A.L. *et al*, *J. Organomet. Chem.*, 1980, **194**, C11 (*synth*)

$C_{10}H_7NO_3Os$ Os-00104

Tricarbonyl[(2,3,4,5-η)-2,4-cyclohexadiene-1-carbonitrile]osmium, 10CI

Tricarbonyl(5-cyano-1,3-cyclohexadiene)osmium

[75535-79-4]

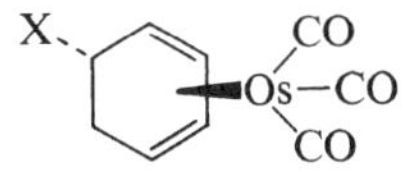

exo-form

X = CN

M 379.370
No phys. props. reported.

Burrows, A.L. *et al*, *J. Organomet. Chem.*, 1980, **194**, C11 (*synth*)

$C_{10}H_{10}O_3Os$ Os-00105

Tricarbonyl[(1,2,3,4-η)-1,3-cycloheptadiene]osmium, 9CI

[59676-07-2]

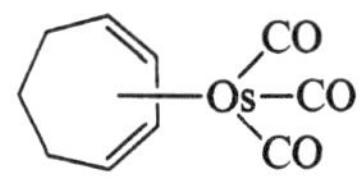

M 368.387
Colourless. $Bp_{0.01}$ 50°.

Knox, S.A.R. *et al*, *J. Chem. Soc., Dalton Trans.*, 1976, 552 (*synth, ir, pmr*)

$C_{10}H_{10}O_4Os$ Os-00106

Dicarbonyl[(1,2,3,4,5-η)-2,4-cyclohexadien-1-yl](methoxycarbonyl)osmium, 10CI

[63022-50-4]

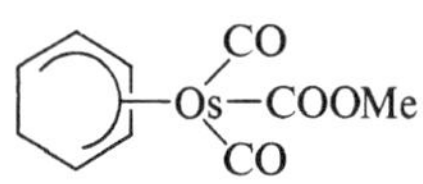

M 384.387
White solid. Thermally dec. to *exo*-form of Tricarbonyl[(1,2,3,4-η)-5-methoxy-1,3-cyclohexadiene]osmium, Os-00107 .

Bryan, E.G. *et al*, *J. Organomet. Chem.*, 1977, **129**, C19 (*synth, pmr, ir*)

$C_{10}H_{10}O_4Os$ Os-00107

Tricarbonyl[(1,2,3,4-η)-5-methoxy-1,3-cyclohexadiene]osmium, 10CI

As Tricarbonyl[(2,3,4,5-η)-2,4-cyclohexadiene-1-carbonitrile]osmium, Os-00104 with

X = OMe

M 384.387

exo-form [63022-51-5]
Characterised spectroscopically.

endo-form [63180-66-5]
Characterised spectroscopically.

Bryan, E.G. *et al*, *J. Organomet. Chem.*, 1977, **129**, C19 (*synth, ms, pmr*)

$C_{10}H_{10}O_6Os$ Os-00108

Tetracarbonyldipropionylosmium, 8CI

As Dibromotetracarbonylosmium, Os-00015 with

$$X = Y = -COCH_2CH_3$$

M 416.385

cis-form [27599-26-4]

White cryst. (pentane).

L'Eplattenier, F. *et al*, *Helv. Chim. Acta*, 1970, **53**, 1091 (*synth, ir, pmr*)

$C_{10}H_{10}Os$ Os-00109

Osmocene, 9CI

Bis(cyclopentadienyl)osmium

[1273-81-0]

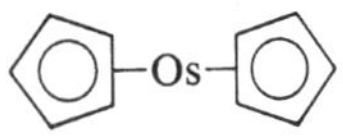

M 320.389

Colourless cryst. (C_6H_6). Mp 229-30°. Sublimes. $FeCl_3$ → $[(C_5H_5)_2Os^{IV}(OH)]^{\oplus}$.

Fischer, E.O. *et al*, *Chem. Ber.*, 1959, **92**, 2302 (*synth*)
Jellinek, F., *Z. Naturforsch., B*, 1959, **14**, 737 (*struct*)
Rausch, M.D. *et al*, *J. Am. Chem. Soc.*, 1960, **82**, 76 (*props*)
Nesmeyanov, A.N. *et al*, *Zh. Strukt. Khim.*, 1972, **13**, 1033 (*cmr*)
Lokshin, B.V. *et al*, *J. Organomet. Chem.*, 1977, **124**, 293 (*ir, raman*)

$C_{10}H_{10}Os^{\oplus}$ Os-00110

Osmocenium, 9CI

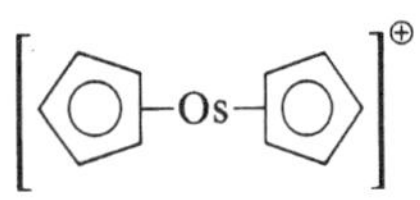

M 320.389 (ion)

Paramagnetic.

Tetrafluoroborate: [54438-62-9].
$C_{10}H_{10}BF_4Os$ M 407.193
Dark-green. Sol. H_2O, insol. Et_2O, EtOH, MeCN, DMF, Py.

Gubin, S.P. *et al*, *J. Organomet. Chem.*, 1971, **30**, 243.
Denisovich, L.I. *et al*, *J. Organomet. Chem.*, 1974, **81**, 207 (*synth, polarog, uv*)

$C_{10}H_{15}IOsP$ Os-00111

(η^6-Benzene)iodomethyl(trimethylphosphine)osmium

[75353-42-3]

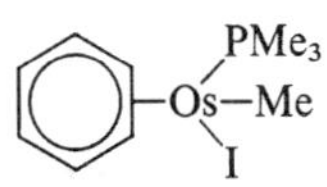

M 483.307

No phys. props. reported.

Werner, H. *et al*, *J. Organomet. Chem.*, 1980, **194**, C7 (*synth*)

$C_{10}H_{16}O_4OsSi_2$ Os-00112

Tetracarbonyl[1,2-ethanediylbis[dimethylsilylene]]osmium, 9CI

[43198-15-8]

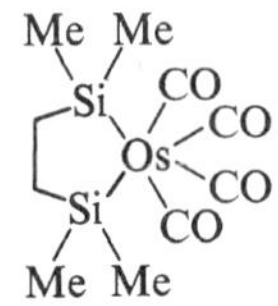

M 446.605

Colourless waxy solid. Mp 98-100°. Subl. at r.t. under reduced press. (0.005 mm).

Vancea, L. *et al*, *Inorg. Chem.*, 1974, **13**, 511 (*synth, ir, pmr, cmr*)

$C_{10}H_{18}F_4O_8OsP_2$ Os-00113

Dicarbonyl(η^2-tetrafluoroethene)bis(trimethyl phosphite-*P*)osmium

Dicarbonylbis(phosphorous acid)(tetrafluoroethylene)-osmium hexamethyl ester, 8CI

[33134-73-5]

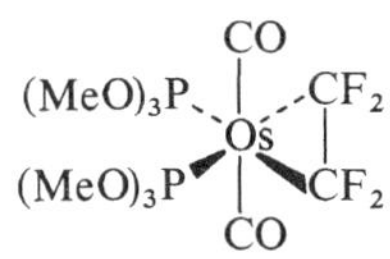

M 594.389

White needles (hexane). Mp 128°.

Cooke, M. *et al*, *J. Chem. Soc. (A)*, 1971, 1200 (*synth, ir, pmr, nmr, ms*)

$C_{10}H_{18}Ge_2O_4Os$ Os-00114

Tetracarbonylbis(trimethylgermyl)osmium, 8CI

[34397-64-3]

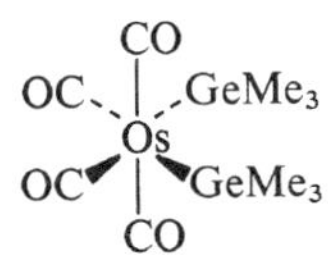

M 537.630

Mixt. of *cis-* and *trans*-isomers. Pale-yellow liq. $Bp_{0.01}$ ca. 60°.

Knox, S.A.R. *et al*, *J. Chem. Soc. (A)*, 1971, 2874 (*synth, ir, pmr*)
Vancea, L. *et al*, *J. Am. Chem. Soc.*, 1976, **98**, 1407 (*cmr*)

$C_{10}H_{18}O_4OsPb_2$ Os-00115

Tetracarbonylbis(trimethylplumbyl)osmium, 9CI

[41553-08-6]

$$(OC)_4Os(PbMe_3)_2$$

M 806.850

Mixt. of *cis-* and *trans*-isomers. Pale-yellow liq. $Bp_{0.01}$ 80-90°.

George, R.D. *et al*, *J. Chem. Soc., Dalton Trans.*, 1973, 972 (*synth, ir, pmr*)
Vancea, L. *et al*, *J. Am. Chem. Soc.*, 1976, **98**, 1407 (*cmr*)

$C_{10}H_{18}O_4OsSi_2$ Os-00116

Tetracarbonylbis(trimethylsilyl)osmium, 8CI

[29994-24-9]

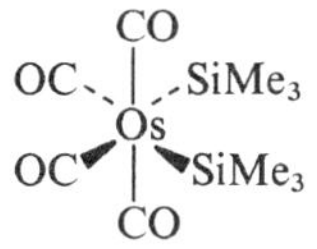

M 448.621

Mixt. of *cis-* and *trans-*isomers. Pale-yellow. Mp 25°.

Knox, S.A.R. *et al*, *J. Chem. Soc. (A)*, 1970, 3147 (*synth, ir*)
Vancea, L. *et al*, *J. Am. Chem. Soc.*, 1976, **98**, 1407 (*cmr*)

$C_{10}H_{18}O_4OsSn_2$ Os-00117

Tetracarbonylbis(trimethylstannyl)osmium, 9CI

[25931-79-7]

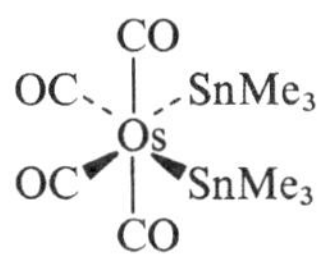

M 629.830

Mixt. of *cis-* and *trans-*isomers. Yellow, air-stable liq. Mp <−23°. $Bp_{0.01}$ 70°.

Knox, S.A.R. *et al*, *J. Chem. Soc. (A)*, 1970, 3147 (*synth, ir*)
Vancea, L. *et al*, *J. Am. Chem. Soc.*, 1976, **98**, 1407 (*cmr*)

$C_{10}I_2O_{10}Os_3$ Os-00118

Decacarbonyldi-μ-iodotriosmium, 8CI

[28109-20-8]

As Di-μ-bromodecacarbonyltriosmium, Os-00090 with

X = I

M 1104.513

Orange cryst. (pentane).

Deeming, A.J. *et al*, *J. Chem. Soc. (A)*, 1970, 897 (*synth, ir*)
Bryan, E.G. *et al*, *J. Chem. Soc., Chem. Commun.*, 1977, 329 (*cmr*)
Vaglio, G.A., *J. Organomet. Chem.*, 1979, **169**, 83 (*ms*)

$C_{10}NO_{11}Os_3^{\ominus}$ Os-00119

Decacarbonyl-μ-nitrosyltriosmate(1−), 10CI

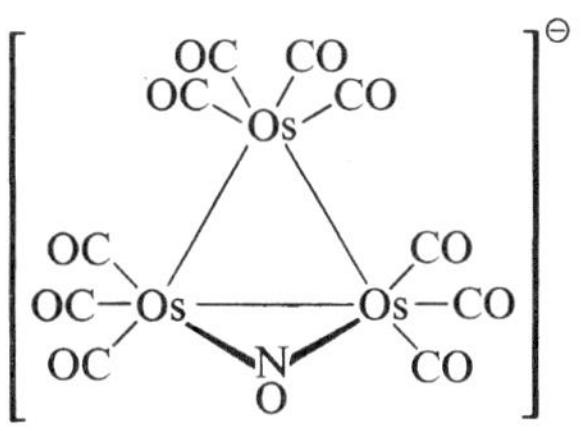

M 880.710 (ion)

Characterised spectroscopically.

Stevens, R.E. *et al*, *J. Am. Chem. Soc.*, 1981, **103**, 4981 (*synth, ir*)

$C_{10}N_2O_{12}Os_3$ Os-00120

Decacarbonyldi-μ-nitrosyltriosmium

[36583-25-2]

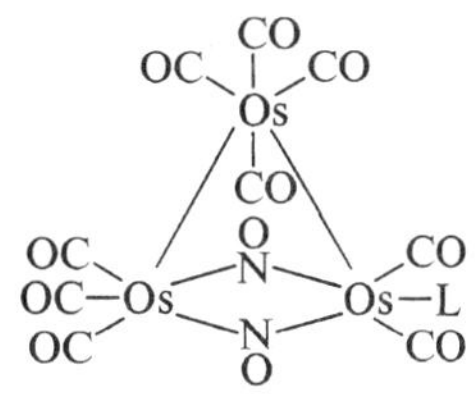

L = CO

M 910.716

Air-stable light-green cryst. (heptane/C_6H_6).

Norton, J. *et al*, *Inorg. Chem.*, 1972, **11**, 382 (*synth, ir, ms*)
Inorg. Synth., 1976, **16**, 40 (*synth, ir*)

$C_{11}HO_{11}Os_3^{\ominus}$ Os-00121

μ-Carbonyldecacarbonyl-μ-hydrotriosmate(1−), 10CI

Undecacarbonylhydrotriosmate(1−)

[61182-07-8]

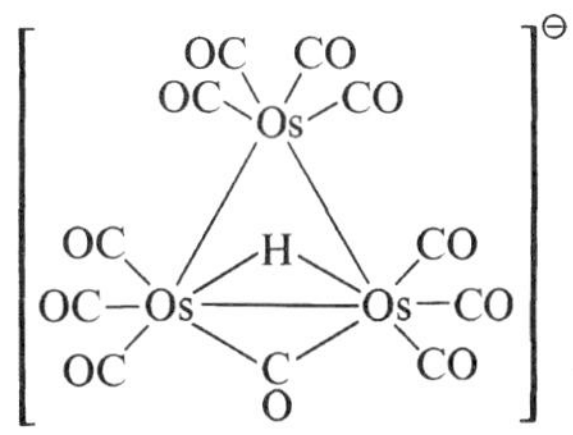

M 879.722 (ion)

Tetramethylammonium salt: [61245-30-5].
$C_{15}H_{16}NO_{11}Os_3$ M 956.892
No phys. props. reported.

Bis(triphenylphosphine)iminium salt: [61182-08-9].
$C_{47}H_{31}NO_{11}Os_3P_2$ M 1418.310
Red cryst.

Eady, C.R. *et al*, *J. Chem. Soc., Dalton Trans.*, 1978, 1358 (*synth, ir, pmr, cmr*)
Johnson, B.F.G. *et al*, *J. Organomet. Chem.*, 1978, **162**, 179 (*synth*)
Burkhardt, E.W. *et al*, *J. Organomet. Chem.*, 1980, **198**, 179 (*synth*)
Braga, D. *et al*, *J. Chem. Soc., Chem. Commun.*, 1982, 419 (*synth*)
Keister, J.B. *et al*, *Organometallics*, 1983, **2**, 219 (*synth, cmr, ir*)

$C_{11}H_2O_9Os_3$ Os-00122

Nonacarbonyl[μ₃-[(1-η:1,2-η:1,2-η)ethynyl]]-μ-hydrotriosmium, 9CI

[55073-02-4]

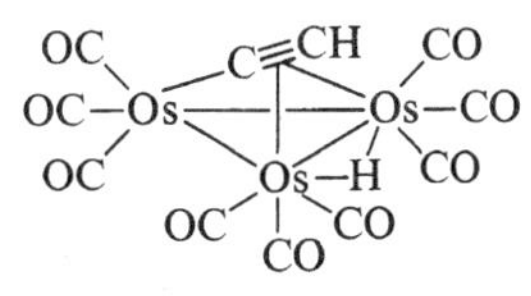

M 848.731

White cryst. Mp 170-4°.

Deeming, A.J. *et al*, *J. Chem. Soc., Dalton Trans.*, 1975, 1614 (*synth, ir, pmr*)
Ferrari, R.P. *et al*, *Gazz. Chim. Ital.*, 1975, **105**, 939 (*synth, ir*)

$C_{11}H_2O_{10}Os_3$ Os-00123

Decacarbonyl-μ-hydro-μ_3-methylidynetriosmium

[83573-03-9]

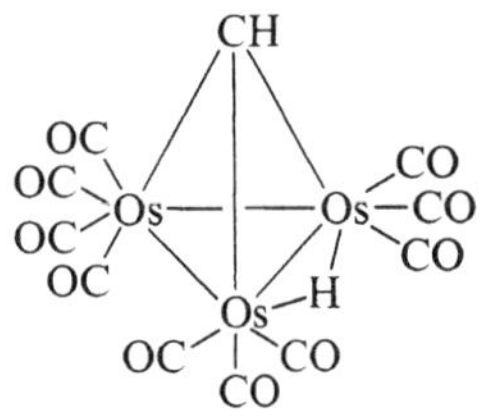

M 864.731

Sl. air-sensitive bright-yellow cryst. (pentane). Rearranges to Nonacarbonyldi-μ-hydro[μ_3-(oxoethenylidene)]triosmium, Os-00124 in refluxing hexane.

Shapley, J.R. *et al*, *J. Am. Chem. Soc.*, 1983, **105**, 140 (*synth, ms, ir, pmr, cmr*)

$C_{11}H_2O_{10}Os_3$ Os-00124

Nonacarbonyldi-μ-hydro[μ_3-(oxoethenylidene)]triosmium

(μ_3-Carbonylmethylidyne)nonacarbonyldi(μ-hydrido)triosmium

[83585-34-6]

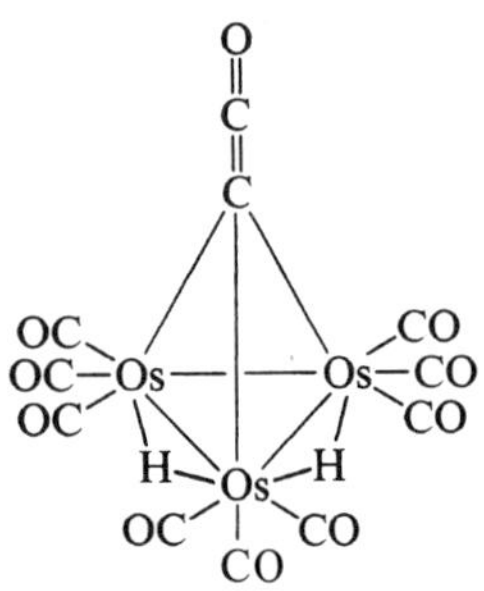

M 864.731

Yellow cryst. (pentane). $Bp_{0.1}$ 85° subl.

Sievert, A.C. *et al*, *Organometallics*, 1982, **1**, 214 (*synth, ir, ms, cmr*)

Shapley, J.R. *et al*, *Organometallics*, 1983, **2**, 185 (*synth, struct*)

$C_{11}H_2O_{11}Os_3$ Os-00125

Undecacarbonyl-μ-hydrohydrotriosmium

Undecacarbonyldihydrotriosmium

[56398-24-4]

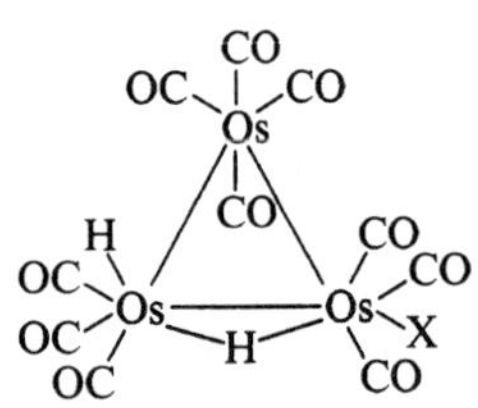

X = CO

M 880.730

H atoms undergo site exchange in soln. Yellow cryst. (hexane). In soln. slowly reverts to Decacarbonyldi-μ-hydrotriosmium, Os-00095 .

Deeming, A.J. *et al*, *J. Organomet. Chem.*, 1975, **88**, C21 (*synth, ir, pmr*)

Shapley, J.R. *et al*, *J. Am. Chem. Soc.*, 1975, **97**, 4145 (*synth, pmr, struct, ms*)

$C_{11}H_4O_9Os_3$ Os-00126

Nonacarbonyl[μ_3-[(1-η:1-η:1,2-η)ethenylidene]]di-μ-hydrotriosmium, 9CI

[42765-74-2]

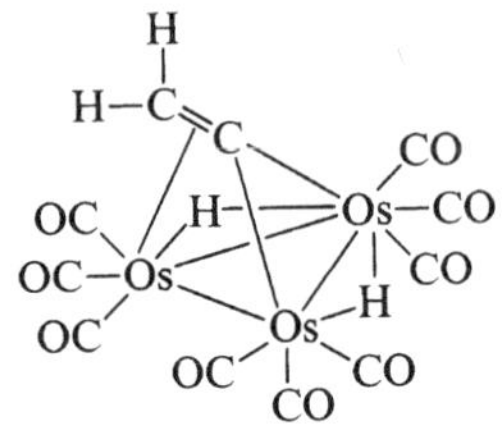

M 850.747

Fluxional molecule. V. pale-yellow cryst. Mp 122-4°. Converted into Nonacarbonyl-μ_3-ethylidynetri-μ-hydrotriosmium, Os-00129 with H_2 in refluxing heptane.

Deeming, A.J. *et al*, *J. Chem. Soc., Dalton Trans.*, 1974, 1415; 1975, 1614 (*synth, ir, pmr, struct*)

Keister, J.B. *et al*, *J. Organomet. Chem.*, 1975, **85**, C29 (*synth, pmr*)

Andrews, J.R. *et al*, *Inorg. Chem.*, 1982, **21**, 2874 (*ir, raman*)

$C_{11}H_4O_{10}Os_3$ Os-00127

Decacarbonyl-μ-hydro-μ-methyltriosmium

Decacarbonyldi-μ-hydro-μ-methylenetriosmium

[64052-01-3]

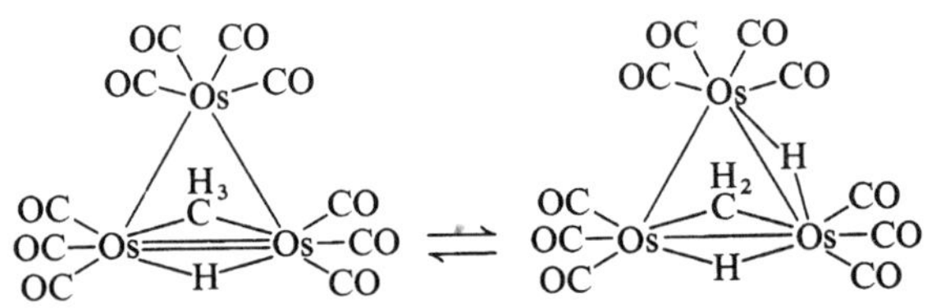

M 866.747

Tautomeric at ambient temps. No phys. props. reported. Decarbonylated to Nonacarbonyltri-μ-hydro-μ_3-methylidynetriosmium, Os-00099 at 110°. Methylene tautomer is less sol. in cold CH_2Cl_2.

Calvert, R.B. *et al*, *J. Am. Chem. Soc.*, 1977, **99**, 5225 (*synth, pmr, cmr*)

Calvert, R.B. *et al*, *J. Am. Chem. Soc.*, 1978, **100**, 7726 (*pmr*)

Jordan, R.F. *et al*, *J. Am. Chem. Soc.*, 1979, **101**, 4853 (*cmr*)

Schultz, A.J. *et al*, *Inorg. Chem.*, 1979, **18**, 319 (*struct*)

$C_{11}H_5NO_{10}Os_3$ Os-00128

Decacarbonyl-μ-hydro[μ-(methanaminato)]triosmium, 10CI

[76581-50-5]

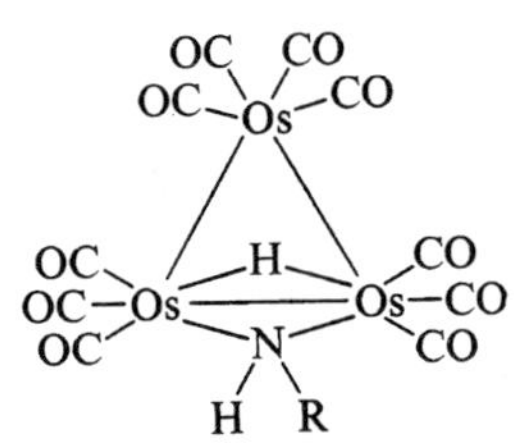

R = Me

M 881.761

No phys. props. reported.

Lin, Y.C. *et al*, *J. Am. Chem. Soc.*, 1981, **103**, 1216 (*synth*)

$C_{11}H_6O_9Os_3$ **Os-00129**

Nonacarbonyl-μ_3-ethylidynetri-μ-hydrotriosmium, 9CI

[51158-87-3]

As [μ_3-(Bromomethylidyne)]nonacarbonyltri-μ-hydrotriosmium, Os-00096 with

$$X = CH_3$$

M 852.763

No phys. props. reported.

Deeming, A.J. *et al*, *J. Chem. Soc., Chem. Commun.*, 1973, 277 (*synth, pmr*)
Yesinowski, J.P. *et al*, *J. Organomet. Chem.*, 1974, **65**, C27 (*struct, pmr*)
Forster, A. *et al*, *J. Organomet. Chem.*, 1976, **104**, 225 (*cmr*)

$C_{11}H_6O_{10}Os_3$ **Os-00130**

Nonacarbonyltri-μ-hydro[μ_3-(methoxymethylidyne)]triosmium, 10CI

[71562-48-6]

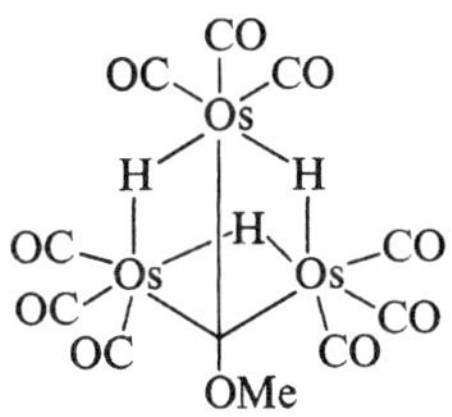

M 868.762

Yellow cryst. (MeOH).

Keister, J.B. *et al*, *Organometallics*, 1983, **2**, 219 (*synth, pmr, cmr, ir, ms*)

$C_{11}H_8O_3Os$ **Os-00131**

Tricarbonyl[(1,2,3,4-η)-1,3,5,7-cyclooctatetraene]osmium

[33309-01-2]

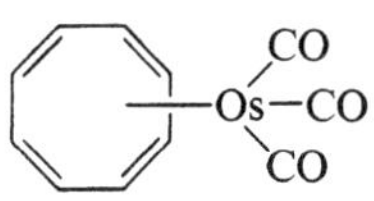

M 378.382

Fluxional molecule. Orange cryst. (hexane). Mp 90.5°.

Bruce, M.I. *et al*, *J. Chem. Soc. (A)*, 1969, 987 (*synth, ms, pmr, ir*)
Cooke, M. *et al*, *J. Chem. Soc., Chem. Commun.*, 1970, 565 (*pmr*)
Harris, P.J. *et al*, *J. Chem. Soc., Dalton Trans.*, 1976, 377 (*synth, pmr, ms*)

$C_{11}H_8O_3Os$ **Os-00132**

Tricarbonyl[(1,2,3,6-η)-1,3,5,7-cyclooctatetraene]osmium

Tricarbonyl-π-σ-cyclooctatrienyleneosmium, 8CI

[12559-97-6]

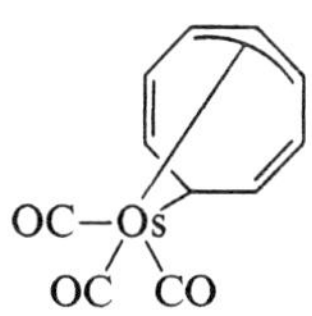

M 378.382

Pale-yellow cryst. (hexane). Mp 69°. Thermally isomerises irreversibly to Tricarbonyl[(1,2,3,4-η)-1,3,5,7-cyclooctatetraene]osmium, Os-00131 .

Bruce, M.I. *et al*, *J. Chem. Soc. (A)*, 1969, 987 (*synth, ir, pmr, ms*)

$C_{11}H_9N_2O_{13}Os_3P$ **Os-00133**

Octacarbonyldinitrosyl(trimethyl phosphite)triosmium

[64972-45-8]

As Nonacarbonyldinitrosyltriosmium, Os-00084 with

$$L = P(OMe)_3$$

M 978.771

Red cryst. (hexane).

Bhaduri, S. *et al*, *J. Chem. Soc., Dalton Trans.*, 1979, 557 (*synth, ir, pmr, cmr, struct*)

$C_{11}H_9O_3Os^{\oplus}$ **Os-00134**

Tricarbonyl[(1,2,5,6,7-η)-1,3,5-cyclooctatrien-1-yl]osmium(1+)

Tricarbonyl-π-cyclooctatrienyliumosmium(1+)

[33309-02-3]

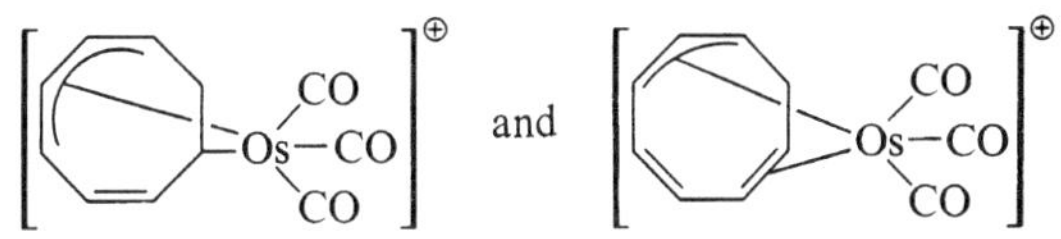

M 379.390 (ion)

Characterised spectroscopically in soln.

Bruce, M. *et al*, *J. Chem. Soc. (A)*, 1969, 987.
Cooke, M. *et al*, *J. Chem. Soc., Chem. Commun.*, 1970, 565.
Cooke, M. *et al*, *J. Chem. Soc., Chem. Commun.*, 1971, 621 (*synth, nmr, struct*)

$C_{11}H_9O_3Os_3^{\oplus}$ **Os-00135**

[(2,3,4,5,6-η)-Bicyclo[5.1.0]octa-3,5-dien-2-yl]tricarbonylosmium(1+)

π-Bicyclo[5.1.0]octadienyliumtricarbonylosmium(1+), 8CI

[33309-03-4]

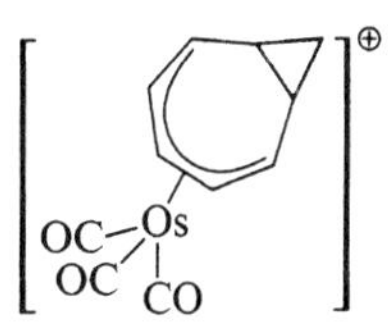

M 759.790 (ion)

Characterised spectroscopically in soln. Isomerises to Tricarbonyl[(1,2,5,6,7-η)-1,3,5-cyclooctatrien-1-yl]osmium(1+), Os-00134 in soln.

Green, M. *et al*, *J. Chem. Soc., Chem. Commun.*, 1971, 621 (*synth*)
Bennett, M.A. *et al*, *Inorg. Chem.*, 1981, **20**, 2353 (*synth, pmr*)

$C_{11}H_{10}O_2Os$ **Os-00136**

Osmocenecarboxylic acid

Carboxyosmocene

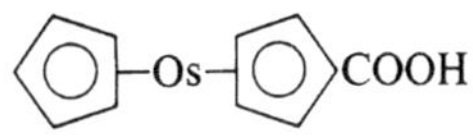

M 364.399

White cryst. by subl. Obt. contaminated with 1,1′-Osmocenedicarboxylic acid, Os-00174 .

Me ester: [55397-90-5]. *Methoxycarbonylosmocene*, 9CI.
$C_{12}H_{12}O_2Os$ M 378.426

Obt. as ^{191}Os deriv.

Rausch, M.D. *et al*, *J. Am. Chem. Soc.*, 1960, **82**, 76 (*synth*)
Langheim, D. *et al*, *Chem. Ber.*, 1975, **108**, 146 (*ester*)

$C_{11}H_{11}Cl_2NOs$ Os-00137

(η^6-Benzene)dichloro(pyridine)osmium, 10CI

[75700-95-7]

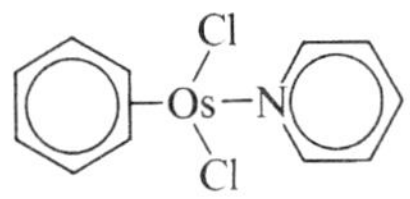

M 418.321

Yellow solid (CH_2Cl_2). Corresponding bromo and iodo analogues also reported.

Hung, Y. *et al*, *Inorg. Chem.*, 1981, **20**, 457 (*synth, uv*)

$C_{11}H_{11}O_3Os^{\oplus}$ Os-00138

Tricarbonyl[(1,2,3,4,5-η)-2,4-cyclooctadien-1-yl]osmium(1+), 9CI

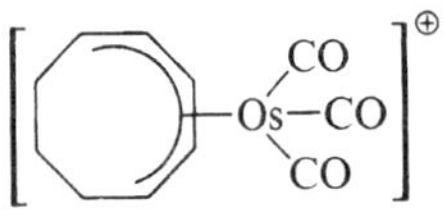

M 381.406 (ion)

Tetrafluoroborate: [55729-67-4].
$C_{11}H_{11}BF_4O_3Os$ M 468.210
White powder.

Deeming, A.J. *et al*, *J. Chem. Soc., Dalton Trans.*, 1974, 2093 (*synth, ir, pmr*)

$C_{11}H_{11}O_3Os^{\oplus}$ Os-00139

Tricarbonyl[(1,2,3,5,6-η)-2,5-cyclooctadien-1-yl]osmium(1+), 9CI

[55763-67-2]

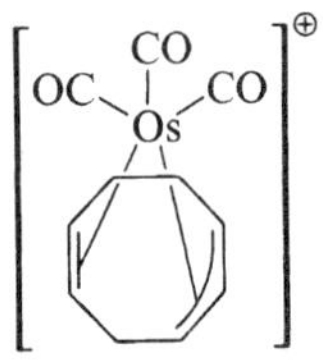

M 381.406 (ion)

Characterised spectroscopically.

Deeming, A.J. *et al*, *J. Chem. Soc., Dalton Trans.*, 1974, 2093 (*synth, ir, pmr*)

$C_{11}H_{11}Os^{\oplus}$ Os-00140

(η^6-Benzene)(η^5-2,4-cyclopentadien-1-yl)osmium(1+)

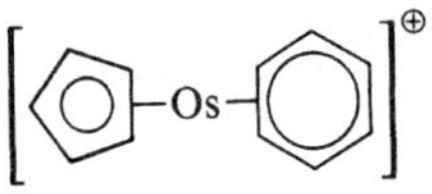

M 333.408 (ion)

Tetraphenylborate:
$C_{35}H_{31}BOs$ M 652.640
Greyish-white ppt. Mp 125-6°.

Robertson, I.W. *et al*, *J. Organomet. Chem.*, 1982, **228**, 171 (*synth, pmr, cmr*)

$C_{11}H_{12}O_3Os$ Os-00141

Tricarbonyl[(1,2,3,4-η)-1,3-cyclooctadiene]osmium, 9CI

[34802-07-8]

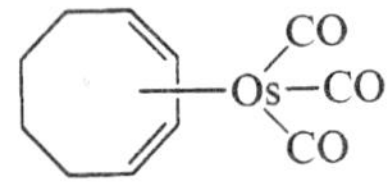

M 382.414

Yellow oil. $Bp_{0.1}$ 70°, $Bp_{0.05}$ 50°.

Deeming, A.J. *et al*, *J. Chem. Soc., Dalton Trans.*, 1974, 2093 (*synth, ir, pmr, ms*)

$C_{11}H_{12}O_3Os$ Os-00142

Tricarbonyl[(1,3,4,5-η)-4-cyclooctene-1,3-diyl]osmium, 9CI

[55938-93-7]

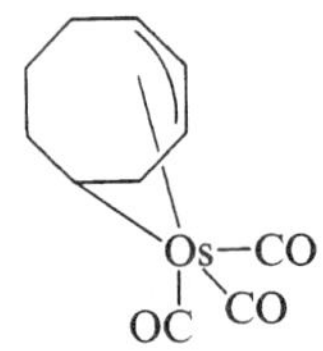

M 382.414

Colourless oil.

Deeming, A.J. *et al*, *J. Chem. Soc., Dalton Trans.*, 1974, 2093 (*synth, ir, pmr*)

$C_{11}H_{18}OOs$ Os-00143

Carbonyltrihydro[(1,2,3,4,5-η)-1,2,3,4,5-pentamethyl-2,4-cyclopentadien-1-yl]osmium

[81753-10-8]

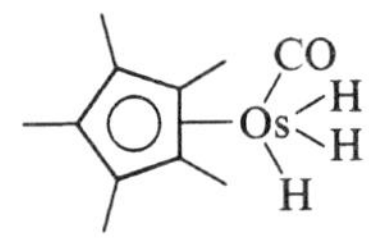

M 356.463

Colourless solid. Mp 25°.

Hoyano, J.K. *et al*, *J. Am. Chem. Soc.*, 1982, **104**, 3722 (*synth, ir, pmr*)

$C_{11}H_{19}IOsP^{\oplus}$ Os-00144

(η^6-Benzene)(η^2-ethene)iodo(trimethylphosphine)osmium(1+), 10CI

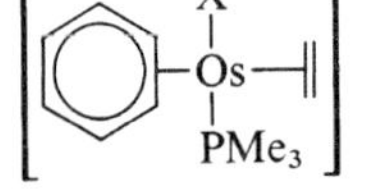

X = I

M 499.349 (ion)

Hexafluorophosphate: [75353-25-2].
$C_{11}H_{19}F_6IOsP_2$ M 644.314
No phys. props. reported.

Werner, H. *et al*, *J. Organomet. Chem.*, 1980, **194**, C7 (*synth*)

$C_{11}H_{19}OsP$ Os-00145

(η^6-Benzene)(η^2-ethene)(trimethylphosphine)osmium, 10CI

(η^6-Benzene)(η^2-ethylene)(trimethylphosphine)osmium

[75353-28-5]

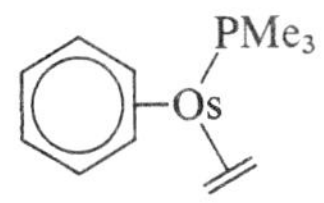

M 372.445

No phys. props. reported.

Werner, H. *et al*, *J. Organomet. Chem.*, 1980, **194**, C7 (*synth, pmr*)

Werner, R. *et al*, *Angew. Chem., Int. Ed. Engl.*, 1981, **20**, 793 (*props*)

$C_{11}H_{20}OsP^{\oplus}$ Os-00146

(η^6-Benzene)(η^2-ethene)hydro(trimethylphosphine)osmium(1+), 10CI

As (η^6-Benzene)(η^2-ethene)iodo(trimethylphosphine)osmium(1+), Os-00144 with

X = H

M 373.453 (ion)

Hexafluorophosphate: [75353-31-0].

$C_{11}H_{20}F_6OsP_2$ M 518.417

No phys. props. reported.

Werner, H. *et al*, *J. Organomet. Chem.*, 1979, **174**, C63; 1980, **194**, C7 (*synth, nmr*)

$C_{11}H_{23}BrO_6OsP_2$ Os-00147

Bromo(η^5-2,4-cyclopentadien-1-yl)bis(trimethyl phosphite-*P*)osmium, 10CI

[71633-32-4]

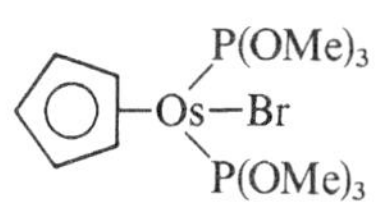

M 583.351

Yellow cryst. (Et_2O/pet. ether). Mp 132-5°.

Ashby, G.S. *et al*, *Aust. J. Chem.*, 1979, **32**, 1003 (*synth, pmr*)

$C_{11}O_{11}Os_3^{\ominus\ominus}$ Os-00148

μ-Carbonyldecacarbonyltriosmate(2−), 10CI

Undecacarbonyltriosmate(2−)

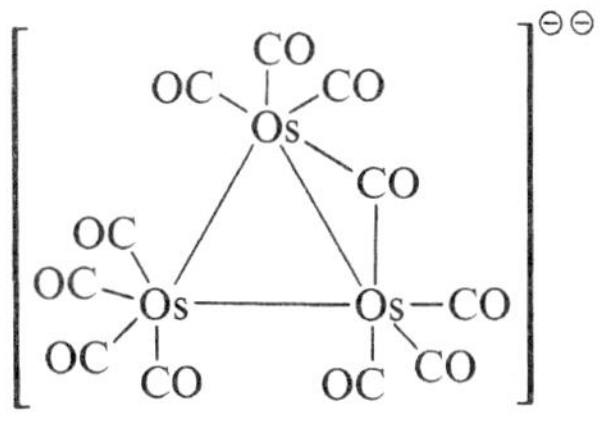

M 878.714 (ion)

Di-K salt: [79949-34-1].

$C_{11}K_2O_{11}Os_3$ M 956.911

Bright yellow-orange powder.

Nagel, C.C. *et al*, *J. Organomet. Chem.*, 1981, **219**, C9 (*synth, ir, cmr, struct*)

U.S.P., 4 349 521, (*1982*); *CA*, **97**, 165494 (*synth*)

$C_{12}Br_2O_{12}Os_3$ Os-00149

Dibromododecacarbonyltriosmium, 10CI

[66749-30-2]

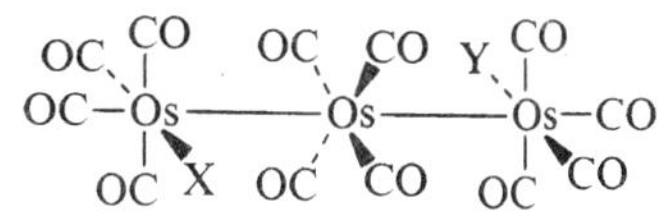

X = Y = Br

M 1066.533

Yellow prisms (CH_2Cl_2/hexane). Mp 145-55° dec. Bp 97° subl. *in vacuo*.

Hartley, D. *et al*, *J. Chem. Soc., Chem. Commun.*, 1968, 493 (*ir, raman*)

Johnson, B.F.G. *et al*, *J. Chem. Soc. (A)*, 1968, 2859 (*synth, ir, ms*)

Moss, J.R. *et al*, *J. Chem. Soc., Dalton Trans.*, 1977, 89 (*synth, ir*)

Vaglio, G.A., *J. Organomet. Chem.*, 1979, **169**, 83 (*ms*)

$C_{12}Cl_2O_{12}Os_3$ Os-00150

Dodecacarbonydichlorotriosmium, 10CI

[21773-70-6]

As Dibromododecacarbonyltriosmium, Os-00149 with

X = Y = Cl

M 977.631

Pale-yellow microprisms (CH_2Cl_2/hexane). Mp 164-70° dec. Bp 129° subl. *in vacuo*.

Hartley, D. *et al*, *J. Chem. Soc., Chem. Commun.*, 1968, 493 (*ir, raman*)

Johnson, B.F.G. *et al*, *J. Chem. Soc. (A)*, 1968, 2859 (*synth, ir, ms*)

Moss, J.R. *et al*, *J. Chem. Soc., Dalton Trans.*, 1977, 89 (*synth, ir*)

$C_{12}FeO_{12}Os_2$ Os-00151

Dodecacarbonylirondiosmium, 8CI

[12560-38-2]

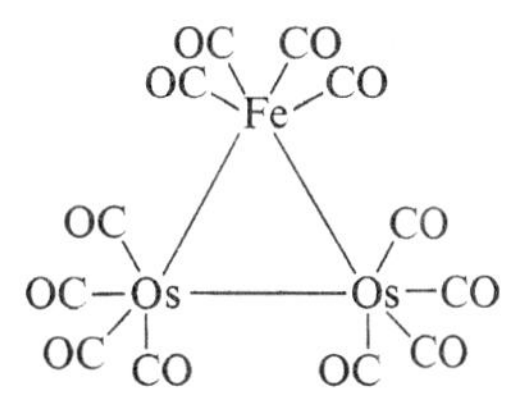

M 772.372

Orange cryst. Mp >150° dec.

Moss, J.R. *et al*, *J. Organomet. Chem.*, 1970, **23**, C23 (*synth, ir*)

$C_{12}Fe_2O_{12}Os$ Os-00152

Tetracarbonyl(di-μ-carbonylhexacarbonyldiiron)osmium

Di-μ-carbonyldecacarbonyldiironosmium, 8CI. *Dodecacarbonyldiironosmium*, 8CI

[32457-60-6]

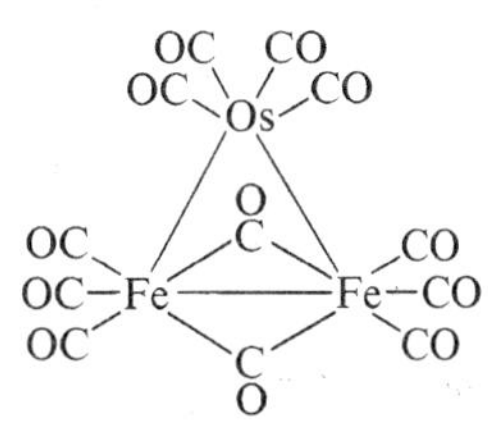

M 638.019

Purple cryst. (hexane). Mp ~130° dec.

Knight, J.A. *et al*, *J. Chem. Soc., Chem. Commun.*, 1970, 1006 (*synth, ir*)
Moss, J.R. *et al*, *J. Organomet. Chem.*, 1970, **23**, C23 (*synth, ir*)

$C_{12}HMnO_{12}Os_3$ Os-00153

Dodecacarbonylhydromanganesediosmium, 8CI

[12560-43-9]

$MnOs_3H(CO)_{12}$

M 962.671

No phys. props. reported.

Knight, J. *et al*, *J. Chem. Soc., Chem. Commun.*, 1971, 62 (*synth, ir*)

$C_{12}HO_{12}Os_2Rh$ Os-00154

Dodecacarbonylhydrorheniumdiosmium, 8CI

[12560-44-0]

$RhOs_2H(CO)_{12}$

M 820.438

No phys. props. reported.

Knight, J. *et al*, *J. Chem. Soc., Chem. Commun.*, 1971, 62 (*synth, ir*)

$C_{12}HO_{12}Os_3^{\oplus}$ Os-00155

Dodecacarbonyl-μ-hydrotriosmium(1+), 8CI

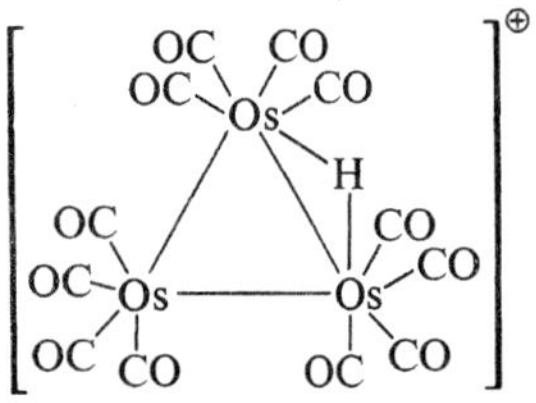

M 907.733 (ion)

Hexafluorophosphate: [26220-99-5].
$C_{12}HF_6O_{12}Os_3P$ M 1052.697
Cream solid.

Knight, J. *et al*, *J. Chem. Soc. (A)*, 1970, 711 (*synth, ir, pmr, ms*)
White, J.W. *et al*, *J. Chem. Soc. (A)*, 1971, 2843 (*struct*)

$C_{12}H_2O_{10}Os_3$ Os-00156

μ-Carbonylnonacarbonyl[μ_3-[(1-η:1,2-η:2-η)-1,2-ethenediyl-]]triosmium, 9CI

μ-Carbonylnonacarbonyl[μ_3-[(1-η:1,2-η:2-η)ethyne-]]triosmium, 9CI

[57373-35-0]

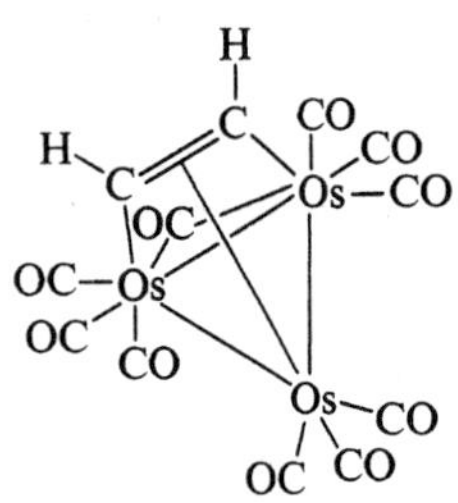

M 876.742

The orange and colourless materials described have identical spectra. Orange or colourless cryst. Mp 120-4°. Thermally decarbonylates to Nonacarbonyl[μ_3-[(1-η:1,2-η:1,2-η)ethynyl]]-μ-hydrotriosmium, Os-00122 and reacts with excess C_2H_2 to give [μ-[(1,2,3,4-η:1,4-η)-1,3-Butadiene-1,4-diyl]]nonacarbonyltriosmium, Os-00216 .

Deeming, A.J. *et al*, *J. Chem. Soc., Dalton Trans.*, 1975, 1614 (*synth, ir, pmr*)
Ferrari, R.P. *et al*, *Gazz. Chim. Ital.*, 1975, **105**, 939 (*synth, ir*)
Bryan, E.G. *et al*, *J. Chem. Soc., Dalton Trans.*, 1977, 1328 (*synth, ir, pmr*)

$C_{12}H_2O_{11}Os_3$ Os-00157

μ-Carbonyldecacarbonyl-μ-methylenetriosmium, 10CI

Undecacarbonyl-μ-methylenetriosmium

[77208-32-3]

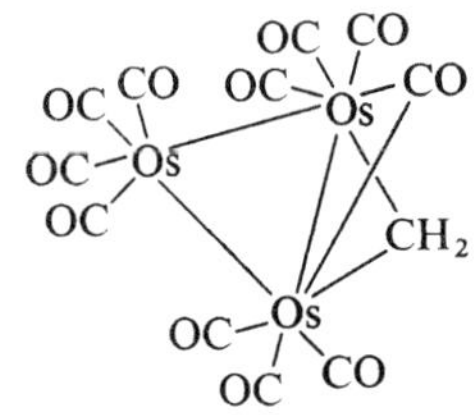

M 892.741

Orange-red air-stable cryst.

Steinmetz, G.R. *et al*, *J. Am. Chem. Soc.*, 1981, **103**, 1278 (*synth, ir, pmr, cmr, ms*)

$C_{12}H_2O_{12}Os_3$ Os-00158

Dodecacarbonyldihydrotriosmium, 10CI

[62863-43-8]

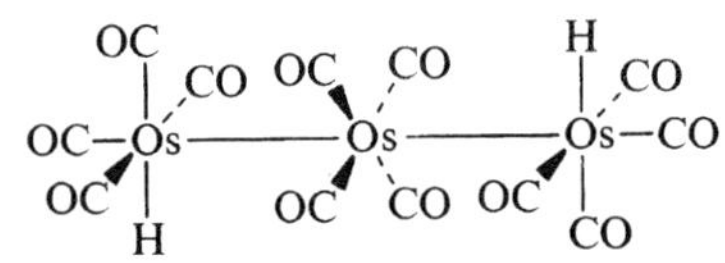

M 908.741

V. pale-yellow or almost colourless cryst. (pentane). Mp 95-8°. $Bp_{0.01}$ 80° subl.

Moss, J.R. *et al*, *Inorg. Chem.*, 1977, **16**, 75 (*synth, ir, pmr, ms*)
Gochin, M. *et al*, *J. Organomet. Chem.*, 1980, **192**, 409 (*ir, raman, cmr*)

$C_{12}H_2O_{12}Os_4^{\ominus\ominus}$ Os-00159

Dodecacarbonyldi-μ-hydrotetraosmate(2−), 10CI

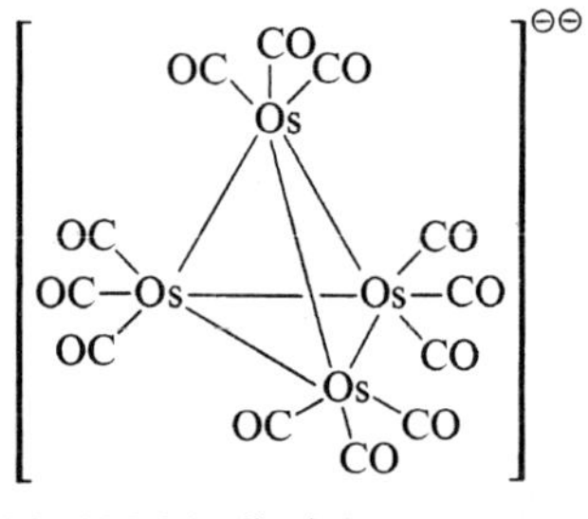

2H atoms not shown

M 1098.941 (ion)

Bis[bis(triphenylphosphine)iminium] salt: [69215-27-6].
$C_{84}H_{62}N_2O_{12}Os_4P_4$ M 2176.115
Orange tablets (Et_2O/CH_2Cl_2). Dec. at 163°.

Johnson, B.F.G. *et al*, *J. Organomet. Chem.*, 1978, **162**, 179 (*synth, ir, pmr, cryst struct*)

$C_{12}H_2O_{12}Os_4S_2$ Os-00160

Dodecacarbonyldi-μ-hydrodi-μ_3-thioxotetraosmium

[72314-28-4]

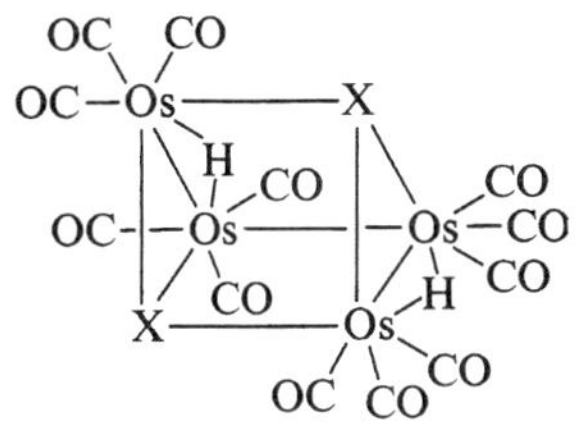

X = S

M 1163.061

Probable positions of H atoms shown. Yellow solid.

Johnson, B.F.G. *et al*, *J. Chem. Soc., Chem. Commun.*, 1979, 719 (*synth, ir*)

$C_{12}H_2O_{12}Os_4Se_2$ Os-00161

Dodecacarbonyldi-μ-hydrodi-μ_3-selenoxotetraosmium

[72268-06-5]

As Dodecacarbonyldi-μ-hydrodi-μ_3-thioxotetraosmium, Os-00160 with

X = Se

M 1256.861

Yellow platelets (hexane).

Johnson, B.F.G. *et al*, *J. Chem. Soc., Chem. Commun.*, 1979, 719 (*synth, ir, struct*)

$C_{12}H_3IO_{12}Os_4$ Os-00162

Dodecacarbonyltri-μ-hydro-μ-iodotetraosmium, 10CI

[67966-16-9]

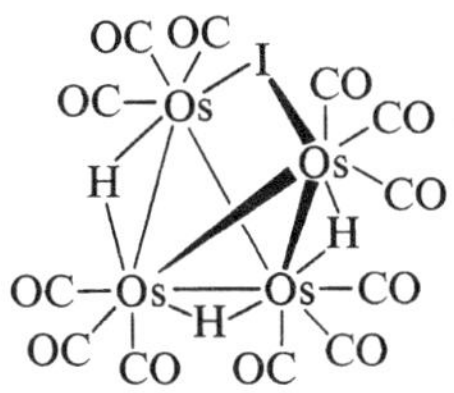

M 1226.853

Yellow cryst. (cyclohexane).

Johnson, B.F.G. *et al*, *J. Chem. Soc., Dalton Trans.*, 1978, 673 (*synth, ir, cryst struct*)

Johnson, B.F.G. *et al*, *J. Chem. Soc., Dalton Trans.*, 1980, 1248 (*neutron struct*)

$C_{12}H_3O_{12}Os_4^{\ominus}$ Os-00163

Dodecacarbonyltri-μ-hydrotetraosmate(1−), 10CI

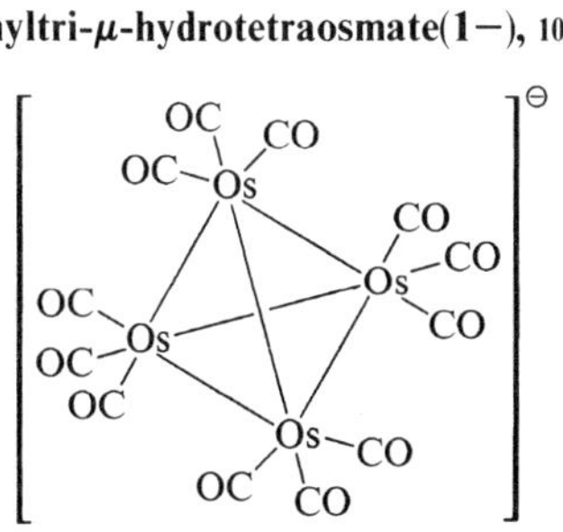

M 1099.949 (ion)

H atoms not shown.

Tetramethylammonium salt: [69074-64-2].

$C_{16}H_{15}NO_{12}Os_4$ M 1174.094

Yellow platelets (CH_2Cl_2/hexane).

Johnson, B.F.G. *et al*, *Acta Crystallogr., Sect. B*, 1978, **34**, 3765 (*cryst struct*)

Johnson, B.F.G. *et al*, *J. Chem. Soc., Dalton Trans.*, 1978, 673 (*synth*)

Johnson, B.F.G. *et al*, *J. Organomet. Chem.*, 1978, **162**, 179 (*synth*)

Constable, E.C. *et al*, *J. Chem. Soc., Chem. Commun.*, 1982, 754 (*nmr*)

$C_{12}H_4O_{10}Os_3$ Os-00164

Decacarbonyl[μ-[(1-η:1,2-η)ethenyl]]-μ-hydrotriosmium, 9CI

[59453-42-8]

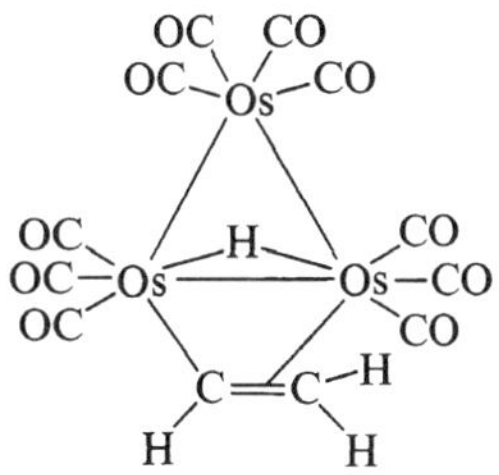

M 878.758

Bridging vinyl group is fluxional. Orange cryst. (pentane). Mp 108-11° dec. Converted into Nonacarbonyl[μ_3-[(1-η:1-η:1,2-η)ethenylidene]]di-μ-hydrotriosmium, Os-00126 at 120°.

Deeming, A.J. *et al*, *J. Chem. Soc., Dalton Trans.*, 1975, 1614 (*synth, ir, pmr*)

Keister, J.B. *et al*, *J. Organomet. Chem.*, 1975, **85**, C29 (*synth, ir, pmr*)

Shapley, J.R. *et al*, *J. Organomet. Chem.*, 1975, **94**, C43 (*cmr*)

Orpen, A.G. *et al*, *Acta Crystallogr., Sect. B*, 1978, **34**, 2466 (*struct*)

Andrews, J.R. *et al*, *Inorg. Chem.*, 1982, **21**, 2874 (*ir, raman*)

$C_{12}H_4O_{11}Os_3$ Os-00165

Decacarbonyl-μ-hydro[μ-(acetyl-*C*:*O*)]triosmium, 10CI

[65908-54-5]

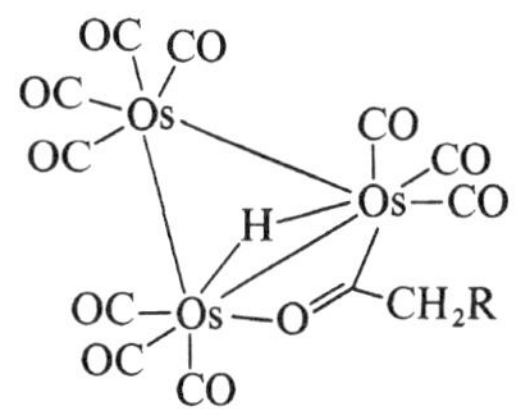

R = H

M 894.757

Yellow cryst. (pentane).

Azam, K.A. *et al*, *J. Chem. Soc., Dalton Trans.*, 1981, 91 (*synth, ir, pmr*)

Johnson, B.F.G. *et al*, *J. Organomet. Chem.*, 1981, **216**, C56 (*synth, struct*)

$C_{12}H_4O_{11}Os_3$ Os-00166

Decacarbonyl-μ-hydro[μ-(methoxymethylidyne)]triosmium, 9CI

[69048-01-7]

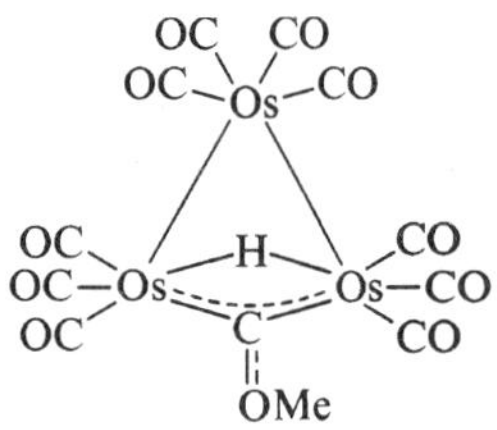

M 894.757

Yellow cryst. (MeOH). Reacts with H_2 at 150° to give $H_3Os_3(\mu COMe)(CO)_9$. Subl. *in vacuo.*

Gavens, P.D. *et al, J. Organomet. Chem.*, 1978, **162**, 389 (*synth, ir, pmr, cmr*)
Keister, J.B. *et al, Organometallics*, 1983, **2**, 219 (*synth, pmr, cmr, ir*)

$C_{12}H_4O_{12}Os_4$ Os-00167

Dodecacarbonyltetra-μ-hydrotetraosmium, 9CI

[12375-04-1]

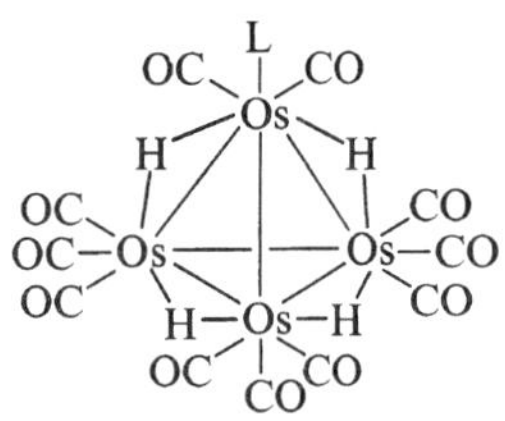

L = CO

M 1100.956

Pale-yellow microprisms (C_6H_6). Mp 232-40° dec. pK_a 12.15 (MeOH, 25°).

Johnson, B.F.G. *et al, J. Chem. Soc., Dalton Trans.*, 1970, 1684 (*ms*)
Kaesz, H.D. *et al, J. Chem. Soc., Chem. Commun.*, 1971, 477 (*synth*)
Johnson, B.F.G. *et al, J. Chem. Soc., Dalton Trans.*, 1978, 673 (*synth, pmr, ir*)
Moss, J.R. *et al, J. Chem. Soc., Dalton Trans.*, 1979, 89 (*synth, ir*)
Walker, H.W. *et al, J. Am. Chem. Soc.*, 1979, **101**, 7428.
Nicholls, J.N. *et al, J. Chem. Soc., Dalton Trans.*, 1982, 1395 (*props, ir*)

$C_{12}H_5O_{12}Os_3Re$ Os-00168

Tricarbonyltri-μ-hydro(nonacarbonyldi-μ-hydrotriosmium-)rhenium, 10CI

Dodecacarbonylpenta-μ-hydrotriosmiumrhenium

[77424-05-6]

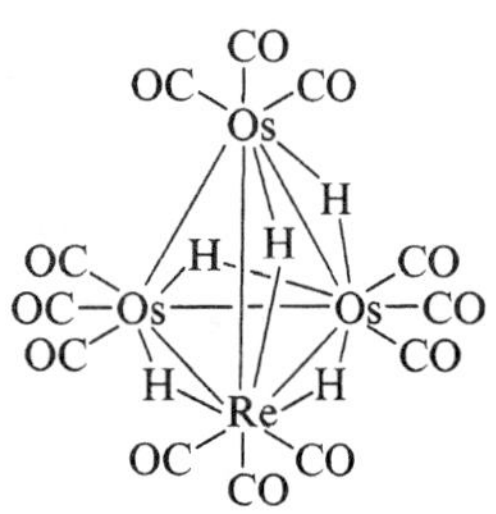

M 1097.971

Yellow cryst.

Churchill, M.R. *et al, J. Am. Chem. Soc.*, 1981, **103**, 2430 (*synth, struct, ir, pmr, ms*)

$C_{12}H_6O_{10}Os_3S$ Os-00169

Decacarbonyl[μ-(ethanethiolato)]-μ-hydrotriosmium, 10CI

[23733-18-8]

As μ-Bromodecacarbonyl-μ-hydrotriosmium, Os-00092 with

X = SEt

M 912.833

Yellow cryst. (C_6H_6). Mp 141-2°.

Crooks, G.R. *et al, J. Chem. Soc. (A)*, 1969, 797 (*synth, ir, pmr*)
Allen, V.F. *et al, J. Organomet. Chem.*, 1977, **140**, 297 (*struct*)
Bryan, E.G. *et al, J. Chem. Soc., Dalton Trans.*, 1978, 196 (*cmr*)

$C_{12}H_6O_{11}Os_3$ Os-00170

Nonacarbonyltri-μ-hydro[μ₃-(2-methoxy-2-oxoethylidyne-)]triosmium

[78697-98-0]

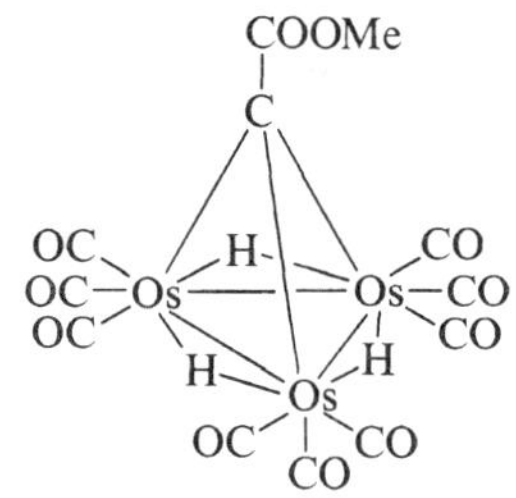

M 896.773

Sievert, A.C. *et al, Organometallics*, 1982, **1**, 214 (*synth, ir, ms, pmr*)

$C_{12}H_8O_9Os_3S$ Os-00171

Nonacarbonyl(η^2-ethene)-μ-hydro[μ-(methanethiolato)]triosmium, 10CI

[68796-11-2]

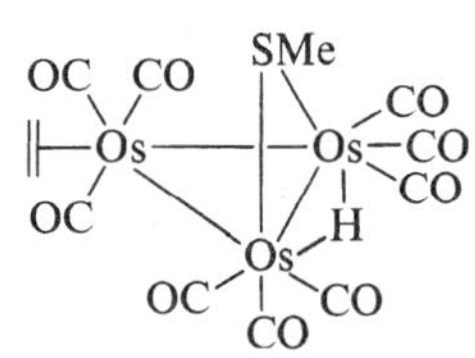

M 898.850

Yellow cryst. (hexane). Generates Nonacarbonyl-μ-hydro[μ_3-(methanethiolato)]triosmium, Os-00100 on heating.

Johnson, B.F.G. *et al, Acta Crystallogr., Sect. B*, 1980, **36**, 703 (*struct*)
Johnson, B.F.G. *et al, J. Organomet. Chem.*, 1981, **213**, 249 (*synth, pmr, ir*)

$C_{12}H_9N_2O_{14}Os_3$ Os-00172

Nonacarbonyldi-μ-nitrosyl(trimethyl phosphite-*P*)triosmium, 10CI

[64972-46-9]

As Decacarbonyldi-μ-nitrosyltriosmium, Os-00120 with

L = $P(OMe)_3$

M 975.808

No phys. props. reported.

Bellard, S. *et al, Acta Crystallogr., Sect. B*, 1980, **36**, 705 (*struct*)

$C_{12}H_9N_3O_{11}Os_3$ Os-00173

Nonacarbonyldi-μ-nitrosyl(trimethylamine)triosmium

*Nonacarbonyl(*N,N*-dimethylmethanamine)di-μ-nitrosyltriosmium*, 10CI

[72910-30-6]

As Decacarbonyldi-μ-nitrosyltriosmium, Os-00120 with

L = Me_3N

M 941.817

Johnson, B.F.G. *et al*, *J. Chem. Soc., Chem. Commun.*, 1979, 916 (*synth, struct, ir, pmr*)

$C_{12}H_{10}O_4Os$ Os-00174

1,1′-Osmocenedicarboxylic acid

1,1′-Dicarboxyosmocene

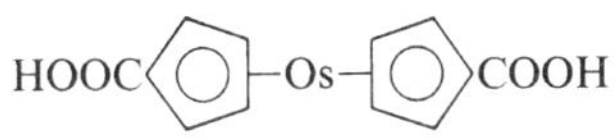

M 408.409

White cryst. by subl. Obt. contaminated by Osmocenecarboxylic acid, Os-00136 .

Rausch, M.D. *et al*, *J. Am. Chem. Soc.*, 1960, **82**, 76 (*synth*)

$C_{12}H_{11}NO_3Os$ Os-00175

Tricarbonyl[(1,2,3,5-η)-4-cyano-2-cyclooctene-1,4-diyl]osmium, 9CI

[55938-92-6]

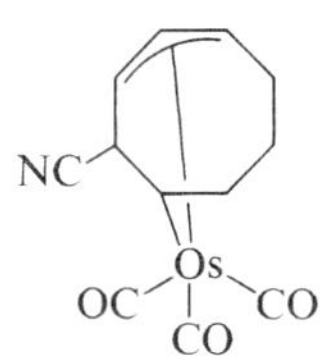

M 407.424

Colourless cryst. Mp 115°. $Bp_{0.1}$ 80°, $Bp_{0.05}$ 60°.

Deeming, A.J. *et al*, *J. Chem. Soc., Dalton Trans.*, 1974, 2093 (*synth, ir, pmr*)

$C_{12}H_{12}Cl_3Os_2^{\oplus}$ Os-00176

Bis(η^6-benzene)tri-μ-chlorodiosmium(1+), 10CI

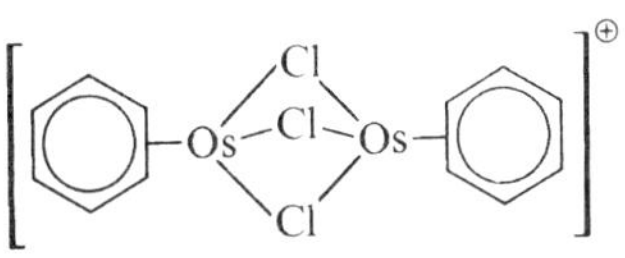

M 642.986 (ion)

Tetrafluoroborate: [70317-01-0].

$C_{12}H_{12}BCl_3F_4Os_2$ M 729.789

No phys. props. reported.

Arthur, T. *et al*, *J. Organomet. Chem.*, 1981, **208**, 369 (*synth, pmr, cmr*)

$C_{12}H_{12}OOs$ Os-00177

Acetylosmocene, 9CI

Methyl osmocenyl ketone

[33039-43-9]

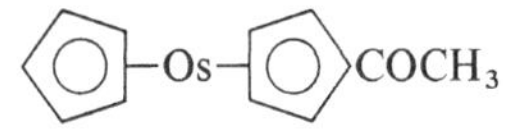

M 362.426

V. pale-yellow needles (heptane). Mp 130.5-131.5°. Subl. at 70-80° *in vacuo*.

Rausch, M.D. *et al*, *Chem. Ind. (London)*, 1958, 756 (*synth*)
Riemschneider, R. *et al*, *Monatsh. Chem.*, 1959, **90**, 658 (*synth*)
Rausch, M.D. *et al*, *J. Am. Chem. Soc.*, 1960, **82**, 76 (*synth, ir*)
Hill, E.A. *et al*, *J. Am. Chem. Soc.*, 1961, **83**, 3840 (*synth*)
Rausch, M.D. *et al*, *J. Org. Chem.*, 1963, **28**, 3225 (*pmr*)

$C_{12}H_{12}O_4Os$ Os-00178

Tricarbonyl[(2,3,4,5-η)-2,3,4,5-tetramethyl-2,4-cyclopentadien-1-one]osmium

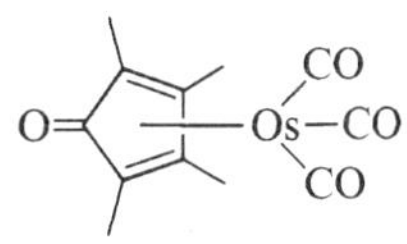

M 410.424

White cryst. Mp 112°. $Bp_{0.1}$ ~20° subl.

Bruce, M.I. *et al*, *J. Chem. Soc. (A)*, 1969, 987 (*synth, ir, ms, pmr*)

$C_{12}H_{12}Os^{\oplus\oplus}$ Os-00179

Bis(η^6-benzene)osmium(2+), 10CI

[76680-98-3]

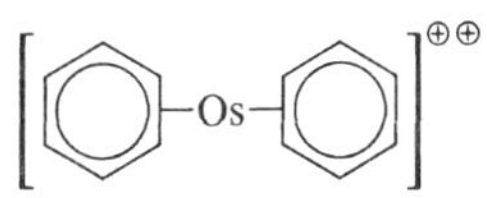

M 346.427 (ion)

Bis(hexafluorophosphate): [76830-28-9].

$C_{12}H_{12}F_{12}OsP_2$ M 636.355

Off-white.

Domaille, P.J. *et al*, *J. Organomet. Chem.*, 1980, **202**, 191 (*synth*)

$C_{12}H_{12}Os$ Os-00180

Vinylosmocene

π-Cyclopentadienyl(vinyl-π-cyclopentadienyl)osmium, 8CI. *Ethenylosmocene*

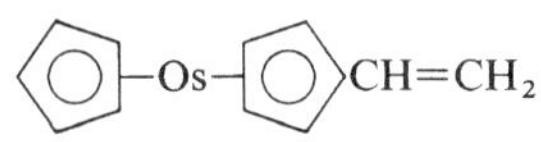

M 346.427

White cryst. by subl. Mp 69.0-70.0°.

Buell, G.R. *et al*, *J. Am. Chem. Soc.*, 1962, **84**, 40 (*synth*)
Rausch, M.D. *et al*, *J. Organomet. Chem.*, 1968, **11**, 317 (*synth, pmr*)

$C_{12}H_{12}OsRu_2$ Os-00181

Tetracarbonyl(octacarbonyldiruthenium)osmium, 10CI

Tetracarbonyl(octacarbonyldiruthenio)osmium, 8CI. *Dodecacarbonylosmiumdiruthenium*

[12389-47-8]

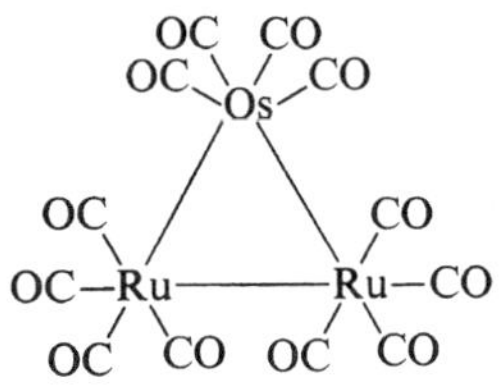

Presumed structure

M 548.567

Shows rapid exchange of axial and equatorial CO groups above 30°.

Johnson, B.F.G. *et al*, *J. Chem. Soc., Chem. Commun.*, 1968, 861 (*synth*)
Lewis, J. *et al*, *Acc. Chem. Res.*, 1968, **1**, 245 (*ms*)
Ferrari, R.P. *et al*, *J. Chem. Soc., Dalton Trans.*, 1978, 1164 (*synth, ms*)
Kettle, S.F.A. *et al*, *Inorg. Chem.*, 1979, **18**, 2749 (*ir, raman*)
Fox, J.R. *et al*, *Inorg. Chem.*, 1980, **19**, 2574 (*synth*)
Koridze, A.A. *et al*, *Izv. Akad. Nauk SSSR, Ser. Khim.*, 1981, 1181; *CA*, **95**, 70500y (*nmr*)

$C_{12}H_{14}OOs$ Os-00182

(1-Hydroxyethyl)osmocene

α-Methylosmocenemethanol. α-Osmocenylethanol. Methylosmocenylcarbinol

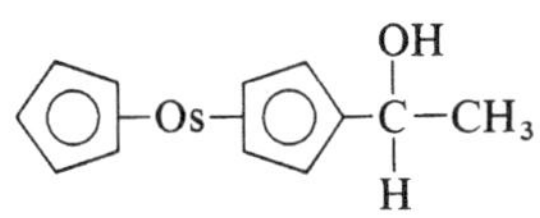

M 364.442

(±)-*form*

White cryst. (hexane). Mp 75-75.2°.

Ac:

$C_{14}H_{16}O_2Os$ M 406.479

White cryst. (isopropyl ether/heptane or by subl.). Mp 67.2-67.7°.

Hill, E.A. *et al*, *J. Am. Chem. Soc.*, 1961, **83**, 3840 (*synth*)
Buell, G.R. *et al*, *J. Am. Chem. Soc.*, 1962, **84**, 40 (*synth*)

$C_{12}H_{14}Os$ Os-00183

(η^6-Benzene)[(1,2,3,4-η)-1,3-Cyclohexadiene]osmium

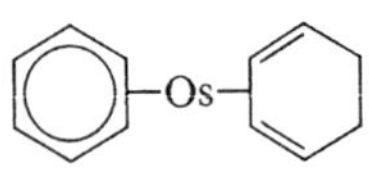

M 348.443

Colourless cryst. (hexane). Mp 114-6°. Bp 60-5° subl. *in vacuo*. Dec. at 170°.

Fischer, E.O. *et al*, *Chem. Ber.*, 1963, **96**, 3217 (*synth, ir, pmr*)

$C_{12}H_{14}Os$ Os-00184

1,1′-Dimethylosmocene, 9CI

[40672-07-9]

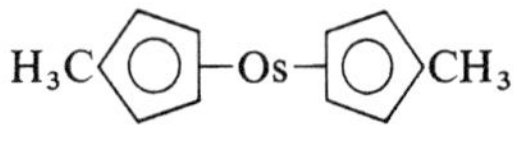

M 348.443

Almost colourless cryst. with camphoraceous odour (pet. ether). Mp 82-3°.

Evans, S. *et al*, *J. Chem. Soc., Faraday Trans. 2*, 1972, **68**, 1847 (*synth, pe*)

$C_{12}H_{14}Os$ Os-00185

Ethylosmocene

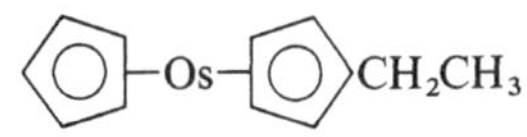

M 348.443

Mp 33-33.5°.

Mark, V. *et al*, *Inorg. Chem.*, 1964, **3**, 1067 (*synth, pmr*)

$C_{12}H_{15}BrO_2Os$ Os-00186

Bromodicarbonyl[(1,2,3,4,5-η)-1,2,3,4,5-pentamethyl-2,4-cyclopentadien-1-yl]osmium

[81554-89-4]

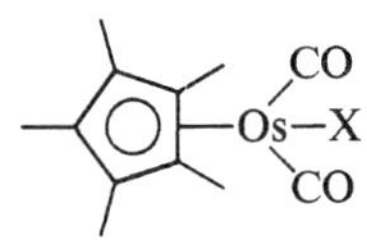

X = Br

M 461.353

Pale-yellow cryst. (CH_2Cl_2/hexane). Mp 185-7°.

Hoyano, J.K. *et al*, *Inorg. Chem.*, 1982, **21**, 3095 (*synth, ir, pmr, ms*)

$C_{12}H_{15}IO_2Os$ Os-00187

Dicarbonyl[(1,2,3,4,5-η)-1,2,3,4,5-pentamethyl-2,4-cyclopentadien-1-yl]iodoosmium

[81554-98-5]

As Bromodicarbonyl[(1,2,3,4,5-η)-1,2,3,4,5-pentamethyl-2,4-cyclopentadien-1-yl]osmium, Os-00186 with

X = I

M 508.354

Yellow cryst. (CH_2Cl_2/heptane). Mp 198-200°.

Hoyano, J.K. *et al*, *Inorg. Chem.*, 1982, **21**, 3095 (*synth, ms, ir, pmr*)

$C_{12}H_{16}O_2Os$ Os-00188

Dicarbonylhydro[(1,2,3,4,5-η)-1,2,3,4,5-pentamethyl-2,4-cyclopentadien-1-yl]osmium

[81554-96-3]

As Bromodicarbonyl[(1,2,3,4,5-η)-1,2,3,4,5-pentamethyl-2,4-cyclopentadien-1-yl]osmium, Os-00186 with

X = H

M 382.457

Colourless solid by subl. Mp 60.5-61.0°. Thermally stable.

Hoyano, J.K. *et al*, *Inorg. Chem.*, 1982, **21**, 3095 (*synth, pmr, ir, ms*)

$C_{12}H_{18}F_6O_8OsP_2$ Os-00189

(Hexafluoro-2-butyne)dicarbonylbis(trimethyl phosphite-*P*)-osmium

Dicarbonyl[bis(trifluoromethyl)vinylene]bis(phosphorous acid)osmium hexamethyl ester, 8CI

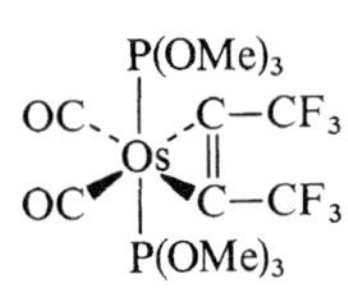

M 656.407

White cryst. (hexane). Mp 131°.

Burt, R. *et al*, *J. Chem. Soc.* (*A*), 1970, 2981 (*synth*, *ir*, *pmr*, *nmr*, *ms*)

$C_{12}H_{18}GeO_2Os$ Os-00190

Dicarbonyl[(1,2,3,4,5-η)-2,4-cycloheptadien-1-yl](trimethylgermyl)osmium, 9CI

[59423-94-8]

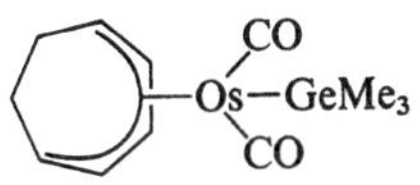

M 457.063

Colourless cryst. (hexane). Mp 48°.

Knox, S.A.R. *et al*, *J. Chem. Soc., Dalton Trans.*, 1976, 552 (*synth*, *ir*, *pmr*)

$C_{12}H_{18}Ge_3O_6Os_2$ Os-00191

Hexacarbonyltris[μ-(dimethylgermylene)]diosmium

[31005-45-5]

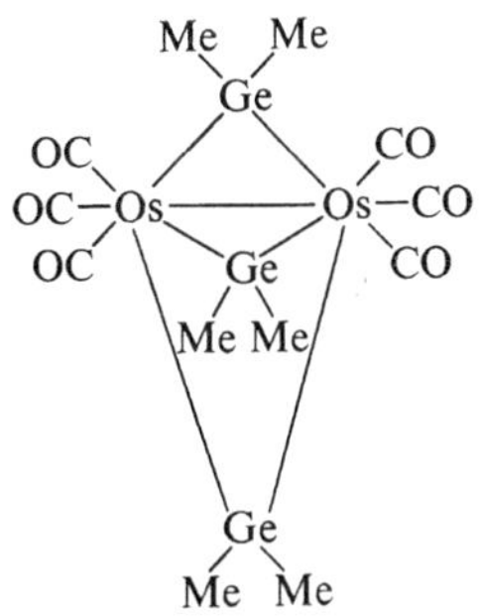

M 856.441

Yellow.

Knox, S.A.R. *et al*, *J. Chem. Soc.* (*A*), 1971, 2874 (*synth*, *ir*)

$C_{12}H_{18}N_6Os^{\oplus\oplus}$ Os-00192

Hexakis(isocyanomethane)osmium(2+), 9CI

Hexakis(methyl isocyanide)osmium(2+)

$Os(CNMe)_6^{\oplus\oplus}$

M 436.514 (ion)

Diperchlorate: [42536-05-0].
$C_{12}H_{18}Cl_2N_6O_8Os$ M 635.416
Long white needles (MeOH). Mp >350°.

Chatt, J. *et al*, *J. Chem. Soc., Dalton Trans.*, 1973, 1433 (*synth*, *ir*, *pmr*)

$C_{12}H_{18}O_2OsSi$ Os-00193

Dicarbonyl[(1,2,3,4,5-η)-2,4-cyclopentadien-1-yl](trimethylsilyl)osmium, 9CI

[59423-89-1]

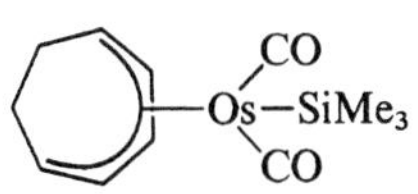

M 412.559

Colourless cryst. (hexane). Mp 44-5°. $Bp_{0.01}$ 30° subl.

Knox, S.A.R. *et al*, *J. Chem. Soc., Dalton Trans.*, 1976, 552 (*synth*, *ir*, *pmr*)

$C_{12}H_{22}OsP^{\oplus}$ Os-00194

(η^6-Benzene)(η^2-ethene)methyl(trimethylphosphine)osmium(1+), 10CI

As (η^6-Benzene)(η^2-ethene)iodo(trimethylphosphine)osmium(1+), Os-00144 with

X = Me

M 387.480 (ion)

Hexafluorophosphate: [75353-35-4].
$C_{12}H_{22}F_6OsP_2$ M 532.444
No phys. props. reported.

Werner, H. *et al*, *J. Organomet. Chem.*, 1979, **174**, C63; 1980, **194**, C7 (*synth*, *nmr*)

$C_{12}H_{24}O_6OsP_2$ Os-00195

(η^6-Benzene)bis(trimethyl phosphite-*P*)osmium, 10CI

[75353-19-4]

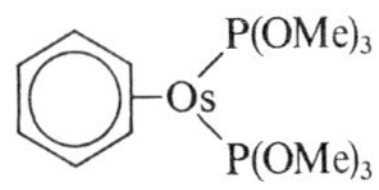

M 516.466

Extremely air-sensitive.

Werner, H. *et al*, *J. Organomet. Chem.*, 1980, **194**, C7 (*synth*, *pmr*)

$C_{12}H_{38}OsP_4$ Os-00196

Dihydrotetrakis(trimethylphosphine)osmium

[84108-59-8]

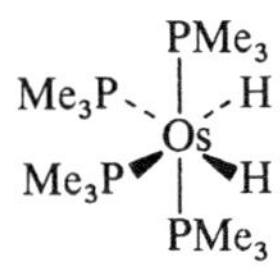

M 496.527

cis-form

Air-stable white cryst. $Bp_{0.01}$ 80° subl.

Werner, H. *et al*, *Organometallics*, 1983, **2**, 547 (*synth*, *nmr*, *pmr*, *ir*)

$C_{12}I_2O_{12}Os_3$ Os-00197

Dodecacarbonyldiiodotriosmium

[66429-64-9]

As Dibromododecacarbonyltriosmium, Os-00149 with

X = Y = I

M 1160.534

Yellow prisms (toluene). Bp 118° subl. *in vacuo*.

Hartley, D. *et al*, *J. Chem. Soc., Chem. Commun.*, 1968, 493 (*ir*, *raman*)

Johnson, B.F.G. *et al*, *J. Chem. Soc.* (*A*), 1968, 2859 (*synth*, *ir*, *ms*)

Cook, N. *et al*, *J. Chem. Soc., Dalton Trans.*, 1977, 1744 (*struct*, *ir*)

Vaglio, G.A., *J. Organomet. Chem.*, 1979, **169**, 83 (*ms*)

$C_{12}MnO_{12}Os_2^{\ominus}$ **Os-00198**

Dodecacarbonylmanganatediosmate(1−), 8CI

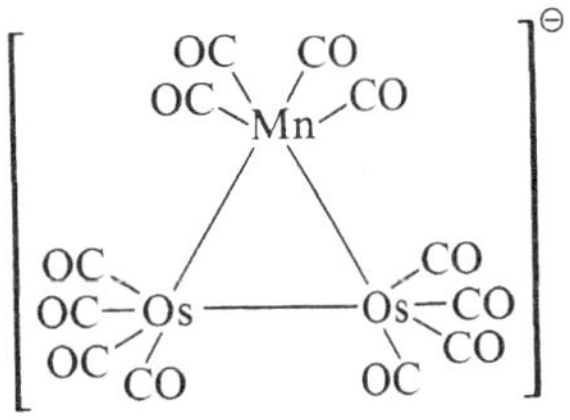

M 771.463 (ion)

Tetramethylammonium salt: [12563-65-4].
$C_{16}H_{12}MnNO_{12}Os_2$ M 845.608
Characterised spectroscopically.

Knight, J. *et al*, *J. Chem. Soc., Chem. Commun.*, 1971, 62 (*synth, ir*)

$C_{12}O_{12}Os_2Re^{\ominus}$ **Os-00199**

Dodecacarbonylrhenatediosmate(1−), 8CI

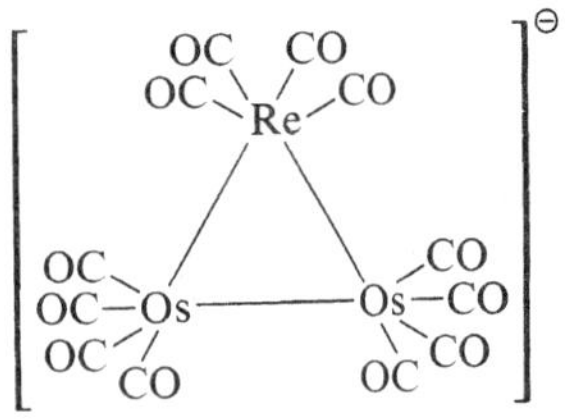

M 902.732 (ion)

Tetramethylammonium salt: [12563-67-6].
$C_{16}H_{12}NO_{12}Os_2Re$ M 976.877
Characterised spectroscopically.

Knight, J. *et al*, *J. Chem. Soc., Chem. Commun.*, 1971, 62 (*synth, ir*)

$C_{12}O_{12}Os_2Ru$ **Os-00200**

Octacarbonyl(tetracarbonylruthenium)diosmium, 10CI

Octacarbonyl[μ-(tetracarbonylruthenio)]diosmium, 8CI.
Dodecacarbonyldiosmiumruthenium

[12389-50-3]

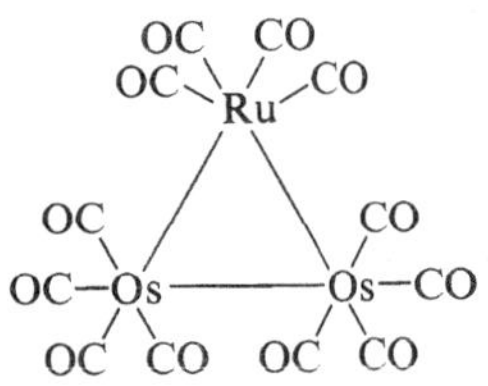

M 817.595

No phys. props. reported. ir ν_{CO} 2066vs, 2032s, 2014w, 2004w cm^{-1}.

Johnson, B.F.G. *et al*, *J. Chem. Soc., Chem. Commun.*, 1968, 861 (*synth*)
Ferrari, R.P. *et al*, *J. Chem. Soc., Dalton Trans.*, 1978, 1164 (*synth, ms*)
Kettle, S.F.A. *et al*, *Inorg. Chem.*, 1979, **18**, 2749 (*raman*)
Battiston, G.A. *et al*, *Inorg. Chem.*, 1980, **19**, 1961 (*ir, raman*)

$C_{12}O_{12}Os_3$ **Os-00201**

Dodecacarbonyltriosmium, 9CI

[15696-40-9]

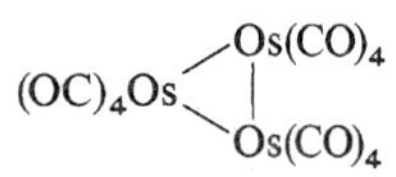

M 906.725

Yellow cryst. (C_6H_6 or $CHCl_3$ or by subl. *in vacuo*). $Bp_{0.01}$ 130° subl.

Quicksall, C.O. *et al*, *Inorg. Chem.*, 1968, **7**, 2365 (*ir, raman*)
Inorg. Synth., 1972, **13**, 92 (*synth, ir*)
Aime, S. *et al*, *Inorg. Chim. Acta*, 1975, **15**, 53 (*cmr*)
Churchill, M.R. *et al*, *Inorg. Chem.*, 1977, **16**, 878 (*struct*)

$C_{12}O_{12}Os_4S$ **Os-00202**

Dodecacarbonyl-μ₃-thioxotetraosmium

[82080-90-8]

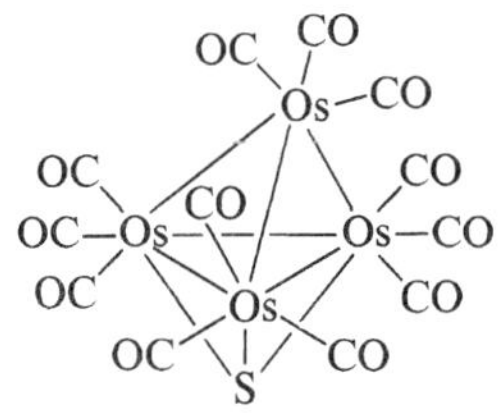

M 1128.985

Adams, R.A. *et al*, *J. Am. Chem. Soc.*, 1982, **104**, 4115 (*synth, ir*)

$C_{12}O_{12}Os_4S_2$ **Os-00203**

Dodecacarbonyldi-μ₃-thioxotetraosmium

[82093-50-3]

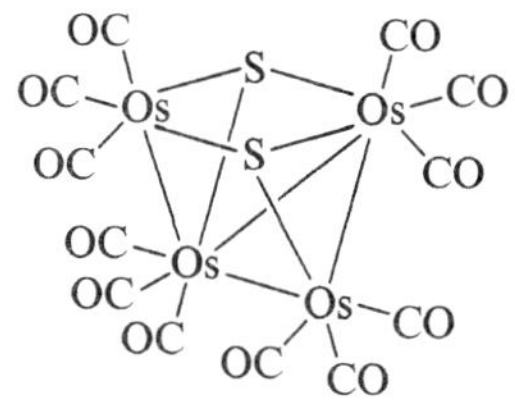

M 1161.045

Adams, R.A. *et al*, *J. Am. Chem. Soc.*, 1982, **104**, 4115 (*synth, struct*)

$C_{12}O_{16}Os_4$ **Os-00204**

Dodecacarbonyltetra-μ₃-oxotetraosmium, 9CI, 8CI

[29994-93-2]

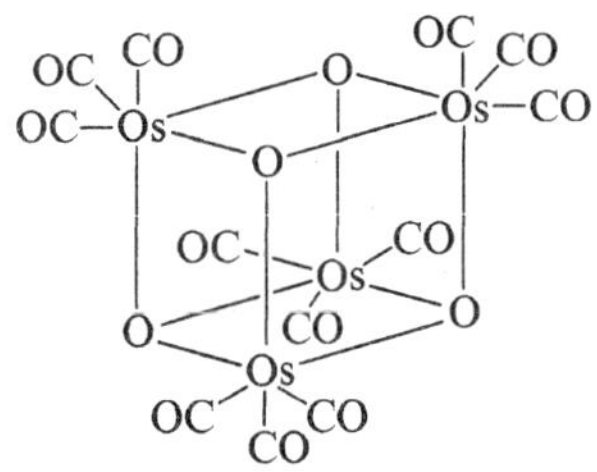

M 1160.922

Yellow-white solid. Sol. org. solvs. $Bp_{0.01}$ 170° subl.

Johnson, B.F.G. *et al*, *J. Chem. Soc., Chem. Commun.*, 1966, 391 (*synth, ms, ir*)
Bradford, C.W. *et al*, *J. Chem. Soc., Chem. Commun.*, 1967, 384 (*synth*)
Bright, D., *J. Chem. Soc., Chem. Commun.*, 1970, 1169 (*struct*)

$C_{13}HO_{13}Os_4^{\ominus}$ Os-00205

μ-Carbonyldodecacarbonyl-μ-hydrotetraosmate(1−), 10CI
Tridecacarbonylhydrotetraosmate(1−)

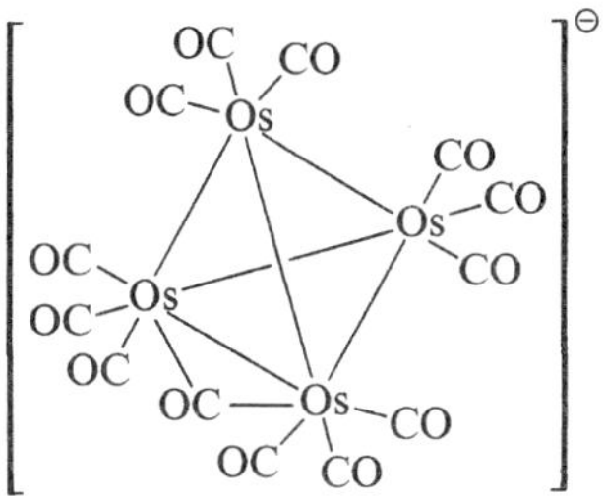

M 1125.943 (ion)
H not shown.
Bis(triphenylphosphine)iminium salt: [76468-65-0]. $C_{49}H_{31}NO_{13}Os_4P_2$ M 1664.530
Dark-red cryst.

Dawson, P.A. *et al*, *J. Chem. Soc., Chem. Commun.*, 1980, 961 (*synth, struct*)

$C_{13}H_2FeO_{13}OsRu_2$ Os-00206

Di-μ-carbonyltricarbonyl(dicarbonyliron)(hexacarbonyl-μ-hydrodiruthenium)-μ-hydroosmium, 10CI
Di-μ-carbonyltricarbonyl(dicarbonyliron)di-μ-hydrobis(tricarbonylruthenium)osmium, 10CI. Tridecacarbonyldihydroirondirutheniumosmium
[61863-84-1]

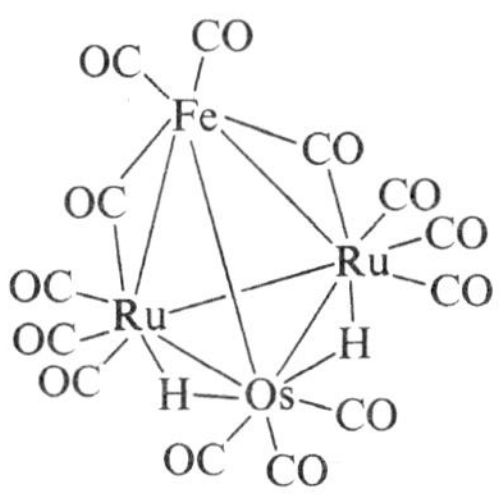

M 814.338
Fluxional molecule, interconverting mixt. of 2 isomers at r.t. The isomer shown predominates, the minor isomer having the Os exchanged with one Ru. Also shows CO group migration. Orange-red.

Geoffroy, G.L. *et al*, *J. Am. Chem. Soc.*, 1977, **99**, 304 (*synth, uv, ir, pmr*)
Gladfelter, W.L. *et al*, *Inorg. Chem.*, 1980, **19**, 2579 (*cmr, pmr*)

$C_{13}H_2FeO_{13}Os_2Ru$ Os-00207

Di-μ-carbonylhexacarbonyl(dicarbonyliron)di-μ-hydro(tricarbonylruthenium)diosmium
Tridecacarbonyldihydroirondiosmiumruthenium
[61816-94-2]

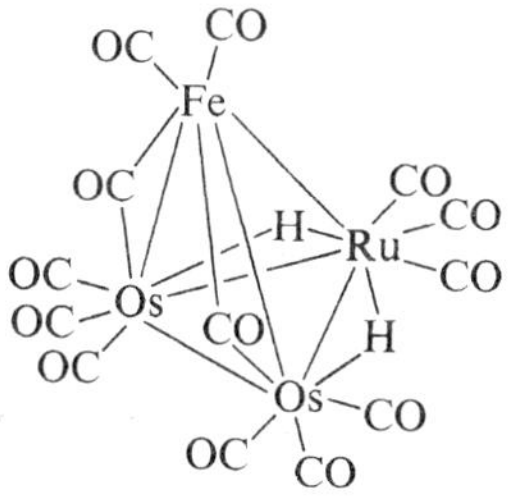

M 903.468
Fluxional molecule showing CO ligand migration and skeletal rearr. having the effect of Os-Ru interchange. Isomer shown predominates. Orange.

Geoffroy, G.L. *et al*, *J. Am. Chem. Soc.*, 1977, **99**, 6775, 7565 (*synth, cmr, ms, ir*)
Gladfelter, W.L. *et al*, *Inorg. Chem.*, 1980, **19**, 2579 (*pmr, cmr*)

$C_{13}H_2FeO_{13}Os_3$ Os-00208

Di-μ-carbonylnonacarbonyl(dicarbonyliron)di-μ-hydrotriosmium, 9CI
Tridecacarbonyldihydrotriosmiumiron, 8CI
[12563-74-5]

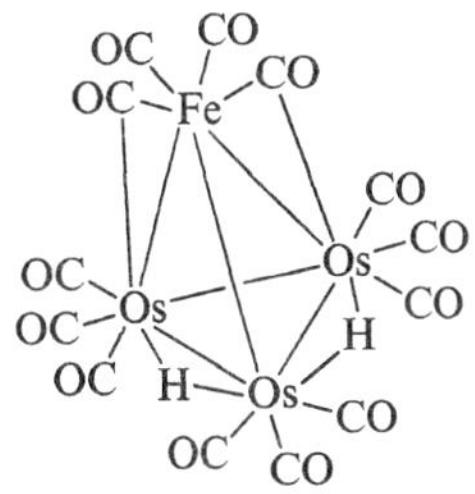

M 992.598
Possesses two semi-bridging CO groups. Mod. air-stable orange cryst. (CH_2Cl_2/hexane). Sol. hexane.

Plotkin, J.S. *et al*, *J. Am. Chem. Soc.*, 1980, **102**, 6157 (*synth, ir, struct, bibl*)
Churchill, M.R. *et al*, *Inorg. Chem.*, 1982, **21**, 1958 (*synth, cmr, cryst struct*)
Inorg. Synth., 1982, **21**, 57 (*synth, spectra*)

$C_{13}H_2O_{13}Os_3Ru$ Os-00209

Di-μ-carbonylnonacarbonyl(dicarbonylruthenium)di-μ-hydrotriosmium, 10CI
Tridecacarbonyldihydrotriosmiumruthenium
[75901-26-7]

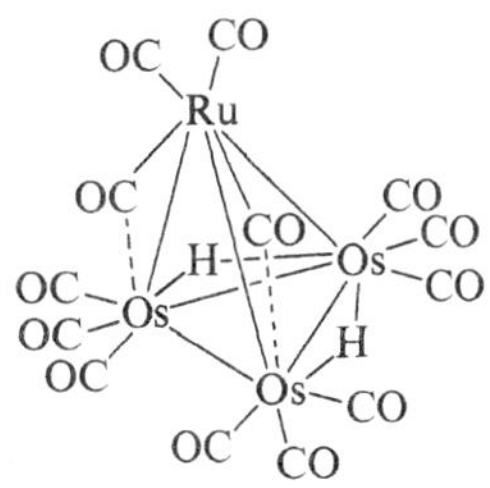

M 1037.821
Probable struct. Orange air-stable cryst. Sol. hexane.

Burkhardt, E.W. *et al*, *J. Organomet. Chem.*, 1980, **198**, 179 (*synth, ir, ms*)
Inorg. Synth., 1982, **21**, 64 (*synth, ir, pmr*)

$C_{13}H_2O_{13}Os_4$ Os-00210

μ_3-Carbonyldodecacarbonyldi-μ-hydrotetraosmium, 8CI
Tridecacarbonyldihydrotetraosmium
[21349-95-1]

$H_2Os_4(CO)_{13}$

M 1126.951
Structure not known. Orange cryst. (pentane).

Johnson, B.F.G. *et al*, *J. Chem. Soc. (A)*, 1968, 2859 (*synth, ir, ms*)

$C_{13}H_3BrO_{12}Os_3$ Os-00211

Bromododecacarbonylmethyltriosmium
[66705-63-3]

As Dibromododecacarbonyltriosmium, Os-00149 with

X = Br, Y = Me

M 1001.664
Yellow cryst.

Kelland, J.W. *et al*, *J. Organomet. Chem.*, 1978, **149**, 185 (*synth, ir, pmr, ms*)

$C_{13}H_3MnO_{13}Os_3$ Os-00212

Tridecacarbonyltrihydromanganesetriosmium, 8CI
[12563-75-6]

$MnOs_3H_3(CO)_{13}$

M 992.697
No phys. props. reported.

Knight, J. *et al*, *J. Chem. Soc., Chem. Commun.*, 1971, 62 (*synth, ir*)

$C_{13}H_3NO_{11}Os_3$ Os-00213

(Acetonitrile)undecacarbonyltriosmium, 10CI
[65702-94-5]

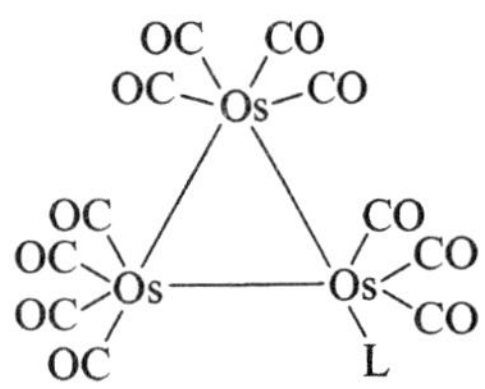

L = MeCN

M 919.767
Intermed. for Os_3 cluster synth. Yellow cryst.

Johnson, B.F.G. *et al*, *J. Organomet. Chem.*, 1978, **145**, C4 (*synth, use*)
Süss-Fink, G., *Z. Naturforsch., B*, 1980, **35**, 454 (*synth, use*)

$C_{13}H_3NO_{12}Os_4$ Os-00214

Dodecacarbonyl[μ_3-[methanaminato(2−)]]tetraosmium
[78770-95-3]

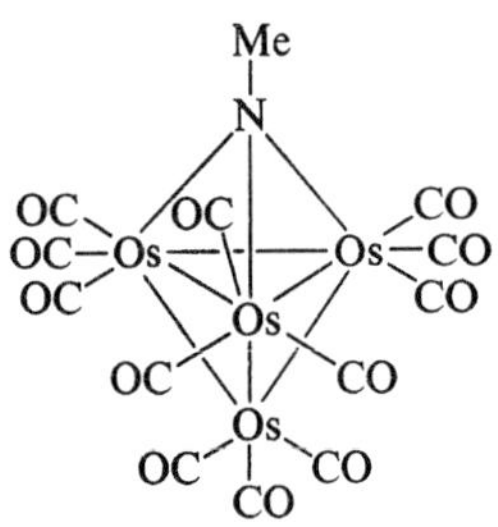

M 1125.966
No phys. props. reported.

Lin, Y.C. *et al*, *J. Organomet. Chem.*, 1981, **213**, C41 (*synth, struct, ir, pmr*)

$C_{13}H_3O_{13}Os_3Re$ Os-00215

Tridecacarbonyltrihydrorheniumtriosmium, 8CI
[12563-76-7]

$ReOs_3H_3(CO)_{13}$

M 1123.966
No phys. props. reported.

Knight, J. *et al*, *J. Chem. Soc., Chem. Commun.*, 1971, 62 (*synth, ir*)

$C_{13}H_4O_9Os_3$ Os-00216

[μ-[(1,2,3,4-η:1,4-η)-1,3-Butadiene-1,4-diyl]]nonacarbonyltriosmium, 9CI
[57373-36-1]

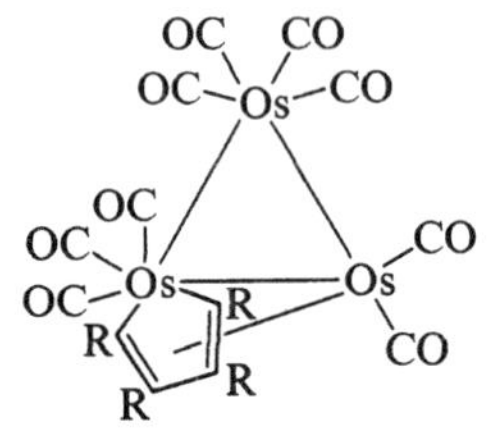

R = H

M 874.769
Pink or purple cryst. Mp 130-4°.

Deeming, A.J. *et al*, *J. Chem. Soc., Dalton Trans.*, 1975, 1614 (*synth, ir, pmr*)
Ferrari, R.P. *et al*, *Gazz. Chim. Ital.*, 1975, **105**, 939 (*synth, ir, ms*)

$C_{13}H_4O_{11}Os_4$ Os-00217

Undecacarbonyl[μ_3-[(1-η:1,2-η:2-η)-1,2-ethenediyl]]di-μ-hydrotetraosmium, 9CI
[60429-09-6]

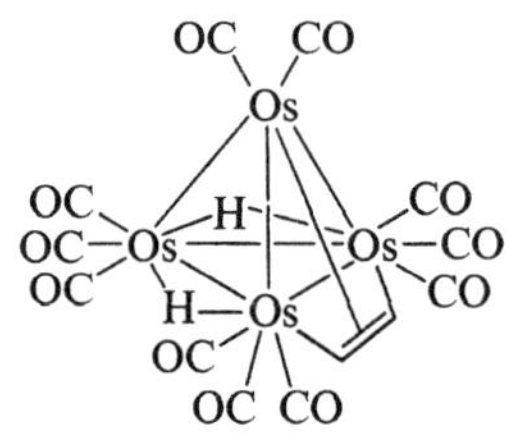

M 1096.968
Suggested struct. shown. Thermolysis (dehydrogenation) product of Undecacarbonyl[μ-(η:η^2-ethenyl)]tri-μ-hydrotetraosmium, Os-00218 . No phys. props. reported.

Bhaduri, S. *et al*, *J. Chem. Soc., Dalton Trans.*, 1979, 562 (*synth, pmr, ir*)

$C_{13}H_6O_{11}Os_4$ Os-00218

Undecacarbonyl[μ-(η:η^2-ethenyl)]tri-μ-hydrotetraosmium, 10CI
[71343-96-9]

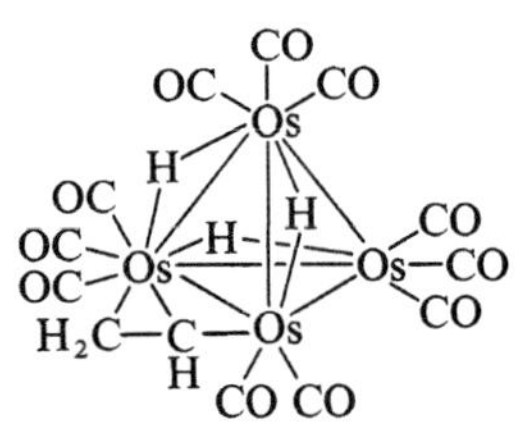

M 1098.984
Yellow-orange tablets (hexane).

Bhaduri, S. *et al*, *J. Chem. Soc., Dalton Trans.*, 1979, 562 (*synth, ir, pmr*)

$C_{13}H_7O_{13}Os_4^{\ominus}$ Os-00219

Undecacarbonyltetra-μ-hydro(methoxycarbonyl)tetraosmate(1−), 10CI

$[H_4Os_4(CO)_{11}(COOMe)]^{\ominus}$

M 1131.991 (ion)
Bis(triphenylphosphine)iminium salt: [67993-17-3].

$C_{49}H_{37}NO_{13}Os_4P_2$ M 1670.578
Pale yellow.

Johnson, B.F.G. *et al*, *J. Chem. Soc., Dalton Trans.*, 1978, 673 (*synth, ir*)

$C_{13}H_{15}O_3Os^{\oplus}$ Os-00220

Tricarbonyl[(1,2,3,4,5-η)-1,2,3,4,5-pentamethyl-2,4-cyclopentadien-1-yl]osmium(1+)

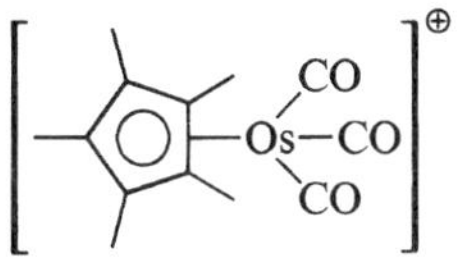

M 409.460 (ion)
Tetrafluoroborate: [81554-87-2].
$C_{13}H_{15}BF_4O_3Os$ M 496.263
Colourless cryst. (CH_2Cl_2). Dec. >310°.

Hoyano, J.K. *et al*, *Inorg. Chem.*, 1982, **21**, 3095 (*synth, ir, pmr*)

$C_{13}H_{18}GeO_2Os$ Os-00221

Dicarbonyl[(1,2,3,3a,6a-η)-1,4,5,6-tetrahydro-1-pentalenyl](trimethylgermyl)osmium, 9CI
[*η*[5]-(*Bicyclo[3.3.0]octa-2,4-dien-1-yl*)]*dicarbonyl*(*trimethylgermyl*)*osmium*
[59584-58-6]

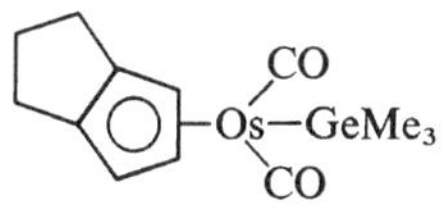

M 469.074
Pale-yellow oil.

Harris, P.J. *et al*, *J. Chem. Soc., Dalton Trans.*, 1976, 377 (*synth, pmr, ir*)

$C_{13}H_{18}O_2OsSi$ Os-00222

Dicarbonyl[(1,2,3,3a,6a-η)-1,4,5,6-tetrahydro-1-pentalenyl](trimethylsilyl)osmium, 9CI
[*η*[5]-(*Bicyclo[3.3.0]octa-2,4-dien-1-yl*)]*dicarbonyl*(*trimethylsilyl*)*osmium*
[59584-57-5]

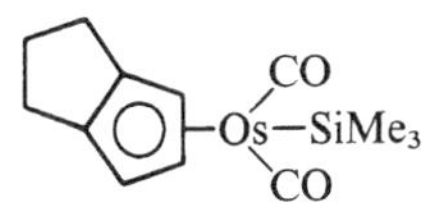

M 424.570
Pale-yellow oil. $Bp_{0.01}$ 50°.

Harris, P.J. *et al*, *J. Chem. Soc., Dalton Trans.*, 1976, 377 (*synth, pmr, ir*)

$C_{13}H_{18}OsSi$ Os-00223

Trimethylsilylosmocene

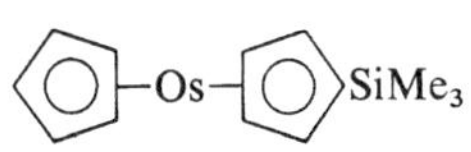

M 392.571

Marr, G. *et al*, *J. Organomet. Chem.*, 1964, **2**, 99.

$C_{14}H_2BrF_6O_{10}Os_3^{\ominus}$ Os-00224

Bromodecacarbonyl[(2,3-η)-1,1,1,4,4,4-hexafluoro-2-butene-]triosmate(1−)

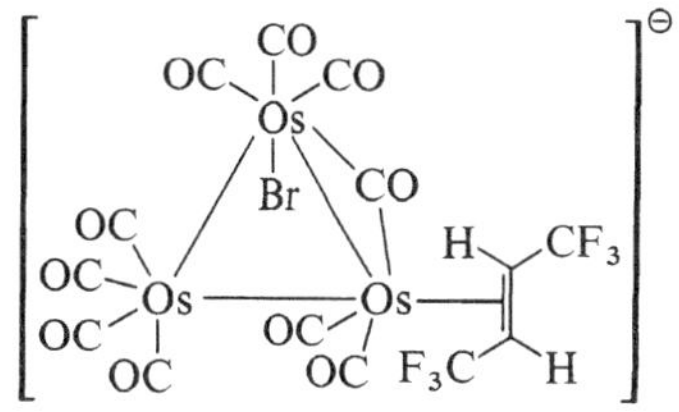

M 1094.658 (ion)
Fluxional.
Bis(triphenylphosphine)iminium salt: [75764-07-7].
$C_{50}H_{32}BrF_6NO_{10}Os_3P_2$ M 1633.245
Bright-amber cubes.

Dawoodi, Z. *et al*, *J. Chem. Soc., Chem. Commun.*, 1980, 641 (*synth, cryst struct, pmr*)

$C_{14}H_2O_{12}Os_4$ Os-00225

Dodecacarbonyl[μ4-[(1-η:1,2-η:1,2-η:2-η)-1,2-ethenediyl]]tetraosmium, 10CI
[75330-60-8]

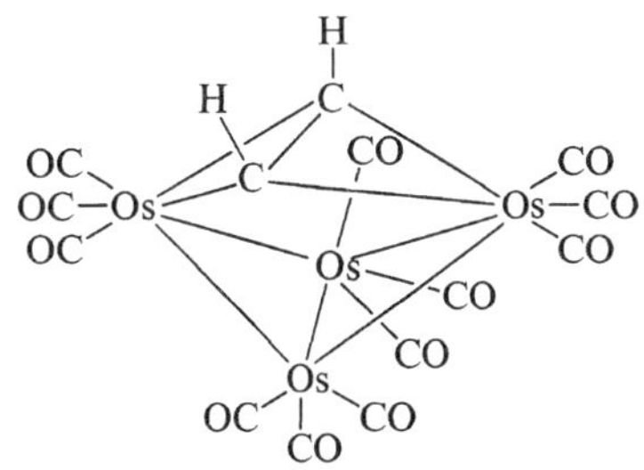

M 1122.963
Orange-yellow cryst. (hexane).

Jackson, R. *et al*, *J. Organomet. Chem.*, 1980, **193**, C1 (*synth, ir, pmr, ms, struct*)

$C_{14}H_6N_2O_{10}Os_3$ Os-00226

Bis(acetonitrile)decacarbonyltriosmium
[61817-93-4]

$Os_3(CO)_{10}(NCCH_3)_2$

M 932.809
Struct. not known. Synthetic intermed. for $Os_3(CO)_{10}$ complexes. Yellow solid, solns. only stable in presence of excess MeCN.

Tachikawa, M. *et al*, *J. Organomet. Chem.*, 1977, **124**, C19 (*synth, ir, pmr*)

$C_{14}H_6O_6Os_2$ Os-00227

Hexacarbonyl[μ-(η[2]:η[4]-1,2-ethenediyl-1,2-phenylene)]diosmium, 9CI
Tricarbonyl(1,1,1-tricarbonylosmaindene)osmium
[59707-01-6]

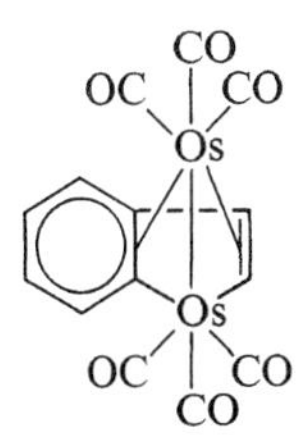

M 650.598

Yellow cryst. Mp 89-91°.

Harris, P.J. *et al*, *J. Chem. Soc., Dalton Trans.*, 1976, 377 (*synth, ir, pmr, struct, ms*)

$C_{14}H_6O_{10}Os_3$ Os-00228

[(1,2,3,4-η)-1,3-Butadiene]decacarbonyltriosmium, 9CI

[60269-72-9]

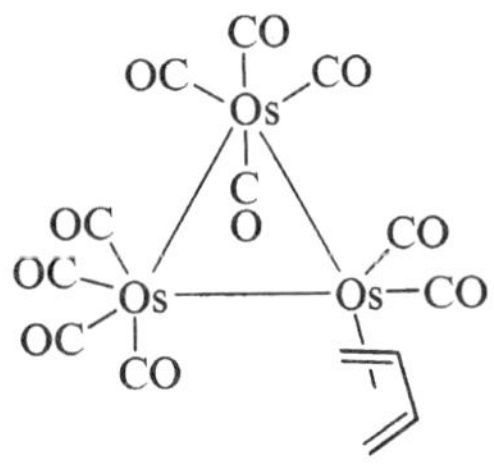

M 904.795

Fluxional molecule.

s-cis-form [60146-16-9]

Yellow cryst. For *s-trans*-form see [μ-[(1,2-η:3,4-η)-1,3-Butadiene]]decacarbonyltriosmium, Os-00229 .

Tachikawa, M. *et al*, *J. Am. Chem. Soc.*, 1976, **98**, 4651 (*synth, ir, pmr, struct*)
Pierpont, C.G., *Inorg. Chem.*, 1978, **17**, 1976 (*cryst struct*)

$C_{14}H_6O_{10}Os_3$ Os-00229

[μ-[(1,2-η:3,4-η)-1,3-Butadiene]]decacarbonyltriosmium, 10CI

[60175-44-2]

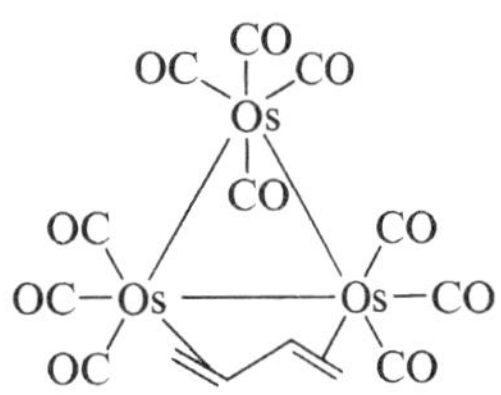

M 904.795

Also descr. as the *s-trans* isomer of [μ-[(1,2,3,4-η)-1,3-Butadiene]]decacarbonyltriosmium, Os-00228. Yellow cryst. ($CHCl_3$).

Tachikawa, M. *et al*, *J. Am. Chem. Soc.*, 1976, **98**, 4651 (*synth*)
Pierpont, C.G., *Inorg. Chem.*, 1978, **17**, 1976 (*cryst struct*)

$C_{14}H_6O_{12}Os_3$ Os-00230

Dodecacarbonyldimethyltriosmium

[60442-72-0]

As Dibromododecacarbonyltriosmium, Os-00149 with

X = Y = Me

M 936.794

White or pale-yellow air-stable cryst.

Evans, J. *et al*, *J. Am. Chem. Soc.*, 1976, **98**, 4000 (*synth, pmr*)
Kelland, J.W. *et al*, *J. Organomet. Chem.*, 1978, **149**, 185 (*synth, ir*)

$C_{14}H_8O_6Os_2$ Os-00231

Hexacarbonyl[μ-[(1,2,3,4,5,8-η)-1,3,5,7-cyclooctatetraen-e]]diosmium

Cyclooctatetraenehexacarbonyldiosmium

[59621-15-7]

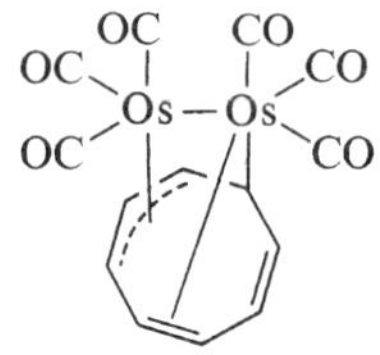

M 652.614

Not well characterised.

Harris, P.J. *et al*, *J. Chem. Soc., Dalton Trans.*, 1976, 377 (*synth, ir*)

$C_{14}H_9O_{14}Os_3P$ Os-00232

Undecacarbonyl(trimethyl phosphite-*P*)triosmium

[66098-55-3]

As (Acetonitrile)undecacarbonyltriosmium, Os-00213 with

L = $P(OMe)_3$

M 1002.790

No phys. props. reported.

Benfield, R.E. *et al*, *Acta Crystallogr., Sect. B*, 1978, **34**, 666 (*struct*)
Fernandez, J.M. *et al*, *J. Chem. Soc., Dalton Trans.*, 1981, 2250.

$C_{14}H_{10}O_4Os_2$ Os-00233

Tetracarbonylbis(η[5]-2,4-cyclopentadien-1-yl)diosmium, 9CI

Tetracarbonyldi-π-cyclopentadienyldiosmium, 8CI. *Bis-(cyclopentadienylosmium dicarbonyl)*

[12131-10-1]

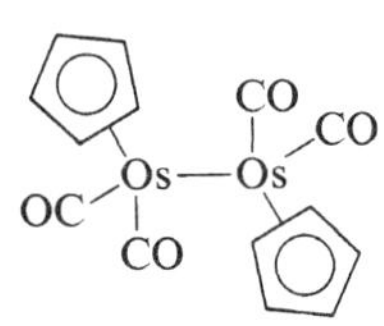

M 622.631

Yellow cryst. by subl. *in vacuo*. Mp 197°. Dipole moment 2.58D at 25°.

Fischer, E.O. *et al*, *Z. Naturforsch., B*, 1962, **17**, 274 (*synth, ir, pmr*)
Fischer, R.D. *et al*, *J. Organomet. Chem.*, 1967, **7**, 135 (*synth, ir*)

$C_{14}H_{11}NO_9Os_3$ Os-00234

Nonacarbonyldi-μ-hydro(2-isocyano-2-methylpropane)triosmium

(tert-*Butylisocyanide*)*nonacarbonyldi-μ-hydrotriosmium*

[70084-45-6]

As Decacarbonyldi-μ-hydrotriosmium, Os-00095 with

X = $(H_3C)_3CNC-$

M 907.842

Red cryst. (hexane). Mp 109-10°. Reacts with CO to produce Decacarbonyl-μ-hydrohydro(2-isocyano-2-methylpropane)triosmium, Os-00259 .

Adams, R.D. *et al*, *Inorg. Chem.*, 1979, **18**, 1909 (*struct*)
Adams, R.D. *et al*, *J. Am. Chem. Soc.*, 1979, **101**, 2579 (*synth, ir, pmr*)

$C_{14}H_{12}O_2Os$ Os-00235

Dicarbonyl(1,3,5-cycloheptatrien-1-yl)(η^5-2,4-cyclopentadien-1-yl)osmium

(η^1-Cycloheptatrienyl)(η^5-cyclopentadienyl)dicarbonylosmium

[81555-01-3]

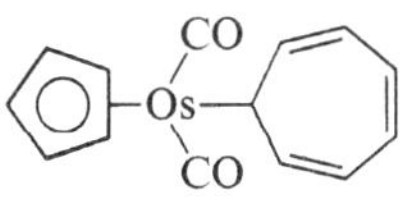

M 402.448

Yellow cryst. by subl. at 50° *in vacuo*. Mp 108-10°.

Hoyano, J.K. *et al*, *Inorg. Chem.*, 1982, **21**, 3095 (*synth, ir, pmr, ms*)

$C_{14}H_{12}O_6Os_2$ Os-00236

Hexacarbonyl[μ-[(1,2,3,4-η:1,4-η)-1,2,3,4-tetramethyl-1,3-butadiene-1,4-diyl]]diosmium, 10CI

Hexacarbonyl[μ-(1,2,3,4-tetramethyl-1,3-butadienylene)]diosmium, 8CI. *Hexacarbonyl[2,3,4,5-tetramethyl(osmacyclopentadiene)]osmium*

[33310-11-1]

As [μ-[(1,2,3,4-η:1,4-η)-1,3-Butadiene-1,4-diyl]]hexacarbonyldiosmium, Os-00098 with

$R = CH_3$

M 656.645

Volatile colourless or yellow cryst. Mp 145-50° dec. $Bp_{0.1}$ 60° subl.

Bruce, M.I. *et al*, *J. Chem. Soc. (A)*, 1969, 987 (*synth, ir, pmr, ms*)
Vaglio, G.A. *et al*, *Org. Mass Spectrom.*, 1971, **5**, 493 (*ms*)
Ferrari, R.P. *et al*, *Gazz. Chim. Ital.*, 1975, **105**, 939 (*synth, ir, ms*)
Farrugia, L.J. *et al*, *J. Chem. Soc., Dalton Trans.*, 1981, 162 (*synth, ir, pmr, ms*)

$C_{14}H_{13}O_2Os^{\oplus}$ Os-00237

Dicarbonyl[(1,2-η)-1,3,5-cycloheptatriene](η^5-2,4-cyclopentadien-1-yl)osmium(1+)

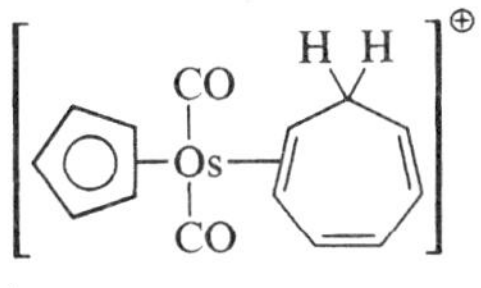

M 403.456 (ion)

Tetrafluoroborate: [81571-98-4].
$C_{14}H_{13}BF_4O_2Os$ M 490.259
White cryst. (CH_2Cl_2/hexane).

Hoyano, J.K. *et al*, *Inorg. Chem.*, 1982, **21**, 3095 (*synth, ir, pmr*)

$C_{14}H_{13}O_{14}Os_4P$ Os-00238

Undecacarbonyltetra-μ-hydro(trimethyl phosphite-*P*)tetraosmium

[78557-65-0]

As Dodecacarbonyltetra-μ-hydrotetraosmium, Os-00167 with

$L = P(OMe)_3$

M 1197.022

Bright-yellow plates (CH_2Cl_2/hexane).

Wei, C-Y. *et al*, *J. Organomet. Chem.*, 1981, **213**, 63 (*struct, synth, ir, pmr*)

$C_{14}H_{14}O_2Os$ Os-00239

1,1′-Diacetylosmocene

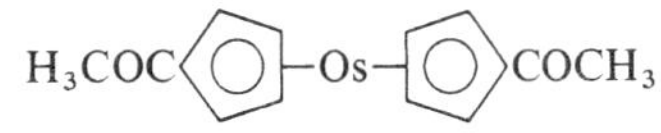

M 404.463

Yellow cryst. (heptane). Mp 148-52°.

Rausch, M.D. *et al*, *J. Org. Chem.*, 1963, **28**, 3225 (*pmr, synth*)

$C_{14}H_{14}Os$ Os-00240

(η^6-Benzene)[(1,2,3,4-η)-1,3,5,7-cyclooctatetraene]osmium, 10CI

[72765-31-2]

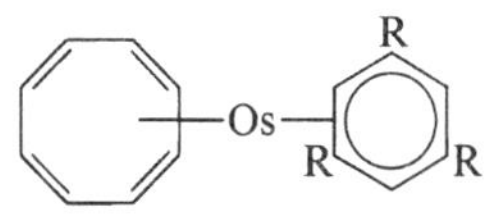

R = H

M 372.465

Orange-yellow cryst.

Bennett, M.A. *et al*, *Inorg. Chem.*, 1980, **19**, 1014 (*synth, pmr, ms*)

$C_{14}H_{16}OOs$ Os-00241

(2-Methyl-1-oxopropyl)osmocene, 9CI

[53540-05-9]

M 390.480

Sokolov, V.I. *et al*, *J. Organomet. Chem.*, 1974, **76**, C15 (*pmr, cmr*)

$C_{14}H_{18}OOs$ Os-00242

(1-Hydroxy-2-methylpropyl)osmocene

(η^5-2,4-Cyclopentadien-1-yl)[(1,2,3,4,5-η)-1-(1-hydroxy-2-methylpropyl)-2,4-cyclopentadien-1-yl]osmium, 9CI

[53886-69-4]

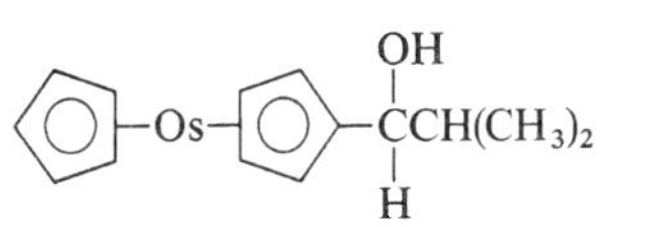

M 392.496

Sokolov, V.I. *et al*, *J. Organomet. Chem.*, 1974, **76**, C15 (*pmr, cmr*)

$C_{14}H_{18}O_8Os_2Si_2$ Os-00243

Octacarbonylbis(trimethylsilyl)diosmium, 9CI, 8CI

[29994-26-1]

$[Me_3SiOs(CO)_4]_2$

M 750.862

Lemon-yellow air-stable cryst. by subl. Sol. hydrocarbons. Mp 106-9°. $Bp_{0.01}$ 70-80° subl.

Monomer: Tetracarbonyl(trimethylsilyl)osmium.
$C_7H_9O_4OsSi$ M 375.431

Unknown.

Knox, S.A.R. *et al*, *J. Chem. Soc.* (*A*), 1970, 3147 (*synth, ir*)

$C_{14}H_{22}Os$ **Os-00244**

[(1,2,5,6-η)-1,5-Cyclooctadiene]bis(η³-2-propenyl)osmium, 9CI

Diallyl(1,5-cyclooctadiene)osmium. Bis(η³-allyl)(1,5-cyclooctadiene)osmium

[53079-97-3]

M 380.528

No phys. props. reported.

Schrock, R.R. *et al*, *J. Chem. Soc., Dalton Trans.*, 1974, 951 (*synth, pmr*)

$C_{14}H_{27}O_7OsP_2^{\oplus}$ **Os-00245**

Carbonyl[(1,2,3,4,5-η)-2,4-cycloheptadien-1-yl]bis(trimethyl phosphite-*P*)osmium(1+), 9CI

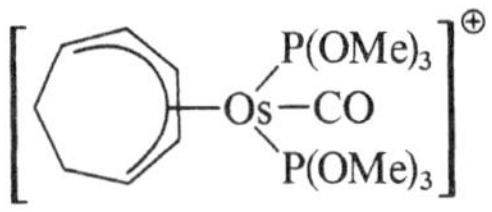

M 559.511 (ion)

Iodide: [59423-97-1].

$C_{14}H_{27}IO_7OsP_2$ M 686.415

Colourless cryst. Mp >250° dec.

Knox, S.A.R. *et al*, *J. Chem. Soc., Dalton Trans.*, 1976, 552 (*synth, ir, pmr, struct*)

$C_{14}Mn_2O_{14}Os$ **Os-00246**

Tetracarbonylbis(pentacarbonylmanganio)osmium, 8CI

Tetradecacarbonyldimanganeseosmium. Dimanganese osmium tetradecacarbonyl

[33292-90-9]

$(OC)_5Mn-Os(CO)_4-Mn(CO)_5$

M 692.222

No phys. props. reported. ir ν_{CO} 2068, 2020, 1987 cm^{-1}.

Abel, E.W. *et al*, *Inorg. Nucl. Chem. Lett.*, 1971, **7**, 587 (*synth, ir*)

$C_{14}O_{14}OsRe_2$ **Os-00247**

Tetracarbonylbis(pentacarbonylrhenio)osmium, 8CI

Tetradecacarbonylosmiumdirhenium. Osmium dirhenium tetradecacarbonyl

[33153-70-7]

$(OC)_5Re-Os(CO)_4-Re(CO)_5$

M 954.760

No phys. props. reported. ir ν_{CO} 2094, 2021, 1982 cm^{-1}.

Abel, E.W. *et al*, *Inorg. Nucl. Chem. Lett.*, 1971, **7**, 587 (*synth, ir*)

$C_{15}HO_{15}Os_5^{\ominus}$ **Os-00248**

Pentadecacarbonyl-μ-hydropentaosmate(1−)

[70804-60-3]

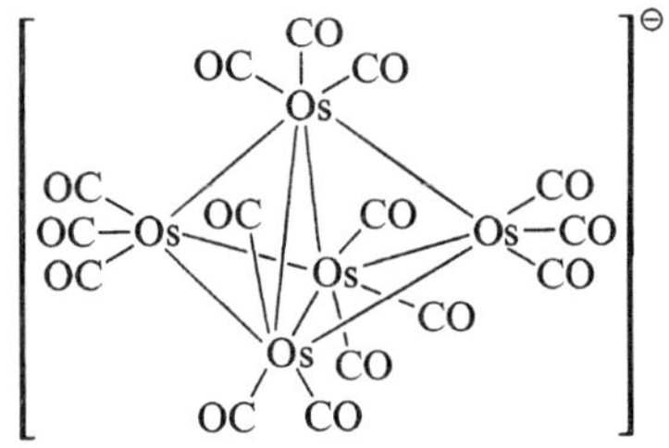

H not shown

M 1372.164 (ion)

Bis(triphenylphosphine)iminium salt: [70804-61-4].

$C_{51}H_{31}NO_{15}Os_5P_2$ M 1910.751

Red-black cryst.

Eady, C.R. *et al*, *J. Chem. Soc., Chem. Commun.*, 1976, 807 (*synth, cmr, struct*)

Guy, J.J. *et al*, *Acta Crystallogr., Sect. B*, 1978, **34**, 1722 (*cryst struct*)

John, G.R. *et al*, *J. Organomet. Chem.*, 1979, **171**, C14 (*synth*)

$C_{15}H_2IO_{15}Os_5^{\ominus}$ **Os-00249**

Pentadecacarbonyldi-μ-hydroiodopentaosmate(1−), 10CI

[71935-55-2]

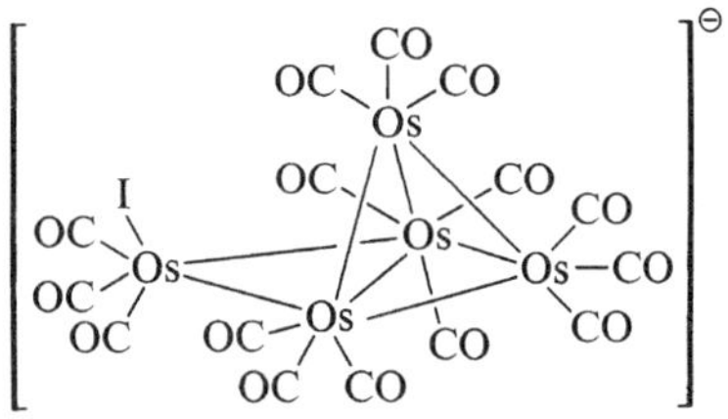

M 1500.076 (ion)

Two bridging H's not shown.

Tetrabutylammonium salt:

$C_{31}H_{38}INO_{15}Os_5$ M 1742.543

Red-brown.

John, G.R. *et al*, *J. Organomet. Chem.*, 1979, **171**, C14 (*synth, ir, struct*)

$C_{15}H_2O_{15}Os_5$ **Os-00250**

Pentadecacarbonyldi-μ-hydropentaosmium

[50648-14-1]

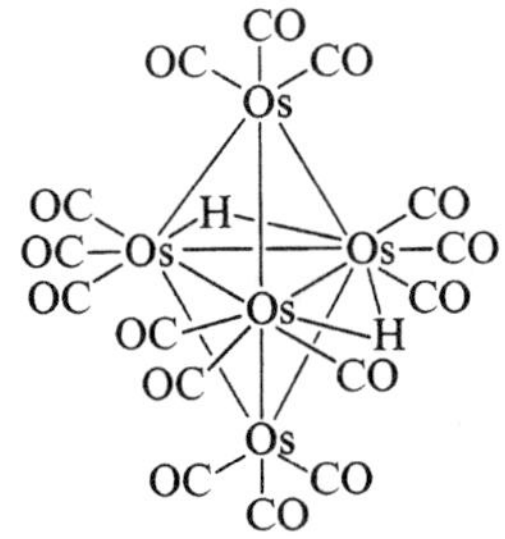

M 1373.172

Location of H atoms uncertain. Characterised spectroscopically. Insol. MeOH.

Eady, C.R. *et al*, *J. Chem. Soc., Dalton Trans.*, 1977, 838 (*synth, ir, pmr*)

Nicholls, J.N. *et al*, *J. Chem. Soc., Dalton Trans.*, 1982, 1395 (*ir*)

$C_{15}H_5NO_{10}Os_3$ Os-00251

Decacarbonyl-μ-hydro[μ-(2-pyridinyl-*C*:*N*)]triosmium, 9CI

[58411-69-1]

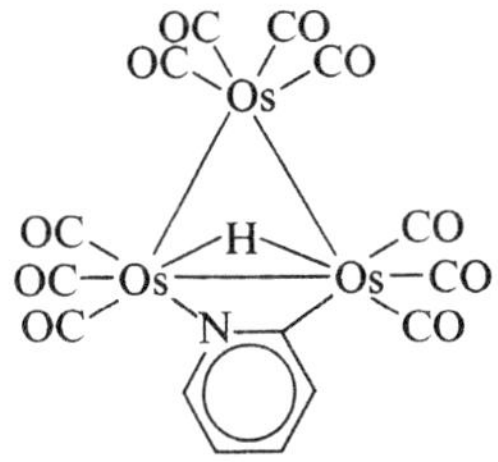

M 929.805

2 Isomers known.

low-melting-form

Yellow cryst. Mp 116-8°.

high-melting-form

Yellow cryst. Mp 144-8°.

Yin, C.C. *et al*, *J. Chem. Soc., Dalton Trans.*, 1975, 2091 (*synth, ir, pmr*)

Calvert, R.B. *et al*, *J. Am. Chem. Soc.*, 1978, **100**, 7726.

$C_{15}H_6O_9Os_3$ Os-00252

Nonacarbonyl[μ₃-[(5-η:5,6-η:6-η)-1,3-cyclohexadien-5-yn-e]]di-μ-hydrotriosmium, 9CI

μ₃-(3,5-Cyclohexadien-1,2-diylidene)-di-μ-hydrido-tris(tricarbonylosmium)

[38979-81-6]

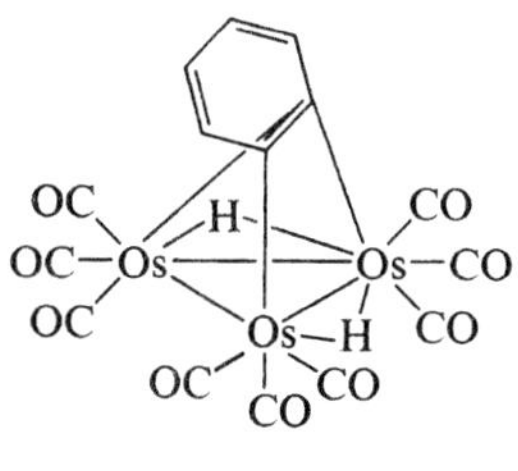

M 900.807

Benzyne complex. Yellow cryst.

Deeming, A.J. *et al*, *J. Chem. Soc., Dalton Trans.*, 1974, 1415 (*synth, ir, pmr*)

$C_{15}H_7NO_9Os_3$ Os-00253

(μ₃-Anilinato)nonacarbonyldi-μ-hydrotriosmium

[*μ₃-[Benzenaminato(2−)]]nonacarbonyldi-μ-hydrotri-osmium*

[81799-11-3]

As Nonacarbonyldi-μ-hydro[μ₃-[methanaminato(2−)]]triosmium, Os-00101 with

X = NPh

M 915.822

Orange cryst. Mp 174-6°.

Yin, C.C. *et al*, *J. Chem. Soc., Dalton Trans.*, 1974, 1013 (*synth, ir, pmr*)

Burgess, K. *et al*, *J. Organomet. Chem.*, 1982, **224**, C40 (*struct*)

The symbol ▷ in Entries highlights hazard or toxicity information

$C_{15}H_7O_9Os_3^{\oplus}$ Os-00254

(Benzene)nonacarbonyl-μ-hydrotriosmium(1+), 9CI

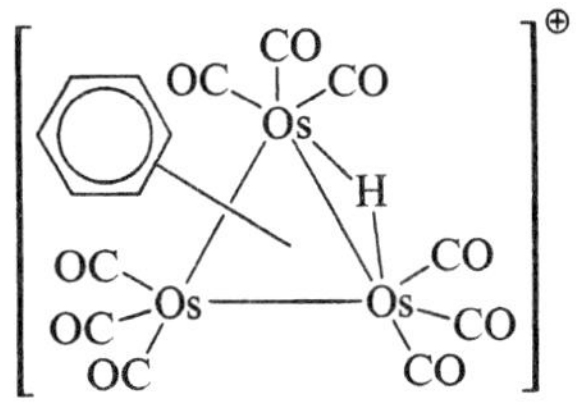

Probable structure

M 901.815 (ion)

Unstable.

Tetrafluoroborate: [60373-24-2].

$C_{15}H_7BF_4O_9Os_3$ M 988.619

No phys. props. reported.

Bryan, E.G. *et al*, *J. Chem. Soc., Chem. Commun.*, 1976, 254 (*synth, ir, pmr*)

$C_{15}H_7O_9Os_3P$ Os-00255

Nonacarbonyldi-μ-hydro[μ₃-(phenylphosphinidene)]trios-mium

[79230-48-1]

As Nonacarbonyldi-μ-hydro[μ₃-[methanaminato-(2−)]]triosmium, Os-00101 with

X = PPh

M 932.789

Pale-yellow solid (cyclohexane).

Iwasaki, F. *et al*, *J. Organomet. Chem.*, 1981, **213**, 185 (*synth, ir, pmr, nmr, ms*)

$C_{15}H_8O_9Os_3$ Os-00256

Nonacarbonyl[μ₃-[(1,2-η:3-η:4,5-η)-2,4-cyclohexadi-en-1-yl]]-μ-hydrotriosmium, 9CI

[60317-82-0]

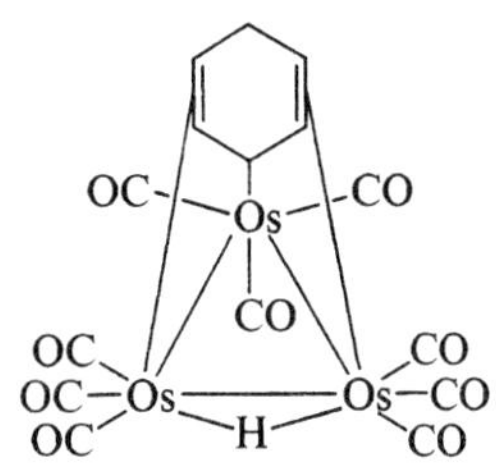

M 902.823

Yellow cryst.

Bryan, E.G. *et al*, *J. Chem. Soc., Chem. Commun.*, 1976, 254 (*synth, ir, pmr, ms, struct*)

$C_{15}H_{10}NO_{10}Os_3^{\ominus}$ Os-00257

[μ-(*tert*-Butylcarbonimidoyl)]decacarbonyl-μ-hydrotrios-mate(1−)

Decacarbonyl[μ-[(1,1-dimethylethyl)carbonimidoyl]]-μ-hydrotriosmate(1−), 10CI

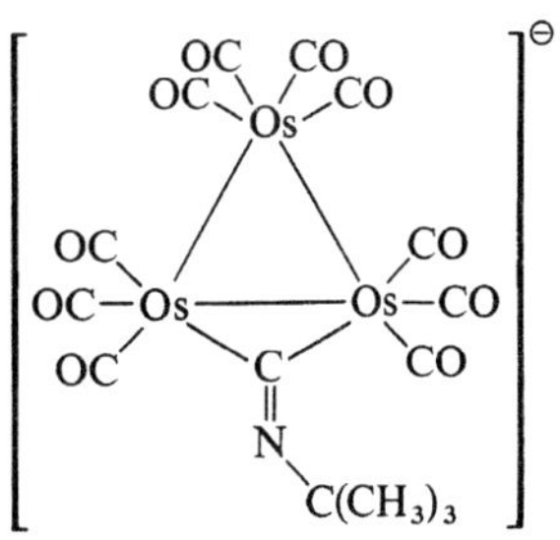

H not shown

M 934.845 (ion)

Hydrogen compd.: [70377-88-7].
$C_{15}H_{11}NO_{10}Os_3$ M 935.853
Yellow. Mp 134-6°.

Adams, R.D. *et al*, *Inorg. Chem.*, 1979, **18**, 2255; *J. Am. Chem. Soc.*, 1979, **101**, 2579 (*cryst struct, synth, ir, pmr*)

$C_{15}H_{10}O_{11}Os_3Si$ Os-00258

μ-Carbonyldecacarbonyl[μ-(trimethylsilyl)methylene]triosmium, 10CI

Undecacarbonyl[μ-(trimethylsilyl)methylene]triosmium

[80030-38-2]

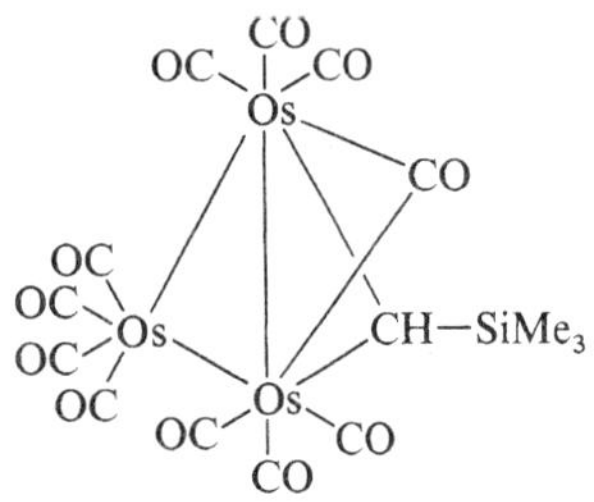

M 964.923
Red, air-stable compd.

Shapley, J.R. *et al*, *J. Am. Chem. Soc.*, 1981, **103**, 6975 (*struct*)

$C_{15}H_{11}NO_{10}Os_3$ Os-00259

Decacarbonyl-μ-hydrohydro(2-isocyano-2-methylpropane-)triosmium, 10CI

(tert-*Butyl isocyanide)decacarbonyldihydrotriosmium*

[69942-19-4]

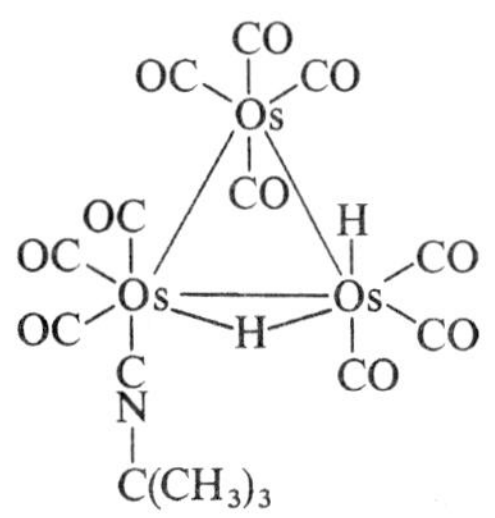

M 935.853
Yellow powder (hexane). Mp 125.5-126°.

Adams, R.D. *et al*, *Inorg. Chem.*, 1979, **18**, 1909 (*struct*)
Adams, R.D. *et al*, *J. Am. Chem. Soc.*, 1979, **101**, 2579 (*synth, ir, pmr*)

$C_{15}H_{12}O_7Os_2$ Os-00260

Hexacarbonyl[μ-[(1,2,4,5-η:1,5-η)-1,2,4,5-tetramethyl-3-oxo-1,4-pentadiene-1,5-diyl]]diosmium, 9CI

[57405-60-4]

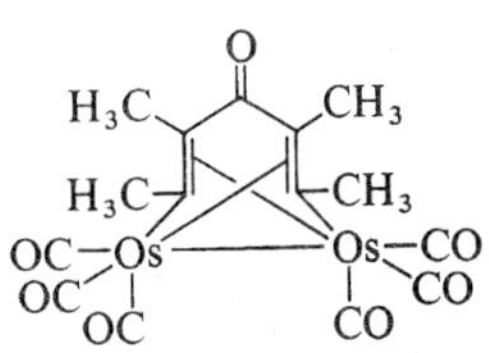

M 684.656
Characterised spectroscopically.

Deeming, A.J. *et al*, *J. Chem. Soc., Dalton Trans.*, 1975, 1614 (*synth, ir, pmr*)

$C_{15}H_{16}GeO_5Os_2$ Os-00261

Pentacarbonyl-μ-2,4,6-cycloheptatrien-1-yl(trimethylgermyl)diosmium, 9CI

[59421-52-2]

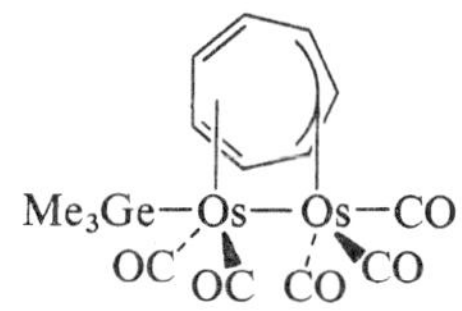

M 729.278
Yellow cryst. Mp 184° dec.

Knox, S.A.R. *et al*, *J. Chem. Soc., Dalton Trans.*, 1976, 552 (*synth, ir, pmr*)

$C_{15}H_{16}O_5Os_2Si$ Os-00262

Pentacarbonyl-μ-2,4,6-cycloheptatrien-1-yl(trimethylsilyl-)diosmium, 9CI

[59421-53-3]

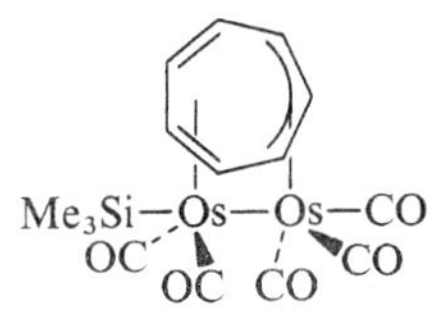

M 684.774
Fluxional molecule. Yellow cryst. (hexane). Mp 176-7°.

Knox, S.A.R. *et al*, *J. Chem. Soc., Dalton Trans.*, 1976, 552 (*synth, ir, pmr*)

$C_{15}H_{18}Ge_3O_9Os_3$ Os-00263

Nonacarbonyltris[μ-(dimethylgermylene)]triosmium, 8CI

[30347-86-5]

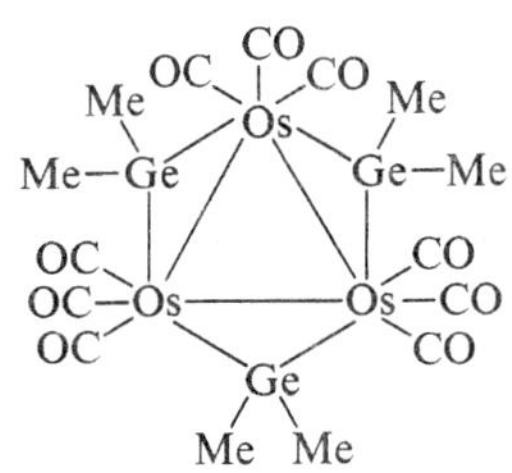

M 1130.672
Yellow cryst. (CH_2Cl_2/hexane).

Knox, S.A.R. *et al*, *J. Chem. Soc. (A)*, 1971, 2874 (*synth, ir, pmr*)

$C_{15}H_{18}O_9Os_3Si_3$ Os-00264

Nonacarbonyltris[μ-(dimethylsilylene)]triosmium

[36632-60-7]

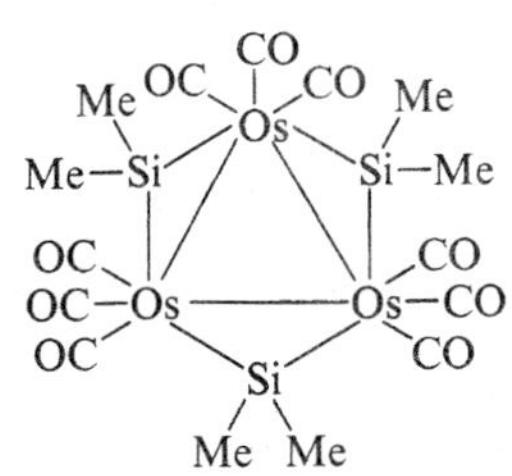

M 997.158
Cream cryst. (hexane). Mp 135° dec.

Brookes, A. *et al*, *J. Chem. Soc. (A)*, 1971, 3469 (*synth, ir, ms, pmr*)

$C_{15}H_{23}N_2Os^{\oplus}$ Os-00265

Bis(acetonitrile)[(1,2,5,6-η)-1,5-cyclooctadiene](η^3-2-propenyl)osmium(1+), 9CI

Bis(acetonitrile)(η^3-allyl)(1,5-cyclooctadiene)osmium

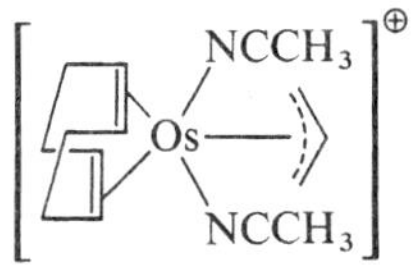

M 421.560 (ion)

Tetrafluoroborate: [53080-00-5].
$C_{15}H_{23}BF_4N_2Os$ M 508.364
White solid.

Schrock, R.R. *et al, J. Chem. Soc., Dalton Trans.*, 1974, 951 (*synth, pmr*)

$C_{15}H_{45}BrO_{15}OsP_5^{\oplus}$ Os-00266

Bromopentakis(trimethyl phosphite-*P*)osmium(1+), 9CI

$BrOs[P(OMe)_3]_5^{\oplus}$

M 890.484 (ion)

Tetraphenylborate: [35083-15-9].
$C_{39}H_{65}BBrO_{15}OsP_5$ M 1209.716
Colourless cryst.

Couch, D.A. *et al, Inorg. Chim. Acta*, 1974, **9**, 39 (*synth*)

$C_{15}H_{45}ClO_{15}OsP_5^{\oplus}$ Os-00267

Chloropentakis(trimethyl phosphite-*P*)osmium(1+), 9CI

$ClOs[P(OMe)_3]_5^{\oplus}$

M 846.033 (ion)

Tetraphenylborate: [35083-14-8].
$C_{39}H_{65}BClO_{15}OsP_5$ M 1165.265
Colourless cryst.

Couch, D.A. *et al, Inorg. Chim. Acta*, 1974, **9**, 39 (*synth*)

$C_{15}H_{45}O_{15}OsP_5$ Os-00268

Pentakis(trimethyl phosphite-*P*)osmium, 10CI

[61880-52-2]

$Os[P(OMe)_3]_5$

M 810.580
Stereochemically nonrigid. White solid (pentane). Not isol. pure.

English, A.D. *et al, J. Am. Chem. Soc.*, 1977, **99**, 117 (*synth, nmr, pmr*)

$C_{15}IO_{15}Os_5^{\ominus}$ Os-00269

Pentadecacarbonyliodopentaosmate(1−)

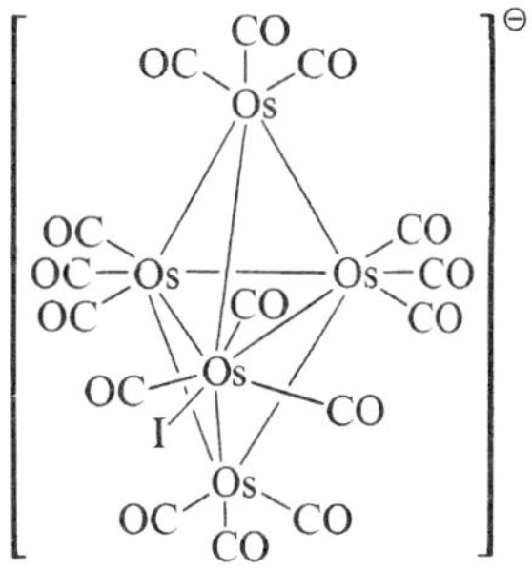

M 1498.061 (ion)

Bis(triphenylphosphine)iminium salt: [68581-83-9].
$C_{51}H_{30}INO_{15}OsP_2$ M 1275.848
Purple cryst.

Eady, C.R. *et al, J. Organomet. Chem.*, 1978, **149**, C43.
Rivera, A.V. *et al, Acta Crystallogr., Sect. B*, 1978, **34**, 3376 (*cryst struct*)

$C_{15}O_{15}Os_5^{\ominus\ominus}$ Os-00270

Pentadecacarbonylpentaosmate(2−), 10CI

[62252-84-0]

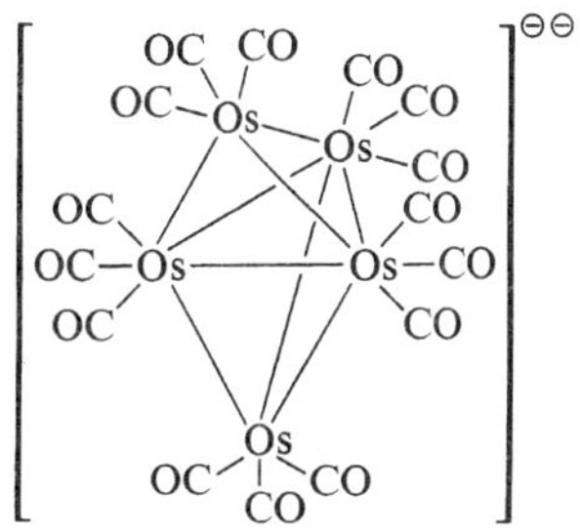

M 1371.156 (ion)

Bis[bis(triphenylphosphine)iminium]salt: [62414-47-5].
$C_{87}H_{60}N_2O_{15}Os_5P_4$ M 2448.330
Deep-purple cryst. (MeCN/Et_2O).

Eady, C.R. *et al, J. Chem. Soc., Chem. Commun.*, 1976, 807 (*synth, cmr, struct*)
John, G.R. *et al, J. Organomet. Chem.*, 1979, **169**, C9; 1979, **171**, C15 (*synth*)

$C_{16}HMnO_{16}Os_3$ Os-00271

Hexadecacarbonylhydromanganesetriosmium, 8CI

[12566-57-3]

$MnOs_3H(CO)_{16}$

M 1074.712
No phys. props. reported.

Knight, J. *et al, J. Chem. Soc., Chem. Commun.*, 1971, 62 (*synth, ir*)

$C_{16}HO_{16}Os_3Re$ Os-00272

Hexadecacarbonylhydrorheniumtriosmium, 8CI

[12566-58-4]

$ReOs_3H(CO)_{16}$

M 1205.981
Yellow solid, not obt. pure.

Knight, J. *et al, J. Chem. Soc., Chem. Commun.*, 1971, 62 (*synth*)

Shapley, J.R. *et al*, *J. Am. Chem. Soc.*, 1977, **99**, 8064 (*synth, ms, pmr*)

$C_{16}H_2O_{16}Os_5$ **Os-00273**

Hexadecacarbonyldi-μ-hydropentaosmium, 9CI

[50648-17-4]

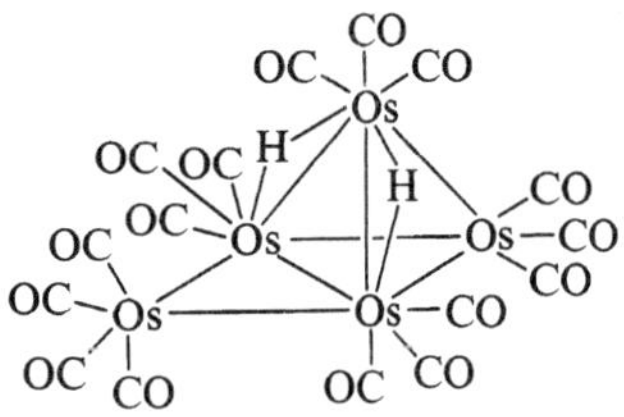

M 1401.182

Green-yellow.

Eady, C.R. *et al*, *J. Chem. Soc., Dalton Trans.*, 1977, 839 (*synth, ir, pmr*)
Guy, J.J. *et al*, *Acta Crystallogr., Sect. B*, 1978, **34**, 1725 (*struct*)
John, G.R. *et al*, *J. Organomet. Chem.*, 1979, **171**, C14 (*synth, ir*)
Nicholls, J.N. *et al*, *J. Chem. Soc., Dalton Trans.*, 1982, 1395 (*props, ir*)

$C_{16}H_3O_{16}Os_5P$ **Os-00274**

Pentadecacarbonyl[μ_4-[methyl phosphinito(2−)-*P*:*P*:*P*:*P*]]-pentaosmium, 10CI

[71191-30-5]

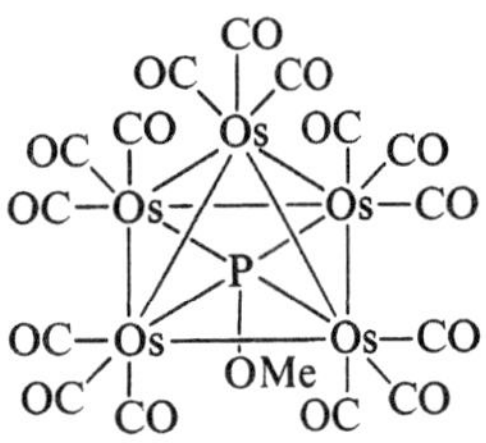

M 1433.164

No phys. props. reported.

Fernandez, J.M. *et al*, *Acta Crystallogr., Sect. B*, 1979, **35**, 1711 (*struct*)

$C_{16}H_6O_{10}Os_3S$ **Os-00275**

[(μ-Benzenethiolato)]decacarbonyl-μ-hydrotriosmium, 8CI

[23733-19-9]

As μ-Bromodecacarbonyl-μ-hydrotriosmium, Os-00092 with

X = SPh

M 960.877

Yellow needles (C_6H_6/MeOH). Mp 121-3°.

Crooks, G.R. *et al*, *J. Chem. Soc. (A)*, 1969, 797 (*synth, ir, pmr*)
Bryan, E.G. *et al*, *J. Chem. Soc., Dalton Trans.*, 1978, 196 (*cmr*)

$C_{16}H_7NO_{10}Os_3$ **Os-00276**

[μ_2-[Benzeneaminato(1−)]]decacarbonyl-μ-hydrotriosmium, 9CI

[53595-03-2]

As Decacarbonyl-μ-hydro[μ-(methanaminato)]triosmium, Os-00128 with

R = Ph

M 943.832

Orange-red cryst. Mp 174-6°.

Yin, C.C. *et al*, *J. Chem. Soc., Dalton Trans.*, 1974, 1013 (*synth, ir, pmr*)
Bryan, E.G. *et al*, *J. Chem. Soc., Dalton Trans.*, 1977, 1328 (*synth*)

$C_{16}H_7O_{10}Os_3^{\oplus}$ **Os-00277**

Decacarbonyl[(1,2,3,4,5-η)-2,4-cyclohexadien-1-yl]triosmium(1+), 9CI

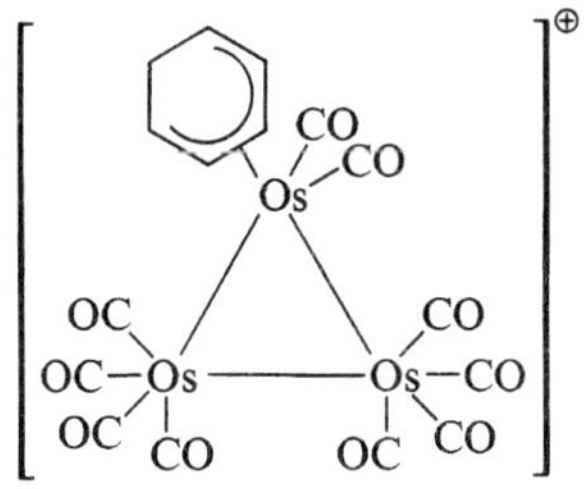

M 929.825 (ion)

Tetrafluoroborate: [60321-20-2].
$C_{16}H_7BF_4O_{10}Os_3$ M 1016.629
Yellow solid.

Bryan, E.G. *et al*, *J. Chem. Soc., Dalton Trans.*, 1977, 1328 (*synth, ir, pmr*)

$C_{16}H_8O_{10}Os_3$ **Os-00278**

Decacarbonyl[(1,2,3,4-η)-1,3-cyclohexadiene]triosmium, 9CI

[60268-95-3]

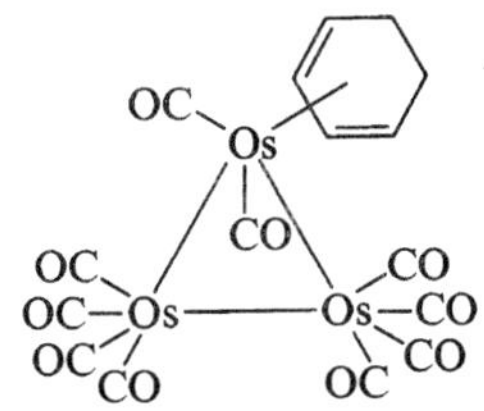

M 930.833

Synthetic intermed. for (μ-H)$Os_3(CO)_{10}$(μ-X). Orange cryst. (hexane).

Tachikawa, M. *et al*, *J. Am. Chem. Soc.*, 1976, **98**, 4651 (*synth, pmr, ir, ms*)
Bryan, E.G. *et al*, *J. Chem. Soc., Dalton Trans.*, 1977, 144 (*cmr, pmr*)
Bryan, E.G. *et al*, *J. Chem. Soc., Dalton Trans.*, 1977, 1328 (*synth*)

$C_{16}H_9N_2O_{12}Os_4^{\oplus}$ **Os-00279**

Bis(acetonitrile)dodecacarbonyltri-μ-hydrotetraosmium(1+)

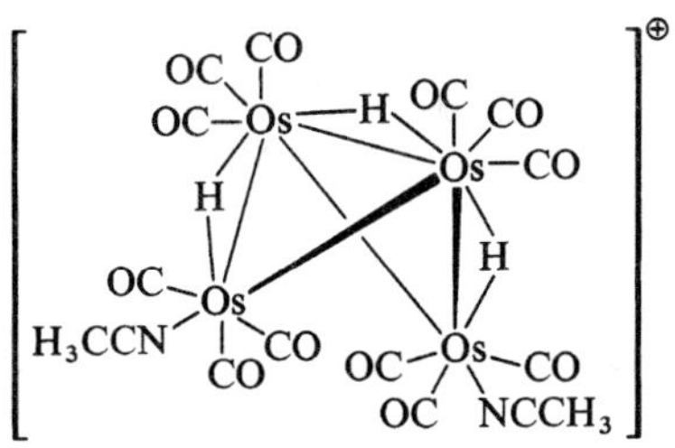

M 1182.053 (ion)

Tetrafluoroborate:
$C_{16}H_9BF_4N_2O_{12}Os_4$ M 1268.857
Yellow block-shaped cryst. (CH_2Cl_2).

Johnson, B.F.G. *et al*, *J. Chem. Soc., Chem. Commun.*, 1982, 611 (*synth, struct, ir, pmr*)

$C_{16}H_{20}N_2O_6Os_2$ Os-00280

Hexacarbonyl[μ-[η^2-N,N′-1,2-ethanediylidenebis[2-methyl-2-propanamine]-N,N′]]diosmium, 10CI

*Hexacarbonyl[glyoxalbis(*tert*-butylimine)]diosmium.*
Hexacarbonyl(2,2,6,6-tetramethyl-3,6-diaza-3,5-octadiene)diosmium

[78192-37-7]

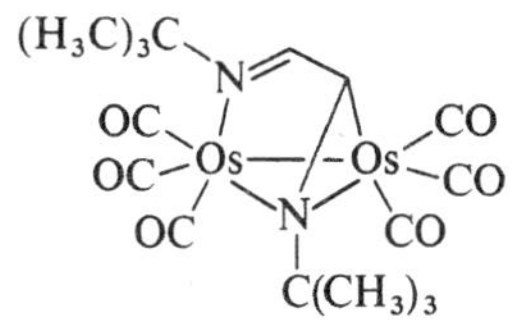

M 716.744

Pale-yellow cryst. (pentane).

Staal, L.H. *et al*, *J. Organomet. Chem.*, 1981, **206**, 99 (*synth, ir, pmr, cmr, ms*)

$C_{16}H_{22}Os$ Os-00281

[(1,2,5,6-η)-1,5-Cyclooctadiene][(1,2,3,4,5,6-η)-1,3,5-cyclooctatriene]osmium, 9CI

[42516-74-5]

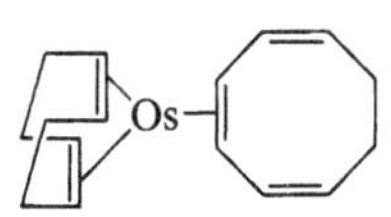

M 404.550

Characterised spectroscopically. Exhibits temp. dependent pmr spectrum.

Fischer, E.O. *et al*, *Chem. Ber.*, 1963, **96**, 3217 (*synth, ir*)
Schrock, R.R. *et al*, *J. Am. Chem. Soc.*, 1973, **95**, 4102 (*synth*)

$C_{16}H_{23}O_3Os^{\oplus}$ Os-00282

Dicarbonyl[(1,2,3,4,5-η)-1,2,3,4,5-pentamethyl-2,4-cyclopentadien-1-yl](tetrahydrofuran)osmium(1+)

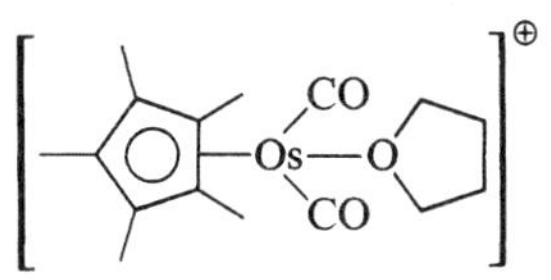

M 453.556 (ion)

Hexafluorophosphate: [81554-80-5].
$C_{16}H_{23}F_6O_3OsP$ M 598.520
Off-white solid (CH_2Cl_2/Et_2O). Dec. at 101-3°.

Hoyano, J.K. *et al*, *Inorg. Chem.*, 1982, **21**, 3095 (*synth, ir, pmr*)

$C_{16}H_{30}Ge_4O_6Os_2$ Os-00283

Hexacarbonylbis[μ-(dimethylgermylene)]bis(trimethylgermyl)diosmium, 8CI

[30854-81-0]

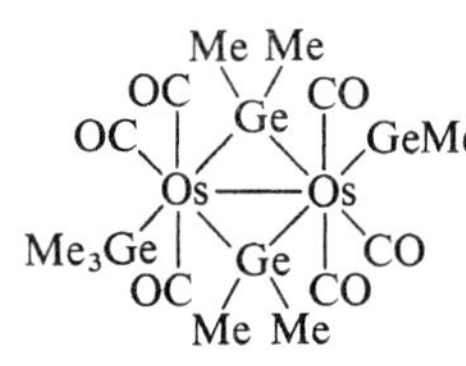

M 989.169

Pale-yellow cryst. (pet. ether). Mp 150-60° dec.

Knox, S.A.R. *et al*, *J. Chem. Soc. (A)*, 1971, 2874 (*synth, ir, pmr*)

$C_{16}H_{30}O_6Os_2Si_4$ Os-00284

Hexacarbonylbis[μ-(dimethylsilylene)]bis(trimethylsilyl)diosmium, 9CI

[36632-59-4]

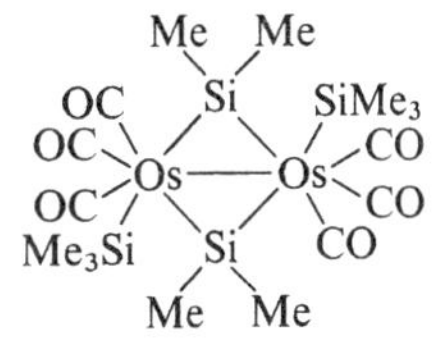

M 811.151

Pale-yellow cryst. (hexane). Mp 130° dec.

Brookes, A. *et al*, *J. Chem. Soc. (A)*, 1971, 3469 (*synth, ir, pmr*)

$C_{16}IO_{15}Os_5^{\ominus}$ Os-00285

[μ_5-(Carbon)]pentadecacarbonyliodopentaosmate(1−)

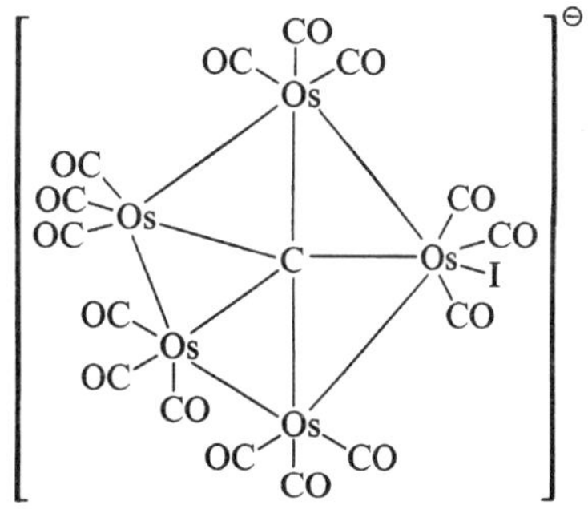

M 1510.072 (ion)

Bis(triphenylphosphine)iminium salt: [76550-98-6].
$C_{52}H_{30}INO_{15}Os_5P_2$ M 2048.659
Orange cryst. ($MeOH/CH_2Cl_2$). Air-stable.

Jackson, P.F. *et al*, *J. Chem. Soc., Chem. Commun.*, 1980, 564 (*synth, cryst struct*)

$C_{16}O_{15}Os_5$ Os-00286

(Carbon)pentadecacarbonylpentaosmium, 9CI

[51205-06-2]

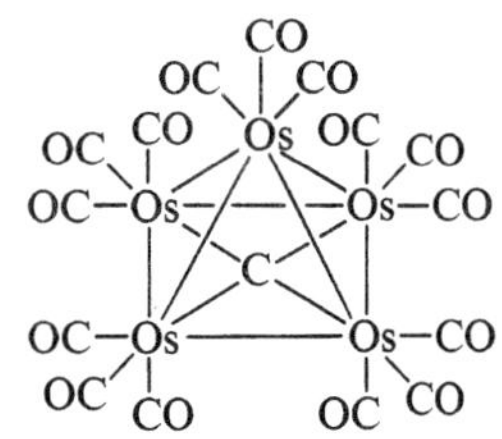

M 1383.167

Yellow or orange cryst. (EtOAc).

Eady, C.R. *et al*, *J. Chem. Soc., Dalton Trans.*, 1975, 2606 (*synth, ir, ms*)
Jackson, P.F. *et al*, *J. Chem. Soc., Chem. Commun.*, 1980, 564 (*struct, ir*)

$C_{16}O_{16}Os_5$ Os-00287

Hexadecacarbonylpentaosmium

Pentaosmium hexadecacarbonyl. Osmium carbonyl[$Os_5(CO)_{16}$]

[60109-68-4]

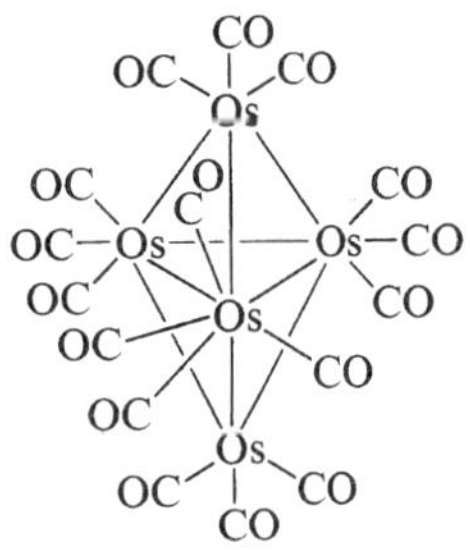

M 1399.166

Pink-red cryst.

Eady, C.R. *et al*, *J. Chem. Soc., Dalton Trans.*, 1975, 2606; 1977, 838 (*synth, ir*)

Reichert, B.E. *et al*, *Acta Crystallogr., Sect. B*, 1977, **33**, 173 (*struct*)

Nicholls, J.N. *et al*, *J. Chem. Soc., Dalton Trans.*, 1982, 1395 (*synth, ir*)

$C_{16}O_{16}Os_6S_2$ Os-00288

Hexadecacarbonyl-μ_3-thioxo-μ_4-thioxohexaosmium

[82190-35-0]

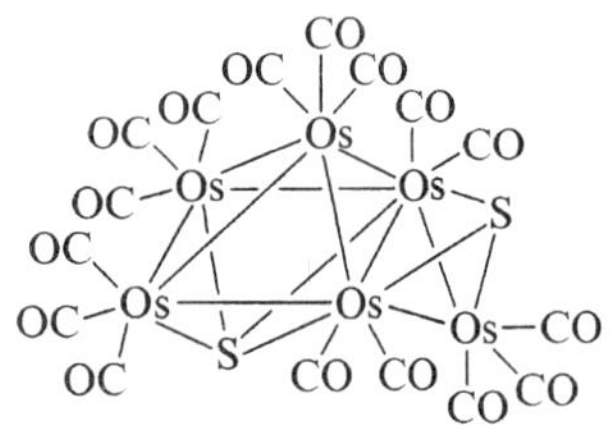

M 1653.486

Adams, R.A. *et al*, *J. Am. Chem. Soc.*, 1982, **104**, 4115 (*synth, struct*)

$C_{16}O_{22}Os_6$ Os-00289

Hexadecacarbonylhexaoxohexaosmium

[11067-28-0]

$Os_6O_6(CO)_{16}$

M 1685.363

Orange-yellow compd. Insol. all common org. solvs., dec. in DMF.

Bradford, C.W. *et al*, *J. Chem. Soc. (A)*, 1971, 2038 (*synth, ms, ir*)

$C_{17}H_4NO_{15}Os_3Re$ Os-00290

[(Acetonitrile)decacarbonyl-μ-hydrotriosmium]pentacarbonylrhenium, 10CI

(Acetonitrile)pentadecacarbonyl-μ-hydrotriosmiumrhenium

[79255-11-1]

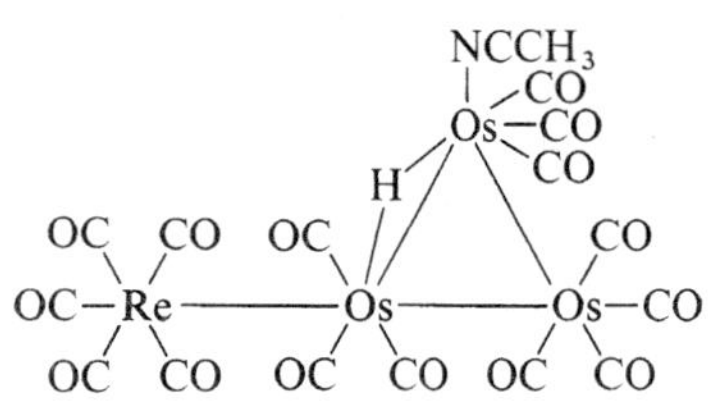

M 1219.023

Previously reported as $HReOs_3(CO)_{15}$. Yellow cryst. (MeCN).

Shapley, J.R. *et al*, *J. Am. Chem. Soc.*, 1977, **99**, 8064 (*synth, ir, ms*)

Churchill, M.R. *et al*, *J. Am. Chem. Soc.*, 1981, **103**, 2430 (*struct*)

$C_{17}H_7O_{17}Os_5P$ Os-00291

[μ_5-(Carbon)]tetradecacarbonyl[μ-(dimethyl phosphito-O'':P)]-μ-hydropentaosmium, 10CI

[67422-61-1]

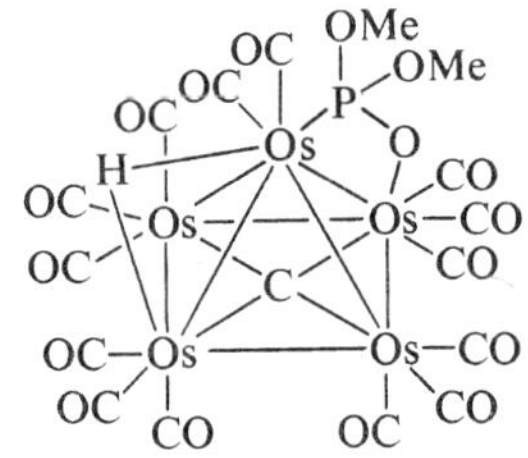

M 1465.206

No phys. props. reported.

Fernandez, J.M. *et al*, *Acta Crystallogr., Sect. B*, 1978, **34**, 1994 (*struct*)

$C_{17}H_8N_2O_{10}Os_3$ Os-00292

Decacarbonyl-μ-hydro[μ-[(4-methylphenyl)azo]]triosmium, 10CI

[76756-17-7]

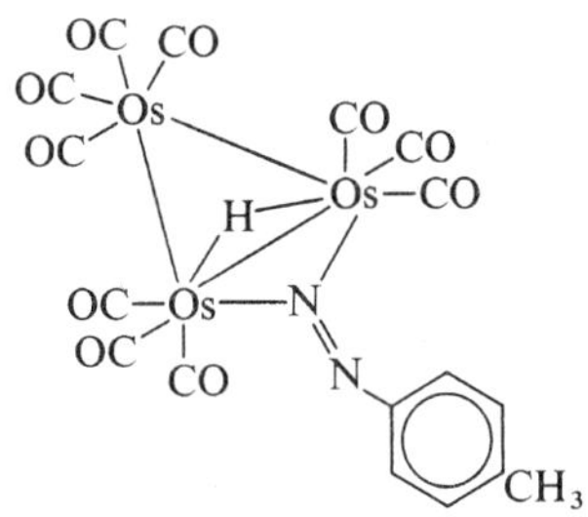

M 970.858

Yellow-brown cryst.

Churchill, M.R. *et al*, *Inorg. Chem.*, 1981, **20**, 1580 (*struct*)

$C_{17}H_8O_9Os_3$ Os-00293

Nonacarbonyldi-μ-hydro[μ_3-[(1-η:1-η:1,2-η)-2-phenylethenylidene]]triosmium

[56214-65-4]

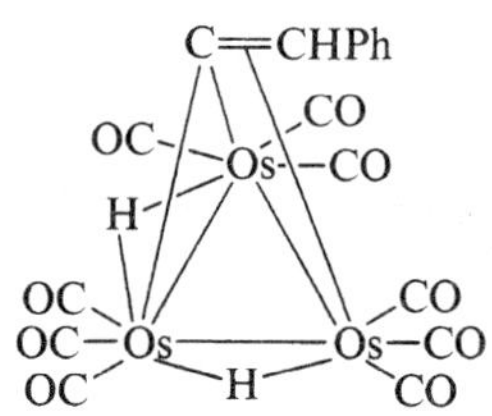

M 926.845

White cryst. Mp 84-5°.

Deeming, A.J. *et al*, *J. Chem. Soc., Dalton Trans.*, 1975, 1614 (*synth, ir, pmr, ms*)

Jackson, W.G. *et al*, *J. Organomet. Chem.*, 1975, **87**, C27.

$C_{17}H_{12}O_{11}Os_4$ Os-00294

Undecacarbonyl[μ-[(1-η:1,2-η)-1-cyclohexen-1-yl]]tri-μ-hydrotetraosmium, 10CI

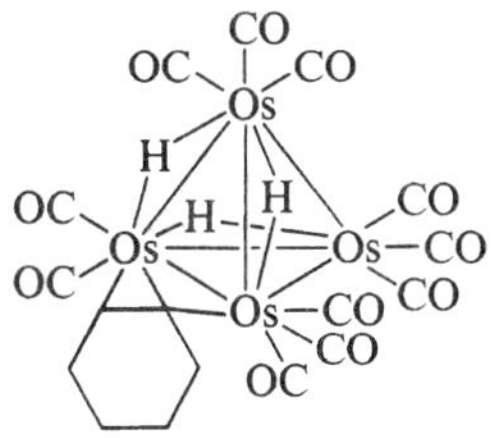

M 1153.075

Yellow-orange tablets (hexane).

Bhaduri, S. *et al, J. Chem. Soc., Dalton Trans.*, 1979, 562 (*synth, ir*)

$C_{17}H_{14}OOs$ Os-00295

Benzoylosmocene

Osmocenylphenylmethanone. Osmocenyl phenyl ketone

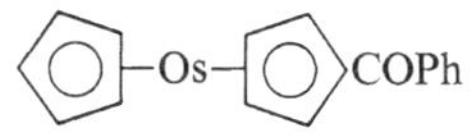

M 424.497

Light-yellow needles by subl. Mp 133.5°.

Rausch, M.D. *et al, Chem. Ind. (London)*, 1958, 756 (*synth*)
Riemschneider, R. *et al, Monatsh. Chem.*, 1959, **90**, 658 (*synth*)
Rausch, M.D. *et al, J. Am. Chem. Soc.*, 1960, **82**, 77 (*synth, ir*)
Rausch, M.D. *et al, J. Org. Chem.*, 1963, **28**, 3225 (*pmr*)
Macdonald, A.C. *et al, Acta Crystallogr.*, 1965, **19**, 1046 (*struct*)

$C_{17}H_{14}O_9Os_3$ Os-00296

Nonacarbonyl[μ₃-[(1-η:1,2-η:2-η)cyclooctyne]]di-μ-hydrotriosmium

[51205-37-9]

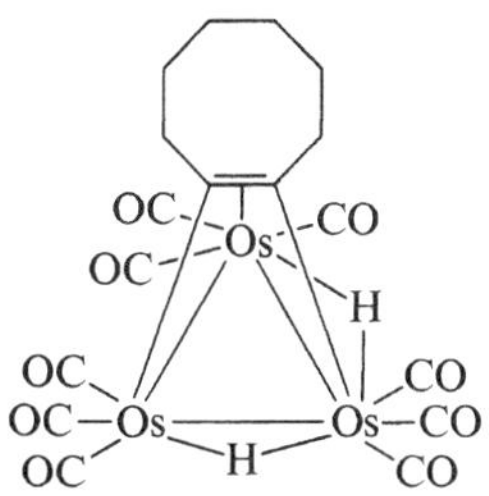

M 932.892

Fluxional molecule. Cream solid (hexane).

Canty, A.J. *et al, J. Chem. Soc., Dalton Trans.*, 1973, 2056 (*synth, ir, pmr*)

$C_{17}H_{16}OOs$ Os-00297

(1-Hydroxy-1-phenylmethyl)osmocene

α-Phenylosmocenemethanol. (α-Hydroxybenzyl)-osmocene

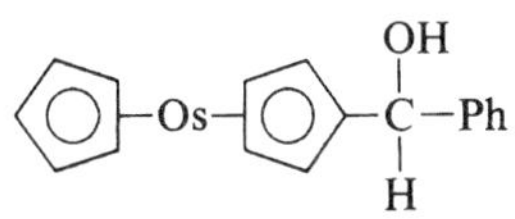

M 426.513

(±)-*form*

White needles (hexane). Mp 115.7-116°.

Bublitz, D.E. *et al, J. Am. Chem. Soc.*, 1962, **84**, 1845 (*synth*)

$C_{17}H_{16}O_7Os_3P_2$ Os-00298

[μ₃-[(1-η:1,2-η:2-η)-Benzene]]heptacarbonylbis[μ-(dimethylphosphino)]triosmium, 9CI

Heptacarbonyl[μ₃-(5-η:5,6-η:6-η)-1,3-cyclohexadien-5-yne]bis[μ-(dimethylphosphine)]triosmium, 9CI

[36835-56-0]

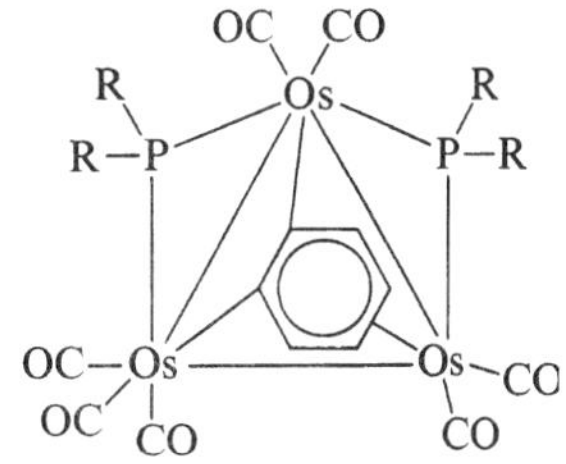

R = Me

M 964.857

Bridging benzyne complex. Red cryst. Mp 190° dec. Corresponding $AsMe_2$ complex also known.

Deeming, A.J. *et al, J. Chem. Soc., Dalton Trans.*, 1973, 2589 (*synth, pmr, ir*)

$C_{17}H_{16}Os$ Os-00299

Benzylosmocene

Phenylmethylosmocene

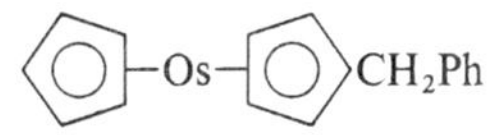

M 410.513

Mp 100-1°.

Mark, V. *et al, Inorg. Chem.*, 1964, **3**, 1067 (*synth, pmr*)

$C_{17}H_{20}Os$ Os-00300

[(1,2,3,4-η)-1,3,5,7-cyclooctatetraene][(1,2,3,4,5,6-η-)-1,3,5-trimethylbenzene]osmium, 10CI

(η^4-Cyclooctatetraene)(η^6-mesitylene)osmium

[72765-30-1]

As (η^6-Benzene)[(1,2,3,4-η)-1,3,5,7-cyclooctatetraene]osmium, Os-00240 with

$R = CH_3$

M 414.545

Fluxional molecule. Orange-yellow cryst. (hexane).

Bennett, M.A. *et al, Inorg. Chem.*, 1980, **19**, 1014 (*synth, ir, pmr, ms, cmr*)

$C_{18}HO_{18}Os_6^{\ominus}$ Os-00301

Octadecacarbonyl-μ₃-hydrohexaosmate(1−)

[60120-83-4]

$HOs_6(CO)_{18}^{\ominus}$

M 1646.395 (ion)

Struct. resembles Octadecacarbonylhexaosmate(2−), Os-00307 .

Bis(triphenylphosphine)iminium salt: [62086-86-6]. $C_{54}H_{31}NO_{18}Os_6P_2$ M 2184.982

Deep-red cryst.

Eady, C.R. *et al, J. Chem. Soc., Chem. Commun.*, 1976, 302 (*synth, cmr*)
McPartlin, M. *et al, J. Chem. Soc., Chem. Commun.*, 1976, 883 (*cryst struct*)
John, G.R. *et al, J. Organomet. Chem.*, 1979, **169**, C23 (*synth*)

Andrews, J.A. *et al*, *Inorg. Chem.*, 1980, **19**, 3033 (*ir*)
Braga, D. *et al*, *J. Chem. Soc., Chem. Commun.*, 1982, 419 (*synth*)

$C_{18}H_2O_{18}Os_6$ Os-00302

Octadecacarbonyldihydrohexaosmium

Octadecacarbonyl-μ-hydro-μ₃-hydrohexaosmium, 9CI

[50648-30-1]

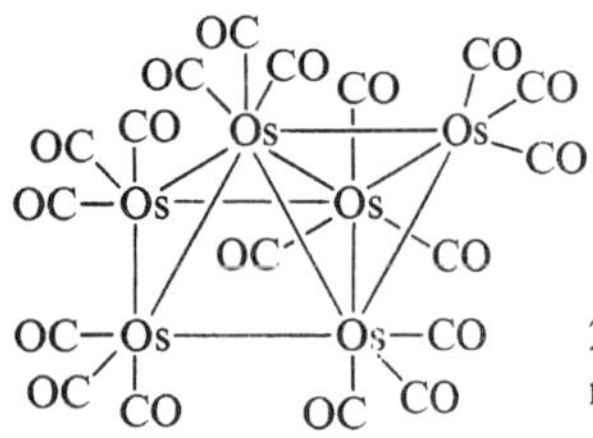

2 H atoms not shown

M 1647.403

Orange-brown cryst. (CH_2Cl_2 solvate).

McPartlin, M. *et al*, *J. Chem. Soc., Chem. Commun.*, 1976, 883 (*struct*)
Eady, C.R. *et al*, *J. Chem. Soc., Dalton Trans.*, 1977, 838 (*synth, ir, pmr*)
Nicholls, J.N. *et al*, *J. Chem. Soc., Dalton Trans.*, 1982, 1395 (*ir*)

$C_{18}H_9NO_{10}Os_3$ Os-00303

Decacarbonyl-μ-hydro[μ-[(methylimino)phenylmethyl-*C*:*N*]]triosmium, 10CI

[63363-66-6]

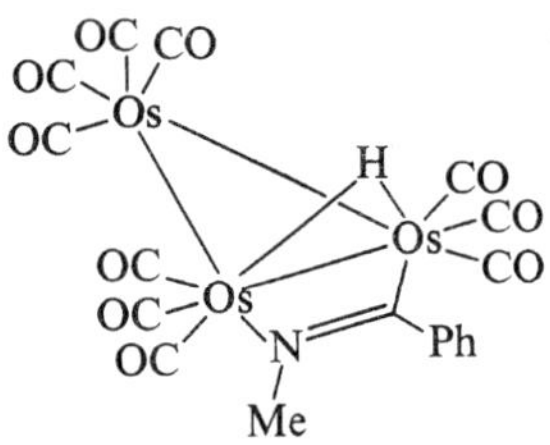

M 969.870

Yellow cryst. (hexane). Mp 152-5°.

Choo Yin, C. *et al*, *J. Organomet. Chem.*, 1977, **133**, 123 (*synth, ir, pmr*)
Adams, R.D. *et al*, *Inorg. Chem.*, 1978, **17**, 1969 (*struct*)

$C_{18}H_{11}O_{18}Os_5P$ Os-00304

Pentadecacarbonyldi-μ-hydro(trimethyl phosphite-*P*)pentaosmium, 10CI

[71935-53-0]

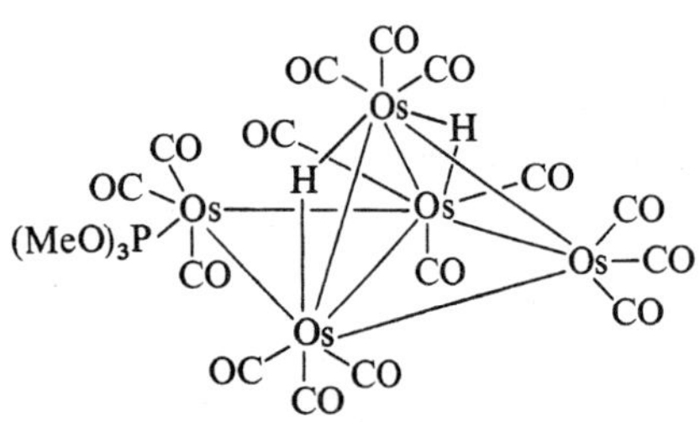

M 1497.248

Orange-red cryst. (CH_2Cl_2).

John, G.R. *et al*, *J. Organomet. Chem.*, 1979, **171**, C14 (*synth, struct, ir, ms, pmr*)

$C_{18}H_{15}Cl_5OsP^{\ominus}$ Os-00305

Pentachloro(triphenylphosphine)osmate(1−), 9CI

$Ph_3POsCl_5^{\ominus}$

M 629.755 (ion)

Tetraphenylarsonium salt: [42594-07-0].
$C_{42}H_{35}AsCl_5OsP$ M 1013.099
Red cryst.

Pawson, D. *et al*, *J. Chem. Soc., Dalton Trans.*, 1975, 417 (*synth, spectra*)

$C_{18}H_{45}Cl_2N_2OsP_3$ Os-00306

Dichloro(dinitrogen)tris(triethylphosphine)osmium, 8CI

[26553-82-2]

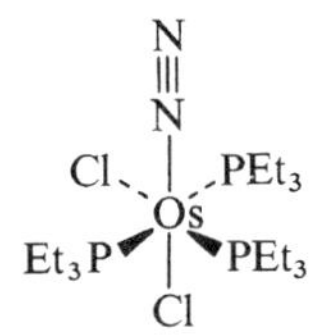

M 643.594

V. pale-pink cryst. (EtOH aq.). Mp 101-4° dec.

Chatt, J. *et al*, *J. Chem. Soc.* (*A*), 1970, 2243 (*synth, ir*)

$C_{18}O_{18}Os_6^{\ominus\ominus}$ Os-00307

Octadecacarbonylhexaosmate(2−)

[60120-85-6]

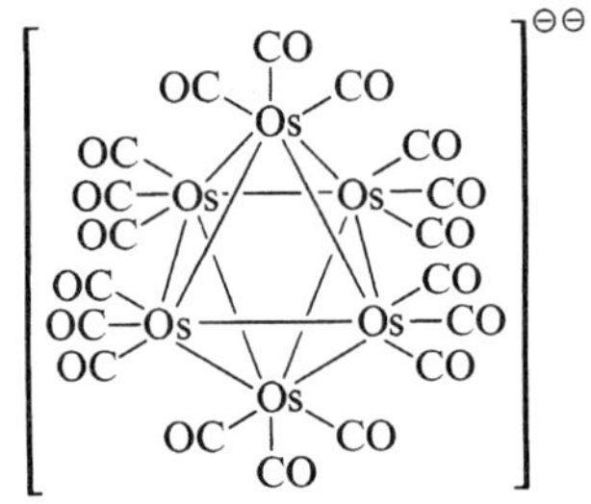

M 1645.387 (ion)

Bis(tetraethylammonium) salt: [82639-15-4].
$C_{34}H_{40}N_2O_{18}Os_6$ M 1905.893
Deep-orange cryst.

Eady, C.R. *et al*, *J. Chem. Soc., Chem. Commun.*, 1976, 302 (*synth, cmr*)
McPartlin, M. *et al*, *J. Chem. Soc., Chem. Commun.*, 1976, 883 (*cryst struct*)
John, G.R. *et al*, *J. Organomet. Chem.*, 1979, **169**, C9 (*synth*)
Hayward, C.-M.T. *et al*, *Inorg. Chem.*, 1982, **21**, 3816 (*synth, struct*)

$C_{18}O_{18}Os_6$ Os-00308

Octadecacarbonylhexaosmium, 9CI
Hexaosmium octadecacarbonyl. Osmium carbonyl[$Os_6(CO)_{18}$]
[37216-50-5]

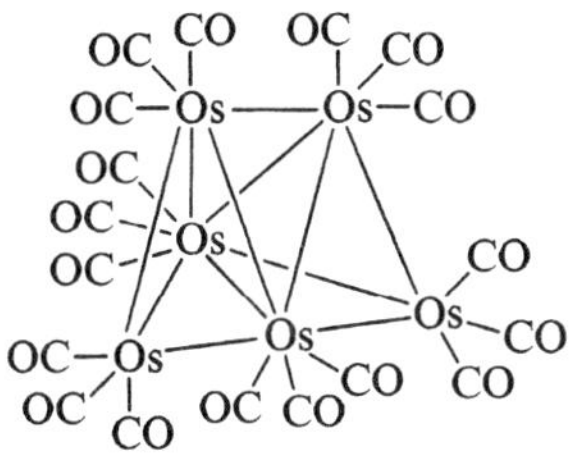

M 1645.387
Dark-brown cryst. (toluene/hexane). Carbonylation gives Nonadecacarbonylpentaosmium, Os-00309 .

Mason, R. *et al*, *J. Am. Chem. Soc.*, 1973, **95**, 3802 (*struct*)
Eady, C.R. *et al*, *J. Chem. Soc., Chem. Commun.*, 1975, 958 (*cmr*)
Eady, C.R. *et al*, *J. Chem. Soc., Dalton Trans.*, 1975, 2606; 1977, 838 (*synth*)
Nicholls, J.N. *et al*, *J. Chem. Soc., Dalton Trans.*, 1982, 1395 (*ir, props*)

$C_{19}O_{19}Os_5$ Os-00309

Nonadecacarbonylpentaosmium
Pentaosmium nonadecacarbonyl. Osmium carbonyl[$Os_5(CO)_{19}$]
[78166-65-1]

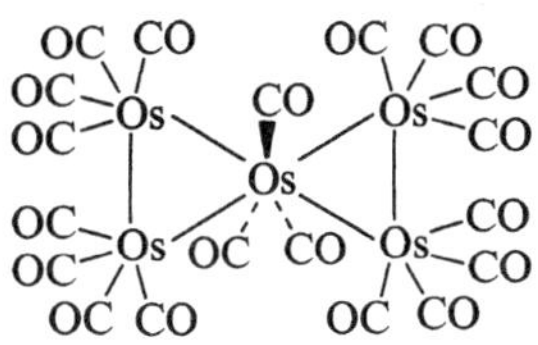

M 1483.198
Orange blocks (CH_2Cl_2). Decarbonylates in boiling octane to Hexadecacarbonylpentaosmium, Os-00287 .

Farrar, D.H. *et al*, *J. Chem. Soc., Chem. Commun.*, 1981, 273 (*synth, struct, ir, ms*)
Nicholls, J.N. *et al*, *J. Chem. Soc., Dalton Trans.*, 1982, 1395 (*ir*)

$C_{20}H_2O_{19}Os_7$ Os-00310

(Carbon)nonadecacarbonyldihydroheptaosmium
[50643-53-3]

$H_2Os_7(C)(CO)_{19}$

M 1877.624
Struct. not known. Green substance.

Eady, C.R. *et al*, *J. Chem. Soc., Dalton Trans.*, 1977, 838 (*synth, ir, ms, pmr*)

$C_{20}H_2O_{20}Os_3Re_2$ Os-00311

Decacarbonyl(decacarbonyldi-μ-hydrotriosmium)diruthenium, 10CI
Icosacarbonyldi-μ-hydrotriosmiumdiruthenium. Eicosacarbonyldi-μ-hydrotriosmiumdiruthenium
[64885-82-1]

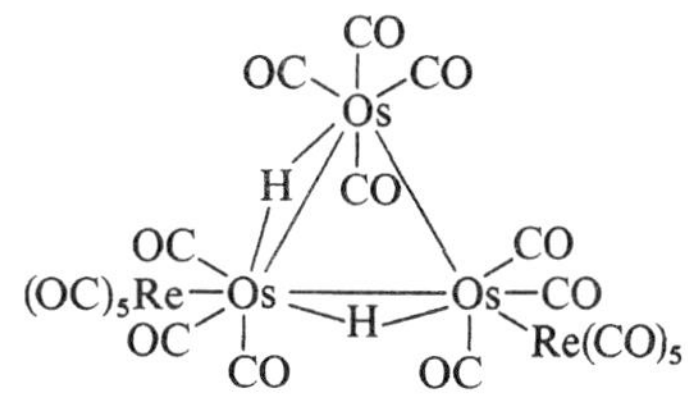

M 1505.238
Clear-yellow plates.

Churchill, M.R. *et al*, *Inorg. Chem.*, 1978, **17**, 3546 (*struct*)
Churchill, M.R. *et al*, *J. Am. Chem. Soc.*, 1981, **103**, 2430 (*synth, ir, pmr, ms*)

$C_{20}H_6O_{16}Os_6$ Os-00312

Hexadecacarbonyl-μ_3-ethylidyne-μ_4-ethylidynehexaosmium, 10CI
[68088-86-8]

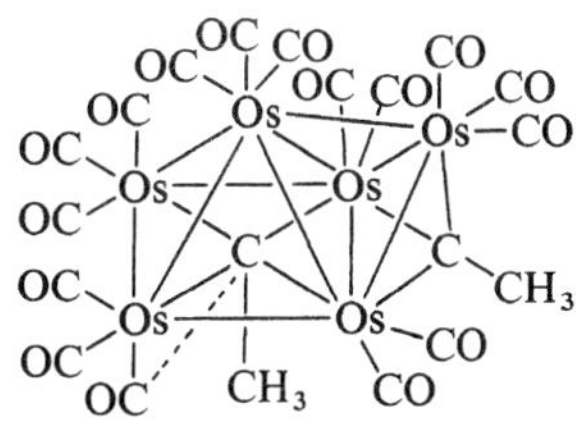

M 1643.458
Green cryst. (EtOAc/cyclohexane).

Eady, C.R. *et al*, *J. Chem. Soc., Chem. Commun.*, 1978, 421 (*synth, struct, pmr, ir*)

$C_{20}H_{10}O_{12}Os_4$ Os-00313

Dodecacarbonyldi-μ-hydro[μ-(2-phenylethylidene)]tetraosmium
[75699-40-0]

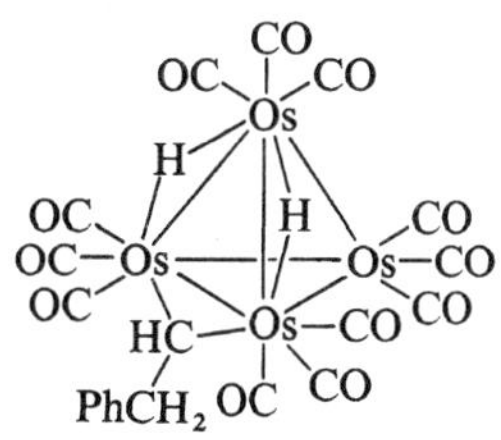

M 1203.092
Cryst. (CH_2Cl_2/hexane).

Johnson, B.F.G. *et al*, *J. Chem. Soc., Chem. Commun.*, 1980, 547 (*synth, struct*)

$C_{20}O_{20}Os_6$ Os-00314

Icosacarbonylhexaosmium

Eicosacarbonylhexaosmium. Hexaosmium icosacarbonyl

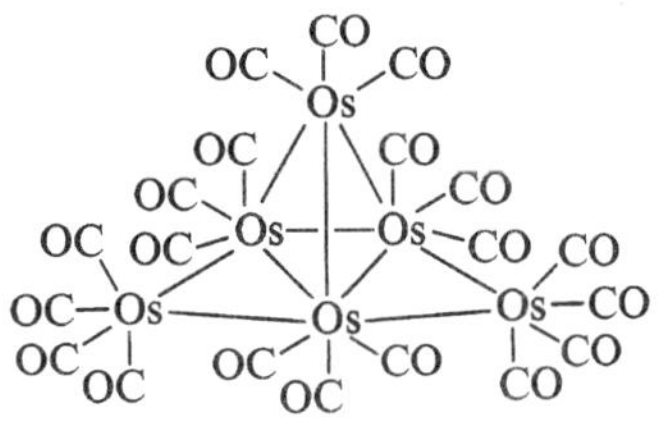

Suggested structure

M 1701.408

Purple solid. Heating generates Octadecacarbonylhexaosmium, Os-00308 .

Farrar, D.H. *et al*, *J. Chem. Soc., Chem. Commun.*, 1981, 273 (*synth, ir, ms*)

$C_{20}O_{20}Os_7^{\ominus\ominus}$ Os-00315

Eicosacarbonylheptaosmate(2−)

Icosacarbonylheptaosmate(2−)

[70379-01-0]

$Os_7(CO)_{20}^{\ominus\ominus}$

M 1891.608 (ion)

Capped octahedral struct. proposed. No phys. props. reported.

John, G.R. *et al*, *J. Organomet. Chem.*, 1979, **169**, C9 (*synth*)

$C_{21}H_6O_{16}Os_6$ Os-00316

Hexadecacarbonyl[μ_3-[(1-η:1,2-η:2-η)-1,2-dimethyl-1,2-ethenediyl]](μ_5-methanetetrayl)hexaosmium, 10CI

[68091-54-3]

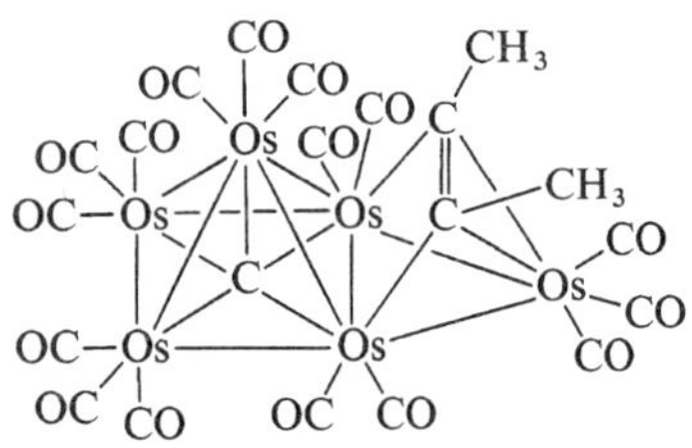

M 1655.469

Red cryst. (EtOAc/cyclohexane).

Eady, C.R. *et al*, *J. Chem. Soc., Chem. Commun.*, 1978, 421 (*synth, struct, ir, pmr*)

$C_{21}H_{15}Br_2O_3OsP$ Os-00317

Dibromotricarbonyl(triphenylphosphine)osmium, 8CI

[16481-98-4]

$Ph_3POsBr_2(CO)_3$

M 696.329

Light-yellow cryst.

Cleare, M.J. *et al*, *Chem. Ind. (London)*, 1967, 1705 (*synth*)

$C_{21}H_{15}Cl_2O_3OsP$ Os-00318

Tricarbonyldichloro(triphenylphosphine)osmium, 8CI

[16520-59-5]

$Ph_3POsCl_2(CO)_3$

M 607.427

Colourless cryst.

Cleare, M.J. *et al*, *Chem. Ind. (London)*, 1967, 1705 (*synth*)

$C_{21}H_{15}I_2O_3OsP$ Os-00319

Tricarbonyldiiodo(triphenylphosphine)osmium

[16481-99-5]

$Ph_3POsI_2(CO)_3$

M 790.330

Yellow cryst.

Bradford, C.W. *et al*, *J. Chem. Soc., Chem. Commun.*, 1967, 384 (*synth*)

$C_{21}H_{16}ClO_3OsP$ Os-00320

Tricarbonylchlorohydro(triphenylphosphine)osmium, 9CI

[59526-98-6]

$Ph_3POsClH(CO)_3$

M 572.982

Buff cryst. (hexane). Mp 185°.

Khan, M.M.T. *et al*, *J. Inorg. Nucl. Chem.*, 1976, **38**, 331 (*synth, ir*)

$C_{21}H_{17}O_3OsP$ Os-00321

Tricarbonyldihydro(triphenylphosphine)osmium, 8CI

[18972-43-5]

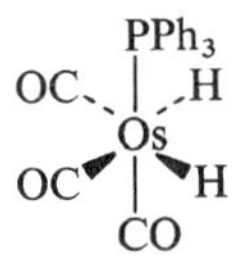

M 538.537

Colourless cryst. (heptane). Mp 148-9°.

L'Eplattenier, F. *et al*, *Inorg. Chem.*, 1967, **6**, 2092; 1968, **7**, 1290 (*synth, ir, pmr*)

$C_{21}H_{18}FeOOs$ Os-00322

(Osmocenylcarbonyl)ferrocene

Ferrocenyl osmocenyl ketone. Ferrocenylosmocenylmethanone. (Ferrocenylcarbonyl)osmocene

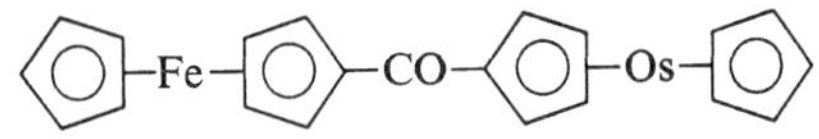

M 532.420

Mp 232° dec.

Gubin, S.P. *et al*, *J. Organomet. Chem.*, 1971, **30**, 243 (*synth, polarog*)

$C_{21}H_{20}FeOs$ Os-00323

Ferrocenylmethylosmocene, 9CI

π-Cyclopentadienyl[(ferrocenylmethyl)-π-cyclopentadienyl]osmium. Osmocenylmethylferrocene

[11105-98-9]

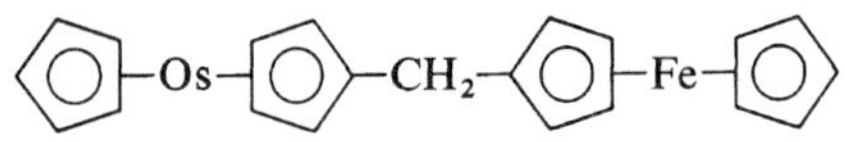

M 518.436

Mp 152-3°.

Gubin, S.P. *et al*, *J. Organomet. Chem.*, 1970, **22**, 183; 1971, **30**, 243 (*polarog*)

Smirnova, S.A. *et al*, *Dokl. Akad. Nauk SSSR, Ser. Sci. Khim.*, 1971, **200**, 605 (*synth, polarog*)

$C_{21}H_{32}OOs_2$ **Os-00324**

μ-Carbonyldi-μ-hydrobis[(1,2,3,4,5-η)-1,2,3,4,5-pentamethyl-2,4-cyclopentadien-1-yl]diosmium

[81740-98-9]

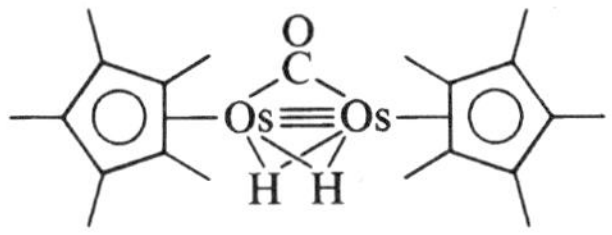

M 680.883

Dark-amber cryst. Mp >160° dec.

Hoyano, J.K. *et al*, *J. Am. Chem. Soc.*, 1982, **104**, 3732 (*synth, ir, pmr, cmr*)

$C_{21}O_{21}Os_7$ **Os-00325**

Heneicosacarbonylheptaosmium, 9CI

Heptaosmium heneicosacarbonyl. Osmium carbonyl $Os_7(CO)_{21}$

[59158-84-8]

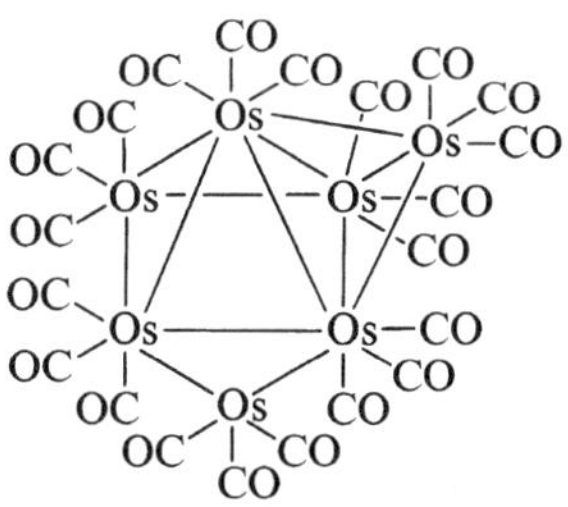

M 1919.618

Orange-brown cryst. (toluene).

Eady, C.R. *et al*, *J. Chem. Soc., Dalton Trans.*, 1975, 2606 (*synth*)

Eady, C.R. *et al*, *J. Chem. Soc., Chem. Commun.*, 1977, 385 (*struct, cmr*)

Nicholls, J.N. *et al*, *J. Chem. Soc., Dalton Trans.*, 1982, 1395 (*ir*)

$C_{22}H_{15}O_4OsP$ **Os-00326**

Tetracarbonyl(triphenylphosphine)osmium, 8CI

[21192-24-5]

$Ph_3POs(CO)_4$

M 564.532

Yellow, light-sensitive in soln. Mp 186° dec.

L'Eplattenier, F. *et al*, *Inorg. Chem.*, 1968, **7**, 1290; **8**, 965 (*synth, ir*)

$C_{22}H_{20}B_2Os$ **Os-00327**

Bis[(1,2,3,4,5,6-η)-1-phenylboratabenzene]osmium, 9CI

Bis(1-phenylborinato)osmium

[60764-69-4]

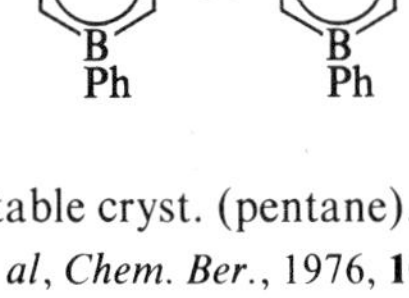

M 496.220

Colourless, air stable cryst. (pentane). Mp 146.5-147°.

Herberich, G.E. *et al*, *Chem. Ber.*, 1976, **109**, 2382 (*synth, pmr, nmr, ms*)

$C_{22}H_{32}O_2Os_2$ **Os-00328**

Dicarbonyldi-μ-hydrobis[(1,2,3,4,5-η)-1,2,3,4,5-pentamethyl-2,4-cyclopentadien-1-yl]diosmium

[81740-97-8]

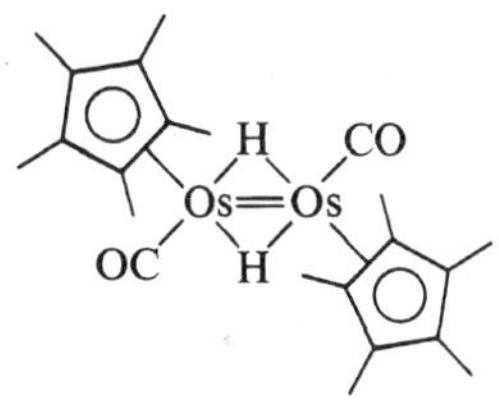

M 708.894

Dark-red cryst. Mp >175° dec.

Hoyano, J.K. *et al*, *J. Am. Chem. Soc.*, 1982, **104**, 3732 (*synth, struct, ir, pmr, cmr*)

$C_{22}O_{21}Os_8$ **Os-00329**

(Carbon)heneicosacarbonyloctaosmium, 9CI

[59113-94-9]

$Os_8(C)(CO)_{21}$

M 2121.829

Struct. not known. Deep-purple cryst. Insol. cyclohexane.

Eady, C.R. *et al*, *J. Chem. Soc., Dalton Trans.*, 1975, 2606; 1977, 838 (*synth, ir, pmr*)

$C_{22}O_{22}Os_8^{\ominus\ominus}$ **Os-00330**

Di-μ-carbonylicosacarbonyloctaosmate(2−)

Docosacarbonyloctaosmate(2−)

[74029-08-6]

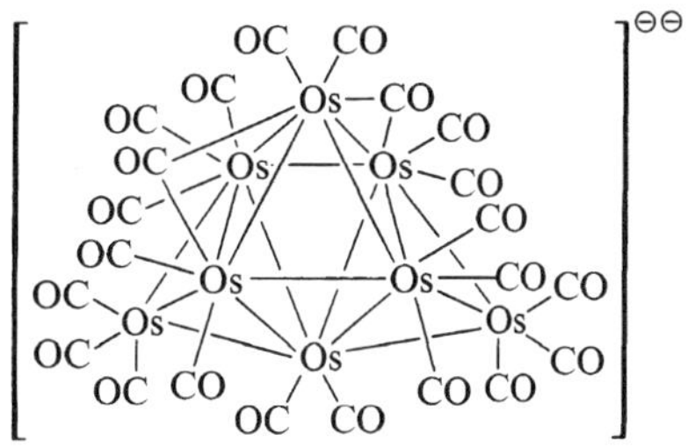

M 2137.829 (ion)

Bis[bis(triphenylphosphine)iminium] salt: [74091-69-3].

$C_{94}H_{60}N_2O_{22}Os_8P_4$ M 3215.003

Dark-red cryst. ($MeOH/CH_2Cl_2$).

Jackson, P.F. *et al*, *J. Chem. Soc., Chem. Commun.*, 1980, 60 (*synth, cryst struct*)

$C_{23}H_{12}N_2O_{10}Os_3$ **Os-00331**

Decacarbonyl[μ-(diphenylmethanone hydrazonato)]-μ-hydrotriosmium, 10CI

[μ-(Benzophenone hydrazonato)]decacarbonyl-μ-hydrotriosmium

[77924-84-6]

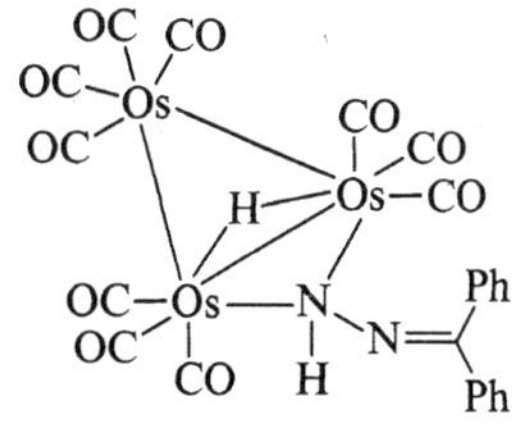

M 1046.955

Yellow microcryst. or amber cryst. (cyclohexane).

Churchill, M.R. *et al*, *Inorg. Chem.*, 1981, **20**, 2905 (*struct*)

$C_{23}H_{21}O_3OsP$ Os-00332

Tricarbonyldimethyl(triphenylphosphine)osmium, 9CI, 8CI

[22639-02-7]

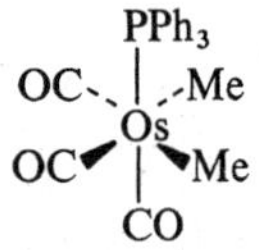

M 566.591

Colourless cryst. (hexane).

L'Eplattenier, F., *Inorg. Chem.*, 1969, **8**, 965 (*synth, ir, pmr*)
Carter, W.J. *et al*, *Inorg. Chem.*, 1982, **21**, 3955 (*synth, ir, pmr*)

$C_{23}H_{39}N_3Os$ Os-00333

[(1,2,5,6-η)-1,5-Cyclooctadiene]tris(2-isocyano-2-methylpropane)osmium, 10CI

*(1,5-Cyclooctadiene)tris(*tert*-butylisocyanide)osmium*

[71744-55-3]

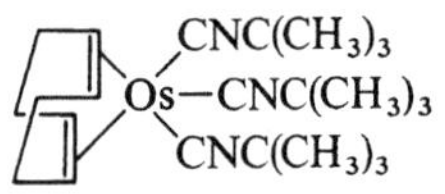

M 547.781

Pale-yellow cryst. (hexane). Mp 102-5°.

Bassett, J-M. *et al*, *J. Chem. Soc., Dalton Trans.*, 1979, 1003 (*synth, ir, pmr, cmr*)

$C_{23}O_{23}Os_8$ Os-00334

Tricosacarbonyloctaosmium

Osmium carbonyl [$Os_8(CO)_{23}$], 10CI. Octaosmium tricosacarbonyl

[37190-66-2]

$Os_8(CO)_{23}$

M 2165.839

Struct. not known. Orange-yellow cryst.

Eady, C.R. *et al*, *J. Chem. Soc., Dalton Trans.*, 1975, 2606 (*synth, ir, ms*)
John, G.R. *et al*, *J. Organomet. Chem.*, 1979, **169**, C9.
Jackson, P.F. *et al*, *J. Chem. Soc., Chem. Commun.*, 1980, 60.

$C_{24}H_{10}O_{10}Os_3$ Os-00335

Decacarbonyl[μ_3-[1,1′-[(1-η:1,2-η:2-η)-1,2-ethynediyl]bis[benzene]]]triosmium, 9CI

[58023-41-9]

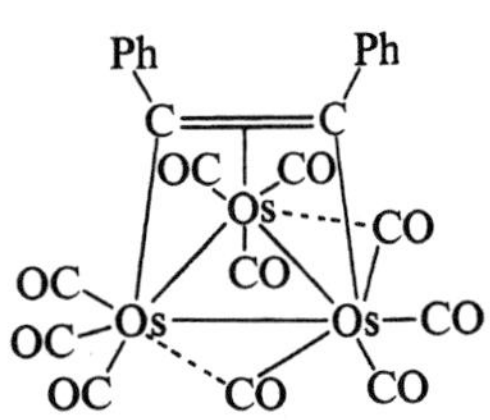

M 1028.937

Possesses semi-bridging CO groups. Yellow cryst. Mp 118-9°.

Tachikawa, M. *et al*, *J. Am. Chem. Soc.*, 1975, **97**, 7172 (*synth, struct, ir, pmr, ms*)
Pierpoint, C.G., *Inorg. Chem.*, 1977, **16**, 639 (*struct*)

$C_{24}H_{20}BrOOsP$ Os-00336

Bromocarbonyl(η^5-2,4-cyclopentadien-1-yl)(triphenylphosphine)osmium

Cyclopentadienylcarbonyl(triphenylphosphine)osmium bromide. (η^5-Cyclopentadienyl)-bromocarbonyl(triphenylphosphine)osmium

[34676-01-2]

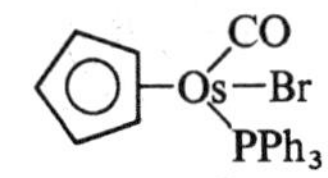

M 625.499

Yellow cryst. (CH_2Cl_2/hexane). Mp 249-51°.

Blackmore, T. *et al*, *J. Chem. Soc. (A)*, 1971, 2376 (*synth, ir, pmr*)

$C_{24}H_{21}I_2OsP$ Os-00337

(η^6-Benzene)diiodo(triphenylphosphine)osmium, 10CI

[75368-02-4]

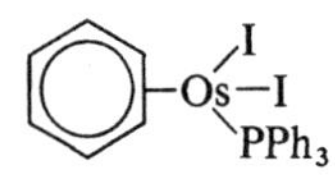

M 784.413

No phys. props. reported.

Werner, H. *et al*, *J. Organomet. Chem.*, 1980, **194**, C7 (*synth*)

$C_{25}H_{10}N_2O_{13}Os_5$ Os-00338

Tridecacarbonyl-μ-hydro[μ_3-[*N*-phenyl-1,2-benzenediaminato(3−)-*N*,*N′*:*N′*,*N′*]]pentaosmium

[76190-51-7]

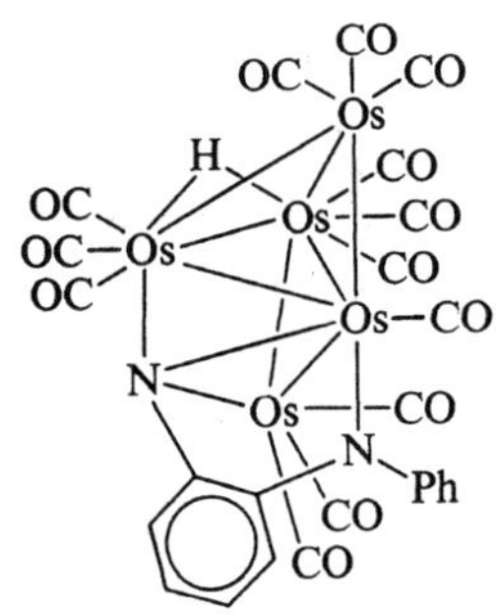

M 1497.360

Green-black needles (hexane).

Dawoodi, Z. *et al*, *J. Chem. Soc., Chem. Commun.*, 1980, 712 (*synth, struct, ir, pmr*)

$C_{25}H_{17}O_7Os_2P$ Os-00339

Heptacarbonyldihydro(triphenylphosphine)diosmium

[62976-01-6]

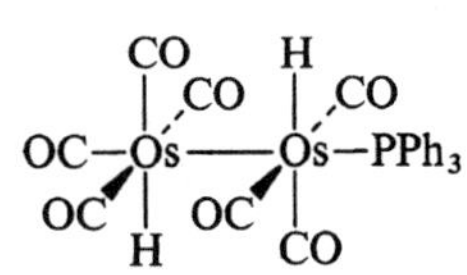

M 840.779

Colourless prisms (hexane). Mp 99-101°.

Moss, J.R. *et al*, *J. Chem. Soc., Dalton Trans.*, 1977, 89 (*synth, ir, pmr*)

$C_{25}IO_{24}Os_{10}^{\ominus}$ Os-00340

[μ_6-(Carbon)]tetracosacarbonyl-μ-iododecaosmate(1−)

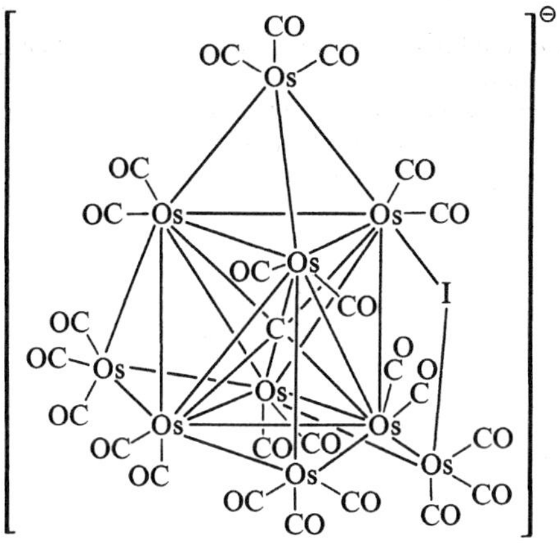

M 2713.165 (ion)

Bis(triphenylphosphine)iminium salt:
$C_{61}H_{30}INO_{24}Os_{10}P_2$ M 3251.752
No props. reported.

Farrar, D.H. *et al*, *J. Chem. Soc., Chem. Commun.*, 1981, 1009 (*synth, struct, ir*)

$C_{25}I_2O_{24}Os_{10}$ Os-00341

(Carbon)tetracosacarbonylbis-μ_2-iododecaosmium

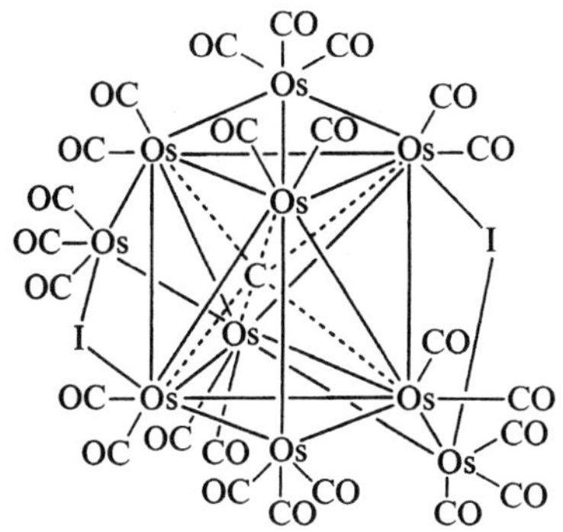

M 2840.070

Farrar, D.H. *et al*, *J. Chem. Soc., Chem. Commun.*, 1981, 1009 (*synth, struct, ir*)

$C_{25}O_{24}Os_{10}^{\ominus\ominus}$ Os-00342

[μ_6-(Carbon)]tetracosacarbonyldecaosmate(2−)
[75026-12-9]

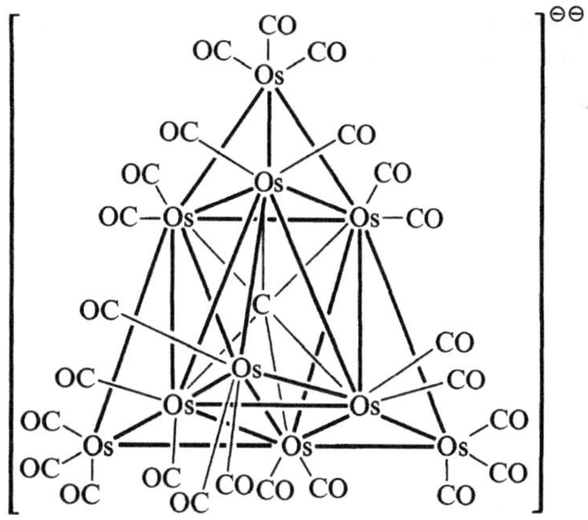

M 2586.261 (ion)

Bis(tetraethylammonium) salt: [82639-16-5].
$C_{41}H_{40}N_2O_{24}Os_{10}$ M 2846.766
Dark red-brown cryst. ($MeOH/CH_2Cl_2$). Air-stable.
Bis[bis(triphenylphosphine)iminium] salt: [75117-74-1].
$C_{97}H_{60}N_2O_{24}Os_{10}P_4$ M 3663.435
No phys. props. reported.

Jackson, P.F. *et al*, *J. Chem. Soc., Chem. Commun.*, 1980, 224 (*synth, cryst struct*)
Oxton, I.A. *et al*, *J. Mol. Struct.*, 1981, **71**, 117 (*ir*)
Hayward, C.-M.T. *et al*, *Inorg. Chem.*, 1982, **21**, 3816 (*synth, ir*)

$C_{27}H_{17}O_9Os_3P$ Os-00343

Nonacarbonyldihydro(triphenylphosphine)triosmium, 10CI
[56093-36-8]

As Decacarbonyldi-μ-hydrotriosmium, Os-00095 with

$$X = PPh_3$$

M 1087.000
Red cryst.

Benfield, R.E. *et al*, *Acta Crystallogr., Sect. B*, 1979, **35**, 2210 (*struct*)

$C_{28}HO_{29}Os_9^{\ominus}$ Os-00344

Heptacosacarbonyl[μ_3-(carboxylato-*C*:*O*:*O'*)]-μ-hydrononaosmate(1−)

[61250-78-0]

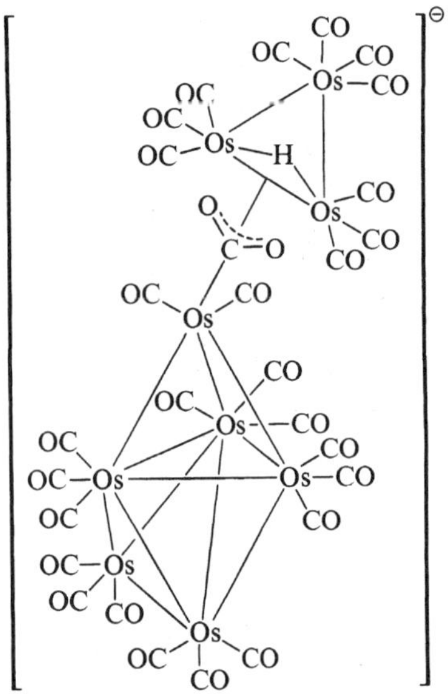

M 2513.099 (ion)

Position of bridging H not certain.

Bis(triphenylphosphine)iminium salt: [61312-93-4].
$C_{64}H_{31}NO_{29}Os_9P_2$ M 3051.686
Brown-black cryst. (CH_2Cl_2/cyclohexane).

Eady, C.R. *et al*, *J. Chem. Soc., Dalton Trans.*, 1978, 1358 (*synth, ir, nmr, cmr, struct*)
Guy, J.J. *et al*, *Acta Crystallogr., Sect. B*, 1978, **34**, 1718 (*cryst struct*)
John, G.R. *et al*, *J. Organomet. Chem.*, 1979, **169**, C23 (*ir, nmr*)

$C_{28}H_{17}O_{10}Os_3P$ Os-00345

Decacarbonyl-μ-hydrohydro(triphenylphosphine)triosmium, 9CI

Decacarbonyldihydro(triphenylphosphine)triosmium

[56398-26-6]

As Undecacarbonyl-μ-hydrohydrotriosmium, Os-00125 with

X = PPh_3

M 1115.010

Yellow cryst. (hexane).

Deeming, A.J. *et al*, *J. Organomet. Chem.*, 1975, **88**, C21 (*synth, pmr, ir*)
Shapley, J.R. *et al*, *J. Am. Chem. Soc.*, 1975, **94**, 4145 (*synth, pmr*)
Churchill, M.R. *et al*, *Inorg. Chem.*, 1977, **16**, 2397 (*struct*)
Aime, S. *et al*, *J. Organomet. Chem.*, 1981, **213**, 207 (*cmr*)

$C_{28}H_{24}I_2O_2OsP_2$ Os-00346

Dicarbonyl[1,2-ethanediylbis[diphenylphosphine]-*P*,*P'*]diiodoosmium

Dicarbonyl[ethylenebis(diphenylphosphine)]diiodoosmium, 8CI. *Bis[1,2-(diphenylphosphino)ethane]dicarbonyldiiodoosmium*

[17633-49-7]

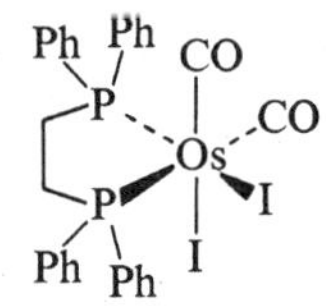

M 898.453

Colourless cryst. (C_6H_6).

Hales, L.A.W. *et al*, *J. Chem. Soc. (A)*, 1967, 1932.

$C_{29}H_{15}AsO_{11}Os_3$ Os-00347

Undecacarbonyl(triphenylarsine)triosmium, 8CI

[30173-89-8]

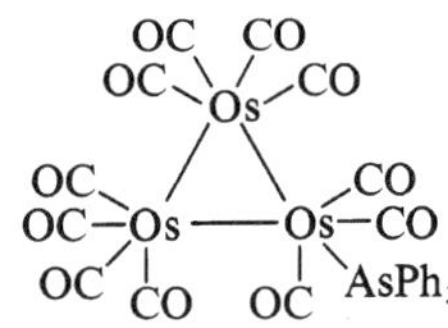

M 1184.953

Yellow cryst. (pet. ether). Mp 173°.

Bradford, C.W. *et al*, *J. Chem. Soc. (A)*, 1970, 2889 (*synth, ir*)

$C_{29}H_{15}O_{11}Os_3P$ Os-00348

Undecacarbonyl(triphenylphosphine)triosmium, 8CI

[30173-88-7]

As (Acetonitrile)undecacarbonyltriosmium, Os-00213 with

L = PPh_3

M 1141.005

Yellow cryst. (pet. ether). Mp 178°.

Bradford, C.W. *et al*, *J. Chem. Soc. (A)*, 1970, 2889 (*synth, ir*)
Tyler, D.R. *et al*, *J. Am. Chem. Soc.*, 1980, **102**, 3022 (*synth, uv*)

$C_{29}H_{36}O_{29}Os_6P_4$ Os-00349

Heptadecacarbonyltetrakis(trimethylphosphite-*P*)hexaosmium

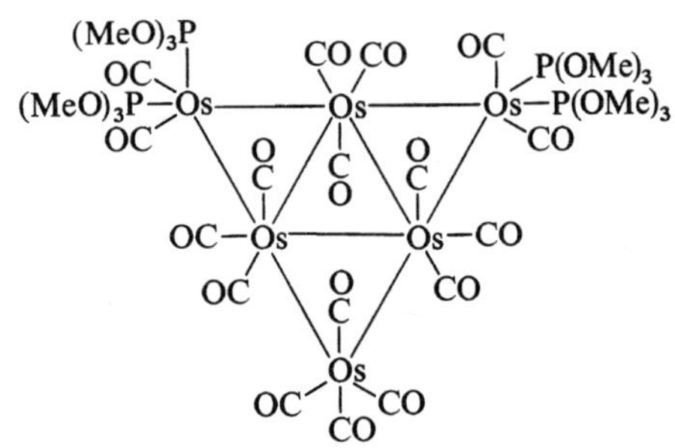

M 2113.681

Dark-blue platelets (CH_2Cl_2/hexane).

Goudsmit, R.J. *et al*, *J. Chem. Soc., Chem. Commun.*, 1982, 640 (*synth, ir, struct*)

$C_{30}H_{75}BrO_{15}OsP_5^{\oplus}$ Os-00350

Bromopentakis(triethyl phosphite-*P*)osmium(1+), 9CI

$BrOs[P(OEt)_3]_5^{\oplus}$

M 1100.886 (ion)

Tetraphenylborate: [35083-17-1].
$C_{54}H_{95}BBrO_{15}OsP_5$ M 1420.118
Colourless cryst. ($EtOH/CHCl_3$).

Couch, D.A. *et al*, *Inorg. Chim. Acta*, 1974, **9**, 39 (*synth, pmr*)

$C_{30}H_{75}ClO_{15}OsP_5^{\oplus}$ Os-00351

Chloropentakis(triethyl phosphite-*P*)osmium(1+), 9CI

$ClOs[P(OEt)_3]_5^{\oplus}$

M 1056.435 (ion)

Tetraphenylborate: [35083-16-0].
$C_{54}H_{95}BClO_{15}OsP_5$ M 1375.667
Colourless cryst.

Couch, D.A. *et al*, *Inorg. Chim. Acta*, 1974, **9**, 39 (*synth*)

$C_{31}HO_{32}Os_{10}^{\ominus}$ Os-00352

Triacontacarbonyl[μ_3-(carboxylato-*C,O,O'*)]-μ-hydrodecaosmate(1−)

[70379-03-2]

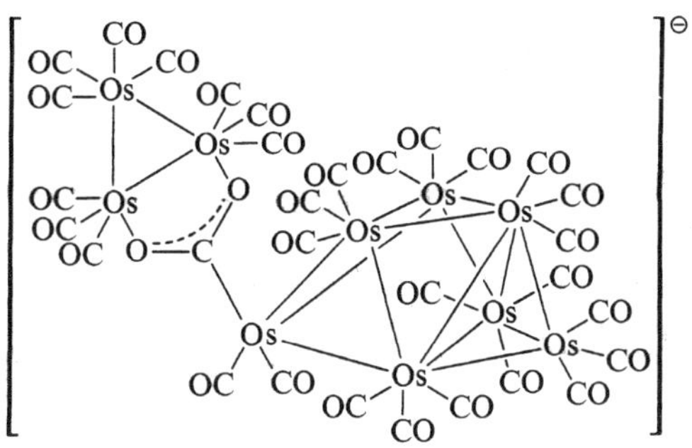

M 2787.330 (ion)

H not located. Dark red-brown.

John, G.R. *et al*, *J. Organomet. Chem.*, 1979, **169**, C23 (*synth, ir, nmr, struct*)

$C_{31}H_{20}O_3Os$ Os-00353

Tricarbonyl[(1,2,3,4-η)-1,2,3,4-tetraphenyl-1,3-cyclobutadiene]osmium

Tricarbonyl[1,1',1'',1'''-(η^4-1,3-cyclobutadiene-1,2,3,4-tetrayl)tetrakis[benzene]]osmium

[42868-19-9]

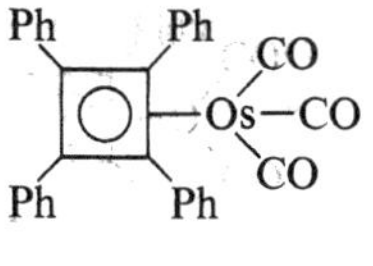

M 630.697

No phys. props. reported.

Vaglio, G.A. *et al*, *Inorg. Chim. Acta*, 1973, **7**, 193 (*synth, ir, ms*)

$C_{31}H_{25}N_2O_{13}Os_5P$ Os-00354

Tridecacarbonyl-μ-hydro[μ_3-[*N*-phenyl-1,2-benzenediaminato(3−)-*N,N':N',N'*]](triethylphosphine)pentaosmium

[79846-24-5]

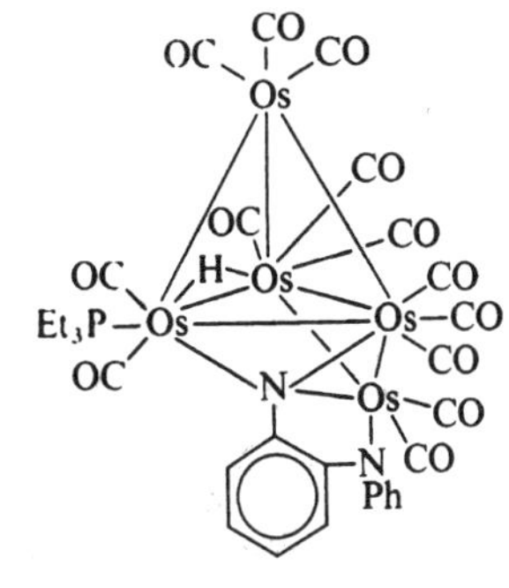

M 1615.518

Square green-black plates.

Dawoodi, Z. *et al*, *J. Chem. Soc., Chem. Commun.*, 1981, 801 (*synth, struct, ir, pmr*)

$C_{31}H_{29}BrOsP_2$ Os-00355

Bromo(η^5-2,4-cyclopentadien-1-yl)[1,2-ethanediylbis[diphenylphosphine]-*P,P'*]osmium, 10CI

[1,2-Bis(diphenylphosphino)ethane]bromo(η^5-cyclopentadienyl)osmium

[71397-13-2]

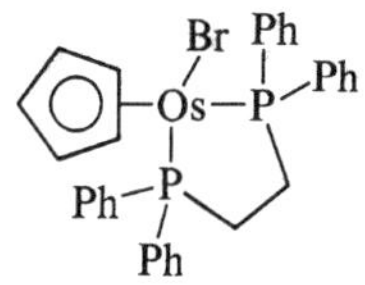

M 733.622

Yellow cryst. + CH_2Cl_2 (CH_2Cl_2/MeOH). Mp >205° dec.

Ashby, G.S. *et al*, *Aust. J. Chem.*, 1979, **32**, 1003 (*synth, pmr*)

$C_{34}H_{20}O_6Os_2$ Os-00356

Hexacarbonyl[μ-[(1,2,3,4-η:1,4-η)-1,2,3,4-tetraphenyl-1,3-butadiene-1,4-diyl]]diosmium, 9CI

Hexacarbonyl[μ-(1,2,3,4-tetraphenyl-1,3-butadienylene)]diosmium

[33310-10-0]

As [μ-[(1,2,3,4-η:1,4-η)-1,3-Butadiene-1,4-diyl]]-hexacarbonyldiosmium, Os-00098 with

R = Ph

M 904.928

Yellow or white cryst. (heptane).

Vaglio, G.A. *et al*, *Org. Mass Spectrom.*, 1971, **5**, 493 (*ms*)
Gambino, O. *et al*, *J. Organomet. Chem.*, 1971, **30**, 381 (*synth, ir, ms*)

$C_{36}H_{20}O_8Os_3$ Os-00357

Octacarbonyl[μ-[(1,2,3,4-η:1,4-η)-1,2,3,4-tetraphenyl-2-butene-1,4-diylidene]]triosmium, 9CI

[38979-93-0]

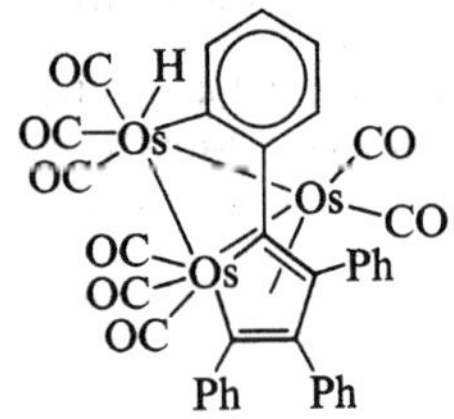

M 1151.149

Probably contains discrete ortho-metallated C_6H_4 ring and Os-H group. Lemon-yellow lamellar cryst. (heptane). Reacts with CO to give Nonacarbonyl[μ-[(1,2,3,4-η:1,4-η)-1,2,3,4-tetraphenyl-1,3-butadiene-1,4-diyl]]triosmium, Os-00368 .

Cetini, G. *et al*, *J. Organomet. Chem.*, 1971, **30**, 381 (*synth, ir, ms*)
Ferrari, R.P. *et al*, *J. Chem. Soc., Dalton Trans.*, 1972, 1998 (*synth, struct, pmr*)
Ferraris, G. *et al*, *J. Chem. Soc., Dalton Trans.*, 1972, 1057 (*struct*)

$C_{36}H_{30}As_2Br_3NOs$ Os-00358

Tribromonitridobis(triphenylarsine)osmium, 9CI

[42531-50-0]

$(Ph_3As)_2OsBr_3(\equiv N)$

M 1056.395

2 Isomers known of undetermined struct.

α-form

Brown cryst. Mp 221°.

β-form

Brown cryst.

Pawson, D. *et al*, *J. Chem. Soc., Dalton Trans.*, 1975, 417 (*prep, ir*)

$C_{36}H_{30}As_2Cl_3NOsS$ Os-00359

Trichlorobis(triphenylarsine)thionitrosylosmium, 9CI

[57902-93-9]

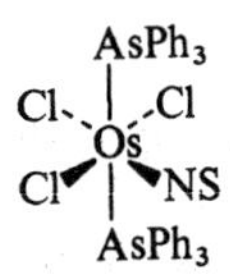

M 955.102

Green cryst. Mp 138-41° dec.

Bishop, M.W. *et al*, *J. Chem. Soc., Chem. Commun.*, 1975, 780 (*synth*)

$C_{36}H_{30}Br_3NOOsP_2$ Os-00360

Tribromonitrosylbis(triphenylphosphine)osmium, 9CI

[29292-11-3]

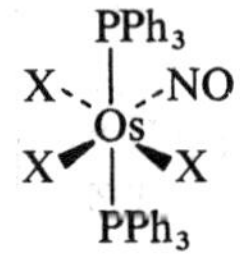

X = Br

M 984.499

Brown cryst. Mp 300° dec.

Robinson, S.D. *et al*, *J. Chem. Soc., Dalton Trans.*, 1972, 1 (*synth, ir*)
Clark, D.T. *et al*, *Inorg. Chem.*, 1977, **16**, 1201 (*uv*)

$C_{36}H_{30}Cl_3HgNOOsP_2$ Os-00361

Dichloro(chloromercurio)nitrosylbis(triphenylphosphine)osmium, 8CI

[30258-60-7]

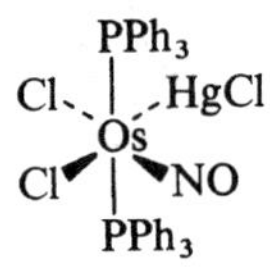

M 1051.736

Yellow cryst. Mp 267-9°.

Bentley, G.A. *et al*, *J. Chem. Soc., Chem. Commun.*, 1970, 998 (*synth, cryst struct*)

$C_{36}H_{30}Cl_3NOOsP_2$ Os-00362

Trichloronitrosylbis(triphenylphosphine)osmium, 9CI, 8CI

[22180-41-2]

As Tribromonitrosylbis(triphenylphosphine)osmium, Os-00360 with

X = Cl

M 851.146

Red-brown microcryst. Mp 265-90° (in air), 330° (sealed tube under N_2).

Robinson, S.D. *et al*, *J. Chem. Soc. (A)*, 1970, 2947 (*synth, ir*)
Inorg. Synth., 1974, **15**, 45 (*synth*)
Pandey, K.K. *et al*, *Indian J. Chem., Sect. A*, 1981, **20**, 370 (*synth, ir*)

$C_{36}H_{30}Cl_3NOsP_2$ Os-00363

Trichloronitridobis(triphenylphosphine)osmium, 9CI

[42531-48-6]

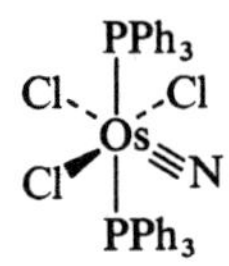

M 835.146

Brown cryst. Mp 145° dec.

Pawson, D. *et al*, *J. Chem. Soc., Dalton Trans.*, 1975, 417 (*synth, spectra*)

$C_{36}H_{30}Cl_4OsP_2^{\ominus}$ Os-00364

Tetrachlorobis(triphenylphosphine)osmate(1–), 9CI

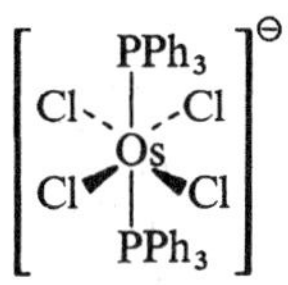

M 856.593 (ion)

Tetraphenylarsonium salt: [55450-03-8].
$C_{60}H_{50}AsCl_4OsP_2$ M 1239.936
Yellow cryst.

Pawson, D. *et al*, *J. Chem. Soc., Dalton Trans.*, 1975, 417 (*synth, ir*)

$C_{36}H_{30}Cl_4OsP_2$ Os-00365

Tetrachlorobis(triphenylphosphine)osmium, 9CI

[42594-06-9]

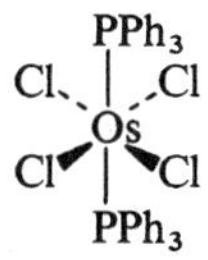

M 856.593

trans-form [60479-38-1]
Yellow. Mp 218-20° dec.

Pawson, D. *et al*, *J. Chem. Soc., Dalton Trans.*, 1975, 417 (*synth, spectra*)
Khan, M.M.T. *et al*, *J. Inorg. Nucl. Chem.*, 1976, **38**, 1135 (*synth, spectra*)

$C_{36}H_{30}N_2O_2OsP_2$ Os-00366

Dinitrosylbis(triphenylphosphine)osmium, 9CI

[30347-88-7]

$(Ph_3P)_2Os(NO)_2$ (C_{2v})

M 774.793

Deep-red prisms + $\frac{1}{2}C_6H_6$ (C_6H_6/hexane).

Haymore, B.L. *et al*, *Inorg. Chem.*, 1975, **14**, 2610 (*synth, ir, cryst struct*)
Bhaduri, S. *et al*, *Transition Met. Chem.*, 1978, **3**, 56 (*use*)

$C_{36}H_{90}O_{18}OsP_6^{\oplus\oplus}$ Os-00367

Hexakis(triethyl phosphite-*P*)osmium(2+), 9CI

$Os[P(OEt)_3]_6^{\oplus\oplus}$

M 1187.139 (ion)

Bis(tetraphenylborate): [53062-03-6].
$C_{84}H_{130}B_2O_{18}OsP_6$ M 1825.603
Colourless cryst.

Couch, D.A. *et al*, *Inorg. Chim. Acta*, 1974, **9**, 39 (*synth*)

$C_{37}H_{20}O_9Os_3$ Os-00368

Nonacarbonyl[μ-[(1,2,3,4-η:1,4-η)-1,2,3,4-tetraphenyl-1,3-butadiene-1,4-diyl]]triosmium, 9CI

Nonacarbonyl[μ-(1,2,3,4-tetraphenyl-1,3-butadienylene)]triosmium, 8CI

[34766-74-0]

As [μ-[(1,2,3,4-η:1,4-η)-1,3-Butadiene-1,4-diyl]]nonacarbonyltriosmium, Os-00216 with

R = Ph

M 1179.160

Amethyst-violet needles or tablets (heptane). Loses CO thermally to give Octacarbonyl[μ-[(1,2,3,4-η:1,4-η)-1,2,3,4-tetraphenyl-2-butene-1,4-diylidene]]-triosmium, Os-00357, reacts with PhC≡CPh to give Heptacarbonyl[μ_3-[1,1′-(η^1:η^1:η^2-1,2-ethynediyl)-bis[benzene]]][μ-(1,2,3,4-η:1,4-η)-1,2,3,4-tetra-phenyl-1,3-butadiene-1,4-diyl]triosmium, Os-00432.

Gambino, O. *et al*, *J. Organomet. Chem.*, 1971, **30**, 381 (*struct, ir, ms, pmr*)
Ferraris, G. *et al*, *J. Chem. Soc., Dalton Trans.*, 1974, 1813 (*struct*)

$C_{37}H_{24}O_7Os_3P_2$ Os-00369

Heptacarbonylbis[μ-(diphenylphosphino)][μ_3-[(1-η:1,2-η:2-η-)-1,3-cyclohexadien-5-yne]]triosmium, 9CI

[36571-25-2]

As [μ_3-[(1-η:1,2-η:2-η)-Benzene]]heptacarbonyl-bis[μ-(dimethylphosphino)]triosmium, Os-00298 with

R = Ph

M 1213.140

Bridging benzyne complex. Red cryst. Mp 238°.

Bradford, C.W. *et al*, *J. Chem. Soc., Dalton Trans.*, 1973, 529 (*synth, struct, ms*)

$C_{37}H_{30}ClNO_2OsP_2$ Os-00370

Carbonylchloronitrosylbis(triphenylphosphine)osmium, 9CI

[22186-54-5]

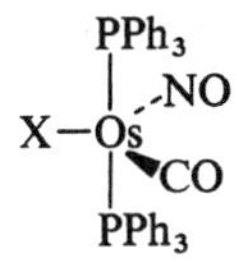

X = Cl

M 808.250

Orange-brown cryst.

Wilson, R.D. *et al*, *Inorg. Chem.*, 1979, **18**, 336 (*synth, cryst struct*)

$C_{37}H_{31}NO_2OsP_2$ Os-00371

Carbonylhydronitrosylbis(triphenylphosphine)osmium, 9CI

[36669-90-6]

As Carbonylchloronitrosylbis(triphenylphosphine)osmium, Os-00370 with

X = H

M 773.805

trans-form [40436-34-8]
Brown cryst.

Johnson, B.F.G. *et al*, *J. Chem. Soc., Dalton Trans.*, 1973, 478 (*synth, ir, pmr*)

$C_{37}H_{32}NO_2OsP_2^{\oplus}$ Os-00372

Carbonyldihydronitrosylbis(triphenylphosphine)osmium(1+), 9CI

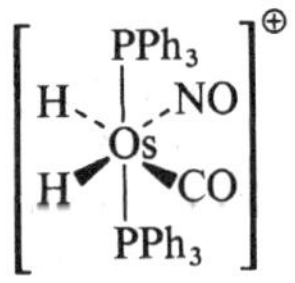

M 774.813 (ion)

Stereochemically nonrigid.

Hexafluorophosphate: [53128-63-5].
$C_{37}H_{32}F_6NO_2OsP_3$ M 919.777
Cream solid (CH_2Cl_2/pentane).

Johnson, B.F.G. *et al*, *J. Chem. Soc., Dalton Trans.*, 1974, 981 (*synth, ir, pmr*)

$C_{37}H_{67}BrOOsP_2$ Os-00373

Bromocarbonylhydrobis(tricyclohexylphosphine)osmium, 9CI

[52359-05-4]

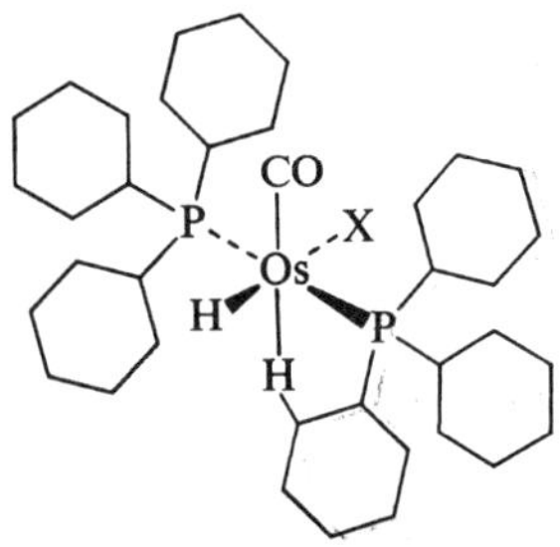

X = Br

M 859.987

Config. uncertain. Brown cryst. Mp 210° dec.

Moers, F.G. *et al*, *J. Organomet. Chem.*, 1974, **65**, 93 (*synth, ir*)

$C_{37}H_{67}ClOOsP_2$ Os-00374

Carbonylchlorohydrobis(tricyclohexylphosphine)osmium, 9CI

[31177-20-5]

As Bromocarbonylhydrobis(tricyclohexylphosphine)osmium, Os-00373 with

X = Cl

M 815.536

Red cryst. or yellow powder. Mp 210° dec. Diamagnetic. Shows H-exchange between Os-H and H's in 2-posn. on cyclohexane rings.

Moers, F.G., *J. Chem. Soc., Chem. Commun.*, 1971, 79 (*synth, ms*)

Moers, F.G. *et al*, *Recl. Trav. Chim. Pays-Bas*, 1972, **91**, 591 (*synth, ir*)

$C_{38}H_{30}As_2Br_2O_2Os$ Os-00375

Dibromodicarbonylbis(triphenylarsine)osmium

Dicarbonylbis(triphenylarsine)dibromoosmium

[17456-75-6]

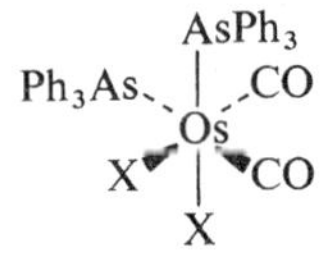

X = Br

M 1018.505

White cryst. ir ν_{CO} 2040, 1971 cm^{-1}.

Hales, L.A.W. *et al*, *J. Chem. Soc. (A)*, 1967, 1932.

$C_{38}H_{30}As_2Cl_2O_2Os$ Os-00376

Dicarbonyldichlorobis(triphenylarsine)osmium

Dicarbonylbis(triphenylarsine)dichloroosmium

[17456-76-7]

As Dibromodicarbonylbis(triphenylarsine)osmium, Os-00375 with

X = Cl

M 929.603

White cryst. (C_6H_6). ir ν_{CO} 2038, 1967 cm^{-1}.

Hales, L.A.W. *et al*, *J. Chem. Soc. (A)*, 1967, 1932 (*synth, ir*)

$C_{38}H_{30}Br_2O_2OsP_2$ Os-00377

Dibromodicarbonylbis(triphenylphosphine)osmium, 8CI

Dicarbonylbis(triphenylphosphine)dibromoosmium

[17456-72-3]

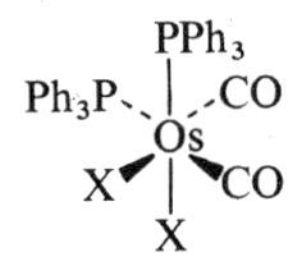

X = Br

M 930.609

Stereochem. not certain, more than one stereoisomer may have been prepd. Colourless cryst. (C_6H_6). Mp 345° (prior dec.). ir ν_{CO} 2043, 1974 (1946) cm^{-1}.

Collman, J.P. *et al*, *J. Am. Chem. Soc.*, 1966, **88**, 3504 (*ir, synth*)

Hales, L.A.W. *et al*, *J. Chem. Soc. (A)*, 1967, 1932 (*synth, ir*)

Cleare, M.J. *et al*, *J. Chem. Soc. (A)*, 1969, 372 (*synth, ir*)

Bradford, C.W. *et al*, *J. Chem. Soc. (A)*, 1970, 2889 (*synth, ir*)

$C_{38}H_{30}Br_2O_2OsSb_2$ Os-00378

Dibromodicarbonylbis(triphenylstibine)osmium

Dicarbonylbis(triphenylstibine)dibromoosmium

[17456-78-9]

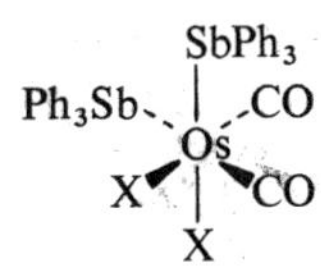

X = Br

M 1112.162

Cryst. (C_6H_6). ir ν_{CO} 2032, 1966 cm^{-1}.

Hales, L.A.W. *et al*, *J. Chem. Soc.* (*A*), 1967, 1932 (*synth*, *ir*)

$C_{38}H_{30}Cl_2OOsP_2S$ Os-00379

(Carbonothioyl)carbonyldichlorobis(triphenylphosphine)osmium, 9CI

Carbonyldichloro(thiocarbonyl)bis(triphenylphosphine)-osmium

[56260-17-4]

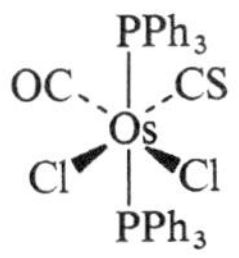

M 857.768

Probable config. Colourless cryst. ir ν_{CO} 2040, 2030, 2020 cm^{-1}, ν_{CS} 1315 cm^{-1}.

Grundy, K.R. *et al*, *J. Organomet. Chem.*, 1975, **90**, C34.
Clark, G.R. *et al*, *J. Am. Chem. Soc.*, 1980, **102**, 1206 (*synth*, *ir*, *struct*)

$C_{38}H_{30}Cl_2OOsP_2Se$ Os-00380

(Carbonoselenoyl)carbonyldichlorobis(triphenylphosphine)-osmium, 10CI

[73249-66-8]

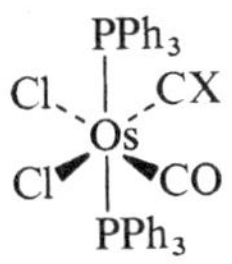

X = Se

M 904.668

Almost colourless cryst.

Clark, G.R. *et al*, *J. Am. Chem. Soc.*, 1980, **102**, 1206 (*synth*, *ir*)

$C_{38}H_{30}Cl_2OOsP_2Te$ Os-00381

(Carbonotelluroyl)carbonyldichlorobis(triphenylphosphine)-osmium, 10CI

[73267-55-7]

As (Carbonoselenoyl)carbonyldichlorobis(triphenylphosphine)osmium, Os-00380 with

X = Te

M 953.308

Air-stable orange cryst. (EtOH). Mp 221-3°.

Clark, G.R. *et al*, *J. Am. Chem. Soc.*, 1980, **102**, 1206 (*synth*, *ir*)

$C_{38}H_{30}Cl_2O_2OsP_2$ Os-00382

Dicarbonyldichlorobis(triphenylphosphine)osmium, 8CI

Dicarbonylbis(triphenylphosphine)dichloroosmium

[16591-90-5]

As Dibromodicarbonylbis(triphenylphosphine)osmium, Os-00377 with

X = Cl

M 841.707

Stereochem. not certain; more than one stereoisomer may have been prepd. Colourless cryst. (C_6H_6). Mp 330° (prior dec.). ir ν_{CO} 2043, 1971 cm^{-1}.

Collman, J.P. *et al*, *J. Am. Chem. Soc.*, 1966, **88**, 3504 (*ir*)
Hales, L.A.W. *et al*, *J. Chem. Soc.* (*A*), 1967, 1932 (*synth*, *ir*)
Cleare, M.J. *et al*, *J. Chem. Soc.* (*A*), 1969, 372 (*synth*, *ir*)

$C_{38}H_{30}Cl_2O_2OsSb_2$ Os-00383

Dicarbonyldichlorobis(triphenylstibine)osmium

Dicarbonylbis(triphenylstibine)dichloroosmium

[18195-25-0]

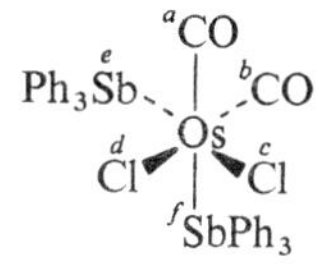

ab-dicarbonyl-*cd*-dichloro-*ef*-distibine-*form*

M 1023.260

ab-dicarbonyl-cd-dichloro-ef-distibine-form

cis-*form*

White cryst. (C_6H_6). Sol. $CHCl_3$, C_6H_6, spar. sol. other solvs. Mp ca. 220-80° dec.

af-dicarbonyl-bd-dichloro-ce-distibine-form

trans-*form*

Yellow cryst. Sol. $CHCl_3$, C_6H_6, spar. sol. other solvs. Mp ca. 220-80° dec.

Araneo, A. *et al*, *Gazz. Chim. Ital.*, 1967, **97**, 885 (*synth*, *ir*)
Hales, L.A.W. *et al*, *J. Chem. Soc.* (*A*), 1967, 1932 (*synth*, *ir*)
Hales, L.A.W. *et al*, *Spectrochim. Acta, Part A*, 1967, **23**, 2981 (*ir*)

$C_{38}H_{30}Cl_4OOsP_2$ Os-00384

Carbonyldichloro(dichloromethylene)bis(triphenylphosphine)-osmium, 10CI

[73249-64-6]

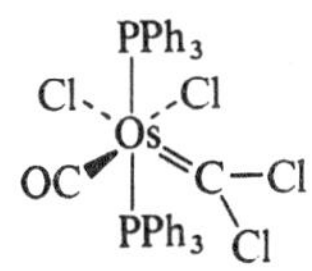

M 896.614

Orange cryst. (C_6H_6/EtOH).

Clark, G.R. *et al*, *J. Am. Chem. Soc.*, 1980, **102**, 1206 (*synth*, *struct*, *ir*, *cmr*)

$C_{38}H_{30}I_2O_2OsP_2$ Os-00385

Dicarbonyldiiodobis(triphenylphosphine)osmium, 8CI

Dicarbonylbis(triphenylphosphine)diiodoosmium

[17456-71-2]

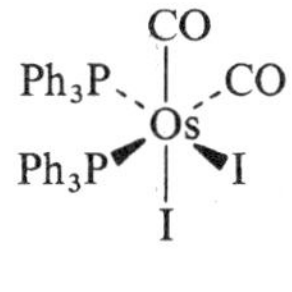

M 1024.610

Colourless cryst. (C_6H_6).

Hales, L.A.W. *et al*, *J. Chem. Soc.* (*A*), 1967, 1932 (*synth*, *ir*)
Hales, L.A.W. *et al*, *Spectrochim. Acta, Part A*, 1967, **23**, 2981 (*ir*)

$C_{38}H_{30}I_2O_2OsSb_2$ Os-00386

Dicarbonyldiiodobis(triphenylstibine)osmium, 8CI
Dicarbonylbis(triphenylstibine)diiodoosmium
[17456-77-8]

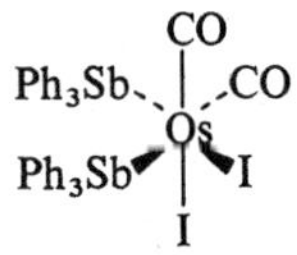

M 1206.163
Cryst. (C_6H_6).

Hales, L.A.W. *et al*, *J. Chem. Soc. (A)*, 1967, 1932 (*synth*, *ir*)
Hales, L.A.W. *et al*, *Spectrochim. Acta*, *Part A*, 1967, **23**, 2981 (*ir*)

$C_{38}H_{30}NO_3OsP_2^{\oplus}$ Os-00387

Dicarbonylnitrosylbis(triphenylphosphine)osmium(1+), 9CI

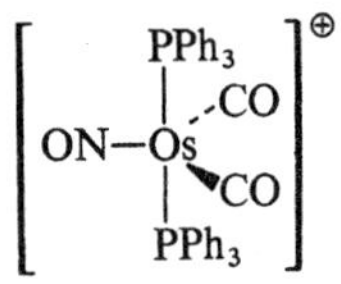

M 800.807 (ion)
Hexafluorophosphate: [40464-73-1].
$C_{38}H_{30}F_6NO_3OsP_3$ M 945.772
Orange cryst. + $1CH_2Cl_2$ ($EtOH/CH_2Cl_2$).
Tetraphenylborate: [40424-93-9].
$C_{62}H_{50}BNO_3OsP_2$ M 1120.039
Cream cryst. ($EtOH/CH_2Cl_2$).

Johnson, B.F.G. *et al*, *J. Chem. Soc.*, *Dalton Trans.*, 1973, 478 (*synth*, *ir*)
Clark, G.R. *et al*, *J. Chem. Soc.*, *Dalton Trans.*, 1975, 2233 (*cryst struct*)

$C_{38}H_{30}O_2OsP_2S_2$ Os-00388

Dicarbonyl(dithioperoxy)bis(triphenylphosphine)osmium, 10CI
[64479-10-3]

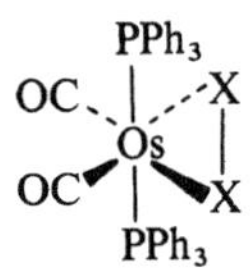

X = S

M 834.921
Orange cryst.

Clark, G.R. *et al*, *J. Organomet. Chem.*, 1977, **136**, C1 (*synth*, *ir*)
Farrar, D.H. *et al*, *J. Am. Chem. Soc.*, 1979, **101**, 6577 (*synth*)

$C_{38}H_{30}O_2OsP_2Se_2$ Os-00389

Dicarbonyl(diseleno)bis(triphenylphosphine)osmium, 10CI
[72155-83-0]
As Dicarbonyl(dithioperoxy)bis(triphenylphosphine)-osmium, Os-00388 with

X = Se

M 928.721
Small red-purple cryst. ($CHCl_3/EtOH$). Mp 192-5°.

Farrar, D.H. *et al*, *J. Am. Chem. Soc.*, 1979, **101**, 6577 (*synth*, *struct*, *ir*)

$C_{38}H_{30}O_4OsP_2$ Os-00390

Dicarbonyl(dioxygen)bis(triphenylphosphine)osmium, 9CI
[35880-57-0]

$Os(O_2)(CO)_2(PPh_3)_2$

M 802.800
Characterised spectroscopically. ir ν_{CO} 1995vs, 1925vs, $\nu_{Os\text{-}O_2}$ 820 cm^{-1}.

Cavitt, B.E. *et al*, *J. Chem. Soc.*, *Chem. Commun.*, 1972, 60 (*synth*, *ir*)

$C_{38}H_{31}ClO_2OsP_2$ Os-00391

Dicarbonylchlorohydrobis(triphenylphosphine)osmium, 9CI
Chlorohydridodicarbonylbis(triphenylphosphine)osmium
[54933-53-8]

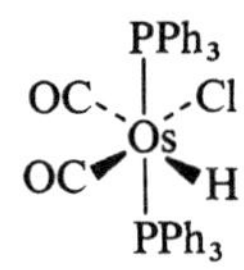

M 807.262
Colourless cryst. ($C_6H_6/EtOH$).

Laing, K.R. *et al*, *J. Chem. Soc. (A)*, 1969, 1889 (*synth*, *ir*)

$C_{38}H_{32}O_2OsP_2$ Os-00392

Dicarbonyldihydrobis(triphenylphosphine)osmium, 9CI
[18974-23-7]

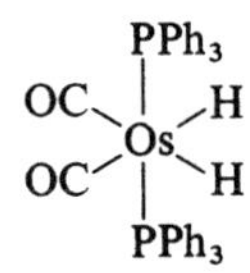

M 772.817
cis,cis,trans-form [25702-68-5]
Creamy-white microcryst. (THF/heptane). Mp 257-8° (sealed under N_2), 243-4° (in air).

L'Eplattenier, F. *et al*, *Inorg. Chem.*, 1968, **7**, 1290 (*synth*, *ir*, *pmr*)
Laing, K.R. *et al*, *J. Chem. Soc. (A)*, 1969, 1889 (*synth*, *ir*)
U.S.P., 3 597 461, (*1971*); *CA*, **75**, 110424 (*synth*)
Ahmad, N. *et al*, *J. Chem. Soc.*, *Dalton Trans.*, 1972, 843 (*synth*, *pmr*)
Inorg. Synth., 1974, **15**, 55 (*synth*, *ir*, *pmr*)

$C_{38}H_{67}BrO_2OsP_2$ Os-00393

Bromodicarbonylhydrobis(tricyclohexylphosphine)osmium, 9CI

[52524-96-6]

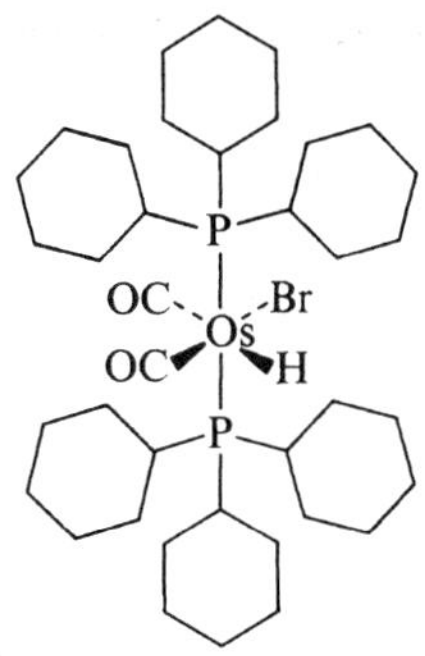

M 887.998

White cryst. Mp 242° dec.

Moers, F.G. *et al*, *J. Organomet. Chem.*, 1974, **65**, 93 (*synth, ir*)

$C_{39}H_{30}ClO_3OsP_2^{\oplus}$ Os-00394

Tricarbonylchlorobis(triphenylphosphine)osmium(1+), 8CI

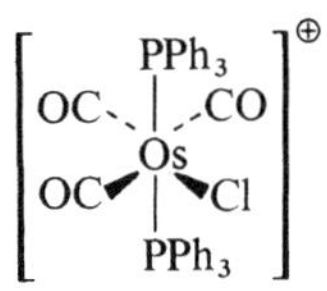

M 834.265 (ion)

trans-form

Chloride: [15928-45-7].
$C_{39}H_{30}Cl_2O_3OsP_2$ M 869.718
White cryst. Dec. at ca. 345°.

Perchlorate: [83114-82-3].
$C_{39}H_{30}Cl_2O_7OsP_2$ M 933.715
White cryst.

Hieber, W. *et al*, *Chem. Ber.*, 1967, **100**, 1961 (*syntn, ir*)
Clark, G.R. *et al*, *J. Organomet. Chem.*, 1982, **231**, 335 (*synth, struct*)

$C_{39}H_{30}O_3OsP_2$ Os-00395

Tricarbonylbis(triphenylphosphine)osmium, 9CI, 8CI

[33635-55-1]

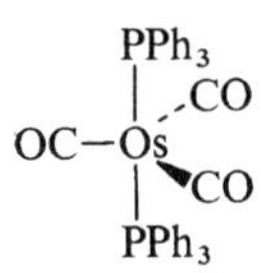

M 798.812

trans-form [20332-51-8]
White cryst. (EtOH/CH_2Cl_2).

L'Eplattenier, F. *et al*, *Inorg. Chem.*, 1968, **7**, 1290 (*synth, ir*)
Piacenti, F. *et al*, *Inorg. Chem.*, 1968, **7**, 1815 (*synth, ir*)
Stalick, J.K. *et al*, *Inorg. Chem.*, 1969, **8**, 419 (*struct*)
Grundy, K.R., *Inorg. Chim. Acta*, 1981, **53**, L225 (*synth*)

$C_{39}H_{31}ClO_2OsP_2S$ Os-00396

Dicarbonylchloro(thioxomethyl)bis(triphenylphosphine)osmium, 10CI

[62793-22-0]

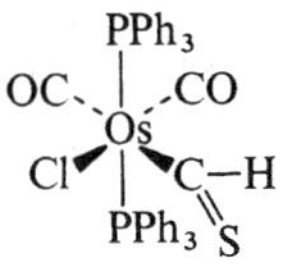

M 851.333

Sky-blue cryst. (hexane). Mp 128-30°. pmr δ 16.39 ppm (C*H*S).

Collins, T.J. *et al*, *J. Chem. Soc., Chem. Commun.*, 1976, 1044; *J. Organomet. Chem.*, 1978, **159**, 73 (*synth, ir, pmr*)

$C_{39}H_{31}O_3OsP_2^{\oplus}$ Os-00397

Tricarbonylhydrobis(triphenylphosphine)osmium(1+), 8CI

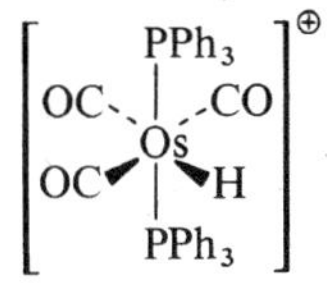

M 799.820 (ion)

Bromide: [25702-65-2].
$C_{39}H_{31}BrO_3OsP_2$ M 879.724
Colourless needles.

Hexafluorophosphate: [25702-66-3].
$C_{39}H_{31}F_6O_3OsP_3$ M 944.784
Colourless cryst. (EtOH).

Laing, K.R. *et al*, *J. Chem. Soc. (A)*, 1969, 1889 (*synth, ir*)

$C_{39}H_{32}NO_2OsP_2^{\oplus}$ Os-00398

Carbonyl(η^2-ethyne)nitrosylbis(triphenylphosphine)osmium(1+), 9CI

Acetylenecarbonylnitrosylbis(triphenylphosphine)osmium(1+)

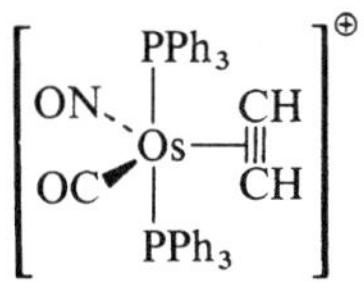

M 798.835 (ion)

Fluxional.

Hexafluorophosphate: [41529-78-6].
$C_{39}H_{32}F_6NO_2OsP_3$ M 943.799
Yellow cryst. + $1CH_2Cl_2$.

Segal, J.A. *et al*, *J. Chem. Soc., Dalton Trans.*, 1975, 1990 (*synth, ir, cmr, pmr*)

$C_{39}H_{32}O_2OsP_2S$ Os-00399

Dicarbonyl(η^2-methanethial)bis(triphenylphosphine)osmium, 10CI

[66926-04-3]

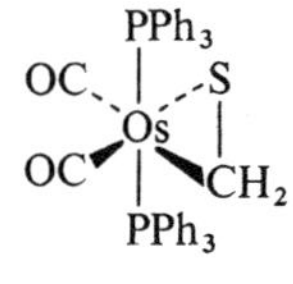

M 816.888

Colourless cryst. +CH_2Cl_2 (CH_2Cl_2/C_6H_6). Mp 193-5°.

Collins, T.J. *et al*, *J. Chem. Soc., Chem. Commun.*, 1977, 901; *J. Organomet. Chem.*, 1978, **159**, 73 (*synth, ir, pmr*)

$C_{39}H_{32}O_3OsP_2$ Os-00400

Dicarbonyl(η^2-formaldehyde)bis(triphenylphosphine)osmium, 10CI

[69558-33-4]

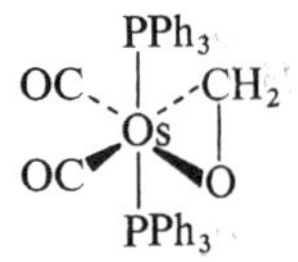

M 800.828

Colourless cryst. (C_6H_6). Rearranges at −75° to $(Ph_3P)_2Os(CO)_2(H)(CHO)$.

Brown, K.L. *et al*, *J. Am. Chem. Soc.*, 1979, **101**, 503 (*synth, struct, ir, pmr*)

$C_{39}H_{33}NO_4OsP_2$ Os-00401

Carbonyl(methoxycarbonyl)nitrosylbis(triphenylphosphine)-osmium, 9CI

[40436-33-7]

As Carbonylchloronitrosylbis(triphenylphosphine)osmium, Os-00370 with

X = −COOMe

M 831.842

Light-brown cryst.

Johnson, B.F.G. *et al*, *J. Chem. Soc., Dalton Trans.*, 1973, 478 (*synth, ir, pmr*)

$C_{40}H_{30}Au_2O_4OsP_2$ Os-00402

Tetracarbonylbis[(triphenylphosphine)gold]osmium

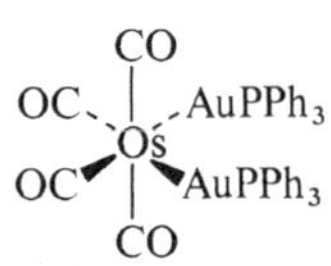

M 1220.755

cis-form [41509-54-0]

Cream cryst. (CH_2Cl_2/hexane). Mp 155-65° dec.

George, R.D. *et al*, *J. Chem. Soc., Dalton Trans.*, 1973, 972 (*synth, ir*)

$C_{40}H_{30}O_4OsP_2^{\oplus\oplus}$ Os-00403

Tetracarbonylbis(triphenylphosphine)osmium(2+), 8CI

Bis(triphenylphosphine)tetracarbonylosmium(2+)

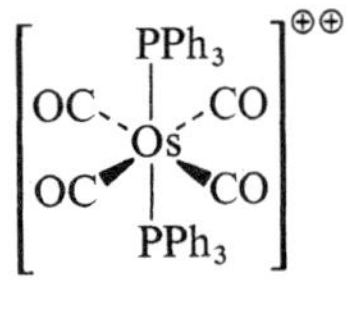

M 826.822 (ion)

Bis(tetrachloroaluminate): [16674-09-2].

$C_{40}H_{30}Al_2Cl_8O_4OsP_2$ M 1164.409

Brownish residue. V. unstable, dec. on washing with Et_2O at −70°.

Hieber, W. *et al*, *Chem. Ber.*, 1967, **100**, 1961 (*synth, ir*)

$C_{40}H_{30}O_4OsSn_2$ Os-00404

Tetracarbonylbis(triphenylstannyl)osmium, 9CI

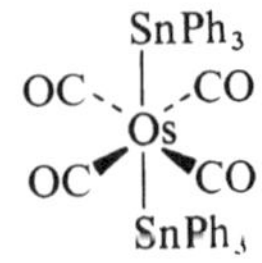

M 1002.255

trans-form [41553-07-5]

Colourless cryst. (CH_2Cl_2/hexane). Mp 228-31°.

George, R.D. *et al*, *J. Chem. Soc., Dalton Trans.*, 1973, 972 (*synth, ir*)

Collman, J.P. *et al*, *Inorg. Chem.*, 1974, **13**, 1 (*synth, struct, ir*)

$C_{40}H_{33}ClO_4OsP_2$ Os-00405

Dicarbonylchloro(methoxycarbonyl)bis(triphenylphosphine)-osmium, 8CI

[16594-57-3]

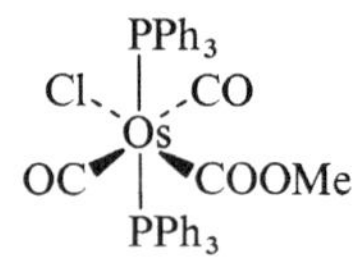

M 865.299

Colourless cryst. (C_6H_6/pet. ether). Mp 272-4° dec.

Hieber, W. *et al*, *Chem. Ber.*, 1967, **100**, 1961 (*synth*)

$C_{40}H_{34}OOsP_2S$ Os-00406

(Carbonothioyl)carbonyl(η^2-ethene)bis(triphenylphosphine)-osmium, 10CI

Carbonyl(ethylene)bis(triphenylphosphine)(thiocarbonyl)osmium

[64883-53-0]

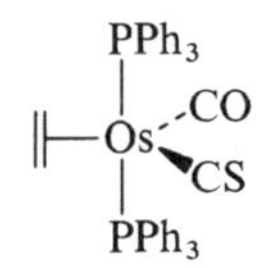

M 814.916

No phys. props. reported.

Collins, T.J. *et al*, *J. Organomet. Chem.*, 1977, **139**, C56 (*synth*)

$C_{40}H_{34}O_2OsP_2$ Os-00407

Dicarbonyl(η^2-ethene)bis(triphenylphosphine)osmium, 10CI

Dicarbonyl(η^2-ethylene)bis(triphenylphosphine)osmium

[79725-76-1]

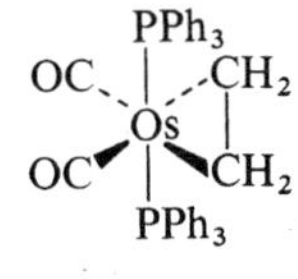

M 798.855

White cryst. (CH_2Cl_2/EtOH). Mp 165-9°.

Grundy, K.R. *et al*, *J. Organomet. Chem.*, 1981, **216**, 255 (*synth, ir, pmr*)

$C_{40}H_{34}O_2OsP_2S_2$ Os-00408

Dicarbonylhydro[(methylthio)thioxomethyl]bis(triphenylphosphine)osmium, 10CI

[63866-74-0]

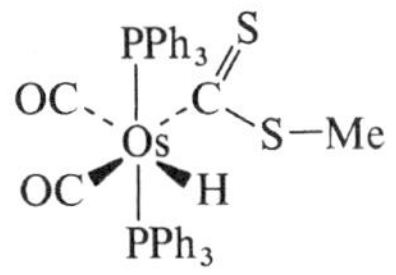

M 862.975

Cryst $+\frac{1}{2}C_6H_6$ (C_6H_6).

Collins, T.J. *et al*, *J. Organomet. Chem.*, 1976, **121**, C41 (*synth, ir, pmr*)

Waters, J.M. *et al*, *Inorg. Chem.*, 1977, **16**, 3273 (*struct*)

$C_{41}H_{35}BrO_6OsP_2$ Os-00409

Bromo(η^5-2,4-cyclopentadien-1-yl)bis(triphenyl phosphite-*P*)osmium, 10CI

[71397-23-4]

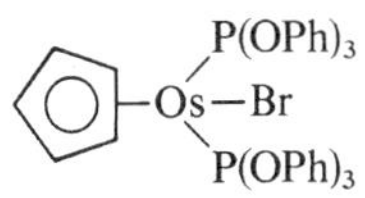

M 955.775

Yellow cryst. (toluene/pet. ether). Mp 155°.

Ashby, G.S. *et al*, *Aust. J. Chem.*, 1979, **32**, 100? (*synth, pmr*)

$C_{41}H_{35}BrOsP_2$ Os-00410

Bromo(η^5-2,4-cyclopentadien-1-yl)bis(triphenylphosphine)osmium

(η^5-Cyclopentadienyl)bromobis(triphenylphosphine)osmium. Cyclopentadienylbis(triphenylphosphine)osmium bromide

[34742-23-9]

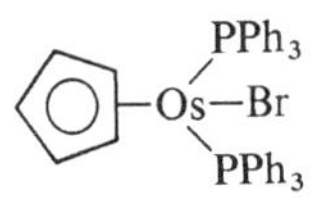

M 859.779

Pale yellow-orange cryst. (CH_2Cl_2/hexane). Mp 182-5° dec.

Blackmore, T. *et al*, *J. Chem. Soc. (A)*, 1971, 2376 (*synth, ir, pmr*)

Bruce, M.I. *et al*, *Aust. J. Chem.*, 1977, **30**, 1601 (*synth*)

$C_{42}H_{32}O_6Os_2P_2$ Os-00411

Hexacarbonyldihydrobis(triphenylphosphine)diosmium

[62793-33-3]

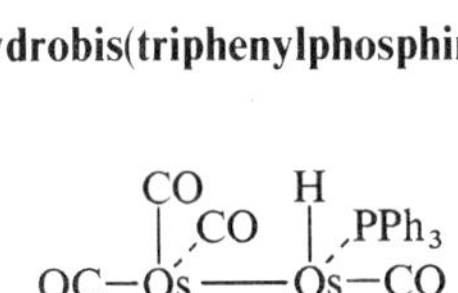

M 1075.059

White plates (CH_2Cl_2/hexane). Mp 207-10°.

Moss, J.R. *et al*, *J. Chem. Soc., Dalton Trans.*, 1977, 89 (*synth, ir, pmr*)

$C_{42}H_{36}IOsP_2^{\oplus}$ Os-00412

(η^6-Benzene)iodobis(triphenylphosphine)osmium(1+), 10CI

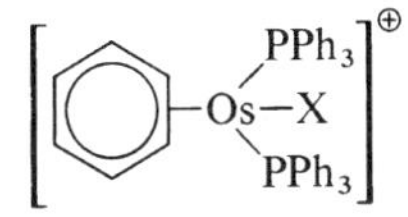

X = I

M 919.798 (ion)

Hexafluorophosphate: [75353-17-2].

$C_{42}H_{36}F_6IOsP_3$ M 1064.763

No phys. props. reported.

Werner, H. *et al*, *J. Organomet. Chem.*, 1980, **194**, C7 (*synth*)

Werner, H., *Pure Appl. Chem.*, 1982, **54**, 177 (*synth*)

$C_{42}H_{36}OsP_2$ Os-00413

(η^6-Benzene)bis(triphenylphosphine)osmium

[75353-18-3]

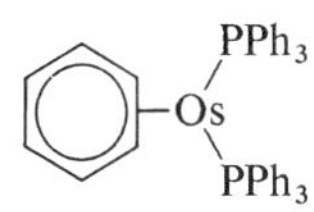

M 792.894

Extremely air-sensitive.

Werner, H. *et al*, *J. Organomet. Chem.*, 1980, **194**, C7 (*synth, pmr*)

$C_{42}H_{37}OsP_2^{\oplus}$ Os-00414

(η^6-Benzene)hydrobis(triphenylphosphine)osmium(1+), 10CI

As (η^6-Benzene)iodobis(triphenylphosphine)osmium(1+), Os-00412 with

X = H

M 793.902 (ion)

Hexafluorophosphate: [75345-68-5].

$C_{42}H_{37}F_6OsP_3$ M 938.866

No phys. props. reported.

Werner, H. *et al*, *J. Organomet. Chem.*, 1980, **194**, C7 (*synth*)

$C_{42}H_{40}O_2OsP_2$ Os-00415

Dicarbonyldiethylbis(triphenylphosphine)osmium, 8CI

[27882-34-4]

$(Ph_3P)_2OsEt_2(CO)_2$

M 828.924

White cryst. (THF/heptane). ir ν_{CO} 2040vs, 1967vs cm^{-1} (THF).

L'Eplattenier, F. *et al*, *Helv. Chim. Acta*, 1970, **53**, 1091.

$C_{42}H_{47}N_2OsP_3$ Os-00416

(Dinitrogen)tris(ethyldiphenylphosphine)dihydroosmium, 9CI

[27497-62-7]

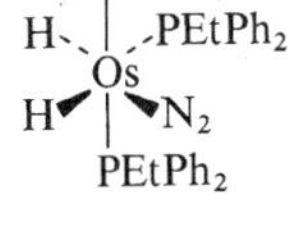

M 862.968

Brown solid. Mp 85-9° dec.

Bell, B. *et al*, *J. Chem. Soc.*, *Dalton Trans.*, 1973, 997 (*synth*, *spectra*)

$C_{43}H_{29}O_7Os_3P_2$ Os-00417

Heptacarbonyl[μ_3-[(5-η:5,6-η:6-η)-1,3-cyclohexadien-5-yne]][μ-(diphenylphosphino)]hydro(triphenylphosphine)triosmium, 9CI

[38960-71-3]

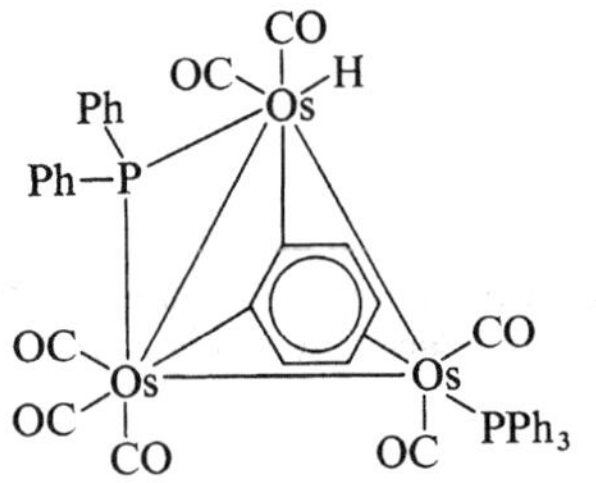

M 1290.245

Bridging benzyne complex. Yellow cryst. Mp 210-20° dec.

Bradford, C.W. *et al*, *J. Chem. Soc.*, *Chem. Commun.*, 1972, 87 (*synth*, *ms*)

Gainsford, G.J. *et al*, *J. Organomet. Chem.*, 1972, **40**, C70 (*struct*)

$C_{43}H_{36}N_2OOsP_2$ Os-00418

Carbonylhydro(phenylazo)bis(triphenylphosphine)osmium, 9CI

Hydridocarbonyl(phenyldiazo)bis(triphenylphosphine)-osmium

[53608-71-2]

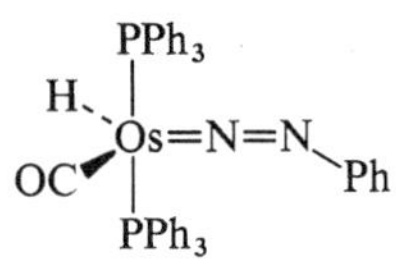

M 848.918

Brown microcryst. powder, bright-orange metastable cryst. or brown cryst. + $1CH_2Cl_2$.

Haymore, B.L. *et al*, *Inorg. Chem.*, 1975, **14**, 2784 (*synth*, *ir*)

$C_{43}H_{38}NOsP_2^{\oplus}$ Os-00419

Acetonitrile(η^5-2,4-cyclopentadien-1-yl)bis(triphenylphosphine)osmium(1+), 10CI

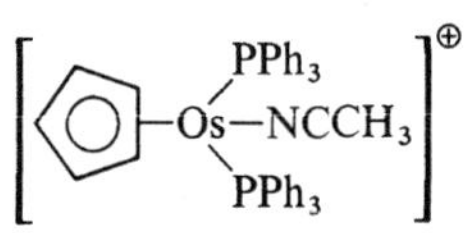

M 820.927 (ion)

Tetraphenylborate: [71397-25-6].

$C_{67}H_{58}BNOsP_2$ M 1140.159

Off-white needles ($MeCN/Et_2O$). Mp 170-3° dec.

Ashby, G.S. *et al*, *Aust. J. Chem.*, 1979, **32**, 1003 (*synth*, *ir*, *pmr*)

$C_{43}H_{39}OsP_2^{\oplus}$ Os-00420

(η^6-Benzene)methylbis(triphenylphosphine)osmium(1+), 10CI

As (η^6-Benzene)iodobis(triphenylphosphine)osmium(1+), Os-00412 with

X = Me

M 807.929 (ion)

Hexafluorophosphate: [75353-21-8].

$C_{43}H_{39}F_6OsP_3$ M 952.893

No phys. props. reported.

Werner, H. *et al*, *J. Organomet. Chem.*, 1980, **194**, C7 (*synth*)

$C_{44}H_{35}BrN_2O_2OsP_2$ Os-00421

Bromodicarbonyl(phenylazo)bis(triphenylphosphine)osmium, 9CI

Bromodicarbonyl(phenydiazo)bis(triphenylphosphine) osmium

[56335-39-8]

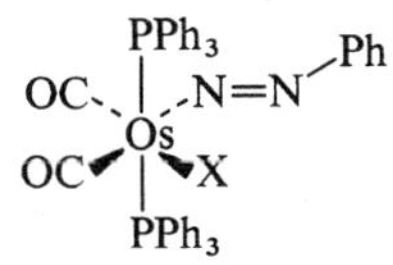

X = Br

M 955.824

Olive-green cryst. + $1Me_2CO$ (Me_2CO).

Haymore, B.L. *et al*, *Inorg. Chem.*, 1975, **14**, 2784 (*synth*, *ir*)

$C_{44}H_{35}ClN_2O_2OsP_2$ Os-00422

Dicarbonylchloro(phenylazo)bis(triphenylphosphine)osmium, 9CI

Chlorodicarbonyl(phenyldiazo)bis(triphenylphosphine)-osmium

[56335-38-7]

As Bromodicarbonyl(phenylazo)bis(triphenylphosphine)-osmium, Os-00421 with

X = Cl

M 911.373

Blue-green cryst. + $1Me_2CO$ (Me_2CO or CH_2Cl_2/Me_2CO).

Haymore, B.L. *et al*, *Inorg. Chem.*, 1975, **14**, 2784 (*synth*, *ir*)

$C_{44}H_{35}IN_2O_2OsP_2$ Os-00423

Dicarbonyliodo(phenylazo)bis(triphenylphosphine)osmium, 9CI

Iododicarbonyl(phenyldiazo)bis(triphenylphosphine)-osmium

[56363-47-4]

As Bromodicarbonyl(phenylazo)bis(triphenylphosphine)-osmium, Os-00421 with

X = I

M 1002.825

Green cryst. + $\frac{1}{2}Me_2CO$ (Me_2CO).

Haymore, B.L. *et al*, *Inorg. Chem.*, 1975, **14**, 2784 (*synth*, *ir*)

$C_{44}H_{35}N_2O_2OsP_2^{\oplus}$ Os-00424

Dicarbonyl(phenylazo)bis(triphenylphosphine)osmium(1+), 9CI

Dicarbonyl(phenyldiazo)bis(triphenylphosphine)osmium(1+)

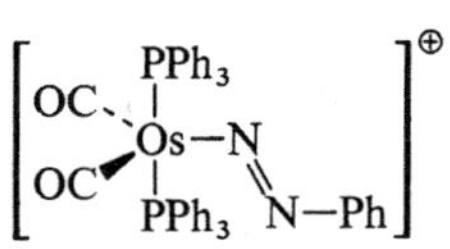

M 875.920 (ion)

Hexafluorophosphate: [56335-36-5].

$C_{44}H_{35}F_6N_2O_2OsP_3$ M 1020.884

Yellow cryst. + CH_2Cl_2 (CH_2Cl_2/Et_2O), yellow cryst. + $\frac{1}{2}C_6H_6$ (C_6H_6). Chemically nonequiv. CO groups due to differing steric relationship with $PF_6^{\ominus}$ counterion.

Haymore, B.L. *et al*, *Inorg. Chem.*, 1975, **14**, 2784 (*synth*)

$C_{44}H_{36}N_2O_2OsP_2$ Os-00425

Dicarbonylhydro(phenylazo)bis(triphenylphosphine)osmium, 9CI

Hydridodicarbonyl(phenyldiazo)bis(triphenylphosphine)osmium

[56335-44-5]

As Bromodicarbonyl(phenylazo)bis(triphenylphosphine)-osmium, Os-00421 with

X = H

M 876.928

Blue-green cryst. powder (C_6H_6/hexane).

Haymore, B.L. *et al*, *Inorg. Chem.*, 1975, **14**, 2784 (*synth, ir*)

$C_{44}H_{38}Cl_2N_2OOsP_2$ Os-00426

(Acetylphenyldiazene-N^2,O^1)dichlorobis(triphenylphosphine)-osmium, 9CI

[58441-24-0]

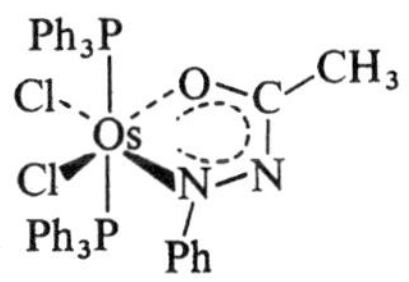

M 933.851

Bright-brown needles (EtOH/CH_2Cl_2). Mp 177-81° dec.

Chatt, J. *et al*, *J. Chem. Soc., Dalton Trans.*, 1975, 2348 (*synth, spectra*)

$C_{45}H_{37}ClOOsP_2$ Os-00427

Carbonylchloro[(4-methylphenyl)methylidyne]bis(triphenylphosphine)osmium, 10CI

[75346-82-6]

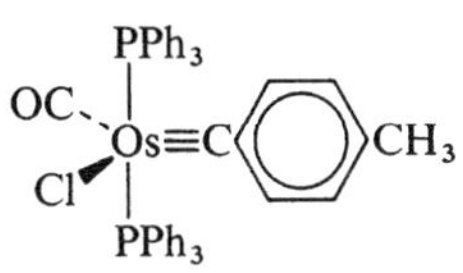

M 881.387

Dark-green cryst.

Clark, G.R. *et al*, *J. Am. Chem. Soc.*, 1980, **102**, 6570 (*synth, struct, ir*)

$C_{46}H_{30}O_{10}Os_3P_2$ Os-00428

Decacarbonylbis(triphenylphosphine)triosmium, 8CI

[41753-97-3]

$Os_3(CO)_{10}(PPh_3)_2$

M 1375.285

Struct. not known. Red cryst. (Me_2CO/pet. ether). Dec. at 195-215°.

Bradford, C.W. *et al*, *J. Chem. Soc. (A)*, 1970, 2889 (*synth, ir*)
Tachikawa, M. *et al*, *J. Organomet. Chem.*, 1977, **124**, C19 (*synth*)
Tyler, D.R. *et al*, *J. Am. Chem. Soc.*, 1980, **102**, 3022 (*synth, uv*)

$C_{46}H_{37}ClNO_2OsP_2^{\oplus}$ Os-00429

Dicarbonylchloro[4-methyl(isocyanobenzene)]bis(triphenylphosphine)osmium(1+), 9CI

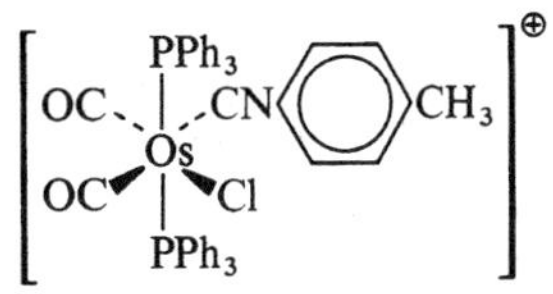

M 923.404 (ion)

Grundy, K.R. *et al*, *J. Organomet. Chem.*, 1976, **113**, C45.

$C_{46}H_{38}O_2OsP_2$ Os-00430

Dicarbonyl[(4-methyl-2,5-cyclohexadien-1-ylidene)methylene]bis(triphenylphosphine)osmium, 10CI

[76451-72-4]

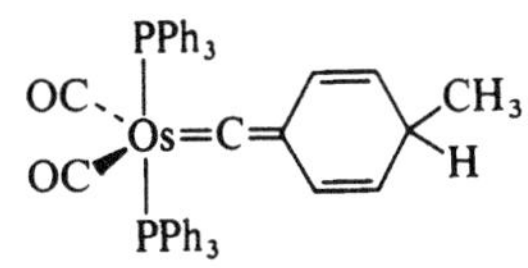

M 874.953

Yellow cryst. (THF/hexane).

Roper, W.R. *et al*, *J. Organomet. Chem.*, 1980, **201**, C27 (*synth, struct, ir, cmr, pmr*)

$C_{48}H_{42}I_4Os_2P_2$ Os-00431

η^6-Benzenediiodo(triphenylphosphine)osmium

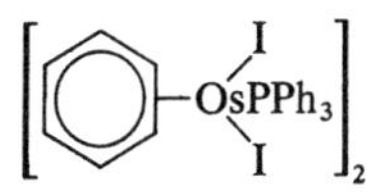

M 1568.825

Dimeric.

Dimer: Bis(benzene)tetraiodobis(triphenylphosphine)-diosmium, 8CI. Red needles (C_6H_6). Mp 220° dec.

Winkhaus, G. *et al*, *Z. Naturforsch., B*, 1966, **21**, 1109 (*synth, pmr*)

$C_{49}H_{30}O_7Os_3$ Os-00432

Heptacarbonyl[μ_3-[1,1′-(η^1:η^1:η^2-1,2-ethynediyl)bis[benzene]]][μ-(1,2,3,4-η:1,4-η)-1,2,3,4-tetraphenyl-1,3-butadiene-1,4-diyl]triosmium, 9CI

[42790-11-4]

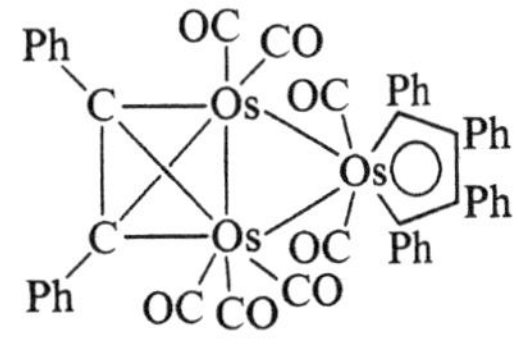

M 1301.372

Red-orange lamellar cryst. (heptane).

Ferrari, R.P. *et al*, *Inorg. Chim. Acta*, 1973, **7**, 193 (*synth, ir, ms, pmr*)
Ferraris, G. *et al*, *J. Chem. Soc., Dalton Trans.*, 1973, 1933 (*struct*)

$C_{49}H_{40}OsP_2$ Os-00433

(η^5-2,4-Cyclopentadien-1-yl)(phenylethynyl)bis(triphenylphosphine)osmium, 10CI

[71957-41-0]

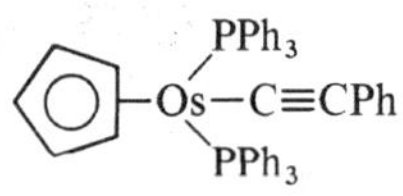

M 881.003

Yellow needles. Mp 217-8° dec.

Bruce, M.I. *et al*, *Aust. J. Chem.*, 1979, **32**, 1471 (*synth, ir, pmr*)

$C_{54}H_{45}Br_2OsP_3$ Os-00434

Dibromotris(triphenylphosphine)osmium, 9CI

[36543-21-2]

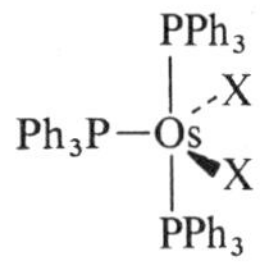

X = Br

M 1136.879

Hoffman, P.R. *et al*, *J. Am. Chem. Soc.*, 1975, **97**, 4221 (*nmr, struct*)

$C_{54}H_{45}Cl_2OsP_3$ Os-00435

Dichlorotris(triphenylphosphine)osmium, 9CI

[40802-32-2]

As Dibromotris(triphenylphosphine)osmium, Os-00434 with

X = Cl

M 1047.977

Green.

Oudeman, A. *et al*, *J. Coord. Chem.*, 1974, **4**, 1 (*synth*)
Hoffman, P.R. *et al*, *J. Am. Chem. Soc.*, 1975, **97**, 4221 (*synth, nmr, struct*)
Contreras, R. *et al*, *J. Organomet. Chem.*, 1981, **215**, C6 (*synth*)

$C_{54}H_{46}NOOsP_3$ Os-00436

Hydronitrosyltris(triphenylphosphine)osmium, 8CI

[33153-18-3]

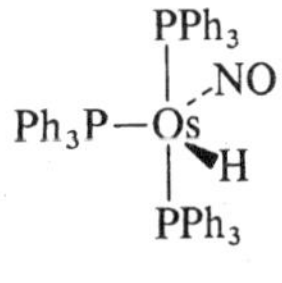

M 1008.085

Deep-brown cryst.

Osborn, J.A. *et al*, *J. Am. Chem. Soc.*, 1971, **93**, 3068 (*synth, ir, pmr*)

$C_{54}H_{49}OsP_3$ Os-00437

Tetrahydrotris(triphenylphosphine)osmium, 9CI, 8CI

[24228-59-9]

$H_4Os(PPh_3)_3$

M 981.102

Creamy-white microcryst. Mp 172-5° (in ar), 219-21° (sealed tube).

Inorg. Synth., 1974, **15**, 45 (*synth*)
Oudeman, A. *et al*, *J. Coord. Chem.*, 1974, **4**, 1 (*synth*)

$C_{55}H_{45}NO_2OsP_3^{\oplus}$ Os-00438

Carbonylnitrosyltris(triphenylphosphine)osmium(1+), 9CI

[36444-68-5]

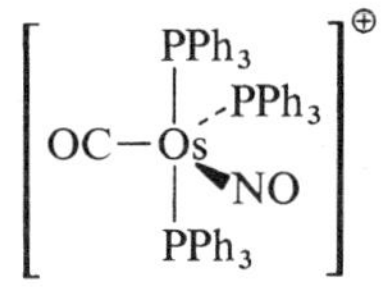

M 1035.087 (ion)

Hexafluorophosphate: [40464-74-2].
$C_{55}H_{45}F_6NO_2OsP_4$ M 1180.051
Light-brown cryst. ($EtOH/CH_2Cl_2$).

Tetraphenylborate: [40424-95-1].
$C_{79}H_{65}BNO_2OsP_3$ M 1354.319
Light-brown cryst. ($EtOH/CH_2Cl_2$ or Et_2O/CH_2Cl_2).

Johnson, B.F.G. *et al*, *J. Chem. Soc., Dalton Trans.*, 1973, 478 (*synth, ir*)

$C_{55}H_{46}As_3ClOOs$ Os-00439

Carbonylchlorohydrotris(triphenylarsine)osmium, 9CI

[55802-01-2]

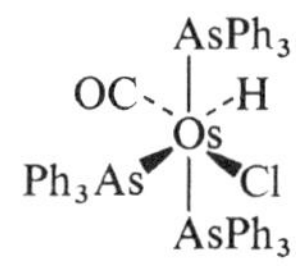

M 1173.386

Colourless cryst. Insol. H_2O, alcohols, spar. sol. other solvs. Mp 200° dec. (in air), 247° dec. (*in vacuo*). Diamagnetic. Air-stable, darkening v. slowly in air over several months.

Vaska, L., *J. Am. Chem. Soc.*, 1964, **86**, 1943 (*synth, ir*)

$C_{55}H_{46}BrOOsP_3$ Os-00440

Bromocarbonylhydrotris(triphenylphosphine)osmium, 9CI

Hydrobromocarbonyltris(triphenylphosphine)osmium.
Hydridobromocarbonyltris(triphenylphosphine)osmium

[16971-32-7]

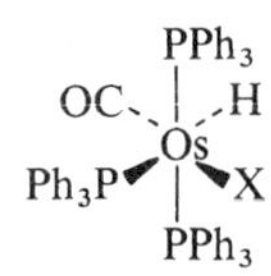

X = Br

M 1085.993

Colourless cryst. Insol. H_2O, alcohols, spar. sol. other solvs. Mp 172° dec. (in air), 276° dec. (*in vacuo*). Diamagnetic. Air-stable, undec. in air after several years in solid state.

Orioli, P.L. *et al*, *Proc. Chem. Soc. London*, 1962, 333 (*cryst struct*)
Vaska, L., *J. Am. Chem. Soc.*, 1964, **86**, 1943 (*synth, ir*)

$C_{55}H_{46}ClOOsP_3$ Os-00441

Carbonylchlorohydrotris(triphenylphosphine)osmium

Hydrochlorocarbonyltris(triphenylphosphine)osmium.
Carbonylhydridochlorotris(triphenylphosphine)osmium

[16971-31-6]

As Bromocarbonylhydrotris(triphenylphosphine)osmium, Os-00440 with

X = Cl

M 1041.542

Isomerisation and hydrogenation catalyst. Colourless cryst. Insol. H_2O, alcohols, spar. sol. other solvs. Mp 179° dec. (in air), 277° dec. (*in vacuo*). Diamagnetic. Air-stable, undec. in air after several years in solid state.

Vaska, L., *J. Am. Chem. Soc.*, 1964, **86**, 1943; 1966, **88**, 4100 (*synth, ir, struct*)
Ahmad, N. *et al, J. Chem. Soc., Dalton Trans.*, 1972, 843 (*pmr*)
Inorg. Synth., 1974, **15**, 45 (*synth*)
Khan, M.M.T. *et al, J. Inorg. Nucl. Chem.*, 1976, **38**, 331 (*synth*)

$C_{55}H_{47}OOsP_3$ Os-00442

Carbonyldihydrotris(triphenylphosphine)osmium, 9CI

[12104-84-6]

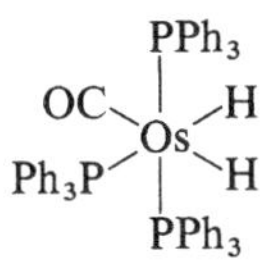

M 1007.097

cis-dihydro-form [36179-61-0]

White microcryst. Mp 245-6° (sealed under N_2), 202-4° (in air). Dipole moment 6.80 D (C_6H_6).

Vaska, L., *J. Am. Chem. Soc.*, 1966, **88**, 4100 (*synth, ir*)
Ahmad, N. *et al, J. Chem. Soc., Dalton Trans.*, 1972, 843 (*synth, pmr*)
Inorg. Synth., 1974, **15**, 54 (*synth, ir, pmr*)

$C_{56}H_{45}O_2OsP_3$ Os-00443

Dicarbonyltris(triphenylphosphine)osmium, 9CI

[35880-55-8]

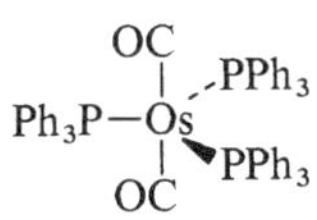

M 1033.092

Bright-yellow needles. Mp 164-6°. ir ν_{CO} 1895vs cm^{-1}.

Cavit, B.E. *et al, J. Chem. Soc., Chem. Commun.*, 1972, 60 (*synth, ir*)
Farrar, D.H. *et al, J. Am. Chem. Soc.*, 1979, **101**, 6577 (*synth, pe*)

$C_{63}H_{45}O_9Os_3P_3$ Os-00444

Nonacarbonyltris(triphenylphosphine)triosmium, 8CI

[16520-58-4]

$Os_3(CO)_9(PPh_3)_3$

M 1609.564

Struct. not known. Orange-red cryst. (Me_2CO/pet. ether). Dec. at 185-95°.

Bradford, C.W. *et al, J. Chem. Soc. (A)*, 1970, 2889 (*synth, ir*)
Tyler, D.R. *et al, J. Am. Chem. Soc.*, 1980, **102**, 3022 (*synth, uv*)

$C_{72}H_{60}Br_2OsP_4$ Os-00445

Dibromotetrakis(triphenylphosphine)osmium, 9CI

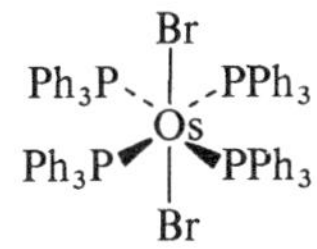

M 1399.169

trans-form [60447-29-2]

Light-grey cryst. Mp 202-5° dec.

Khan, M.M.T. *et al, J. Inorg. Nucl. Chem.*, 1976, **38**, 1135 (*synth, spectra*)

$F_{12}HOsP_4^{\ominus}$ Os-00446

Hydrotetrakis(phosphorous trifluoride)osmate(1–)

[32236-82-1]

$HOs(PF_3)_4^{\ominus}$

M 543.084 (ion)

Stereochemically nonrigid. Characterised spectroscopically.

Kruck, T. *et al, Z. Anorg. Allg. Chem.*, 1969, **371**, 1 (*synth, pmr, nmr*)
Meakin, P. *et al, J. Am. Chem. Soc.*, 1971, **93**, 1797 (*nmr*)

$F_{12}H_2OsP_4$ Os-00447

Dihydrotetrakis(phosphorous trifluoride)osmium

[24899-54-5]

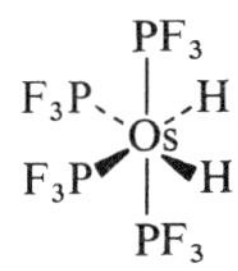

M 544.092

Colourless v. volatile liq. Mp −72°. Dec. at 340°.

Kruck, T. *et al, Z. Anorg. Allg. Chem.*, 1969, **371**, 1 (*synth, ir, pmr, nmr*)
Meakin, P. *et al, J. Am. Chem. Soc.*, 1971, **93**, 1797 (*pmr, nmr*)

$F_{12}OsP_4^{\ominus\ominus}$ Os-00448

Tetrakis(phosphorous trifluoride)osmate(2–), 8CI

$Os(PF_3)_4^{\ominus\ominus}$

M 542.076 (ion)

Dihydrogen compd.: see Dihydrotetrakis(phosphorous trifluoride)osmium, Os-00447

Di-K salt: [26876-74-4].

$F_{12}K_2OsP_4$ M 620.272

Colourless air-sensitive solid.

Kruck, T. *et al, Z. Anorg. Allg. Chem.*, 1969, **371**, 1 (*synth, ir, nmr*)

$F_{15}OsP_5$ **Os-00449**

Pentakis(phosphorous trifluoride)osmium, 9CI

[19702-31-9]

$$Os(PF_3)_5$$

M 630.045

Stereochemically nonrigid. Extremely volatile, colourless cryst

Kruck, T. *et al*, *Z. Anorg. Allg. Chem.*, 1968, **356**, 118 (*synth, ir*)

Meakin, P. *et al*, *J. Am. Chem. Soc.*, 1972, **94**, 5271 (*nmr*)

Ru Ruthenium

G. R. Knox

Ruthénium (Fr.), Ruthenium (Ger.), Rutenio (Sp., Ital.), Рутений (Rutenii) (Russ.), ルテニウム (Japan.)

Atomic Number. 44

Atomic Weight. 101.07

Electronic Configuration. [Kr] $4d^7\ 5s^1$

Oxidation States. Usually −2, −1, 0, +1 or +2 in organometallics.

Coordination Number. Commonly 5 and 6 in organometallics.

Colour. Ruthenocene derivatives are usually pale yellow whilst simple, diamagnetic ruthenium carbonyl complexes are usually bright yellow, orange or red.

Availability. The material sold as ruthenium trichloride hydrate, corresponding approximately to $RuCl_3{\cdot}3H_2O$, is the commonest organometallic precursor; ruthenocene, $Ru_3(CO)_{12}$, $[Ru(CO)_3Cl_2]_2$ are widely available. Ruthenium derivatives are relatively expensive.

Handling. Ruthenocene derivatives are amenable to the usual purification techniques available to organic chemists, column and thin-layer chromatography of the coloured materials being particularly suitable; other neutral organometallics of ruthenium are usually purified by chromatography and/or sublimation, but may be air-sensitive in solution.

Toxicity. Organoruthenium compounds are not generally regarded as being especially toxic.

Isotopic Abundance. ^{96}Ru, 5.51%; ^{98}Ru, 1.87%; ^{99}Ru, 12.72%; ^{100}Ru, 12.62%; ^{101}Ru, 17.07%; ^{102}Ru, 31.61%; ^{104}Ru, 18.58%. Values of the isotopic composition of ruthenium were determined by measuring the mass spectral pattern of ruthenocene.

Spectroscopy. Of the naturally occurring isotopes of ruthenium, only ^{99}Ru ($I = \frac{5}{2}$) and ^{101}Ru ($I = \frac{5}{2}$) are, in principle, amenable to nmr spectroscopy, but their quadrupole moments make direct observation difficult.

Analysis. See references listed in the introduction to the *Sourcebook*.

References. Various aspects of the organometallic chemistry of ruthenium are described in reviews; some are listed in the introduction to the *Sourcebook* and others, more specialized below:

Chemistry of the Iron Group Metallocenes; Ferrocene, Ruthenocene, Osmocene, Rosenblum, M. Ed., Wiley, 1965.

Transition Metal Clusters, Johnson, B. F. G. Ed., Wiley, 1980.

Johnson, B. F. G. *et al.*, *Adv. Inorg. Chem. Radiochem.*, 1981, **24**, 225.

Gmelins Handbuch der Anorganischen Chemie, System No. 63 (Ruthenium) details ruthenocene and derivatives together with many carbonyl and nitrosyl complexes.

$Ru(CO)Br_3$

Ru-00001

$Br_5RuCO^{\ominus\ominus}$

Ru-00002

$Ru(CO)Cl_3$

Ru-00003

$Cl_5RuCO^{\ominus\ominus}$

Ru-00004

$(F_3P)_4RuCO$

Ru-00005

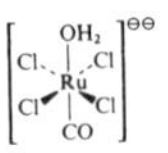

Ru-00006

$Ru(CO)I_3$

Ru-00007

$Ru(CO)_2Br_2$

Ru-00008

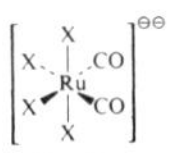

X = Br

Ru-00009

$Ru(CO)_2Cl_2$

Ru-00010

As Ru-00009 with
X = Cl

Ru-00011

$(F_3P)_3Ru(CO)_2$

Ru-00012

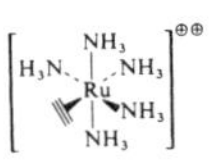

Ru-00013

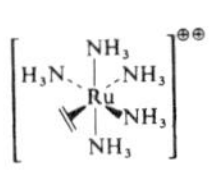

Ru-00014

$Ru(CO)_2I_2$

Ru-00015

As Ru-00009 with
X = I

Ru-00016

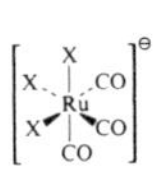

X = Br

Ru-00017

As Ru-00017 with
X = Cl

Ru-00018

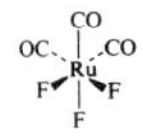

Ru-00019

$(F_3P)_2Ru(CO)_3$

Ru-00020

As Ru-00017 with
X = I

Ru-00021

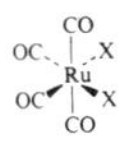

X = Br

Ru-00022

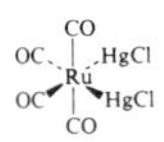

Ru-00023

As Ru-00022 with
X = Cl

Ru-00024

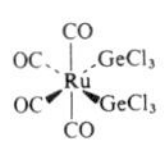

Ru-00025

$F_3PRu(CO)_4$

Ru-00026

As Ru-00022 with
X = H

Ru-00027

As Ru-00022 with
X = I

Ru-00028

$Ru(CO)_4^{\ominus\ominus}$

Ru-00029

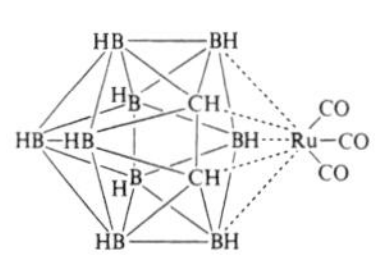

Ru-00030

$Ru(CO)_5$

Ru-00031

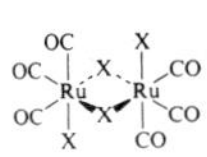

X = Br

Ru-00032

As Ru-00032 with
X = Cl

Ru-00033

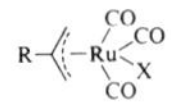

X = Br, R = H

Ru-00034

As Ru-00034 with
X = Cl, R = H

Ru-00035

As Ru-00034 with
X = I, R = H

Ru-00036

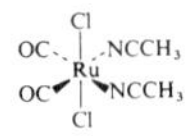

Ru-00037

As Ru-00237 with
X = Cl

Ru-00038

Ru-00039

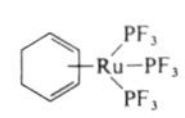

Ru-00040

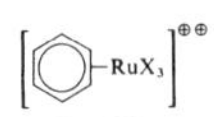

$X = OH_2$

Ru-00041

As Ru-00041 with
$X = NH_3$

Ru-00042

As Ru-00032 with
X = I

Ru-00043

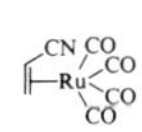

Ru-00044

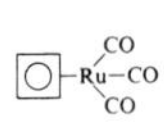

Ru-00045

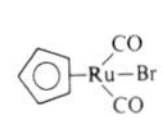

Ru-00046

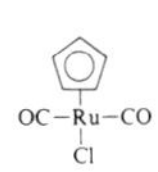

Ru-00047

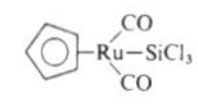

Ru-00048

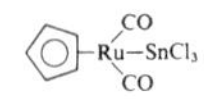

Ru-00049

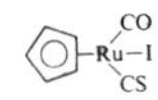

Ru-00050

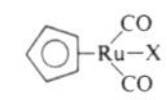

X = I

Ru-00051

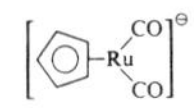

Ru-00052

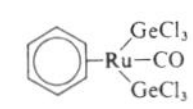

Ru-00053

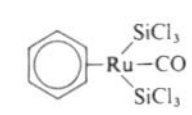

Ru-00054

As Ru-00051 with
X = H

Ru-00055

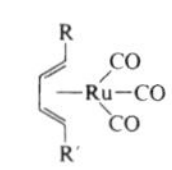

R = R′ = H

Ru-00056

As Ru-00034 with
$X = Cl, R = CH_3$

Ru-00057

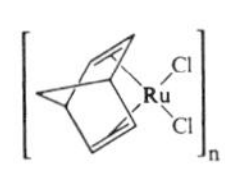

Ru-00058

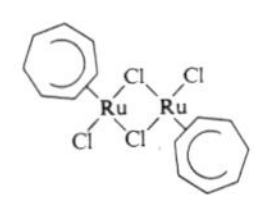

Ru-00059

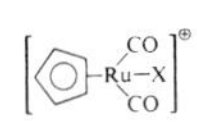

$X = NH_3$

Ru-00060

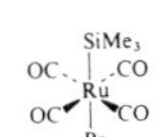

Ru-00061

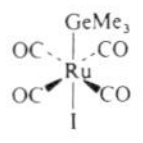

Ru-00062

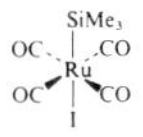

Ru-00063

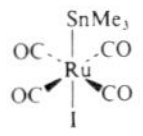

Ru-00064

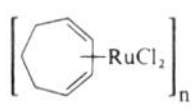

Ru-00065

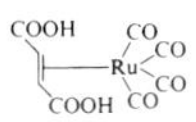

(E)-form

Ru-00066

As Ru-00051 with
X = CN

Ru-00067

As Ru-00051 with
X = –NCS

Ru-00068

As Ru-00051 with
X = –SCN

Ru-00069

As Ru-00051 with
X = –NCO

Ru-00070

As Ru-00060 with
X = CS

Ru-00071

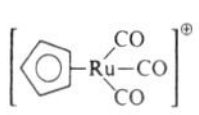

Ru-00072

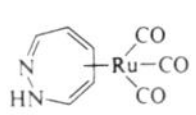

Ru-00073

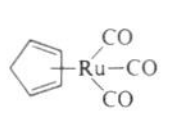

Ru-00074

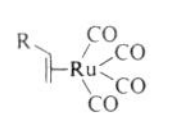

R = –COCH$_3$

Ru-00075

As Ru-00075 with
R = –COOMe

Ru-00076

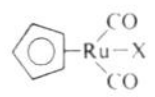

X = CONH$_2$

Ru-00077

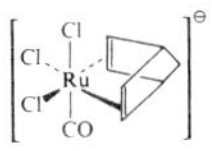

Ru-00078

As Ru-00051 with
X = Me

Ru-00079

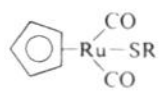

R = Me

Ru-00080

As Ru-00056 with
R = CH$_3$, R′ = H

Ru-00081

As Ru-00051 with
X = –SO$_2$Me

Ru-00082

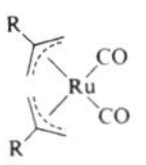

R = H

Ru-00083

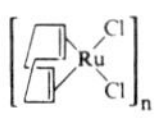

Ru-00084

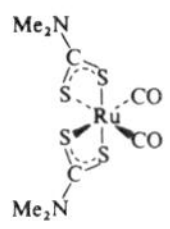

Ru-00085

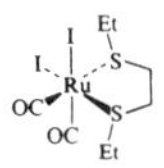

Ru-00086

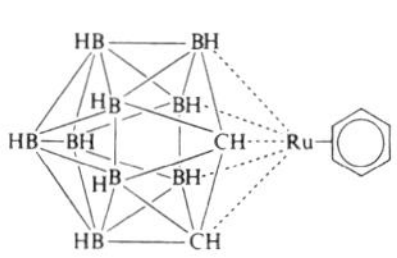

Ru-00087

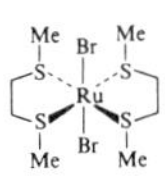

Ru-00088

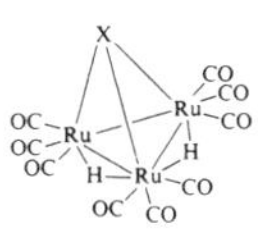

X = S

Ru-00089

As Ru-00089 with
X = Se

Ru-00090

As Ru-00089 with
X = Te

Ru-00091

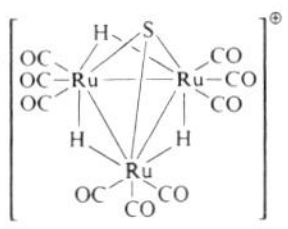

Ru-00092

As Ru-00051 with
X = –COCF$_3$

Ru-00093

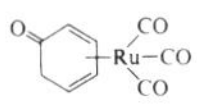

Ru-00094

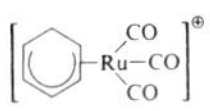

Ru-00095

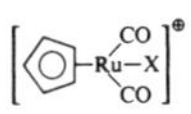

X = NCMe

Ru-00096

As Ru-00060 with
X = MeNC–

Ru-00097

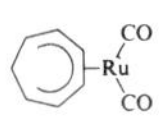

Ru-00098

As Ru-00077 with
X = –C(S)SMe

Ru-00099

As Ru-00051 with
X = Ac

Ru-00100

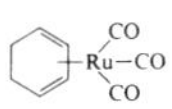

Ru-00101

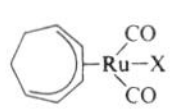

X = Br

Ru-00102

As Ru-00102 with
X = I

Ru-00103

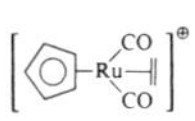

Ru-00104

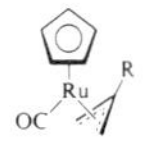

exo-form

R = H

Ru-00105

As Ru-00051 with
X = Et

Ru-00106

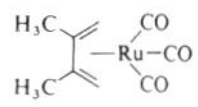

Ru-00107

As Ru-00056 with
R = R′ = CH$_3$

Ru-00108

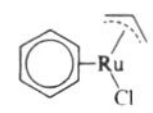

Ru-00109

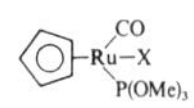

X = Br

Ru-00110

As Ru-00110 with
X = Cl

Ru-00111

As Ru-00110 with
X = I

Ru-00112

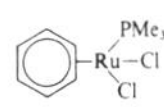

Ru-00113

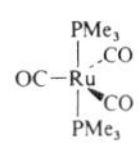

Ru-00114

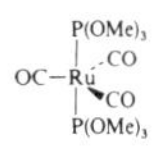

Ru-00115

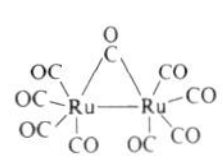

Ru-00116

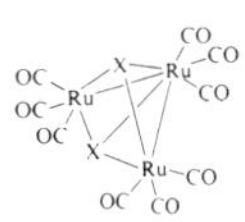

X = S

Ru-00117

As Ru-00117 with
X = Se

Ru-00118

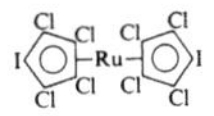

Ru-00119

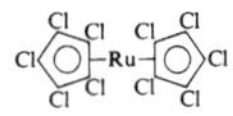

Ru-00120

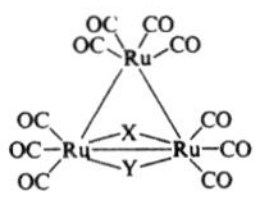

X = H, Y = NO

Ru-00121

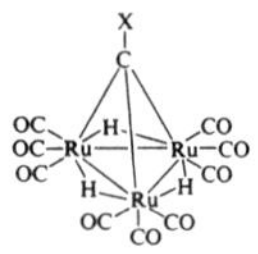

X = Br

Ru-00122

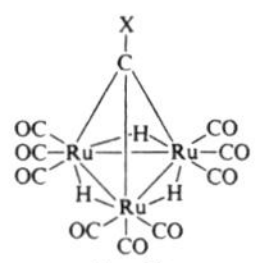

X = Cl

Ru-00123

As Ru-00123 with
X = H

Ru-00124

PhCNRuCl$_2$(CO)$_3$

Ru-00125

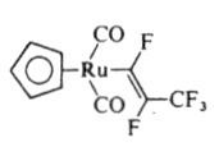

Ru-00126

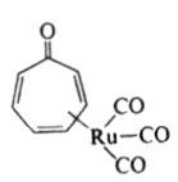

Ru-00127

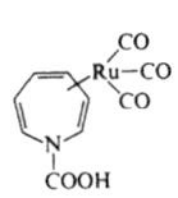

Ru-00128

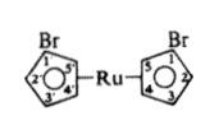

Ru-00129

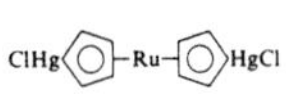

Ru-00130

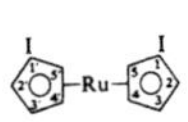

Ru-00131

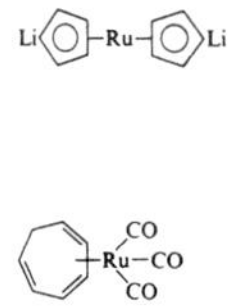

Ru-00132

Ru-00133

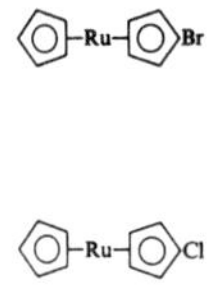

Ru-00134

Ru-00135

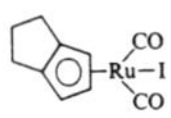

Ru-00136

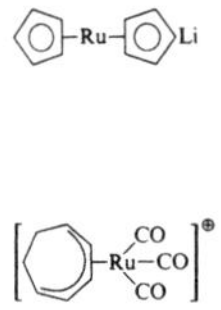

Ru-00138

Ru-00139

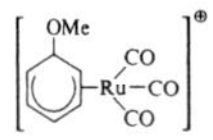

Ru-00140

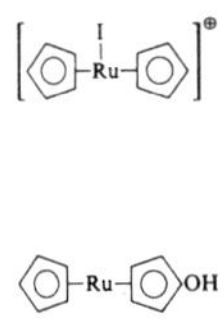

Ru-00141

Ru-00142

As Ru-00051 with
X = $-CH_2CH{=}CH_2$

Ru-00143

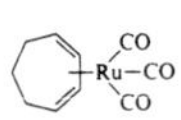

Ru-00144

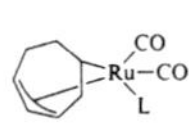

L = CO

Ru-00145

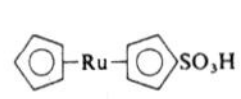

Ru-00146

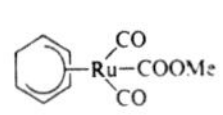

Ru-00147

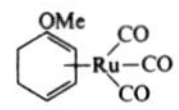

Ru-00148

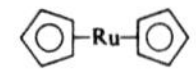

Ru-00149

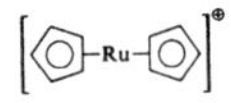

Ru-00150

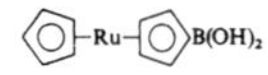

Ru-00151

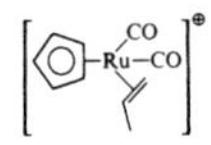

Ru-00152

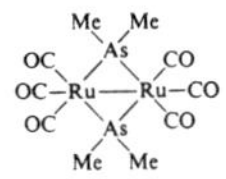

Ru-00153

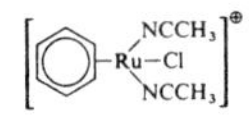

Ru-00154

As Ru-00105 with
R = CH$_3$

Ru-00155

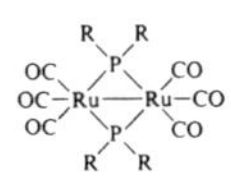

R = Me

Ru-00156

As Ru-00096 with
X = PMe$_3$

Ru-00157

As Ru-00083 with
R = CH$_3$

Ru-00158

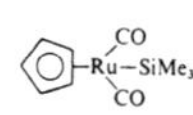

Ru-00159

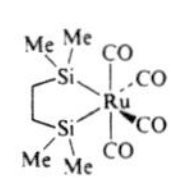

Ru-00160

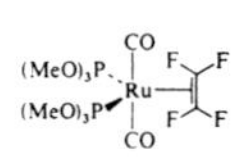

Ru-00161

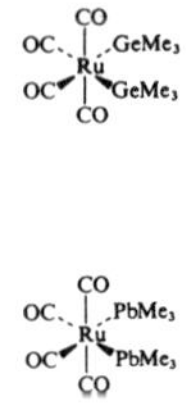

Ru-00162

Ru-00163

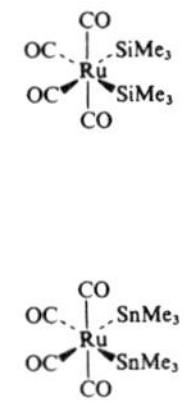

Ru-00164

Ru-00165

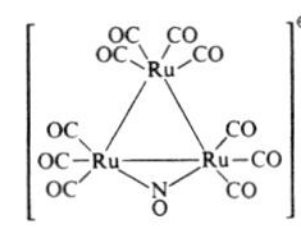

Ru-00166

As Ru-00121 with
X = Y = NO

Ru-00167

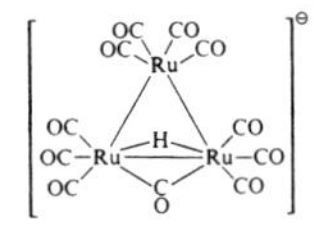

Ru-00168

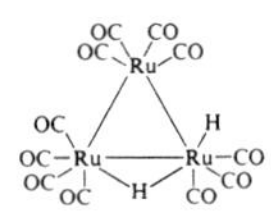

Ru-00169

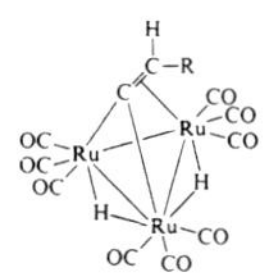

R = H

Ru-00170

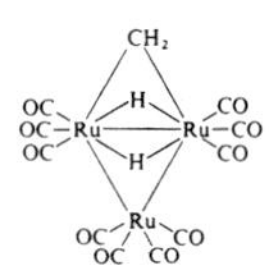

Ru-00171

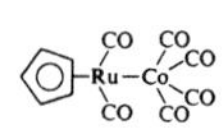

Ru-00172

As Ru-00123 with
X = CH$_3$

Ru-00173

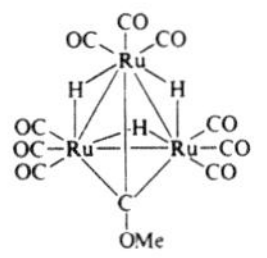

Ru-00174

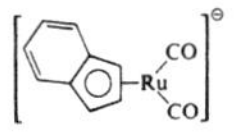

Ru-00175

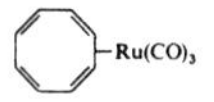

Ru-00176

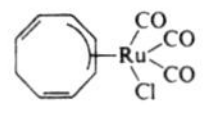

Ru-00177

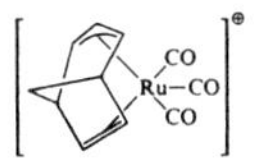

Ru-00178

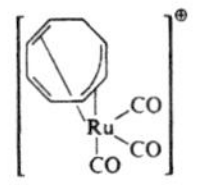

Ru-00179

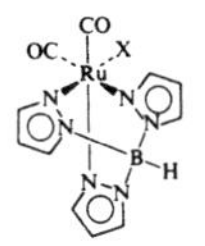

X = Br

Ru-00180

As Ru-00180 with
X = Cl

Ru-00181

As Ru-00180 with
X = I

Ru-00182

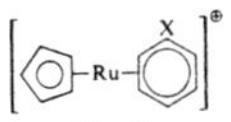

X = Cl

Ru-00183

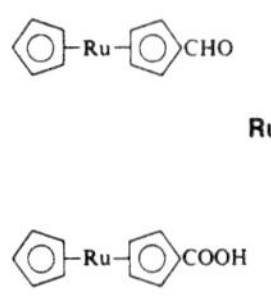

Ru-00184

Ru-00185

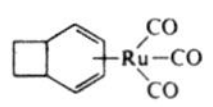

Ru-00186

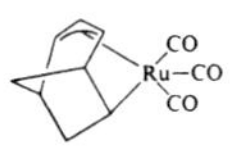

Ru-00187

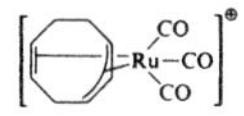

Ru-00188

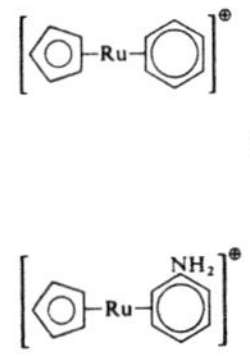

Ru-00189

Ru-00190

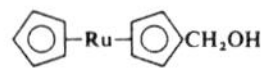

Ru-00191

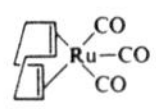

Ru-00192

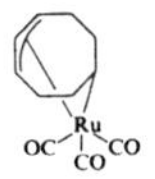

Ru-00193

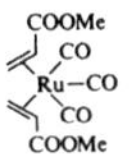

Ru-00194

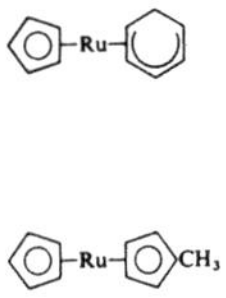

Ru-00195

Ru-00196

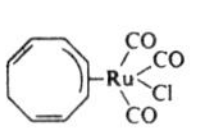

Ru-00197

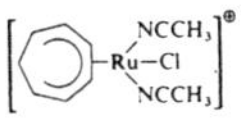

Ru-00198

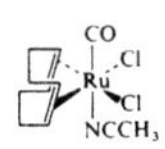

Ru-00199

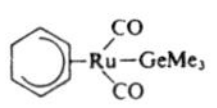

Ru-00200

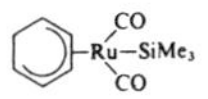

Ru-00201

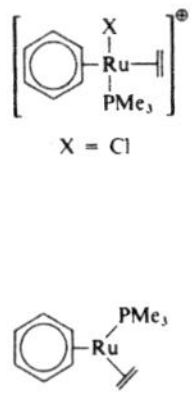

X = Cl

Ru-00202

Ru-00203

As Ru-00202 with
X = H

Ru-00204

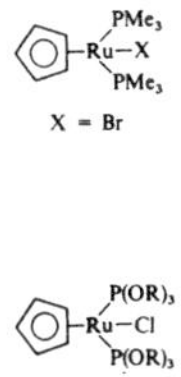

X = Br

Ru-00205

R = Me

Ru-00206

As Ru-00205 with
X = Cl

Ru-00207

As Ru-00205 with
X = I

Ru-00208

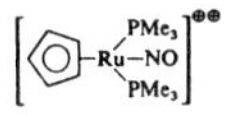

Ru-00209

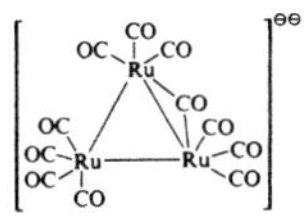

Ru-00210

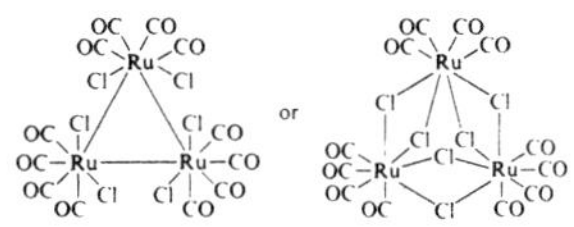

Ru-00211

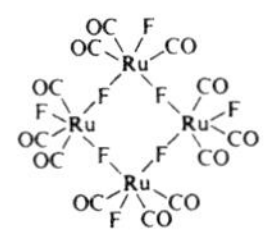

Ru-00212

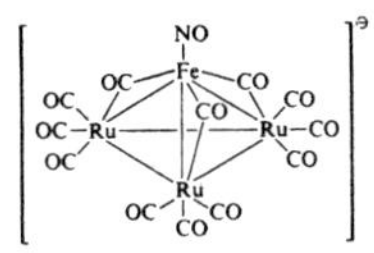

Ru-00213

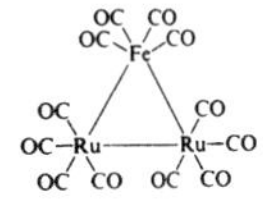

Ru-00214

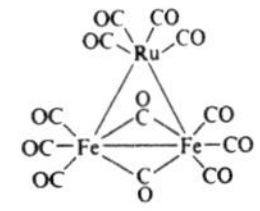

Ru-00215

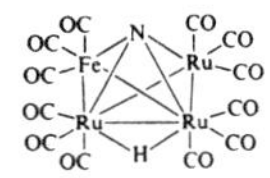

Ru-00216

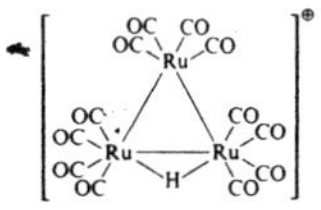

Ru-00217

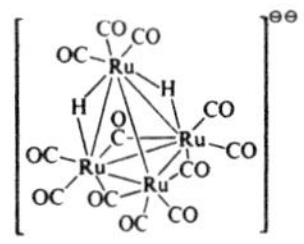

Ru-00218

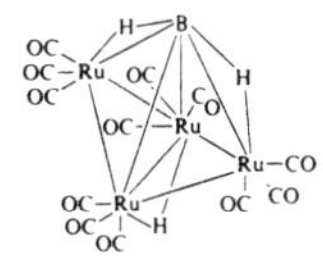

Ru-00219

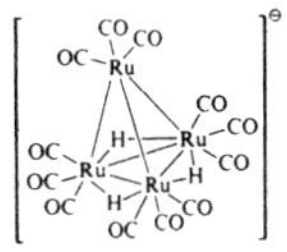

Ru-00220

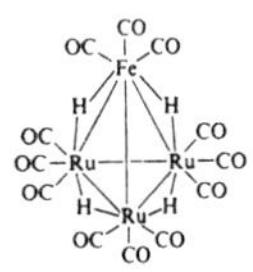

Ru-00221

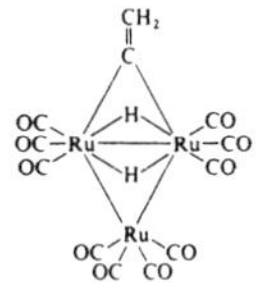

Possible structure

Ru-00222

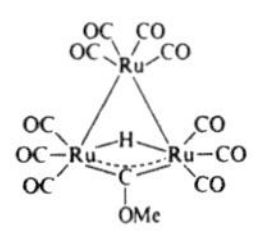

Ru-00223

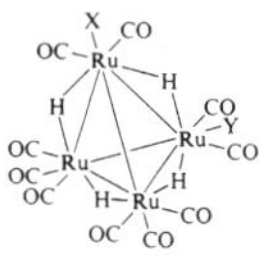

X = Y = CO

Ru-00224

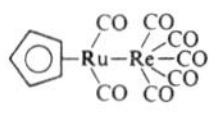

Ru-00225

As Ru-00121 with
X = H, Y = SEt

Ru-00226

As Ru-00122 with
X = COOMe

Ru-00227

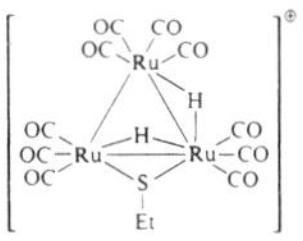

Ru-00228

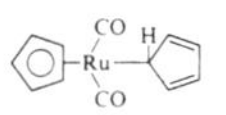

Ru-00229

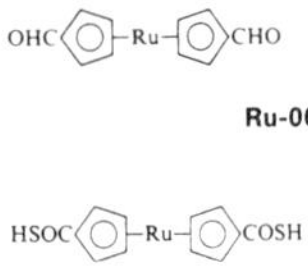

Ru-00230

Ru-00231

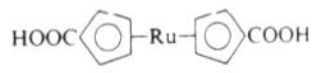

Ru-00232

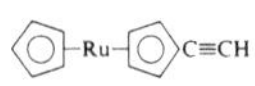

Ru-00233

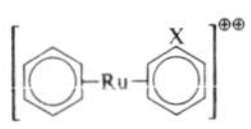

X = Cl

Ru-00234

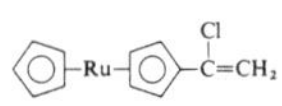

Ru-00235

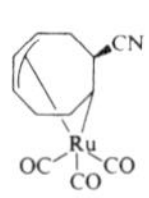

exo-form

Ru-00236

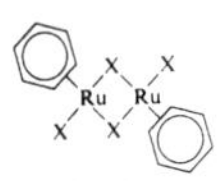

X = Br

Ru-00237

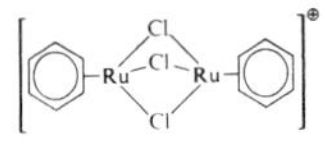

Ru-00238

As Ru-00237 with
X = I

Ru-00239

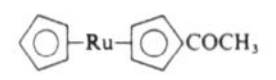

Ru-00240

As Ru-00234 with
X = OH

Ru-00241

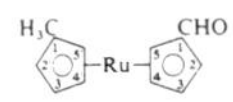

Ru-00242

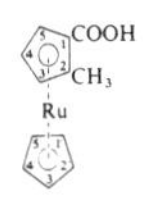

(1*S*)-*form*

Ru-00246

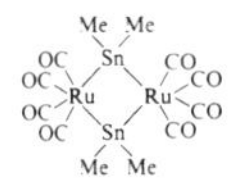

Ru-00248

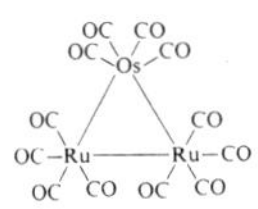

Presumed structure

Ru-00249

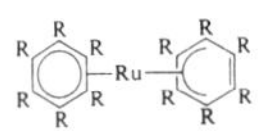

R = H

Ru-00250

As Ru-00234 with
X = H

Ru-00251

Ru-00252

As Ru-00183 with
X = OMe

Ru-00253

As Ru-00183 with
$R = CH_3$

Ru-00254

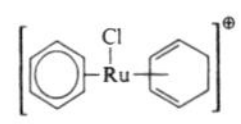

Ru-00255

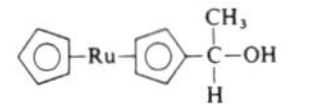

Ru-00256

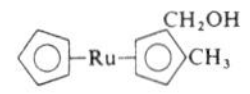

Ru-00257

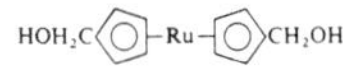

Ru-00258

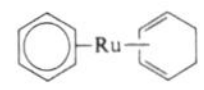

Ru-00259

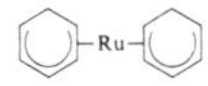

Ru-00260

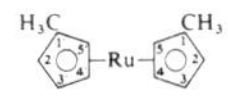

Ru-00261

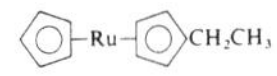

Ru-00264

As Ru-00041 with
$X = -NCCH_3$

Ru-00265

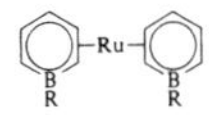

R = Me

Ru-00266

Ru-00267

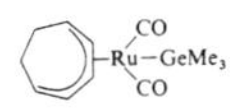

Ru-00268

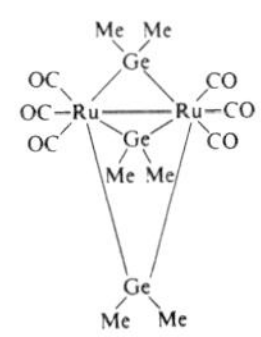

Ru-00269

$Ru(CNMe)_6^{\oplus\oplus}$

Ru-00270

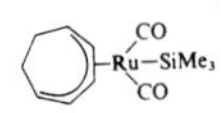

Ru-00271

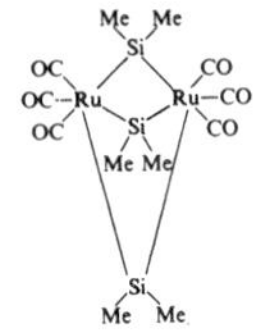

Ru-00272

As Ru-00202 with
R = Me

Ru-00273

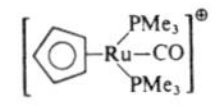

Ru-00274

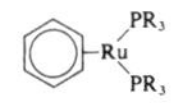

R = OMe

Ru-00275

As Ru-00275 with
R = Me

Ru-00276

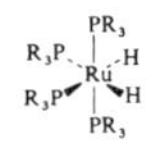

R = OMe

Ru-00277

As Ru-00277 with
R = Me

Ru-00278

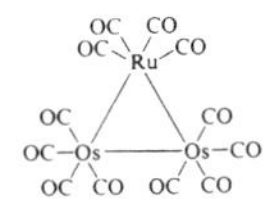

Ru-00279

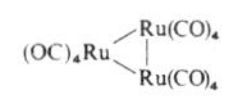

Ru-00280

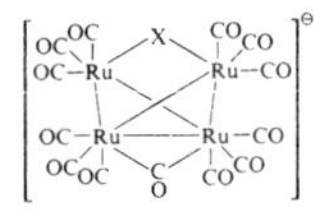

X = Cl

Ru-00281

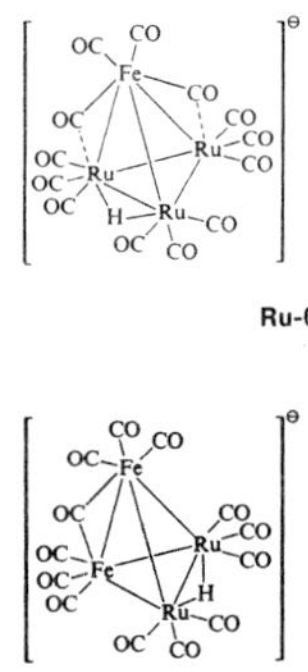

Ru-00282

Ru-00283

As Ru-00281 with
X = H

Ru-00284

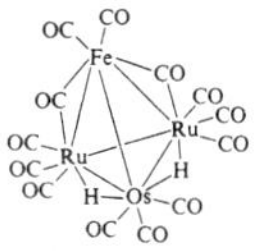

Ru-00285

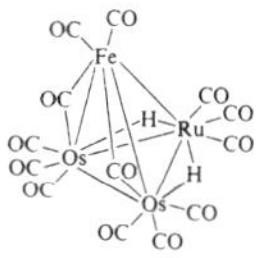

Ru-00286

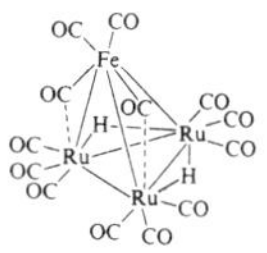

Ru-00287

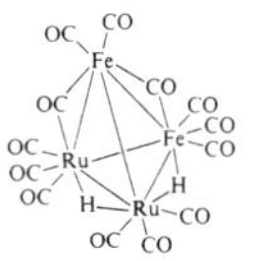

Ru-00289

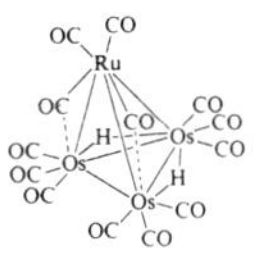

Ru-00290

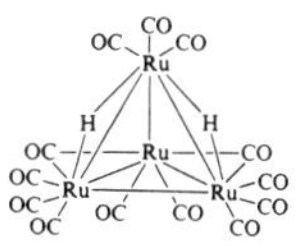

Ru-00291

As Ru-00051 with
X = C_6F_5

Ru-00292

As Ru-00080 with
R = C_6F_5

Ru-00293

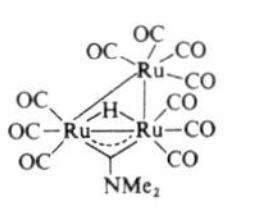

Ru-00294

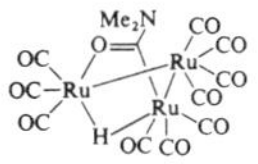

Ru-00295

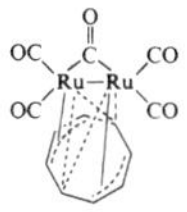

Ru-00296

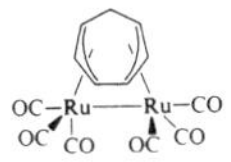

Ru-00297

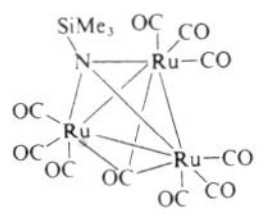

Ru-00298

As Ru-00051 with
X = Ph

Ru-00299

As Ru-00080 with
R = Ph

Ru-00300

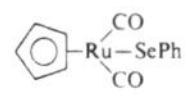

Ru-00301

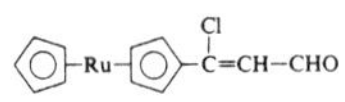

Ru-00302

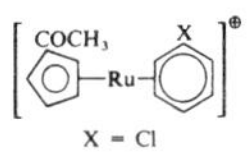

Ru-00303

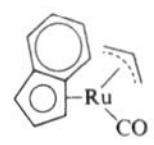

Ru-00304

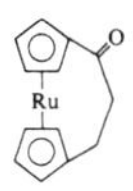

Ru-00305

As Ru-00234 with
X = COOH

Ru-00306

Ru-00307

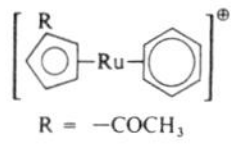

Ru-00308

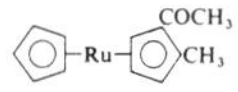

Ru-00309

As Ru-00234 with
X = OMe

Ru-00310

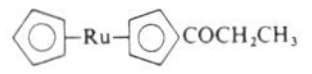

Ru-00311

Ru-00312

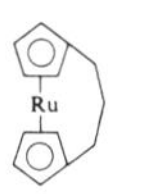

Ru-00313

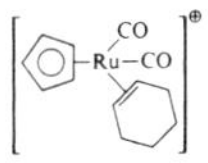

Ru-00314

As Ru-00308 with
R = CH_2CH_3

Ru-00315

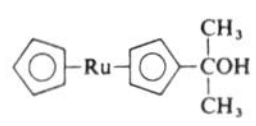

Ru-00316

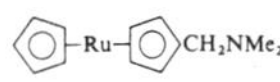

Ru-00317

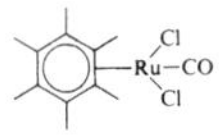

Ru-00318

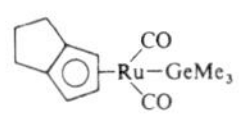

Ru-00319

$[(H_3C)_3CNC]_2Ru(CO)_3$

Ru-00320

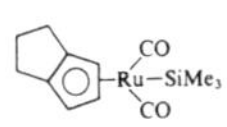

Ru-00321

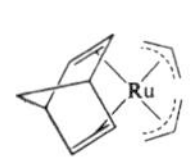

Ru-00322

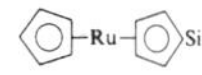

Ru-00323

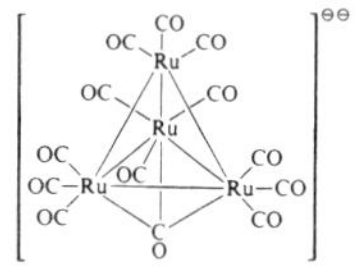

Ru-00324

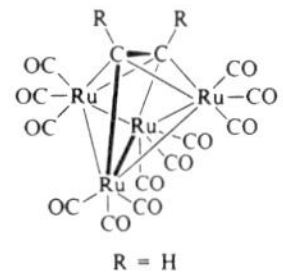

Proposed structure

Ru-00325

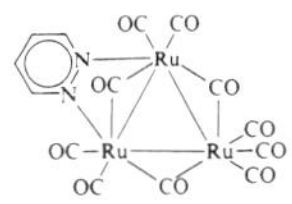

Ru-00326

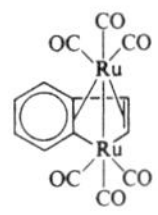

Ru-00327

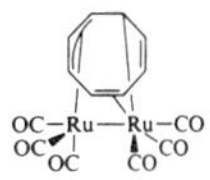

Ru-00328

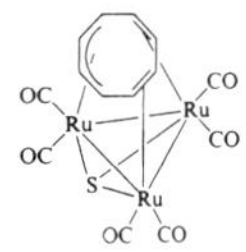

Ru-00329

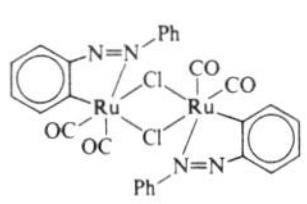

Ru-00330

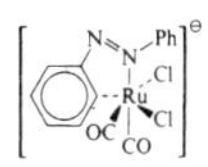

Ru-00331

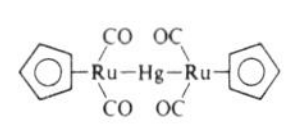

Ru-00332

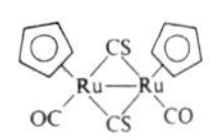

cis-form

Ru-00333

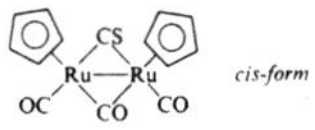

Ru-00334

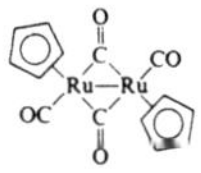

Ru-00335

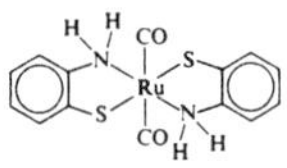

Ru-00336

As Ru-00051 with
X = $-CH_2Ph$

Ru-00337

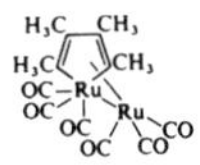

Ru-00338

As Ru-00224 with
X = $P(OMe)_3$, Y = CO

Ru-00339

$Ru(PhNH_2)_2(CO)_2Br_2$

Ru-00340

$Ru(PhNH_2)_2(CO)_2Cl_2$

Ru-00341

As Ru-00234 with
X = $COCH_3$

Ru-00342

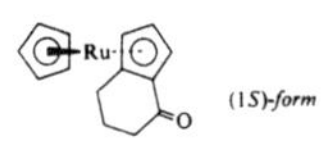

Ru-00343

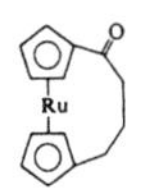

Ru-00344

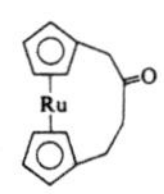

Ru-00345

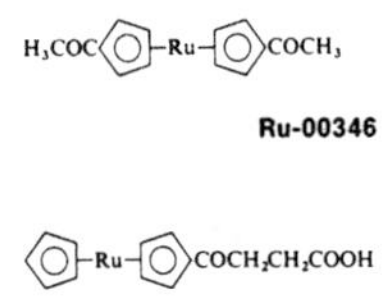

Ru-00346

Ru-00347

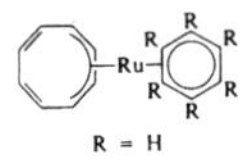

Ru-00348

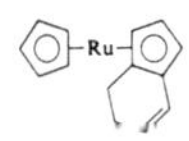

Ru-00349

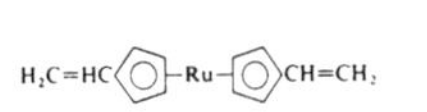

Ru-00350

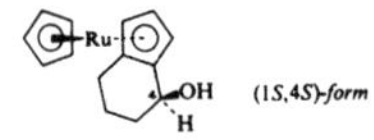

Ru-00351

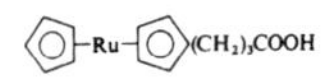

Ru-00352

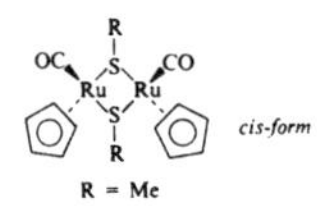

Ru-00353

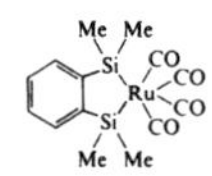

Ru-00354

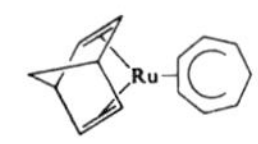

Ru-00355

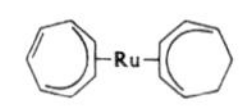

Ru-00356

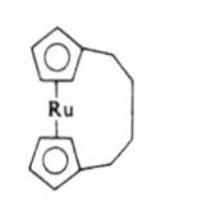

Ru-00357

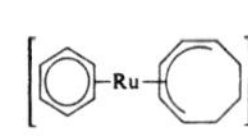

Ru-00358

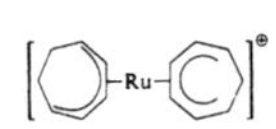

Ru-00359

$Me_3Ge-Ru(CO)_4-Ru(CO)_4-GeMe_3$
(*diax-*)

Ru-00360

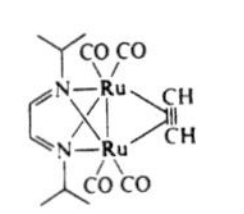

Ru-00361

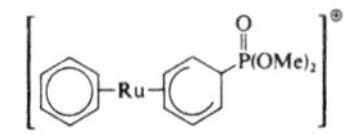

Ru-00362

$Me_3Si-Ru(CO)_4-Ru(CO)_4-SiMe_3$
(*diax*)

Ru-00363

$Me_3Sn-Ru(CO)_4-Ru(CO)_4-SnMe_3$
(*diax-*)

Ru-00364

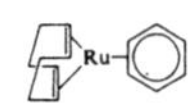

Ru-00365

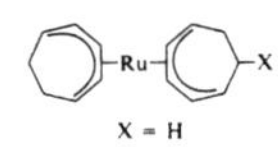

Ru-00366

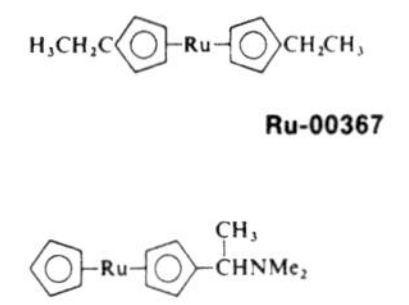

Ru-00367

Ru-00368

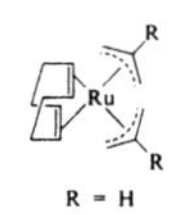

Ru-00369

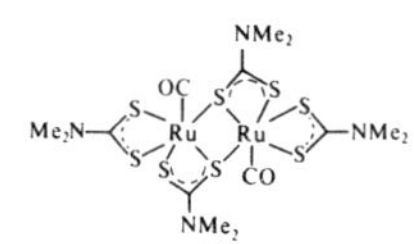

Ru-00370

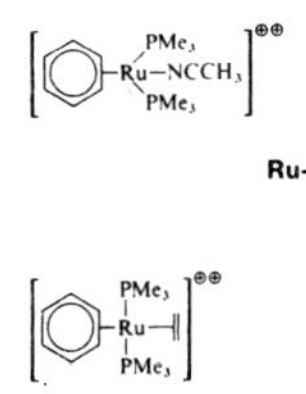

Ru-00371

Ru-00372

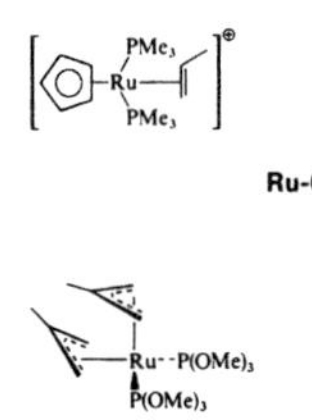

Ru-00373

Ru-00374

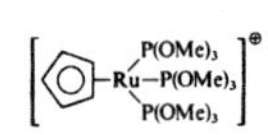

Ru-00375

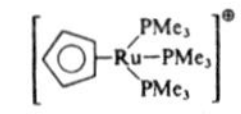

Ru-00376

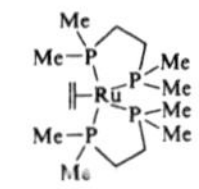

Ru-00377

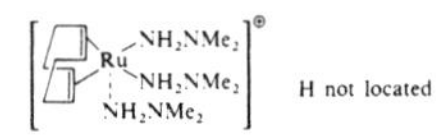

Ru-00378

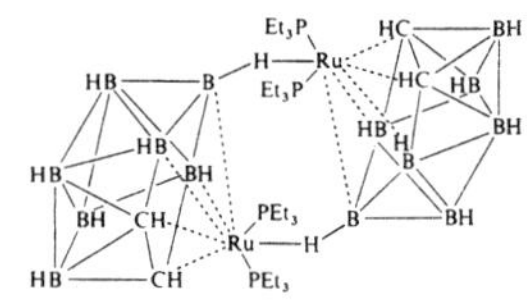

Ru-00379

$(OC)_5Mn-Ru(CO)_4-Mn(CO)_5$

Ru-00380

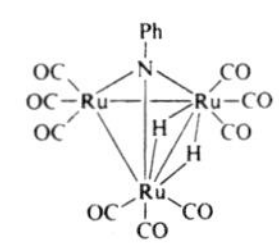

Ru-00381

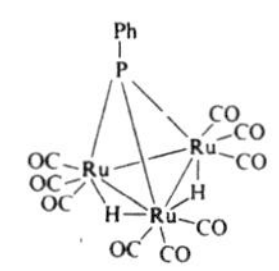

Ru-00382

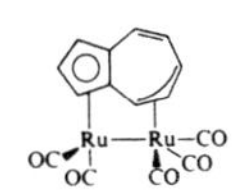

Ru-00383

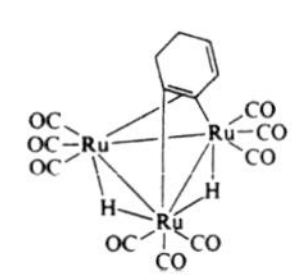

Ru-00384

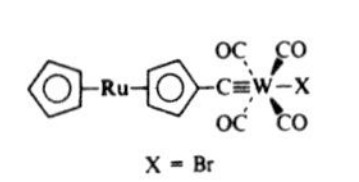

Ru-00385

As Ru-00385 with
X = Cl

Ru-00386

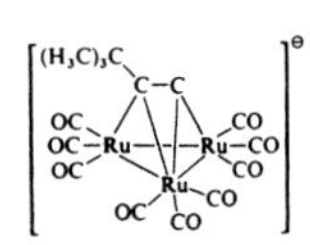

Ru-00387

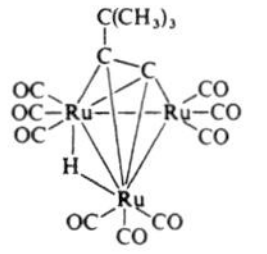

Ru-00388

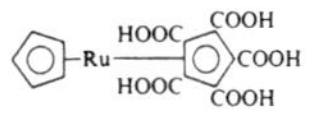

Ru-00389

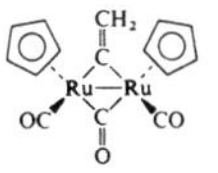

Ru-00390

As Ru-00170 with
R = $-C(CH_3)_3$

Ru-00391

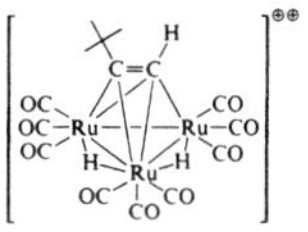

Ru-00392

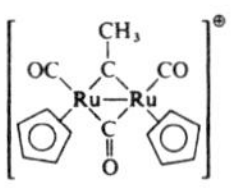

cis-form

Ru-00393

As Ru-00123 with
X = $-CH_2C(CH_3)_3$

Ru-00394

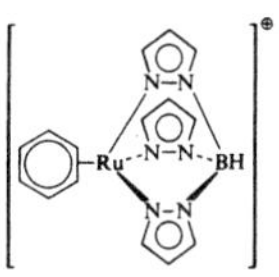

Ru-00395

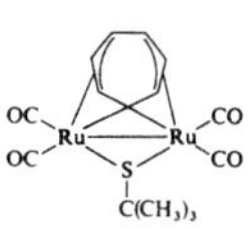

Ru-00396

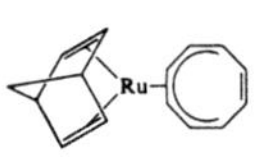

Ru-00397

As Ru-00366 with
X = CN

Ru-00398

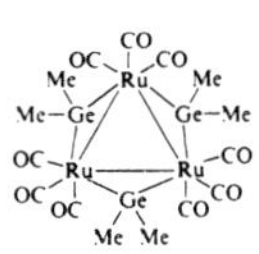

Ru-00399

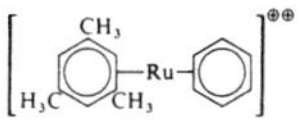

Ru-00400

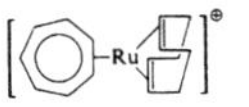

Ru-00401

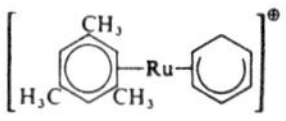

Ru-00402

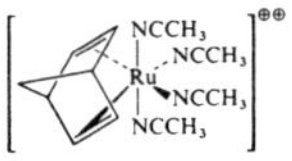

Ru-00403

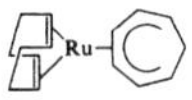

Ru-00404

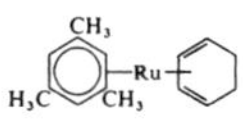

Ru-00405

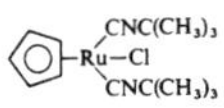

Ru-00406

As Ru-00041 with
X = PMe_3

Ru-00407

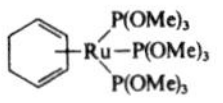

Ru-00408

$BrRu[P(OMe)_3]_5^{\oplus}$

Ru-00409

$ClRu[P(OMe)_3]_5^{\oplus}$

Ru-00410

$Ru[P(OMe)_3]_5$

Ru-00411

$HRu[P(OMe)_3]_5^{\oplus}$

Ru-00412

$Ru_5(C)(CO)_{15}Cl^{\ominus}$

Ru-00413

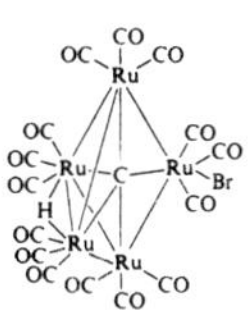

Ru-00414

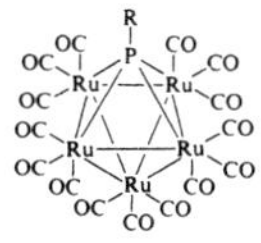

R = Me

Ru-00415

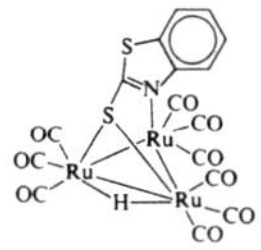

Ru-00416

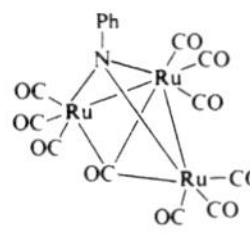

Ru-00417

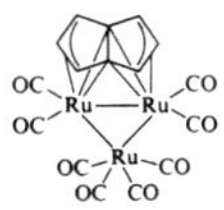

Ru-00418

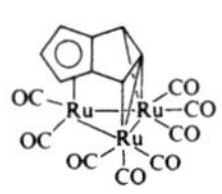

Ru-00419

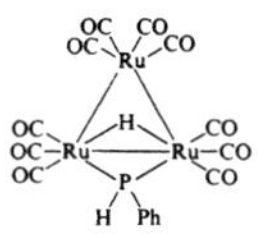

Ru-00420

$(PhCN)_2RuCl_2(CO)_2$

Ru-00421

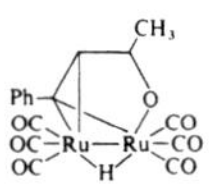

Ru-00422

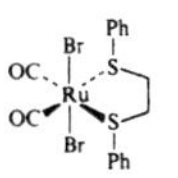

Ru-00423

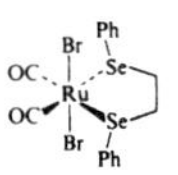

Ru-00424

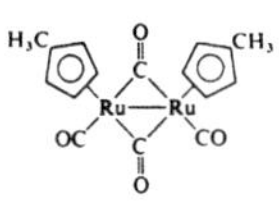

Ru-00425

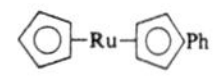

Ru-00426

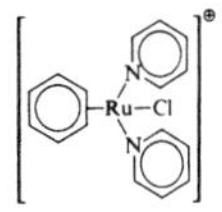

Ru-00427

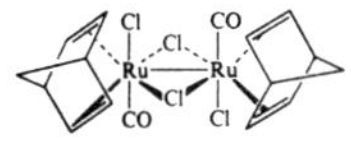

Ru-00428

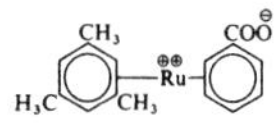

Ru-00429

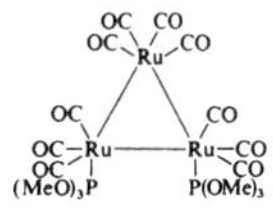

Ru-00430

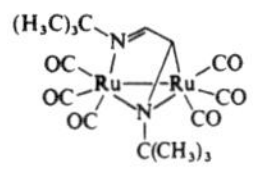

Ru-00431

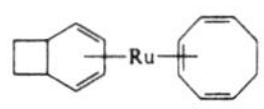

Ru-00432

As Ru-00224 with
X = Y = $P(OMe)_3$

Ru-00433

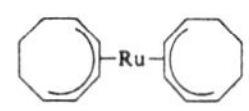

Ru-00434

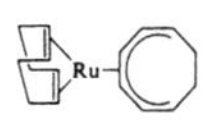

Ru-00435

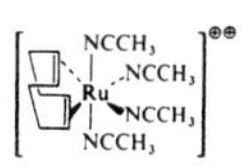

Ru-00436

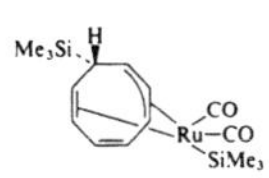

Ru-00437

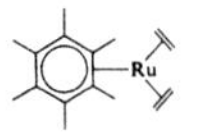

Ru-00438

As Ru-00369 with
R = CH_3

Ru-00439

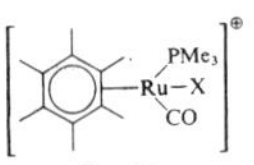

X = Cl

Ru-00440

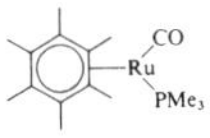

Ru-00441

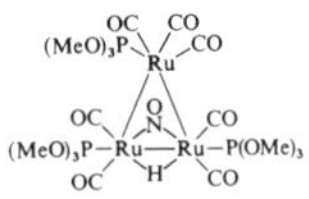

Ru-00442

As Ru-00440 with
X = H

Ru-00443

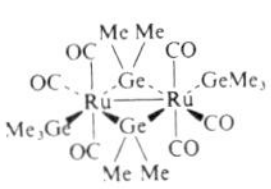

Ru-00444

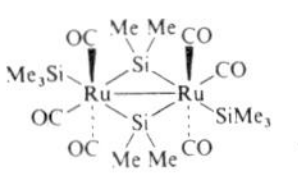

Ru-00445

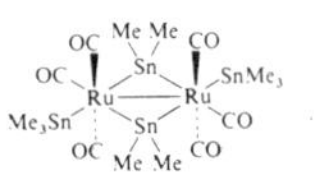

Ru-00446

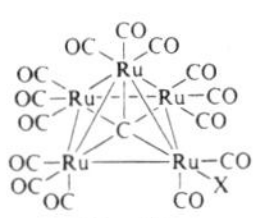

X = CO

Ru-00447

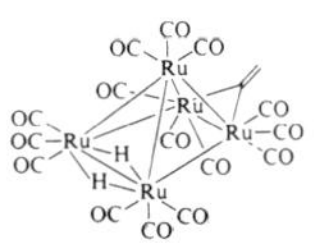

Ru-00449

As Ru-00415 with
R = Et

Ru-00450

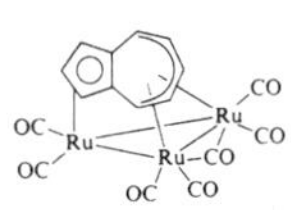

Ru-00451

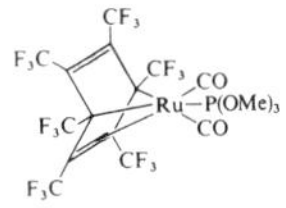

Ru-00452

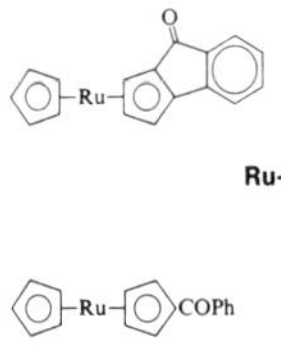

Ru-00453

Ru-00454

As Ru-00183 with
X = Ph

Ru-00455

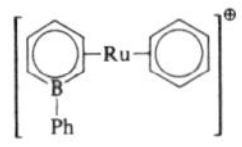

Ru-00456

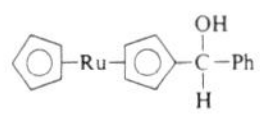

Ru-00457

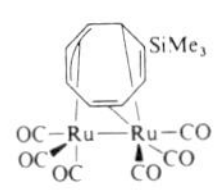

Ru-00458

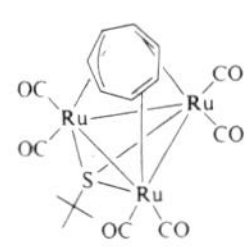

Ru-00459

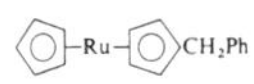

Ru-00460

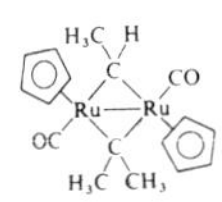

Ru-00461

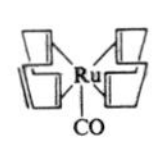

Ru-00462

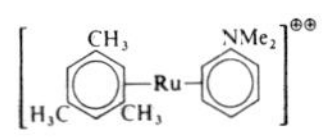

Ru-00463

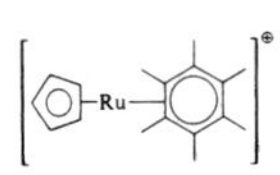

Ru-00464

As Ru-00440 with
X = Me

Ru-00465

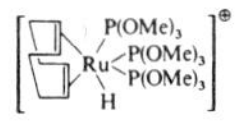

Ru-00466

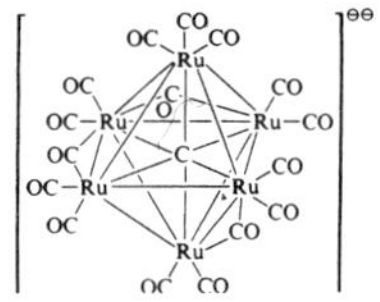

Ru-00467

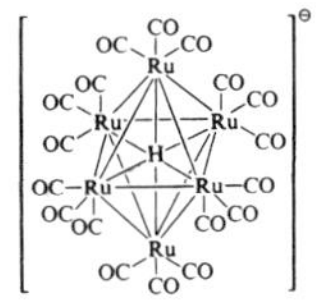

Ru-00468

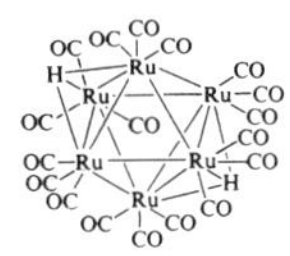

Ru-00469

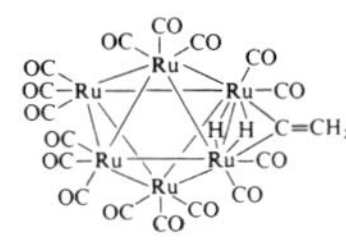

Ru-00470

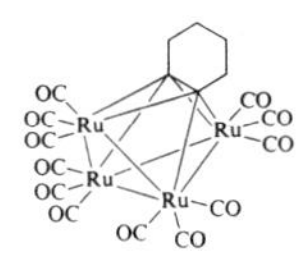

Ru-00471

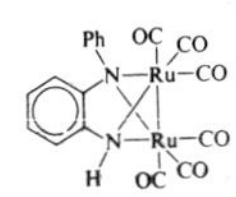

Ru-00472

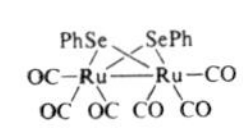

Ru-00473

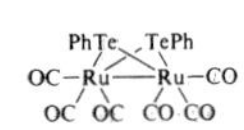

Ru-00474

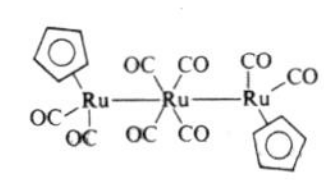

Ru-00475

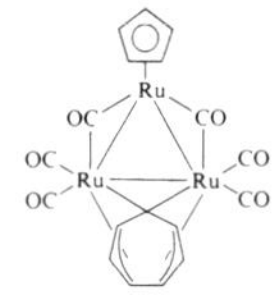

Ru-00476

$(C_8H_{12})Ru_3(CO)_{10}$

Ru-00477

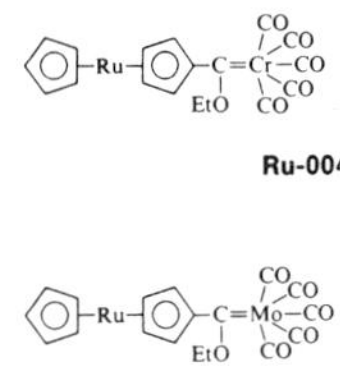

Ru-00478

Ru-00479

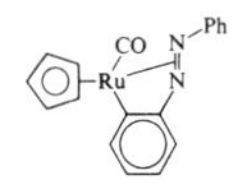

Ru-00480

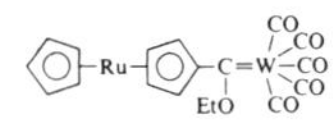

Ru-00481

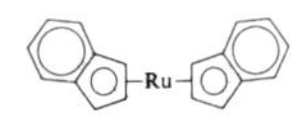

Ru-00482

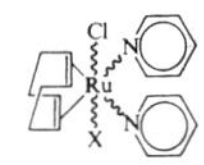

X = Cl

Ru-00483

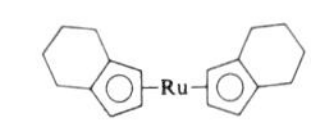

Ru-00484

As Ru-00483 with
X = H

Ru-00485

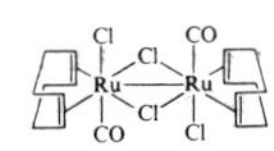

Ru-00486

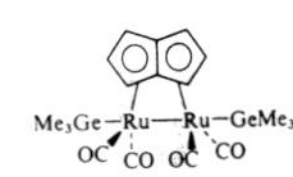

Ru-00487

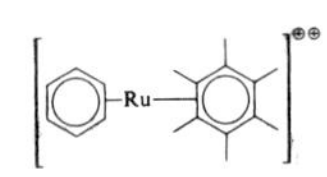

Ru-00488

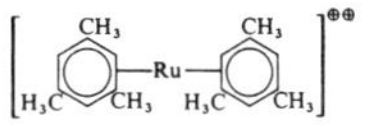

Ru-00489

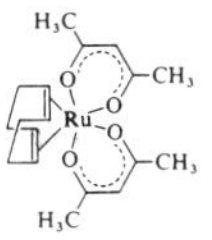

Ru-00490

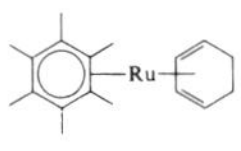

Ru-00491

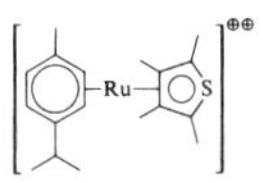

Ru-00492

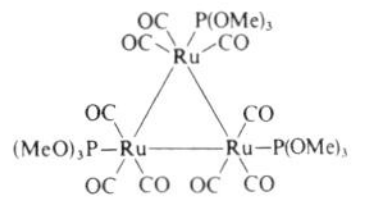

Ru-00493

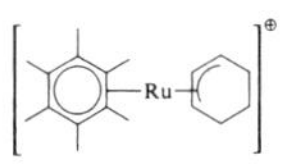

Ru-00494

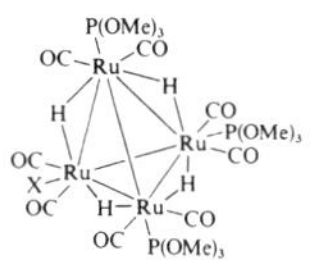

X = CO

Ru-00495

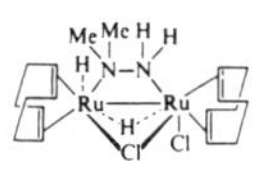

Ru-00496

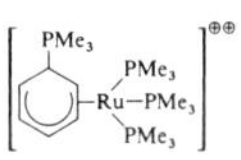

Ru-00497

$Ru[P(OMe)_3]_6^{\oplus\oplus}$

Ru-00498

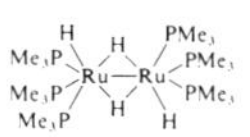

Ru-00499

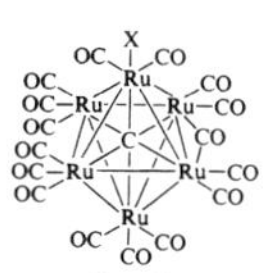

X = CO

Ru-00500

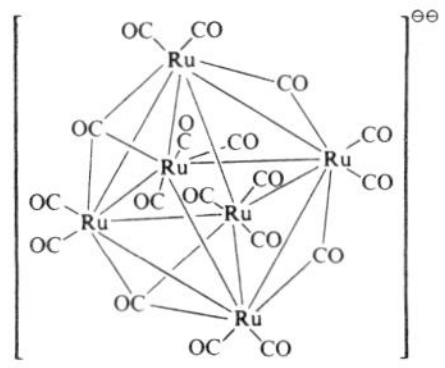

Ru-00501

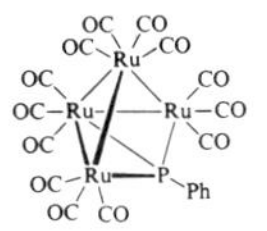

Ru-00502

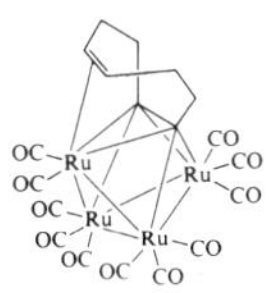

Ru-00503

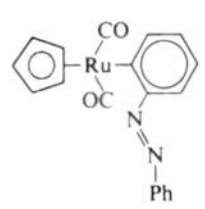

Ru-00504

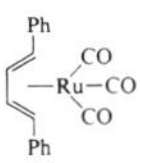

Ru-00505

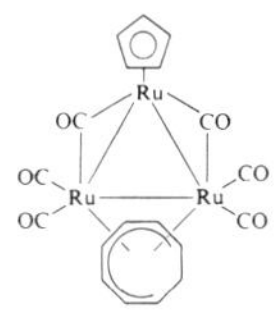

Ru-00506

As Ru-00051 with
X = $-PPh_2$

Ru-00507

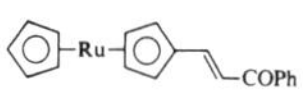

Ru-00508

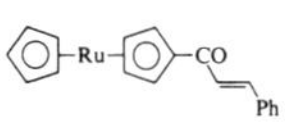

Ru-00509

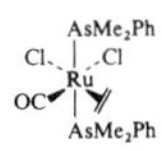

Ru-00510

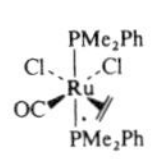

Ru-00511

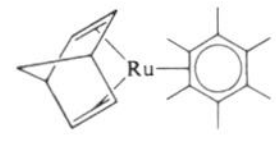

Ru-00512

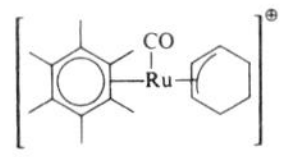

Ru-00513

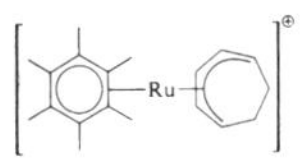

Ru-00514

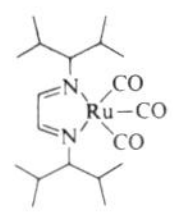

Ru-00515

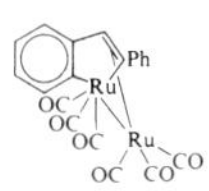

Possible structure

Ru-00516

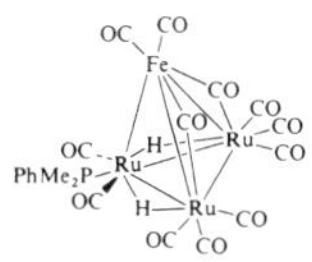

Ru-00517

$Ph_3PRu(CO)_2I_3^{\ominus}$

Ru-00518

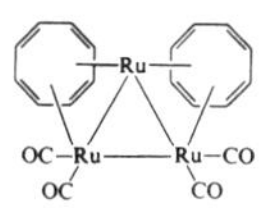

Ru-00519

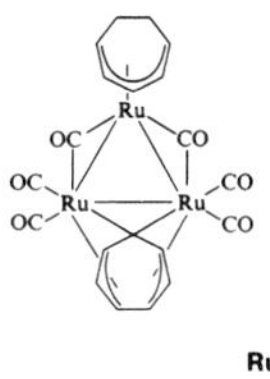

Ru-00520

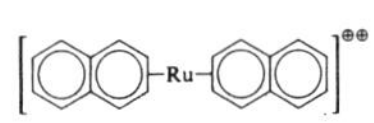

Ru-00521

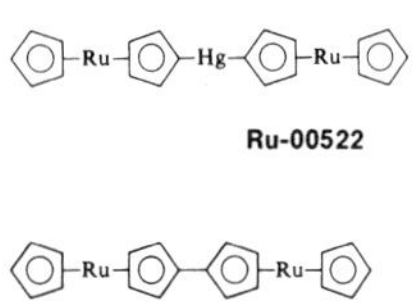

Ru-00522

Ru-00523

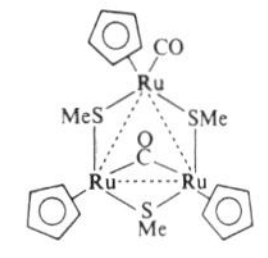

Ru-00524

As Ru-00348 with
R = CH_3

Ru-00525

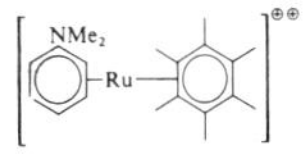

Ru-00526

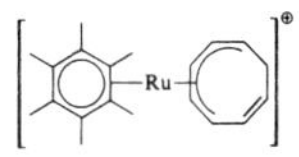

Ru-00527

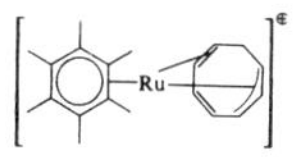

Ru-00528

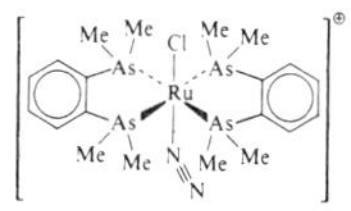

Ru-00529

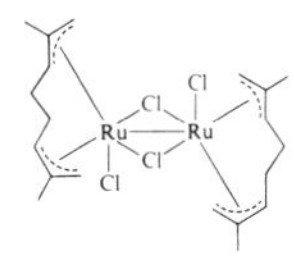

Ru-00530

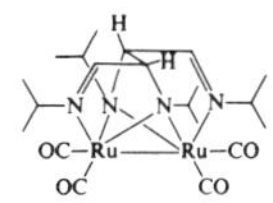

Ru-00531

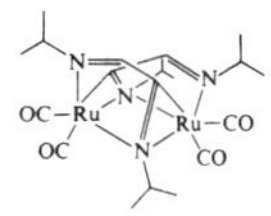

Ru-00532

As Ru-00495 with
X = $P(OMe)_3$

Ru-00533

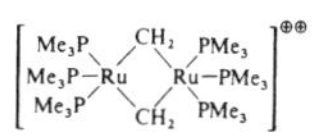

Ru-00534

As Ru-00415 with
R = Ph

Ru-00535

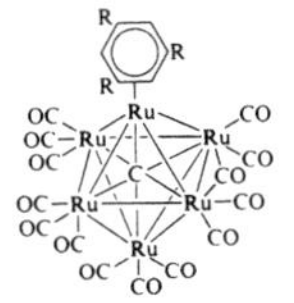

R = H

Ru-00536

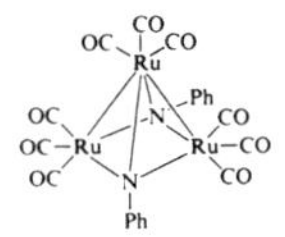

Ru-00537

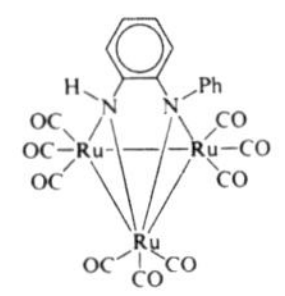

Ru-00538

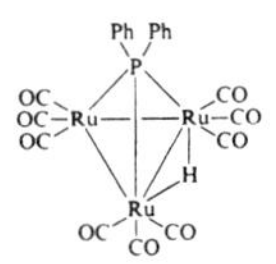

Ru-00539

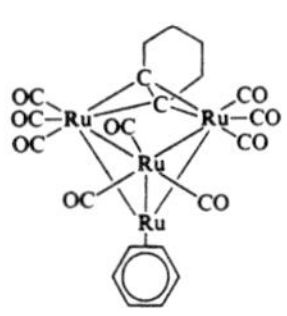

Ru-00540

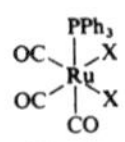

X = Br

Ru-00541

As Ru-00541 with
X = Cl

Ru-00542

As Ru-00541 with
X = I

Ru-00543

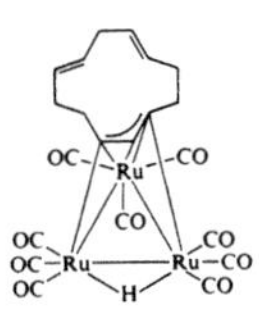

Ru-00544

$H_2Ru(CO)_3(PPh_3)$

Ru-00545

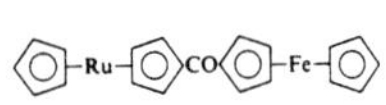

Ru-00546

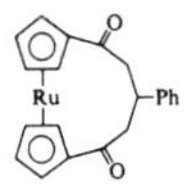

Ru-00547

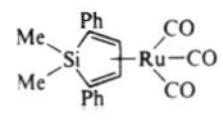

Ru-00548

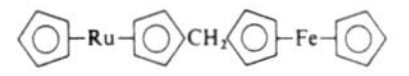

Ru-00549

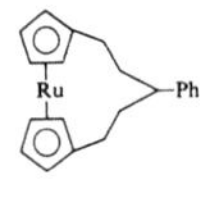

Ru-00550

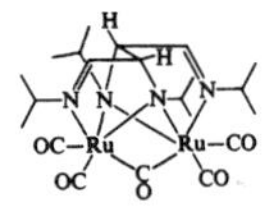

Ru-00551

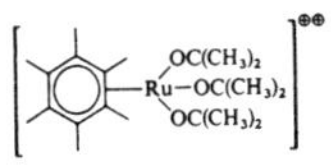

Ru-00552

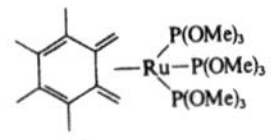

Ru-00553

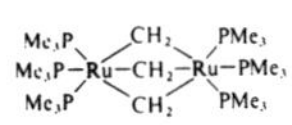

Ru-00554

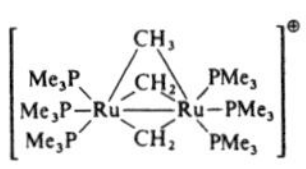

Ru-00555

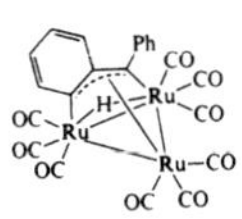

Ru-00556

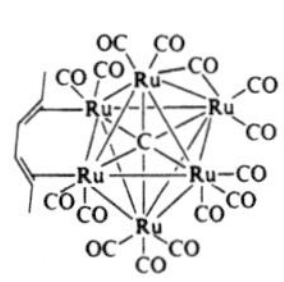

Ru-00557

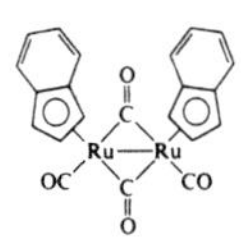

Ru-00558

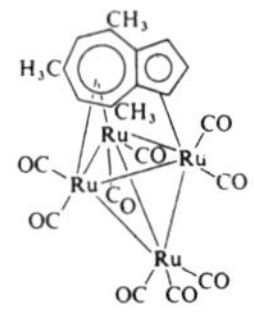

Ru-00559

$Ph_3PRu(CO)_4$

Ru-00560

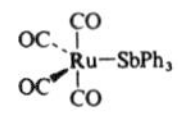

Ru-00561

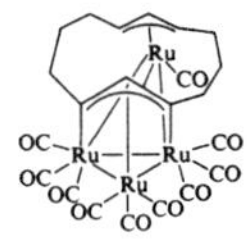

Ru-00562

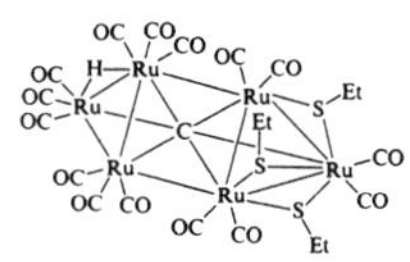

Ru-00563

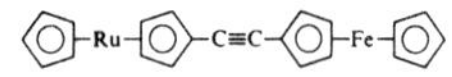

Ru-00564

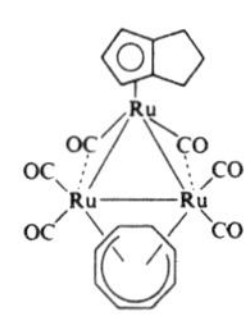

Ru-00565

As Ru-00266 with
R = Ph

Ru-00566

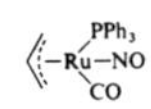

Ru-00567

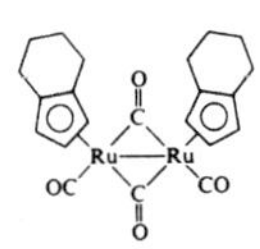

Ru-00568

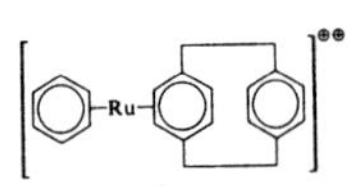

Ru-00569

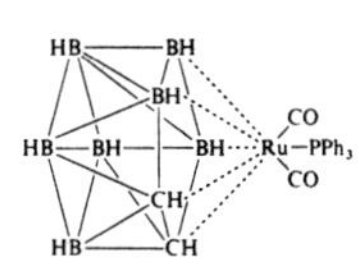

Ru-00570

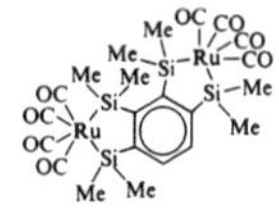

Ru-00571

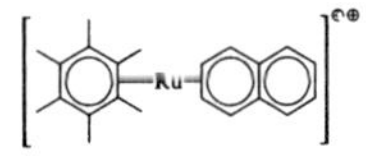

Ru-00573

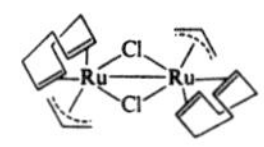

Ru-00574

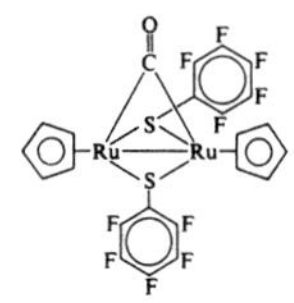

Ru-00575

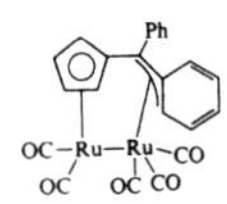

Ru-00576

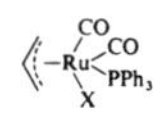

X = Br

Ru-00577

As Ru-00577 with
X = Cl

Ru-00578

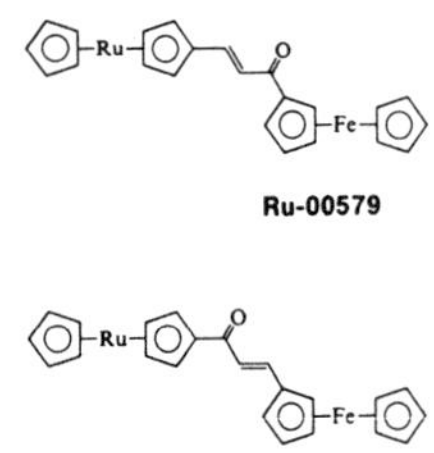

Ru-00579

Ru-00580

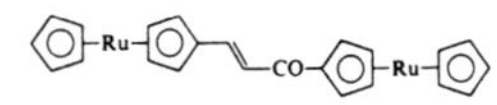

Ru-00581

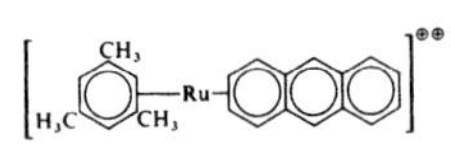

Ru-00582

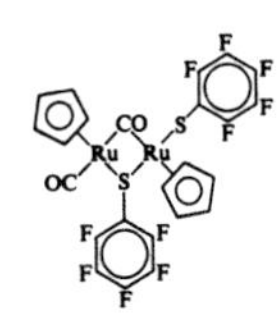

Ru-00583

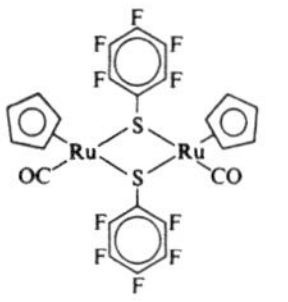

Ru-00584

As Ru-00536 with
R = CH_3

Ru-00585

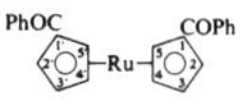

Ru-00586

Ru-00588

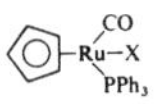

X = Br

Ru-00589

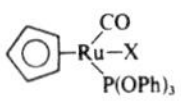

X = Br

Ru-00590

As Ru-00589 with
X = Cl

Ru-00591

As Ru-00590 with
X = Cl

Ru-00592

As Ru-00589 with
X = I

Ru-00593

As Ru-00590 with
X = I

Ru-00594

As Ru-00353 with
R = Ph

Ru-00595

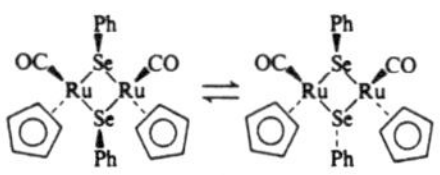

cis-form

Ru-00596

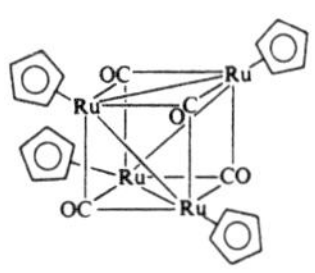

Ru-00597

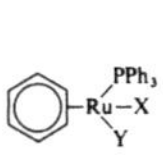

X = Y = Br

Ru-00598

As Ru-00598 with
X = Y = Cl

Ru-00599

As Ru-00598 with
X = Y = I

Ru-00600

As Ru-00589 with
X = H

Ru-00601

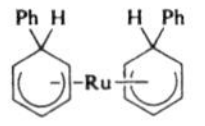

Ru-00602

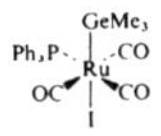

Ru-00603

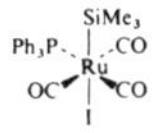

Ru-00604

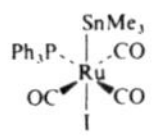

Ru-00605

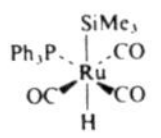

Ru-00606

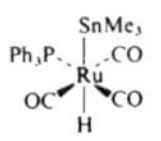

Ru-00607

$Ru_6(CO)_{13}(\mu\text{-}CO)(\mu_6\text{-}C)[P(OMe)_3]_3$

Ru-00608

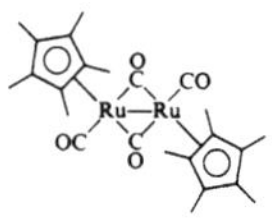

Ru-00609

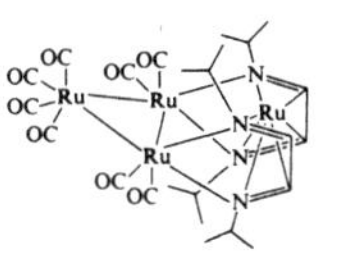

Ru-00610

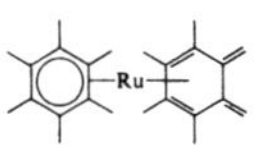

Ru-00611

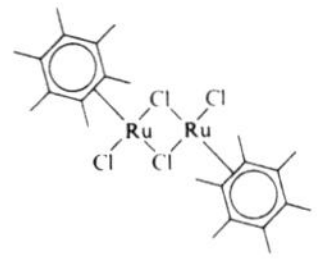

Ru-00612

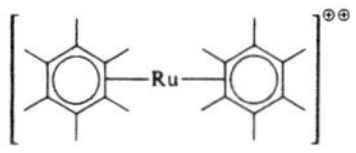

Ru-00613

As Ru-00250 with
R = CH_3

Ru-00614

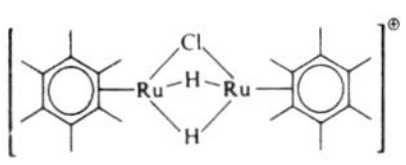

Ru-00615

$H_2Ru[P(OEt)_3]_4$

Ru-00616

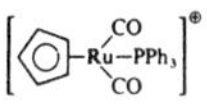

Ru-00617

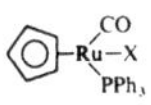

X = $CONH_2$

Ru-00618

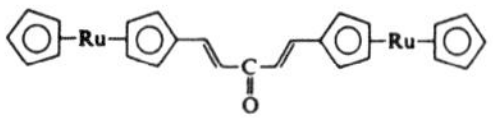

Ru-00619

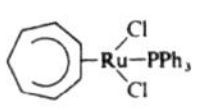

Ru-00620

As Ru-00618 with
X = Me

Ru-00621

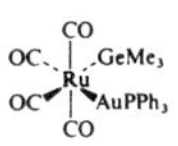

Ru-00622

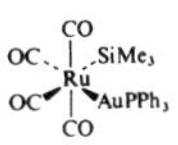

Ru-00623

As Ru-00598 with
X = Cl, Y = Me

Ru-00624

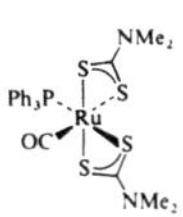

Ru-00625

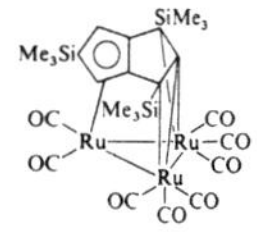

Ru-00626

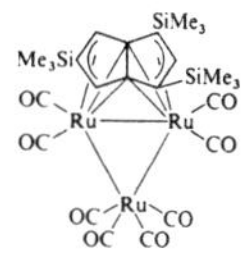

Ru-00627

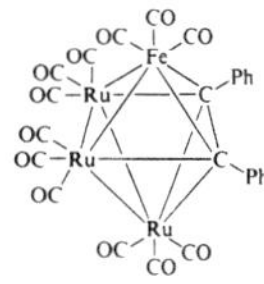

axial-form

Ru-00628

As Ru-00325 with
R = Ph

Ru-00629

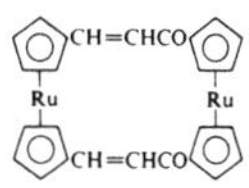

Ru-00630

As Ru-00589 with
X = Ac

Ru-00631

As Ru-00589 with
X = Et

Ru-00632

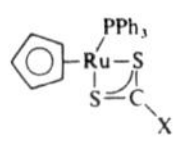

X = OEt

Ru-00633

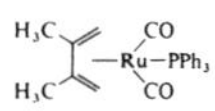

Ru-00634

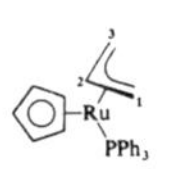

trans-form

Ru-00635

As Ru-00598 with
X = Y = Me

Ru-00636

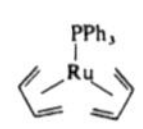

Ru-00637

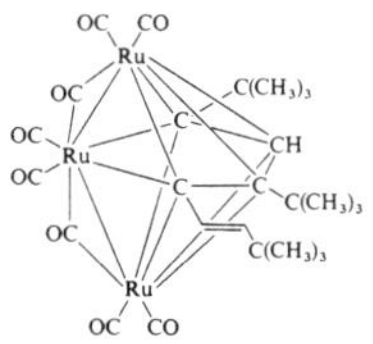

Ru-00638

$Ru_6(CO)_{12}(\mu\text{-}CO)(\mu_6\text{-}C)[P(OMe)_3]_4$

Ru-00639

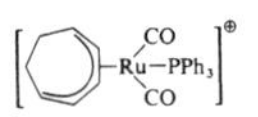

Ru-00640

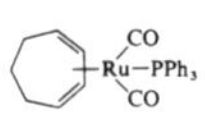

Ru-00641

As Ru-00145 with
L = PPh_3

Ru-00642

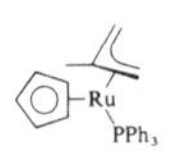

Ru-00643

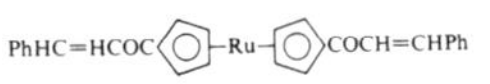

Ru-00644

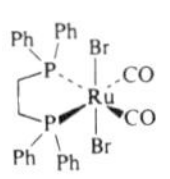

Ru-00645

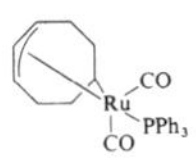

Ru-00646

As Ru-00633 with
X = NEt_2

Ru-00647

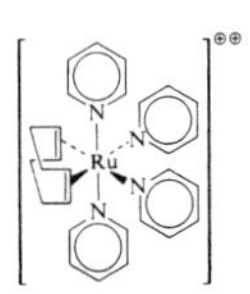

Ru-00648

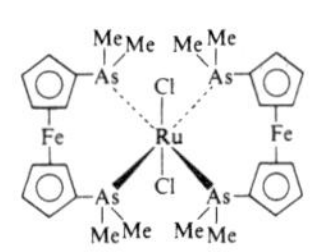

Ru-00649

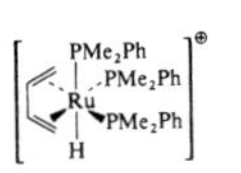

Ru-00650

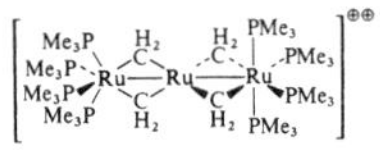

Ru-00651

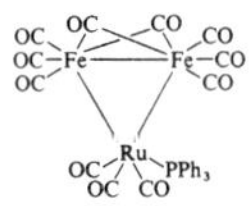

Ru-00652

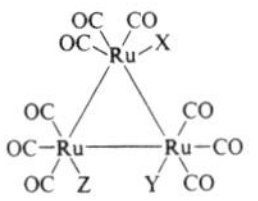

X = PPh_3, Y = Z = CO

Ru-00653

As Ru-00224 with
X = PPh_3, Y = CO

Ru-00654

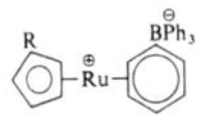

R = H

Ru-00655

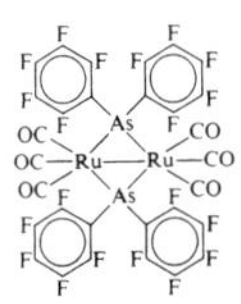

Ru-00656

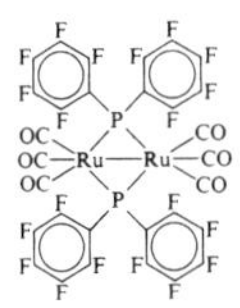

Ru-00657

As Ru-00156 with
R = Ph

Ru-00658

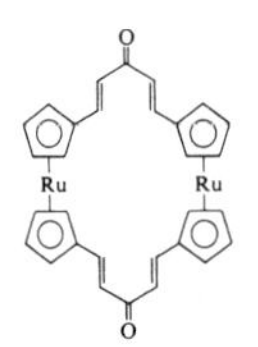

Ru-00659

As Ru-00598 with
X = Cl, Y = Ph

Ru-00660

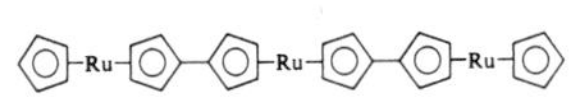

Ru-00661

As Ru-00655 with
R = CH_3

Ru-00662

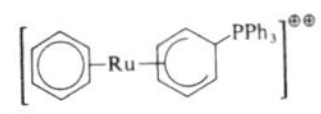

Ru-00663

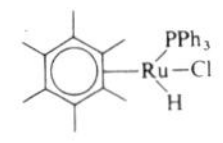

Ru-00664

$BrRu[P(OEt)_3]_5^{\oplus}$

Ru-00665

$ClRu[P(OEt)_3]_5^{\oplus}$

Ru-00666

$HRu[P(OEt)_3]_5^{\oplus}$

Ru-00667

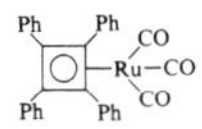

Ru-00668

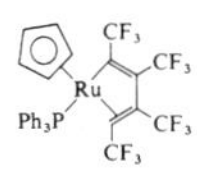

Ru-00669

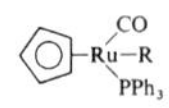

R = CH_2Ph

Ru-00670

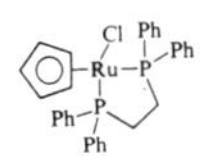

Ru-00671

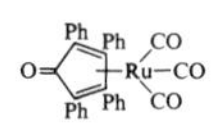

Ru-00672

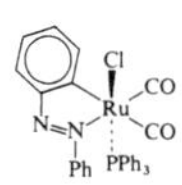

Ru-00673

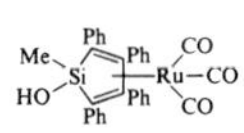

Ru-00674

As Ru-00670 with
R = —C≡CPh

Ru-00675

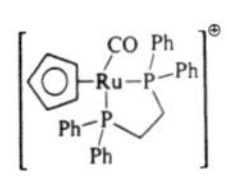

Ru-00676

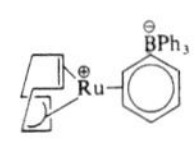

Ru-00677

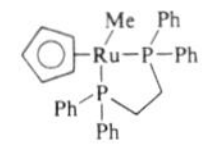

Ru-00678

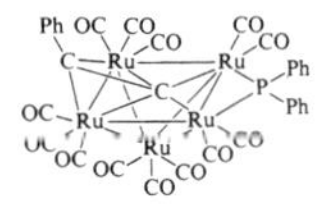

Ru-00679

As Ru-00447 with
X = PPh_3

Ru-00680

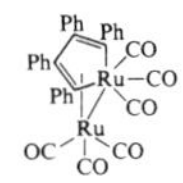

Ru-00681

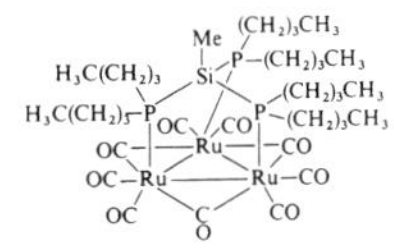

Ru-00682

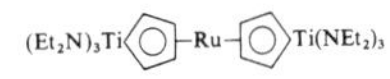

Ru-00683

As Ru-00500 with
X = PPh_3

Ru-00684

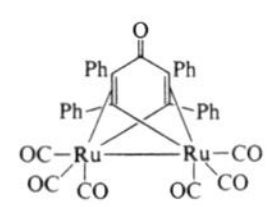

Possible structure

Ru-00685

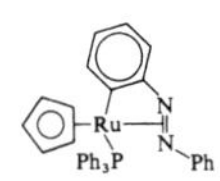

Ru-00686

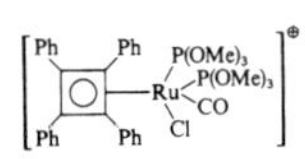

Ru-00687

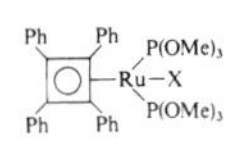

X = CO

Ru-00688

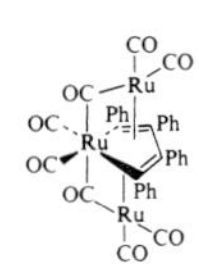

Ru-00689

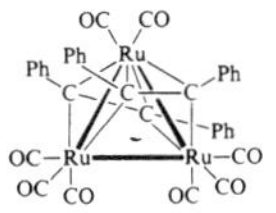

Ru-00690

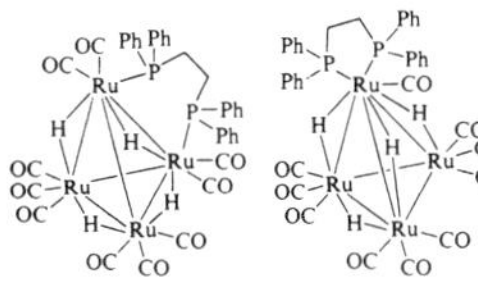

vicinal-form *geminal-form*

Ru-00691

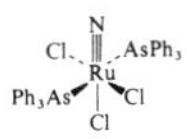

Ru-00692

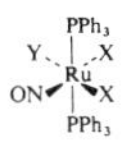

X = Y = Br

Ru-00693

As Ru-00693 with
X = I, Y = Cl

Ru-00694

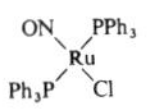

Ru-00695

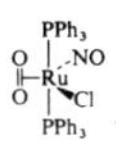

Ru-00696

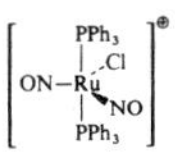

Ru-00697

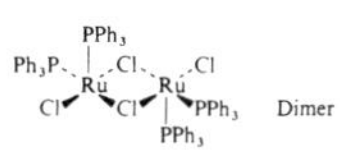

Ru-00698

As Ru-00693 with
X = Y = Cl

Ru-00699

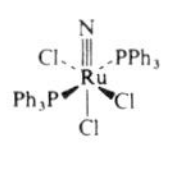

Ru-00700

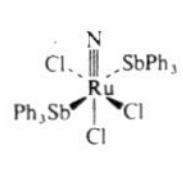

Ru-00701

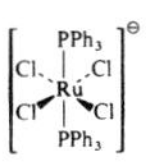

Ru-00702

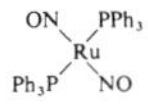

Ru-00703

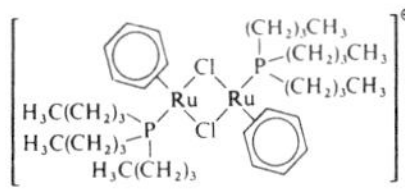

Ru-00704

$Ru[P(OEt)_3]_6^{\oplus\oplus}$

Ru-00705

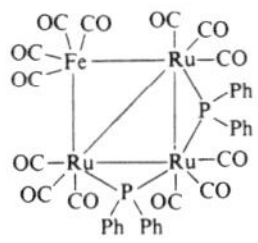

Ru-00706

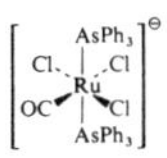

Ru-00707

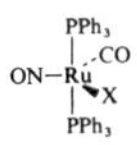

X = Br

Ru-00708

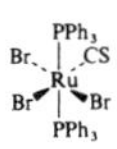

Ru-00709

As Ru-00708 with
X = Cl

Ru-00710

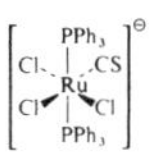

Ru-00711

As Ru-00708 with
X = I

Ru-00712

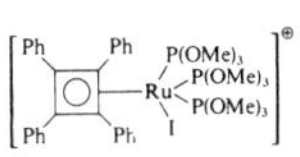

Ru-00713

As Ru-00688 with
$X = P(OMe)_3$

Ru-00714

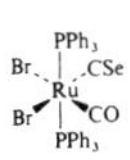

Ru-00715

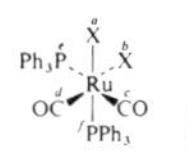

ab-dibromo-*cd*-dicarbonyl-*ef*-diphosphine-*form*

X = Br

Ru-00716

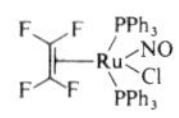

Ru-00717

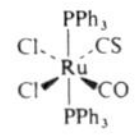

Ru-00718

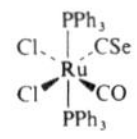

Ru-00719

As Ru-00716 with
X = Cl

Ru-00720

$Ru(CO)_2Cl_2[P(OPh)_3]_2$

Ru-00721

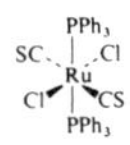

Ru-00722

As Ru-00716 with
X = I

Ru-00723

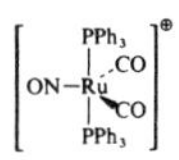

Ru-00724

As Ru-00726 with
X = S

Ru-00725

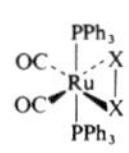

X = O

Ru-00726

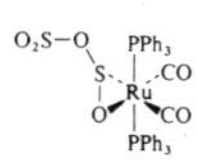

SO_2 complex

Ru-00727

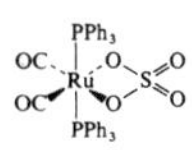

Ru-00728

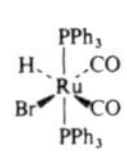

Ru-00729

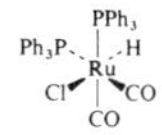

Ru-00730

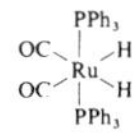

Ru-00731

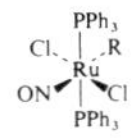

R = Ac

Ru-00732

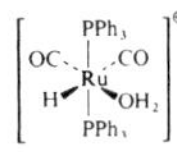

Ru-00733

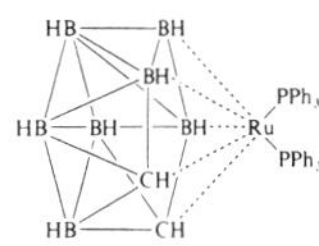

Ru-00734

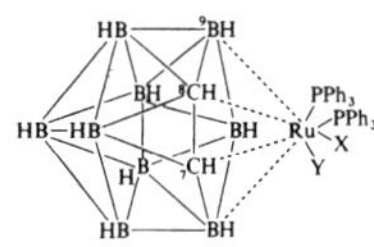

X = Cl, Y = H

Ru-00735

As Ru-00735 with
X = Y = H

Ru-00737

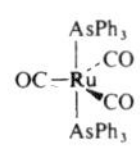

Ru-00739

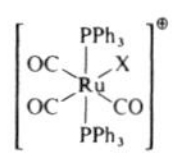

X = Cl

Ru-00740

As Ru-00740 with
X = I

Ru-00741

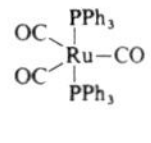

Ru-00742

$[(PhO)_3P]_2Ru(CO)_3$

Ru-00743

As Ru-00740 with
X = H

Ru-00744

$(H_3CCSe_2)RuCl(CO)(PPh_3)_2$

Ru-00745

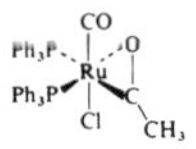

Ru-00746

$(H_3CCS_2)RuCl(CO)(PPh_3)_2$

Ru-00747

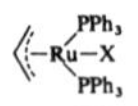

X = NO

Ru-00748

As Ru-00735 with
XY = CO

Ru-00749

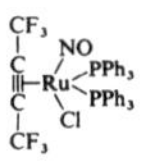

Ru-00751

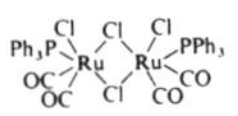

Ru-00752

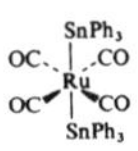

Ru-00753

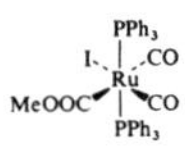

Ru-00754

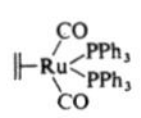

Ru-00755

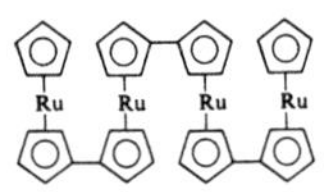

Ru-00756

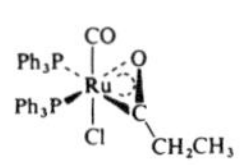

Ru-00757

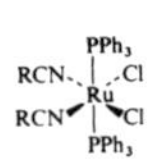

R = H_3C–

Ru-00758

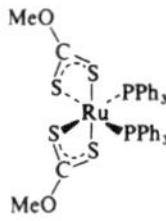

Ru-00759

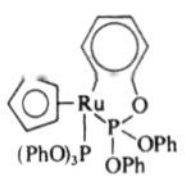

Ru-00760

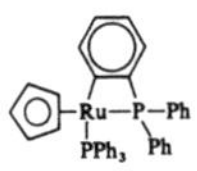

Ru-00761

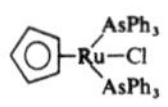

Ru-00762

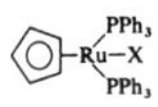

X = Br

Ru-00763

As Ru-00206 with
R = Ph

Ru-00764

As Ru-00763 with
X = Cl

Ru-00765

As Ru-00763 with
X = F

Ru-00766

As Ru-00763 with
X = I

Ru-00767

As Ru-00763 with
X = H

Ru-00768

As Ru-00748 with
X = H_3CCN—

Ru-00769

As Ru-00763 with
X = BH_4

Ru-00770

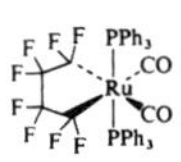

Ru-00771

As Ru-00763 with
X = —NCO

Ru-00772

As Ru-00763 with
X = CN

Ru-00773

As Ru-00763 with
X = —NCS

Ru-00774

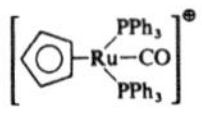

Ru-00775

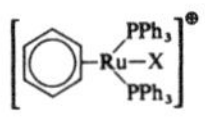

X = Cl

Ru-00776

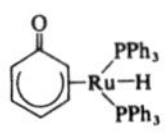

Ru-00777

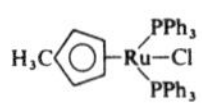

Ru-00778

As Ru-00776 with
X = H

Ru-00779

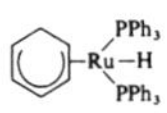

Ru-00780

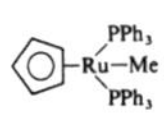

Ru-00781

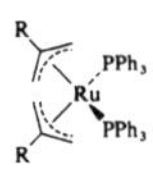

R = H

Ru-00782

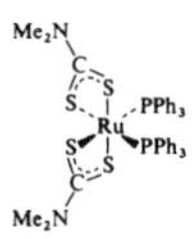

Ru-00783

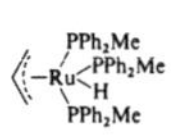

Ru-00784

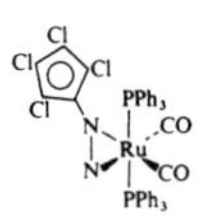

Ru-00785

As Ru-00732 with
R = COPh

Ru-00786

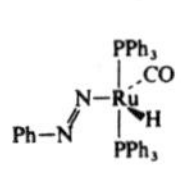

Ru-00787

As Ru-00732 with
R = $-CH_2Ph$

Ru-00788

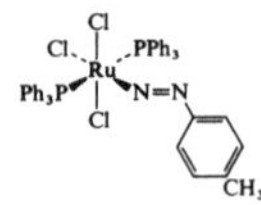

Ru-00789

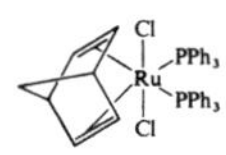

Ru-00790

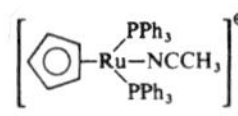

Ru-00791

As Ru-00763 with
X = –SC(S)OMe

Ru-00792

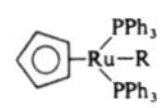

R = Et

Ru-00793

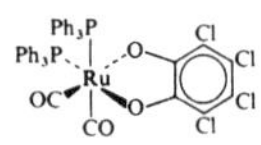

Ru-00794

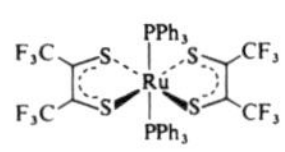

Ru-00795

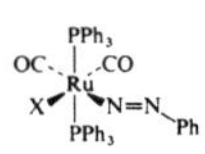

X = Br

Ru-00796

As Ru-00796 with
X = I

Ru-00797

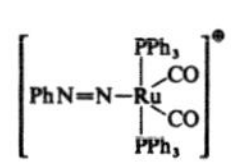

Ru-00798

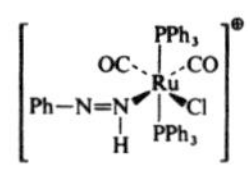

Ru-00799

As Ru-00796 with
X = H

Ru-00800

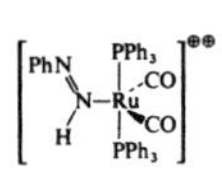

Ru-00801

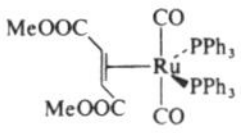

Ru-00802

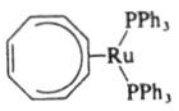

Ru-00803

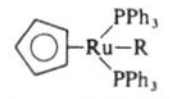

R = −C≡CCH$_3$

Ru-00804

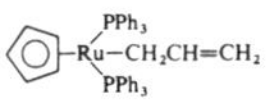

Ru-00805

As Ru-00793 with
R = $-CH_2CH_2CH_3$

Ru-00806

As Ru-00782 with
R = CH_3

Ru-00807

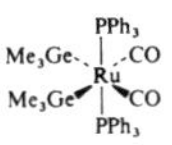

Ru-00808

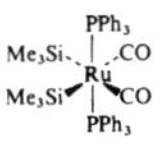

Ru-00809

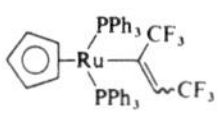

Ru-00810

As Ru-00793 with
R = $-(CH_2)_3CH_3$

Ru-00811

As Ru-00653 with
X = Y = PPh_3, Z = CO

Ru-00812

As Ru-00224 with
X = Y = PPh_3

Ru-00813

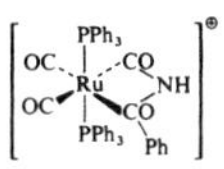

Ru-00814

As Ru-00763 with
X = C_6F_5

Ru-00815

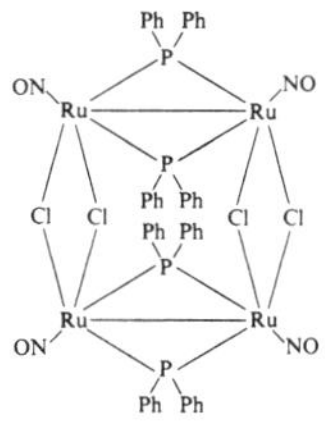

Ru-00816

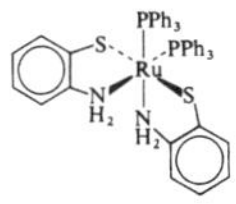

Ru-00817

As Ru-00804 with
R = $-CH_2Ph$

Ru-00818

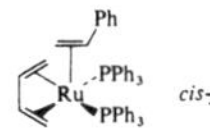

cis-form

Ru-00819

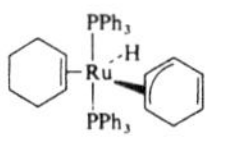

diapical-diphosphine-form

Ru-00820

As Ru-00804 with
R = −C≡CPh

Ru-00821

As Ru-00758 with
R = Ph

Ru-00822

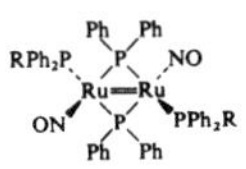

R = Me

Ru-00823

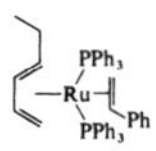

Ru-00824

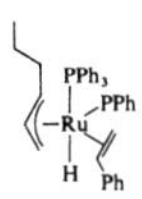

Ru-00825

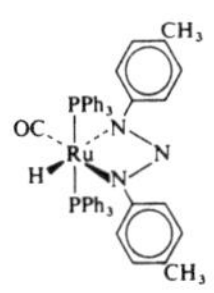

Ru-00826

$Ru_6(CO)_{14}(\mu\text{-}CO)(\mu_6\text{-}C)(PPh_3)_2$

Ru-00827

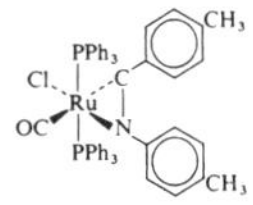

Ru-00828

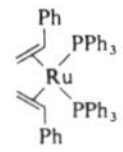

Ru-00829

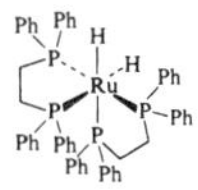

Ru-00830

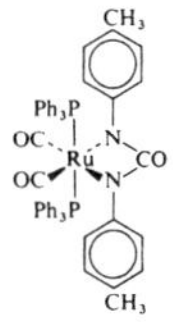

Ru-00831

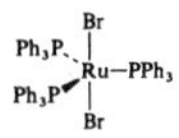

Ru-00832

$(Ph_3P)_3RuCl_2$

Ru-00833

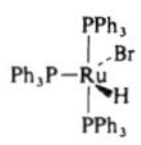

Ru-00834

$HRu(PPh_3)_3Cl$

Ru-00835

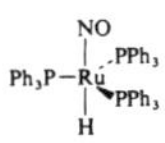

Ru-00836

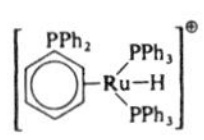

Ru-00837

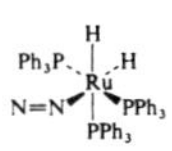

Ru-00838

$H_4Ru(PPh_3)_3$

Ru-00839

$[(PhO)_3P]_3Ru(CO)Br_2$

Ru-00840

$(Ph_3P)_3Ru(CO)Cl_2$

Ru-00841

$[(PhO)_3P]_3Ru(CO)Cl_2$

Ru-00842

$[(PhO)_3P]_3Ru(CO)I_2$

Ru-00843

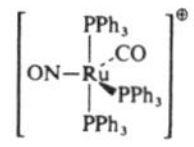

Ru-00844

$(Ph_3P)_3RuH(CO)Br$

Ru-00845

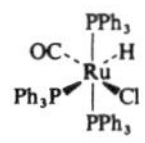

Ru-00846

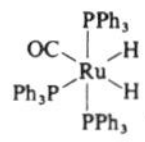

White isomer

Ru-00847

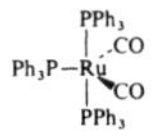

Ru-00848

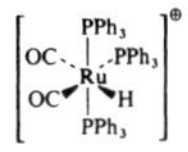

Ru-00849

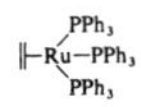

Ru-00850

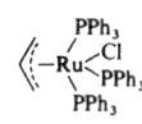

Ru-00851

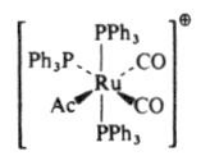

Ru-00852

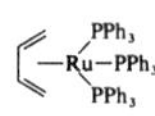

Ru-00853

As Ru-00823 with
R = Ph

Ru-00854

As Ru-00653 with
X = Y = Z = PPh_3

Ru-00855

$Ru_4H_4(CO)_9(PPh_3)_3$

Ru-00856

As Ru-00858 with
R = OPh

Ru-00857

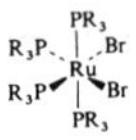

R = Ph

Ru-00858

$Ru(PPh_3)_4Cl_2$

Ru-00859

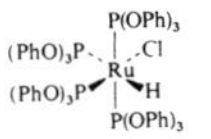

Ru-00860

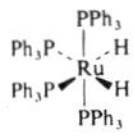

Ru-00861

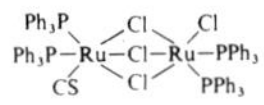

Ru-00862

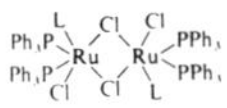

L = CO

Ru-00863

As Ru-00863 with
L = CS

Ru-00864

$Ru_4H_4(CO)_8(PPh_3)_4$

Ru-00865

$Cl_5Ru(NO)^{\ominus\ominus}$

Ru-00866

$Ru(PF_3)_5$

Ru-00867

$HRu(PF_3)_4^{\ominus}$

Ru-00868

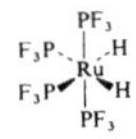

Ru-00869

$Ru(PF_3)_4^{\ominus\ominus}$

Ru-00870

CBr_3RuO **Ru-00001**

Tribromocarbonylruthenium, 8CI

[32073-81-7]

$Ru(CO)Br_3$

M 368.792

Deep-red. Compd. not isol. pure.

Colton, R. *et al*, *Aust. J. Chem.*, 1971, **24**, 903 (*ir, synth*)

$CBr_5ORu^{\ominus\ominus}$ **Ru-00002**

Pentabromocarbonylruthenate(2−), 9CI, 8CI

$Br_5RuCO^{\ominus\ominus}$

M 528.600 (ion)

Paramagnetic.

Di-Cs salt: [30191-89-0].

CBr_5Cs_2ORu M 794.411

Dark-blue cryst. or red solid.

Cleare, M.J. *et al*, *J. Chem. Soc. (A)*, 1970, 2788 (*synth*)
Colton, R. *et al*, *Aust. J. Chem.*, 1971, **24**, 903 (*synth, ir*)
Cleare, M.J. *et al*, *Spectrochim. Acta, Part A*, 1972, **28**, 2019 (*synth, ir, raman*)

CCl_3ORu **Ru-00003**

Carbonyltrichlororuthenium, 9CI, 8CI

[32073-80-6]

$Ru(CO)Cl_3$

M 235.439

Deep-red. Compd. not isol. pure because of ready reduction to $Ru(CO)(H_2O)Cl_2$.

Colton, R. *et al*, *Aust. J. Chem.*, 1971, **24**, 903 (*ir*)
Berch, M.L. *et al*, *J. Inorg. Nucl. Chem.*, 1973, **35**, 3763 (*synth*)

$CCl_5ORu^{\ominus\ominus}$ **Ru-00004**

Carbonylpentachlororuthenate(2−), 8CI

[51418-48-5]

$Cl_5RuCO^{\ominus\ominus}$

M 306.345 (ion)

Paramagnetic.

Di-K salt: [66115-12-6].

CCl_5K_2ORu M 384.542

Brick-red cryst. $+H_2O$ (HCl aq.).

Di-Cs salt:

CCl_5Cs_2ORu M 572.156

Red solid.

Cleare, M.J. *et al*, *J. Chem. Soc. (A)*, 1969, 372 (*synth, ir, raman*)
Colton, R. *et al*, *Aust. J. Chem.*, 1971, **24**, 903 (*synth, ir*)
Fergusson, J.E. *et al*, *Aust. J. Chem.*, 1978, **31**, 497 (*synth, ir*)
Gould, R.O. *et al*, *J. Chem. Res. (S)*, 1978, 254 (*struct*)

$CF_{12}OP_4Ru$ **Ru-00005**

Carbonyltetrakis(phosphorous trifluoride)ruthenium, 9CI

[36542-23-1]

$(F_3P)_4RuCO$

M 480.956

Yellow liq.

Udovich, C.A. *et al*, *J. Organomet. Chem.*, 1972, **36**, 355 (*synth, ir, nmr, ms*)

$CH_2Cl_4O_2Ru^{\ominus\ominus}$ **Ru-00006**

Aquacarbonyltetrachlororuthenate(2−)

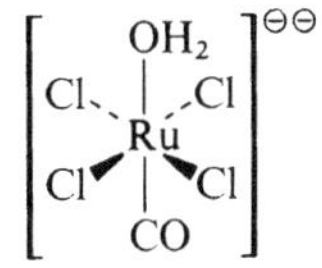

M 288.908 (ion)

Di-NH_4 salt: [14912-60-8].

$CH_{10}Cl_4N_2O_2Ru$ M 324.984

Pale-green cryst.

Di-Cs salt: [17568-72-8].

$CH_2Cl_4Cs_2O_2Ru$ M 554.718

Green solid. Dec. at ca. 250°. Air-stable.

Halpern, J. *et al*, *J. Am. Chem. Soc.*, 1966, **88**, 5142 (*synth, ir*)
Cleare, M.J. *et al*, *Chem. Ind. (London)*, 1967, 1705 (*synth*)
Colton, R. *et al*, *Aust. J. Chem.*, 1967, **22**, 2011; 1971, **24**, 903 (*synth, ir*)
Stanko, J.A. *et al*, *J. Am. Chem. Soc.*, 1970, **92**, 5580 (*cryst struct*)

CI_3ORu **Ru-00007**

Carbonyltriiodoruthenium, 8CI

[32073-82-8]

$Ru(CO)I_3$

M 509.794

Deep-red. Characterised spectroscopically.

Colton, R. *et al*, *Aust. J. Chem.*, 1969, **22**, 2011; 1971, **24**, 903 (*ir, synth*)

$C_2Br_2O_2Ru$ **Ru-00008**

Dibromodicarbonylruthenium, 9CI, 8CI

[16369-41-8]

$Ru(CO)_2Br_2$

M 316.899

Monomeric and polymeric forms known.

Polymer: [28986-71-2]. Cream solid.

Colton, R. *et al*, *Aust. J. Chem.*, 1967, **20**, 1283 (*synth, ir*)
Sandhu, S.S. *et al*, *Indian J. Chem.*, 1972, **10**, 946.

$C_2Br_4O_2Ru^{\ominus\ominus}$ **Ru-00009**

Tetrabromodicarbonylruthenate(2−), 8CI

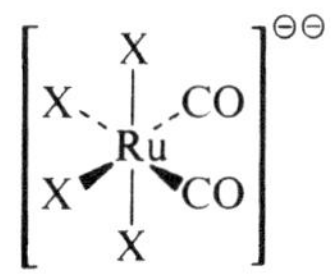

X = Br

M 476.707 (ion)

cis-form

Di-Cs salt: [16369-43-0].

$C_2Br_4Cs_2O_2Ru$ M 742.518

Deep-yellow cryst.

Cleare, M.J. *et al*, *J. Chem. Soc. (A)*, 1969, 372 (*synth, ir, raman*)
Colton, R. *et al*, *Aust. J. Chem.*, 1971, **24**, 903 (*ir*)

$C_2Cl_2O_2Ru$ **Ru-00010**

Dicarbonyldichlororuthenium, 9CI

$Ru(CO)_2Cl_2$

M 227.997
Monomeric and polymeric forms known.
Monomer: [16369-40-7]. Fine yellow cryst.

Cleare, M.J., *Platinum Metals Rev.*, 1967, **11**, 148 (*synth*)
Colton, R. *et al*, *Aust. J. Chem.*, 1967, **20**, 1283 (*synth*, *ir*)
Berch, M.L. *et al*, *J. Inorg. Nucl. Chem.*, 1973, **35**, 3763 (*synth*)

$C_2Cl_4O_2Ru^{\ominus\ominus}$ Ru-00011

Dicarbonyltetrachlororuthenate(2−), 9CI, 8CI
[60105-91-1]
As Tetrabromodicarbonylruthenate(2−), Ru-00009 with

X = Cl

M 298.903 (ion)
***cis*-form**
Di-Cs salt: [17439-05-3].
$C_2Cl_4Cs_2O_2Ru$ M 564.714
Bright-yellow cryst.
Bis(tetraphenylarsonium) salt: [51982-86-6].
$C_{50}H_{40}As_2Cl_4O_2Ru$ M 1065.590
Yellow cryst. (CH_2Cl_2/pet. ether).

Halpern, J. *et al*, *J. Am. Chem. Soc.*, 1966, **88**, 5142 (*synth*, *spectra*)
Cleare, M.J. *et al*, *J. Chem. Soc. (A)*, 1969, 372 (*synth*, *ir*, *raman*)
Colton, R. *et al*, *Aust. J. Chem.*, 1971, **24**, 903 (*synth*, *ir*)
Stephenson, T.A. *et al*, *J. Chem. Soc., Dalton Trans.*, 1973, 2112 (*synth*)

$C_2F_9O_2P_3Ru$ Ru-00012

Dicarbonyltris(phosphorous trifluoride)ruthenium, 9CI
[36430-00-9]

$(F_3P)_3Ru(CO)_2$

M 420.998
Light-yellow liq.

Udovich, C.A. *et al*, *J. Organomet. Chem.*, 1972, **36**, 355 (*synth*, *ir*, *nmr*, *ms*)

$C_2H_{17}N_5Ru^{\oplus\oplus}$ Ru-00013

Pentaammine(η^2-ethyne)ruthenium(2+)
η^2-Acetylene(pentaamine)ruthenium(2+)

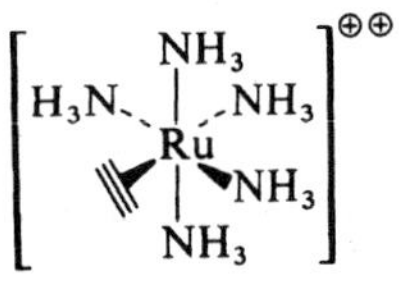

M 212.260 (ion)
Dithionate: [64799-85-5]. Brown microcryst.

Sullivan, B.P. *et al*, *J. Am. Chem. Soc.*, 1977, **99**, 7368 (*synth*)
Lehmann, H. *et al*, *J. Am. Chem. Soc.*, 1979, **101**, 6197 (*synth*, *uv*)

$C_2H_{19}N_5Ru^{\oplus\oplus}$ Ru-00014

Pentaamine(η^2-ethene)ruthenium(2+)
Pentammine(ethylene)ruthenium(2+)

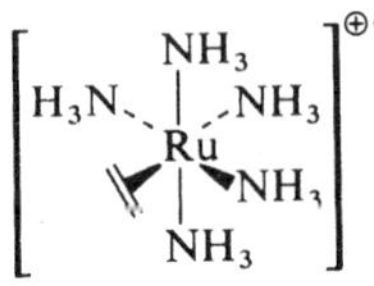

M 214.276 (ion)
Dithionate: [72098-73-8]. No phys. description recorded.

Lehmann, H. *et al*, *J. Am. Chem. Soc.*, 1979, **101**, 6197 (*synth*, *uv*)

$C_2I_2O_2Ru$ Ru-00015

Dicarbonyldiiodoruthenium
[16369-42-9]

$Ru(CO)_2I_2$

M 410.900
Monomeric and polymeric forms known.
Monomer: [28986-72-3]. Yellow.
Polymer: [29088-68-4]. Dark-orange-tan solid.

Colton, R. *et al*, *Aust. J. Chem.*, 1967, **20**, 1283 (*synth*, *ir*)
Abrahamson, H.B. *et al*, *J. Am. Chem. Soc.*, 1979, **101**, 4123 (*synth*)

$C_2I_4O_2Ru^{\ominus\ominus}$ Ru-00016

Dicarbonyltetraiodoruthenate(2−), 8CI
As Tetrabromodicarbonylruthenate(2−), Ru-00009 with

X = I

M 664.709 (ion)
***cis*-form**
Di-Cs salt: [16369-44-1].
$C_2Cs_2I_4O_2Ru$ M 930.520
Orange cryst.

Cleare, M.J. *et al*, *J. Chem. Soc. (A)*, 1969, 372 (*synth*, *ir*, *raman*)
Colton, R. *et al*, *Aust. J. Chem.*, 1971, **24**, 903 (*ir*)

$C_3Br_3O_3Ru^{\ominus}$ Ru-00017

Tribromotricarbonylruthenate(1−), 9CI, 8CI
[32423-82-8]

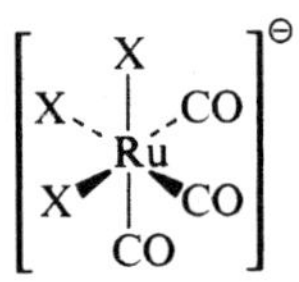

X = Br

M 424.813 (ion)
Cs salt: [22594-82-7].
$C_3Br_3CsO_3Ru$ M 557.719
V. pale-yellow cryst.

Cleare, M.J. *et al*, *J. Chem. Soc. (A)*, 1969, 372 (*synth*, *ir*, *raman*)
Colton, R. *et al*, *Aust. J. Chem.*, 1971, **24**, 903 (*synth*, *ir*)

$C_3Cl_3O_3Ru^{\ominus}$ Ru-00018

Tricarbonyltrichlororuthenate(1−), 9CI, 8CI
[32423-81-7]

As Tribromotricarbonylruthenate(1−), Ru-00017 with

X = Cl

M 291.460 (ion)

Cs salt: [22594-81-6].
$C_3Cl_3CsO_3Ru$ M 424.366
Pale-yellow cryst. (Me_2CO/pet. ether).

Cleare, M.J. *et al*, *J. Chem. Soc. (A)*, 1969, 372 (*synth, ir, raman*)
Colton, R. *et al*, *Aust. J. Chem.*, 1971, **24**, 903 (*synth, ir*)
Berch, M.L. *et al*, *J. Inorg. Nucl. Chem.*, 1973, **35**, 3763 (*synth*)
Teulon, P. *et al*, *Z. Anorg. Allg. Chem.*, 1981, **483**, 219 (*synth*)

$C_3F_3O_3Ru$ Ru-00019

Tricarbonyltrifluororuthenium, 9CI
[59559-91-0]

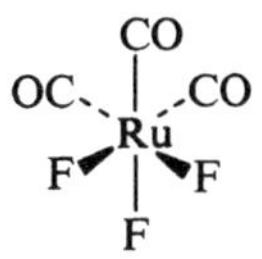

M 242.096
Buff, air-sensitive.

Hewitt, A.J. *et al*, *J. Chem. Soc., Dalton Trans.*, 1976, 579 (*prep, ir, esr*)

$C_3F_6O_3P_2Ru$ Ru-00020

Tricarbonylbis(phosphorous trifluoride)ruthenium, 9CI
[36429-99-9]

$(F_3P)_2Ru(CO)_3$

M 361.039
Light-yellow liq., identified spectroscopically.

Udovich, C.A. *et al*, *J. Organomet. Chem.*, 1972, **36**, 355 (*synth, ir*)

$C_3I_3O_3Ru^{\ominus}$ Ru-00021

Tricarbonyltriiodoruthenate(1−), 8CI
[32423-83-9]
As Tribromotricarbonylruthenate(1−), Ru-00017 with

X = I

M 565.815 (ion)

Cs salt: [22594-83-8].
$C_3CsI_3O_3Ru$ M 698.720
Yellow cryst.

Tetraethylammonium salt: [83721-25-9].
$C_{11}H_{20}I_3NO_3Ru$ M 696.067
Yellow cryst. (MeOH).

Cleare, M.J. *et al*, *J. Chem. Soc. (A)*, 1969, 372 (*synth, ir, raman*)
Colton, R. *et al*, *Aust. J. Chem.*, 1971, **24**, 903 (*synth, ir*)
Dombek, B.D. *et al*, *J. Am. Chem. Soc.*, 1981, **103**, 6508 (*synth, use*)
Dykstra, R.W. *et al*, *J. Magn. Reson.*, 1982, **46**, 338 (*nmr*)
Hidai, M. *et al*, *Organometallics*, 1983, **2**, 292 (*synth*)

$C_4Br_2O_4Ru$ Ru-00022

Dibromotetracarbonylruthenium, 9CI, 8CI
[15663-29-3]

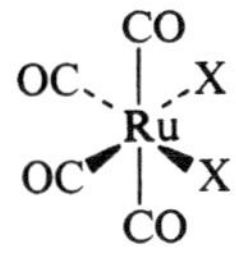

X = Br

M 372.920

cis-form [21439-16-7]
Yellow cryst.

Calderazzo, F. *et al*, *Inorg. Chem.*, 1967, **6**, 1220 (*synth*)
Cotton, J.D. *et al*, *J. Chem. Soc. (A)*, 1968, 2162 (*synth*)
Johnson, B.F.G. *et al*, *J. Chem. Soc. (A)*, 1969, 792 (*synth, ir*)
Benedetti, E. *et al*, *J. Organomet. Chem.*, 1972, **37**, 361 (*synth, ir*)

$C_4Cl_2Hg_2O_4Ru$ Ru-00023

Tetracarbonylbis(chloromercurio)ruthenium, 8CI

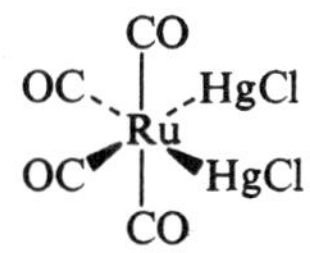

M 685.198

cis-form [21710-53-2]
Pale-yellow solid. Mp 170-5° dec.

Bradford, C.W. *et al*, *J. Chem. Soc. (A)*, 1968, 2456 (*synth, ir*)

$C_4Cl_2O_4Ru$ Ru-00024

Tetracarbonyldichlororuthenium, 8CI
As Dibromotetracarbonylruthenium, Ru-00022 with

X = Cl

M 284.018

cis-form [21475-91-2]
Yellow.

Allegra, G. *et al*, *Tetrahedron Lett.*, 1965, 4187 (*synth*)
Johnson, B.F.G. *et al*, *J. Chem. Soc. (A)*, 1969, 792 (*synth, ir*)
Bruce, M.I. *et al*, *J. Chem. Soc., Chem. Commun.*, 1970, 1325 (*synth, ir*)

$C_4Cl_6Ge_2O_4Ru$ Ru-00025

Tetracarbonylbis(trichlorogermyl)ruthenium, 9CI
[37834-28-9]

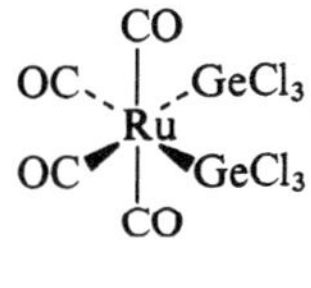

M 571.010

cis-form [34397-63-2]
White cryst. (CH_2Cl_2/hexane). Mp 90-100° dec.

Knox, S.A.R. *et al*, *J. Chem. Soc. (A)*, 1971, 2874 (*synth, ir*)
Ball, R. *et al*, *Inorg. Chem.*, 1972, **11**, 1806 (*cryst struct*)
Vancea, L. *et al*, *J. Am. Chem. Soc.*, 1976, **98**, 1407 (*cmr*)

$C_4F_3O_4PRu$ Ru-00026

Tetracarbonyl(phosphorous trifluoride)ruthenium, 9CI

[36429-98-8]

$F_3PRu(CO)_4$

M 301.081

Light-yellow liq.

Udovich, C.A. *et al*, *J. Organomet. Chem.*, 1972, **36**, 355 (*synth, ir*)

$C_4H_2O_4Ru$ Ru-00027

Tetracarbonyldihydroruthenium, 9CI, 8CI

[21029-23-2]

As Dibromotetracarbonylruthenium, Ru-00022 with

X = H

M 215.127

cis-form [42781-58-8]

V. volatile colourless liq. Mp ca. −63°.

Cotton, J.D. *et al*, *J. Chem. Soc.* (*A*), 1968, 2162 (*synth, ir*)
Whyman, R., *J. Organomet. Chem.*, 1973, **56**, 339 (*ir*)
Bradley, G.F. *et al*, *J. Chem. Soc., Chem. Commun.*, 1975, 325 (*struct, raman*)

$C_4I_2O_4Ru$ Ru-00028

Tetracarbonyldiiodoruthenium

[15663-30-6]

As Dibromotetracarbonylruthenium, Ru-00022 with

X = I

M 466.921

cis-form [18475-75-7]

Yellow cryst. *trans-form* not found (cf. Os analogue).

Calderazzo, F. *et al*, *Inorg. Chem.*, 1967, **6**, 1220 (*synth, ir*)
Johnson, B.F.G. *et al*, *J. Chem. Soc.* (*A*), 1969, 792 (*synth, ir*)
Pankowski, M. *et al*, *J. Organomet. Chem.*, 1969, **19**, 393 (*synth, ir*)
Vancea, L. *et al*, *J. Organomet. Chem.*, 1977, **134**, 219 (*synth, cmr*)

$C_4O_4Ru^{\ominus\ominus}$ Ru-00029

Tetracarbonylruthenate(2−), 9CI

$Ru(CO)_4^{\ominus\ominus}$

M 213.112 (ion)

Di-Na salt: [57398-60-4].

$C_4Na_2O_4Ru$ M 259.091

Pale-fawn solid. V. air-sensitive.

Cotton, J.D. *et al*, *J. Chem. Soc.* (*A*), 1968, 2162 (*synth, ir*)
Green, M. *et al*, *J. Chem. Soc., Dalton Trans.*, 1975, 939 (*use*)

The symbol ▷ in Entries highlights hazard or toxicity information

$C_5H_{11}B_9O_3Ru$ Ru-00030

Tricarbonyl[(7,8,9,10,11-η)undecahydro-7,8-dicarbaundecaborato(2−)]ruthenium, 9CI

[56465-06-6]

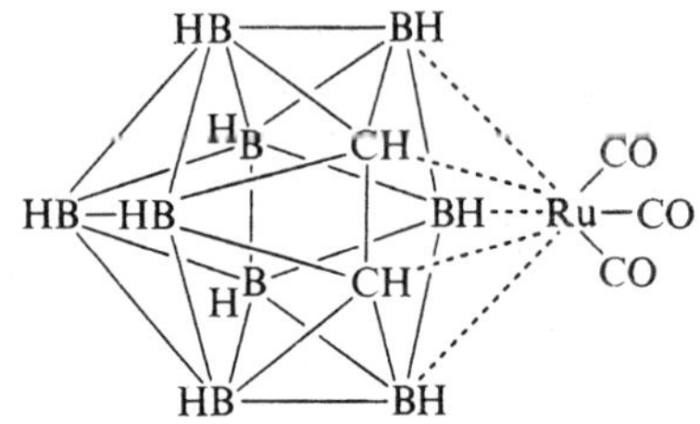

M 317.500

Yellow cryst. + $\frac{1}{2}C_6H_6$ (C_6H_6/cyclohexane). Mp >350°.

Siedle, A.R., *J. Organomet. Chem.*, 1975, **90**, 249 (*synth, ir, uv, nmr, pmr, ms*)

C_5O_5Ru Ru-00031

Pentacarbonylruthenium

Ruthenium pentacarbonyl

[16406-48-7]

$Ru(CO)_5$

M 241.122

Colourless, light-sensitive liq. Mp −17° to −16°.

Rushman, P. *et al*, *Organometallics*, 1983, **2**, 693 (*synth, ir, cmr*)

$C_6Br_4O_6Ru_2$ Ru-00032

Di-μ-bromodibromohexacarbonyldiruthenium, 10CI, 9CI

Tetrabromohexacarbonyldiruthenium

[22594-70-3]

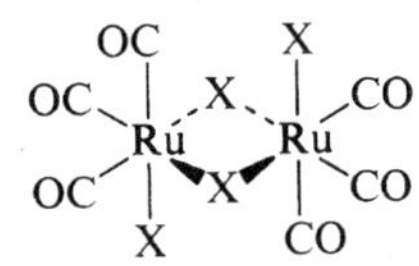

X = Br

M 689.818

Pale-yellow cryst. ($CHCl_3$/pet. ether). $Bp_{0.001}$ 150° subl. Forms stable yellow crystalline THF solvate. Two isomers present, separable chromatographically.

Monomer: Dibromotricarbonylruthenium.

$C_3Br_2O_3Ru$ M 344.909

Unknown.

Bruce, M.I., *Int. J. Mass Spectrom. Ion Phys.*, 1968, **1**, 141 (*ms*)
Merlino, S. *et al*, *Acta Crystallogr., Sect. B*, 1968, **24**, 424 (*struct*)
Braca, G. *et al*, *Chim. Ind.* (*Milan*), 1969, **51**, 1245 (*synth*)
Cleare, M.J. *et al*, *J. Chem. Soc.* (*A*), 1969, 372 (*synth, ir, raman*)
Johnson, B.F.G. *et al*, *J. Chem. Soc.* (*A*), 1969, 792 (*synth, ir, ms*)
Trovati, A. *et al*, *Inorg. Chem.*, 1970, **9**, 671 (*synth, ir*)
Colton, R. *et al*, *Aust. J. Chem.*, 1971, **24**, 903 (*synth, ir*)
Benedetti, E. *et al*, *J. Organomet. Chem.*, 1972, **37**, 361 (*ir*)

$C_6Cl_4O_6Ru_2$ Ru-00033

Hexacarbonyldi-μ-chlorodichlorodiruthenium, 10CI, 9CI, 8CI

Hexacarbonyltetrachlorodiruthenium

[22594-69-0]

As Di-μ-bromodibromohexacarbonyldiruthenium, Ru-00032 with

X = Cl

M 512.014

White solid; yellow when impure. Sl. sol. $CHCl_3$; sol. MeOH, THF. Turns orange-brown at 215°; dec. >315°. Two isomers present, separable chromatographically. Chloride bridges readily cleaved by THF, PPh_3 etc.

Trichlorostibine adduct: [79056-45-4].
$C_6Cl_7O_6Ru_2Sb$ M 740.123
Pale-yellow cryst. (CH_2Cl_2). Stable in dry air. Dec. by moisture.

THF adduct: Valuable synthetic intermed.

Monomer: Tricarbonyldichlororuthenium.
$C_3Cl_2O_3Ru$ M 256.007
Unknown.

Merlino, S. *et al*, *Acta Crystallogr., Sect. B*, 1965, **24**, 424 (*struct*)
Bruce, M.I. *et al*, *J. Chem. Soc. (A)*, 1967, 1238 (*synth, ir, use*)
Bruce, M.I., *Int. J. Mass Spectrom. Ion Phys.*, 1968, **1**, 141 (*ms*)
Cotton, J.D. *et al*, *J. Chem. Soc. (A)*, 1968, 2162 (*synth, isom, ir*)
Benedetti, E. *et al*, *J. Organomet. Chem.*, 1972, **37**, 361 (*ir*)
Inorg. Synth., 1976, **16**, 51 (*synth*)
Teulon, P. *et al*, *J. Organomet. Chem.*, 1981, **214**, 391 (*struct, deriv*)

$C_6H_5BrO_3Ru$ Ru-00034

Bromotricarbonyl(η^3-2-propenyl)ruthenium, 9CI

π-Allylbromotricarbonylruthenium

[31781-74-5]

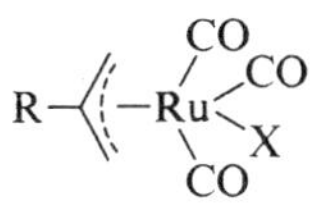

X = Br, R = H

M 306.078

Yellow cryst. (2,2,4-trimethylpentane at −50°). Mp 89-91°.

Sbrana, G. *et al*, *J. Organomet. Chem.*, 1968, **13**, 240 (*synth, ir, pmr*)
Sbrana, G. *et al*, *J. Chem. Soc., Dalton Trans.*, 1975, 754 (*props*)
Carlini, C. *et al*, *J. Mol. Catal.*, 1977, **2**, 379 (*props*)

$C_6H_5ClO_3Ru$ Ru-00035

Tricarbonylchloro(η^3-2-propenyl)ruthenium, 9CI

π-Allyltricarbonylchlororuthenium, 8CI

[31781-73-4]

As Bromotricarbonyl(η^3-2-propenyl)ruthenium, Ru-00034 with

X = Cl, R = H

M 261.627

Pale-yellow cryst. (2,2,4-trimethylpentane at −50°). Mp 75-7°.

Sbrana, G. *et al*, *J. Organomet. Chem.*, 1968, **13**, 240 (*synth, ir, pmr*)
Carlini, C. *et al*, *J. Mol. Catal.*, 1977, **2**, 379 (*synth*)
Merlino, S. *et al*, *Inorg. Chim. Acta*, 1978, **27**, 233 (*props*)

$C_6H_5IO_3Ru$ Ru-00036

Tricarbonyliodo(η^3-2-propenyl)ruthenium

π-Allyltricarbonyliodoruthenium

[31781-75-6]

As Bromotricarbonyl(η^3-2-propenyl)ruthenium, Ru-00034 with

X = I, R = H

M 353.078

Orange cryst. (2,2,4-trimethylpentane at −50°). Mp 73-4°.

Sbrana, G. *et al*, *J. Organomet. Chem.*, 1968, **13**, 240 (*synth, ir, pmr*)

$C_6H_6Cl_2N_2O_2Ru$ Ru-00037

Bis(acetonitrile)dicarbonyldichlororuthenium, 9CI

[30066-39-8]

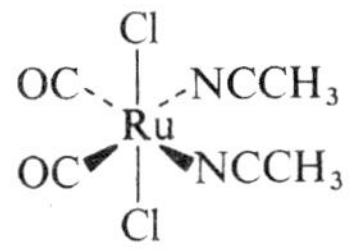

M 310.102

Yellow cryst. Mp 210° dec.

Wilkinson, G. *et al*, *J. Chem. Soc. (A)*, 1970, 2765 (*synth, ir*)
Newton, W.E. *et al*, *Inorg. Chim. Acta*, 1973, **7**, 349 (*synth, ir*)

$C_6H_6Cl_2Ru$ Ru-00038

(η^6-Benzene)dichlororuthenium, 10CI, 9CI

[51831-98-2]

As Bis(η^6-benzene)di-μ-bromodibromodiruthenium, Ru-00237 with

X = Cl

M 250.089

Dimeric or polymeric.

Dimer: [35796-48-6]. *Bis(η^6-benzene)di-μ-chlorodichlorodiruthenium, 9CI. Bis(η^6-benzene)tetrachlorodiruthenium.*
$C_{12}H_{12}Cl_4Ru_2$ M 500.179
Hydrogenation catalyst. Red solid. Sol. DMSO, insol. most solvs. Mp 249° dec. Halide bridges cleaved by donor solvs., eg. MeCN, DMSO.

Polymer: [12628-24-9]. Brown solid. Mp 242°.

DMSO complex: [69058-50-0]. *(η^6-Benzene)dichloro[sulfinylbis[methane]-O]ruthenium, 10CI. (Benzene)dichloro(methylsulfoxide)ruthenium, 8CI.*
$C_8H_{12}Cl_2ORuS$ M 328.218
Formed when the parent compd. is diss. in DMSO. Hydrogenation catalyst precursor. Red needles (EtOH aq.). Mp >180° dec.

Ogata, I. *et al*, *Tetrahedron Lett.*, 1970, 3011.
Zelonka, R.A. *et al*, *Can. J. Chem.*, 1972, **50**, 3063 (*synth, ir, pmr*)
Bennett, M.A. *et al*, *J. Chem. Soc., Dalton Trans.*, 1974, 233 (*synth, ir, pmr*)
Games, D.E. *et al*, *J. Organomet. Chem.*, 1980, **193**, 229 (*ms*)

$C_6H_6Cl_3Ru^{\ominus}$ Ru-00039

(η^6-Benzene)trichlororuthenate(1−), 10CI, 9CI

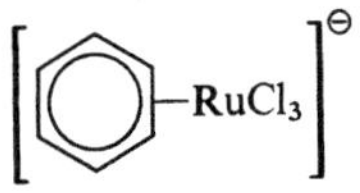

M 285.542 (ion)

Cs salt: [60829-94-9].
$C_6H_6Cl_3CsRu$ M 418.448
Orange solid. Mp 270° dec.

Robertson, D.R. *et al*, *J. Organomet. Chem.*, 1978, **162**, 121 (*synth, ir, pmr*)

$C_6H_8F_9P_3Ru$ Ru-00040

[(1,2,3,4-η)-1,3-Cyclohexadiene]tris(phosphorous trifluoride)-ruthenium

(η^4-1,3-Cyclohexadiene)tris(trifluorophosphine)ruthenium

[76786-24-8]

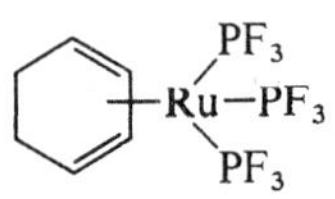

M 445.106

Off-white, glassy solid.

Al-Ohaly, A.-R. *et al*, *J. Organomet. Chem.*, 1980, **202**, 297 (*synth, ir, pmr, nmr*)

$C_6H_{12}O_3Ru^{\oplus\oplus}$ Ru-00041

Triaqua(η^6-benzene)ruthenium(2+), 10CI

[75701-06-3]

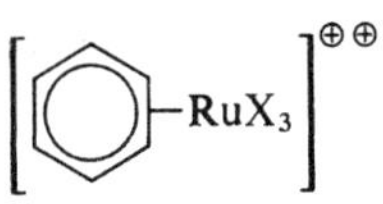

X = OH_2

M 233.229 (ion)

Intermed. for (η^6-benzene) complexes.

Robertson, D.R. *et al*, *J. Organomet. Chem.*, 1978, **157**, C47 (*synth*)
Hung, Y. *et al*, *Inorg. Chem.*, 1981, **20**, 457 (*spectra*)

$C_6H_{15}N_3Ru^{\oplus\oplus}$ Ru-00042

Triammine(η^6-benzene)ruthenium(2+), 10CI

As Triaqua(η^6-benzene)ruthenium(2+), Ru-00041 with

X = NH_3

M 230.275 (ion)

Bis(hexafluorophosphate): [68193-71-5].
$C_6H_{15}F_{12}N_3P_2Ru$ M 520.203
Yellow.

Robertson, D.R. *et al*, *J. Organomet. Chem.*, 1978, **157**, C47 (*synth*)

$C_6I_4O_6Ru_2$ Ru-00043

Hexacarbonyldi-μ-iododiiododiruthenium, 9CI

Hexacarbonyltetraiododiruthenium

[32423-80-6]

As Di-μ-bromodibromohexacarbonyldiruthenium, Ru-00032 with

X = I

M 877.820

Pale-yellow cryst. (CH_2Cl_2/hexane). Two isomers present, separable chromatographically.

Monomer: Tricarbonyldiiodoruthenium.
$C_3I_2O_3Ru$ M 438.910
Unknown.

Bruce, M.I. *et al*, *Int. J. Mass Spectrom. Ion Phys.*, 1968, **1**, 141 (*ms*)
Bruce, M.I. *et al*, *J. Organomet. Chem.*, 1968, **13**, 227 (*synth, ir*)
Cotton, J.D. *et al*, *J. Chem. Soc.* (*A*), 1968, 2162 (*synth, ir, isom*)
Johnson, B.F.G. *et al*, *J. Chem. Soc.* (*A*), 1969, 792 (*synth, ir, ms*)
Trovati, A. *et al*, *Inorg. Chem.*, 1970, **9**, 671 (*synth, ir*)

$C_7H_3NO_4Ru$ Ru-00044

Tetracarbonyl[(2,3-η)-2-propenenitrile]ruthenium

(Acrylonitrile)tetracarbonylruthenium. Tetracarbonyl-(cyanoethylene)ruthenium

[78319-39-8]

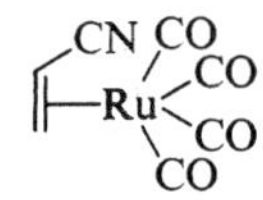

M 266.175

White cryst. (hexane containing free olefin), unstable in soln.

Grevels, F.W. *et al*, *J. Am. Chem. Soc.*, 1981, **103**, 4069 (*synth, ir*)

$C_7H_4O_3Ru$ Ru-00045

Tricarbonyl(η^4-1,3-cyclobutadiene)ruthenium

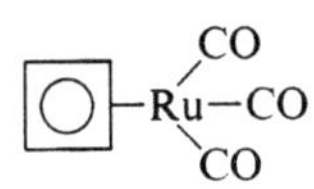

M 237.177

Phys. description not recorded. Stable for prolonged periods in an inert atmosphere.

Amiet, R.G. *et al*, *J. Chem. Soc., Chem. Commun.*, 1967, 1208 (*synth, pmr, ir*)

$C_7H_5BrO_2Ru$ Ru-00046

Bromodicarbonyl(η^5-2,4-cyclopentadien-1-yl)ruthenium, 9CI

η^5-Cyclopentadienylbromodicarbonylruthenium. Cyclopentadienyldicarbonylruthenium bromide

[32613-75-5]

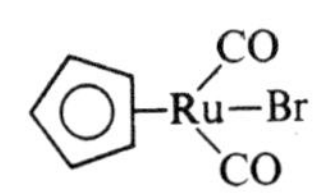

M 302.089

Yellow cryst. (CH_2Cl_2/pet. ether). Mp 94.5° (88.5-88.8°).

Bruce, M.I., *Int. J. Mass Spectrom. Ion Phys.*, 1968, **1**, 141 (*ms*)
Haines, R.J. *et al*, *J. Chem. Soc., Dalton Trans.*, 1972, 944 (*synth, ir, pmr*)
Gibson, D.H. *et al*, *J. Organomet. Chem.*, 1981, **208**, 89 (*synth*)

$C_7H_5ClO_2Ru$ Ru-00047

Dicarbonylchloro(η^5-2,4-cyclopentadien-1-yl)ruthenium, 9CI

Cyclopentadienyldicarbonylchlororuthenium. η^5-Cyclopentadienyldicarbonylruthenium chloride

[32611-12-4]

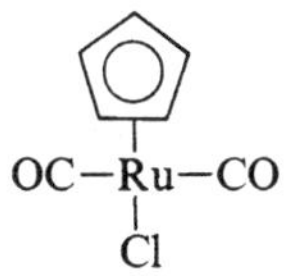

M 257.638

Pale-yellow cryst. (pet. ether/Et_2O). Mp 102-4°.

Blackmore, T. *et al*, *J. Chem. Soc.* (*A*), 1968, 2931 (*synth, ir, pmr*)
Bruce, M.I., *Int. J. Mass Spectrom. Ion Phys.*, 1968, **1**, 141 (*ms*)
Wnuk, T.A. *et al*, *Inorg. Chem.*, 1977, **16**, 1173 (*synth*)

$C_7H_5Cl_3O_2RuSi$ Ru-00048

Dicarbonyl(η^5-2,4-cyclopentadien-1-yl)(trichlorosilyl)ruthenium

(Dicarbonyl-π-cyclopentadienylruthenium)trichlorosilane

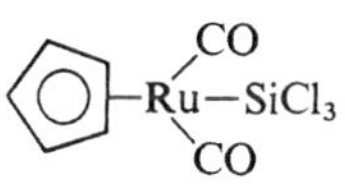

M 356.630

White cryst. Mp 60-3°. $Bp_{0.1}$ 70° subl.

Blackmore, T. *et al*, *J. Chem. Soc.* (*A*), 1968, 2931 (*synth, ir, ms*)

$C_7H_5Cl_3O_2RuSn$ Ru-00049

Dicarbonyl(η^5-2,4-cyclopentadien-1-yl)(trichlorostannyl)ruthenium

(Dicarbonyl-π-cyclopentadienylruthenium)trichlorostannane. (Cyclopentadienyldicarbonylruthenium)tin trichloride

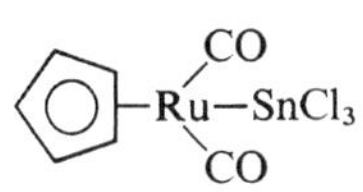

M 447.234

Mustard-yellow cryst. (C_6H_6). Mp 175-6°.

Blackmore, T. *et al*, *J. Chem. Soc.* (*A*), 1968, 2931 (*synth, ir, ms*)

C_7H_5IORuS Ru-00050

(Carbonothioyl)carbonyl(η^5-2,4-cyclopentadien-1-yl)iodoruthenium, 10CI

[61930-82-3]

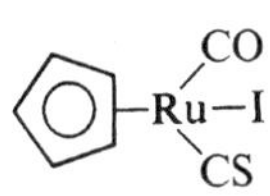

M 365.150

Dark red-black.

Wnuk, T.A. *et al*, *Inorg. Chem.*, 1977, **16**, 1173 (*synth, ir, pmr, ms*)

$C_7H_5IO_2Ru$ Ru-00051

Dicarbonyl(η^5-2,4-cyclopentadien-1-yl)iodoruthenium, 9CI

η^5-Cyclopentadienyldicarbonyliodoruthenium. Cyclopentadienyldicarbonylruthenium iodide

[31781-83-6]

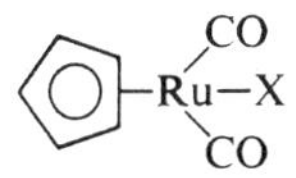

X = I

M 349.090

Orange cryst. (CH_2Cl_2/pet. ether). Mp 121-2°. Bp_1 100° subl.

Davison, A. *et al*, *J. Chem. Soc.*, 1963, 1133 (*synth*)
Bruce, M.I., *Int. J. Mass Spectrom. Ion Phys.*, 1968, **1**, 141 (*ms*)
Kruse, A.E. *et al*, *J. Organomet. Chem.*, 1970, **24**, 231 (*synth, ms, ir*)
Haines, R.J. *et al*, *J. Chem. Soc., Dalton Trans.*, 1972, 944 (*synth, ir, pmr*)

$C_7H_5O_2Ru^{\ominus}$ Ru-00052

Dicarbonyl(η^5-2,4-cyclopentadien-1-yl)ruthenate(1−), 9CI

(η^5-Cyclopentadienyl)dicarbonylruthenium(1−)

[12128-15-3]

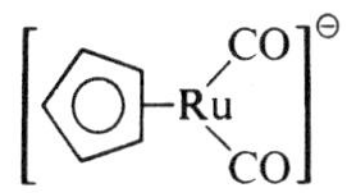

M 222.185 (ion)

Prep. and used *in situ*.

Na salt: [42802-20-0].
$C_7H_5NaO_2Ru$ M 245.175
Not isol.

Dessy, R.E. *et al*, *J. Am. Chem. Soc.*, 1966, **88**, 5121 (*synth*)
Blackmore, T. *et al*, *J. Chem. Soc.* (*A*), 1968, 2158 (*synth*)
Gibson, D.H. *et al*, *J. Organomet. Chem.*, 1981, **215**, 379 (*synth, use*)

$C_7H_6Cl_6Ge_2ORu$ Ru-00053

(η^6-Benzene)carbonylbis(trichlorogermyl)ruthenium, 9CI

[54932-36-4]

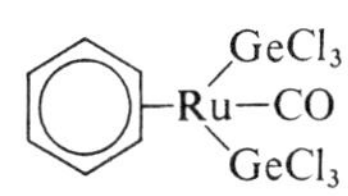

M 565.092

Pale-yellow needles (C_6H_6). Mp >300°.

Chan, L.Y.Y. *et al*, *Inorg. Chem.*, 1975, **14**, 1778 (*struct*)
Pomeroy, R.K. *et al*, *Can. J. Chem.*, 1975, **53**, 2985 (*synth, ir, pmr*)

$C_7H_6Cl_6ORuSi_2$ Ru-00054

(η^6-Benzene)carbonylbis(trichlorosilyl)ruthenium, 9CI

[58369-31-6]

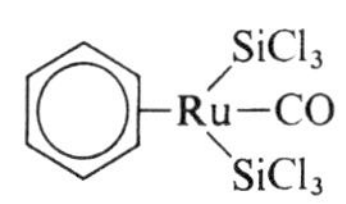

M 476.083

White cryst. (CH_2Cl_2/heptane). Mp 203° dec.

Pomeroy, R.K. *et al*, *Can. J. Chem.*, 1975, **53**, 2985 (*synth, ir, pmr*)

$C_7H_6O_2Ru$ Ru-00055

Dicarbonyl(η^5-2,4-cyclopentadien-1-yl)hydroruthenium

Dicarbonyl-π-cyclopentadienylhydridoruthenium. Hydrodicarbonyl-π-cyclopentadienylruthenium. Cyclopentadienyldicarbonylruthenium hydride

[57349-59-4]

As Dicarbonyl(η^5-2,4-cyclopentadien-1-yl)-iodoruthenium, Ru-00051 with

X = H

M 223.193

Colourless, v. air-sensitive volatile liq. with obnoxious odour.

Davison, A. *et al*, *J. Chem. Soc.*, 1963, 1133 (*synth, ir, pmr*)
Humphries, A.P. *et al*, *J. Chem. Soc., Dalton Trans.*, 1975, 1710 (*synth*)
Gibson, D.H. *et al*, *J. Organomet. Chem.*, 1981, **206**, C17 (*synth*)

$C_7H_6O_3Ru$ Ru-00056

(η^4-Butadiene)tricarbonylruthenium, 10CI

[62883-45-8]

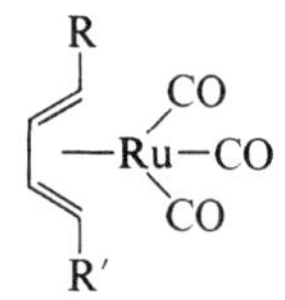

R = R′ = H

M 239.193

Colourless oil. Mp ca. 10°.

Ruh, S. *et al*, *J. Organomet. Chem.*, 1977, **127**, C59 (*synth, ms, ir, pmr*)
Zobl-Ruh, S. *et al*, *Helv. Chim. Acta*, 1980, **63**, 773 (*cmr*)

$C_7H_7ClO_3Ru$ Ru-00057

Tricarbonylchloro[(1,2,3-η)-2-methyl-2-propenyl]ruthenium

Tricarbonylchloro(2-methyl-π-allyl)ruthenium

[31781-76-7]

As Bromotricarbonyl(η^3-2-propenyl)ruthenium, Ru-00034 with

X = Cl, R = CH_3

M 275.654

Can be used in conjunction with iodide promotors as a catalyst for carbonylation and homologation. Pale-yellow cryst. (2,2,4-trimethylpentane at −50°). Mp 94-6°.

Sbrana, G. *et al*, *J. Organomet. Chem.*, 1968, **13**, 240 (*synth, ir, pmr*)

$C_7H_8Cl_2Ru$ Ru-00058

[(2,3,5,6-η)-Bicyclo[2.2.1]hepta-2,5-diene]dichlororuthenium, 9CI

Dichloro(norbornadiene)ruthenium

[48107-17-1]

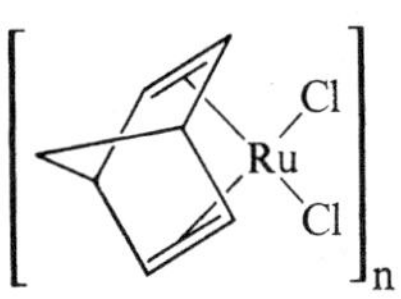

M 264.116

Polymeric. Red solid. Insol. H_2O, all common solvs. Diamagnetic.

Abel, E.W. *et al*, *J. Chem. Soc.*, 1959, 3178 (*synth, ir*)
Schrock, R.R. *et al*, *J. Chem. Soc., Dalton Trans.*, 1974, 951.
Sullivan, B.P. *et al*, *J. Am. Chem. Soc.*, 1977, **99**, 7368.

$C_7H_8Cl_2Ru$ Ru-00059

Dichloro[(1,2,3,4-η)-1,3,5-cycloheptatriene]ruthenium, 9CI

[54866-29-4]

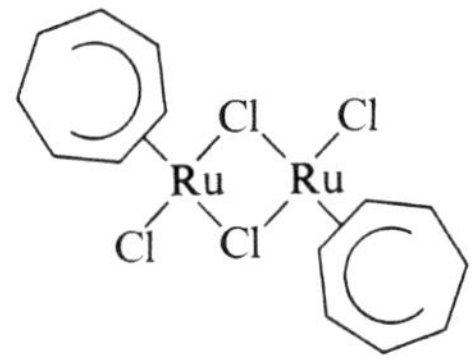

M 264.116

Chloride bridged dimer. Pmr spectrum inconsistent with 1,3-bound olefin, in spite of 9CI name.

Dimer: [63630-81-9]. *Di-μ-chlorodichlorobis(1,3,5-cycloheptatriene)diruthenium. Tetrachlorobis(1,3,5-cycloheptatriene)diruthenium.*
$C_{14}H_{16}Cl_4Ru_2$ M 528.232
Brown solid. Sol. CF_3COOH, insol. $CHCl_3$, C_6H_6, sol. DMSO with dec.

Winkhaus, G. *et al*, *J. Organomet. Chem.*, 1967, **7**, 487 (*synth, ir, pmr*)
Müller, J. *et al*, *Chem. Ber.*, 1975, **108**, 273 (*props*)
Johnson, B.F.G. *et al*, *J. Chem. Soc., Dalton Trans.*, 1977, 719 (*synth, struct, pmr*)

$C_7H_8NO_2Ru^{\oplus}$ Ru-00060

Amminedicarbonyl(η^5-2,4-cyclopentadien-1-yl)ruthenium(1+)

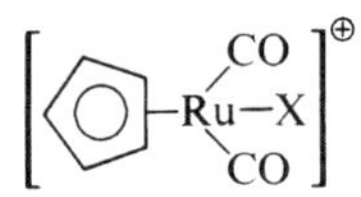

X = NH_3

M 239.216 (ion)

Hexafluorophosphate: [72763-42-9].
$C_7H_8F_6NO_2PRu$ M 384.180
Colourless cryst. (Me_2CO/Et_2O).

Tetraphenylborate: [68663-43-4].
$C_{31}H_{28}BNO_2Ru$ M 558.448
Cryst. (Me_2CO/Et_2O). Mp 225° dec.

Jungbauer, A. *et al*, *Z. Naturforsch., B*, 1978, **33**, 1083, **34**, 1477 (*synth, ir, pmr*)

$C_7H_9BrO_4RuSi$ Ru-00061

Bromotetracarbonyl(trimethylsilyl)ruthenium, 8CI

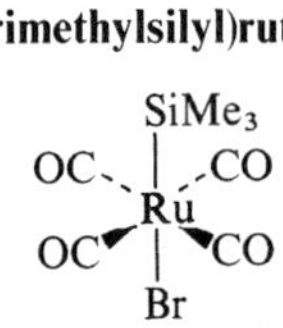

M 366.205

trans-form [30848-49-8]

V. unstable yellow cryst.

Stone, F.G.A. *et al*, *J. Chem. Soc. (A)*, 1971, 458 (*synth, ir, pmr*)

$C_7H_9GeIO_4Ru$ Ru-00062

Tetracarbonyliodo(trimethylgermyl)ruthenium, 8CI

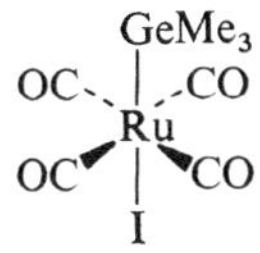

M 457.710

***trans*-form** [30846-57-2]
Orange solid. Mp 63-5°.

Ash, M.J. *et al*, *J. Chem. Soc. (A)*, 1971, 458 (*synth, ir, pmr*)

$C_7H_9IO_4RuSi$ Ru-00063

Tetracarbonyliodo(trimethylsilyl)ruthenium, 9CI, 8CI

[30848-50-1]

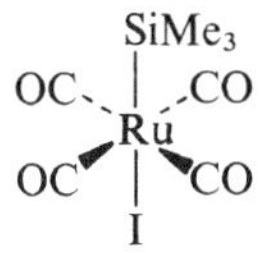

M 413.206

***trans*-form** [49564-81-0]
Yellow cryst. Mp 85°.

Stone, F.G.A. *et al*, *J. Chem. Soc. (A)*, 1971, 458 (*synth, ir, pmr*)

$C_7H_9IO_4RuSn$ Ru-00064

Tetracarbonyliodo(trimethylstannyl)ruthenium, 8CI

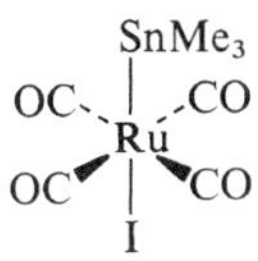

M 503.810

***trans*-form** [30846-58-3]
Orange. Not obt. pure.

Stone, F.G.A. *et al*, *J. Chem. Soc. (A)*, 1971, 458 (*synth, ir, pmr*)

$C_7H_{10}Cl_2Ru$ Ru-00065

Dichloro[(1,2,3,4-η)-1,3-cycloheptadiene]ruthenium, 9CI

[54866-27-2]

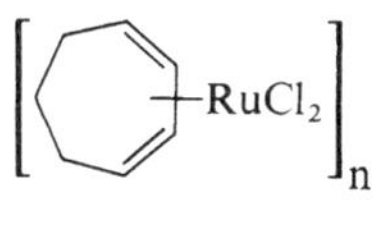

M 266.132

Polymeric. Brown solid.

Müller, J. *et al*, *Chem. Ber.*, 1975, **108**, 273 (*synth*)

$C_8H_4O_8Ru$ Ru-00066

[(2,3-η)-2-Butenedioic acid]tetracarbonylruthenium

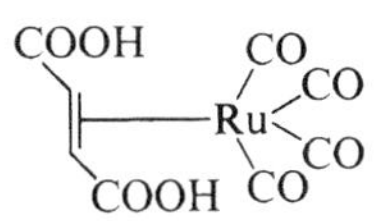

(*E*)-*form*

M 329.185

(*E*)-form

(*Fumaric acid*)*tetracarbonylruthenium*

Di-Me ester: (*Dimethyl fumarate*)*tetracarbonylruthenium.*
$C_{10}H_8O_8Ru$ M 357.238
White cryst. (hexane containing free olefin). Mp 101-2°.

(*Z*)-form

(*Maleic acid*)*tetracarbonylruthenium*

Di-Me ester: [78319-36-5]. (*Dimethyl maleate*)-*tetracarbonylruthenium.* White cryst. (hexane containing free olefin). Mp 45-45.5°.

Anhydride: [78319-40-1]. *Tetracarbonyl*[(*3,4-η*)-*2,5-furandione*]*ruthenium.* (*Maleic anhydride*)*tetracarbonylruthenium.*
$C_8H_2O_7Ru$ M 311.170
White cryst. (hexane containing free olefin).

Kruczynski, L. *et al*, *J. Organomet. Chem.*, 1974, **80**, C9 (*synth, ir, pmr, cmr, ms*)
Grevels, F.W. *et al*, *J. Am. Chem. Soc.*, 1981, **103**, 4069 (*synth, ir, pmr, cmr, ms, uv*)

$C_8H_5NO_2Ru$ Ru-00067

Dicarbonyl(cyano-*C*)(η^5-2,4-cyclopentadien-1-yl)ruthenium, 9CI

Cyano-η^5-cyclopentadienyldicarbonylruthenium. Cyclopentadienyldicarbonylruthenium cyanide

[36180-35-5]

As Dicarbonyl(η^5-2,4-cyclopentadien-1-yl)-iodoruthenium, Ru-00051 with

X = CN

M 248.203

White cryst. (CH_2Cl_2/pet. ether).

Haines, R.J. *et al*, *J. Chem. Soc., Dalton Trans.*, 1972, 944 (*synth, ir, pmr*)
Haines, R.J. *et al*, *J. Organomet. Chem.*, 1975, **84**, 357 (*synth, ir, pmr*)

$C_8H_5NO_2RuS$ Ru-00068

Dicarbonyl(η^5-2,4-cyclopentadien-1-yl)(thiocyanato-*N*)ruthenium

Dicarbonyl-π-cyclopentadienyl(isothiocyanato)ruthenium, 8CI. (*η^5-Cyclopentadienyldicarbonyl(isothiocyanato)ruthenium. Cyclopentadienyldicarbonylruthenium isothiocyanate*

[31741-75-0]

As Dicarbonyl(η^5-2,4-cyclopentadien-1-yl)-iodoruthenium, Ru-00051 with

X = –NCS

M 280.263

Yellow cryst. ($CHCl_3$/hexane). Mp 54-6°.

Kruse, A.E. *et al*, *J. Organomet. Chem.*, 1970, **24**, 231 (*synth, ir, ms*)
Wnuk, T.A. *et al*, *Inorg. Chem.*, 1977, **16**, 1173 (*synth, ir, pmr, ms*)

$C_8H_5NO_2RuS$ Ru-00069

Dicarbonyl(η^5-2,4-cyclopentadien-1-yl)(thiocyanato-*S*)ruthenium, 9CI

η^5-Cyclopentadienyldicarbonyl(thiocyanato)ruthenium. Cyclopentadienyldicarbonylruthenium thiocyanate

[31741-74-9]

As Dicarbonyl(η^5-2,4-cyclopentadien-1-yl)-iodoruthenium, Ru-00051 with

X = –SCN

M 280.263
Yellow cryst. (C_6H_6/pet. ether).

Haines, R.J. *et al*, *J. Chem. Soc., Dalton Trans.*, 1972, 944 (*synth, ir, pmr*)
Wnuk, T.A. *et al*, *Inorg. Chem.*, 1977, **16**, 1173 (*synth, ir, pmr, ms*)

$C_8H_5NO_3Ru$ Ru-00070

Dicarbonyl(cyanato-*N*)(η^5-2,4-cyclopentadien-1-yl)ruthenium

Dicarbonyl-π-cyclopentadienyl(isocyanato)ruthenium, 8CI. *η^5-Cyclopentadienyldicarbonyl(isocyanato)ruthenium. Cyclopentadienyldicarbonylruthenium isocyanate*

[31741-73-8]

As Dicarbonyl(η^5-2,4-cyclopentadien-1-yl)iodoruthenium, Ru-00051 with

X = –NCO

M 264.202
Yellow or green cryst. ($CHCl_3/Et_2O$). Mp 145-50° dec.

Kruse, A.E. *et al*, *J. Organomet. Chem.*, 1970, **24**, 231 (*synth, ms, ir*)

$C_8H_5O_2RuS^{\oplus}$ Ru-00071

(Carbonothioyl)dicarbonyl(η^5-2,4-cyclopentadien-1-yl)ruthenium(1+), 10CI

As Amminedicarbonyl(η^5-2,4-cyclopentadien-1-yl)ruthenium(1+), Ru-00060 with

X = CS

M 266.256 (ion)

Trifluoromethanesulfonate: [61930-84-5].
$C_9H_5F_3O_5RuS_2$ M 415.321
Tan powder (Me_2CO/Et_2O).

Wnuk, T.A. *et al*, *Inorg. Chem.*, 1977, **16**, 1173 (*synth, ir, pmr*)

$C_8H_5O_3Ru^{\oplus}$ Ru-00072

Tricarbonyl(η^5-2,4-cyclopentadien-1-yl)ruthenium(1+), 9CI

(η^5-Cyclopentadienyl)tricarbonylruthenium(1+)

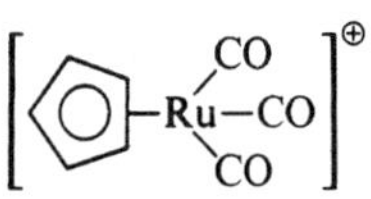

M 250.196 (ion)

Hexafluorophosphate: [31741-71-6].
$C_8H_5F_6O_3PRu$ M 395.160
White cryst. (Me_2CO/Et_2O).

Tetrafluoroborate: [57327-07-8].
$C_8H_5BF_4O_3Ru$ M 336.999
White cryst. Mp 202° dec.

Tetraphenylborate: [33113-17-6].
$C_{32}H_{25}BO_3Ru$ M 569.428
White cryst. (Me_2CO). Mp 195-7° dec.

Blackmore, T. *et al*, *J. Chem. Soc. (A)*, 1968, 2931 (*synth, ir*)
Kruse, A.E. *et al*, *J. Organomet. Chem.*, 1970, **24**, 231 (*synth, ir*)
Haines, R.J. *et al*, *J. Organomet. Chem.*, 1975, **84**, 357 (*synth, ir*)
Humphries, A.P. *et al*, *J. Chem. Soc., Dalton Trans.*, 1975, 1710 (*synth, ir, pmr*)
Jungbauer, A. *et al*, *Z. Naturforsch., B*, 1978, **33**, 1083 (*ir, pmr*)

$C_8H_6N_2O_3Ru$ Ru-00073

Tricarbonyl[(4,5,6,7-η)-1*H*-1,2-diazepine]ruthenium

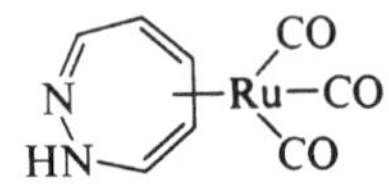

M 279.217

1-Ac: [34822-84-9]. *(1-Acetyl-1H-1,2-diazepine)tricarbonylruthenium*, 8CI.
$C_{10}H_8N_2O_4Ru$ M 321.254
Pale-yellow needles (heptane). Mp 77-8°.

Carty, A.J. *et al*, *J. Organomet. Chem.*, 1971, **32**, 241 (*synth, pmr, ir, uv, ms*)

$C_8H_6O_3Ru$ Ru-00074

Tricarbonyl[(1,2,3,4-η)-1,3-cyclopentadiene]ruthenium, 9CI

[42442-15-9]

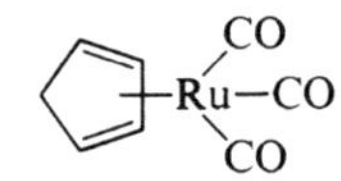

M 251.204
Yellow liq., dec. on distillation. $Bp_{0.1}$ 25°.

Humphries, A.P. *et al*, *J. Chem. Soc., Dalton Trans.*, 1975, 1710 (*synth, ir, pmr*)

$C_8H_6O_5Ru$ Ru-00075

[(3,4-η)-3-Buten-2-one]tetracarbonylruthenium

[78319-38-7]

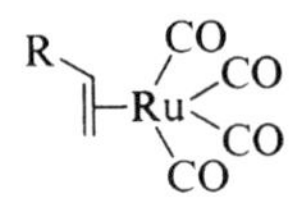

R = $-COCH_3$

M 283.202
Prepd. in hexane soln. and not isolated.

Grevels, F.W. *et al*, *J. Am. Chem. Soc.*, 1981, **103**, 4069 (*synth, ir*)

$C_8H_6O_6Ru$ Ru-00076

Tetracarbonyl[(2,3-η)-methyl 2-propenoate]ruthenium

Tetracarbonyl(methyl acrylate)ruthenium

[78319-35-4]

As [(3,4-η)-3-Buten-2-one]tetracarbonylruthenium, Ru-00075 with

R = –COOMe

M 299.202
White cryst. (hexane/methyl acrylate). Stable indefinitely under Ar at low temps. Sol. toluene. Dec. $>-10°$.

Grevels, F.W. *et al*, *J. Am. Chem. Soc.*, 1981, **103**, 4069 (*synth, ir, pmr*)

$C_8H_7NO_3Ru$ Ru-00077

(Aminocarbonyl)dicarbonyl(η^5-2,4-cyclopentadien-1-yl)ruthenium, 10CI

[72477-51-1]

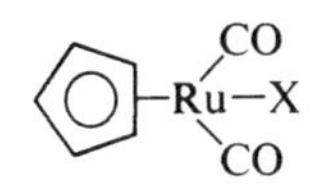

X = $CONH_2$

M 266.218

Colourless cryst. (C_6H_6). Mp 128°.

Behrens, H. *et al*, *Z. Naturforsch., B*, 1979, **34**, 1477 (*synth, ms, pmr, ir*)

Wagner, H. *et al*, *Z. Naturforsch., B*, 1979, **34**, 1487 (*struct*)

$C_8H_8Cl_3ORu^{\ominus}$ Ru-00078

[(2,3,5,6-η)-Bicyclo[2.2.1]hepta-2,5-diene]carbonyltrichlororuthenate(1–)

Carbonylnorbornadienyltrichlororuthenate(1–)

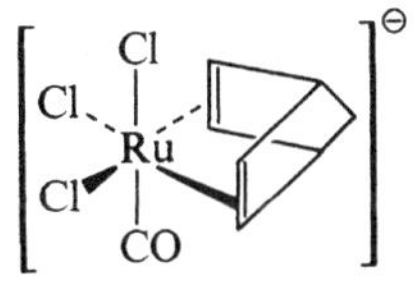

M 327.580 (ion)

Cs salt: [34779-38-9].

$C_8H_8Cl_3CsORu$ M 460.485

Yellow powder (MeOH). Mp ca. 270° dec.

Benzyltriphenylphosphonium salt: [34852-76-1].

$C_{33}H_{30}Cl_3OPRu$ M 681.002

Orange cryst. (CH_2Cl_2/heptane). Mp 177-80°.

Stephenson, T.A. *et al*, *J. Chem. Soc., Dalton Trans.*, 1973, 2112 (*synth, ir, pmr*)

Gould, R.O. *et al*, *J. Chem. Res. (S)*, 1978, 254 (*synth, cryst struct*)

Howard, J. *et al*, *Spectrochim. Acta, Part A*, 1978, **34**, 807 (*ir, raman*)

$C_8H_8O_2Ru$ Ru-00079

Dicarbonyl(η^5-2,4-cyclopentadien-1-yl)methylruthenium

η^5-Cyclopentadienyldicarbonylmethylruthenium. σ-Methyl-π-cyclopentadienylruthenium dicarbonyl

[53449-90-4]

As Dicarbonyl(η^5-2,4-cyclopentadien-1-yl)-iodoruthenium, Ru-00051 with

X = Me

M 237.220

White volatile cryst. Mp 39-40°. $Bp_{0.1}$ 40° subl.

Davison, A. *et al*, *J. Chem. Soc.*, 1963, 1133 (*synth, ir, pmr*)

Howell, J.A.S. *et al*, *J. Chem. Soc., Dalton Trans.*, 1980, 1845 (*cmr*)

$C_8H_8O_2RuS$ Ru-00080

Dicarbonyl(η^5-2,4-cyclopentadien-1-yl)(methanethiolato)ruthenium, 10CI

Dicarbonyl-π-cyclopentadienyl(methylthio)ruthenium. Cyclopentadienyldicarbonylruthenium methylmercaptide

[53556-39-1]

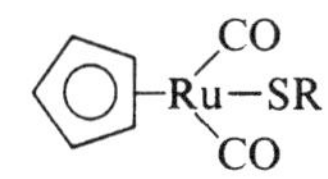

R = Me

M 269.280

Orange cryst. (Et_2O/pet. ether). Mp 60° (55-6°). Unstable in soln.

Knox, G.R. *et al*, *J. Organomet. Chem.*, 1974, **74**, 105 (*synth, ir, pmr*)

Killops, S.D. *et al*, *J. Chem. Soc., Dalton Trans.*, 1978, 1260 (*synth, ir, pmr*)

$C_8H_8O_3Ru$ Ru-00081

Tricarbonyl[(1,2,3,4-η)-1,3-pentadiene]ruthenium

[75756-65-9]

As (η^4-Butadiene)tricarbonylruthenium, Ru-00056 with

R = CH_3, R′ = H

M 253.219

No phys. props. recorded.

Zobl-Ruh, S. *et al*, *Helv. Chim. Acta*, 1980, **63**, 773 (*synth, cmr*)

$C_8H_8O_4RuS$ Ru-00082

Dicarbonyl(η^5-2,4-cyclopentadien-1-yl)(methylsulfonyl)ruthenium, 9CI

Dicarbonyl-η^5-cyclopentadienyl(methyl-S-sulfinato)ruthenium

[53449-92-6]

As Dicarbonyl(η^5-2,4-cyclopentadien-1-yl)-iodoruthenium, Ru-00051 with

X = $-SO_2Me$

M 301.279

No phys. props. reported.

Jacobson, S.E. *et al*, *J. Organomet. Chem.*, 1974, **72**, 113 (*synth, ir*)

$C_8H_{10}O_2Ru$ Ru-00083

Dicarbonylbis(η^3-2-propenyl)ruthenium

Di-π-allyldicarbonylruthenium

[33307-38-9]

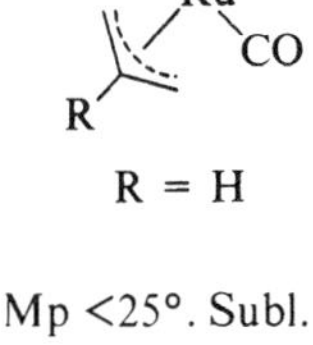

R = H

M 239.236

Pale-yellow cryst. Mp <25°. Subl. *in vacuo*.

Abel, E.W. *et al*, *Angew. Chem., Int. Ed. Engl.*, 1971, **10**, 339 (*synth*)

Cooke, M. *et al*, *J. Chem. Soc. (A)*, 1971, 16 (*synth, ir*)

$C_8H_{12}Cl_2Ru$ Ru-00084

Dichloro[(1,2,5,6-η)-1,5-cyclooctadiene]ruthenium

[50982-13-3]

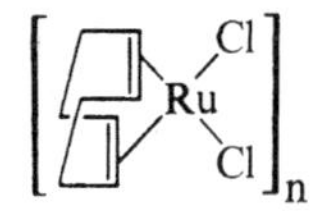

M 280.159

Polymeric. Light-brown solid. Almost insol. org. solvs. Dec. >200°.

Bennett, M.A. *et al*, *Chem. Ind. (London)*, 1959, 1516 (*synth*)
Schrock, R.R. *et al*, *J. Am. Chem. Soc.*, 1973, **95**, 4102 (*props*)
Ashworth, T.V. *et al*, *J. Organomet. Chem.*, 1974, **77**, C31 (*props*)
Potvin, C. *et al*, *Bull. Soc. Chim. Fr.*, 1974, 783 (*props*)

$C_8H_{12}N_2O_2RuS_4$ Ru-00085

Dicarbonylbis(dimethylcarbamodithioato-*S,S'*)ruthenium, 9CI

Dicarbonylbis(dimethyldithiocarbamato)ruthenium, 8CI

[15393-11-0]

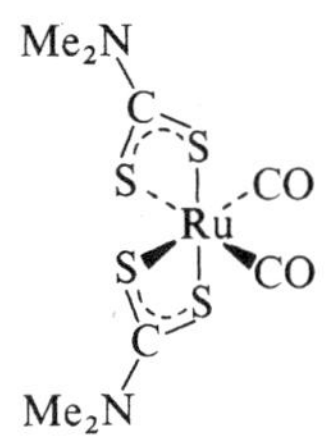

M 397.505

cis-form

Yellow cryst. (EtOH). Mp 230-1°. Bp 170° subl.

Kingston, J.V. *et al*, *J. Inorg. Nucl. Chem.*, 1966, **28**, 2709 (*synth*)
Cole-Hamilton, D.J. *et al*, *J. Chem. Soc., Dalton Trans.*, 1974, 739 (*synth, pmr*)

$C_8H_{14}I_2O_2RuS_2$ Ru-00086

[1,2-Bis(ethylthio)ethane-*S,S'*]dicarbonyldiiodoruthenium

[29099-76-1]

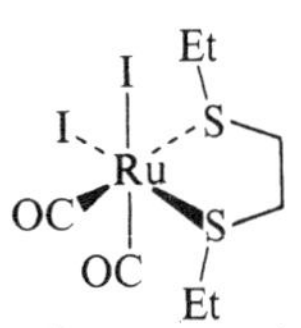

M 561.196

Orange. Mp 144°, 180°.

John, P. *et al*, *Chem. Ber.*, 1970, **103**, 2161, 2178 (*synth, struct*)

$C_8H_{17}B_9Ru$ Ru-00087

(η^6-Benzene)[(7,8,9,10,11-η)-undecahydro-7,8-dicarbaundecaborato(2−)]ruthenium

3,1,2-(η^6-Benzene)-$RuC_2B_9H_{11}$

[82704-20-9]

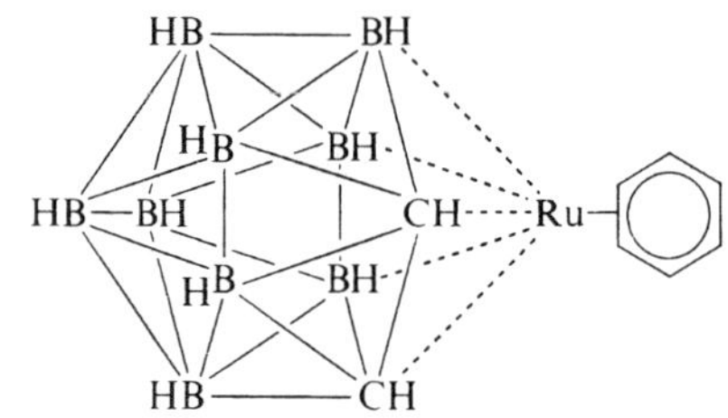

M 311.582

Canary-yellow needles (Me_2CO/hexane). Mp 308-9° dec.

Hanusa, T. *et al*, *Polyhedron*, 1982, **1**, 77 (*synth, pmr, ir, nmr, ms*)

$C_8H_{20}Br_2RuS_4$ Ru-00088

Bis[1,2-bis(methylthio)ethane-*S,S'*]dibromoruthenium

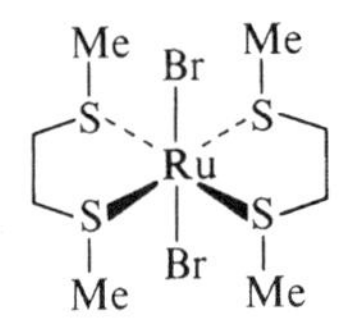

M 505.364

trans-form [32648-27-4]

Lilac cryst. (butanol). Mp >250°.

Chatt, J. *et al*, *J. Chem. Soc. (A)*, 1971, 1380 (*synth, ir*)

$C_9H_2O_9Ru_3S$ Ru-00089

Nonacarbonyldi-μ-hydro-μ_3-thioxotriruthenium

[11127-38-1]

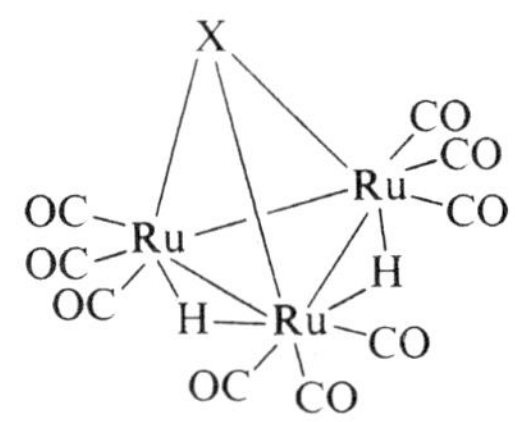

X = S

M 589.379

Air-sensitive yellow cryst. (C_6H_6). Mp 118°.

Forster, A. *et al*, *J. Organomet. Chem.*, 1976, **104**, 225 (*cmr*)
Johnson, B.F.G. *et al*, *J. Chem. Soc., Chem. Commun.*, 1979, 719 (*synth, ir, pmr*)
Cresswell, T.A. *et al*, *J. Chem. Soc., Dalton Trans.*, 1981, 2220 (*synth*)
Adams, R.D. *et al*, *Organometallics*, 1982, **1**, 53 (*synth, bibl, cryst struct*)

$C_9H_2O_9Ru_3Se$ Ru-00090

Nonacarbonyldi-μ-hydro-μ_3-selenoxotriruthenium, 9CI

[72268-07-6]

As Nonacarbonyldi-μ-hydro-μ_3-thioxotriruthenium, Ru-00089 with

X = Se

M 636.279

Orange-yellow. Mp 110° dec.

Sappa, E. *et al*, *J. Organomet. Chem.*, 1972, **35**, 375 (*synth, ir, pmr, ms*)
Johnson, B.F.G. *et al*, *J. Chem. Soc., Chem. Commun.*, 1979, 719 (*synth*)

$C_9H_2O_9Ru_3Te$ Ru-00091

Nonacarbonyldi-μ-hydro-μ$_3$-telluroxotriruthenium, 9CI

[72268-08-7]

As Nonacarbonyldi-μ-hydro-μ$_3$-thioxotriruthenium, Ru-00089 with

X = Te

M 684.919
Yellow. Mp 135° dec.

Sappa, E. *et al*, *J. Organomet. Chem.*, 1972, **35**, 375 (*synth, ir, pmr, ms*)
Johnson, B.F.G. *et al*, *J. Chem. Soc., Chem. Commun.*, 1979, 719 (*synth*)

$C_9H_3O_9Ru_3S^{\oplus}$ Ru-00092

Nonacarbonyltri-μ-hydro-μ$_3$-thioxotriruthenium(1+)

[32591-91-6]

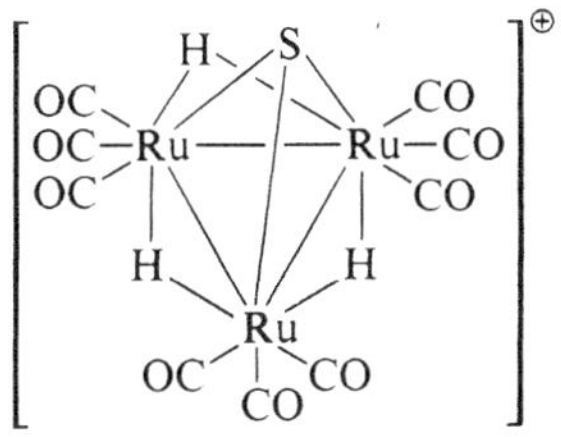

M 590.387 (ion)
Prepared in H_2SO_4 soln. only, no salts isolated.

Lewis, J. *et al*, *J. Chem. Soc. (A)*, 1971, 1797 (*synth, ir*)

$C_9H_5F_3O_3Ru$ Ru-00093

Dicarbonyl(η^5-2,4-cyclopentadien-1-yl)(trifluoroacetyl)ruthenium

η^5-Cyclopentadienyldicarbonyl(trifluoroacetyl)ruthenium

As Dicarbonyl(η^5-2,4-cyclopentadien-1-yl)-iodoruthenium, Ru-00051 with

X = $-COCF_3$

M 319.202
Colourless liq.

Blackmore, T. *et al*, *J. Chem. Soc. (A)*, 1968, 2158 (*synth, ir, nmr*)

$C_9H_6O_4Ru$ Ru-00094

Tricarbonyl[(2,3,4,5-η)-2,4-cyclohexadien-1-one]ruthenium, 9CI

[38903-39-8]

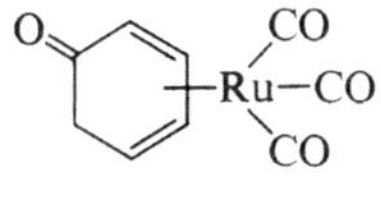

M 279.214
White cryst. by subl. $Bp_{0.01}$ 30° subl.

Cowles, R.J.H. *et al*, *J. Chem. Soc., Dalton Trans.*, 1972, 1768 (*synth, ir, ms*)

$C_9H_7O_3Ru^{\oplus}$ Ru-00095

Tricarbonyl[(1,2,3,4,5-η)-2,4-cyclohexadien-1-yl]ruthenium(1+), 9CI

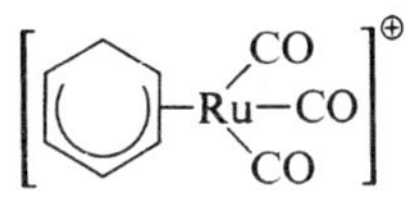

M 264.223 (ion)

Tetrafluoroborate: [43089-98-1].
$C_9H_7BF_4O_3Ru$ M 351.026
White solid. Sol. $MeNO_2$. Mp 175° dec.

Kane-Maguire, L.A.P. *et al*, *J. Chem. Soc., Chem. Commun.*, 1973, 540 (*props*)
Sosinsky, B.A. *et al*, *J. Chem. Soc., Dalton Trans.*, 1975, 1633 (*synth, pmr*)

$C_9H_8NO_2Ru^{\oplus}$ Ru-00096

(Acetonitrile)dicarbonyl(η^5-2,4-cyclopentadien-1-yl)ruthenium(1+)

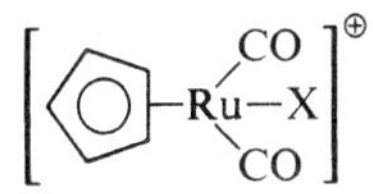

X = NCMe

M 263.238 (ion)

Hexafluorophosphate: [68663-47-8].
$C_9H_8F_6NO_2PRu$ M 408.202
Yellow cryst. (Me_2CO/Et_2O). Mp 149°.

Jungbauer, A. *et al*, *Z. Naturforsch., B*, 1978, **33**, 1083 (*synth, ir, pmr*)

$C_9H_8NO_2Ru^{\oplus}$ Ru-00097

Dicarbonyl(η^5-2,4-cyclopentadien-1-yl)(isocyanomethane)-ruthenium(1+), 10CI

Dicarbonyl(η^5-2,4-cyclopentadienyl)(methyl isocyanide)ruthenium(1+)

As Amminedicarbonyl(η^5-2,4-cyclopentadien-1-yl)-ruthenium(1+), Ru-00060 with

X = MeNC–

M 263.238 (ion)

Hexafluorophosphate: [68643-64-1].
$C_9H_8F_6NO_2PRu$ M 408.202
Mp 149°.

Jungbauer, A. *et al*, *Z. Naturforsch., B*, 1978, **33**, 1083 (*synth, ir, pmr*)

$C_9H_8O_2Ru$ Ru-00098

Dicarbonyl[(1,2,3,4,5,6-η)-1,3,5-cycloheptatriene]ruthenium, 9CI

[51508-72-6]

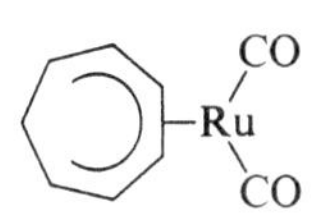

M 249.231
Air-sensitive pale-yellow solid. Sol. hexane.

Burt, J.C. *et al*, *J. Chem. Soc., Dalton Trans.*, 1975, 731 (*synth, ir, pmr*)
Deganello, G. *et al*, *J. Organomet. Chem.*, 1977. **135**, 215 (*synth*)

$C_9H_8O_2S_2Ru$ Ru-00099

Dicarbonyl(η^5-2,4-cyclopentadien-1-yl)[(methylthio)thioxomethyl]ruthenium, 10CI

[61930-80-1]

As (Aminocarbonyl)dicarbonyl(η^5-2,4-cyclopentadien-1-yl)ruthenium, Ru-00077 with

X = –C(S)SMe

M 313.351

Prescursor for thiocarbonyl complexes. Dark orange-red oil.

Wnuk, T.A. *et al*, *Inorg. Chem.*, 1977, **16**, 1173 (*synth*, *ir*, *pmr*, *ms*)

$C_9H_8O_3Ru$ Ru-00100

Acetyldicarbonyl(η^5-2,4-cyclopentadien-1-yl)ruthenium

η^5-Cyclopentadienyldicarbonylacetylruthenium

[32610-56-3]

As Dicarbonyl(η^5-2,4-cyclopentadien-1-yl)-iodoruthenium, Ru-00051 with

X = Ac

M 265.230

White cryst. (pet. ether). Mp 44°. Bp 70-80° subl. *in vacuo*.

Green, M.L.H. *et al*, *J. Chem. Soc.* (*A*), 1971, 794 (*synth*, *pmr*, *ir*)

$C_9H_8O_3Ru$ Ru-00101

Tricarbonyl[(1,2,3,4-η)-1,3-cyclohexadiene]ruthenium, 9CI

[12108-25-7]

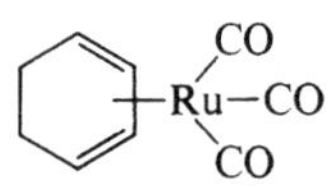

M 265.230

Colourless liq. Mp ca. 20°.

Johnson, B.F.G. *et al*, *Nature* (*London*), 1967, **213**, 901 (*synth*, *ir*)

Cowles, R.J.H. *et al*, *J. Chem. Soc.*, *Chem. Commun.*, 1969, 392 (*synth*)

Whitesides, T.H. *et al*, *Inorg. Chem.*, 1975, **14**, 664 (*synth*, *ir*, *pmr*)

Sosinksy, B.A. *et al*, *J. Chem. Soc.*, *Dalton Trans.*, 1975, 1633 (*synth*, *ir*, *pmr*)

Green, J.C. *et al*, *J. Chem. Soc.*, *Dalton Trans.*, 1976, 1974 (*pe*)

Zobl-Ruh, S. *et al*, *Helv. Chim. Acta*, 1980, **63**, 773 (*cmr*)

$C_9H_9BrO_2Ru$ Ru-00102

Bromodicarbonyl[(1,2,3,4,5-η)-2,4-cycloheptadien-1-yl]ruthenium, 9CI

[51508-74-8]

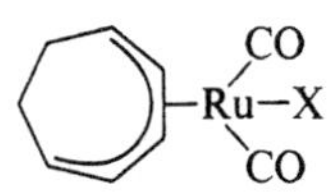

X = Br

M 330.143

Yellow solid, solns. are air-sensitive. Sol. hexane. Mp 114-6°.

Burt, J.C. *et al*, *J. Chem. Soc.*, *Dalton Trans.*, 1975, 731 (*synth*, *ir*, *pmr*)

$C_9H_9IO_2Ru$ Ru-00103

Dicarbonyl[(1,2,3,4,5-η)-2,4-cycloheptadien-1-yl]iodoruthenium, 9CI

[51508-73-7]

As Bromodicarbonyl[(1,2,3,4,5-η)-2,4-cycloheptadien-1-yl]ruthenium, Ru-00102 with

X = I

M 377.143

Yellow solid. Sol. hexane. Mp 108-10°.

Burt, J.C. *et al*, *J. Chem. Soc.*, *Dalton Trans.*, 1975, 731 (*synth*, *ir*, *pmr*)

Sosinsky, B.A. *et al*, *J. Chem. Soc.*, *Dalton Trans.*, 1975, 1633 (*props*)

$C_9H_9O_2Ru^{\oplus}$ Ru-00104

Dicarbonyl(η^5-2,4-cyclopentadien-1-yl)(η^2-ethene)ruthenium(1+), 9CI

Dicarbonyl-π-cyclopentadienyl(ethylene)ruthenium(1+). (η^5-Cyclopentadienyl)(η^2-ethene)-dicarbonylruthenium(1+)

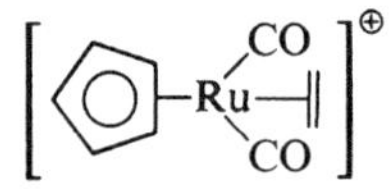

M 250.239 (ion)

Tetrafluoroborate: [56285-09-7].

$C_9H_9BF_4O_2Ru$ M 337.043

Colourless cryst. ($MeCN/Et_2O$).

Hexafluorophosphate: [68643-61-8].

$C_9H_9F_6O_2Ru$ M 364.229

Colourless cryst.

Fischer, E.O. *et al*, *Z. Naturforsch*, *B.*, 1962, **17**, 421 (*synth*)

Faller, J.W. *et al*, *J. Organomet. Chem.*, 1975, **88**, 101 (*synth*, *pmr*, *ir*)

Jungbauer, A. *et al*, *Z. Naturforsch*, *B*, 1978, **33**, 1083 (*pmr*, *ir*)

$C_9H_{10}ORu$ Ru-00105

Carbonyl(η^5-2,4-cyclopentadien-1-yl)(η^3-2-propenyl)ruthenium

π-Allylcarbonyl-π-cyclopentadienylruthenium, 8CI. (η^3-2-Propenyl)(η^5-cyclopentadienyl)carbonylrutheniu-m

[31741-51-2]

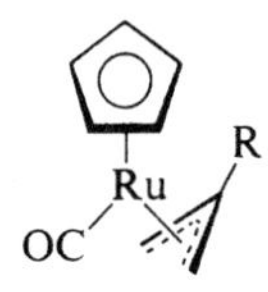

exo-form

R = H

M 235.247

exo-form [77983-18-7]

Pale-yellow needles (pentane). Mp 52-3°. $Bp_{0.1}$ 40° subl.

endo-form [77983-17-6]

Yellow needles (pentane). Mp 71-71.5°. Isomerises thermally to *exo*-form.

King, R.B. *et al*, *Inorg. Chim. Acta*, 1970, **4**, 258 (*synth*, *pmr*, *ir*, *ms*)

Faller, J.W. *et al*, *J. Organomet. Chem.*, 1974, **65**, 395 (*synth*, *ir*, *nmr*)

Gibson, D.H. *et al*, *J. Organomet. Chem.*, 1981, **208**, 89 (*synth*, *ir*, *pmr*, *cmr*, *isom*)

Worley, S.D. *et al*, *Organometallics*, 1982, **1**, 134 (*pe*)

$C_9H_{10}O_2Ru$ Ru-00106

Dicarbonyl(η^5-2,4-cyclopentadien-1-yl)ethylruthenium

η^5-Cyclopentadienyldicarbonylethylruthenium. σ-Ethyl-π-cyclopentadienylruthenium dicarbonyl

[72042-46-7]
As Dicarbonyl(η^5-2,4-cyclopentadien-1-yl)iodoruthenium, Ru-00051 with

X = Et

M 251.247
Colourless volatile oil. Mp ca. −5°.

Davison, A. *et al*, *J. Chem. Soc.*, 1963, 1133 (*synth, ir, pmr*)

$C_9H_{10}O_3Ru$ Ru-00107

Tricarbonyl[(1,2,3,4-η)-2,3-dimethyl-1,3-butadiene]ruthenium, 9CI
[52649-53-3]

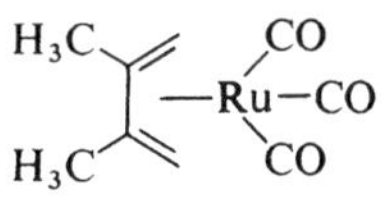

M 267.246
No phys. props. recorded. Sol. CCl_4.

Gambino, O. *et al*, *Inorg. Chim. Acta*, 1974, **8**, 71 (*synth, ms, ir, pmr*)

$C_9H_{10}O_3Ru$ Ru-00108

Tricarbonyl(2,4-hexadiene)ruthenium, 9CI
[52504-75-3]
As (η^4-Butadiene)tricarbonylruthenium, Ru-00056 with

R = R′ = CH_3

M 267.246
Yellow solid. Sol. CCl_4.

Gambino, O. *et al*, *Inorg. Chim. Acta*, 1974, **8**, 71 (*synth, ir, pmr, ms*)

$C_9H_{11}ClRu$ Ru-00109

(η^6-Benzene)chloro(η^3-2-propenyl)ruthenium, 9CI
η^3-Allyl(η^6-benzene)chlororuthenium
[41500-34-9]

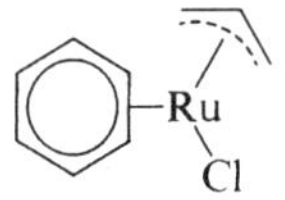

M 255.709
Yellow platelets (MeCN/Et_2O or C_6H_6/heptane). Mp 176° dec. Substituted allyl complexes also prepd.

Zelonka, R.A. *et al*, *J. Organomet. Chem.*, 1972, **44**, 383 (*synth, pmr, ir*)
Ivanov, A.S. *et al*, *Izv. Akad. Nauk SSSR, Ser. Khim.*, 1974, 931 (*synth, pmr, ir*)

$C_9H_{14}BrO_4PRu$ Ru-00110

Bromocarbonyl(η^5-2,4-cyclopentadien-1-yl)(trimethylphosphite-*P*)ruthenium
Bromocarbonyl-π-cyclopentadienyl(phosphorous acid)-ruthenium trimethyl ester, 8CI
[32611-26-0]

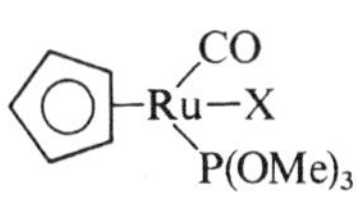

X = Br

M 398.155
Yellow-orange cryst. (hexane). Mp 65-7°.

Brown, D.A. *et al*, *Inorg. Chim. Acta*, 1970, **4**, 621 (*synth, ir*)

$C_9H_{14}ClO_4PRu$ Ru-00111

Carbonylchloro(η^5-2,4-cyclopentadien-1-yl)(trimethylphosphite-*P*)ruthenium
Carbonylchloro-π-cyclopentadienyl(phosphorous acid)-ruthenium trimethyl ester, 8CI
[32611-30-6]
As Bromocarbonyl(η^5-2,4-cyclopentadien-1-yl)(trimethylphosphite-*P*)ruthenium, Ru-00110 with

X = Cl

M 353.704
Yellow-orange cryst. (hexane). Mp 91-8°.

Brown, D.A. *et al*, *Inorg. Chim. Acta*, 1970, **4**, 621 (*synth, ir*)

$C_9H_{14}IO_4PRu$ Ru-00112

Carbonyl(η^5-2,4-cyclopentadien-1-yl)iodo(trimethyl phosphite-*P*)ruthenium
Carbonyl-π-cyclopentadienyliodo(phosphorous acid)-ruthenium trimethyl ester, 8CI
[31960-26-6]
As Bromocarbonyl(η^5-2,4-cyclopentadien-1-yl)(trimethylphosphite-*P*)ruthenium, Ru-00110 with

X = I

M 445.155
Yellow-orange cryst. (hexane). Mp 77-8°.

Brown, D.A. *et al*, *Inorg. Chim. Acta*, 1970, **4**, 428, 621 (*synth, ir*)

$C_9H_{15}Cl_2PRu$ Ru-00113

(η^6-Benzene)dichloro(trimethylphosphine)ruthenium, 10CI
[67612-85-5]

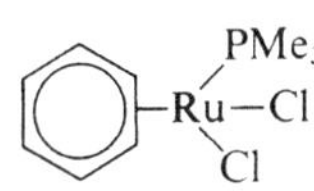

M 326.167
Hydrogenation catalyst precursor. Orange cryst. Mp 151°.

Singer, H. *et al*, *J. Organomet. Chem.*, 1979, **178**, C13 (*synth, pmr*)

$C_9H_{18}O_3P_2Ru$ Ru-00114

Tricarbonylbis(trimethylphosphine)ruthenium, 10CI

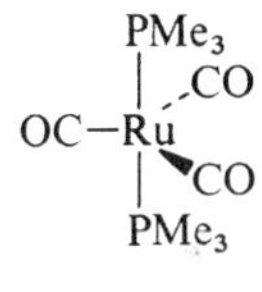

M 337.257
trans-form [75687-45-5]
Colourless prisms (toluene).

Jones, R.A. *et al*, *J. Chem. Soc., Dalton Trans.*, 1980, 1771 (*synth, struct, pmr, nmr, ir*)

$C_9H_{18}O_9P_2Ru$ Ru-00115

Tricarbonylbis(trimethyl phosphite-*P*)ruthenium

Tricarbonylbis(phosphorous acid)ruthenium hexamethyl ester, 8CI

[31541-94-3]

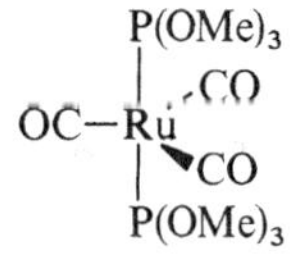

M 433.253

trans-form [30759-74-1]

White cryst. (hexane). Mp 54.5°.

Cooke, M. *et al*, *J. Chem. Soc. (A)*, 1971, 16 (*synth, ir, pmr, ms*)

Graaf, J.L. *et al*, *Organometallics*, 1982, **1**, 837 (*synth, ir*)

$C_9O_9Ru_2$ Ru-00116

μ-Carbonyloctacarbonyldiruthenium, 9CI

Nonacarbonyldiruthenium. Diruthenium nonacarbonyl. Diruthenium enneacarbonyl

[63128-11-0]

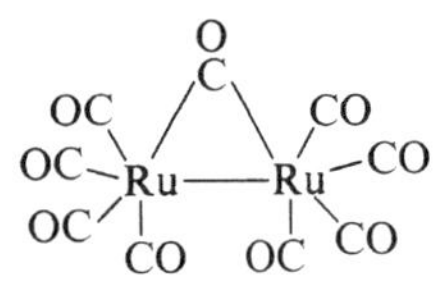

M 454.234

Orange-brown. Unstable.

Moss, J.R. *et al*, *J. Chem. Soc., Dalton Trans.*, 1977, 95 (*synth, ir, ms*)

$C_9O_9Ru_3S_2$ Ru-00117

Nonacarbonyldi-μ_3-thioxotriruthenium, 10CI

[72282-38-3]

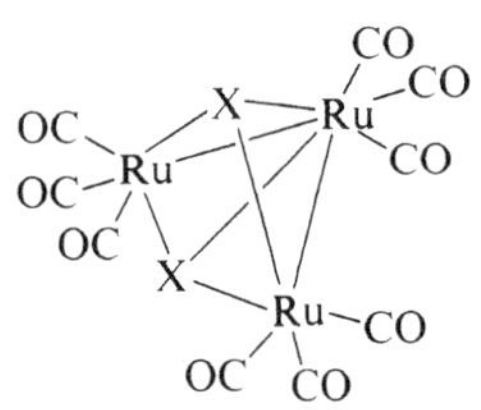

X = S

M 619.424

No phys. data reported.

Johnson, B.F.G. *et al*, *J. Chem. Soc., Chem. Commun.*, 1979, 719 (*synth, ir*)

$C_9O_9Ru_3Se_2$ Ru-00118

Nonacarbonyldi-μ_3-selenoxotriruthenium, 10CI

[72282-39-4]

As Nonacarbonyldi-μ_3-thioxotriruthenium, Ru-00117 with

X = Se

M 713.224

No phys. data reported.

Johnson, B.F.G. *et al*, *J. Chem. Soc., Chem. Commun.*, 1979, 719 (*synth, ir*)

$C_{10}Cl_8I_2Ru$ Ru-00119

1,1′,2,2′,3,3′,4,4′-Octachloro-5,5′-diiodoruthenocene, 9CI

[37298-24-1]

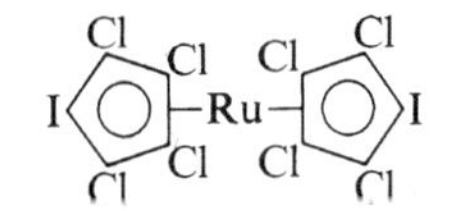

M 758.613

Cryst. (hexane). Mp >325° (darkens). Subl. >300°.

Hedberg, F.L. *et al*, *J. Am. Chem. Soc.*, 1973, **95**, 870 (*synth, pmr*)

$C_{10}Cl_{10}Ru$ Ru-00120

Decachlororuthenocene, 9CI

Perchlororuthenocene

[12679-33-3]

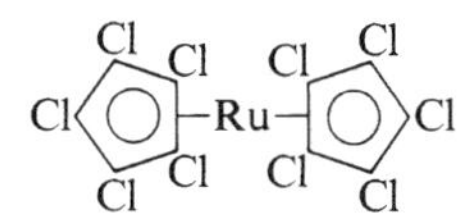

M 575.710

Cryst. (heptane). Mp 360-5° dec.

Brown, G.M. *et al*, *J. Chem. Soc., Chem. Commun.*, 1972, 5 (*struct*)

Hedberg, F.L. *et al*, *J. Am. Chem. Soc.*, 1973, **95**, 870 (*synth, pmr, ms*)

$C_{10}HNO_{11}Ru_3$ Ru-00121

Decacarbonyl-μ-hydro-μ-nitrosyltriruthenium, 10CI

[73230-19-0]

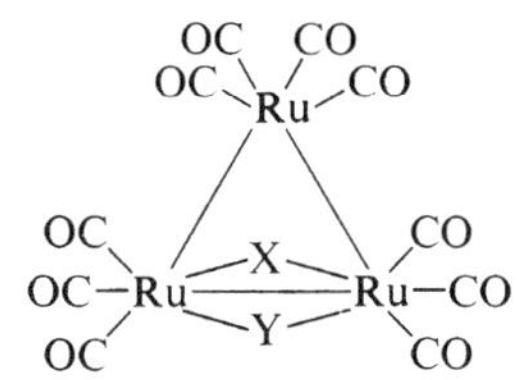

X = H, Y = NO

M 614.328

Sl. air-sensitive dark-red cryst. (hexane).

Johnson, B.F.G. *et al*, *J. Chem. Soc., Dalton Trans.*, 1980, 99 (*synth, ir, pmr, cmr*)

Stevens, R.E. *et al*, *J. Am. Chem. Soc.*, 1981, **103**, 4981 (*synth*)

$C_{10}H_3BrO_9Ru_3$ Ru-00122

[μ_3-(Bromomethylidyne)]nonacarbonyltri-μ-hydrotriruthenium, 10CI

[73746-95-9]

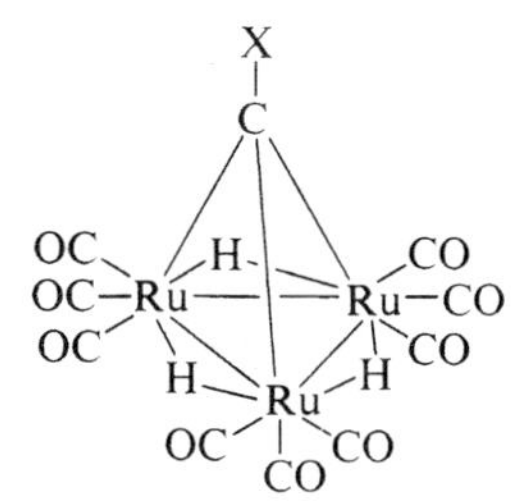

X = Br

M 650.242

Important precursor for μ_3-methylidyne complexes.
Bright-yellow cryst. With $AlCl_3$ reacts as an electrophile towards aromatic substrates.

Keister, J.B. *et al*, *Inorg. Chem.*, 1980, **19**, 2304.

$C_{10}H_3ClO_9Ru_3$ — Ru-00123

Nonacarbonyl[μ_3-(chloromethylidyne)]tri-μ-hydrotriruthenium

Nonacarbonyltris(μ-hydrido)(μ-chloromethylidyne)triruthenium

[73746-97-1]

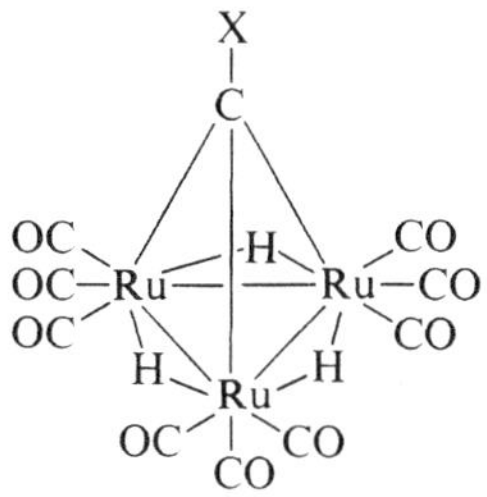

X = Cl

M 605.791

Bright-yellow.

Keister, J.B. *et al*, *Inorg. Chem.*, 1980, **19**, 2304 (*synth, ir, pmr*)
Sherwood, D.E. *et al*, *Organometallics*, 1982, **1**, 1519 (*pe*)
Zhu, N.J. *et al*, *Acta Crystallogr., Sect. B*, 1982, **38**, 1286 (*cryst struct*)

$C_{10}H_4O_9Ru_3$ — Ru-00124

Nonacarbonyltri-μ-hydro-μ_3-methylidynetriruthenium

Nonacarbonyltris(μ-hydrido)(μ-methylidyne)triruthenium

[63280-43-3]

As Nonacarbonyl[μ_3-(chloromethylidyne)]tri-μ-hydrotriruthenium, Ru-00123 with

X = H

M 571.346

Yellow.

Eady, C.R. *et al*, *J. Chem. Soc., Dalton Trans.*, 1977, 477 (*ir, nmr, ms*)
Keister, J.B. *et al*, *Inorg. Chem.*, 1980, **19**, 2304 (*synth, ir, pmr*)
Sherwood, D.E. *et al*, *Organometallics*, 1982, **1**, 1519 (*pe*)

$C_{10}H_5Cl_2NO_3Ru$ — Ru-00125

(Benzonitrile)tricarbonyldichlororuthenium, 9CI

Dichloro(cyanobenzene)tricarbonylruthenium. Dichlorotricarbonyl(phenyl cyanide)ruthenium

[36502-87-1]

$PhCNRuCl_2(CO)_3$

M 359.130

Pale-yellow cryst. (CH_2Cl_2/pentane). Mp 146-8°.

Benedetti, E. *et al*, *J. Organomet. Chem.*, 1972, **37**, 361 (*synth, ir, pmr*)

$C_{10}H_5F_5O_2Ru$ — Ru-00126

Dicarbonyl(η^5-2,4-cyclopentadien-1-yl)(1,2,3,3,3-pentafluoro-1-propenyl)ruthenium

σ-(Pentafluoro-1-propenyl)-π-cyclopentadienylruthenium dicarbonyl

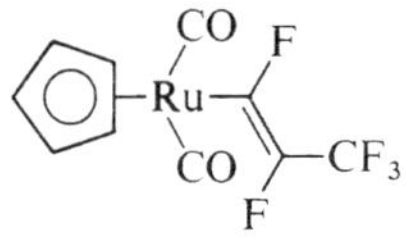

M 353.210

(Z)-form

trans-*form*

White needles (pet. ether). Mp 70-1°. $Bp_{0.2}$ 50° subl.

Blackmore, T. *et al*, *J. Chem. Soc. (A)*, 1968, 2158 (*synth, ir, nmr*)
Bruce, M.I., *Org. Mass Spectrom.*, 1969, **2**, 63 (*ms*)

$C_{10}H_6O_4Ru$ — Ru-00127

Tricarbonyl[(2,3,4,5-η)-2,4,6-cycloheptatrien-1-one]ruthenium, 9CI

Troponeruthenium tricarbonyl

[41550-69-0]

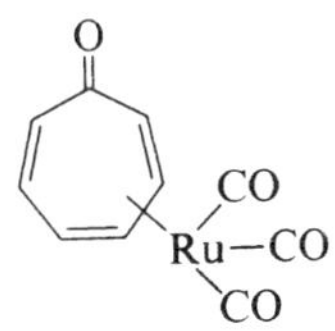

M 291.225

Orange oil, mod. stable under N_2.

Domingos, A.J.P. *et al*, *J. Chem. Soc., Dalton Trans.*, 1975, 2288 (*synth, ir, pmr*)

$C_{10}H_7NO_5Ru$ — Ru-00128

[(2,3,4,5-η)-1*H*-Azepine-1-carboxylic acid]tricarbonylruthenium

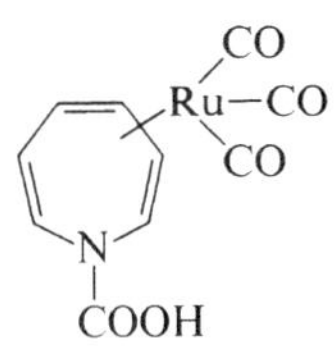

M 322.239

Free acid not known.

Me ester: [62745-23-7]. *Tricarbonyl[(2,3,4,5-η)methyl-1H-azepine-1-carboxylate]ruthenium*, 9CI. *Tricarbonyl(η-N-methoxycarbonyl-1H-azepine)ruthenium.* $C_{11}H_9NO_5Ru$ M 336.266
Yellow cryst. (hexane). Mp 109-10°.

Green, M. *et al*, *J. Chem. Soc., Dalton Trans.*, 1977, 204 (*synth, ir, pmr, ms*)

$C_{10}H_8Br_2Ru$ — Ru-00129

1,1'-Dibromoruthenocene

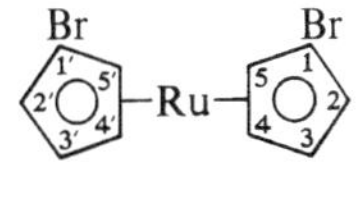

M 389.051

Mp 117-8°.

Bednarik, L. *et al*, *J. Organomet. Chem.*, 1979, **168**, C8 (*synth*)

$C_{10}H_8Cl_2Hg_2Ru$ Ru-00130

1,1′-Bis(chloromercuri)ruthenocene

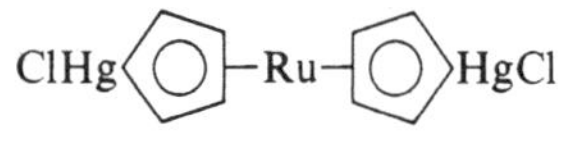

M 701.329

Long yellow needles (butanol). Mp 217-8°. Contaminated with monosubstitution prod.

Rausch, M.D. *et al*, *J. Am. Chem. Soc.*, 1960, **82**, 76 (*synth*)
Nesmeyanov, A.N. *et al*, *Izv. Akad. Nauk SSSR, Ser. Khim.*, 1972, 1823.

$C_{10}H_8I_2Ru$ Ru-00131

1,1′-Diiodoruthenocene

[70210-63-8]

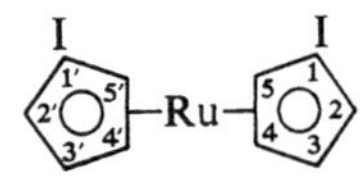

M 483.052

Pale-yellow cryst. (hexane). Mp 87-8°.

Bednarik, L. *et al*, *J. Organomet. Chem.*, 1979, **168**, C8 (*synth, ir, pmr*)

$C_{10}H_8Li_2Ru$ Ru-00132

1,1′-Dilithioruthenocene

μ-1,1′-Ruthenocenediyldilithium, 9CI

[60898-13-7]

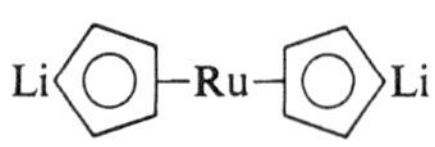

M 243.125

Prepared and used *in situ*.

Rausch, M.D. *et al*, *J. Am. Chem. Soc.*, 1960, **82**, 76 (*synth*)

$C_{10}H_8O_3Ru$ Ru-00133

Tricarbonyl[(1,2,3,4-η)-1,3,5-cycloheptatriene]ruthenium, 9CI

[41550-66-7]

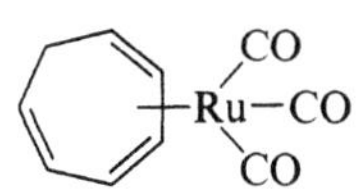

M 277.241

Pale-yellow liq. $Bp_{0.1}$ 30°.

Burt, J.C. *et al*, *J. Chem. Soc., Dalton Trans.*, 1975, 731 (*synth, ir, pmr*)
Domingos, A.J.P. *et al*, *J. Chem. Soc., Dalton Trans.*, 1975, 2288 (*synth, ir, pmr*)
Deganello, G. *et al*, *J. Organomet. Chem.*, 1977, **135**, 215 (*synth*)

$C_{10}H_9BrRu$ Ru-00134

Bromoruthenocene, 9CI

[38816-65-8]

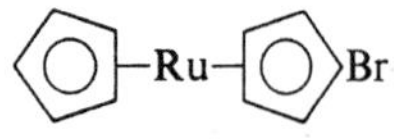

M 310.155

Cryst. (EtOH). Mp 66.5-67.5°.

Nesmeyanov, A.N. *et al*, *Izv. Akad. Nauk SSSR, Ser. Khim.*, 1972, 1823 (*synth*)
Bednarik, L. *et al*, *J. Organomet. Chem.*, 1979, **168**, C8 (*synth*)

$C_{10}H_9ClRu$ Ru-00135

Chlororuthenocene

(η^5-1-Chloro-2,4-cyclopentadien-1-yl)(η^5-2,4-cyclopentadien-1-yl)ruthenium, 9CI. *(Chloro-π-cyclopentadienyl)-π-cyclopentadienylruthenium*

[38959-07-8]

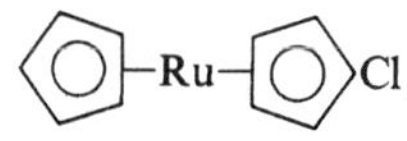

M 265.704

Cryst. (hexane). Mp 81-2°.

Nesmeyanov, A.N. *et al*, *Izv. Akad. Nauk SSSR, Ser. Khim.*, 1972, 1823 (*synth*)

$C_{10}H_9IO_2Ru$ Ru-00136

Dicarbonyliodo[(1,2,3,3*a*,6*a*-η)-1,4,5,6-tetrahydro-1-pentalenyl]ruthenium

[69251-43-0]

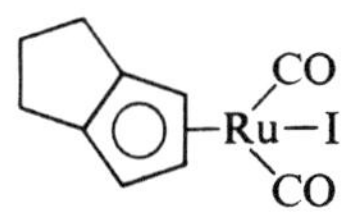

M 389.154

Orange cryst. Sol. $CHCl_3$. Mp 50-54°.

Humphries, A.P. *et al*, *J. Chem. Soc., Dalton Trans.*, 1978, 1523 (*synth, ir, pmr*)

$C_{10}H_9IRu$ Ru-00137

Iodoruthenocene

[70210-62-7]

M 357.156

Pale-yellow needles (hexane). Mp 89°.

Bednarik, L. *et al*, *J. Organomet. Chem.*, 1979, **168**, C8 (*synth, ir, pmr*)

$C_{10}H_9LiRu$ Ru-00138

Lithioruthenocene

Ruthenocenyllithium

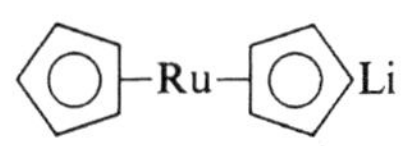

M 237.192

Prepared and used *in situ*. Ruthenocene is more reactive than ferrocene towards metalation.

Rausch, M.D. *et al*, *J. Am. Chem. Soc.*, 1960, **82**, 76 (*synth*)
Nesmeyanov, A.N. *et al*, *Izv. Akad. Nauk SSSR, Ser. Khim.*, 1972, 1823 (*synth*)

$C_{10}H_9O_3Ru^{\oplus}$ Ru-00139

Tricarbonyl[(1,2,3,4,5-η)-2,4-cycloheptadien-1-yl]ruthenium(1+), 9CI

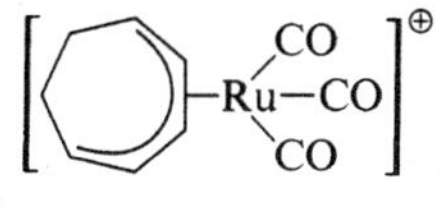

M 278.249 (ion)

Tetrafluoroborate: [57659-27-5].
$C_{10}H_9BF_4O_3Ru$ M 365.053
White cryst. Sol. $MeNO_2$. Mp 225-50° dec.
Hexafluorophosphate: [58320-36-8].
$C_{10}H_9F_6O_3PRu$ M 423.213

Air-stable white cryst. (liq. SO_2).

Sosinsky, B.A. *et al*, *J. Chem. Soc., Dalton Trans.*, 1975, 1633 (*synth, pmr*)
Domingos, A.J.P. *et al*, *J. Chem. Soc., Dalton Trans.*, 1975, 2288 (*synth*)

$C_{10}H_9O_4Ru^{\oplus}$ Ru-00140

Tricarbonyl[(2,3,4,5,6-η)-1-methoxy-2,4-cyclohexadien-1-yl]ruthenium(1+)

Tricarbonyl(methoxycyclohexadienylium)-ruthenium(1+), 8CI

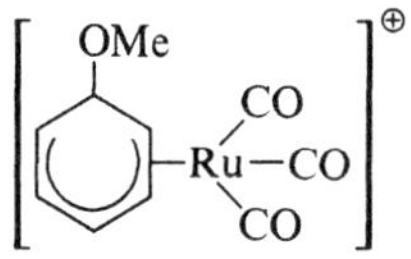

M 294.249 (ion)

Tetrafluoroborate: [12116-59-5].
$C_{10}H_9BF_4O_4Ru$ M 381.052
No phys. description recorded.

Johnson, B.F.G. *et al*, *Nature (London)*, 1967, **213**, 901 (*synth, ir*)

$C_{10}H_{10}IRu^{\oplus}$ Ru-00141

Bis(η[5]-2,4-cyclopentadien-1-yl)iodoruthenium(1+), 9CI

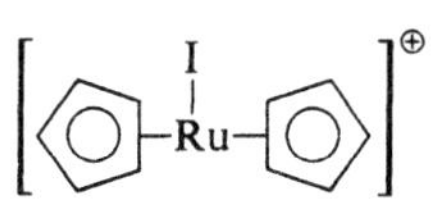

M 358.164 (ion)

Diamagnetic. The Cl analogue is also known.

Triiodide: [39427-31-1].
$C_{10}H_{10}I_4Ru$ M 738.877
Dark-red needles ($MeNO_2$).

Nesmeyanov, A.N. *et al*, *Izv. Akad. Nauk SSSR, Ser. Khim.*, 1967, 935.
Sohn, Y.S. *et al*, *Inorg. Chem.*, 1974, **13**, 301 (*synth, struct, ir*)

$C_{10}H_{10}ORu$ Ru-00142

Hydroxyruthenocene

Ruthenocenol. (π-Cyclopentadienyl)(π-hydroxycyclopentadienyl)ruthenium

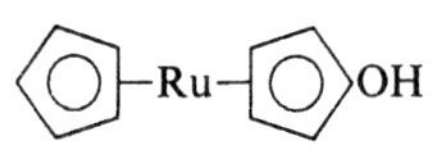

M 247.258

Me ether: [38833-39-5]. *Methoxyruthenocene. Methyl ruthenocenyl ether.*
$C_{11}H_{12}ORu$ M 261.285
Cryst. (pentane). Mp 49°.

Ac: [38816-67-0]. *Acetoxyruthenocene.*
$C_{12}H_{12}O_2Ru$ M 289.296
Cryst. (MeOH). Mp 56°.

Nesmeyanov, A.N. *et al*, *Izv. Akad. Nauk SSSR, Ser. Khim.*, 1972, 1823.

$C_{10}H_{10}O_2Ru$ Ru-00143

Dicarbonyl(η[5]-2,4-cyclopentadien-1-yl)-2-propenylruthenium

Allyldicarbonyl-π-cyclopentadienylruthenium, 8CI.
σ-Allyl-π-cyclopentadienylruthenium dicarbonyl.
η[1]-Allyl(η[5]-cyclopentadienyl)dicarbonylruthenium

[31811-43-5]

As Dicarbonyl(η[5]-2,4-cyclopentadien-1-yl)-iodoruthenium, Ru-00051 with

$$X = -CH_2CH{=}CH_2$$

M 263.258

Yellow liq.

King, R.B. *et al*, *Inorg. Chim. Acta*, 1970, **4**, 258 (*synth, pmr, ir, ms*)

$C_{10}H_{10}O_3Ru$ Ru-00144

Tricarbonyl[(1,2,3,4-η)-1,3-cycloheptadiene]ruthenium, 9CI

[41550-67-8]

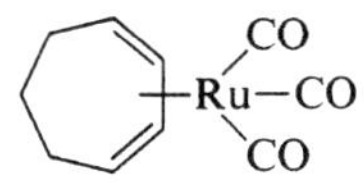

M 279.257

Pale-yellow oil. Sol. hexane. $Bp_{0.1}$ 35°.

Burt, J.C. *et al*, *J. Chem. Soc., Dalton Trans.*, 1975, 731 (*synth, ir, pmr*)
Domingos, A.J.P. *et al*, *J. Chem. Soc., Dalton Trans.*, 1975, 2288 (*synth*)
Whitesides, T.H. *et al*, *Inorg. Chem.*, 1975, **14**, 664 (*synth, ir, ms, pmr*)

$C_{10}H_{10}O_3Ru$ Ru-00145

Tricarbonyl[(1,3,4,5-η)-4-cycloheptene-1,3-diyl]ruthenium, 9CI

[57603-32-4]

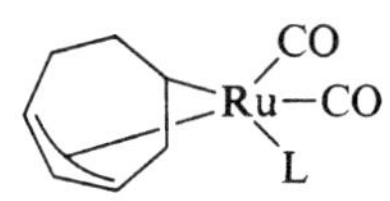

L = CO

M 279.257

Yellow solid. Mp ca. 30°. $Bp_{0.1}$ 65°. Obt. 93% pure.

Sosinsky, B.A. *et al*, *J. Chem. Soc., Dalton Trans.*, 1975, 1633 (*synth, ir, pmr*)

$C_{10}H_{10}O_3RuS$ Ru-00146

Ruthenocenesulfonic acid

Sulforuthenocene

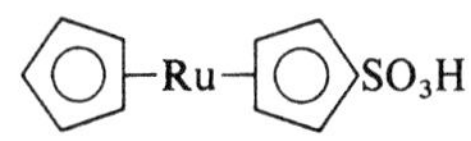

M 311.317

Isol. as Pb^{2+} salt.

Et ester:
$C_{12}H_{14}O_3RuS$ M 339.371
White needles (hexane). Mp 118-9°.

Chloride: [38816-61-4]. *Ruthenocenesulfonyl chloride. (Chlorosulfonyl)ruthenocene.*
$C_{10}H_9ClO_2RuS$ M 329.763
Yellow cryst. (pet. ether). Mp 80° dec.

Nesmeyanov, A.N. *et al*, *Izv. Akad. Nauk SSSR, Ser. Khim.*, 1972, 1823 (*synth*)

$C_{10}H_{10}O_4Ru$ Ru-00147

Dicarbonyl[(1,2,3,4,5-η)-2,4-cyclohexadien-1-yl](methoxycarbonyl)ruthenium

[12116-81-3]

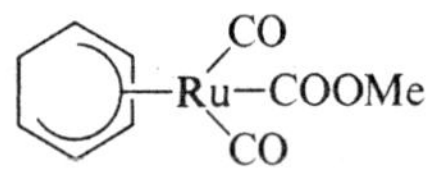

M 295.257

Characterised spectroscopically.

Cowles, R.J.H. *et al*, *J. Chem. Soc., Chem. Commun.*, 1969, 392 (*synth*)

$C_{10}H_{10}O_4Ru$ Ru-00148

Tricarbonyl[(1,2,3,4-η)-1-methoxy-1,3-cyclohexadiene]ruthenium

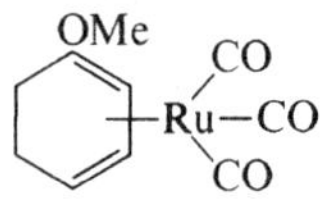

M 295.257

Yellow oil. Obt. as a mixt. with the 2-methoxy isomer.

Cowles, R.J.H. *et al*, *J. Chem. Soc., Dalton Trans.*, 1972, 1768 (*synth*)

$C_{10}H_{10}Ru$ Ru-00149

Ruthenocene, 9CI

Di-π-cyclopentadienylruthenium, 8CI.

Bis(cyclopentadienyl)ruthenium

[1287-13-4]

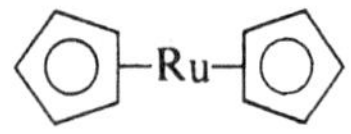

M 231.259

Light-yellow cryst. (CCl_4). Mp 199-200°. Sublimes *in vacuo*. Less reactive than ferrocene towards electrophiles.

Wilkinson, G. *et al*, *J. Am. Chem. Soc.*, 1952, **74**, 6146 (*synth*)
Piper, T.S. *et al*, *J. Inorg. Nucl. Chem.*, 1956, **3**, 104 (*pmr*)
Fischer, E.O. *et al*, *Chem. Ber.*, 1959, **92**, 2302 (*synth*)
Rausch, M.D. *et al*, *J. Am. Chem. Soc.*, 1960, **82**, 76 (*synth*)
Org. Synth., 1961, **41**, 96 (*synth*)
Aleksanyan, V.T. *et al*, *J. Organomet. Chem.*, 1977, **124**, 293 (*ir*, *raman*)
Seiler, P. *et al*, *Acta Crystallogr., Sect. B*, 1980, **36**, 2946 (*struct*)
Senda, Y. *et al*, *Bull. Chem. Soc. Jpn.*, 1980, **53**, 3381 (*cmr*)

$C_{10}H_{10}Ru^{\oplus}$ Ru-00150

Ruthenocenium

[54538-51-1]

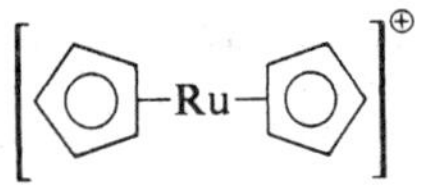

M 231.259 (ion)

Paramagnetic. Yellow. Simple cation not isol.

Nesmeyanov, A.N. *et al*, *Izv. Akad. Nauk SSSR, Ser. Khim.*, 1967, 935.
Gubin, S.P. *et al*, *J. Organomet. Chem.*, 1971, **30**, 243.
Denisovich, L.I. *et al*, *J. Organomet. Chem.*, 1974, **81**, 207 (*synth*, *polarog*, *uv*)

$C_{10}H_{11}BO_2Ru$ Ru-00151

Ruthenoceneboronic acid

Boronoruthenocene, 9CI

[38816-64-7]

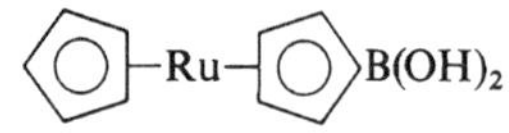

M 275.076

White cryst. Sol. Et_2O.

Nesmeyanov, A.N. *et al*, *Izv. Akad. Nauk SSSR, Ser. Khim.*, 1972, 1823 (*synth*)

$C_{10}H_{11}O_2Ru^{\oplus}$ Ru-00152

Dicarbonyl(η^5-2,4-cyclopentadien-1-yl)[(1,2-η)-1-propene]ruthenium(1+), 10CI

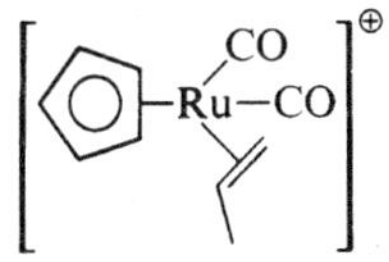

M 264.266 (ion)

Hexafluorophosphate: [68643-65-2].
$C_{10}H_{11}F_6O_2PRu$ M 409.230
Mp 210° dec.

Jungbauer, A. *et al*, *Z. Naturforsch., B*, 1978, **33**, 1083 (*synth*, *ir*, *pmr*)

$C_{10}H_{12}As_2O_6Ru_2$ Ru-00153

Hexacarbonylbis[μ-(dimethylarsino)]diruthenium, 9CI

[74095-85-5]

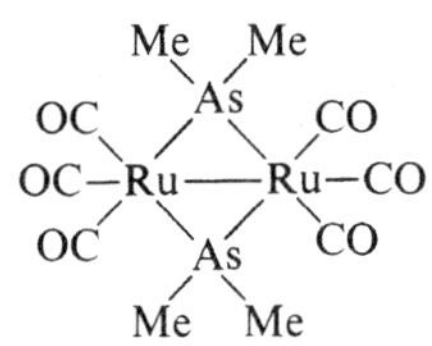

M 580.184

Yellow cryst. (pet. ether). Mp 129° dec.

Roland, E. *et al*, *Chem. Ber.*, 1980, **113**, 1799 (*synth*, *ir*, *pmr*)

$C_{10}H_{12}ClN_2Ru^{\oplus}$ Ru-00154

Bis(acetonitrile)(η^6-benzene)chlororuthenium(1+), 9CI

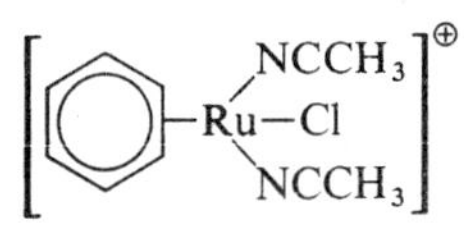

M 296.741 (ion)

Intermed. for (η^6-benzene) Ru complexes.

Trichloromercurate: [35796-51-1].
$C_{10}H_{12}Cl_4HgN_2Ru$ M 603.690
Yellow-orange cryst. Dec. at 118°.

Zelonka, R.A. *et al*, *Can. J. Chem.*, 1972, **50**, 3063 (*synth*, *pmr*)

$C_{10}H_{12}ORu$ Ru-00155

Carbonyl(η^5-2,4-cyclopentadien-1-yl)[(1,2,3-η)-2-methyl-2-propenyl]ruthenium

π-Cyclopentadienyl-π-(2-methylallyl)carbonylruthenium. (η^3-2-Methyl-2-propenyl)(η^5-cyclopentadienyl)carbonylruthenium

As Carbonyl(η^5-2,4-cyclopentadien-1-yl)(η^3-2-propenyl)ruthenium, Ru-00105 with

$R = CH_3$

M 249.274

***exo*-form** [78037-33-9]

Pale-yellow plates (pentane). Mp 50-50.5°.

***endo*-form** [77933-78-9]

Yellow plates (pentane). Mp 65°. Isom. thermally to *exo*-form.

Gibson, D.H. *et al*, *J. Organomet. Chem.*, 1981, **208**, 89 (*synth*, *ir*, *pmr*, *cmr*, *isom*)

Worley, S.D. *et al*, *Organometallics*, 1982 **1**, 134 (*pe*)

$C_{10}H_{12}O_6P_2Ru_2$ Ru-00156

Hexacarbonylbis[μ-(dimethylphosphino)]diruthenium, 9CI

[36568-57-7]

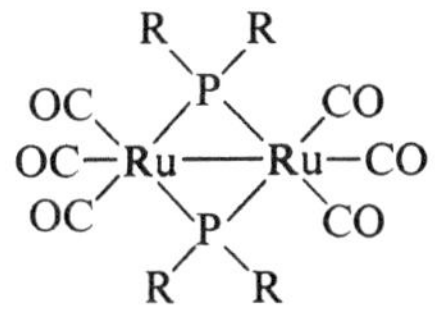

R = Me

M 492.289

Orange cryst. (C_6H_6/hexane). Subl. *in vacuo*.

Dessy, R.E. *et al*, *J. Am. Chem. Soc.*, 1972, **94**, 746 (*synth*, *pmr*, *props*)

$C_{10}H_{14}O_2PRu^{\oplus}$ Ru-00157

Dicarbonyl(η^5-2,4-cyclopentadien-1-yl)(trimethylphosphine)-ruthenium(1+)

As (Acetonitrile)dicarbonyl(η^5-2,4-cyclopentadien-1-yl)-ruthenium(1+), Ru-00096 with

X = PMe_3

M 298.263 (ion)

Hexafluorophosphate: [75182-23-9].

$C_{10}H_{14}F_6O_2P_2Ru$ M 443.227

Colourless cryst. (Me_2CO/pet. ether). Mp >300°.

Treichel, P.M. *et al*, *Synth, React. Inorg. Metal-Org. Chem.*, 1980, **10**, 205 (*synth*, *ir*, *pmr*)

$C_{10}H_{14}O_2Ru$ Ru-00158

Dicarbonylbis[(1,2,3-η)-2-methyl-2-propenyl]ruthenium

Dicarbonylbis(2-methyl-π-allyl)ruthenium, 8CI

[32677-88-6]

As Dicarbonylbis(η^3-2-propenyl)ruthenium, Ru-00083 with

R = CH_3

M 267.289

Characterised spectroscopically, not isol.

Green, M. *et al*, *J. Chem. Soc. (A)*, 1971, 16 (*ir*)

$C_{10}H_{14}O_2RuSi$ Ru-00159

Dicarbonyl(η^5-2,4-cyclopentadien-1-yl)(trimethylsilyl)ruthenium, 9CI

η^5-Cyclopentadienyldicarbonyl(trimethylsilyl)ruthenium. (Cyclopentadienyldicarbonylruthenium)-trimethylsilane

[43184-26-5]

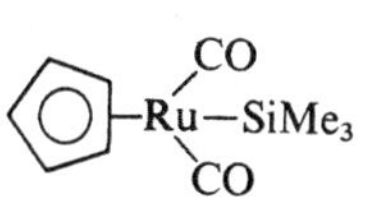

M 295.375

Liq.

Abel, E.W. *et al*, *J. Chem. Soc., Dalton Trans.*, 1973, 1706 (*synth*)

$C_{10}H_{16}O_4RuSi_2$ Ru-00160

Tetracarbonyl[1,2-ethanediylbis[dimethylsilylene]]ruthenium

[1,2-Bis(dimethylsilyl)ethane]tetracarbonylruthenium

[43198-14-7]

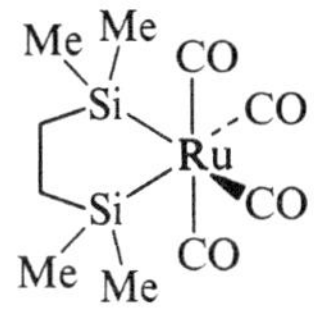

M 357.475

Colourless waxy solid. Sol. $CHCl_3$, heptane. Mp 91-2°. $Bp_{0.005}$ ~20° subl.

Vancea, L. *et al*, *Inorg. Chem.*, 1974, **13**, 511 (*synth*, *ir*, *pmr*, *nmr*)

Edwards, J.D. *et al*, *J. Chem. Soc., Dalton Trans.*, 1980, 545 (*props*)

$C_{10}H_{18}F_4O_8P_2Ru$ Ru-00161

Dicarbonyl(tetrafluoroethene)bis(trimethyl phosphite-*P*)-ruthenium

Dicarbonylbis(phosphorous acid)(tetrafluoroethylene)-ruthenium hexamethyl ester, 8CI

[22588-62-1]

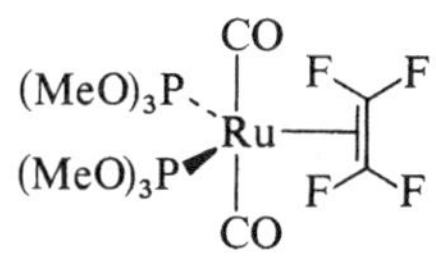

M 505.259

Low-melting white cryst. (hexane at −80°).

Cooke, M. *et al*, *J. Chem. Soc. (A)*, 1969, 651 (*synth*, *ms*, *nmr*, *ir*)

Green, M. *et al*, *J. Chem. Soc. (A)*, 1970, 2975 (*synth*)

$C_{10}H_{18}Ge_2O_4Ru$ Ru-00162

Tetracarbonylbis(trimethylgermyl)ruthenium

[30854-78-5]

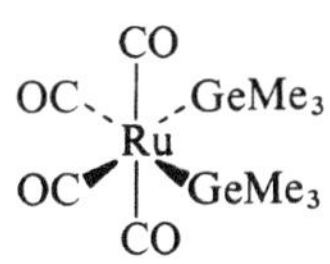

M 448.500

***cis*-form** [34397-77-8]

Yellow liq. $Bp_{0.01}$ 75°.

Knox, S.A.R. *et al*, *J. Chem. Soc. (A)*, 1971, 2874 (*synth*, *ir*, *pmr*)

Vancea, L. *et al*, *J. Am. Chem. Soc.*, 1976, **98**, 1407 (*cmr*)

$C_{10}H_{18}O_4Pb_2Ru$ Ru-00163

Tetracarbonylbis(trimethylplumbyl)ruthenium

[21868-07-5]

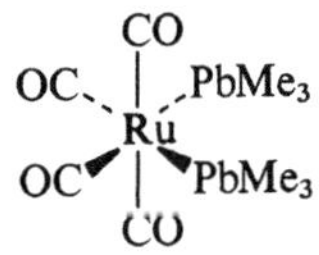

M 717.720

***cis*-form**

Pale-yellow solid.

Cotton, J.D. *et al*, *J. Chem. Soc. (A)*, 1968, 2758 (*synth, ir*)
Vancea, L. *et al*, *J. Am. Chem. Soc.*, 1976, **98**, 1407 (*cmr*)

$C_{10}H_{18}O_4RuSi_2$ Ru-00164

Tetracarbonylbis(trimethylsilyl)ruthenium

[25929-58-2]

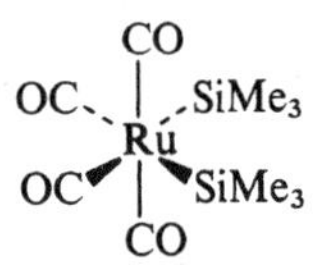

M 359.491

***cis*-form** [53228-31-2]

Yellow liq. $Bp_{0.01}$ 50°.

Knox, S.A.R. *et al*, *J. Chem. Soc. (A)*, 1969, 2559 (*synth, ir*)
Vancea, L. *et al*, *J. Am. Chem. Soc.*, 1976, **98**, 1407 (*cmr*)

$C_{10}H_{18}O_4RuSn_2$ Ru-00165

Tetracarbonylbis(trimethylstannyl)ruthenium

[21439-09-8]

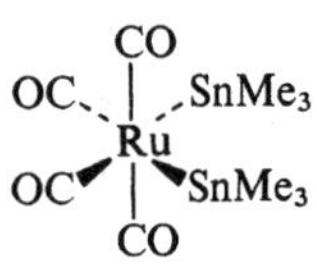

M 540.700

***cis*-form** [58073-96-4]

Yellow liq. $Bp_{0.01}$ ca. 80°.

Cotton, J.D. *et al*, *J. Chem. Soc. (A)*, 1968, 2162, 2758 (*synth, ir*)
Churchill, M.R. *et al*, *J. Am. Chem. Soc.*, 1975, **97**, 7158 (*synth*)
Vancea, L. *et al*, *J. Am. Chem. Soc.*, 1976, **98**, 1407 (*cmr*)

$C_{10}NO_{11}Ru_3^{\ominus}$ Ru-00166

Decacarbonyl-μ-nitrosyltriruthenate(1−), 10CI

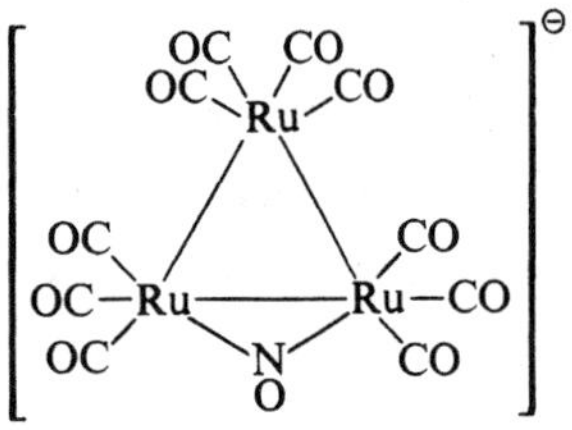

M 613.320 (ion)

Probable struct.

Bis(triphenylphosphine)iminium salt: [79085-63-5].
$C_{46}H_{30}N_2O_{11}P_2Ru_3$ M 1151.907
Green cryst. (Et_2O).

Stevens, R.E. *et al*, *J. Am. Chem. Soc.*, 1981, **103**, 4981 (*synth, ir*)

$C_{10}N_2O_{12}Ru_3$ Ru-00167

Decacarbonyldi-μ-nitrosyltriruthenium, 9CI

Decacarbonyldinitrosyltriruthenium

[36583-24-1]

As Decacarbonyl-μ-hydro-μ-nitrosyltriruthenium, Ru-00121 with

X = Y = NO

M 643.326

Golden-green cryst. (heptane/pentane).

Turner, J.J. *et al*, *J. Chem. Soc. (A)*, 1971, 654 (*ir*)
Norton, J.R. *et al*, *Inorg. Chem.*, 1972, **11**, 382 (*synth, ir, ms, cryst struct*)
Forster, A. *et al*, *J. Chem. Soc., Chem. Commun.*, 1974, 1042 (*cmr*)
Inorg. Synth., 1976, **16**, 39 (*synth, ir*)

$C_{11}HO_{11}Ru_3^{\ominus}$ Ru-00168

μ-Carbonyldecacarbonyl-μ-hydrotriruthenate(1−), 10CI

Undecacarbonylhydrotriruthenate(1−)

[60496-59-5]

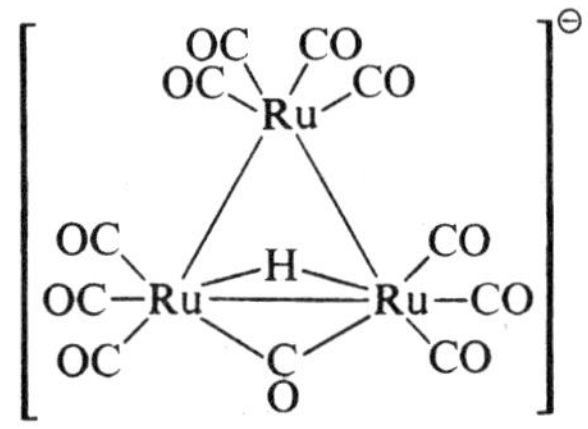

M 612.332 (ion)

Tetraethylammonium salt: [71936-71-5].
$C_{19}H_{21}NO_{11}Ru_3$ M 742.585
Dark purplish-red cryst. (CH_2Cl_2/pet. ether). Sl. air-sensitive. Dec. under N_2 at 325-30°.

Bis(triphenylphosphine)iminium salt: [71936-70-4].
$C_{47}H_{31}NO_{11}P_2Ru_3$ M 1150.920
Red-brown platelets (MeOH). Sl. air-sensitive.

Johnson, B.F.G. *et al*, *J. Chem. Soc., Dalton Trans.*, 1979, 1356 (*synth, cmr, ir, struct*)
Ungerman, C. *et al*, *J. Am. Chem. Soc.*, 1979, **101**, 5922 (*cmr*)
Keister, J.B. *et al*, *Organometallics*, 1983, **2**, 219 (*synth, ir, cmr*)

$C_{11}H_2O_{11}Ru_3$ Ru-00169

Undecacarbonyl-μ-hydrohydrotriruthenium

Undecacarbonyldihydrotriruthenium

[74790-26-4]

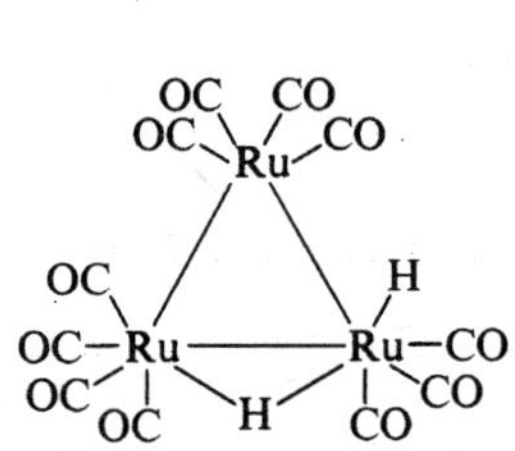

M 613.340

Stable only <−30°. Formed by protonation of $HRu_3(CO)_{11}^{\ominus}$. Characterised spectroscopically.

Keister, J.B., *J. Organomet. Chem.*, 1980, **190**, C36 (*synth, pmr*)

$C_{11}H_4O_9Ru_3$ Ru-00170

Nonacarbonyl[μ-[(1-η:1-η:1,2-η)-ethenylidene]]di-μ-hydrotriruthenium, 9CI

[55802-24-9]

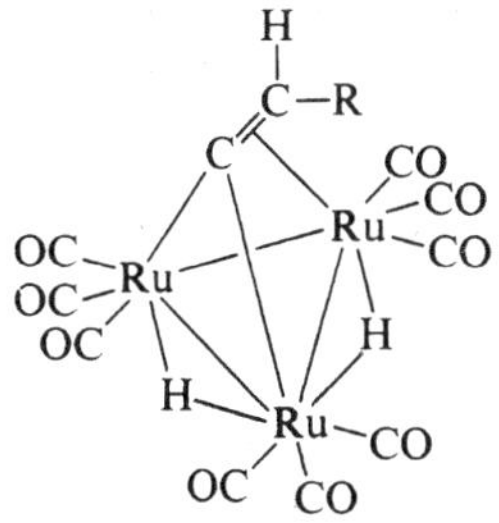

R = H

M 583.357

Characterised spectroscopically.

Deeming, A.J. *et al*, *J. Chem. Soc., Chem. Commun.*, 1974, 807 (*synth, nmr*)
Eady, C.R. *et al*, *J. Chem. Soc., Dalton Trans.*, 1977, 477 (*synth, ir, pmr, ms*)

$C_{11}H_4O_{10}Ru_3$ Ru-00171

Decacarbonyldi-μ-hydro-μ-methylenetriruthenium

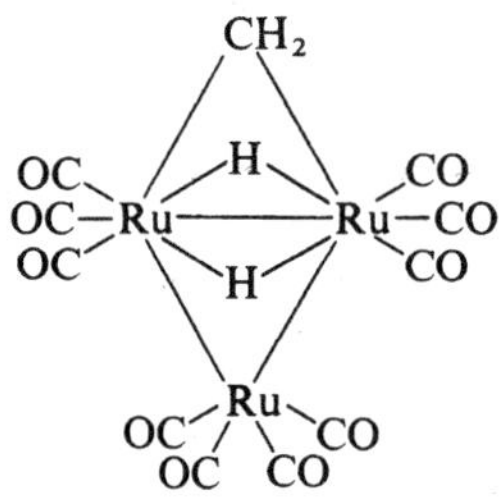

M 599.357

Not isolated spectroscopically pure.

Eady, C.R. *et al*, *J. Chem. Soc., Dalton Trans.*, 1977, 477 (*synth, nmr, ms*)

$C_{11}H_5CoO_6Ru$ Ru-00172

Dicarbonyl(η^5-2,4-cyclopentadien-1-yl)(tetracarbonylcobalt)-ruthenium

Cyclopentadienylrutheniumdicarbonyltetracarbonylcobalt. (η^5-Cyclopentadienyl)-dicarbonylrutheniumtetracarbonylcobalt

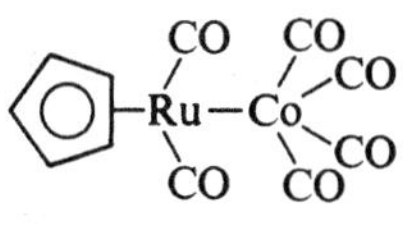

M 393.160

Light-orange cryst. (pentane). Mp 60-2°. Sublimes.

Blackmore, T. *et al*, *J. Chem. Soc. (A)*, 1968, 2931 (*synth, ms, ir*)
Bruce, M.I., *Int. J. Mass Spectrom. Ion Phys.*, 1968, **1**, 141 (*ms*)
Manning, A.R., *J. Chem. Soc. (A)*, 1971, 2321 (*ir*)

$C_{11}H_6O_9Ru_3$ Ru-00173

Nonacarbonyl-μ_3-ethylidynetri-μ-hydrotriruthenium, 9CI

[41517-70-8]

As Nonacarbonyl[μ_3-(chloromethylidyne)]tri-μ-hydrotriruthenium, Ru-00123 with

X = CH_3

M 585.373

Yellow cryst.

Carty, A.J. *et al*, *J. Chem. Soc., Chem. Commun.*, 1972, 1331 (*synth, pmr*)
Sheldrick, G.M. *et al*, *J. Chem. Soc., Dalton Trans.*, 1975, 873 (*cryst struct*)
Forster, A. *et al*, *J. Organomet. Chem.*, 1976, **104**, 225 (*cmr*)
Eady, C.R. *et al*, *J. Chem. Soc., Dalton Trans.*, 1977, 477 (*ir, nmr, ms*)

$C_{11}H_6O_{10}Ru_3$ Ru-00174

Nonacarbonyltri-μ-hydro[μ_3-(methoxymethylidyne)]triruthenium, 10CI

[71562-47-5]

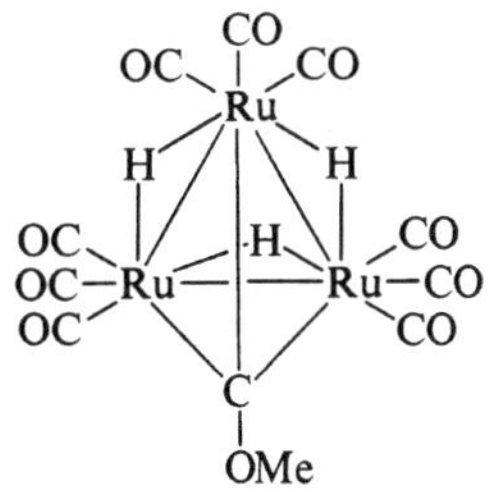

M 601.372

Bright-orange cryst. (MeOH).

Keister, J.B. *et al*, *Organometallics*, 1983, **2**, 219 (*synth, pmr, ir, ms*)

$C_{11}H_7O_2Ru^{\ominus}$ Ru-00175

Dicarbonyl[(1,2,3,3*a*,7*a*-η)-1-*H*-inden-1-yl]ruthenate(1–), 9CI

(η^5-Indenyl)dicarbonylruthenium(1–)

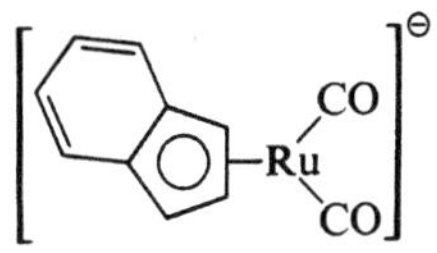

M 272.245 (ion)

Na salt: [52326-31-5].

$C_{11}H_{17}NaO_2Ru$ M 305.314

Not isolated; generated and used *in situ*.

Faller, J.W. *et al*, *J. Organomet. Chem.*, 1974, **65**, 395 (*synth*)

$C_{11}H_8O_3Ru$ Ru-00176

Tricarbonyl[(1,2,3,4-η)-1,3,5,7-cyclooctatetraene]ruthenium, 9CI

Cyclooctatetraenetricarbonylruthenium

[12282-26-7]

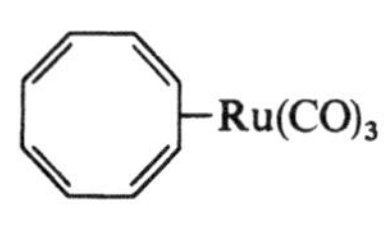

M 289.252

Exhibits valence tautomerism in soln. at r.t. Orange cryst. Mp 75-6°. $Bp_{0.05}$ 47° subl.

Bruce, M.I., *Int. J. Mass Spectrom. Ion Phys.*, 1969, **2**, 349 (*ms*)
Cotton, F.A. *et al*, *J. Am. Chem. Soc.*, 1969, **91**, 6591, 6598 (*synth, struct, pmr, ir*)
Cooke, M. *et al*, *J. Chem. Soc., Chem. Commun.*, 1971, 621.
Cotton, F.A. *et al*, *J. Am. Chem. Soc.*, 1976, **98**, 1413 (*cmr*)

$C_{11}H_9ClO_3Ru$ Ru-00177

Tricarbonylchloro[(1,2,3-η)-2,4,7-cyclooctatrien-1-yl]ruthenium, 10CI

[61127-38-6]

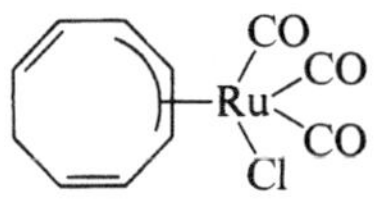

M 325.713

Pale-yellow cryst. (Et_2O).

Charles, A.D. *et al*, *J. Organomet. Chem.*, 1976, **116**, C25 (*synth, ir, pmr, ms*)

$C_{11}H_9O_3Ru^{\oplus}$ Ru-00178

[(2,3,4,6,7-η)-Bicyclo[3.2.1]octa-3,6-dien-2-yl]tricarbonylruthenium(1+), 9CI

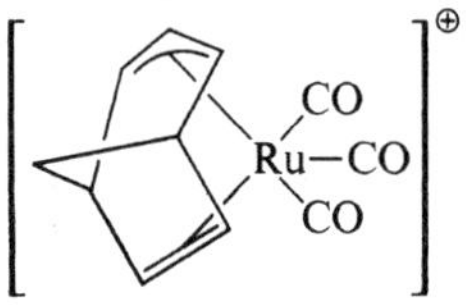

M 290.260 (ion)

Tetrafluoroborate:

$C_{11}H_9BF_4O_3Ru$ M 377.064

Air-stable white cryst. (CH_2Cl_2/Et_2O or liq. SO_2).

Domingos, A.J.P. *et al*, *J. Chem. Soc., Dalton Trans.*, 1974, 145 (*synth, pmr, ir*)

$C_{11}H_9O_3Ru^{\oplus}$ Ru-00179

Tricarbonyl[(1,2,3,6,7-η)-2,4,6-cyclooctatrien-1-yl]ruthenium(1+)

Tricarbonyl-π-cyclooctatrienyliumruthenium(1+)

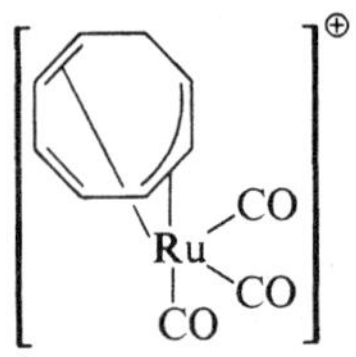

M 290.260 (ion)

Previously assigned a bicyclo[5.1.0]octa-3,5-dien-2-yl structure.

Hexafluorophosphate: [33306-40-0].

$C_{11}H_9F_6O_3PRu$ M 435.224

White cryst.

Tetrafluoroborate: [33306-42-2].

$C_{11}H_9BF_4O_3Ru$ M 377.064

Solid (Ac_2O/Et_2O).

Green, M. *et al*, *J. Chem. Soc., Chem. Commun.*, 1971, 621 (*synth, ir, pmr*)

Bennett, M.A. *et al*, *Inorg. Chem.*, 1981, **20**, 2353 (*synth, pmr, nmr*)

$C_{11}H_{10}BBrN_6O_2Ru$ Ru-00180

Bromodicarbonyl[hydrotris(pyrazolato)borato(1−)]ruthenium, 8CI

[33849-21-7]

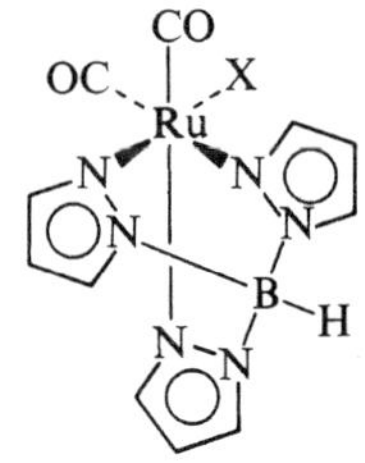

X = Br

M 450.023

Yellow microcryst. (MeOH). Mp 256°.

Bruce, M.I. *et al*, *J. Organomet. Chem.*, 1971, **31**, 269 (*synth, pmr, ir*)

$C_{11}H_{10}BClN_6O_2Ru$ Ru-00181

Dicarbonylchloro[hydrotris(pyrazolato)borato(1−)]ruthenium, 8CI

[33849-22-8]

As Bromodicarbonyl[hydrotris(pyrazolato)borato(1−)]-ruthenium, Ru-00180 with

X = Cl

M 405.572

White cryst. (MeOH). Mp 261-2°.

Bruce, M.I. *et al*, *J. Organomet. Chem.*, 1971, **31**, 269 (*synth, pmr, ir*)

$C_{11}H_{10}BIN_6O_2Ru$ Ru-00182

Dicarbonyl[hydrotris(pyrazolato)borato(1−)]iodoruthenium, 8CI

[33849-20-6]

As Bromodicarbonyl[hydrotris(pyrazolato)borato(1−)]-ruthenium, Ru-00180 with

X = I

M 497.024

Bright-yellow cryst. (MeOH). Mp 243°.

Bruce, M.I. *et al*, *J. Organomet. Chem.*, 1971, **31**, 269 (*synth, pmr, ir*)

$C_{11}H_{10}ClRu^{\oplus}$ Ru-00183

($η^6$-Chlorobenzene)($η^5$-2,4-cyclopentadien-1-yl)ruthenium(1+), 10CI

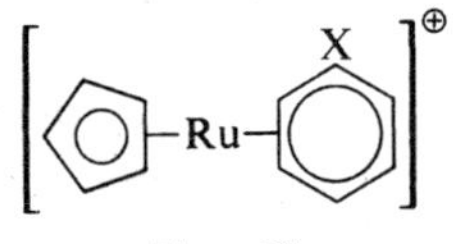

X = Cl

M 278.723 (ion)

Air-stable.

Tetrafluoroborate: [72804-61-6].

$C_{11}H_{10}BClF_4Ru$ M 365.527

Light-yellow cryst. (EtOH). Mp 160-2°.

Hexafluorophosphate: [78618-96-9].

$C_{11}H_{10}ClF_6PRu$ M 423.687

Light-yellow cryst. (EtOH).

Nesmeyanov, I.N. *et al*, *J. Organomet. Chem.*, 1979, **182**, C36; *Dokl. Akad. Nauk SSSR, Ser. Sci. Khim.*, 1981, **258**, 193 (*ir, pmr, synth*)

$C_{11}H_{10}ORu$ Ru-00184

Ruthenocenecarboxaldehyde

π-Cyclopentadienyl(formyl-π-cyclopentadienyl)ruthenium. Formylruthenocene

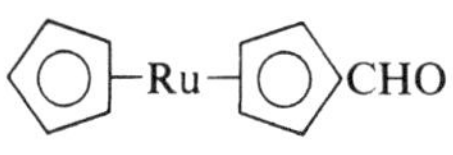

M 259.269

Yellow needles (hexane). Mp 100.2-100.8°.

2,4-Dinitrophenylhydrazone: Mp 220-2° dec.

Semicarbazone: Light-yellow leaflets (EtOH aq.). Mp 208.0-208.4°.

Bublitz, D.E. *et al*, *J. Am. Chem. Soc.*, 1962, **84**, 1845 (*synth*)
Hofer, O. *et al*, *J. Organomet. Chem.*, 1968, **13**, 443.

$C_{11}H_{10}O_2Ru$ Ru-00185

Ruthenocenecarboxylic acid

(Carboxy-π-cyclopentadienyl)-π-cyclopentadienylruthenium, 8CI. *Carboxyruthenocene*

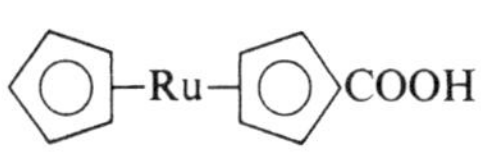

M 275.269

Fine yellow cryst. (C_6H_6). Mp 237-8° dec. pK_a 5.43 (20°, 30% EtOH aq.).

Me ester: Methoxycarbonylruthenocene.
$C_{12}H_{12}O_2Ru$ M 289.296
White leaflets (MeOH aq.). Mp 111° (107-107.5°).

Amide: [55411-35-3]. *(Aminocarbonyl)ruthenocene*, 9CI. *Ruthenocenecarboxamide.*
$C_{11}H_{11}NORu$ M 274.284
Mp 216°.

Nitrile: [12245-07-7]. *(Cyano-π-cyclopentadienyl)-π--cyclopentadienylruthenium*, 8CI. *Cyanoruthenocene. Ruthenocenenitrile.*
$C_{11}H_9NRu$ M 256.269
Yellow cryst. (heptane). Mp 118-9°.

Anilide:
$C_{17}H_{15}NORu$ M 350.382
Light-yellow cryst. (EtOH). Mp 210°.

Rausch, M.D. *et al*, *J. Am. Chem. Soc.*, 1960, **82**, 76 (*synth*, *ir*)
Nesmeyanov, A.N. *et al*, *Izv. Akad. Nauk SSSR, Ser. Khim.*, 1972, 1823 (*synth*)
Rapić, V. *et al*, *J. Organomet. Chem.*, 1975, **94**, 87 (*synth*)

$C_{11}H_{10}O_3Ru$ Ru-00186

[(2,3,4,5-η)-Bicyclo[4.2.0]octa-2,4-diene]tricarbonylruthenium, 9CI

[41348-18-9]

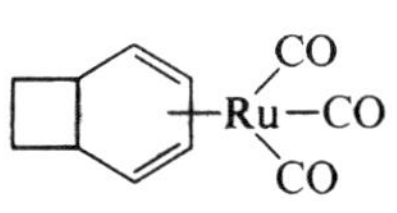

M 291.268

Yellow liq. $Bp_{0.1}$ 30°.

Szary, A.C. *et al*, *J. Chem. Soc., Dalton Trans.*, 1974, 662 (*synth*)
Domingos, A.J.P. *et al*, *J. Chem. Soc., Dalton Trans.*, 1975, 2288 (*synth*, *pmr*)

$C_{11}H_{10}O_3Ru$ Ru-00187

[(2,3,4,6-η)Bicyclo[3.2.1]oct-3-ene-2,6-diyl]tricarbonylruthenium, 9CI

[41685-13-6]

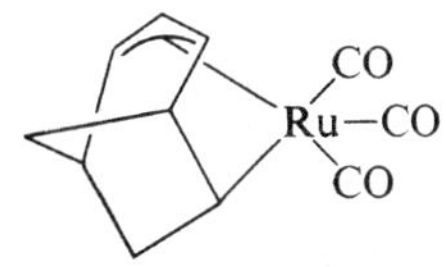

M 291.268

Colourless oil at r.t.; white cryst. below −10°. Mp ca. −10°. Bp_1 ~20°.

Domingos, A.J.P. *et al*, *J. Chem. Soc., Dalton Trans.*, 1974, 145 (*synth*, *pmr*, *ir*)

$C_{11}H_{11}O_3Ru^{\oplus}$ Ru-00188

Tricarbonyl[(1,2,3,5,6-η)-2,5-cyclooctadien-1-yl]ruthenium(1+), 9CI

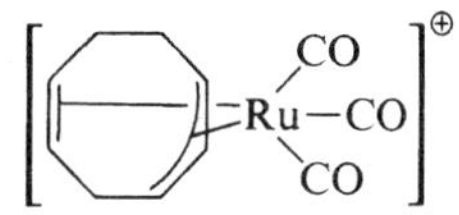

M 292.276 (ion)

Tetrafluoroborate: [55729-62-9].
$C_{11}H_{11}BF_4O_3Ru$ M 379.080
Air-stable white cryst. (CH_2Cl_2). Dec. at 209° without melting.

Deeming, A.J. *et al*, *J. Chem. Soc., Dalton Trans.*, 1974, 2093 (*synth*, *ir*, *pmr*)

$C_{11}H_{11}Ru^{\oplus}$ Ru-00189

(η^6-Benzene)(η^5-2,4-cyclopentadien-1-yl)ruthenium(1+), 9CI

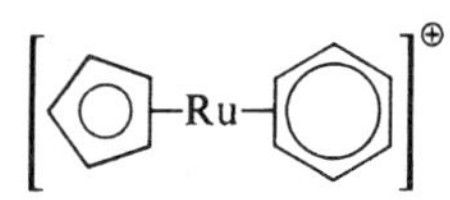

M 244.278 (ion)

Chloride: [39358-58-2].
$C_{11}H_{11}ClRu$ M 279.731
Pale-brown cryst. (MeCN/Et_2O). Dec. at 161°.

Tetraphenylborate: [72804-54-7].
$C_{35}H_{31}BRu$ M 563.510
Mp 248-50°.

Zelonka, R.A. *et al*, *J. Organomet. Chem.*, 1972, **35**, C43; **44**, 383 (*synth*, *ir*, *pmr*)
Nesmeyanov, A.N. *et al*, *J. Organomet. Chem.*, 1979, **182**, C36; *Dokl. Akad. Nauk SSSR, Ser. Sci. Khim.*, 1981, **258**, 193 (*synth*, *ir*, *pmr*)

$C_{11}H_{12}NRu^{\oplus}$ Ru-00190

(η^6-Aniline)(η^5-cyclopentadienyl)ruthenium(1+)

(η^6-Benzeneamine)(η^5-2,4-cyclopentadien-1-yl)ruthenium(1+), 10CI

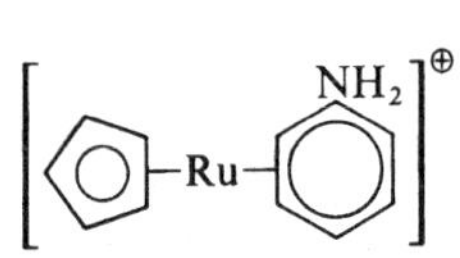

M 259.293 (ion)

Air-stable.

Tetrafluoroborate: [77170-78-6].
$C_{11}H_{12}BF_4NRu$ M 346.096
Darkens on heating, does not melt <230°.

Nesmeyanov, A.N. *et al*, *Dokl. Akad. Nauk SSSR, Ser. Sci. Khim.*, 1980, **254**, 480 (*synth*, *ir*, *pmr*)

$C_{11}H_{12}ORu$ Ru-00191

Ruthenocenemethanol

π-Cyclopentadienyl[(hydroxymethyl)-π-cyclopentadienyl]ruthenium, 8CI. *Hydroxymethylruthenocene*

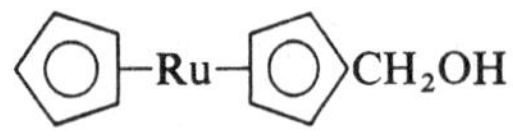

M 261.285

Hofer, O. *et al*, *J. Organomet. Chem.*, 1968, **13**, 443.

$C_{11}H_{12}O_3Ru$ Ru-00192

Tricarbonyl[(1,2,5,6-η)-1,5-cyclooctadiene]ruthenium, 9CI

[32874-17-2]

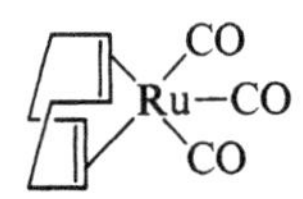

M 293.284

White cryst. Unstable even at −20° under N_2, but may be kept for 1 month in frozen C_6H_6. $Bp_{0.005}$ 0° subl.

Lewis, J. *et al*, *J. Am. Chem. Soc.*, 1971, **93**, 4624 (*synth*)
Deeming, A.J. *et al*, *J. Chem. Soc., Dalton Trans.*, 1974, 2093 (*synth, ir, pmr*)
Inorg. Synth., 1976, **16**, 105 (*synth, ir, pmr*)

$C_{11}H_{12}O_3Ru$ Ru-00193

Tricarbonyl[(1,4,5,6-η)-5-cyclooctene-1,4-diyl]ruthenium, 9CI

[52247-30-0]

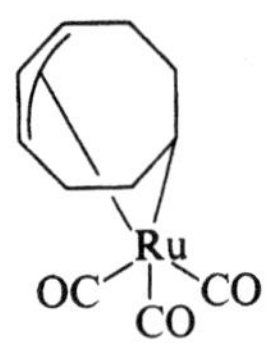

M 293.284

White cryst. Mp ca. 30°.

Cotton, F.A. *et al*, *J. Am. Chem. Soc.*, 1971, **93**, 4624 (*synth*)
Deeming, A.J. *et al*, *J. Chem. Soc., Dalton Trans.*, 1974, 2093 (*synth, ir, pmr*)
Johnson, B.F. *et al*, *J. Chem. Soc., Dalton Trans.*, 1974, 241 (*props*)

$C_{11}H_{12}O_7Ru$ Ru-00194

Tricarbonylbis[(2,3-η)-2-methyl-2-propenoate]ruthenium

[77495-49-9]

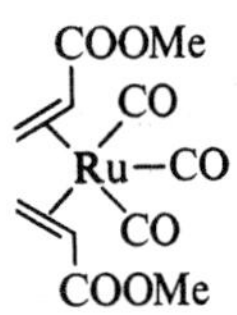

M 357.282

White cryst. Stable in solid state at ambient temperatures, but of limited stability in soln. Mp 36-8°.

Grevels, F.W. *et al*, *Angew. Chem., Int. Ed. Engl.*, 1981, **20**, 452 (*synth, ir, pmr, nmr, ms*)

$C_{11}H_{12}Ru$ Ru-00195

[(1,2,3,4,5-η)-2,4-Cyclohexadien-1-yl]($η^5$-2,4-cyclopentadien-1-yl)ruthenium

[65630-15-1]

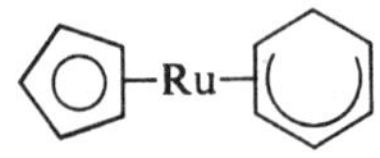

M 245.286

Yellow solid (pet. ether).

Robertson, I.W. *et al*, *J. Organomet. Chem.*, 1982, **228**, 171 (*synth, pmr, ms*)

$C_{11}H_{12}Ru$ Ru-00196

Methylruthenocene

π-Cyclopentadienyl(methyl-π-cyclopentadienyl)ruthenium, 8CI

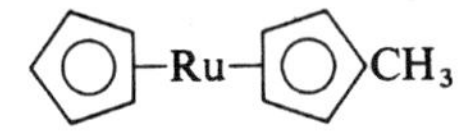

M 245.286

Yellow cryst. Mp 41-2°.

Hofer, O. *et al*, *J. Organomet. Chem.*, 1968, **13**, 443 (*synth, pmr*)

$C_{11}H_{13}ClO_3Ru$ Ru-00197

Tricarbonylchloro[(1,2,3-η)-2,4,7-cycloctatrien-1-yl]ruthenium

[61127-38-6]

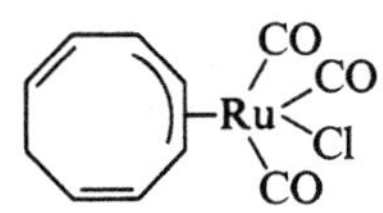

M 329.745

Pale-yellow cryst.

Charles, A.D. *et al*, *J. Organomet. Chem.*, 1976, **116**, C25 (*synth, ir, pmr*)

$C_{11}H_{14}ClN_2Ru^{\oplus}$ Ru-00198

Bis(acetonitrile)chloro[(1,2,3,4,5,6-η)-1,3,5-cycloheptatriene]ruthenium(1+), 10CI

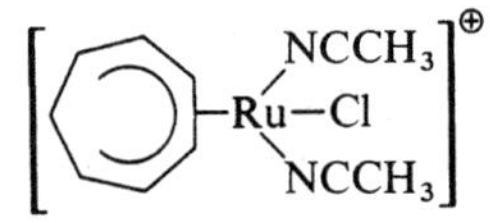

M 310.768 (ion)

Tetrafluoroborate: [63625-18-3].
$C_{11}H_{14}BClF_4N_2Ru$ M 397.572
Pale-yellow solid. Not obt. pure.

Johnson, B.F.G. *et al*, *J. Chem. Soc., Dalton Trans.*, 1977, 719 (*synth, pmr*)

$C_{11}H_{15}Cl_2NORu$ Ru-00199

(Acetonitrile)carbonyldichloro[(1,2,5,6-η)-1,5-cyclooctadiene]ruthenium, 10CI

[62795-62-4]

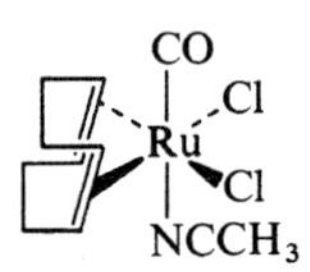

M 349.222

Yellow-orange cryst. Insol. org. solvs. Mp 180° dec. Too insol. and involatile for nmr, ms, or MW studies.

Gould, R.O. *et al*, *J. Chem. Soc.*, *Dalton Trans.*, 1977, 129 (*synth*, *ir*, *struct*)

$C_{11}H_{16}GeO_2Ru$ Ru-00200

Dicarbonyl[(1,2,3,4,5-η)-2,4-cyclohexadien-1-yl](trimethylgermyl)ruthenium, 9CI

[57650-20-1]

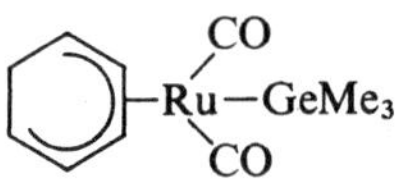

M 353.906

Colourless oil. Sol. hexane. $Bp_{0.1}$ 50°.

Sosinsky, B.A. *et al*, *J. Chem. Soc.*, *Dalton Trans.*, 1975, 1633 (*synth*, *ir*, *pmr*)

$C_{11}H_{16}O_2RuSi$ Ru-00201

Dicarbonyl[(1,2,3,4,5-η)-2,4-cyclohexadien-1-yl](trimethylsilyl)ruthenium, 9CI

[57659-31-1]

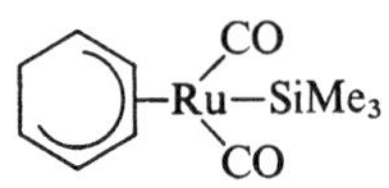

M 309.402

Pale-yellow liq.

Sosinsky, B.A. *et al*, *J. Chem. Soc.*, *Dalton Trans.*, 1975, 1633 (*synth*, *ir*)

$C_{11}H_{19}ClPRu^{\oplus}$ Ru-00202

(η^6-Benzene)chloro(η^2-ethene)(trimethylphosphine)ruthenium(1+), 10CI

(η^6-Benzene)chloro(η^2-ethylene)(trimethylphosphine)ruthenium

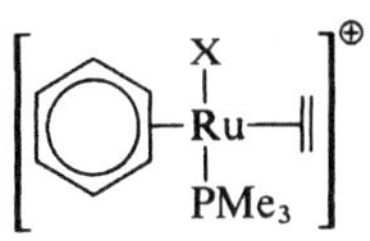

X = Cl

M 318.768 (ion)

Hexafluorophosphate: [71685-25-1].
$C_{11}H_{19}ClF_6P_2Ru$ M 463.732
No phys. props. reported.

Werner, H. *et al*, *J. Organomet. Chem.*, 1979, **174**, C63 (*synth*, *pmr*)

$C_{11}H_{19}PRu$ Ru-00203

(η^6-Benzene)(η^2-ethene)(trimethylphosphine)ruthenium, 10CI

(η^6-Benzene)ethylene(trimethylphosphine)ruthenium

[71685-17-1]

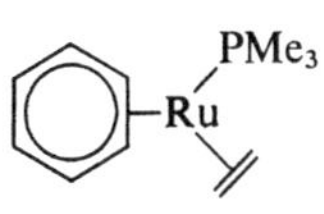

M 283.315

Characterised spectroscopically.

Werner, H. *et al*, *J. Organomet. Chem.*, 1979, **174**, C63 (*synth*, *pmr*)

$C_{11}H_{20}PRu^{\oplus}$ Ru-00204

(η^6-Benzene)(η^2-ethene)hydro(trimethylphosphine)ruthenium(1+), 10CI

(η^6-Benzene)(η^2-ethylene)hydro(trimethylphosphine)ruthenium(1+)

As (η^6-Benzene)chloro(η^2-ethene)(trimethylphosphine)ruthenium(1+), Ru-00202 with

X = H

M 284.323 (ion)

Hexafluorophosphate: [71685-19-3].
$C_{11}H_{20}F_6P_2Ru$ M 429.287
No phys. props. reported.

Werner, H. *et al*, *J. Organomet. Chem.*, 1979, **174**, C63 (*synth*, *pmr*)

$C_{11}H_{23}BrP_2Ru$ Ru-00205

Bromo(η^5-2,4-cyclopentadien-1-yl)bis(trimethylphosphine)ruthenium, 10CI

[75182-16-0]

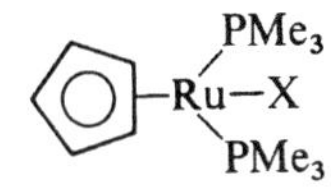

X = Br

M 398.224

Feathery-yellow cryst. (Me_2CO/pet. ether). Mp 144° dec.

Treichel, P.M. *et al*, *Synth. React. Inorg. Metal-Org. Chem.*, 1980, **10**, 205 (*synth*, *pmr*)

$C_{11}H_{23}ClO_6P_2Ru$ Ru-00206

Chloro(η^5-2,4-cyclopentadien-1-yl)bis(trimethyl phosphite-*P*)ruthenium, 10CI

Chloro-π-cyclopentadienylbis(phosphorous acid)ruthenium hexamethyl ester, 8CI. *Cyclopentadienylbis(trimethyl phosphite)ruthenium chloride*

[34822-87-2]

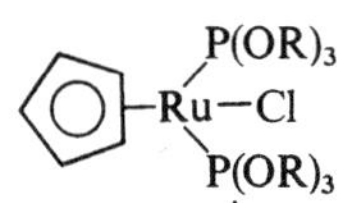

R = Me

M 449.770

Yellow cryst. (hexane). Mp 126-9°.

Blackmore, T. *et al*, *J. Chem. Soc.* (*A*), 1971, 2376 (*synth*, *pmr*)
Ashby, G.S. *et al*, *Aust. J. Chem.*, 1979, **32**, 1003 (*synth*, *ir*, *pmr*)

$C_{11}H_{23}ClP_2Ru$ Ru-00207

Chloro(η^5-2,4-cyclopentadien-1-yl)bis(trimethylphosphine)ruthenium, 10CI

[74558-74-0]

As Bromo(η^5-2,4-cyclopentadien-1-yl)bis(trimethylphosphine)ruthenium, Ru-00205 with

X = Cl

M 353.773

Somewhat air-sensitive orange needles (pet. ether). Mp 130°, 143-4°.

Treichel, P.M. *et al*, *Synth. React. Inorg. Metal-Org. Chem.*, 1980, **10**, 205 (*synth*, *pmr*)
Bruce, M.I. *et al*, *J. Chem. Soc.*, *Dalton Trans.*, 1981, 1398 (*synth*, *struct*, *pmr*, *cmr*)

$C_{11}H_{23}IP_2Ru$ Ru-00208

(η^5-2,4-Cyclopentadien-1-yl)iodobis(trimethylphosphine)ruthenium

[75182-17-1]

As Bromo(η^5-2,4-cyclopentadien-1-yl)bis(trimethylphosphine)ruthenium, Ru-00205 with

X = I

M 445.225

Orange cryst. (Me_2CO/pet. ether). Mp 130° dec.

Treichel, P.M. *et al*, *Synth. React. Inorg. Metal-Org. Chem.*, 1980, **10**, 205 (*synth, ir, pmr*)

$C_{11}H_{23}NOP_2Ru^{⊕⊕}$ Ru-00209

(η^5-2,4-Cyclopentadien-1-yl)nitrosylbis(trimethylphosphine)-ruthenium(2+)

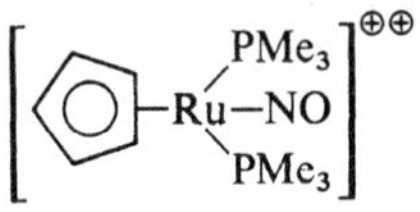

M 348.326 (ion)

Bis(hexafluorophosphate): [75182-21-7].
$C_{11}H_{23}F_{12}NOP_4Ru$ M 638.255
Bright-yellow cryst. (Me_2CO/Et_2O). Mp 245° dec.

Treichel, P.M. *et al*, *Synth. React. Inorg. Metal-Org. Chem.*, 1980, **10**, 205 (*synth, ir, pmr*)

$C_{11}O_{11}Ru_3^{⊖⊖}$ Ru-00210

μ-Carbonyldecacarbonyltriruthenate(2−), 10CI

Undecacarbonyltriruthenate(2−)

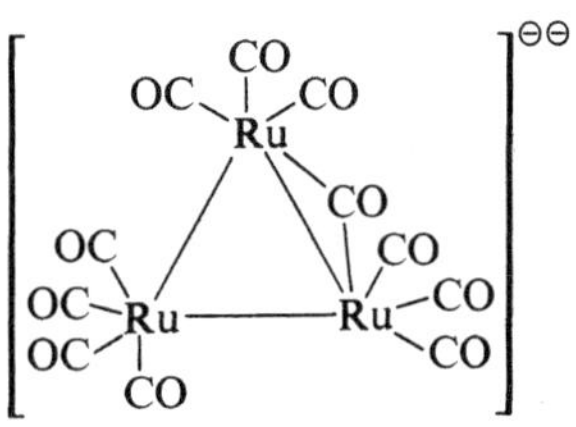

M 611.324 (ion)

Ca salt: [79949-33-0].
$C_{11}O_{11}CaRu$ M 449.264
Orange powder. Air- and water-sensitive.

Nagel, C.C. *et al*, *J. Organomet. Chem.*, 1981, **219**, C9 (*synth, ir*)

$C_{12}Cl_6O_{12}Ru_3$ Ru-00211

Dodecacarbonylhexachlorotriruthenium

[12109-37-4]

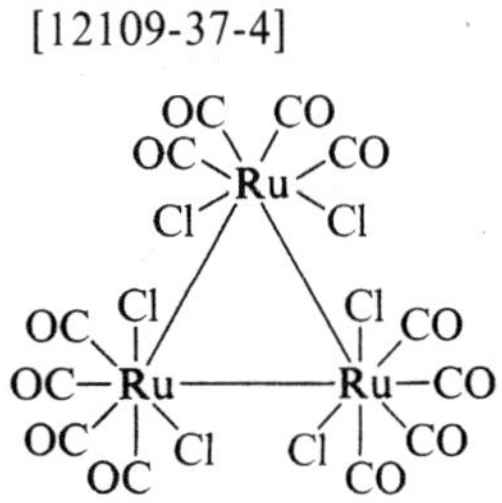

or

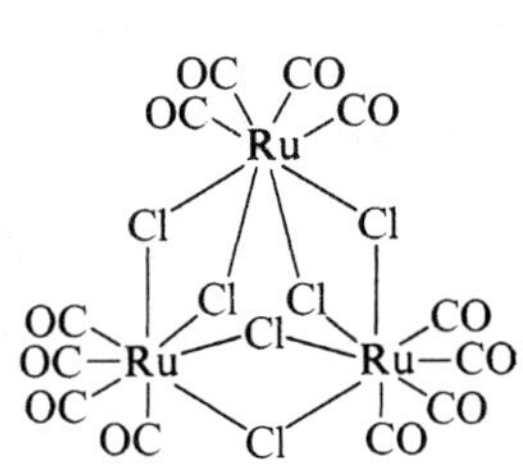

M 852.053

Struct. uncertain, two possibilities discussed. Pale-yellow.

Johnson, B.F.G. *et al*, *J. Chem. Soc. (A)*, 1969, 792 (*synth, ir, struct*)

$C_{12}F_8O_{12}Ru_4$ Ru-00212

Dodecacarbonyltetra-μ-fluorotetrafluorotetraruthenium, 9CI

Dodecacarbonyloctafluorotetraruthenium

[31252-76-3]

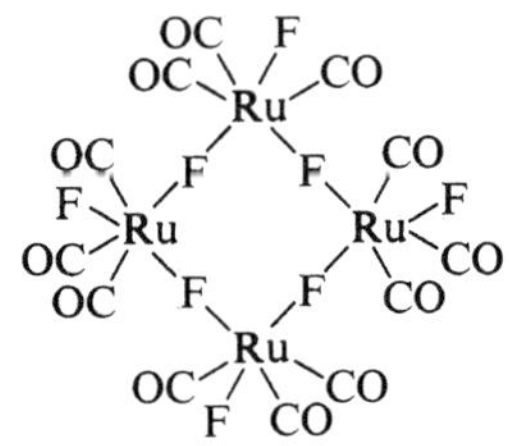

M 892.392

Air-sensitive involatile pale-yellow cryst.

Monomer: Tricarbonyldifluororuthenium.
$C_3F_2O_3Ru$ M 223.098
Unknown.

Peacock, R.D. *et al*, *J. Chem. Soc., Chem. Commun.*, 1970, 1643 (*synth, cryst struct*)
Hewitt, A.J. *et al*, *J. Chem. Soc., Dalton Trans.*, 1976, 579 (*synth, ir*)

$C_{12}FeNO_{13}Ru_3^{⊖}$ Ru-00213

Tri-μ-carbonylnonacarbonyl(nitrosylferrate)triruthenate(1−)

Dodecacarbonyl(nitrosylferrate)triruthenate(1−)

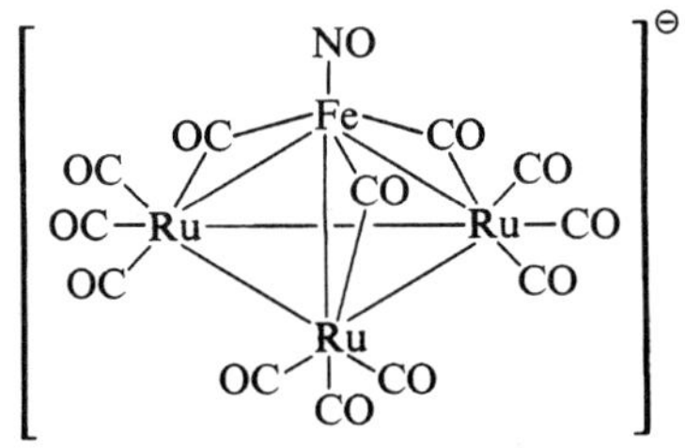

M 725.188 (ion)

Bis(triphenylphosphine)iminium salt:
$C_{48}H_{30}FeN_2O_{13}P_2Ru_3$ M 1263.775
No phys. props. reported.

Fjare, D.E. *et al*, *J. Am. Chem. Soc.*, 1981, **103**, 1572 (*synth, struct*)

$C_{12}FeO_{12}Ru_2$ Ru-00214

Octacarbonyl(tetracarbonyliron)diruthenium

Octacarbonyl[μ-(tetracarbonylferrio)]diruthenium, 8CI.
Tetracarbonyl(octacarbonyldiruthenio)iron, 8CI.
Dodecacarbonylirondiruthenium

[12388-68-0]

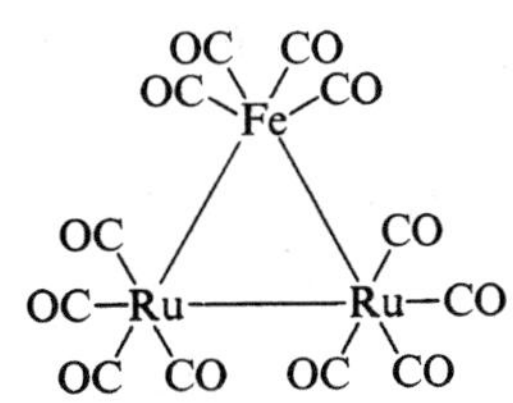

M 594.112

Thought to be isostructural with $Ru_3(CO)_{12}$. Brown cryst.

Knight, J. *et al*, *Chem. Ind. (London)*, 1968, 1159 (*synth, ir*)
Yawney, D.B.W. *et al*, *J. Chem. Soc. (A)*, 1969, 502 (*synth, ir, uv, ms*)
Tyler, D.R. *et al*, *J. Am. Chem. Soc.*, 1978, **100**, 7888 (*uv*)
Baev, A.K. *et al*, *J. Organomet. Chem.*, 1981, **213**, 151 (*props*)

$C_{12}Fe_2O_{12}Ru$ Ru-00215

Tetracarbonyl(di-μ-carbonylhexacarbonyldiiron)ruthenium, 9CI

Di-μ-carbonylhexacarbonyl[μ-(tetracarbonylruthenio-)]diiron, 8CI. *Tetracarbonyl(di-μ-carbonylhexacarbonyl-diferrio)ruthenium*, 8CI. *Dodecacarbonyldiironruthenium*

[20468-34-2]

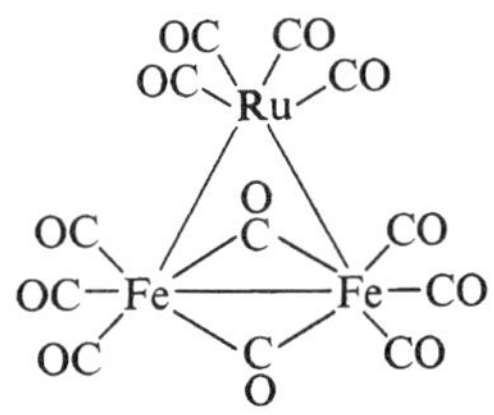

M 548.889

Violet cryst. Mp >125° dec.

Knight, J.A. *et al*, *Chem. Ind. (London)*, 1968, 1159 (*synth*)
Yawney, D.B.W. *et al*, *J. Chem. Soc. (A)*, 1969, 502 (*synth, ir, uv, ms*)
Knight, J.A. *et al*, *J. Chem. Soc., Chem. Commun.*, 1970, 1006 (*ir*)
Foster, A. *et al*, *J. Chem. Soc., Chem. Commun.*, 1974, 1042 (*cmr*)
Tyler, D.R. *et al*, *J. Am. Chem. Soc.*, 1978, **100**, 7888 (*struct*)

$C_{12}HFeNO_{12}Ru_3$ Ru-00216

Nonacarbonyl-μ-hydro-μ₄-nitrido(tricarbonyliron)triruthenium

Dodecacarbonylhydronitridoirontriruthenium

[76791-95-2]

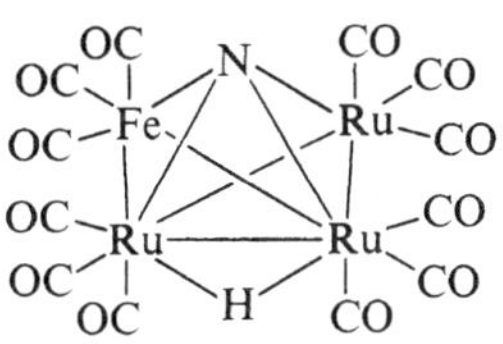

M 710.196

Probable struct. ir ν_{CO} 2104w, 2071s, 2062s, 2043s, 2029s, 2000m, 1991m, 1968w, 1955w cm^{-1} (hexane).

Fjare, D.E. *et al*, *J. Am. Chem. Soc.*, 1981, **103**, 1572 (*synth*)

$C_{12}HO_{12}Ru_3^{\oplus}$ Ru-00217

Dodecacarbonyl-μ-hydrotriruthenium(1+)

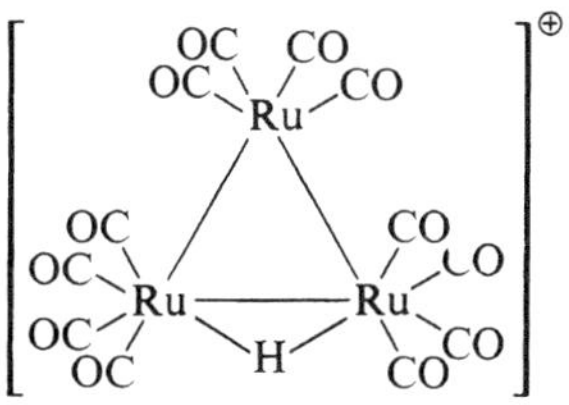

M 640.343 (ion)

Hexafluorophosphate: [26220-98-4].
$C_{12}HF_6O_{12}PRu_3$ M 785.307
Bright-yellow solid. Sol. $MeNO_2$, Me_2CO.

Knight, J. *et al*, *J. Chem. Soc. (A)*, 1970, 711 (*synth, ir, pmr*)

$C_{12}H_2O_{12}Ru_4^{\ominus\ominus}$ Ru-00218

Tri-μ-carbonylnonacarbonyldi-μ-hydrotetraruthenate(2−), 10CI

Dodecacarbonyldihydrotetraruthenate(2−)

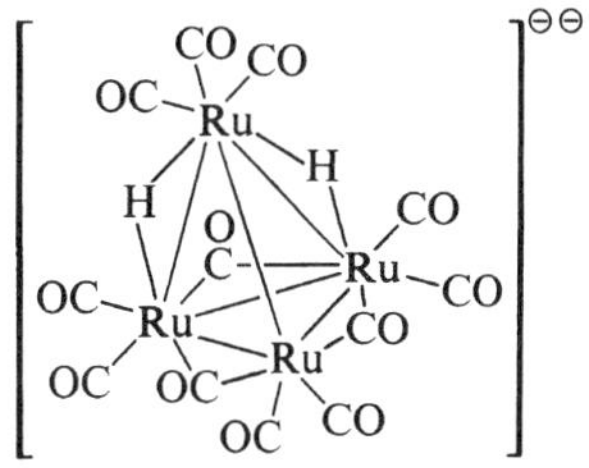

M 742.421 (ion)

Proposed struct. shown.

Di-K salt: [67269-69-6].
$C_{12}H_2K_2O_{12}Ru_4$ M 820.617
Yellow powder.

Inkrott, K.E. *et al*, *Inorg. Chem.*, 1979, **18**, 2817 (*synth, ir, pmr, cmr*)

$C_{12}H_3BO_{12}Ru_4$ Ru-00219

[μ₄-(Boryl-*B*:*B*:*H*:*H'*)]dodecacarbonyl-μ-hydrotetraruthenium, 10CI

[63374-84-5]

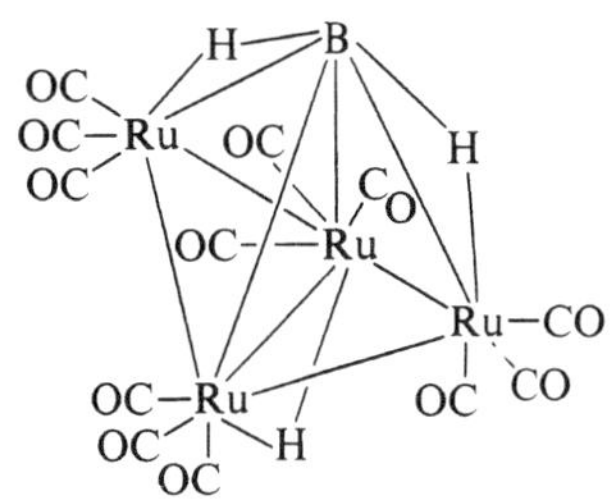

M 754.239

Air-sensitive yellow solid.

Eady, C.R. *et al*, *J. Chem. Soc., Dalton Trans.*, 1977, 477 (*synth, ms, ir, pmr*)

$C_{12}H_3O_{12}Ru_4^{\ominus}$ Ru-00220

Dodecacarbonyltrihydrotetraruthenate(1−), 10CI

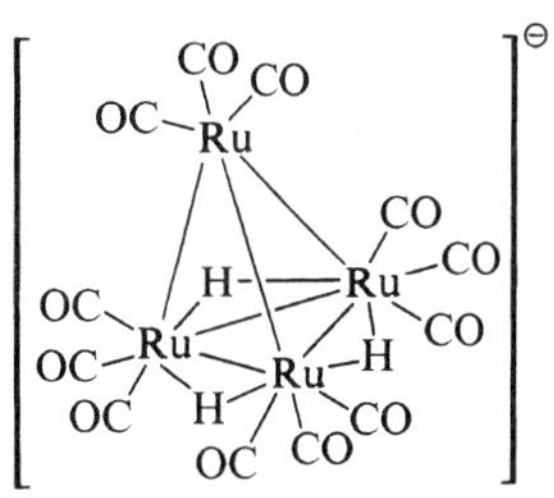

M 743.429 (ion)

Tetraphenylarsonium salt: [56669-40-0].
$C_{36}H_{23}AsO_{12}Ru_4$ M 1126.772
No phys. props. reported.

Bis(triphenylphosphine)iminium salt: [70073-17-5].
$C_{48}H_{33}NO_{12}P_2Ru_4$ M 1282.016
Yellow-orange solid.

Koepke, J.W. *et al*, *J. Am. Chem. Soc.*, 1975, **97**, 3947 (*synth, ir, raman*)
Jackson, P.F. *et al*, *J. Chem. Soc., Chem. Commun.*, 1978, 920 (*cryst struct*)
Inkrott, K.E. *et al*, *Inorg. Chem.*, 1979, **18**, 2817 (*synth, cmr*)

$C_{12}H_4FeO_{12}Ru_3$ Ru-00221

Dodecacarbonyltetrahydro(iron)triruthenium, 10CI, 8CI

[11064-20-3]

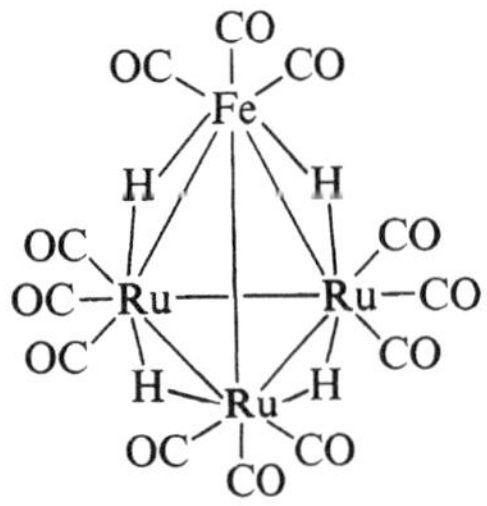

M 699.213

Orange-red or dark-red powder (CH_2Cl_2/hexane). pK_a 13.4 (MeOH at 25°).

Knox, S.A.R. *et al*, *J. Am. Chem. Soc.*, 1975, **97**, 3942 (*synth, ir, pmr, ms*)
Walker, H.W. *et al*, *J. Am. Chem. Soc.*, 1979, **101**, 7428.
Foley, H.C. *et al*, *J. Am. Chem. Soc.*, 1981, **103**, 7176.

$C_{12}H_4O_{10}Ru_3$ Ru-00222

Decacarbonyl-μ-ethenylidenedi-μ-hydrotriruthenium, 9CI

[63322-99-6]

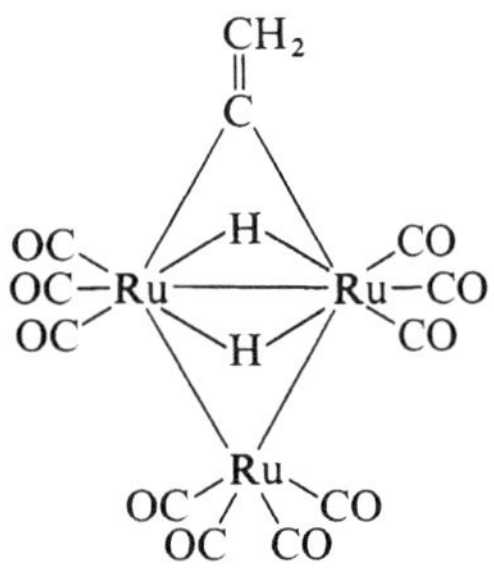

Possible structure

M 611.368

Not isolated spectroscopically pure.

Eady, C.R. *et al*, *J. Chem. Soc., Dalton Trans.*, 1977, 477 (*synth, pmr, ms*)

$C_{12}H_4O_{11}Ru_3$ Ru-00223

Decacarbonyl-μ-hydro[μ-(methoxymethylidyne)]triruthenium, 10CI

[71737-42-3]

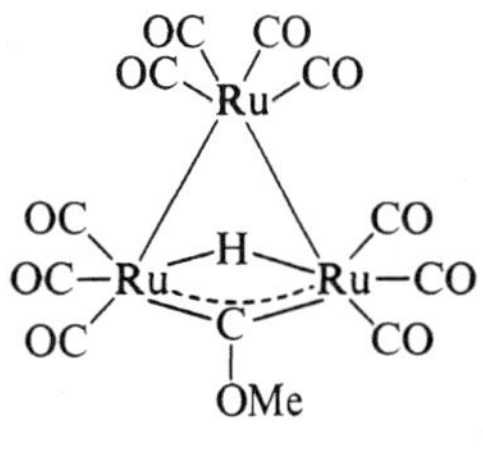

M 627.367

Reacts with H_2 at 80° to give H_3Ru_3 (μ_3-$COCH_3$)-$(CO)_9$. Yellow cryst. (MeOH).

Johnson, B.F.G. *et al*, *J. Organomet. Chem.*, 1979, **173**, 187 (*synth, cmr, cryst struct*)
Keister, J.B. *et al*, *Organometallics*, 1983, **2**, 219 (*synth, ir, pmr, cmr*)

$C_{12}H_4O_{12}Ru_4$ Ru-00224

Dodecacarbonyltetra-μ-hydrotetraruthenium, 8CI

[34438-91-0]

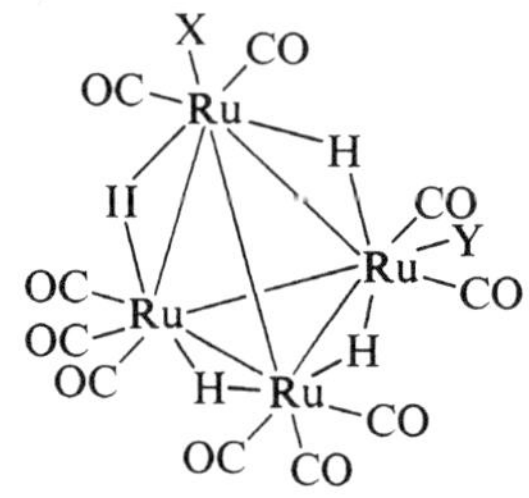

X = Y = CO

M 744.436

Water gas shift reaction and hydroformylation catalyst. Yellow-orange microcryst. Mod. stable to air and moisture, slowly dec. in soln. Mod. sol. hydrocarbons, sol. CH_2Cl_2. pK_a 11.8 (MeOH, 25°). 2 forms formerly thought to exist, later proved to have one only.

Johnson, B.F.G. *et al*, *J. Chem. Soc. (A)*, 1968, 2856 (*synth, ir, pmr*)
Lewis, J. *et al*, *Acc. Chem. Res.*, 1968, **1**, 245 (*ms*)
Kaesz, H.D. *et al*, *J. Chem. Soc., Chem. Commun.*, 1971, 477 (*synth*)
Kaesz, H.D. *et al*, *J. Am. Chem. Soc.*, 1971, **93**, 4594 (*pmr*)
Piacenti, F. *et al*, *Inorg. Chem.*, 1971, **10**, 2759 (*synth*)
Banditelli, P. *et al*, *Thermochim. Acta*, 1975, **12**, 277.
Knox, S.A.R. *et al*, *J. Am. Chem. Soc.*, 1975, **97**, 3942 (*ir, pmr, ms, raman*)
Wilson, R.D. *et al*, *Inorg. Chem.*, 1978, **17**, 1271 (*cryst struct*)
Walker, H.W. *et al*, *J. Am. Chem. Soc.*, 1979, **101**, 7428.

$C_{12}H_5O_7ReRu$ Ru-00225

Dicarbonyl(η^5-2,4-cyclopentadien-1-yl)(pentacarbonylrhenio)ruthenium

Cyclopentadienylrutheniumdicarbonylpentacarbonylrhenium. (η^5-Cyclopentadienyl)dicarbonylrutheniumpentacarbonylrhenium. Heptacarbonyl(η^5-cyclopentadienyl)rheniumruthenium

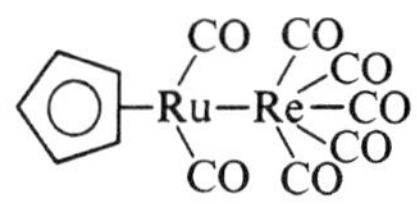

M 548.444

Yellow solid. $Bp_{0.1}$ 60° subl.

Blackmore, T. *et al*, *J. Chem. Soc. (A)*, 1968, 2931 (*synth, ir*)
Bruce, M.I., *Int. J. Mass Spectrom. Ion Phys.*, 1968, **1**, 141 (*ms*)

$C_{12}H_6O_{10}Ru_3S$ Ru-00226

Decacarbonyl[μ-(ethanethiolato)]-μ-hydrotriruthenium

[23733-20-2]

As Decacarbonyl-μ-hydro-μ-nitrosyltriruthenium, Ru-00121 with

X = H, Y = SEt

M 645.443

Orange cryst. (pentane). Mp 110-3° dec.

Crooks, G.R. *et al*, *J. Chem. Soc. (A)*, 1969, 797 (*synth, ir, pmr*)

$C_{12}H_6O_{11}Ru_3$ Ru-00227

Nonacarbonyltri-μ-hydro[μ_3-(2-methoxy-2-oxoethylidyne)]triruthenium, 10CI

[73746-98-2]

As [μ_3-(Bromomethylidyne)]nonacarbonyltri-μ-hydrotriruthenium, Ru-00122 with

X = COOMe

M 629.383

Bright-yellow cryst.

Keister, J.B. *et al*, *Inorg. Chem.*, 1980, **19**, 2304 (*synth, ir, pmr, ms*)

$C_{12}H_7O_{10}Ru_3S^{\oplus}$ Ru-00228

Decacarbonyl(μ-ethanethiolato)di-μ-hydrotriruthenium(1+), 8CI

[32574-34-8]

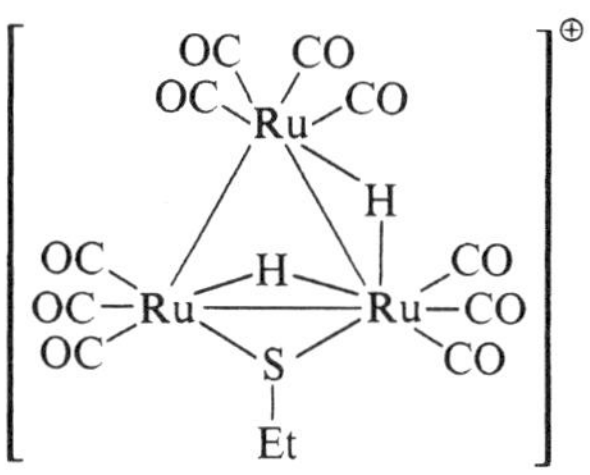

M 646.451 (ion)

Prepd. in 98% H_2SO_4 soln. only, no salts isolated.

Lewis, J. *et al*, *J. Chem. Soc. (A)*, 1971, 1797 (*synth, pmr*)

$C_{12}H_{10}O_2Ru$ Ru-00229

Dicarbonyl-2,4-cyclopentadien-1-yl(η^5-2,4-cyclopentadien-1-yl)ruthenium

σ-Cyclopentadienyl-π-cyclopentadienyldicarbonylruthenium. Dicarbonylbis(cyclopentadienyl)ruthenium

[31870-71-0]

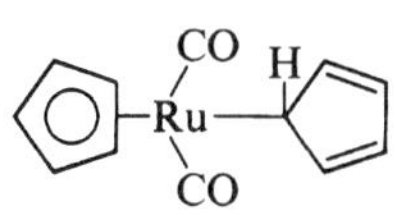

M 287.280

Fluxional σ-bonded C_5H_5 ring. Yellow cryst. (2-methylpentane). Mp 52-3°.

Cotton, F.A. *et al*, *J. Am. Chem. Soc.*, 1969, **91**, 7523 (*pmr*)
Campbell, C.H. *et al*, *J. Chem. Soc. (A)*, 1970, 1318 (*synth, ir, pmr*)

$C_{12}H_{10}O_2Ru$ Ru-00230

1,1′-Ruthenocenedicarboxaldehyde

1,1′-Diformylruthenocene, 10CI

[79354-80-6]

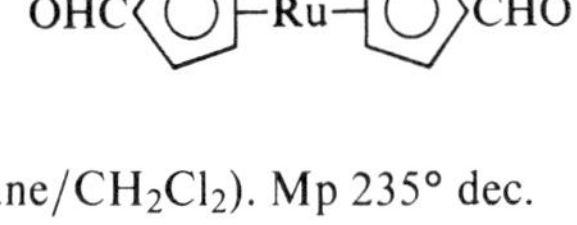

M 287.280

Cryst. (hexane/CH_2Cl_2). Mp 235° dec.

Kamiyama, S. *et al*, *Bull. Chem. Soc. Jpn.*, 1981, **54**, 2079 (*synth, ir, pmr, ms*)

$C_{12}H_{10}O_2RuS_2$ Ru-00231

1,1′-Ruthenocenebis(thioic acid)

Bis[(thiocarboxy)-π-cyclopentadienyl]ruthenium, 8CI. *1,1′-Di(thiocarboxy)ruthenocene*

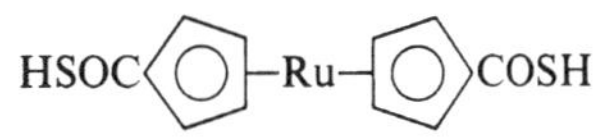

M 351.400

U.S.P., 3 387 009, (*1968*); *CA*, **69**, 87191

$C_{12}H_{10}O_4Ru$ Ru-00232

1,1′-Ruthenocenedicarboxylic acid

1,1′-Dicarboxyruthenocene. Bis(carboxycyclopentadienyl)ruthenium

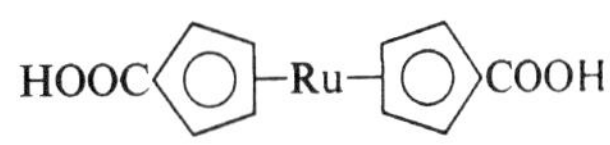

M 319.279

Yellow cryst. (AcOH). Dec. at ca. 325°.

Rausch, M.D. *et al*, *J. Am. Chem. Soc.*, 1960, **82**, 76 (*synth, ir*)
Mulay, L.N. *et al*, *Naturwissenschaften*, 1962, **49**, 446 (*pmr*)
Rapić, V. *et al*, *J. Organomet. Chem.*, 1975, **94**, 87 (*synth*)

$C_{12}H_{10}Ru$ Ru-00233

Ethynylruthenocene

(π-Cyclopentadienyl)(ethynyl-π-cyclopentadienyl)ruthenium, 8CI. *Ruthenocenylacetylene*

[12337-22-3]

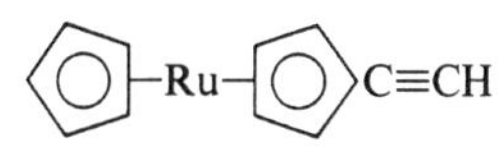

M 255.281

Pale-yellow cryst. (pentane). Mp 73-4°.

Hofer, O. *et al*, *J. Organomet. Chem.*, 1968, **13**, 443 (*synth, pmr, ir*)
Rausch, M.D. *et al*, *J. Org. Chem.*, 1969, **34**, 1974 (*synth, pmr, ir*)

$C_{12}H_{11}ClRu^{\oplus\oplus}$ Ru-00234

(η^6-Benzene)(η^6-chlorobenzene)ruthenium(2+), 10CI

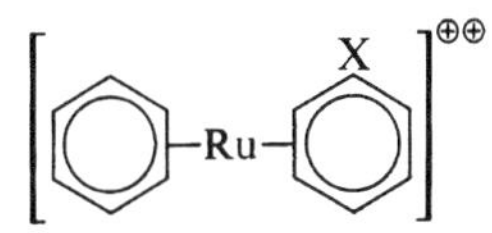

X = Cl

M 291.742 (ion)

Air-stable.

Bis(tetrafluoroborate): [71825-94-0].
$C_{12}H_{11}B_2ClF_8Ru$ M 465.349
Colour and Mp not reported.

Bennett, M.A. *et al*, *J. Organomet. Chem.*, 1979, **175**, 87 (*synth, pmr*)

$C_{12}H_{11}ClRu$ Ru-00235

(1-Chloroethenyl)ruthenocene

[(1-Chlorovinyl)-π-cyclopentadienyl]-π-cyclopentadienylruthenium, 8CI. *(1-Chlorovinyl)ruthenocene. 1-Chloro-1-ruthenocenylethylene*

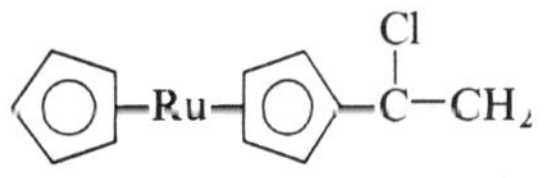

M 291.742

Mp 70-1°.

Hofer, O. *et al*, *J. Organomet. Chem.*, 1968, **13**, 443 (*synth, ir*)

$C_{12}H_{11}NO_3Ru$ Ru-00236

Tricarbonyl[(1,4,5,6-η)-2-cyano-5-cyclooctene-1,4-diyl]ruthenium

Tricarbonyl(2-cyano-π,σ-5-cycloocten-1,4-ylene)ruthenium, 8CI. *[(η-1,2,3,6)-5-Cyanocyclooctadienyl]ruthenium tricarbonyl*

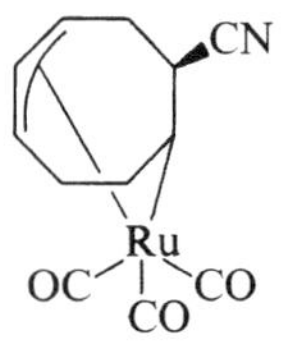

exo-form

M 318.294

exo-form [34738-50-6]

Air-stable orthorhombic cryst.

Green, M. *et al*, *J. Chem. Soc., Chem. Commun.*, 1971, 621 (*synth*)

Cotton, F.A. *et al*, *J. Am. Chem. Soc.*, 1971, **93**, 4626 (*synth, cryst struct*)

$C_{12}H_{12}Br_4Ru_2$ Ru-00237

Bis(η⁶-benzene)di-μ-bromodibromodiruthenium, 9CI

Bis(η⁶-benzene)tetrabromodiruthenium

[37362-03-1]

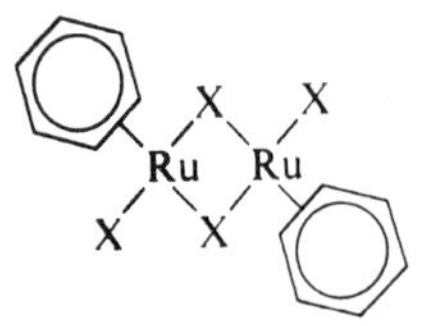

X = Br

M 677.983

Bridged dimer. Red. Dec. >250°.

Monomer: (η⁶-Benzene)dibromoruthenium.

$C_6H_6Br_2Ru$ M 338.991

Unknown.

Zelonka, R.A. *et al*, *Can. J. Chem.*, 1972, **50**, 3063 (*synth, pmr*)

Bennett, M.A. *et al*, *J. Chem. Soc., Dalton Trans.*, 1974, 233 (*synth, ir, pmr*)

Games, D.C. *et al*, *J. Organomet. Chem.*, 1980, 193, 229 (*ms*)

$C_{12}H_{12}Cl_3Ru_2^{\oplus}$ Ru-00238

Bis(η⁶-benzene)tri-μ-chlorodiruthenium(1+), 9CI

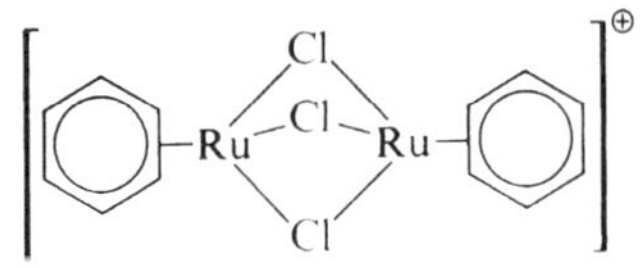

M 464.726 (ion)

Hexafluorophosphate: [52462-47-2].

$C_{12}H_{12}Cl_3F_6Ru_2$ M 578.716

Orange solid. Mp 245°.

Tetrafluoroborate:

$C_{12}H_{12}BCl_3F_4Ru_2$ M 551.529

Orange solid. Mp 270° dec.

Bennett, M.A. *et al*, *J. Chem. Soc., Dalton Trans.*, 1974, 233 (*synth, ir*)

Stephenson, T.A. *et al*, *J. Organomet. Chem.*, 1981, **208**, 369 (*synth, pmr, cmr*)

$C_{12}H_{12}I_4Ru_2$ Ru-00239

Bis(η⁶-benzene)di-μ-iododiiododiruthenium, 9CI

Bis(η⁶-benzene)tetraiododiruthenium

[37375-79-4]

As Bis(η⁶-benzene)di-μ-bromodibromodiruthenium, Ru-00237 with

X = I

M 865.985

Bridged dimer. Violet. Mp 255°.

Monomer: (η⁶-Benzene)diiodoruthenium.

$C_6H_6I_2Ru$ M 432.992

Unknown.

Zelonka, R.A. *et al*, *Can. J. Chem.*, 1972, **50**, 3063 (*synth, pmr*)

Bennett, M.A. *et al*, *J. Chem. Soc., Dalton Trans.*, 1974, 233 (*synth, ir, pmr*)

Games, D.E. *et al*, *J. Organomet. Chem.*, 1980, **193**, 229 (*ms*)

$C_{12}H_{12}ORu$ Ru-00240

Acetylruthenocene

(Acetyl-π-cyclopentadienyl)-π-cyclopentadienylruthenocene, 8CI. *1-Ruthenocenylethanone. Methyl ruthenocenyl ketone*

[1316-92-3]

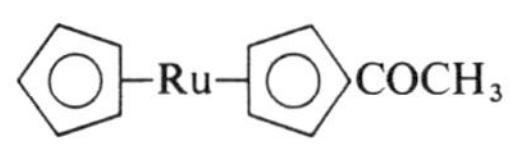

M 273.296

Light-yellow cryst. (hexane). Mp 113.5-114°.

Rausch, M.D. *et al*, *J. Am. Chem. Soc.*, 1960, **82**, 76 (*synth, ir*)

Hill, E.A. *et al*, *J. Am. Chem. Soc.*, 1961, **83**, 3840 (*synth*)

Rausch, M.D. *et al*, *J. Org. Chem.*, 1963, **28**, 3225 (*nmr*)

$C_{12}H_{12}ORu^{\oplus\oplus}$ Ru-00241

(η⁶-Benzene)(η⁶-phenol)ruthenium(2+), 10CI

As (η⁶-Benzene)(η⁶-chlorobenzene)ruthenium(2+), Ru-00234 with

X = OH

M 273.296 (ion)

Air-stable.

Bis(tetrafluoroborate): [71825-96-2].

$C_{12}H_{12}B_2F_8ORu$ M 446.903

Colour and Mp not reported.

Me ether: see (η⁶-Benzene)[(1,2,3,4,5,6-η)-methoxybenzene]ruthenium(2+), Ru-00310

Bennett, M.A. *et al*, *J. Organomet. Chem.*, 1979, **175**, 87 (*synth, pmr*)

$C_{12}H_{12}ORu$ Ru-00242

1′-Methylruthenocenecarboxaldehyde

(Formyl-π-cyclopentadienyl)(methyl-π-cyclopentadienyl)ruthenium, 8CI. *1-Formyl-1′-methylruthenocene*

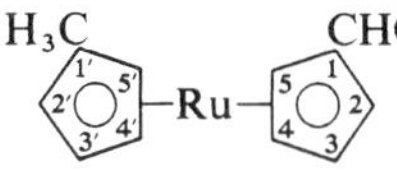

M 273.296

Hofer, O. *et al*, *J. Organomet. Chem.*, 1968, **13**, 443 (*synth, pmr, ir*)

$C_{12}H_{12}ORu$ Ru-00243

2-Methylruthenocenecarboxaldehyde

π-Cyclopentadienyl(1-formyl-2-methyl-π-cyclopentadienyl)ruthenium, 8CI. *1-Formyl-2-methylruthenocene*

M 273.296

(+)-***form***

Mp 70-3°. $[\alpha]_D$ +62° (EtOH).

Hofer, O. *et al*, *J. Organomet. Chem.*, 1968, **13**, 443, 457 (*synth, pmr, ir, resoln*)

$C_{12}H_{12}ORu$ Ru-00244

3-Methylruthenocenecarboxaldehyde

π-Cyclopentadienyl(1-formyl-3-methyl-π-cyclopentadienyl)ruthenium, 8CI. *1-Formyl-3-methylruthenocene*

M 273.296

Hofer, O. *et al*, *J. Organomet. Chem.*, 1968, **13**, 443 (*synth, pmr, ir*)

$C_{12}H_{12}O_2Ru$ Ru-00245

1′-Methylruthenocenecarboxylic acid

(Carboxy-π-cyclopentadienyl)(methyl-π-cyclopentadienyl)ruthenium, 8CI. *1-Carboxy-1′-methylruthenocene*

M 289.296

Mp 195-7°.

Me ester:

$C_{13}H_{14}O_2Ru$ M 303.322

Yellow cryst. Mp 79-83°.

Nitrile: (Cyano-π-cyclopentadienyl)(methyl-π-cyclopentadienyl)ruthenium, 8CI. *1-Cyano-1′-methylruthenocene.*

$C_{12}H_{11}NRu$ M 270.296

Yellow oil.

Hofer, O. *et al*, *J. Organomet. Chem.*, 1968, **13**, 443 (*synth, pmr, ir*)

Handle all chemicals with care

$C_{12}H_{12}O_2Ru$ Ru-00246

2-Methylruthenocenecarboxylic acid

(1-Carboxy-2-methyl-π-cyclopentadienyl)π-cyclopentadienylruthenium, 8CI. *1-Carboxy-2-methylruthenocene*

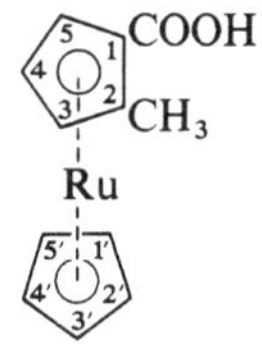

(1*S*)-*form*

M 289.296

(*1S*)-***form***

Mp 191-7°. $[\alpha]_D$ +47°.

Me ester:

$C_{13}H_{14}O_2Ru$ M 303.322

Yellow oil. $[\alpha]_D$ +21° (EtOH).

(±)-***form***

Mp 171-5°.

Me ester: Pale-yellow oil.

Nitrile: (1-Cyano-2-methyl-π-cyclopentadienyl)(π-cyclopentadienyl)ruthenium, 8CI. *1-Cyano-2-methylruthenocene.*

$C_{12}H_{11}NRu$ M 270.296

Yellow oil.

Hofer, O. *et al*, *J. Organomet. Chem.*, 1968, **13**, 443 (*synth, pmr, ir, uv*)

Hofer, O. *et al*, *J. Organomet. Chem.*, 1968, **13**, 457 (*synth, resoln, ord*)

$C_{12}H_{12}O_2Ru$ Ru-00247

3-Methylruthenocenecarboxylic acid

(1-Carboxy-3-methyl-π-cyclopentadienyl)-π-cyclopentadienylruthenium, 8CI. *1-Carboxy-3-methylruthenocene*

M 289.296

Mp 176-9°.

Me ester:

$C_{13}H_{14}O_2Ru$ M 303.322

Pale-yellow oil.

Nitrile: (1-Cyano-3-methyl-π-cyclopentadienyl)(π-cyclopentadienyl)ruthenium, 8CI. *1-Cyano-3-methylruthenocene.*

$C_{12}H_{11}NRu$ M 270.296

Yellow oil.

Hofer, O. *et al*, *J. Organomet. Chem.*, 1968, **13**, 443 (*synth, pmr, ir*)

$C_{12}H_{12}O_8Ru_2Sn_2$ Ru-00248

Octacarbonylbis[μ-(dimethylstannylene)]diruthenium

[25158-72-9]

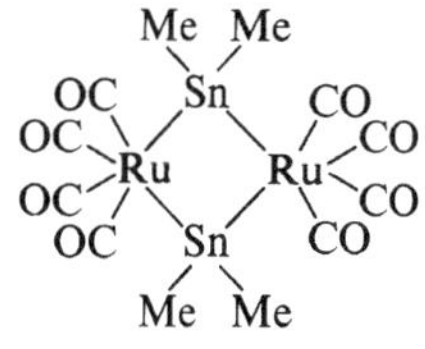

M 723.742

Yellow-orange solid. Mp 130°.

Knox, S.A.R. *et al*, *J. Chem. Soc. (A)*, 1969, 2559 (*synth, ir*)

$C_{12}H_{12}OsRu_2$ **Ru-00249**

Tetracarbonyl(octacarbonyldiruthenium)osmium, 10CI

Tetracarbonyl(octacarbonyldiruthenio)osmium, 8CI. *Dodecacarbonylosmiumdiruthenium*

[12389-47-8]

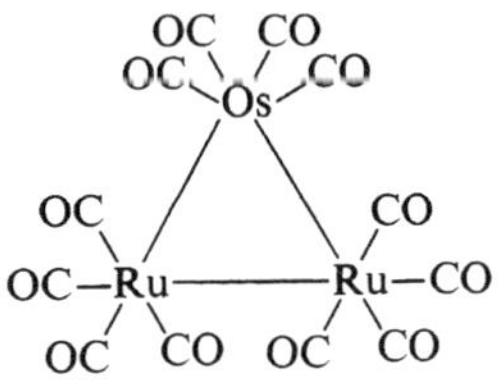

Presumed structure

M 548.567

Shows rapid exchange of axial and equatorial CO groups above 30°.

Johnson, B.F.G. *et al*, *J. Chem. Soc., Chem. Commun.*, 1968, 861 (*synth*)
Lewis, J. *et al*, *Acc. Chem. Res.*, 1968, **1**, 245 (*ms*)
Ferrari, R.P. *et al*, *J. Chem. Soc., Dalton Trans.*, 1978, 1164 (*synth*, *ms*)
Kettle, S.F.A. *et al*, *Inorg. Chem.*, 1979, **18**, 2749 (*ir*, *raman*)
Fox, J.R. *et al*, *Inorg. Chem.*, 1980, **19**, 2574 (*synth*)
Koridze, A.A. *et al*, *Izv. Akad. Nauk SSSR, Ser. Khim.*, 1981, 1181; *CA*, **95**, 70500y (*nmr*)

$C_{12}H_{12}Ru$ **Ru-00250**

[(1,2,3,4-η)-Benzene](η^6-benzene)ruthenium, 10CI

Dibenzeneruthenium

[69191-10-2]

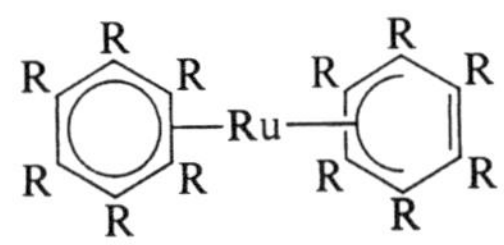

R = H

M 257.297

Thermally unstable orange cryst. Mp <0° dec.

Timms, P.L. *et al*, *J. Chem. Soc., Chem. Commun.*, 1978, 898 (*synth*, *pmr*)

$C_{12}H_{12}Ru^{\oplus\oplus}$ **Ru-00251**

Bis(η^6-Benzene)ruthenium(2+), 10CI

As (η^6-Benzene)(η^6-chlorobenzene)ruthenium(2+), Ru-00234 with

X = H

M 257.297 (ion)

Bis(hexafluorophosphate): [76687-31-5].
$C_{12}H_{12}F_{12}P_2Ru$ M 547.225
Brown.

Diperchlorate:
$C_{12}H_{12}Cl_2O_8Ru$ M 456.198
Pale-yellow cryst. Air-stable.

▷Explodes at high temps.

Bennett, M.A. *et al*, *J. Organomet. Chem.*, 1979, **175**, 87 (*synth*, *pmr*)
Domaille, P.J. *et al*, *J. Organomet. Chem.*, 1980, **202**, 191 (*synth*)

$C_{12}H_{12}Ru$ **Ru-00252**

Vinylruthenocene

π-Cyclopentadienyl(vinyl-π-cyclopentadienyl)ruthenium, 8CI. *Ethenylruthenocene*

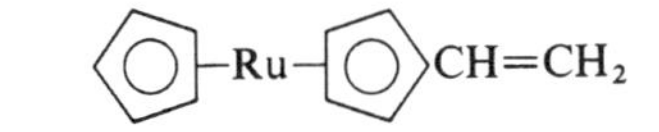

M 257.297

Mp 53.5-54.5°.

Buell, G.R. *et al*, *J. Am. Chem. Soc.*, 1962, **84**, 40 (*synth*)
Rausch, M.D. *et al*, *J. Organomet. Chem.*, 1968, **11**, 317 (*synth*, *pmr*)

$C_{12}H_{13}ORu^{\oplus}$ **Ru-00253**

(η^5-2,4-Cyclopentadien-1-yl)[(1,2,3,4,5,6-η)-methoxybenzene]ruthenium(1+), 10CI

(η^6-Anisole)(η^5-cyclopentadienyl)ruthenium(1+)

As (η^6-Chlorobenzene)(η^5-2,4-cyclopentadien-1-yl)-ruthenium(1+), Ru-00183 with

X = OMe

M 274.304 (ion)

Tetrafluoroborate: [77170-74-2].
$C_{12}H_{13}BF_4ORu$ M 361.108
Mp 131-4°. Dec. >300°.

Tetraphenylborate: [82466-20-4].
$C_{36}H_{33}BORu$ M 593.536
Greyish-white solid. Mp 123-5°.

Nesmeyanov, A.N. *et al*, *Dokl. Akad. Nauk SSSR, Ser. Sci. Khim.*, 1980, **254**, 480 (*synth*, *ir*, *pmr*)
Robertson, I.W. *et al*, *J. Organomet. Chem.*, 1982, **228**, 171 (*synth*, *pmr*, *cmr*)

$C_{12}H_{13}Ru^{\oplus}$ **Ru-00254**

(η^5-2,4-Cyclopentadien-1-yl)[(1,2,3,4,5,6-η)-methylbenzene]ruthenium(1+), 10CI

(η^5-Cyclopentadienyl)(η^6-toluene)ruthenium(1+)

As (η^6-Chlorobenzene)(η^5-2,4-cyclopentadien-1-yl)-ruthenium(1+), Ru-00183 with

R = CH_3

M 258.305 (ion)

Hexafluorophosphate: [78618-95-8].
$C_{12}H_{13}F_6PRu$ M 403.269
Yellow cryst. (EtOH). Dec. at 220-6°.

Tetraphenylborate: [72804-56-9].
$C_{36}H_{33}BRu$ M 577.537
Yellow. Mp ca. 260°.

Nesmeyanov, A.N. *et al*, *J. Organomet. Chem.*, 1979, **182**, C36 (*synth*, *pmr*)
Nesmeyanov, A.N. *et al*, *Dokl. Chem. (Engl. Transl.)*, 1981, **258**, 193 (*synth*, *pmr*)

$C_{12}H_{14}ClRu^{\oplus}$ **Ru-00255**

(η^6-Benzene)chloro[(1,2,3,4-η)-1,3-cyclohexadiene]ruthenium(1+), 9CI

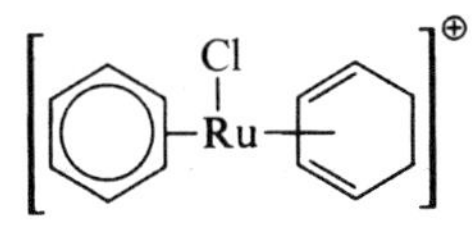

M 294.766 (ion)

Tetraphenylborate: [39732-39-3].

$C_{36}H_{34}BClRu$ M 613.998
Too unstable to purify.

Zelonka, R.A. *et al*, *Can. J. Chem.*, 1972, **50**, 3063 (*synth*)

$C_{12}H_{14}ORu$ Ru-00256

(1-Hydroxyet.yl)ruthenocene

1-Ruthenocenylethanol. α-Methylruthenocenemethanol

[12089-68-8]

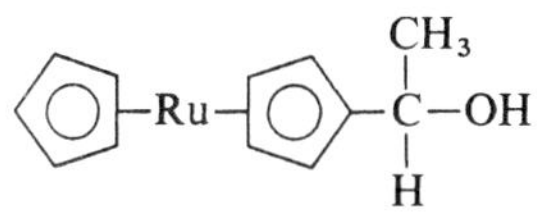

M 275.312

(+)-form

$[\alpha]_D^{25}$ +20.7° (c, 1.5 in C_6H_6).

(−)-form

$[\alpha]_D^{25}$ −13.1° (c, 1.5 in C_6H_6).

(±)-form

Colourless cryst. (hexane). Mp 63-4°.

Ac:

$C_{14}H_{16}O_2Ru$ M 317.349
White cryst. (isopropyl ether or by subl.). Mp 64-64.5°.

Hill, E.A. *et al*, *J. Am. Chem. Soc.*, 1961, **83**, 3840, 4216 (*synth*)
Buell, G.R. *et al*, *J. Am. Chem. Soc.*, 1962, **83**, 40 (*synth*)
Turbitt, T.D. *et al*, *J. Chem. Soc., Perkin Trans. 2*, 1974, 177 (*synth, resoln*)

$C_{12}H_{14}ORu$ Ru-00257

1-(Hydroxymethyl)-2-methylruthenocene

π-Cyclopentadienyl[1-(hydroxymethyl)-2-methyl-π-cyclopentadienyl]ruthenium, 8CI. *2-Methylruthenocenemethanol*

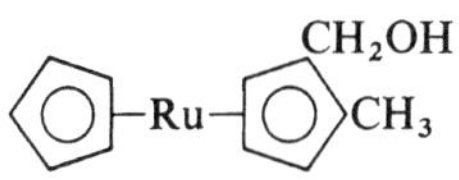

M 275.312

(+)-form

$[\alpha]_D$ +5.4° (EtOH).

Hofer, O. *et al*, *J. Organomet. Chem.*, 1968, **13**, 457 (*synth, ir*)

$C_{12}H_{14}O_2Ru$ Ru-00258

1,1′-Bis(hydroxymethyl)ruthenocene, 10CI

1,1′-Ruthenocenedimethanol

[79354-83-9]

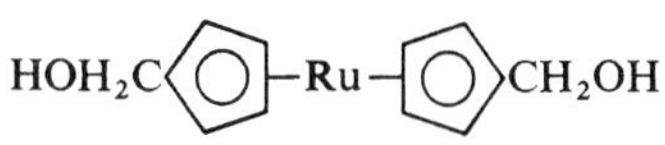

M 291.311

Yellow cryst. ($CHCl_3$). Mp 125-7°.

Kamiyama, S. *et al*, *Bull. Chem. Soc. Jpn.*, 1981, **54**, 2079 (*synth, pmr, ir, ms*)

$C_{12}H_{14}Ru$ Ru-00259

(η^6-Benzene)[(1,2,3,4-η)-1,3-cyclohexadiene]ruthenium, 10CI

[12215-07-5]

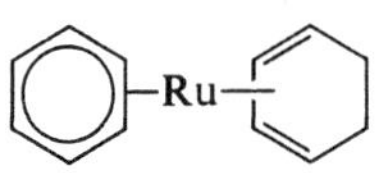

M 259.313

Air-sensitive green cryst. or light-yellow cryst. (pentane). Mp 118-9° (114-5°). Bp 50-60° subl. *in vacuo*.

Jones, D. *et al*, *J. Chem. Soc.*, 1962, 4458 (*synth, ir, pmr*)
Fischer, E.O. *et al*, *Chem. Ber.*, 1963, **96**, 3217 (*synth, ir, pmr*)
Müller, J. *et al*, *Chem. Ber.*, 1975, **108**, 273 (*synth*)
Pertici, P. *et al*, *J. Chem. Soc., Dalton Trans.*, 1980, 1961 (*synth, pmr, cmr*)

$C_{12}H_{14}Ru$ Ru-00260

Bis(η^5-2,4-cyclohexadien-1-yl)ruthenium

Dicyclohexadienylruthenium

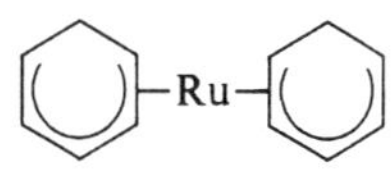

M 259.313

Yellow cryst. Isomerises to (η^6-Benzene)[(1,2,3,4-η)-1,3-cyclohexadiene]ruthenium, Ru-00259 on heating.

Jones, D. *et al*, *J. Chem. Soc.*, 1962, 4458 (*synth, pmr, ir*)

$C_{12}H_{14}Ru$ Ru-00261

1,1′-Dimethylruthenocene, 9CI

Bis(methyl-π-cyclopentadienyl)ruthenium, 8CI

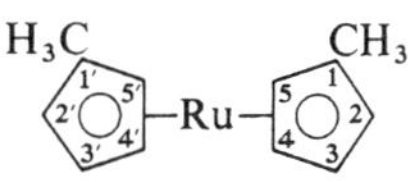

M 259.313

Mp 61-3°.

Hofer, O. *et al*, *J. Organomet. Chem.*, 1968, **13**, 443 (*synth, pmr, ir*)
Evans, S. *et al*, *J. Chem. Soc., Faraday Trans. 2*, 1972, **68**, 1847 (*synth, pe*)

$C_{12}H_{14}Ru$ Ru-00262

1,2-Dimethylruthenocene

(π-Cyclopentadienyl)(1,2-dimethyl-π-cyclopentadienyl)ruthenium, 8CI

M 259.313

Yellow cryst. Mp 46-7°.

Hofer, O. *et al*, *J. Organomet. Chem.*, 1968, **13**, 443 (*synth, pmr*)

$C_{12}H_{14}Ru$ Ru-00263

1,3-Dimethylruthenocene

(π-Cyclopentadienyl)(1,3-dimethyl-π-cyclopentadienyl)ruthenium, 8CI

M 259.313

Yellow oil.

Hofer, O. *et al*, *J. Organomet. Chem.*, 1968, **13**, 443 (*synth, pmr*)

$C_{12}H_{14}Ru$ Ru-00264

Ethylruthenocene

π-Cyclopentadienyl(ethyl-π-cyclopentadienyl)ruthenium

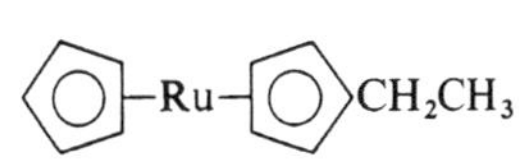

M 259.313

Mp 12-12.5°.

Mark, V. *et al*, *Inorg. Chem.*, 1964, **3**, 1067 (*synth, pmr*)

$C_{12}H_{15}N_3Ru^{\oplus\oplus}$ **Ru-00265**

Tris(acetonitrile)(η^6-benzene)ruthenium(2+), 10CI

[52462-53-0]

As Triaqua(η^6-benzene)ruthenium(2+), Ru-00041 with

$$X = -NCCH_3$$

M 302.341 (ion)

Intermed. for η^6-benzene complexes.

Bis(tetrafluoroborate): [52462-54-1].

$C_{12}H_{15}B_2F_8N_3Ru$ M 475.948

Yellow-orange cryst.

Bennett, M.A. *et al*, *J. Chem. Soc., Dalton Trans.*, 1974, 233 (*synth*)

Werner, H. *et al*, *J. Organomet. Chem.*, 1981, **210**, C11 (*props*)

$C_{12}H_{16}B_2Ru$ **Ru-00266**

Bis[(1,2,3,4,5,6-η)-1-methylboratabenzene]ruthenium, 9CI

Bis(1-methylborinato)ruthenium

[60552-37-6]

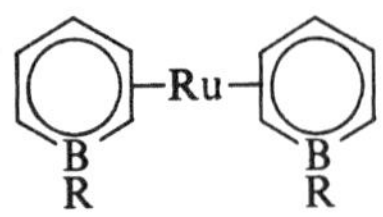

R = Me

M 282.948

Air stable, pale-yellow cryst. (pentane). Mp 64-5°. $Bp_{0.01}$ 60° subl.

Herberich, G.E. *et al*, *Chem. Ber.*, 1976, **109**, 2382 (*synth, pmr, nmr, ms*)

$C_{12}H_{18}Cl_2Ru$ **Ru-00267**

Dichloro[(1,2,3,6,7,10,11,12-η)-2,6,10-dodecatriene-1,12-diyl]ruthenium, 9CI

Dichloro(3-hexenylenedi-π-allyl)ruthenium, 8CI. *Dichloro-π_2-2,6,10-dodecatrienyleneruthenium*

[12170-97-7]

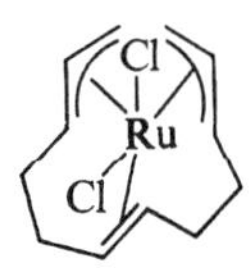

M 334.250

Monomeric in C_6H_6. Catalyst for the polymerisation of butadiene. Together with diamines, catalyst for hydrogenation of C=C and C≡C bonds. Yellow-bronze or yellow-green dichroic cryst. Sol. $CHCl_3$. Mp 210-20° dec.

Nicholson, J.K. *et al*, *J. Chem. Soc. (A)*, 1966, 807 (*synth, pmr*)

Lydon, J.E. *et al*, *J. Chem. Soc. (A)*, 1968, 362 (*struct*)

Hiraki, K. *et al*, *Macromolecules*, 1970, **3**, 382 (*use*)

$C_{12}H_{18}GeO_2Ru$ **Ru-00268**

Dicarbonyl[(1,2,3,4,5-η)-2,4-cycloheptadien-1-yl](trimethylgermyl)ruthenium, 9CI

[57650-19-8]

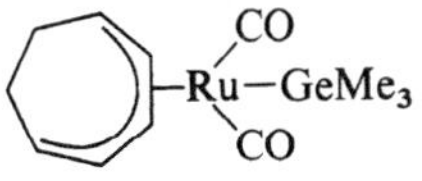

M 367.933

Pale-yellow liq. $Bp_{0.1}$ 75°.

Sosinsky, B.A. *et al*, *J. Chem. Soc., Dalton Trans.*, 1975, 1633 (*synth, ir, pmr*)

$C_{12}H_{18}Ge_3O_6Ru_2$ **Ru-00269**

Hexacarbonyltris[μ-(dimethylgermylene)]diruthenium, 8CI

[34307-40-9]

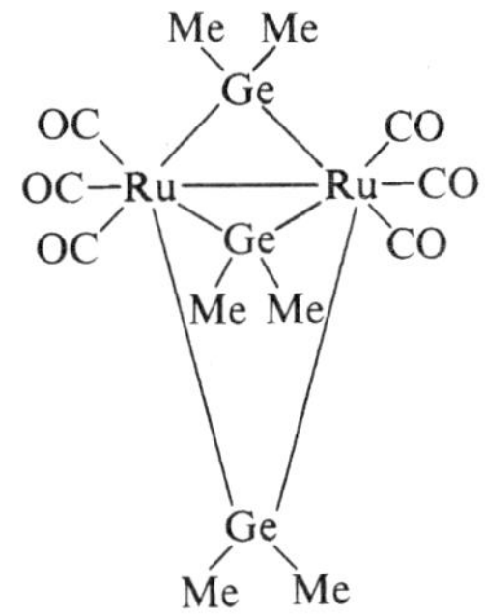

M 678.181

Lemon-yellow cryst. Mp 150-60° dec.

Knox, S.A.R. *et al*, *J. Chem. Soc. (A)*, 1971, 2874 (*synth, ir, pmr*)

$C_{12}H_{18}N_6Ru^{\oplus\oplus}$ **Ru-00270**

Hexakis(isocyanomethane)ruthenium(2+), 9CI

Hexakis(methyl isocyanide)ruthenium(2+)

[42572-84-9]

$$Ru(CNMe)_6^{\oplus\oplus}$$

M 347.384 (ion)

Bis(tetrafluoroborate): [49631-68-7].

$C_{12}H_{18}B_2F_8N_6Ru$ M 520.992

White cryst. (MeOH).

Bis(hexafluorophosphate): [49631-67-6].

$C_{12}H_{18}F_{12}N_6P_2Ru$ M 637.313

White cryst. (MeCN).

Doonan, D.J. *et al*, *Inorg. Chem.*, 1974, **13**, 921 (*synth, ir*)

Dykstra, R.W. *et al*, *J. Magn. Reson.*, 1982, **46**, 338 (*nmr*)

$C_{12}H_{18}O_2RuSi$ **Ru-00271**

Dicarbonyl[(1,2,3,4,5-η)-2,4-cycloheptadien-1-yl](trimethylsilyl)ruthenium, 9CI

[57659-28-6]

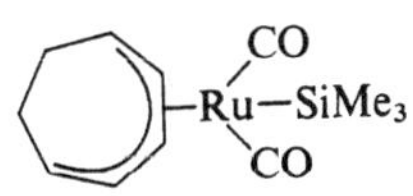

M 323.429

Pale-yellow liq. Sol. C_6H_6. $Bp_{0.1}$ 70°.

Sosinsky, B.A. *et al*, *J. Chem. Soc., Dalton Trans.*, 1975, 1633 (*synth, ir, pmr*)

$C_{12}H_{18}O_6Ru_2Si_3$ **Ru-00272**

Hexacarbonyltris[μ-(dimethylsilyl)]diruthenium, 9CI

Hexacarbonyltris[μ-(dimethylsilylene)]diruthenium

[36632-57-2]

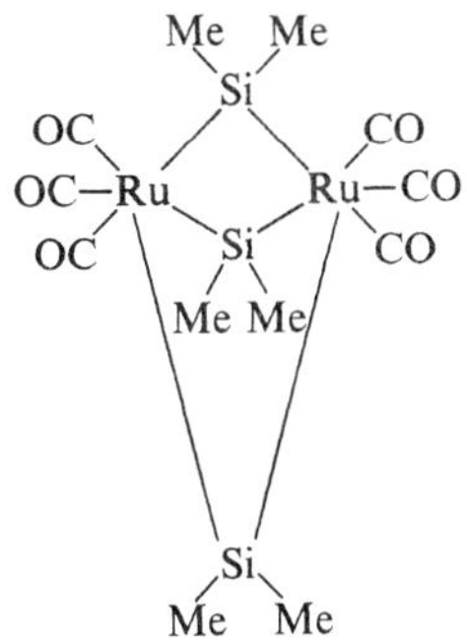

M 544.667

Cream powder (hexane). Mp 140° dec.

Brookes, A. *et al*, *J. Chem. Soc. (A)*, 1971, 3469 (*synth*, *ir*)

$C_{12}H_{22}PRu^{\oplus}$ **Ru-00273**

(η[6]-Benzene)(η[2]-ethene)methyl(trimethylphosphine)ruthenium(1+), 10CI

As (η[6]-Benzene)chloro(η[2]-ethene)(trimethylphosphine)-ruthenium(1+), Ru-00202 with

X = Me

M 298.350 (ion)

Hexafluorophosphate: [71685-21-7].
$C_{12}H_{22}F_6P_2Ru$ M 443.314
No phys. props. reported.

Werner, H. *et al*, *J. Organomet. Chem.*, 1979, **174**, C63 (*synth*, *pmr*)

$C_{12}H_{23}OP_2Ru^{\oplus}$ **Ru-00274**

Carbonyl(η[5]-2,4-cyclopentadien-1-yl)bis(trimethylphosphine)-ruthenium(1+), 10CI

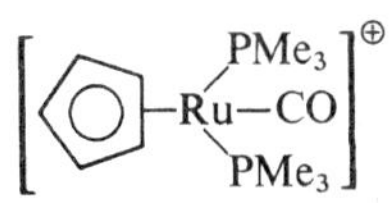

M 346.331 (ion)

Hexafluorophosphate: [75182-25-1].
$C_{12}H_{23}F_6OP_3Ru$ M 491.295
White solid (CH_2Cl_2/Et_2O). Mp >300° dec.

Treichel, P.M. *et al*, *Synth. React. Inorg. Metal-Org. Chem.*, 1980, **10**, 205 (*synth*, *ir*, *pmr*)
Bruce, M.I. *et al*, *J. Chem. Soc., Dalton Trans.*, 1981, 1398 (*synth*, *ir*, *pmr*, *cmr*)

$C_{12}H_{24}O_6P_2Ru$ **Ru-00275**

(η[6]-Benzene)bis(trimethyl phosphite-*P*)ruthenium, 10CI

[67612-94-6]

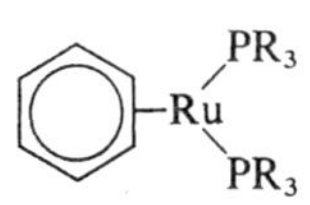

R = OMe

M 427.336

Low melting air-sensitive orange cryst. (pentane). Mp <20°.

Werner, H. *et al*, *Angew. Chem., Int. Ed. Engl.*, 1978, **17**, 683 (*synth*, *pmr*)

$C_{12}H_{24}P_2Ru$ **Ru-00276**

(η[6]-Benzene)bis(trimethylphosphine)ruthenium, 10CI

[67612-92-4]

As (η[6]-Benzene)bis(trimethyl phosphite-*P*)ruthenium, Ru-00275 with

R = Me

M 331.339

Air-sensitive orange cryst. (pentane). Sol. aprotic solvs. Mp <20°.

Werner, H. *et al*, *Angew. Chem., Int. Ed. Engl.*, 1978, **17**, 683 (*synth*, *pmr*)

$C_{12}H_{38}O_{12}P_4Ru$ **Ru-00277**

Dihydrotetrakis(trimethyl phosphite-*P*)ruthenium, 10CI

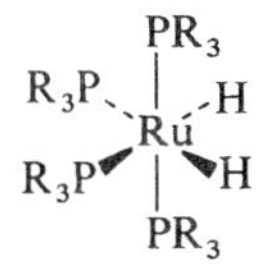

R = OMe

M 599.390

cis-form [74219-48-0]

White cryst. (MeOH).

Gerlach, D.H. *et al*, *J. Am. Chem. Soc.*, 1972, **94**, 4545 (*synth*)
Meakin, P. *et al*, *J. Am. Chem. Soc.*, 1973, **95**, 75 (*nmr*, *pmr*)
Pertici, P. *et al*, *Inorg. Chim. Acta*, 1979, **37**, L521 (*synth*)

$C_{12}H_{38}P_4Ru$ **Ru-00278**

Dihydrotetrakis(trimethylphosphine)ruthenium, 10CI

[76171-49-8]

As Dihydrotetrakis(trimethyl phosphite-*P*)ruthenium, Ru-00277 with

R = Me

M 407.397

cis-form

White needles (pentane). Mp 160-4°. $Bp_{0.001}$ 100° subl.

Jones, R.A. *et al*, *J. Chem. Soc., Dalton Trans.*, 1980, 2480 (*synth*, *pmr*, *nmr*, *ir*)

$C_{12}O_{12}Os_2Ru$ **Ru-00279**

Octacarbonyl(tetracarbonylruthenium)diosmium, 10CI

Octacarbonyl[μ-(tetracarbonylruthenio)]diosmium, 8CI.
Dodecacarbonyldiosmiumruthenium

[12389-50-3]

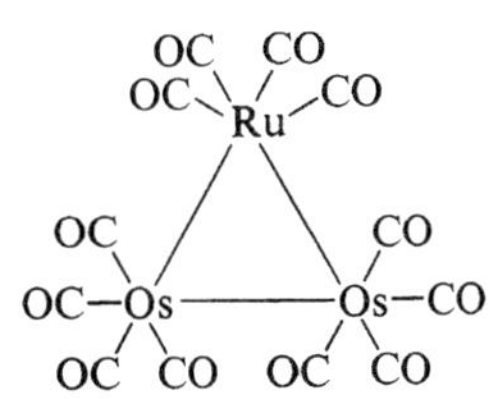

M 817.595

No phys. props. reported. ir ν_{CO} 2066vs, 2032s, 2014w, 2004w cm^{-1}.

Johnson, B.F.G. *et al*, *J. Chem. Soc., Chem. Commun.*, 1968, 861 (*synth*)
Ferrari, R.P. *et al*, *J. Chem. Soc., Dalton Trans.*, 1978, 1164 (*synth, ms*)
Kettle, S.F.A. *et al*, *Inorg. Chem.*, 1979, **18**, 2749 (*raman*)
Battiston, G.A. *et al*, *Inorg. Chem.*, 1980, **19**, 1961 (*ir, raman*)

$C_{12}O_{12}Ru_3$ Ru-00280

Dodecacarbonyltriruthenium, 9CI
Triruthenium dodecacarbonyl. Ruthenium carbonyl
[$Ru_3(CO)_{12}$]
[15243-33-1]

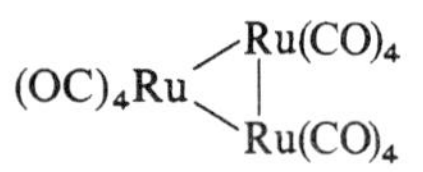

M 639.335
Orange cryst. (hexane or C_6H_6). Mp 154-5° dec.

Mason, R. *et al*, *J. Chem. Soc. (A)*, 1968, 778 (*struct*)
Quicksall, C.O. *et al*, *Inorg. Chem.*, 1968, **7**, 2365 (*ir, raman*)
Dawes, J.L. *et al*, *Inorg. Nucl. Chem. Lett.*, 1971, **7**, 847.
Herlan, A., *J. Organomet. Chem.*, 1971, **28**, 423 (*ms*)
Inorg. Synth., 1972, **13**, 92; 1976, **16**, 45, 47 (*synth*)
Forster, A. *et al*, *J. Chem. Soc., Chem. Commun.*, 1974, 1042 (*cmr*)
U.S.P., 3 786 132, (*1974*); *CA*, **80**, 122898 (*synth*)
Churchill, M.R. *et al*, *Inorg. Chem.*, 1977, **16**, 2655 (*cryst struct*)
Bruce, M.I. *et al*, *Angew. Chem., Int. Ed. Engl.*, 1978, **7**, 427 (*rev*)

$C_{13}ClO_{13}Ru_4^{\ominus}$ Ru-00281

μ-Carbonyldodecacarbonyl-μ-chlorotetraruthenate(1−)
Tridecacarbonylchlorotetraruthenate(1−)

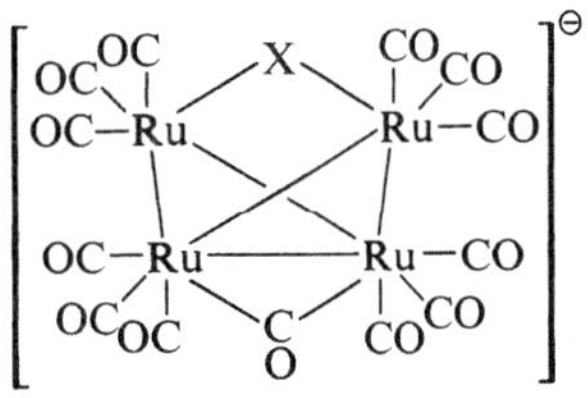

X = Cl

M 803.868 (ion)
Bis(triphenylphosphine)iminium salt: [74231-12-2].
$C_{49}H_{30}ClNO_{13}P_2Ru_4$ M 1342.455
Red microcryst. Slightly air-sensitive.

Steimetz, G.R. *et al*, *Inorg. Chem.*, 1980, **19**, 2985 (*synth, cryst struct*)

$C_{13}HFeO_{13}Ru_3^{\ominus}$ Ru-00282

Di-μ-carbonylnonacarbonyl(dicarbonylferrate)-μ-hydrotriruthenate(1−)
Tridecacarbonylhydroirontriruthenium(1−)
[74128-10-2]

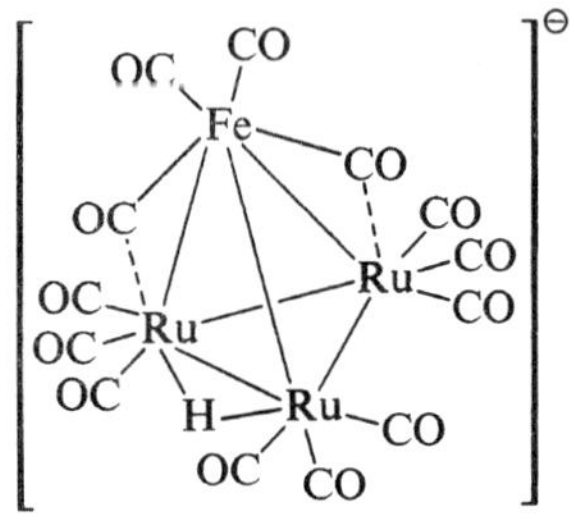

M 724.200 (ion)
Air-stable.
Bis(triphenylphosphine)iminium salt: [78571-90-1].
$C_{49}H_{31}FeNO_{13}P_2Ru_3$ M 1262.787
Air-stable black cryst. (Et_2O/hexane). Sol. CH_2Cl_2, Et_2O, insol. hexane.

Takusagawa, F. *et al*, *Inorg. Chem.*, 1981, **20**, 3823 (*synth, cryst struct*)
Inorg. Synth., 1982, **21**, 60 (*synth, ir, pmr*)

$C_{13}HFe_2O_{13}Ru_2^{\ominus}$ Ru-00283

Hexacarbonyl(μ-carbonylhexacarbonyldiferrate)-μ-hydrodiruthenate(1−), 10CI

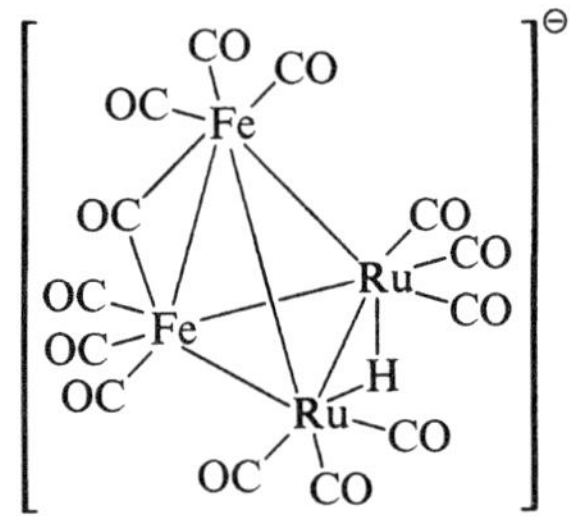

M 678.977 (ion)
Bis(triphenylphosphine)iminium salt: [78571-92-3].
$C_{49}H_{31}Fe_2NO_{13}P_2Ru_2$ M 1217.564
Black cryst. (Et_2O/hexane). Sol. Et_2O, THF.

Takusagawa, F. *et al*, *Inorg. Chem.*, 1981, **20**, 3823 (*synth, ir, cryst struct*)

$C_{13}HO_{13}Ru_4^{\ominus}$ Ru-00284

μ_3-Carbonyldodecacarbonylhydrotetraruthenate(1−)
Tridecacarbonylhydridotetraruthenate(1−)
[76917-52-7]
As
μ-Carbonyldodecacarbonyl-μ-chlorotetraruthenate(1−), Ru-00281 with

X = H

M 769.423 (ion)
K salt: [76917-54-9].
$C_{13}HKO_{13}Ru_4$ M 808.521
Red.
Bis(triphenylphosphine)iminium salt: [76917-53-8].
$C_{49}H_{31}NO_{13}P_2Ru_4$ M 1308.010
Light-red-brown powder (CH_2Cl_2/Et_2O).

Nagel, C.C. *et al*, *J. Chem. Soc., Chem. Commun.*, 1980, 530 (*synth, ir, nmr*)

U.S.P., 4 349 521, (*1982*); *CA*, **97**, 165494 (*synth*)

$C_{13}H_2FeO_{13}OsRu_2$ Ru-00285

Di-μ-carbonyltricarbonyl(dicarbonyliron)(hexacarbonyl-μ-hydrodiruthenium)-μ-hydroosmium, 10CI

Di-μ-carbonyltricarbonyl(dicarbonyliron)di-μ-hydrobis(tricarbonylruthenium)osmium, 10CI. Tridecacarbonyldihydroirondirutheniumosmium

[61863-84-1]

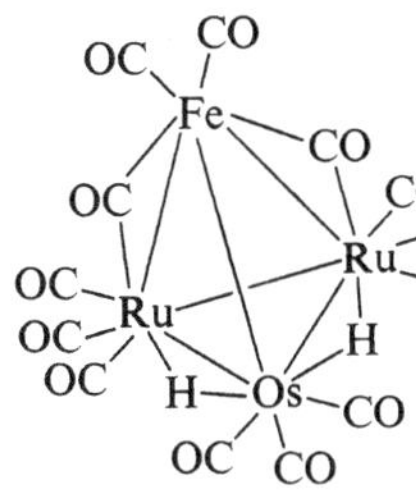

M 814.338

Fluxional molecule, interconverting mixt. of 2 isomers at r.t. The isomer shown predominates, the minor isomer having the Os exchanged with one Ru. Also shows CO group migration. Orange-red.

Geoffroy, G.L. *et al*, *J. Am. Chem. Soc.*, 1977, **99**, 304 (*synth, uv, ir, pmr*)
Gladfelter, W.L. *et al*, *Inorg. Chem.*, 1980, **19**, 2579 (*cmr, pmr*)

$C_{13}H_2FeO_{13}Os_2Ru$ Ru-00286

Di-μ-carbonylhexacarbonyl(dicarbonyliron)di-μ-hydro(tricarbonylruthenium)diosmium

Tridecacarbonyldihydroirondiosmiumruthenium

[61816-94-2]

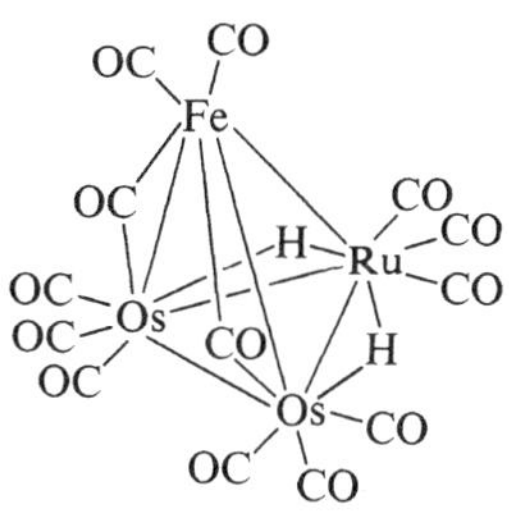

M 903.468

Fluxional molecule showing CO ligand migration and skeletal rearr. having the effect of Os-Ru interchange. Isomer shown predominates. Orange.

Geoffroy, G.L. *et al*, *J. Am. Chem. Soc.*, 1977, **99**, 6775, 7565 (*synth, cmr, ms, ir*)
Gladfelter, W.L. *et al*, *Inorg. Chem.*, 1980, **19**, 2579 (*pmr, cmr*)

$C_{13}H_2FeO_{13}Ru_3$ Ru-00287

Di-μ-carbonyldicarbonyldi-μ-hydrotris[μ_3-(tricarbonylruthenio)]iron, 10CI

Tridecacarbonyldihydroirontriruthenium, 9CI. Tridecacarbonyldihydrotriruthenium iron, 8CI

[12375-24-5]

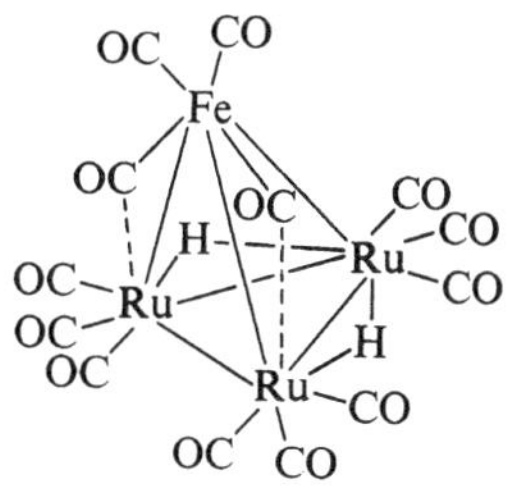

M 725.208

Fluxional molecule showing CO group migration and skeletal rearr. Orange or red needles. pK_a 14.3 (MeOH, 25°). Air-stable.

Knight, J. *et al*, *Chem. Ind.* (*London*), 1968, 115 (*synth, ir*)
Yawney, D.B.W. *et al*, *J. Chem. Soc.* (*A*), 1969, 502 (*synth*)
Gilmore, C.J. *et al*, *J. Chem. Soc.* (*A*), 1971, 3453 (*cryst struct*)
Geoffroy, G.L. *et al*, *J. Am. Chem. Soc.*, 1977, **99**, 7565 (*synth*)
Walker, H.W. *et al*, *J. Am. Chem. Soc.*, 1979, **101**, 7428.
Gladfelter, W.L. *et al*, *Inorg. Chem.*, 1980, **19**, 2579 (*pmr, cmr*)
Geoffroy, G.L. *et al*, *ACS Symp. Ser.*, 1981, **155**, 111 (*rev*)
Inorg. Synth., 1982, **21**, 57 (*synth, bibl*)

$C_{13}H_2Fe_2O_{13}Ru_2$ Ru-00289

μ-Carbonylhexacarbonyl(μ-carbonylpentacarbonyldiiron)di-μ-hydrodiruthenium

Tridecacarbonyldihydrodiirondiruthenium

[65724-47-2]

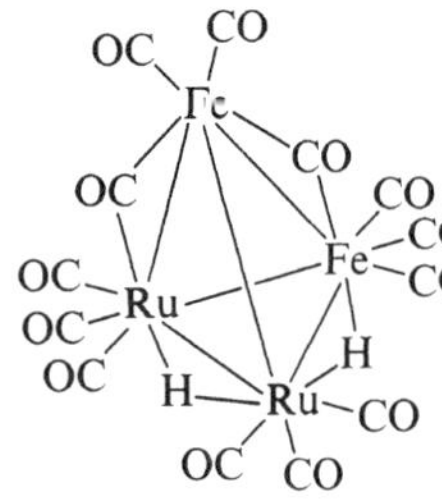

M 679.985

Obt. only as mixt. with Tridecacarbonyldihydroirontriruthenium, Ru-00288 .

Geoffroy, G.L. *et al*, *J. Am. Chem. Soc.*, 1977, **99**, 7565 (*synth, ms, ir*)

$C_{13}H_2O_{13}Os_3Ru$ Ru-00290

Di-μ-carbonylnonacarbonyl(dicarbonylruthenium)di-μ-hydrotriosmium, 10CI

Tridecacarbonyldihydrotriosmiumruthenium

[75901-26-7]

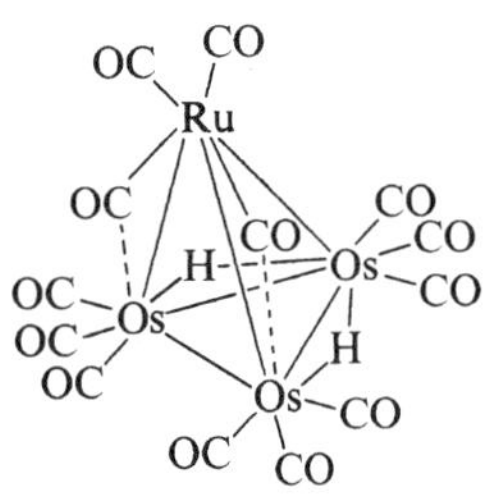

M 1037.821

Probable struct. Orange air-stable cryst. Sol. hexane.

Burkhardt, E.W. *et al*, *J. Organomet. Chem.*, 1980, **198**, 179 (*synth, ir, ms*)

Inorg. Synth., 1982, **21**, 64 (*synth, ir, pmr*)

$C_{13}H_2O_{13}Ru_4$ Ru-00291

Di-μ-carbonylundecacarbonyldi-μ-hydrotetraruthenium, 9CI

Tridecacarbonyldihydrotetraruthenium, 8CI

[21077-76-9]

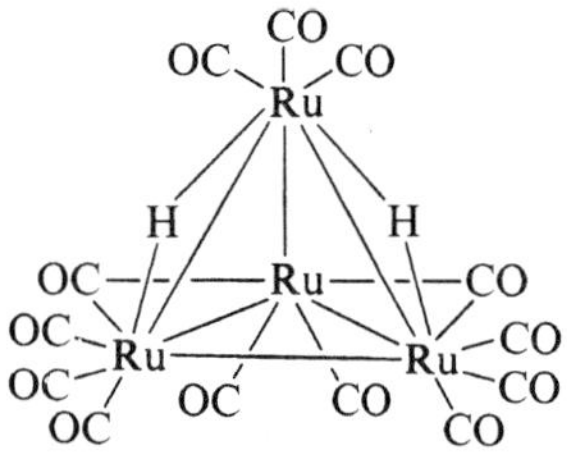

M 770.431

Suggested struct. shown. Deep-red rhombic cryst. Stable indefinitely as solid, but slowly dec. in soln. Sol. org. solvs. Bp_1 ca. 60°. Volatile.

Johnson, B.F.G. *et al*, *J. Chem. Soc. (A)*, 1968, 2856, 2865 (*synth, ir, pmr*)

Yawney, D.B.W. *et al*, *J. Chem. Soc. (A)*, 1969, 502 (*synth, ms, ir*)

Johnson, B.F.G. *et al*, *J. Chem. Soc. (A)*, 1970, 901 (*synth*)

Yawney, D.B.W. *et al*, *Inorg. Chem.*, 1972, **11**, 838 (*cryst struct*)

Eady, C.R. *et al*, *J. Chem. Soc., Dalton Trans.*, 1977, 477 (*pmr*)

Aime, S. *et al*, *Inorg. Chim. Acta*, 1978, **29**, L211 (*ir, cmr*)

Fox, J.R. *et al*, *Inorg. Chem.*, 1980, **19**, 2574 (*props*)

$C_{13}H_5F_5O_2Ru$ Ru-00292

Dicarbonyl(η^5-2,4-cyclopentadien-1-yl)(pentafluorophenyl)ruthenium

η^5-Cyclopentadienyldicarbonyl(pentafluorophenyl)ruthenium. σ-(Pentafluorophenyl)-π-cyclopentadienylruthenium dicarbonyl

As Dicarbonyl(η^5-2,4-cyclopentadien-1-yl)iodoruthenium, Ru-00051 with

$$X = C_6F_5$$

M 389.243

White cryst. Mp 159-60°. $Bp_{0.2}$ 65° subl.

Blackmore, T. *et al*, *J. Chem. Soc. (A)*, 1968, 2158 (*synth, ir, nmr*)

Bruce, M.I. *et al*, *Org. Mass Spectrom.*, 1968, **1**, 835 (*ms*)

$C_{13}H_5F_5O_2RuS$ Ru-00293

Dicarbonyl(η^5-2,4-cyclopentadien-1-yl)(pentafluorobenzenethiolato)ruthenium, 9CI

Dicarbonyl-π-cyclopentadienyl(pentafluorophenylthio)ruthenium. Cyclopentadienyldicarbonylruthenium pentafluorophenylmercaptide

[53556-38-0]

As Dicarbonyl(η^5-2,4-cyclopentadien-1-yl)(methanethiolato)ruthenium, Ru-00080 with

$$R = C_6F_5$$

M 421.303

Yellow solid. Mp 90-2°.

Knox, G.R. *et al*, *J. Organomet. Chem.*, 1974, **74**, 103 (*synth, ir, pmr, nmr*)

$C_{13}H_7NO_{10}Ru_3$ Ru-00294

Decacarbonyl[μ-[(dimethyliminio)methylene]]-μ-hydrotriruthenium, 9CI

[59532-85-3]

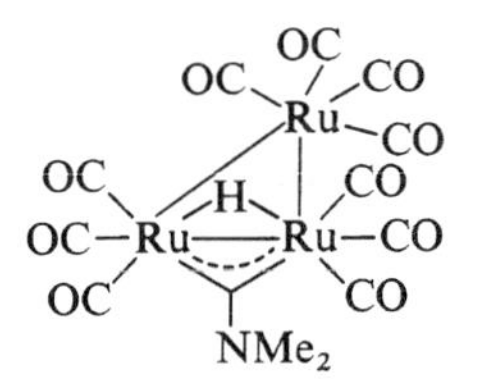

M 640.409

Orange cryst. (hexane). Mp 112° dec.

Churchill, M.R. *et al*, *J. Am. Chem. Soc.*, 1975, **97**, 7158 (*struct*)

Abel, E.W. *et al*, *J. Inorg. Nucl. Chem.*, 1980, **42**, 541 (*synth, ir, pmr*)

$C_{13}H_7NO_{11}Ru_3$ Ru-00295

Decacarbonyl[μ-[(dimethylamino)carbonyl-*C*:*O*]]-μ-hydrotriruthenium

[74325-95-4]

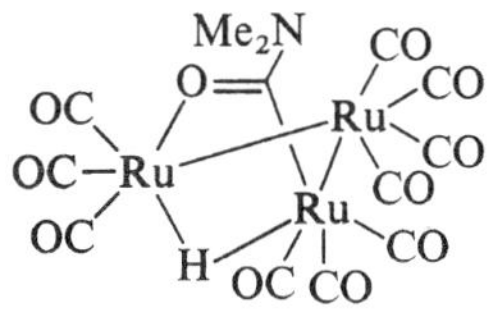

M 656.408

Red-orange or dark-red cryst. (hexane/CH_2Cl_2).

Szostak, R. *et al*, *J. Organomet. Chem.*, 1980, **191**, 243 (*synth, ir, raman, pmr, cmr, struct*)

$C_{13}H_8O_5Ru_2$ Ru-00296

μ-Carbonyltetracarbonyl[μ-[(1,2,3,4,5,6,7,8-η)-1,3,5,7-cyclooctatetraene]]diruthenium, 8CI

(Cyclooctatetraene)pentacarbonyldiruthenium

[12289-05-3]

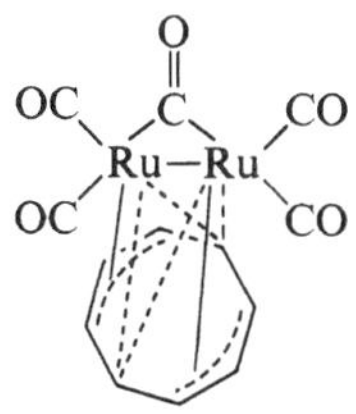

M 446.343

Fluxional molecule. Orange cryst. (toluene). Mp ~160° dec.

Cotton, F.A. *et al*, *J. Am. Chem. Soc.*, 1969, **91**, 6598 (*synth, pmr, ir*)

$C_{13}H_8O_6Ru_2$ Ru-00297

Hexacarbonyl[μ-[(1,2,3-η:4,5,6-η)-1,3,5-cycloheptatriene]]diruthenium, 9CI

[57383-59-2]

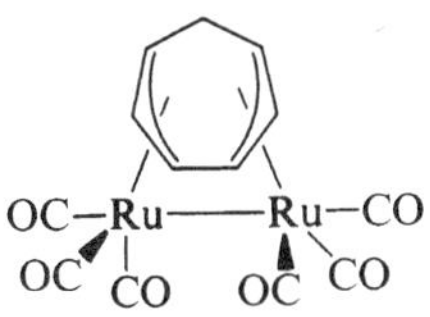

M 462.343

Yellow.

Burt, J.C. *et al*, *J. Chem. Soc., Dalton Trans.*, 1975, 731 (*synth, ir, pmr*)

$C_{13}H_9NO_{10}Ru_3Si$ Ru-00298

μ_3-Carbonylnonacarbonyl[μ_3-[1,1,1-trimethylsilanaminato(2−)]]triruthenium, 9CI

Decacarbonyl(μ_3-trimethylsilylaminato)triruthenium

[53652-60-1]

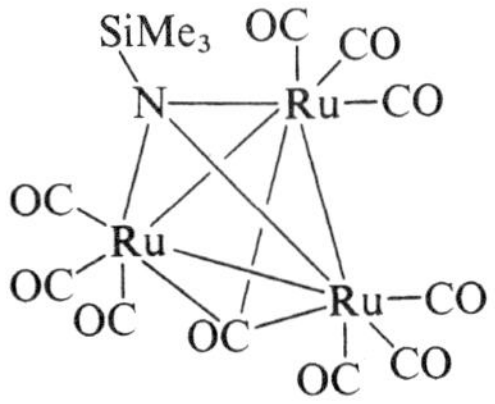

M 670.510

Sl. air-sensitive yellow-orange cryst. (hexane). Mp 160° dec.

Abel, E.W. *et al*, *Inorg. Nucl. Chem. Lett.*, 1974, **10**, 941 (*synth, ir, pmr, ms*)

$C_{13}H_{10}O_2Ru$ Ru-00299

Dicarbonyl(η^5-2,4-cyclopentadien-1-yl)phenylruthenium, 9CI

η^5-Cyclopentadienyldicarbonylphenylruthenium. σ-Phenyl-π-cyclopentadienylruthenium dicarbonyl

[36180-37-7]

As Dicarbonyl(η^5-2,4-cyclopentadien-1-yl)-iodoruthenium, Ru-00051 with

X = Ph

M 299.291

Orange or yellow oil.

Haines, R.J. *et al*, *J. Chem. Soc., Dalton Trans.*, 1972, 944 (*synth, ir, pmr*)

Haines, R.J. *et al*, *J. Organomet. Chem.*, 1975, **84**, 357 (*synth, ir, pmr*)

$C_{13}H_{10}O_2RuS$ Ru-00300

(Benzenethiolato)dicarbonyl(η^5-2,4-cyclopentadien-1-yl)ruthenium, 10CI

Cyclopentadienyldicarbonylruthenium phenylmercaptide. Dicarbonyl-π-cyclopentadienyl(phenylthio)ruthenium

[69047-82-1]

As Dicarbonyl(η^5-2,4-cyclopentadien-1-yl)-(methanethiolato)ruthenium, Ru-00080 with

R = Ph

M 331.351

Yellow cryst. (CH_2Cl_2/hexane). Mp 100-1°.

Killops, S.D. *et al*, *J. Chem. Soc., Dalton Trans.*, 1978, 1260 (*synth, ir, pmr*)

$C_{13}H_{10}O_2RuSe$ Ru-00301

(Benzeneselenolato)dicarbonyl(η^5-2,4-cyclopentadien-1-yl)-ruthenium, 10CI

Cyclopentadienyldicarbonylruthenium phenylselenide

[69047-83-2]

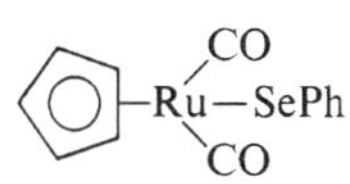

M 378.251

Red cryst. (CH_2Cl_2/hexane). Mp 205-6°. V. unstable in soln.

Killops, S.D. *et al*, *J. Chem. Soc., Dalton Trans.*, 1978, 1260 (*synth, ir, pmr*)

$C_{13}H_{11}ClORu$ Ru-00302

3-Chloro-3-ruthenocenyl-2-propen-1-al

[(1-Chloro-2-formylvinyl)-π-cyclopentadienyl]-π-cyclopentadienylruthenium, 8CI. β-Ruthenocenyl-β-chloroacrolein

[12337-23-4]

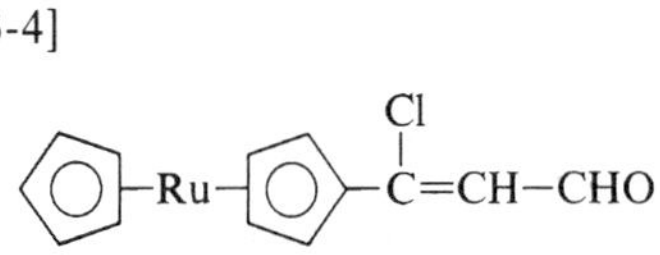

M 319.752

Mp 112-4°.

Hofer, O. *et al*, *J. Organomet. Chem.*, 1968, **13**, 443 (*synth, ir*)
Rausch, M.D. *et al*, *J. Org. Chem.*, 1969, **34**, 1974 (*synth*)

$C_{13}H_{12}ClORu^{\oplus}$ Ru-00303

[(1,2,3,4,5-η)-1-Acetyl-2,4-cyclopentadien-1-yl](η⁶-chlorobenzene)ruthenium(1+), 10CI

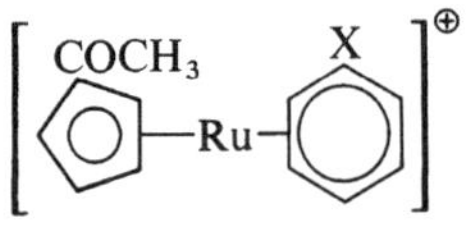

X = Cl

M 320.760 (ion)
Air-stable.

Hexafluorophosphate: [78643-30-8].
$C_{13}H_{12}ClF_6OPRu$ M 465.724
No phys. props. reported.

Nesmeyanov, A.N. *et al*, *Dokl. Akad. Nauk SSSR, Ser. Sci. Khim.*, 1981, **258**, 193 (*synth, ir, pmr*)

$C_{13}H_{12}ORu$ Ru-00304

Carbonyl[(1,2,3,3*a*,7*a*-η)-1*H*-inden-1-yl](η³-2-propenyl)ruthenium, 9CI

(*π-Allyl*)*carbonyl*(*π-indenyl*)*ruthenium*
[43184-28-7]

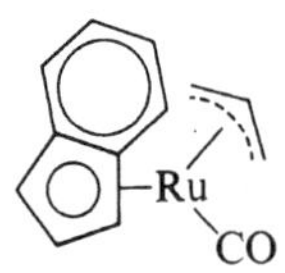

M 285.307
Air-sensitive, pale-yellow solid. Mp 69°.

Abel, E.W. *et al*, *J. Chem. Soc., Dalton Trans.*, 1973, 1706 (*synth*)
Faller, J.W. *et al*, *J. Organomet. Chem.*, 1974, **65**, 395 (*synth, pmr, ir*)

$C_{13}H_{12}ORu$ Ru-00305

[3]Ruthenocenophan-1-one

1,1′-(1-Oxo-1,3-propanediyl)ruthenocene, 10CI
[66507-20-8]

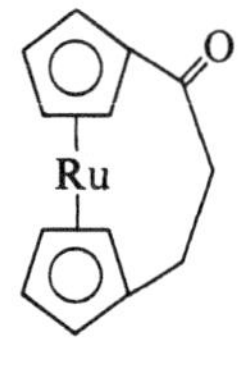

M 285.307
Cryst. (Me_2CO aq.). Mp 175-6°.

Kamiyama, S. *et al*, *Bull. Chem. Soc. Jpn.*, 1978, **51**, 909 (*synth, pmr, ir, uv*)
Senda, Y. *et al*, *Bull. Chem. Soc. Jpn.*, 1980, **53**, 3381 (*cmr*)

$C_{13}H_{12}O_2Ru^{\oplus\oplus}$ Ru-00306

(η⁶-Benzene)[(1,2,3,4,5,6-η)-benzoic acid]ruthenium(2+)

As (η⁶-Benzene)(η⁶-chlorobenzene)ruthenium(2+), Ru-00234 with

X = COOH

M 301.307 (ion)

Me ester, bis(tetrafluoroborate): [71826-04-5]. (*η⁶-Benzene*)[(*1,2,3,4,5,6-η*)-*methylbenzoate*]*ruthenium*(*2+*) *bis*(*tetrafluoroborate*), 10CI. (*η⁶-Benzene*)(*η⁶-methoxycarbonylbenzene*)*ruthenium*(*2+*) *bis*(*tetrafluoroborate*).
$C_{14}H_{14}B_2F_8O_2Ru$ M 488.941
Colour and Mp not reported. Air-stable.

Bennett, M.A. *et al*, *J. Organomet. Chem.*, 1979, **175**, 87 (*synth, pmr*)

$C_{13}H_{12}O_2Ru$ Ru-00307

3-Ruthenocenyl-2-propenoic acid

(*2-Carboxyethenyl*)*ruthenocene*, 10CI. *3-Ruthenocenylacrylic acid*
[66507-26-4]

M 301.307
Cryst. (dioxan aq.). Mp 228-9°.

Kamiyama, S. *et al*, *Bull. Chem. Soc. Jpn.*, 1978, **51**, 909 (*synth, pmr, ir, uv*)
Senda, Y. *et al*, *Bull. Chem. Soc. Jpn.*, 1980, **53**, 3381 (*cmr*)

$C_{13}H_{13}ORu^{\oplus}$ Ru-00308

[(1,2,3,4,5-η)-1-Acetyl-2,4-cyclopentadien-1-yl](η⁶-benzene)ruthenium(1+), 10CI

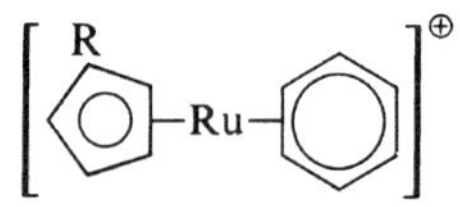

R = —$COCH_3$

M 286.315 (ion)

Hexafluorophosphate: [78618-91-4].
$C_{13}H_{13}F_6OPRu$ M 431.279
Yellow solid. Not isolated in a pure state.

Nesmeyanov, A.N. *et al*, *Dokl. Chem.* (*Engl. Transl.*), 1981, **258**, 193 (*synth, pmr*)

$C_{13}H_{14}ORu$ Ru-00309

1-Acetyl-2-methylruthenocene

(*1-Acetyl-2-methyl-π-cyclopentadienyl*)-*π-cyclopentadienylruthenium*, 8CI

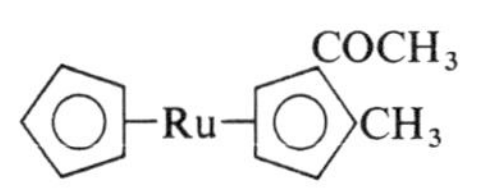

M 287.323

(−)-form

Cryst. Mp 57-65°. $[\alpha]_D$ −181° (EtOH).

Hofer, O. *et al*, *J. Organomet. Chem.*, 1968, **13**, 457 (*synth, ir, uv, cd, ord, pmr*)

$C_{13}H_{14}ORu^{\oplus\oplus}$ Ru-00310

(η⁶-Benzene)[(1,2,3,4,5,6-η)-methoxybenzene]ruthenium(2+), 10CI

(*η⁶-Anisole*)(*η⁶-benzene*)*ruthenium*(*2+*)

As (η⁶-Benzene)(η⁶-chlorobenzene)ruthenium(2+), Ru-00234 with

X = OMe

M 287.323 (ion)
Air-stable.

Bis(tetrafluoroborate): [71825-88-2].

$C_{13}H_{14}B_2F_8ORu$ M 460.930
No phys. props. reported.

Bennett, M.A. *et al*, *J. Organomet. Chem.*, 1979, **175**, 87 (*synth, pmr*)

$C_{13}H_{14}ORu$ Ru-00311

Propanoylruthenocene

π-Cyclopentadienyl(propionyl-π-cyclopentadienyl)ruthenium, 8CI. *Ethyl ruthenocenyl ketone. 1-Ruthenocenylpropanone*

[12512-44-6]

M 287.323

Yellow cryst. Mp 70-1°.

McDonnell, J.J. *et al*, *J. Org. Chem.*, 1971, **36**, 2092 (*synth, pmr*)

$C_{13}H_{14}O_2Ru$ Ru-00312

3-Ruthenocenylpropanoic acid

(2-Carboxyethyl)ruthenocene, 10CI

[66560-23-4]

M 303.322

Cryst. (pet. ether). Mp 99-100°.

Kamiyama, S. *et al*, *Bull. Chem. Soc. Jpn.*, 1978, **51**, 909 (*synth, pmr, ir, uv*)
Senda, Y. *et al*, *Bull. Chem. Soc. Jpn.*, 1980, **53**, 3381 (*cmr*)

$C_{13}H_{14}Ru$ Ru-00313

[3]Ruthenocenophane

1,1′-(1,3-Propanediyl)ruthenocene, 10CI

[66507-21-9]

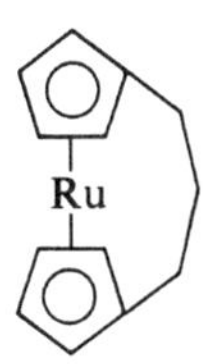

M 271.324

Cryst. (hexane). Mp 123-4°.

Kamiyama, S. *et al*, *Bull. Chem. Soc. Jpn.*, 1978, **51**, 909 (*synth, uv, pmr*)
Senda, Y. *et al*, *Bull. Chem. Soc. Jpn.*, 1980, **53**, 3381 (*cmr*)

$C_{13}H_{15}O_2Ru^{\oplus}$ Ru-00314

Dicarbonyl[(1,2-η)-cyclohexene](η⁵-2,4-cyclopentadien-1-yl-)ruthenium(1+), 10CI

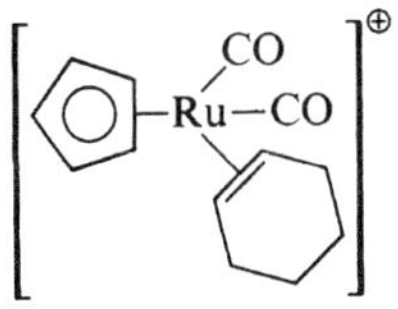

M 304.330 (ion)

Hexafluorophosphate: [68643-67-4].
$C_{13}H_{15}F_6O_2PRu$ M 449.294
Mp 155° dec.

Jungbauer, A. *et al*, *Z. Naturforsch., B*, 1978, **33**, 1083 (*synth, ir, pmr*)

$C_{13}H_{15}Ru^{\oplus}$ Ru-00315

(η^6-Benzene)[(1,2,3,4,5-η)-1-ethyl-2,4-cyclopentadien-1-yl-]ruthenium(1+), 10CI

As [(1,2,3,4,5-η)-1-Acetyl-2,4-cyclopentadien-1-yl](η^6-benzene)ruthenium(1+), Ru-00308 with

$$R = CH_2CH_3$$

M 272.332 (ion)

Hexafluorophosphate: [78618-87-8].
$C_{13}H_{15}F_6PRu$ M 417.296
Yellow cryst. (EtOH). Dec. at 132-42°.

Nesmeyanov, A.N. *et al*, *Dokl. Chem. (Engl. Transl.)*, 1981, **258**, 193 (*synth, ir, pmr*)

$C_{13}H_{16}ORu$ Ru-00316

(1-Hydroxy-1-methylethyl)ruthenocene, 9CI

2-Ruthenocenyl-2-propanol.

[52537-43-6]

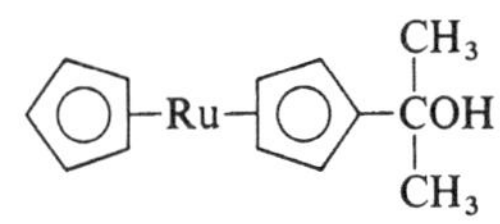

M 289.339

Pale-yellow solid. Mp 80-2°.

Turbitt, T.D. *et al*, *J. Chem. Soc., Perkin Trans. 2*, 1974, 189 (*synth, pmr*)

$C_{13}H_{17}NRu$ Ru-00317

(N,N-Dimethylaminomethyl)ruthenocene

π-Cyclopentadienyl[[(dimethylamino)methyl]-π-cyclopentadienyl]ruthenium, 8CI. N,N-*Dimethylruthenocenemethylamine*

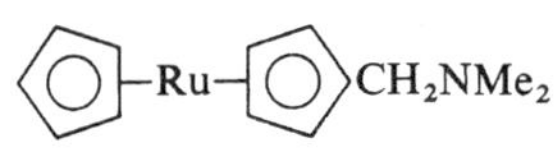

M 288.354

Bright-yellow oil or cryst. Mp 39-42°.

B,MeI:
$C_{14}H_{20}INRu$ M 430.293
Mp 177-80°.

Hofer, O. *et al*, *J. Organomet. Chem.*, 1968, **13**, 443 (*synth, pmr, ir*)

$C_{13}H_{18}Cl_2ORu$ Ru-00318

Carbonyldichloro[(1,2,3,4,5,6-η)-hexamethylbenzene]ruthenium

[71652-17-0]

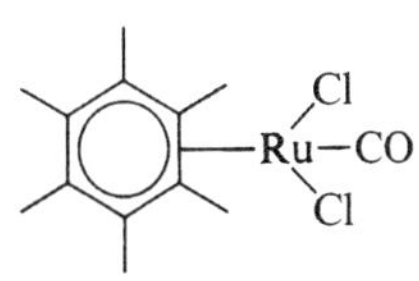

M 362.261

Characterised spectroscopically.

Werner, H. *et al*, *J. Organomet. Chem.*, 1979, **174**, C67 (*synth, pmr, ir*)

$C_{13}H_{18}GeO_2Ru$ Ru-00319

Dicarbonyl[(1,2,3,3a,6a-η)-1,4,5,6-tetrahydro-1-pentalenyl-](trimethylgermyl)ruthenium, 9CI

[41357-40-8]

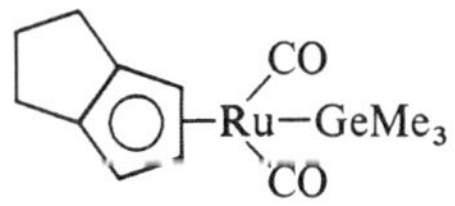

M 379.944

Yellow liq. Sol. $CHCl_3$. $Bp_{0.1}$ ca. 60°.

Knox, S.A.R. *et al*, *J. Chem. Soc., Dalton Trans.*, 1974, 658 (*synth, ir, pmr, ms*)

$C_{13}H_{18}N_2O_3Ru$ Ru-00320

Bis(2-isocyano-2-methylpropane)tricarbonylruthenium

Bis(tert-*butyl isocyanide*)*tricarbonylruthenium*, 8CI

[31532-12-4]

$[(H_3C)_3CNC]_2Ru(CO)_3$

M 351.367

No phys. props. reported.

Ger. Pat., 2 026 926, (*1971*); *CA*, **74**, 124822

$C_{13}H_{18}O_2RuSi$ Ru-00321

Dicarbonyl[(1,2,3,3a,6a-η)-1,4,5,6-tetrahydro-1-pentalenyl-](trimethylsilyl)ruthenium, 9CI

Tetrahydro-π-pentalenyldicarbonyl(trimethylsilyl)ruthenium

[53265-34-2]

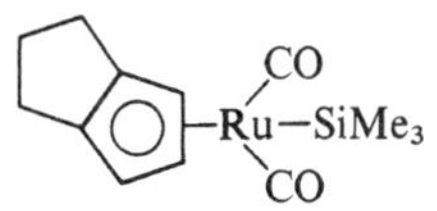

M 335.440

Yellow. $Bp_{0.01}$ ca. 60°.

Knox, S.A.R. *et al*, *J. Chem. Soc., Dalton Trans.*, 1974, 658 (*synth, ms, pmr, ir*)

$C_{13}H_{18}Ru$ Ru-00322

[(2,3,5,6-η)-Bicyclo[2.2.1]hepta-2,5-diene]bis(η^3-2-propenyl-)ruthenium, 9CI

Di-π-allyl-π-2,5-norbornadienylruthenium

[12289-15-5]

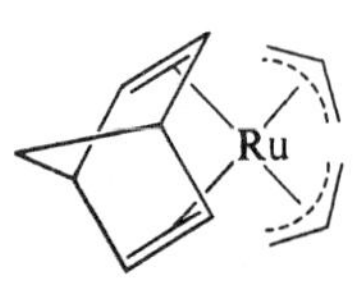

M 275.355

White prisms (MeOH/pet. ether). Spar. sol. alcohols, sol. pet. ether, C_6H_6, dec. by chlorinated solvs. Mp 105-10°.

Powell, J. *et al*, *J. Chem. Soc. (A)*, 1968, 159 (*synth*)

Schrock, R.R. *et al*, *J. Chem. Soc., Dalton Trans.*, 1974, 951 (*props*)

$C_{13}H_{18}RuSi$ Ru-00323

Trimethylsilylruthenocene

π-Cyclopentadienyl-(trimethylsilyl-π-cyclopentadienyl-)ruthenium. Trimethyl(ruthenocenyl)silane

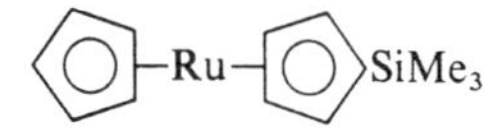

M 303.441

Marr, G. *et al*, *J. Organomet. Chem.*, 1964, **2**, 99.

$C_{13}O_{13}Ru_4^{\ominus\ominus}$ Ru-00324

μ_3-Carbonyldodecacarbonyltetraruthenate(2−)

Tridecacarbonyltetraruthenate(2−)

[75215-45-1]

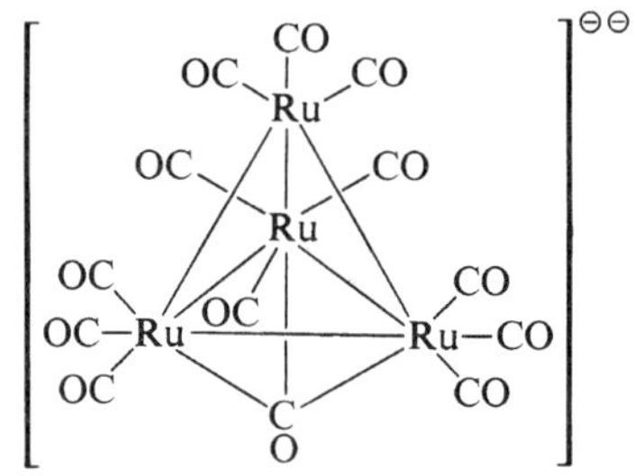

M 768.415 (ion)

Di-K salt: [75215-44-0].

$C_{13}K_2O_{13}Ru_4$ M 846.612

Red.

Bis[bis(triphenylphosphine)iminium]salt: [75215-46-2].

$C_{85}H_{60}N_2O_{13}P_4Ru_4$ M 1845.590

Light red-brown powder (CH_2Cl_2/Et_2O).

Nagel, C.C. *et al*, *J. Chem. Soc., Chem. Commun.*, 1980, 530 (*synth, ir, cmr*)

U.S.P., 4 349 521, (*1982*); *CA*, **97**, 165494 (*synth*)

$C_{14}H_2O_{12}Ru_4$ Ru-00325

Dodecacarbonyl[μ_4-(1-η:η^2:η^2:2-η-1,2-ethenediyl)]tetraruthenium, 10CI

[63374-85-6]

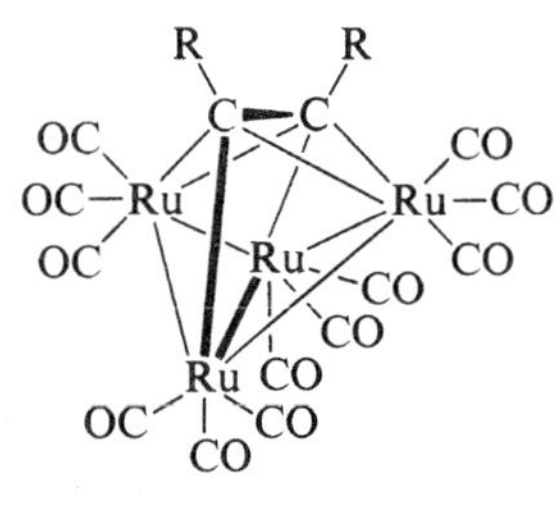

Proposed structure

R = H

M 766.443

Air-stable yellow solid. Not well characterised.

Eady, C.R. *et al*, *J. Chem. Soc., Dalton Trans.*, 1977, 477 (*synth, ir*)

$C_{14}H_4N_2O_{10}Ru_3$ Ru-00326

Tri-μ-carbonylheptacarbonyl[μ-(pyridazine-N^1:N^2)]triruthenium, 9CI

*Decacarbonyl[μ-(pyridazine-*N[1],N[2]*)]triruthenium*

[60767-64-8]

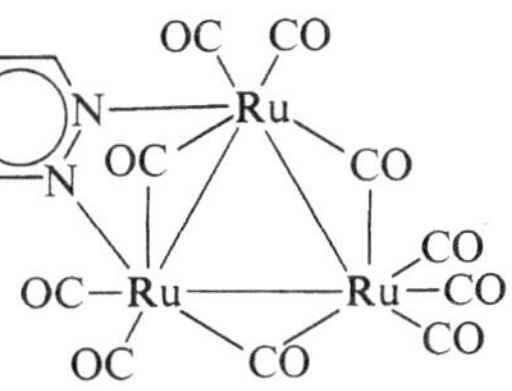

M 663.403

Maroon cryst. (THF/hexane).

Cotton, F.A. *et al*, *J. Am. Chem. Soc.*, 1977, **99**, 6588 (*synth, ir, cmr, cryst struct*)

$C_{14}H_6O_6Ru_2$ Ru-00327

Hexacarbonyl[μ-(η^2:η^4-1,2-ethenediyl-1,2-phenylene)]diruthenium, 9CI

Tricarbonyl(1,1,1-tricarbonylruthenaindene)ruthenium

[59687-85-3]

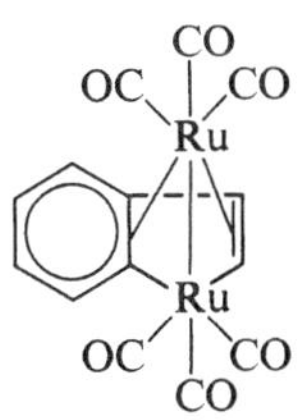

M 472.338

Pale-yellow cryst. Mp 91-92.5°.

Harris, P.J. *et al*, *J. Chem. Soc., Dalton Trans.*, 1976, 377 (*synth, ir*)

$C_{14}H_8O_6Ru_2$ Ru-00328

Hexacarbonyl[μ-[(1,2,3,4-η:1,4,5,8-η)-1,3,5,7-cyclooctatetraene]]diruthenium, 9CI

Hexacarbonyl(μ-1,3,5,7-cyclooctatetraene)diruthenium

[74333-23-6]

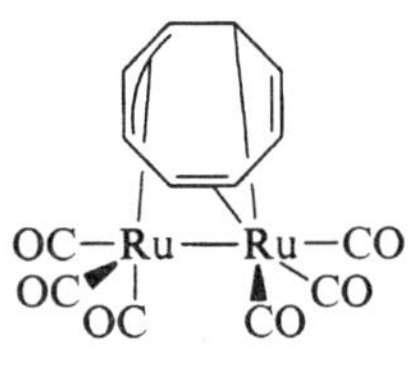

M 474.354

Fluxional in soln. Pale-yellow air-stable cryst. (hexane). Mp 101-4° (95-9°).

Bruce, M.I. *et al*, *J. Organomet. Chem.*, 1968, **13**, 227 (*synth, ir, pmr*)

Cotton, F.A. *et al*, *J. Am. Chem. Soc.*, 1968, **90**, 5412 (*cryst struct*)

Cotton, F.A. *et al*, *J. Am. Chem. Soc.*, 1969, **91**, 6598 (*synth, pmr, ir*)

Edwards, J.D. *et al*, *J. Chem. Soc., Dalton Trans.*, 1980, 552 (*pmr*)

$C_{14}H_8O_6Ru_3S$ Ru-00329

Hexacarbonyl[μ_3-[(1,2,3,4-η:5,6-η:7,8-η)-1,3,5,7-cyclooctatetraene]]-μ_3-thioxotriruthenium, 10CI

[73085-12-8]

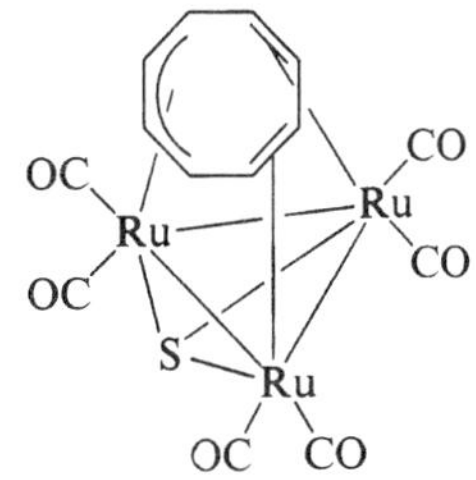

M 607.484

No phys. props. reported.

Howard, J.A.K. *et al*, *J. Chem. Soc., Chem. Commun.*, 1979, 839 (*synth, cmr*)

$C_{14}H_9ClN_2O_2Ru$ Ru-00330

Dicarbonylchloro[2-(phenylazo)phenyl]ruthenium, 8CI

*Dicarbonyl[2-*C,N′*-(phenylazo)phenyl]chlororuthenium. (Azobenzene-*C,N′*)dicarbonylchlororuthenium*

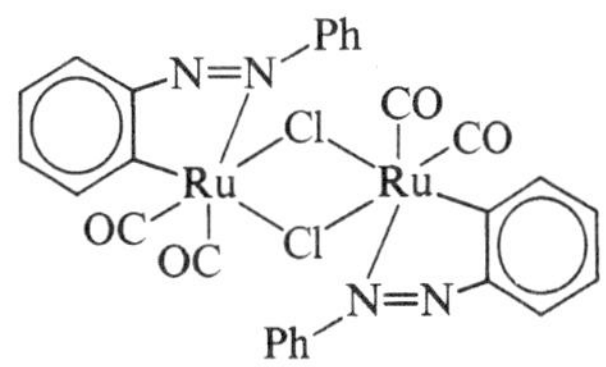

M 373.760

Dimeric.

Dimer: [34269-35-7]. *Tetracarbonyldi-μ-chlorobis[2-(phenylazo)phenyl-*C,N′*]diruthenium.*

$C_{28}H_{18}Cl_2N_4O_4Ru_2$ M 747.521

Orange-yellow cryst. (pet. ether). Mp 249-50°.

Bruce, M. *et al*, *J. Chem. Soc. (A)*, 1971, 2820 (*synth, ir, pmr*)

$C_{14}H_9Cl_2N_2O_2Ru^{\ominus}$ Ru-00331

Dicarbonyldichloro[2-(phenylazo)phenyl]ruthenate(1−), 8CI

*[(Azobenzene)-*C,N′*]dicarbonyldichlororuthenate(1−)*

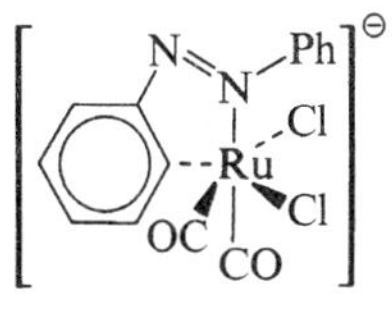

M 409.213 (ion)

Tetramethylammonium salt: [30754-01-9].

$C_{18}H_{21}Cl_2N_3O_2Ru$ M 483.359

Orange needles (Me_2CO/pet. ether). Mp 112-3°.

Bruce, M.I. *et al*, *J. Chem. Soc. (A)*, 1971, 2820 (*synth, ir, pmr*)

$C_{14}H_{10}HgO_4Ru_2$ Ru-00332

Tetracarbonylbis(η^5-2,4-cyclopentadien-1-yl)(mercury)diruthenium

Tetracarbonyldi-π-cyclopentadienyl-μ-mercuriodiruthenium, 8CI

[12131-08-7]

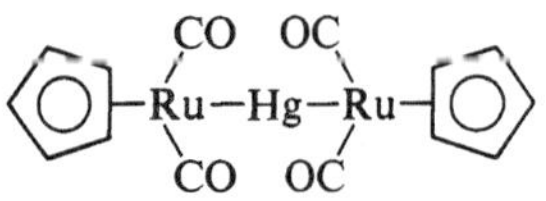

M 644.961

Fine-yellow cryst. (pet. ether). Mp 176° (173-5°).

Fischer, R.D. *et al*, *J. Organomet. Chem.*, 1967, **7**, 135 (*synth, ir*)
Blackmore, T. *et al*, *J. Chem. Soc. (A)*, 1968, 2931 (*synth, ir, ms*)

$C_{14}H_{10}O_2Ru_2S_2$ Ru-00333

Bis[μ-(carbonothioyl)]dicarbonylbis(η^5-2,4-cyclopentadien-1-yl)diruthenium, 10CI

Dicarbonylbis(η^5-2,4-cyclopentadien-1-yl)bis[μ-(thiocarbonyl)]diruthenium

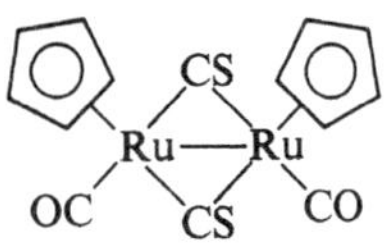

cis-form

M 476.492

Cis- and *trans*-forms interconvert in soln. Dark-red air-stable solid (CS_2/hexane).

cis-form [61930-78-7]
Not separately characterised.

trans-form [61989-65-9]
Not separately characterised.

Wnuk, T.A. *et al*, *Inorg. Chem.*, 1977, **16**, 1173 (*synth, pmr, ir, ms*)

$C_{14}H_{10}O_3Ru_2S$ Ru-00334

[μ-(Carbonothioyl)]-μ-carbonyldicarbonylbis(η^5-2,4-cyclopentadien-1-yl)diruthenium, 10CI

μ-(Carbonothioyl)tricarbonyl(η^5-cyclopentadienyl)diruthenium. μ-Carbonyldicarbonylbis(η^5-2,4-cyclopentadien-1-yl)[μ-(thiocarbonyl)]diruthenium

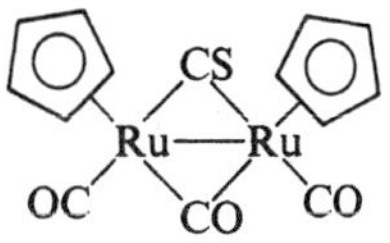

cis-form

M 460.431

Cis- and *trans*-forms interconvert in soln. Air-stable orange-red solid.

cis-form [61930-77-6]
Not separately characterised.

trans-form [61930-76-5]
Not separately characterised.

Wnuk, T.A. *et al*, *Inorg. Chem.*, 1977, **16**, 1173 (*synth, ir, pmr, ms*)

$C_{14}H_{10}O_4Ru_2$ Ru-00335

Di-μ-carbonyldicarbonylbis(η^5-2,4-cyclopentadien-1-yl)diruthenium, 9CI

Bis(cyclopentadienyldicarbonylruthenium). Bis(cyclopentadienyl)tetracarbonyldiruthenium

[12132-87-5]

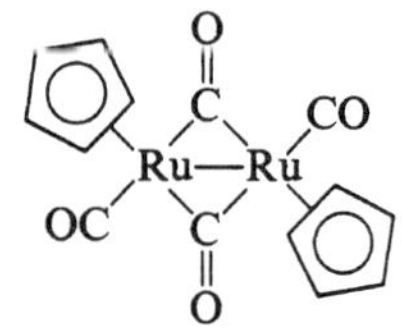

M 444.371

Struct. of solid as shown; in soln. compd. exists as equil. mixt. with *cis* and non-CO bridged forms.

cis-form [31725-27-6]
Not isol.

trans-form [31725-28-7]
Orange cryst. (Et_2O). Mp 185° dec.

Fischer, E.O. *et al*, *Z. Naturforsch., B*, 1962, **17**, 420 (*synth, pmr, ir*)
Cotton, F.A. *et al*, *Inorg. Chem.*, 1967, **6**, 15 (*ir*)
Fischer, R.D. *et al*, *J. Organomet. Chem.*, 1967, **7**, 135 (*cryst struct, ir*)
Noack, K., *J. Organomet. Chem.*, 1967, **7**, 151 (*ir*)
Blackmore, T. *et al*, *J. Chem. Soc. (A)*, 1968, 2158 (*synth*)
King, R.B., *Org. Mass Spectrom.*, 1969, **2**, 657 (*ms*)
McArdle, P. *et al*, *J. Chem. Soc. (A)*, 1970, 2128 (*ir*)
Bullitt, J.G. *et al*, *Inorg. Chem.*, 1972, **11**, 671 (*pmr, ir*)
Harris, D.C. *et al*, *J. Chem. Soc., Dalton Trans.*, 1974, 2398 (*cmr*)
Humpries, A.P. *et al*, *J. Chem. Soc., Dalton Trans.*, 1975, 1710 (*synth*)
Onaka, S. *et al*, *Inorg. Chem.*, 1976, **15**, 915 (*raman*)

$C_{14}H_{12}N_2O_2RuS_2$ Ru-00336

Bis(2-aminothiophenol)dicarbonylruthenium

Bis(o-aminobenzenethiolato)dicarbonylruthenium, 8CI. *Dicarbonylbis(2-aminothiophenol)ruthenium*

[14978-16-6]

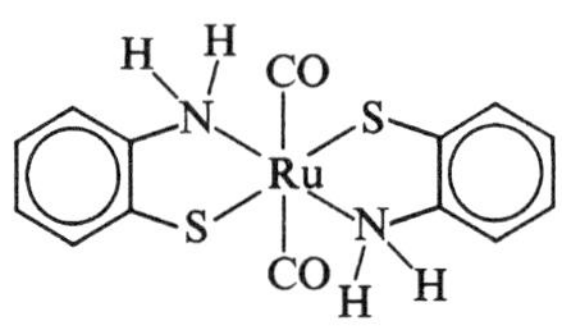

M 405.451

Orange-brown. Mp 191-3°. ir ν_{CO} 2035, 1970 cm^{-1}. Diamagnetic, nonelectrolyte in $PhNO_2$.

Kingston, J.V. *et al*, *J. Inorg. Nucl. Chem.*, 1967, **29**, 133 (*synth, ir*)

$C_{14}H_{12}O_2Ru$ Ru-00337

Benzyldicarbonyl(η^5-2,4-cyclopentadien-1-yl)ruthenium

Dicarbonyl(η^5-2,4-cyclopentadien-1-yl)phenylmethylruthenium, 9CI. *η^5-Cyclopentadienyldicarbonylbenzylruthenium. σ-Benzyl-π-cyclopentadienylruthenium dicarbonyl*

[53449-91-5]

As Dicarbonyl(η^5-2,4-cyclopentadien-1-yl)iodoruthenium, Ru-00051 with

$$X = -CH_2Ph$$

M 313.318

Not isol. pure, charcterised spectroscopically.

Jacobson, S.E. *et al*, *J. Organomet. Chem.*, 1974, **72**, 113 (*synth, ir*)

$C_{14}H_{12}O_6Ru_2$ Ru-00338

Hexacarbonyl[μ-[(1,2,3,4-η:1,4-η)-1,2,3,4-tetramethyl-1,3-butadiene-1,4-diyl]]diruthenium

[33310-09-7]

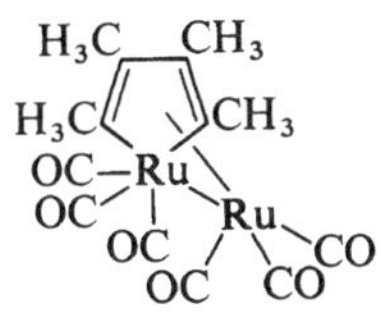

M 478.385

No phys. props. reported.

Vaglio, G.A. *et al*, *Org. Mass. Spectrom.*, 1971, **5**, 493 (*ms*)

$C_{14}H_{13}O_{14}PRu_4$ Ru-00339

Undecacarbonyltetra-μ-hydro(trimethyl phosphite-*P*)tetraruthenium

Undecacarbonyltetra-μ-hydro(phosphorous acid)tetraruthenium trimethyl ester, 8CI

[34438-92-1]

As Dodecacarbonyltetra-μ-hydrotetraruthenium, Ru-00224 with

$$X = P(OMe)_3, Y = CO$$

M 840.502

Catalyst precursor for olefin isomerisation and hydrogenation. Air-stable orange cryst. Sol. hydrocarbons, halocarbons.

Kaesz, H.D. *et al*, *J. Am. Chem. Soc.*, 1971, **93**, 4594 (*synth, pmr*)

Graff, J.L. *et al*, *J. Am. Chem. Soc.*, 1980, **102**, 2123 (*use*)

$C_{14}H_{14}Br_2N_2O_2Ru$ Ru-00340

Bis(aniline)dibromodicarbonylruthenium, 8CI

Bis(benzeneamine)dibromodicarbonylruthenium

[14911-97-8]

$$Ru(PhNH_2)_2(CO)_2Br_2$$

M 503.155

Mixt. of *cis*- and *trans*-forms. Green cryst. Mp >250°. Diamagnetic.

Kingston, J.V. *et al*, *J. Inorg. Nucl. Chem.*, 1967, **29**, 133 (*synth, ir*)

$C_{14}H_{14}Cl_2N_2O_2Ru$ Ru-00341

Bis(aniline)dicarbonyldichlororuthenium, 8CI

Bis(benzenamine)dicarbonyldichlororuthenium, 9CI

[14911-96-7]

$$Ru(PhNH_2)_2(CO)_2Cl_2$$

M 414.253

Mixt. of *cis*- and *trans*-forms. Blue solid. Mp 185-7°. Diamagnetic.

Kingston, J.V. *et al*, *J. Inorg. Nucl. Chem.*, 1967, **29**, 133 (*synth, ir*)

$C_{14}H_{14}ORu^{\oplus\oplus}$ Ru-00342

(η^6-Benzene)[1-(η^6-phenyl)ethanone]ruthenium(2+), 10CI

(η^6-Acetophenone)(η^6-benzene)ruthenium(2+). *(η^6-Acetylbenzene)(η^6-benzene)ruthenium(2+)*

As (η^6-Benzene)(η^6-chlorobenzene)ruthenium(2+), Ru-00234 with

$$X = COCH_3$$

M 299.334 (ion)

Air-stable.

Bis(tetrafluoroborate): [71861-33-1].

$C_{14}H_{14}B_2F_8ORu$ M 472.941

Colour and Mp not reported.

Bennett, M.A. *et al*, *J. Organomet. Chem.*, 1979, **175**, 87 (*synth, pmr*)

$C_{14}H_{14}ORu$ Ru-00343

π-Cyclopentadienyl(4,5,6,7-tetrahydro-4-oxo-π-indenyl)ruthenium, 8CI

1,2-(α-Oxotetramethylene)ruthenocene

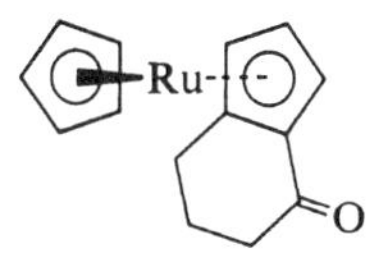

(1*S*)-*form*

M 299.334

Cryst. (C_6H_6/pet. ether). Mp 104-7°.

(1S)-form

Cryst. (hexane). Mp 117-9°. $[\alpha]_D$ +375° (EtOH).

Menthylhydrazone: Cryst. (EtOH). Mp 208-10°. $[\alpha]_D$ +645° (EtOH).

Hofer, O. *et al*, *J. Organomet. Chem.*, 1968, **13**, 443 (*synth, ir, uv*)

Hofer, O. *et al*, *J. Organomet. Chem.*, 1968, **13**, 457 (*resoln, ord, cd*)

$C_{14}H_{14}ORu$ Ru-00344

[4]-Ruthenocenophan-1-one

1,1′-(1-Oxo-1,4-butanediyl)ruthenocene, 10CI

[66507-23-1]

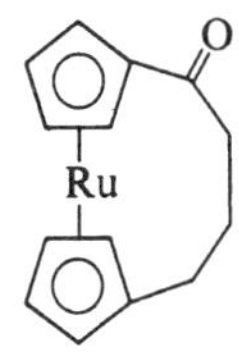

M 299.334

Not isol. pure.

Kamiyama, S-I. *et al*, *Bull. Chem. Soc. Jpn.*, 1978, **51**, 909 (*synth*)

$C_{14}H_{14}ORu$ Ru-00345

[4]Ruthenocenophan-2-one

1,1′-(2-Oxo-1,4-butanediyl)ruthenocene, 10CI

[66507-22-0]

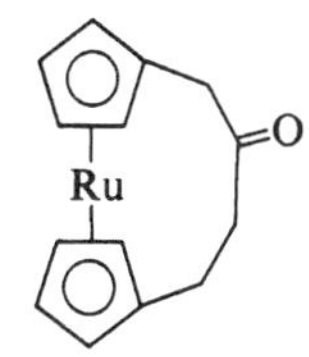

M 299.334

Cryst. (hexane/CH_2Cl_2). Mp 165-6°.

Kamiyama, S. *et al*, *Bull. Chem. Soc. Jpn.*, 1978, **51**, 909 (*synth, pmr, uv, ir*)

Senda, Y. *et al*, *Bull. Chem. Soc. Jpn.*, 1980, **53**, 3381 (*cmr*)

$C_{14}H_{14}O_2Ru$ **Ru-00346**

1,1′-Diacetylruthenocene, 9CI

Bis(acetyl-π-cyclopentadienyl)ruthenium

[54628-71-6]

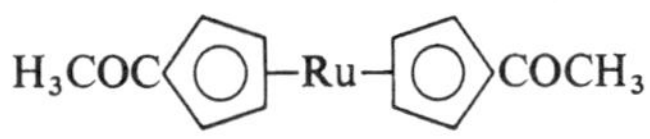

M 315.333

Long yellow needles (C_6H_6/pet. ether). Mp 149-50°. Subl. *in vacuo*.

Rausch, M.D. *et al*, *J. Am. Chem. Soc.*, 1960, **82**, 76 (*synth, ir*)
Rausch, M.D. *et al*, *J. Org. Chem.*, 1963, **28**, 3225 (*pmr*)
Trotter, J., *Acta Crystallogr.*, 1963, **16**, 571 (*struct*)

$C_{14}H_{14}O_3Ru$ **Ru-00347**

γ-Oxoruthenocenebutanoic acid

[(3-Carboxypropionyl)-π-cyclopentadienyl]-π-cyclopentadienylruthenium, 8CI

M 331.333

Mp 201-2°.

Hofer, O. *et al*, *Tetrahedron Lett.*, 1967, 3485 (*synth*)
Hofer, O. *et al*, *J. Organomet. Chem.*, 1968, **13**, 443 (*synth*)

$C_{14}H_{14}Ru$ **Ru-00348**

(η^6-Benzene)[(1,2,3,4-η)-1,3,5,7-cyclooctatetraene]ruthenium, 10CI

[62422-34-8]

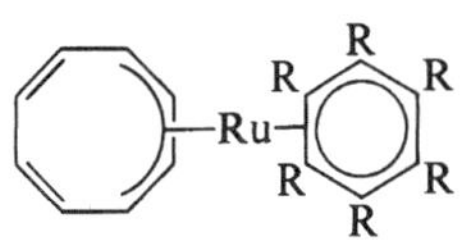

R = H

M 283.335

Air-sensitive cryst. (hexane). Fluxional C_8H_8 ring.

Bennett, M.A. *et al*, *J. Organomet. Chem.*, 1976, **121**, C18.
Bennett, M.A. *et al*, *Inorg. Chem.*, 1980, **19**, 1014 (*synth, ir, pmr*)

$C_{14}H_{14}Ru$ **Ru-00349**

π-Cyclopentadienyl(4,5-dihydro-π-indenyl)ruthenium

1,2-(1-Butenylene)ruthenocene

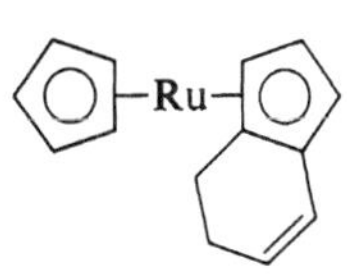

M 283.335

(−)-*form*

Mp 55-64°. $[\alpha]_D$ −920° (EtOH).

Hofer, O. *et al*, *J. Organomet. Chem.*, 1968, **13**, 457 (*synth, ir, uv, ord*)

$C_{14}H_{14}Ru$ **Ru-00350**

1,1′-Divinylruthenocene

Bis(vinyl-π-cyclopentadienyl)ruthenium, 8CI. *1,1′-Bis(ethenyl)ruthenocene. 1,1′-Diethenylruthenocene*

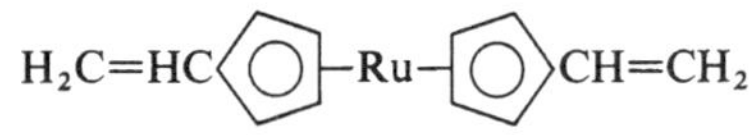

M 283.335

Yellow cryst. by subl. Mp 55-7°.

Rausch, M.D. *et al*, *J. Organomet. Chem.*, 1968, **11**, 317 (*synth, pmr*)

$C_{14}H_{16}ORu$ **Ru-00351**

π-Cyclopentadienyl(4,5,6,7-tetrahydro-4-hydroxy-π-indenyl)ruthenium

α-Hydroxy-1,2-tetramethyleneruthenocene

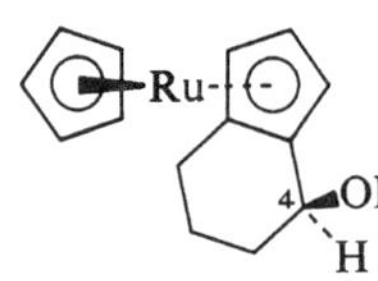

(1*S*,4*S*)-*form*

M 301.350

(*1S,4S*)-*form*

(+)-exo-*form*

$[\alpha]_D$ +100° (EtOH).

(*1S,4R*)-*form*

(+)-endo-*form*

$[\alpha]_D$ +110° (EtOH).

Hofer, O. *et al*, *J. Organomet. Chem.*, 1968, **13**, 457 (*synth, resoln, ir*)

$C_{14}H_{16}O_2Ru$ **Ru-00352**

Ruthenocenebutanoic acid

[(3-Carboxypropyl)-π-cyclopentadienyl]-π-cyclopentadienylruthenium, 8CI

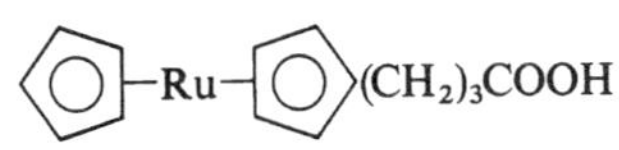

M 317.349

Cryst. (pet. ether). Mp 115-6°.

Hofer, O. *et al*, *J. Organomet. Chem.*, 1968, **13**, 443 (*synth*)

$C_{14}H_{16}O_2Ru_2S_2$ **Ru-00353**

Dicarbonylbis(η^5-2,4-cyclopentadien-1-yl)bis[μ-(methanethiolato)]diruthenium

Dicarbonyldi-π-cyclopentadienylbis(methanethiolato-)diruthenium

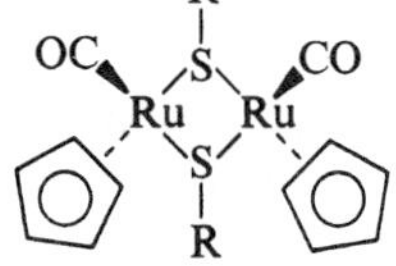

cis-form

R = Me

M 482.539

One *cis*- and one *trans*-form known. Further isomerism caused by *cis*, *trans*-orientation of the R groups has not been demonstrated (contrast homologues with larger R groups).

cis-form [53537-18-1]

Orange cryst. (CH_2Cl_2/hexane). Mp 168-70°. Exhibits one ν_{CO} ir band in soln. Thermodynamically more stable form, converted to *trans*-form by uv irradiation.

***trans*-form** [53585-80-1]
Red cryst. (CH_2Cl_2/hexane). Mp 134-6°. Exhibits two ν_{CO} ir bands in soln. Thermodynamically less stable than *cis*-form. Undergoes rapid ring inversion.

Knox, G.R. *et al*, *J. Organomet. Chem.*, 1974, **74**, 105 (*synth*, *ir*, *pmr*)
Killops, S.D. *et al*, *J. Chem. Soc.*, *Dalton Trans.*, 1978, 1260 (*synth*, *ir*, *pmr*)

$C_{14}H_{16}O_4RuSi_2$ Ru-00354

Tetracarbonyl[1,2-phenylenebis[dimethylsilylene]]ruthenium, 9CI
1,1,3,3-Tetramethyl-2,2,2,2-tetracarbonyl-1,3-dihydrobenzo[d][*2,1,3*]*ruthenadisilole*
[59327-83-2]

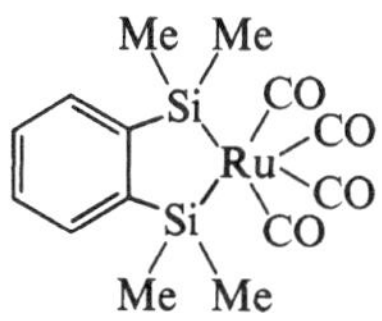

M 405.519
Mp 53.5°. $Bp_{0.005}$ 40°.

Fink, W., *Helv. Chim. Acta*, 1976, **59**, 606 (*synth*, *pmr*, *ms*)

$C_{14}H_{16}Ru$ Ru-00355

(Bicyclo[2.2.1]hepta-2,5-diene)(1,3,5-cycloheptatriene)ruthenium
(*1,3,5-Cycloheptatriene*)(*2,5-norbornadiene*)*ruthenium*

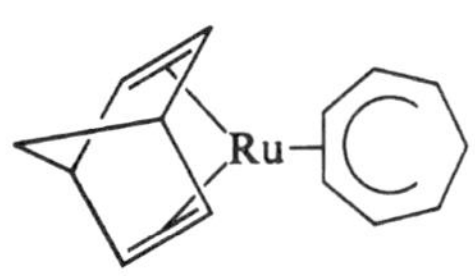

M 285.350
Yellow cryst. by subl. *in vacuo*. Mp 47-9°. Dec. >115°.

Müller, J. *et al*, *J. Organomet. Chem.*, 1966, **5**, 275 (*synth*, *pmr*, *ir*)

$C_{14}H_{16}Ru$ Ru-00356

[(1,2,3,4,5-η)-2,4-Cycloheptadien-1-yl][(1,2,3,4,5-η)-2,4,6-cycloheptatrien-1-yl]ruthenium, 9CI
[54873-27-7]

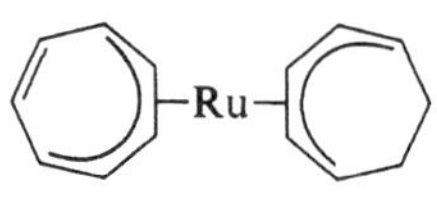

M 285.350
Yellow cryst. Mod. air-stable in solid state but air-sensitive in soln. V. sol. pentane, C_6H_6, Et_2O, mod. sol. EtOH, insol. H_2O. Mp 73°.

Müller, J. *et al*, *Chem. Ber.*, 1975, **108**, 273 (*synth*, *pmr*)
Müller, J. *et al*, *J. Organomet. Chem.*, 1975, **97**, 275 (*props*)
Schmid, H. *et al*, *Chem. Ber.*, 1976, **109**, 125 (*cryst struct*)

$C_{14}H_{16}Ru$ Ru-00357

[4]Ruthenocenophane
1,1′-(1,4-Butanediyl)ruthenocene, 10CI
[66507-24-2]

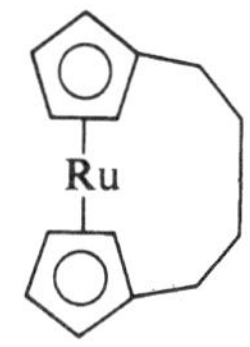

M 285.350
Cryst. (C_6H_6). Mp 73-4°.

Kamiyama, S. *et al*, *Bull. Chem. Soc. Jpn.*, 1978, **51**, 909 (*synth*, *pmr*, *uv*)
Senda, Y. *et al*, *Bull. Chem. Soc. Jpn.*, 1980, **53**, 3381 (*cmr*)

$C_{14}H_{17}Ru^{\oplus}$ Ru-00358

(η^6-Benzene)[(1,2,3,4,5-η)-2,4,6-cyclooctatrien-1-yl]ruthenium(1+), 10CI

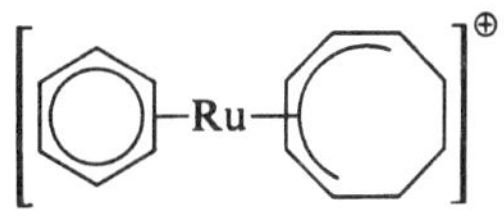

M 286.358 (ion)
Hexafluorophosphate: [77357-79-0].
$C_{14}H_{17}F_6PRu$ M 431.322
Not isol. in pure form.

Bennett, M.A. *et al*, *Inorg. Chem.*, 1981, **20**, 2353 (*synth*)

$C_{14}H_{17}Ru^{\oplus}$ Ru-00359

[(1,2,3,4,5-η)-2,4-Cycloheptadien-1-yl][(1,2,3,4,5,6-η)-1,3,5-cycloheptatriene]ruthenium(1+), 9CI

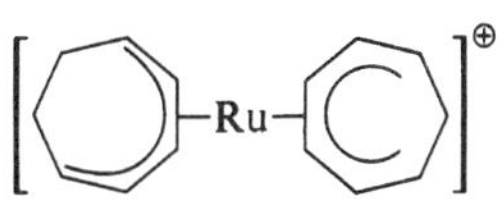

M 286.358 (ion)
Hexafluorophosphate: [57603-50-6].
$C_{14}H_{17}F_6PRu$ M 431.322
Light-yellow solid. Sol. H_2O, CH_2Cl_2, Me_2CO, MeCN, insol. hydrocarbons, Et_2O.
Tetrafluoroborate: [57603-49-3].
$C_{14}H_{17}BF_4Ru$ M 373.162
Light-yellow cryst. (Me_2CO/Et_2O). Sol. H_2O, CH_2Cl_2, Me_2CO, MeCN, insol. Et_2O, hydrocarbons.

Müller, J. *et al*, *J. Organomet. Chem.*, 1975, **97**, 275 (*synth*, *ir*, *pmr*)

$C_{14}H_{18}Ge_2O_8Ru_2$ Ru-00360

Octacarbonylbis(trimethylgermyl)diruthenium, 9CI
[30854-80-9]

$Me_3Ge{-}Ru(CO)_4{-}Ru(CO)_4{-}GeMe_3$ (*diax-*)

M 661.611
Yellow cryst. Mp 126-7°.

Knox, S.A.R. *et al*, *J. Chem. Soc.* (*A*), 1971, 2874 (*synth*, *ir*, *nmr*)

$C_{14}H_{18}N_2O_4Ru_2$ Ru-00361

Tetracarbonyl[μ-[η^4-*N,N'*-1,2-ethanediylidenebis[2-propanamine]-*N,N'*]]-μ-1,2-ethenediyldiruthenium, 10CI

Tetracarbonyl(μ^2-acetylene)[glyoxalbis(isopropylimine)]diruthenium

[75963-10-9]

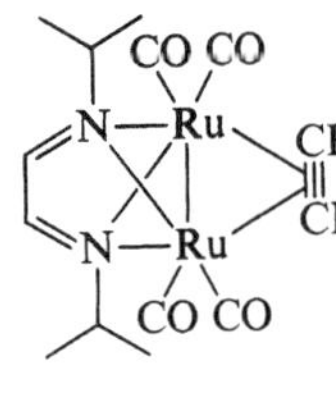

M 480.447

Yellow cryst. (CH_2Cl_2).

Staal, L.H. *et al*, *Inorg. Chem.*, 1981, **20**, 1830 (*synth, cryst struct, pmr, ir*)
Staal, L.H. *et al*, *Inorg. Chim. Acta*, 1981, **50**, 205 (*ms*)

$C_{14}H_{18}O_3PRu^{\oplus}$ Ru-00362

(η^6-Benzene)[(1,2,3,4,5,6-η)-6-(dimethoxyphosphinyl-)-2,4-cyclohexadien-1-yl]ruthenium(1+), 10CI

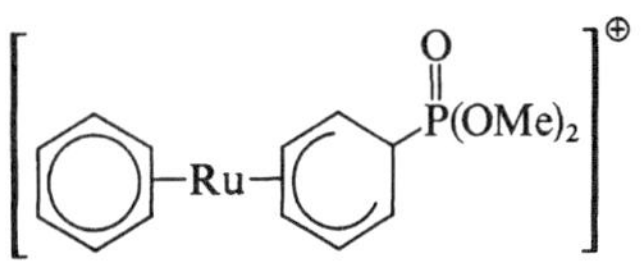

M 366.338 (ion)

Catalyses Arbuzov rearrangements.

Hexafluorophosphate: [76722-36-6].
$C_{14}H_{18}F_6O_3P_2Ru$ M 511.302
White solid (MeCN aq.).

Sweigert, D., *J. Chem. Soc., Chem. Commun.*, 1980, 1159 (*synth, pmr*)

$C_{14}H_{18}O_8Ru_2Si_2$ Ru-00363

Octacarbonylbis(trimethylsilyl)diruthenium, 9CI, 8CI

[51267-41-5]

$Me_3Si-Ru(CO)_4-Ru(CO)_4-SiMe_3$ (*diax-*)

M 572.602

Bright-yellow needles (pentane). Mp 129-31°.

Knox, S.A.R. *et al*, *J. Chem. Soc. (A)*, 1969, 2559 (*synth, ir, pmr*)
Knox, S.A.R. *et al*, *J. Chem. Soc., Dalton Trans.*, 1980, 235 (*use*)

$C_{14}H_{18}O_8Ru_2Sn_2$ Ru-00364

Octacarbonylbis(trimethylstannyl)diruthenium, 9CI

[58280-32-3]

$Me_3Sn-Ru(CO)_4-Ru(CO)_4-SnMe_3$ (*diax-*)

M 753.811

Yellow prisms. $Bp_{0.01}$ ca. 80° subl.

Cotton, J.D. *et al*, *J. Chem. Soc. (A)*, 1968, 2758 (*synth, ir*)
Howard, J.A.K. *et al*, *J. Chem. Soc., Dalton Trans.*, 1975, 2332 (*cryst struct*)

$C_{14}H_{18}Ru$ Ru-00365

(η^6-Benzene)[(1,2,5,6-η)-1,5-cyclooctadiene]ruthenium, 9CI

[54913-13-2]

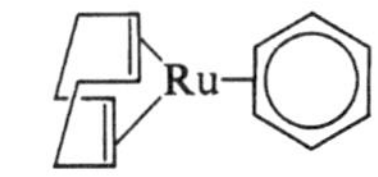

M 287.366

Yellow needles (pentane). Mp 134-135.5°.

Müller, J. *et al*, *Chem. Ber.*, 1975, **108**, 273 (*synth*)
Schmid, H. *et al*, *Chem. Ber.*, 1976, **109**, 132 (*struct*)
Muetterties, E.L. *et al*, *J. Organomet. Chem.*, 1979, **178**, 197 (*synth, pmr, ms*)

$C_{14}H_{18}Ru$ Ru-00366

Bis[(1,2,3,4,5-η)-2,4-cycloheptadien-1-yl]ruthenium, 9CI

[54873-26-6]

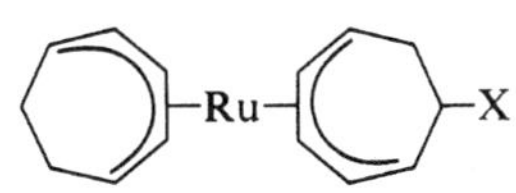

X = H

M 287.366

Yellow cryst. Mp 43-4°.

Müller, J. *et al*, *Chem. Ber.*, 1975, 273 (*synth*)
Müller, J. *et al*, *J. Organomet. Chem.*, 1975, **97**, 275 (*props*)
Schmid, H. *et al*, *Chem. Ber.*, 1976, **109**, 125 (*cryst struct*)
Pertici, P. *et al*, *J. Chem. Soc., Dalton Trans.*, 1980, 1961 (*synth, pmr*)

$C_{14}H_{18}Ru$ Ru-00367

1,1'-Diethylruthenocene

Bis(ethyl-π-cyclopentadienyl)ruthenium, 8CI

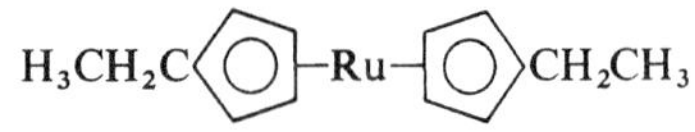

M 287.366

Characterised by pmr.

Gauthier, G., *J. Chem. Soc., Chem. Commun.*, 1969, 690 (*synth, pmr*)

$C_{14}H_{19}NRu$ Ru-00368

1-[(Dimethylamino)ethyl]ruthenocene

N,N-*Dimethyl-1-ruthenocenylethylamine.* α,N,N-*Trimethylruthenocenemethanamine*

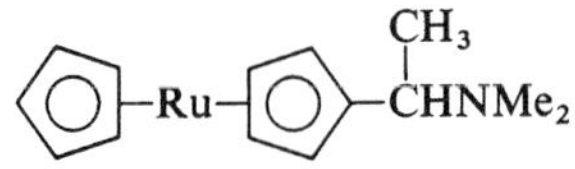

M 302.381

(+)-*form*

$[\alpha]_D^{25}$ +8.0° (c, 1.0 in EtOH).

(−)-*form*

$[\alpha]_D^{25}$ −12.0° (c, 1.0 in EtOH).

(±)-*form*

Yellow liq.

Turbitt, T.D. *et al*, *J. Chem. Soc., Perkin Trans. 2*, 1974, 177 (*synth, pmr*)

$C_{14}H_{22}Ru$ Ru-00369

[(1,2,5,6-η)-1,5-Cyclooctadiene]bis(η³-2-propenyl)ruthenium, 9CI

Bis(π-allyl)(1,5-cyclooctadiene)ruthenium

[12289-52-0]

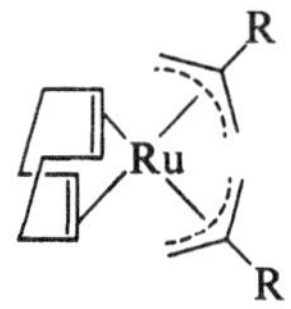

R = H

M 291.398

White microprisms (MeOH/pet. ether). Sol. C_6H_6, pet. ether, spar. sol. alcohols. Dec. by chlorinated solvs. Mp 130-40° dec.

Powell, J. *et al*, *J. Chem. Soc. (A)*, 1968, 159 (*synth, pmr*)
Schrock, R.R. *et al*, *J. Chem. Soc., Dalton Trans.*, 1974, 951.

$C_{14}H_{24}N_4O_2Ru_2S_8$ Ru-00370

Dicarbonylbis[μ-(dimethylcarbamodithioato-*S*:*S*,*S'*)]bis(dimethylcarbamodithioato-*S*,*S'*)-diruthenium

[15393-10-9]

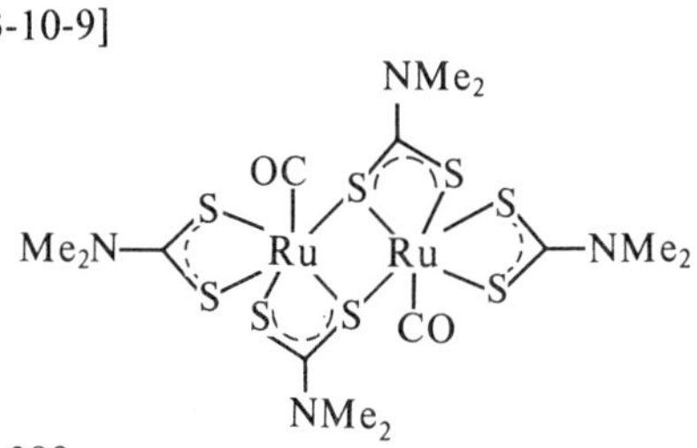

M 738.989

Yellow cryst.

Kingston, J.V. *et al*, *J. Inorg. Nucl. Chem.*, 1966, **28**, 2709 (*synth, ir*)
Raston, C.L. *et al*, *J. Chem. Soc., Dalton Trans.*, 1975, 2418 (*struct*)

$C_{14}H_{27}NP_2Ru^{\oplus\oplus}$ Ru-00371

(Acetonitrile)(η⁶-benzene)bis(trimethylphosphine)ruthenium(2+), 10CI

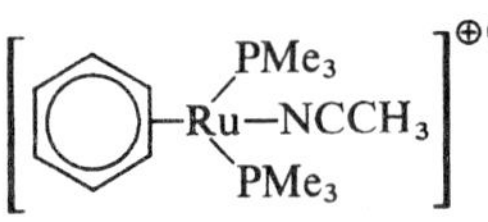

M 372.392 (ion)

Bis(hexafluorophosphate): [78343-13-2].
$C_{14}H_{27}F_{12}NP_4Ru$ M 662.320
No phys. props. reported.

Werner, H. *et al*, *J. Organomet. Chem.*, 1981, **210**, C11 (*synth, pmr, nmr*)

$C_{14}H_{28}P_2Ru^{\oplus\oplus}$ Ru-00372

(η⁶-Benzene)(η²-ethene)bis(trimethylphosphine)ruthenium(2+), 10CI

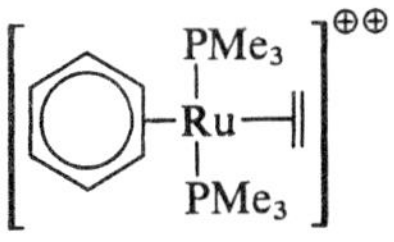

M 359.393 (ion)

Bis(hexafluorophosphate): [73046-96-5].
$C_{14}H_{28}F_{12}P_4Ru$ M 649.321
No phys. props. reported.

Werner, H. *et al*, *J. Organomet. Chem.*, 1979, **181**, C7 (*synth, pmr, nmr*)

$C_{14}H_{29}P_2Ru^{\oplus}$ Ru-00373

(η⁵-2,4-Cyclopentadien-1-yl)[(1,2-η)-1-propene]bis(trimethylphosphine)ruthenium(1+), 10CI

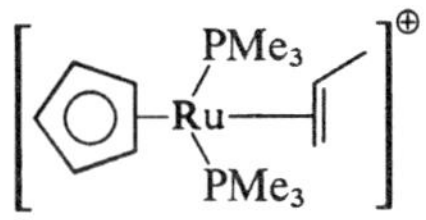

M 360.401 (ion)

Hexafluorophosphate: [76297-42-2].
$C_{14}H_{29}F_6P_3Ru$ M 505.365
White cryst. (CH_2Cl_2/pet. ether). Mp 276° dec.

Treichel, P.M. *et al*, *Inorg. Chim. Acta*, 1980, **42**, 277 (*synth, pmr*)
Bruce, M.I. *et al*, *J. Organomet. Chem.*, 1981, **210**, C5.

$C_{14}H_{32}O_6P_2Ru$ Ru-00374

Bis[(1,2,3-η)-2-methyl-2-propenyl]bis(trimethyl phosphite-*P*)ruthenium, 9CI

Bis(2-methyl-π-allyl)bis(phosphorous acid)ruthenium hexamethyl ester, 8CI

[41654-97-1]

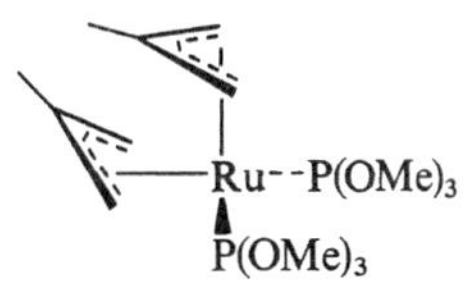

M 459.421

cis-form [32677-89-7]
Well-formed colourless needles (hexane). Slowly dec. in air. Mp 125°.

Green, M. *et al*, *J. Chem. Soc. (A)*, 1971, 16 (*synth, pmr*)
Marsh, R.A. *et al*, *J. Chem. Soc., Dalton Trans.*, 1973, 778 (*cryst struct*)

$C_{14}H_{32}O_9P_3Ru^{\oplus}$ Ru-00375

(η⁵-2,4-Cyclopentadien-1-yl)tris(trimethyl phosphite-*P*)ruthenium(1+)

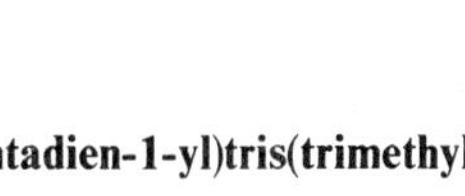

M 538.393 (ion)

Hexafluorophosphate: [71397-30-3].
$C_{14}H_{32}F_6O_9P_4Ru$ M 683.357
Cream cryst. (CH_2Cl_2/Et_2O or $CHCl_3$/pet. ether).

Ashby, G.S. *et al*, *Aust. J. Chem.*, 1979, **32**, 1003 (*synth, pmr, ir*)

$C_{14}H_{32}P_3Ru^{\oplus}$ Ru-00376

(η⁵-2,4-Cyclopentadien-1-yl)tris(trimethylphosphine)ruthenium(1+), 10CI

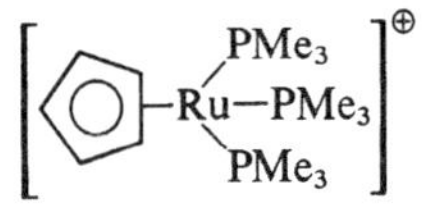

M 394.398 (ion)

Hexafluorophosphate: [75182-27-3].
$C_{14}H_{32}F_6P_4Ru$ M 539.362
White solid (CH_2Cl_2/Et_2O). Mp >250° dec.

Treichel, P.M. *et al*, *Synth. React. Inorg. Metal-Org. Chem.*, 1980, **10**, 205 (*synth, ir, pmr*)
Bruce, M.I. *et al*, *J. Chem. Soc., Dalton Trans.*, 1981, 1398 (*synth, ir, pmr, cmr*)

$C_{14}H_{36}P_4Ru$ Ru-00377

Bis[1,2-ethanediylbis[dimethylphosphine]-*P,P'*](η^2-ethene)-ruthenium

Bis[1,2-bis(dimethylphosphino)ethane](η^2-ethylene)-ruthenium

[67421-25-4]

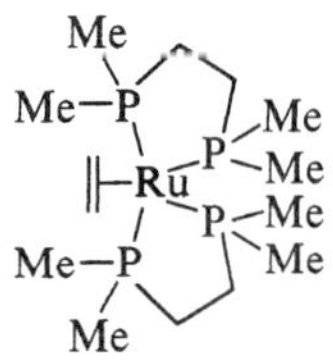

M 429.403

No phys. description recorded. Sol. toluene.

Tolman, C.A. *et al*, *J. Am. Chem. Soc.*, 1978, **100**, 4080 (*nmr*)

$C_{14}H_{37}N_6Ru^{\oplus}$ Ru-00378

[(1,2,5,6-η)-1,5-Cyclooctadiene]tris(1,1-dimethylhydrazine-N^2)hydroruthenium(1+)

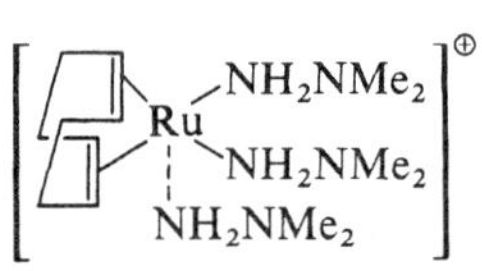

H not located

M 390.557 (ion)

Hexafluorophosphate: [64722-25-4].
$C_{14}H_{37}F_6N_6PRu$ M 535.521
White solid. Mp 115° dec.

Tetraphenylborate:
$C_{38}H_{57}BN_6Ru$ M 709.789
Colourless plates (MeOH or H_2O). Mp 148° dec.

Ashworth, T.V. *et al*, *J. Chem. Soc., Dalton Trans.*, 1977, 1809 (*synth, ir, pmr*)

Ashworth, T.V. *et al*, *J. Chem. Soc., Dalton Trans.*, 1978, 1040 (*cryst struct*)

$C_{14}H_{39}B_7P_2Ru$ Ru-00379

[η^6-Nonahydrodicarbanonaborato(2−)-*B,H*]bis(triethylphosphine)ruthenium, 10CI

hyper-closo-*6,6-Bis(triethylphosphine)-6,2,3-RuC$_2$B-$_7$H$_9$*

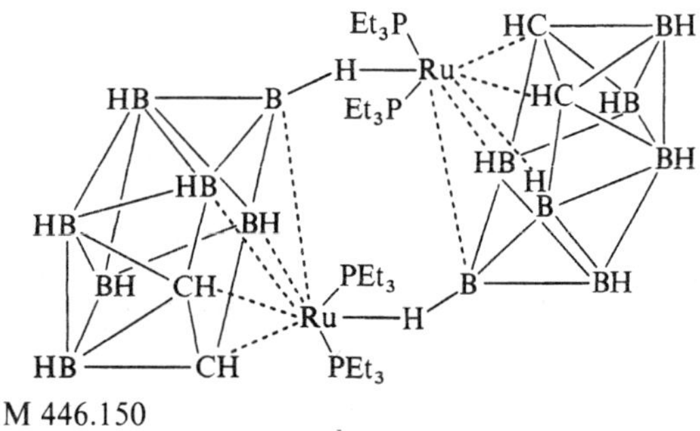

M 446.150

Dimeric.

Dimer: [76798-39-5]. *Bis[μ-[η^6-nonahydrodicarbanonaborato(2−)-B,H]]tetrakis(triethylphosphine)diruthenium, 10CI.*
$C_{28}H_{78}B_{14}P_4Ru_2$ M 892.299
Large blue-black or small bright-green cryst. (toluene/pentane).

Jung, C.W. *et al*, *J. Am. Chem. Soc.*, 1981, **103**, 810 (*synth, nmr, pmr, ir*)

$C_{14}Mn_2O_{14}Ru$ Ru-00380

Tetracarbonylbis(pentacarbonylmanganio)ruthenium, 8CI

Tetradecacarbonyldimanganeseruthenium

[33152-40-8]

$(OC)_5Mn—Ru(CO)_4—Mn(CO)_5$

M 603.092

Yellow.

Abel, E.W. *et al*, *Inorg. Nucl. Chem. Lett.*, 1971, **7**, 587 (*synth, ir*)

$C_{15}H_7NO_9Ru_3$ Ru-00381

[μ_3-[Benzenaminato(2−)]]nonacarbonyldi-μ-hydrotriruthenium, 9CI

μ_3-Anilinatononacarbonyldihydrotriruthenium

[51185-97-8]

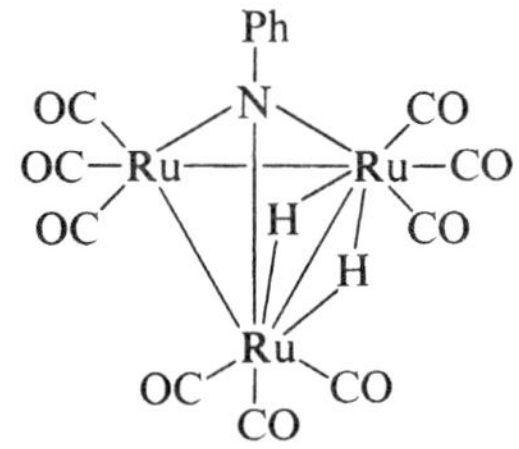

M 648.432

Yellow cryst. Mp 130-40° dec.

Sappa, E. *et al*, *J. Organomet. Chem.*, 1973, **61**, 383 (*synth, ir, ms, pmr*)

$C_{15}H_7O_9PRu_3$ Ru-00382

Nonacarbonyldi-μ-hydro[μ_3-(phenylphosphinidene)]triruthenium, 9CI

Di-μ-hydrido-μ_3-(phenylphosphiniden)-triangulo-*tris-(tricarbonylruthenium)*

[79230-49-2]

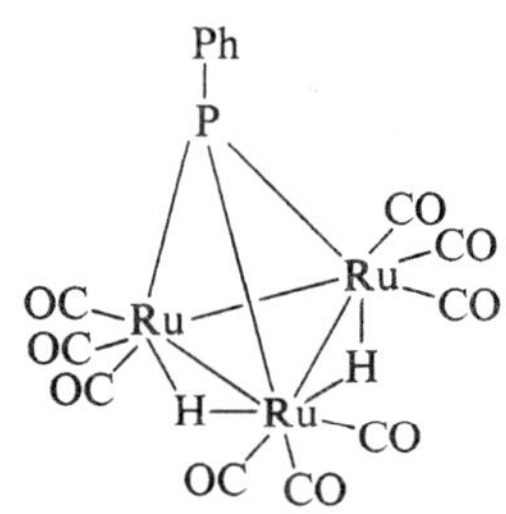

M 665.399

Yellow cryst. (hexane/toluene). Mp 117° dec.

Iwasaki, F. *et al*, *J. Organomet. Chem.*, 1981, **213**, 185 (*synth, ir, nmr, cryst struct*)

Carty, A.J., *Pure Appl. Chem.*, 1982, **54**, 113 (*synth, struct*)

MacLaughlin, S.A. *et al*, *Can. J. Chem.*, 1982, **60**, 87 (*synth*)

Schneider, J. *et al*, *Chem. Ber.*, 1982, **115**, 989 (*synth, pmr, nmr, ms*)

$C_{15}H_8O_5Ru_2$ Ru-00383

[μ-[(1,2,3,3*a*,8*a*-η:4,5,6-η)-azulene]]pentacarbonyldiruthenium, 9CI

[57903-36-3]

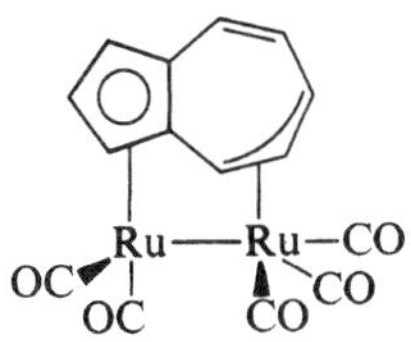

M 470.365

Dark-red cryst. Mp 168-9°.

Knox, S.A.R. *et al*, *J. Chem. Soc., Dalton Trans.*, 1975, 1647 (*synth*)
Cotton, F.A. *et al*, *Inorg. Chem.*, 1977, **16**, 89 (*pmr*)

$C_{15}H_8O_9Ru_3$ Ru-00384

Nonacarbonyl[μ_3-[(1-η:1,2-η:2-η)-1,3-cyclohexadiene-1,2-diyl]]di-μ-hydrotriruthenium, 9CI

[70772-42-8]

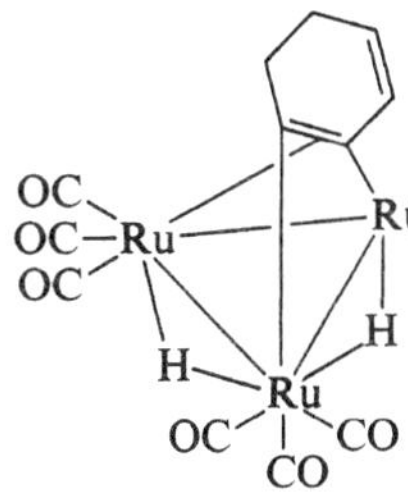

M 635.433

Pale-yellow.

Aime, S. *et al*, *Inorg. Chim. Acta*, 1979, **34**, 49 (*synth, ir, ms*)

$C_{15}H_9BrO_4RuW$ Ru-00385

Bromotetracarbonyl(ruthenocenylmethylidyne)tungsten, 10CI

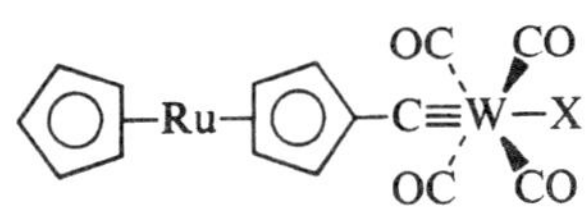

X = Br

M 618.058

***trans*-form** [75293-20-8]

Bright-yellow powder. Mp >−5° dec.

Fischer, E.O. *et al*, *J. Organomet. Chem.*, 1980, **191**, 261 (*synth, ir, ms, cmr, pmr*)

$C_{15}H_9ClO_4RuW$ Ru-00386

Tetracarbonylchloro(ruthenocenomethylidyne)tungsten, 10CI

As Bromotetracarbonyl(ruthenocenylmethylidyne)tungsten, Ru-00385 with

X = Cl

M 573.607

***trans*-form**

Bright-yellow.

Fischer, E.O. *et al*, *J. Organomet. Chem.*, 1980, **191**, 261.

$C_{15}H_9O_9Ru_3^{\ominus}$ Ru-00387

Nonacarbonyl[μ_3-[(1-η:1,2-η:1,2-η)-3,3-dimethyl-1-butynyl-]]triruthenate(1−)

Nonacarbonyl(tert-*butylacetylene*)*triruthenate*(*1*−)

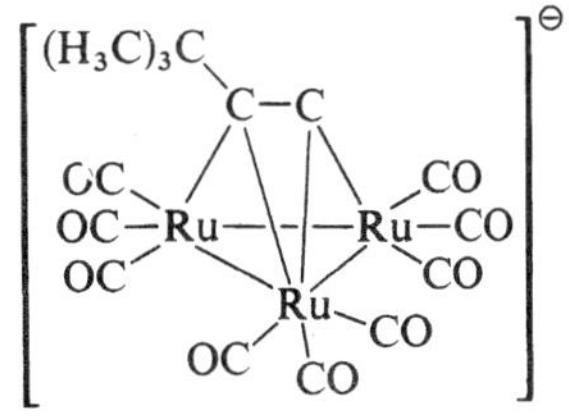

M 636.441 (ion)

Exhibits CO scrambling at ambient temps. Obt. from Nonacarbonyl[μ_3-[(1-η:1,2-η:1,2-η)-3,3-dimethyl-1-butynyl]]-μ-hydrotriruthenium, Ru-00388 .

Tetraphenylarsonium salt: [76741-75-8].
$C_{39}H_{29}AsO_9Ru_3$ M 1019.784
Orange-yellow cryst. (EtOH).

Barner-Thorsen, C. *et al*, *Inorg. Chem.*, 1981, **20**, 4306 (*synth, cryst struct, cmr, pmr*)

$C_{15}H_{10}O_9Ru_3$ Ru-00388

Nonacarbonyl[μ_3-[(1-η:1,2-η:1,2-η)-3,3-dimethyl-1-butynyl]]-μ-hydrotriruthenium, 9CI

Nonacarbonyl(tert-*butylacetylene*)*hydrotriruthenium*

[57673-31-1]

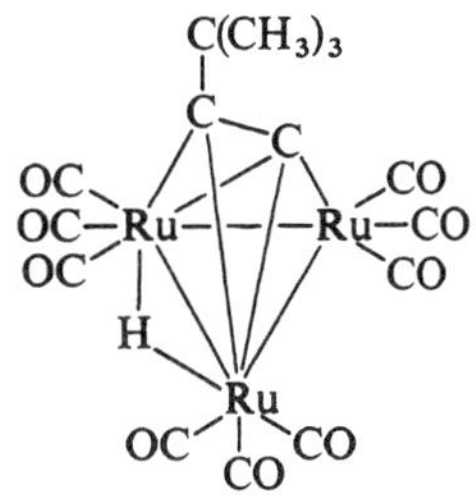

M 637.449

Exhibits CO scrambling at ambient temp. Yellow cryst. (heptane). Deprotonation yields Nonacarbonyl[μ_3-[(1-η:1,2-η:1,2-η)-3,3-dimethyl-1-butynyl]]triruthenate(1−), Ru-00387 .

Rosenberg, E. *et al*, *Inorg. Chim. Acta*, 1975, **15**, 53 (*cmr*)
Catti, M. *et al*, *J. Chem. Soc., Dalton Trans.*, 1977, 2260 (*cryst struct*)
Ermer, S. *et al*, *J. Organomet. Chem.*, 1980, **187**, 81 (*synth, ir, pmr*)
Jangala, C. *et al*, *Inorg. Chem.*, 1980, **19**, 1571 (*synth*)
Castiglioni, M. *et al*, *Inorg. Chim. Acta*, 1981, **49**, 217 (*synth*)
Carty, A.J., *Pure Appl. Chem.*, 1982, **54**, 113 (*cmr*)

$C_{15}H_{10}O_{10}Ru$ Ru-00389

1,2,3,4,5-Ruthenocenepentacarboxylic acid

1,2,3,4,5-Pentacarboxyruthenocene

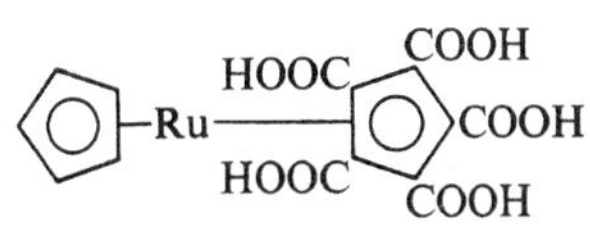

M 451.308

Penta-Me ether: [79151-49-8]. *1,2,3,4,5-Pentakis-(methoxycarbonyl)ruthenocene*, 10CI.
$C_{20}H_{20}O_{10}Ru$ M 521.442
Pale-yellow cryst.

Bruce, M. I. *et al*, *J. Chem. Soc., Chem. Commun.*, 1981, 428 (*synth, pmr, cmr, struct*)

$C_{15}H_{12}O_3Ru_2$ Ru-00390

μ-Carbonyldicarbonylbis(η^5-2,4-cyclopentadien-1-yl-)-μ-ethenylidenediruthenium, 10CI

Tricarbonylbis(η^5-cyclopentadienyl)-μ-ethenylidenediruthenium

[75952-49-7]

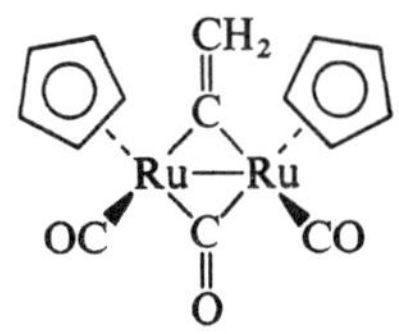

M 442.398

cis-form

Yellow cryst. Mp 178-80°. Slowly converts in soln. at r.t. to an equilibrium mixt. of *cis-* and *trans*-forms.

Davies, D.L. *et al*, *J. Organomet. Chem.*, 1980, **198**, C43 (*synth, cryst struct, ir, pmr, cmr*)

$C_{15}H_{12}O_9Ru_3$ Ru-00391

Nonacarbonyl[μ_3-[(1-η:1-η:1,2-η)-3,3-dimethyl-1-butenylidene]]di-μ-hydrotriruthenium, 9CI

[79061-97-5]

As Nonacarbonyl[μ-[(1-η:1-η:1,2-η)-ethenylidene]]di-μ-hydrotriruthenium, Ru-00170 with

$$R = -C(CH_3)_3$$

M 639.464

Dark-yellow cryst.

Castiglioni, M. *et al*, *Inorg. Chim. Acta*, 1981, **49**, 217 (*synth, pmr, ms*)

$C_{15}H_{12}O_9Ru_3^{\oplus\oplus}$ Ru-00392

Nonacarbonyl[μ_3-[(1-η:1,2-η:2-η)-1-(1,1-dimethylethyl)-1,2-ethenediyl]]di-μ-hydrotriruthenium(2+)

Nonacarbonyl(3,3-dimethyl-1-butene)dihydrotriruthenium(2+). (tert-*Butylethylene*)-*nonacarbonyldihydrotriruthenium(2+)*

[76721-83-0]

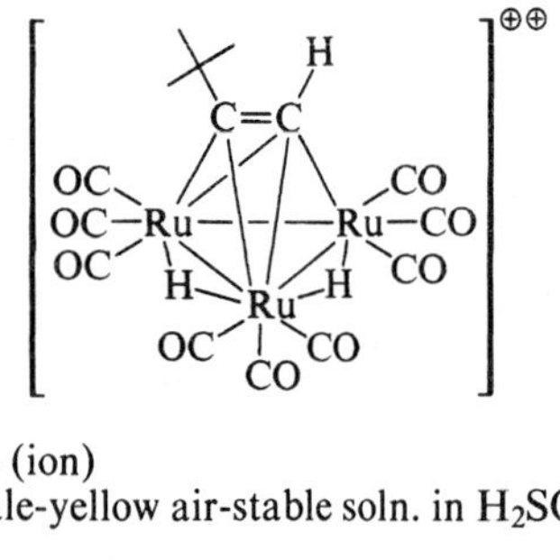

M 639.464 (ion)

Forms a pale-yellow air-stable soln. in H_2SO_4; no salts isol.

Barner-Thorsen, C. *et al*, *Inorg. Chem.*, 1981, **20**, 1592 (*synth, pmr*)

$C_{15}H_{13}O_3Ru_2^{\oplus}$ Ru-00393

μ-Carbonyldicarbonylbis(η^5-2,4-cyclopentadien-1-yl-)(μ-ethylidiyne)diruthenium(1+), 10CI

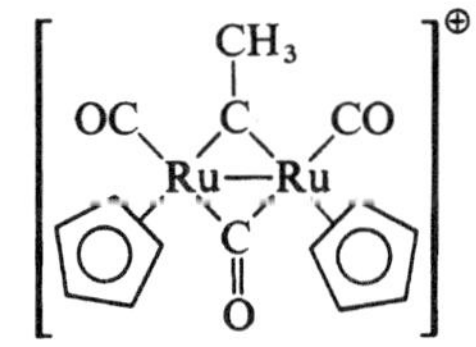

cis-form

M 443.406 (ion)

cis-form [75952-48-6]

Tetrafluoroborate:

$C_{15}H_{13}BF_4O_3Ru_2$ M 530.210

Yellow cryst. Mp 170-2° dec. Deprotonated by water to μ-Carbonyldicarbonylbis(η^5-2,4-cyclopentadien-1-yl)-μ-ethenylidenediruthenium, Ru-00390 .

Davies, D.L. *et al*, *J. Organomet. Chem.*, 1980, **198**, C43 (*synth, struct, ir, pmr, cmr*)

$C_{15}H_{14}O_9Ru_3$ Ru-00394

Nonacarbonyl[μ_3-(3,3-dimethylbutylidyne)]tri-μ-hydrotriruthenium, 10CI

[79075-59-5]

As Nonacarbonyl[μ_3-(chloromethylidyne)]tri-μ-hydrotriruthenium, Ru-00123 with

$$X = -CH_2C(CH_3)_3$$

M 641.480

Green-yellow cryst. ($CHCl_3$/heptane).

Castiglioni, M. *et al*, *Inorg. Chim. Acta*, 1981, **49**, 217 (*synth, ir, pmr, cryst struct*)

$C_{15}H_{16}BN_6Ru^{\oplus}$ Ru-00395

(η^6-Benzene)[hydrotris(1*H*-pyrazolato-*N'*)borato(1−)-N^2,$N^{2'}$,$N^{2''}$]ruthenium(1+), 9CI

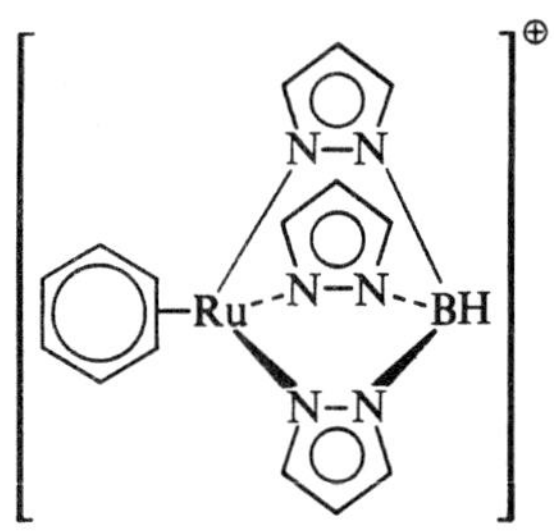

M 392.212 (ion)

Hexafluorophosphate: [52015-87-9].

$C_{15}H_{16}BF_6N_6PRu$ M 537.176

Yellow cryst.

O'Sullivan, D.J. *et al*, *J. Organomet. Chem.*, 1973, **57**, C58 (*synth*)

$C_{15}H_{16}O_4Ru_2S$ **Ru-00396**

Tetracarbonyl[μ-(1,2,3,4-η:1,5,6,7-η)-2,4,6-cyclohepta-trien-1-yl][μ-(2-methyl-2-propanethiolato)]diruthenium, 10CI

[73107-22-9]

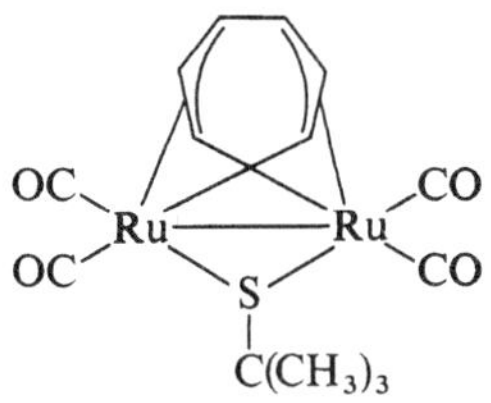

M 494.489

Yellow solid. Mp 124°.

Howard, J.A.K. *et al*, *J. Chem. Soc., Chem. Commun.*, 1979, 839 (*synth, pmr, cmr*)

$C_{15}H_{16}Ru$ **Ru-00397**

[(2,3,5,6-η)-Bicyclo[2.2.1]hepta-2,5-dien-e][(1,2,3,4,5,6-η)-1,3,5,7-cyclooctatetraene]ruthenium

[42516-70-1]

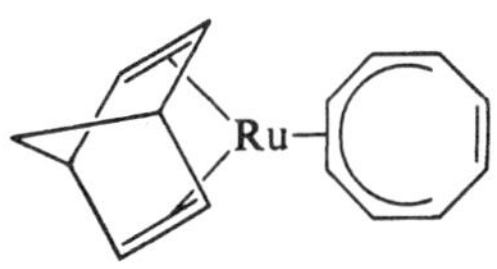

M 297.361

Orange plates. V. air-sensitive in soln. and mildly so in solid state. May be stored indefinitely at −30° under N_2. Dec. on subl.

Schrock, R.R. *et al*, *J. Am. Chem. Soc.*, 1973, **95**, 4102 (*synth, ir, ms*)

Cotton, F.A. *et al*, *J. Organomet. Chem.*, 1976, **107**, 113 (*pmr, cmr*)

$C_{15}H_{17}NRu$ **Ru-00398**

[(1,2,3,4,5-η)-6-Cyano-2,4-cycloheptadi-en-1-yl][(1,2,3,4,5-η)-2,4-cycloheptadien-1-yl]ruthenium

[57603-52-8]

As Bis[(1,2,3,4,5-η)-2,4-cycloheptadien-1-yl]ruthenium, Ru-00366 with

X = CN

M 312.376

Yellow cryst. (pentane).

Müller, J. *et al*, *J. Organomet. Chem.*, 1975, **97**, 275 (*synth, ir, ms*)

$C_{15}H_{18}Ge_3O_9Ru_3$ **Ru-00399**

Nonacarbonyltris[μ-(dimethylgermylene)]triruthenium

[30347-85-4]

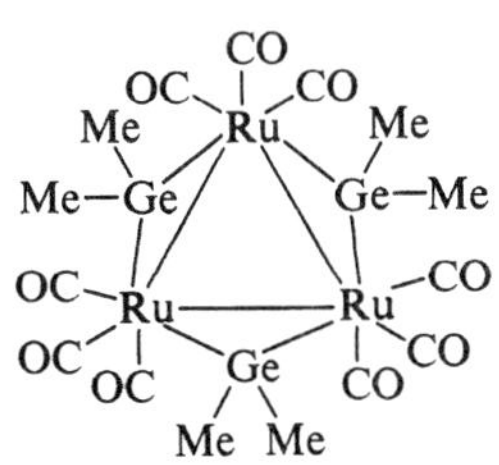

M 863.282

Small golden hexagonal plates. Mp 140-50° dec.

Howard, J. *et al*, *J. Chem. Soc. (A)*, 1971, 3648 (*cryst struct*)

Knox, S.A.R. *et al*, *J. Chem. Soc. (A)*, 1971, 2874 (*synth, ir, pmr*)

$C_{15}H_{18}Ru^{\oplus\oplus}$ **Ru-00400**

(η⁶-Benzene)[(1,2,3,4,5,6-η)-1,3,5-trimethylbenzene]ruthen-ium(2+)

π-Benzene-π-mesityleneruthenium(2+)

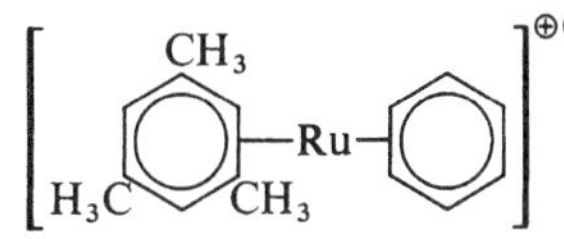

M 299.377 (ion)

Bis(tetraphenylborate):

$C_{63}H_{58}B_2Ru$ M 937.841

White or pale-yellow solid (Me_2CO/Et_2O). Mp 120-2° dec.

Rybinskaya, M.I. *et al*, *J. Organomet. Chem.*, 1982, **235**, 215 (*synth*)

$C_{15}H_{19}Ru^{\oplus}$ **Ru-00401**

(η⁷-Cycloheptatrienylium)[(1,2,5,6-η)-1,5-cyclooctadiene]ru-thenium(1+), 9CI

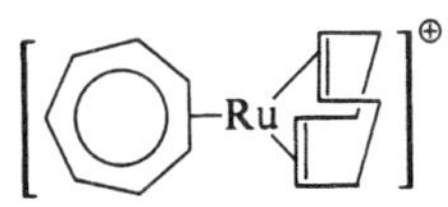

M 300.385 (ion)

Tetrafluoroborate: [57603-47-1].

$C_{15}H_{19}BF_4Ru$ M 387.189

Light-orange cryst. (Me_2CO/Et_2O). Sol. H_2O, CH_2Cl_2, Me_2CO, MeCN. Insol. hydrocarbons, Et_2O.

Müller, J. *et al*, *J. Organomet. Chem.*, 1975, **97**, 275 (*synth, ir, pmr*)

$C_{15}H_{19}Ru^{\oplus}$ **Ru-00402**

(η⁵-2,4-Cyclohexadien-1-yl)[(1,2,3,4,5,6-η)-1,3,5-trimethyl-benzene]ruthenium(1+)

(η⁵-2,4-Cyclohexadienyl)(η⁶-mesitylene)-ruthenium(1+)

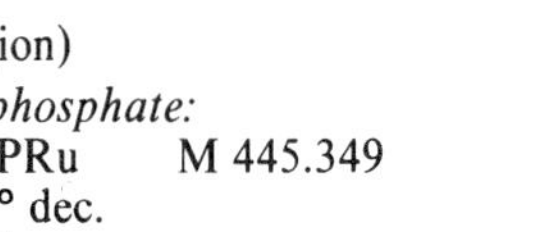

M 300.385 (ion)

Hexafluorophosphate:

$C_{15}H_{19}F_6PRu$ M 445.349

Mp 227-8° dec.

Tetraphenylborate:

$C_{39}H_{39}BRu$ M 619.617

Pale-yellow fine cryst. (Me_2CO/Et_2O). Mp 150-60° dec.

Rybinskaya, M.I. *et al*, *J. Organomet. Chem.*, 1982, **235**, 215 (*synth*)

$C_{15}H_{20}N_4Ru^{\oplus\oplus}$ Ru-00403

Tetrakis(acetonitrile)[(2,3,5,6-η)bicyclo[2.2.1]hepta-2,5-diene]ruthenium(2+), 9CI

Tetrakis(acetonitrile)norbornadieneruthenium(2+)

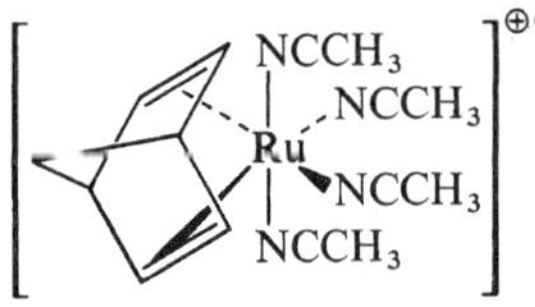

M 357.420 (ion)

Diperchlorate: [53080-27-6].

$C_{15}H_{20}Cl_2N_4O_8Ru$ M 556.321

Yellow needles.

Schrock, R.R. *et al*, *J. Chem. Soc., Dalton Trans.*, 1974, 951 (*synth, ir*)

$C_{15}H_{20}Ru$ Ru-00404

[(1,2,3,4,5,6-η)-1,3,5-Cycloheptatriene][(1,2,5,6-η)-1,5-cyclooctadiene]ruthenium, 9CI

[12216-20-5]

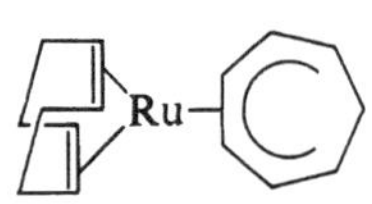

M 301.393

Orange cryst. Mp 96-8°. Subl. at 60-70° *in vacuo*. Dec. at 150-60°.

Müller, J. *et al*, *J. Organomet. Chem.*, 1966, **5**, 275; 1975, **99**, 275 (*synth, ir, pmr*)

Pertici, P. *et al*, *J. Chem. Soc., Chem. Commun.*, 1977, 132 (*synth*)

$C_{15}H_{20}Ru$ Ru-00405

(η^4-1,3-Cyclohexadiene)[(1,2,3,4,5,6-η)-1,3,5-trimethylbenzene]ruthenium

(η^4-1,3-Cyclohexadiene)(η^6-mesitylene)ruthenium

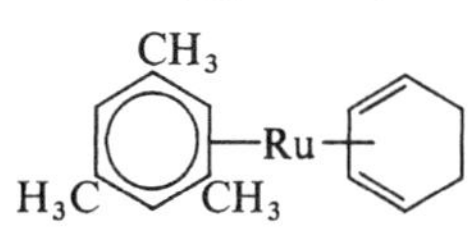

M 301.393

Pale-yellow cryst. (pentane). Mp 53-4°.

Rybinskaya, M.I. *et al*, *J. Organomet. Chem.*, 1982, **235**, 215 (*synth, ms*)

$C_{15}H_{23}ClN_2Ru$ Ru-00406

Chloro(η^5-2,4-cyclopentadien-1-yl)bis(2-isocyano-2-methylpropane)ruthenium, 10CI

*Chloro-π-cyclopentadienylbis(*tert*-butylisonitrile)ruthenium*

[79234-38-1]

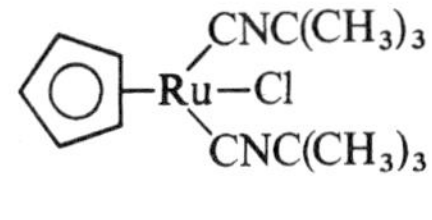

M 367.883

Unstable white solid.

Bruce, M.I. *et al*, *Aust. J. Chem.*, 1981, **34**, 209 (*synth, pmr, cmr*)

$C_{15}H_{33}P_3Ru^{\oplus\oplus}$ Ru-00407

(η^6-Benzene)tris(trimethylphosphine)ruthenium(2+), 10CI

As Triaqua(η^6-benzene)ruthenium(2+), Ru-00041 with

$X = PMe_3$

M 407.417 (ion)

Bis(hexafluorophosphate): [78343-09-6].

$C_{15}H_{33}F_{12}P_5Ru$ M 697.345

No phys. props. reported.

Werner, H. *et al*, *J. Organomet. Chem.*, 1981, **210**, C11 (*synth, pmr, nmr*)

$C_{15}H_{35}O_9P_3Ru$ Ru-00408

[(1,2,3,4-η)-1,3-Cyclohexadiene]tris(trimethyl phosphite-*P*)-ruthenium

[76786-22-6]

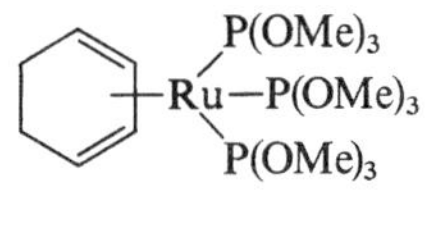

M 553.427

Pale-yellow oil.

Al-Ohaly, A.-R. *et al*, *J. Organomet. Chem.*, 1980, **202**, 297 (*synth, pmr, nmr*)

$C_{15}H_{45}BrO_{15}P_5Ru^{\oplus}$ Ru-00409

Bromopentakis(trimethyl phosphite-*P*)ruthenium(1+), 9CI

$BrRu[P(OMe)_3]_5^{\oplus}$

M 801.354 (ion)

Tetraphenylborate: [35083-11-5].

$C_{39}H_{65}BBrO_{15}P_5Ru$ M 1120.586

White rectangular plates ($MeOH/CH_2Cl_2$).

Couch, D.A. *et al*, *Inorg. Chim. Acta*, 1974, **9**, 39 (*synth, pmr*)

$C_{15}H_{45}ClO_{15}P_5Ru^{\oplus}$ Ru-00410

Chloropentakis(trimethyl phosphite-*P*)ruthenium(1+), 9CI

$ClRu[P(OMe)_3]_5^{\oplus}$

M 756.903 (ion)

Tetraphenylborate: [35083-10-4].

$C_{39}H_{65}BClO_{15}P_5Ru$ M 1076.135

Colourless cryst.

Couch, D.A. *et al*, *Inorg. Chim. Acta*, 1974, **9**, 39 (*synth, pmr*)

$C_{15}H_{45}O_{15}P_5Ru$ Ru-00411

Pentakis(trimethyl phosphite-*P*)ruthenium, 10CI

[61839-26-7]

$Ru[P(OMe)_3]_5$

M 721.450

Stereochemically nonrigid. Air-sensitive white cryst. (pentane). Mp 190°. Protonated by alcohols.

English, A.D. *et al*, *J. Am. Chem. Soc.*, 1977, **99**, 117 (*synth, pmr, nmr*)

Inorg. Synth., 1980, **20**, 79 (*synth, nmr*)

$C_{15}H_{46}O_{15}P_5Ru^{\oplus}$ Ru-00412

Hydropentakis(trimethyl phosphite-*P*)ruthenium(1+), 9CI

$HRu[P(OMe)_3]_5^{\oplus}$

M 722.458 (ion)

Tetraphenylborate: [35083-12-6].

$C_{39}H_{66}BO_{15}P_5Ru$ M 1041.690

White cryst. ($MeOH/CH_2Cl_2$).

Couch, D.A. *et al*, *Inorg. Chim. Acta*, 1974, **9**, 39 (*synth, ir, pmr*)

$C_{16}ClO_{15}Ru_5^{\ominus}$ Ru-00413

[μ_5-(Carbon)]pentadecacarbonylchloropentaruthenate(1−)

[79845-73-1]

$Ru_5(C)(CO)_{15}Cl^{\ominus}$

M 972.970 (ion)

Struct. unknown. No phys. props. reported.

Farrar, D.H. *et al*, *J. Chem. Soc., Chem. Commun.*, 1981, 415 (*synth*)

$C_{16}HBrO_{15}Ru_5$ Ru-00414

Bromopentadecacarbonyl-μ-hydro-μ_5-methanetetraylpentaruthenium

[78919-48-9]

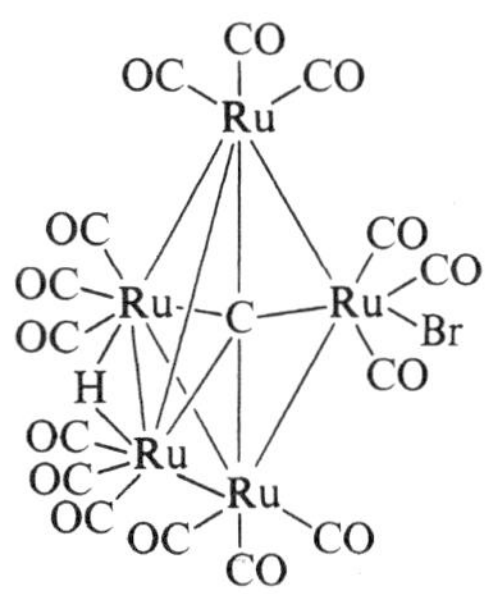

M 1018.429

Yellow solid (CH_2Cl_2). Sol. CH_2Cl_2. Corresponding Cl compound also known.

Oxton, I.A. *et al*, *Inorg. Chem.*, 1981, **20**, 4302 (*synth, ir, pmr, struct*)

$C_{16}H_3O_{15}PRu_5$ Ru-00415

Pentadecacarbonyl[μ_4-(methylphosphoranetetrayl)]pentaruthenium

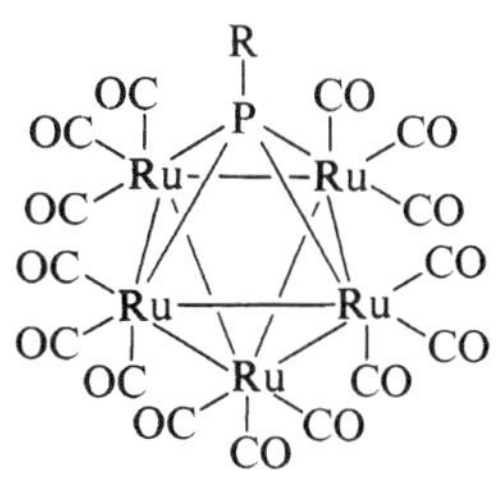

R = Me

M 971.514

Green cryst. Mp 135-40°.

Natarajan, K. *et al*, *J. Organomet. Chem.*, 1981, **209**, 85.

$C_{16}H_5NO_9Ru_3S_2$ Ru-00416

[μ_3-(2(3*H*)-Benzothiazolethionato-N^3:S^2:S^2)]nonacarbonyl-μ-hydrotriruthenium, 9CI

[66515-97-7]

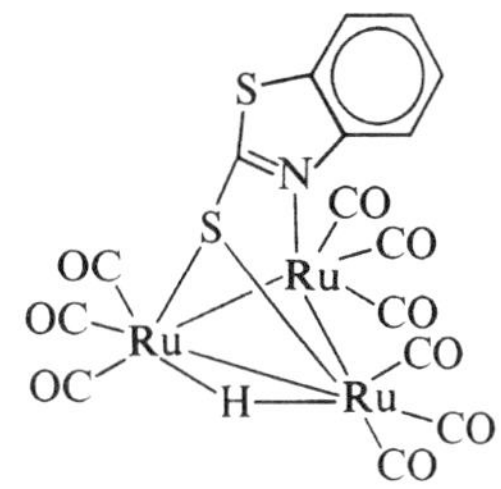

M 722.547

Red cryst.

Jeannin, S. *et al*, *Inorg. Chem.*, 1978, **17**, 2103 (*synth, ir, cryst struct*)

$C_{16}H_5NO_{10}Ru_3$ Ru-00417

[μ_3-[Benzenaminato(2−)]]-μ_3-carbonylnonacarbonyltriruthenium, 9CI

μ_3-Anilinatodecacarbonyltriruthenium

[51185-99-0]

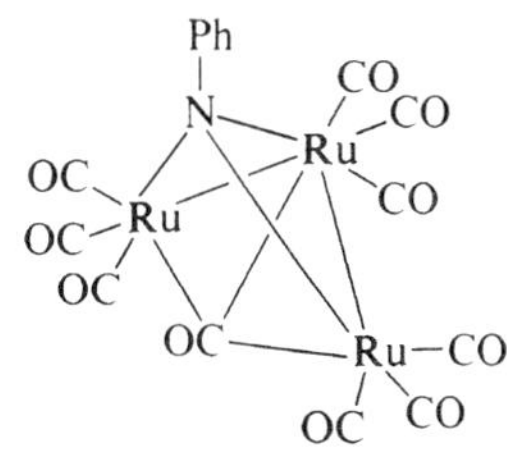

M 674.426

Lemon-yellow cryst. Mp 120° dec.

Sappa, E. *et al*, *J. Organomet. Chem.*, 1973, **61**, 383 (*synth, ir, pmr*)

$C_{16}H_6O_8Ru_3$ Ru-00418

Octacarbonyl[(1,2,3,3*a*,6*a*-η;3*a*,4,5,6,6*a*-η)pentalene]triruthenium

[54327-52-5]

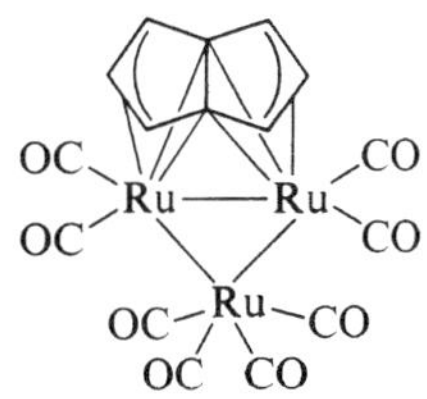

M 629.429

Edge-bonded isomer, equilibrates with Octacarbonyl[μ_3-[(1,2-η:2,3-η:3*a*,4,5,6,6*a*-η)-pentalene]]triruthenium, Ru-00419 in soln. Yellow. Fluxional.

Howard, J.A.K. *et al*, *J. Chem. Soc., Chem. Commun.*, 1974, 452 (*cryst struct*)

Knox, S.A.R. *et al*, *J. Chem. Soc., Dalton Trans.*, 1979, 1801 (*synth, ir, pmr, ms*)

$C_{16}H_6O_8Ru_3$ Ru-00419

Octacarbonyl[μ_3-[(1,2-η:2,3-η:3*a*,4,5,6,6*a*-η)-pentalene]]triruthenium

[72643-21-1]

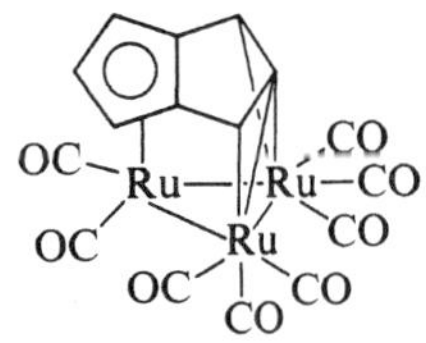

M 629.429

Face bonded isomer. Equilibrates with Octacarbonyl[(1,2,3,3*a*,6*a*-η;3*a*,4,5,6,6*a*-η)pentalene]triruthenium, Ru-00418 in soln. Orange solid. Mp 220°.

Howard, J.A.K. *et al*, *J. Chem. Soc., Chem. Commun.*, 1974, 452 (*cryst struct*)

Knox, S.A.R. *et al*, *J. Chem. Soc., Dalton Trans.*, 1979, 1801 (*synth, ir, pmr, ms*)

$C_{16}H_7O_{10}PRu_3$ Ru-00420

Decacarbonyl-μ-hydro[μ-(phenylphosphino)]triruthenium, 10CI

[79235-61-3]

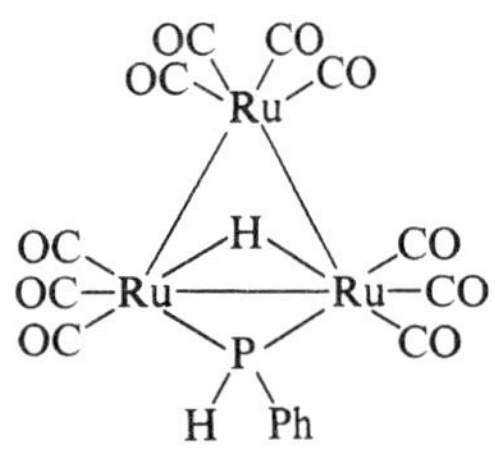

M 693.409

Orange-red.

Iwasaki, F. *et al*, *J. Organomet. Chem.*, 1981, **213**, 185 (*synth, ir, pmr, ms, cryst struct*)

Field, J.S. *et al*, *J. Organomet. Chem.*, 1982, **224**, C49 (*synth*)

$C_{16}H_{10}Cl_2N_2O_2Ru$ Ru-00421

Bis(benzonitrile)dicarbonyldichlororuthenium, 8CI

Dicarbonyldichlorobis(cyanobenzene)ruthenium

[30066-41-2]

$(PhCN)_2RuCl_2(CO)_2$

M 434.243

Yellow cryst. Insol. Et_2O. Mp 125-9°.

Gilbert, J.D. *et al*, *J. Chem. Soc. (A)*, 1970, 2765 (*synth, ir*)

$C_{16}H_{10}O_7Ru_2$ Ru-00422

Hexacarbonyl-μ-hydro[μ-[(*O*,1-η:1,2-η)-3-oxo-1-phenyl-1-butenyl]]diruthenium, 9CI

μ-Hydrido-μ-(benzylideneacetonato)hexacarbonyldiruthenium

[52675-73-7]

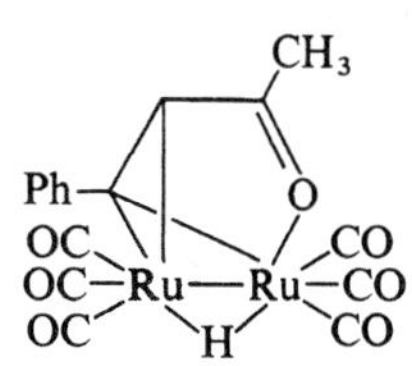

M 516.391

Air-stable orange cryst.

Domingos, A.J.P. *et al*, *J. Chem. Soc., Chem. Commun.*, 1973, 912 (*synth, pmr, cryst struct*)

$C_{16}H_{14}Br_2O_2RuS_2$ Ru-00423

[1,2-Bis(benzenethiolato)ethane-*S,S'*]dibromodicarbonylruthenium

[1,2-Bis(phenylthio)ethane]dibromodicarbonylruthenium, 8CI

[32648-41-2]

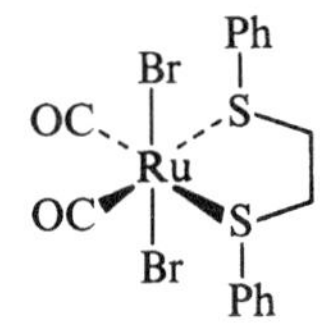

M 563.283

Yellow cryst. Mp 175-8°.

Chatt, J. *et al*, *J. Chem. Soc. (A)*, 1971, 1380 (*synth, ir, uv*)

$C_{16}H_{14}Br_2O_2RuSe_2$ Ru-00424

[1,2-Bis(benzeneseleno)ethane-*Se,Se'*]dibromodicarbonylruthenium

[1,2-Bis(phenylseleno)ethane]dibromodicarbonylruthenium, 8CI

[29079-55-8]

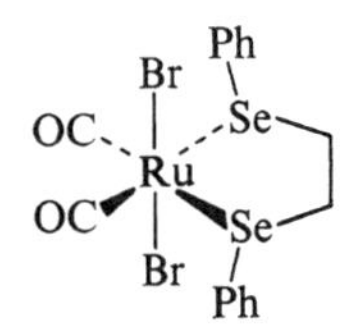

M 657.083

Lemon-yellow cryst. Mp ca. 185°.

John, P. *et al*, *Chem. Ber.*, 1970, **103**, 2161, 2178 (*synth, struct*)

$C_{16}H_{14}O_4Ru_2$ Ru-00425

Di-μ-carbonyldicarbonylbis[(1,2,3,4,5-η)-1-methyl-2,4-cyclopentadien-1-yl]diruthenium, 9CI

Tetracarbonylbis(methyl-π-cyclopentadienyl)diruthenium. Bis(methylcyclopentadienyldicarbonylruthenium)

[60687-58-3]

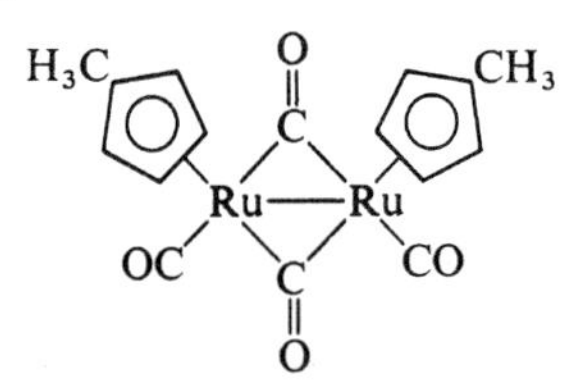

M 472.424

Exists in soln. as mixt. of bridged and non-bridged forms. Yellow cryst. (C_6H_6/hexane). Mp 113-5° dec.

McArdle, P. *et al*, *J. Chem. Soc. (A)*, 1970, 2128 (*synth, ir*)

Gansow, O.A. *et al*, *J. Am. Chem. Soc.*, 1976, **98**, 5817 (*synth, cmr, pmr*)

$C_{16}H_{14}Ru$ **Ru-00426**

Phenylruthenocene, 9CI

[38816-66-9]

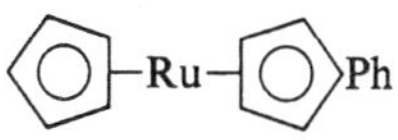

M 307.357

Cryst. (pet. ether). Mp 98°.

Rausch, M.D. *et al*, *J. Am. Chem. Soc.*, 1960, **82**, 76.
Nesmeyanov, A.N. *et al*, *Izv. Akad. Nauk SSSR, Ser. Khim.*, 1972, 1823 (*synth*)

$C_{16}H_{16}ClN_2Ru^{\oplus}$ **Ru-00427**

(η^6-Benzene)chlorobis(pyridine)ruthenium(1+), 9CI

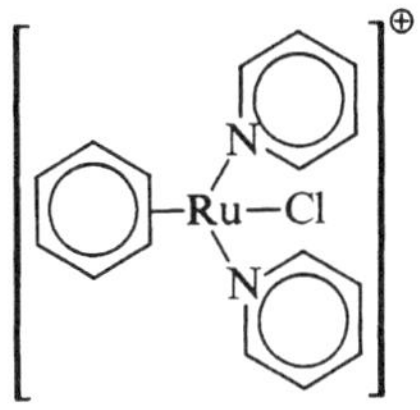

M 372.839 (ion)

Hexafluorophosphate: [60879-64-3].
$C_{16}H_{16}ClF_6N_2PRu$ M 517.803
Yellow. Mp 227° dec.

Arthur, T. *et al*, *J. Organomet. Chem.*, 1981, **208**, 369 (*synth, pmr*)

$C_{16}H_{16}Cl_4O_2Ru_2$ **Ru-00428**

Bis(bicyclo[2.2.1]hepta-2,5-diene)dicarbonyltetrachlorodiruthenium, 9CI

Tetrachlorodicarbonylbis(2,5-norbornadiene)diruthenium

[60843-19-8]

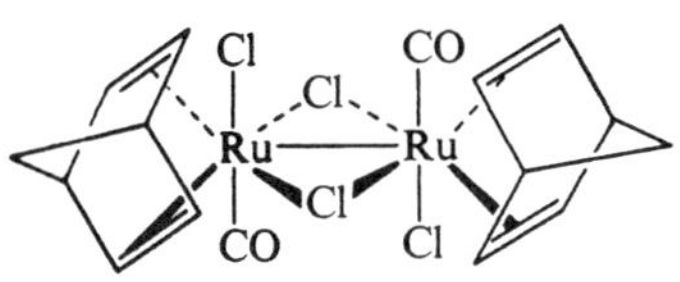

M 584.253

Bridged dimer. Brick-red cryst. Insol. common solvs. Dec. >240°.

Monomer: Bicyclo[2.2.1]hepta-2,5-dienecarbonyldichlororuthenium. Norbornadienerutheniumcarbonyl dichloride.
$C_8H_8Cl_2ORu$ M 292.127
Unknown.

Robinson, S.D. *et al*, *J. Chem. Soc. (A)*, 1966, 300 (*synth, ir*)
Stephenson, T.A. *et al*, *J. Chem. Soc., Dalton Trans.*, 1973, 2112.
Gould, R.O. *et al*, *J. Chem. Res. (S)*, 1978, 254.

$C_{16}H_{18}O_2Ru^{\oplus\oplus}$ **Ru-00429**

(η^6-Benzoic acid)(η^6-1,3,5-trimethylbenzene)ruthenium(2+)

(η^6-Benzoic acid)(η^6-mesitylene)ruthenium(2+)

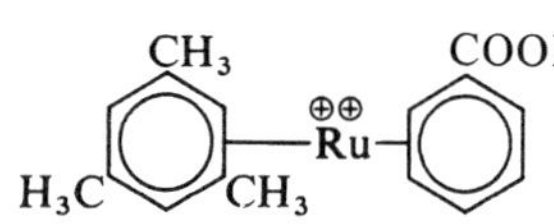

M 343.387 (ion)

Bis(tetrafluoroborate): [71825-64-4]. *[(1,2,3,4,5,6-η)-Benzoato][(1,2,3,4,5,6-η)-1,3,5-trimethylbenzene]ruthenium(1+) hydrogen tetrafluoroborate(1:1:2),* 10CI.
$C_{16}H_{18}B_2F_8O_2Ru$ M 516.994
No phys. props. reported. Air-stable.

Bennett, M.A. *et al*, *J. Organomet. Chem.*, 1979, **175**, 87 (*synth, pmr*)

$C_{16}H_{18}O_{16}P_2Ru_3$ **Ru-00430**

Decacarbonylbis(trimethyl phosphite-*P*)triruthenium, 10CI

[78168-08-8]

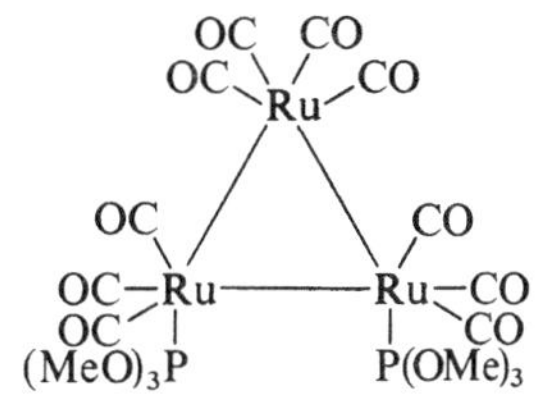

M 831.466

Air-stable yellow cryst. (CH_2Cl_2/hexane). Mp 93-4°.

Jones, D.F. *et al*, *Inorg. Chem.*, 1981, **20**, 3247 (*synth, ir, ms*)

$C_{16}H_{20}N_2O_6Ru_2$ **Ru-00431**

Hexacarbonyl[μ-[η^2-*N,N'*-1,2-ethanediylidenebis[2-methyl-2-propanamine]-*N,N'*]]diruthenium, 10CI

*Hexacarbonyl[glyoxalbis(*tert*-butylimine)]diruthenium. Hexacarbonyl(2,2,6,6-tetramethyl-3,6-diaza-3,5-octadiene)diruthenium*

[74552-68-4]

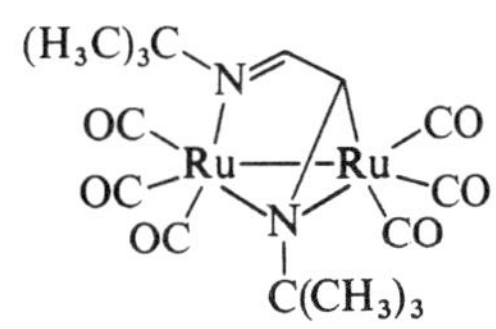

M 538.484

Orange cryst. (pentane).

Staal, L.H. *et al*, *Inorg. Chim. Acta*, 1979, **37**, L485 (*ir, pmr*)
Staal, L.H. *et al*, *Inorg. Chem.*, 1980, **19**, 3343 (*synth, ir, pmr, uv, ms*)
Staal, L.H. *et al*, *J. Organomet. Chem.*, 1981, **204**, 101; **206**, 99 (*ir, pmr, cmr*)
Gast, C.H. *et al*, *J. Organomet. Chem.*, 1981, **208**, 225 (*hplc*)

$C_{16}H_{20}Ru$ **Ru-00432**

[(2,3,4,5-η)Bicyclo[4.2.0]octa-2,4-diene][(1,2,3,4,5,6-η)-1,3,5-cyclooctatriene]ruthenium, 9CI

[12212-72-5]

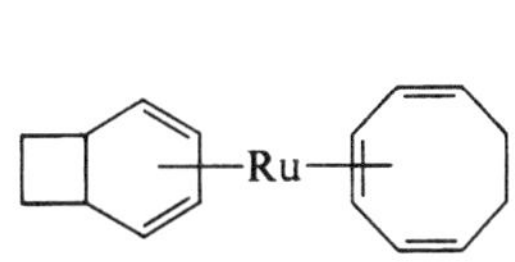

M 313.404

Yellow cryst. (hexane). Mp 99-101°. Bp 65-80° subl. *in vacuo*. Dec. at 130-60°.

Müller, J. *et al*, *J. Organomet. Chem.*, 1966, **5**, 275 (*synth, pmr, ir*)
Müller, J. *et al*, *Chem. Ber.*, 1975, **108**, 273 (*synth*)

$C_{16}H_{22}O_{16}P_2Ru_4$ **Ru-00433**

Decacarbonyltetra-μ-hydrobis(trimethyl phosphite-*P*)tetraruthenium

Decacarbonyltetra-μ-hydrobis(phosphorous acid)tetraruthenium hexamethyl ester, 8CI

[34438-93-2]

As Dodecacarbonyltetra-μ-hydrotetraruthenium, Ru-00224 with

$$X = Y = P(OMe)_3$$

M 936.568

Air-stable orange cryst. Sol. hydrocarbons, halocarbons.

Kaesz, H.D. *et al*, *J. Am. Chem. Soc.*, 1971, **93**, 4594 (*synth, ir, pmr*)

$C_{16}H_{22}Ru$ Ru-00434

Bis[(1,2,3,4,5-η)-2,4-cyclooctadien-1-yl]ruthenium, 10CI

[63395-36-8]

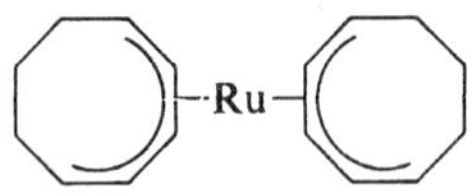

M 315.420

Yellow cryst. Mp 89-90°.

Pertici, P. *et al*, *J. Chem. Soc., Dalton Trans.*, 1980, 1961 (*synth, pmr*)

$C_{16}H_{22}Ru$ Ru-00435

[(1,2,5,6-η)-1,5-Cyclooctadiene][(1,2,3,4,5,6-η)-1,3,5-cyclooctatriene]ruthenium, 9CI

[42516-72-3]

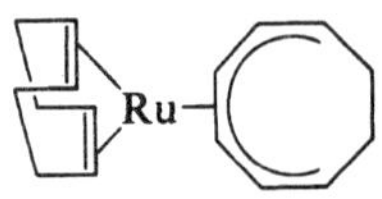

M 315.420

Yellow cryst. Mp 92-4°.

Fischer, E.O. *et al*, *Chem. Ber.*, 1963, **96**, 3217 (*synth*)
Pertici, P. *et al*, *J. Chem. Soc., Dalton Trans.*, 1980, 1961 (*synth, pmr*)

$C_{16}H_{24}N_4Ru^{\oplus\oplus}$ Ru-00436

Tetrakis(acetonitrile)[(1,2,5,6-η)-1,5-cyclooctadiene]ruthenium(2+), 9CI

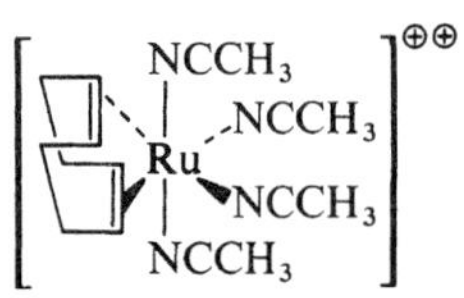

M 373.462 (ion)

Bis(hexafluorophosphate): [54071-76-0].
$C_{16}H_{24}F_{12}N_4P_2Ru$ M 663.391
No phys. description recorded. Sol. $MeNO_2$.

Ashworth, T.V. *et al*, *J. Organomet. Chem.*, 1974, **77**, C31 (*synth*)
Ashworth, T.V. *et al*, *Inorg. Chim. Acta*, 1977, **25**, L33 (*pmr*)

$C_{16}H_{26}O_2RuSi_2$ Ru-00437

Dicarbonyl(trimethylsilyl)[(1,2,3,6,7-η)-8-(trimethylsilyl)-2,4,6-cyclooctatrien-1-yl]ruthenium

[58355-41-2]

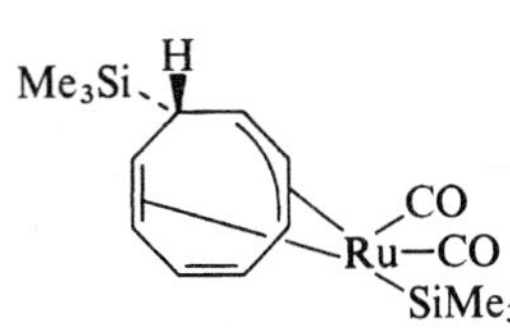

M 407.621

Fluxional. White cryst. (hexane at −78°). Mp 78°.

Harris, P.J. *et al*, *J. Chem. Soc., Dalton Trans.*, 1978, 403 (*synth, ir, pmr, struct, cmr*)

$C_{16}H_{26}Ru$ Ru-00438

Bis(η^2-ethene)[(1,2,3,4,5,6-η)-hexamethylbenzene]ruthenium, 10CI

Bis(ethylene)(hexamethylbenzene)ruthenium

[67420-77-3]

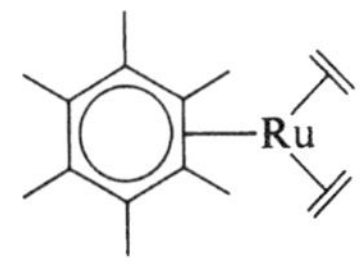

M 319.451

Tan or off-white, air-sensitive needles (hexane).

Bennett, M.A. *et al*, *J. Organomet. Chem.*, 1978, **153**, C25 (*synth, pmr*)
Bennett, M.A. *et al*, *J. Chem. Soc., Chem. Commun.*, 1979, 312 (*synth*)
Inorg. Synth., 1982, **21**, 76 (*synth, ir, pmr*)

$C_{16}H_{26}Ru$ Ru-00439

[(1,2,5,6-η)-1,5-Cyclooctadiene]bis[(1,2,3-η)-2-methyl-2-propenyl]ruthenium, 9CI

π-1,5-Cyclooctadienylbis(2-methyl-π-allyl)ruthenium

[12289-94-0]

As [(1,2,5,6-η)-1,5-Cyclooctadiene]bis(η^3-2-propenyl)ruthenium, Ru-00369 with

$$R = CH_3$$

M 319.451

White prisms (MeOH/pet. ether). Sol. C_6H_6, pet. ether, spar. sol. alcohols. Dec. by chlorinated solvs. Mp 80-5°.

Powell, J. *et al*, *J. Chem. Soc. (A)*, 1968, 159 (*synth, pmr*)
Schrock, R.R. *et al*, *J. Chem. Soc., Dalton Trans.*, 1974, 951.

$C_{16}H_{27}ClOPRu^{\oplus}$ Ru-00440

Carbonylchloro[(1,2,3,4,5,6-η)-hexamethylbenzene](trimethylphosphine)ruthenium(1+), 10CI

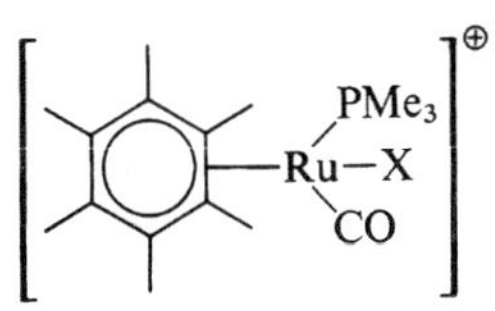

X = Cl

M 402.885 (ion)

Hexafluorophosphate: [71652-19-2].
$C_{16}H_{27}ClF_6OP_2Ru$ M 547.850
No phys. props. reported.

Werner, H. *et al*, *J. Organomet. Chem.*, 1979, **174**, C67 (*synth, pmr*)

$C_{16}H_{27}OPRu$ Ru-00441

Carbonyl[(1,2,3,4,5,6-η)-hexamethylbenzene](trimethylphosphine)ruthenium

[71652-20-5]

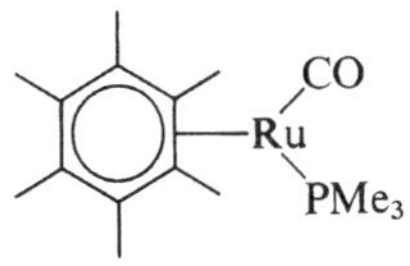

M 367.432

Yellow air-sensitive cryst. Sol. aprotic solvs.

Werner, H. *et al*, *J. Organomet. Chem.*, 1979, **174**, C67 (*synth, pmr, ir*)

$C_{16}H_{28}NO_{17}P_3Ru_3$ Ru-00442

Heptacarbonyl-μ-hydro-μ-nitrosyltris(trimethyl phosphite-*P*)triruthenium

[73230-18-9]

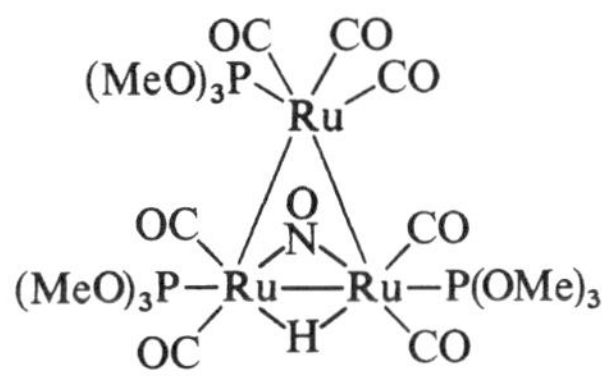

M 902.525

Deep-red blocks (hexane).

Johnson, B.F.G. *et al*, *J. Chem. Soc., Dalton Trans.*, 1980, 99 (*synth, struct, cmr, ir*)

$C_{16}H_{28}OPRu^{\oplus}$ Ru-00443

Carbonyl[(1,2,3,4,5,6-η)-hexamethylbenzene]hydro(trimethylphosphine)ruthenium(1+), 10CI

As Carbonylchloro[(1,2,3,4,5,6-η)-hexamethylbenzene](trimethylphosphine)ruthenium(1+), Ru-00440 with

X = H

M 368.440 (ion)

Hexafluorophosphate: [71652-22-7].
$C_{16}H_{28}F_6OP_2Ru$ M 513.405
No phys. props. reported.

Werner, H. *et al*, *J. Organomet. Chem.*, 1979, **174**, C67 (*synth, pmr*)

$C_{16}H_{30}Ge_4O_6Ru_2$ Ru-00444

Hexacarbonylbis[μ-(dimethylgermylene)]bis(trimethylgermyl)diruthenium, 8CI

[34307-41-0]

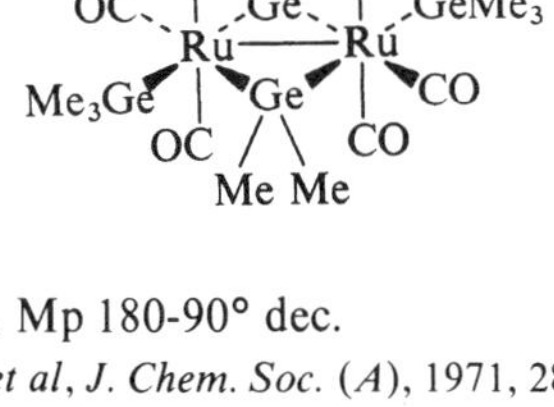

M 810.909

Yellow cryst. Mp 180-90° dec.

Knox, S.A.R. *et al*, *J. Chem. Soc. (A)*, 1971, 2874 (*synth, ir, pmr*)

$C_{16}H_{30}O_6Ru_2Si_4$ Ru-00445

Hexacarbonylbis[μ-(dimethylsilylene)]bis(trimethylsilyl)diruthenium, 9CI

[39945-24-9]

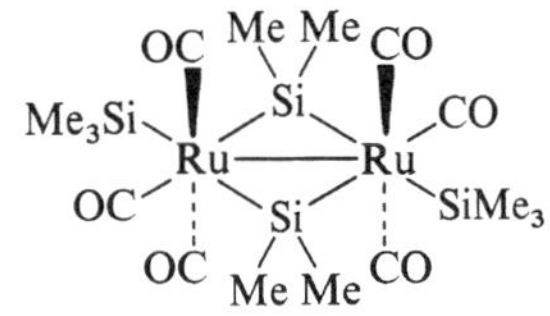

M 632.891

Golden-yellow plates. Mp 140° dec.

Brookes, A. *et al*, *J. Chem. Soc. (A)*, 1971, 3469 (*synth, ms, ir, pmr*)

Crozat, M.M. *et al*, *J. Chem. Soc., Dalton Trans.*, 1972, 2512 (*cryst struct*)

$C_{16}H_{30}O_6Ru_2Sn_4$ Ru-00446

Hexacarbonylbis[μ-(dimethylstannylene)]bis(trimethylstannyl)diruthenium

[21868-05-3]

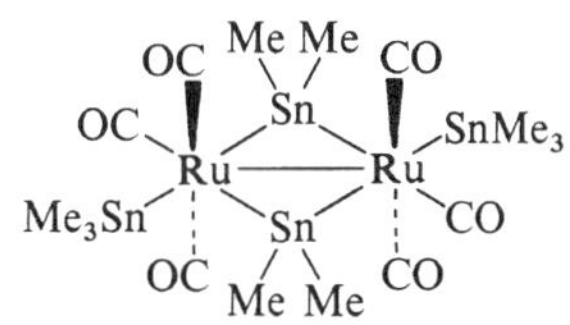

M 995.309

Yellow cryst. (cyclohexane). Mp 190-2°.

Cotton, J.D. *et al*, *J. Chem. Soc. (A)*, 1968, 2758 (*synth, ir, pmr*)

Watkins, S.F., *J. Chem. Soc. (A)*, 1969, 1552 (*cryst struct*)

$C_{16}O_{15}Ru_5$ Ru-00447

(Carbon)pentadecacarbonylpentaruthenium, 9CI

[51205-07-3]

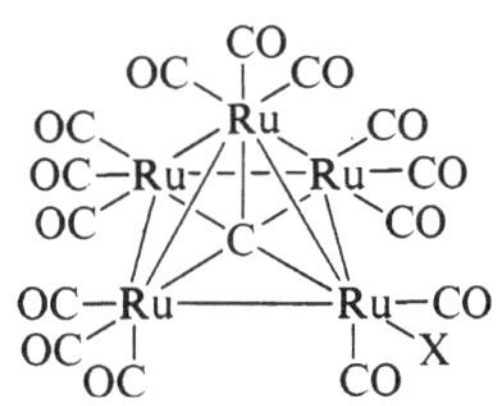

X = CO

M 937.517

No phys. details recorded.

Eady, C.R. *et al*, *J. Organomet. Chem.*, 1973, **57**, C82 (*synth, ir*)

Farrar, D.H. *et al*, *J. Chem. Soc., Chem. Commun.*, 1981, 415 (*synth, cryst struct*)

$C_{17}H_2O_{16}Ru_6$ Ru-00448

[μ₆-(Carbon)]hexadecacarbonyldihydrohexaruthenium

M 1068.613

Struct. probably like [μ_6-(Carbon)]tri-μ-carbonyltridecacarbonylhexaruthenate(2−), Ru-00467 H atoms not located. Red cryst.

Johnson, B.F.G. *et al*, *J. Organomet. Chem.*, 1980, **191**, C3 (*synth, ir, pmr*)

$C_{17}H_4O_{15}Ru_5$ Ru-00449

Pentadecacarbonyl-μ-ethenylidenedi-μ-hydropentaruthenium

[63322-98-5]

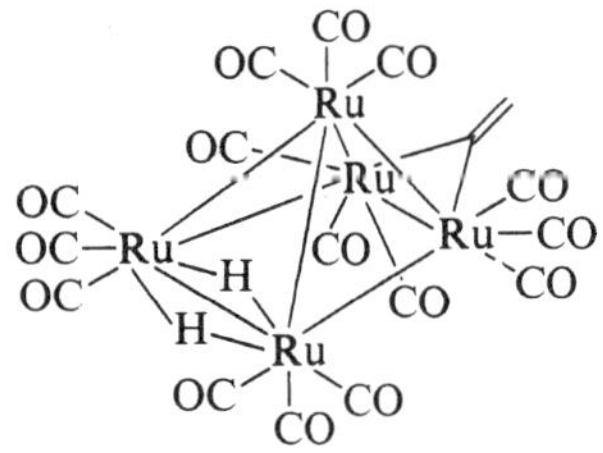

M 953.560

Air-stable solid.

Eady, C.R. *et al*, *J. Chem. Soc., Dalton Trans.*, 1977, 477 (*synth, ir, ms*)

$C_{17}H_5O_{15}PRu_5$ Ru-00450

Pentadecacarbonyl[μ$_4$-(ethylphosphoranetetrayl)]pentaruthenium

[78330-37-7]

As Pentadecacarbonyl[μ$_4$-(methylphosphoranetetrayl)]-pentaruthenium, Ru-00415 with

R = Et

M 985.541

Green cryst. (CH_2Cl_2/pentane), stable in air for weeks. Solns. dec. on long standing. Mp 210-20° dec.

Natarajan, K. *et al*, *J. Organomet. Chem.*, 1981, **209**, 85 (*synth, ir, pmr, ms, cryst struct*)

$C_{17}H_8O_7Ru_3$ Ru-00451

[μ-[(1,2,3,3*a*,8*a*-η:4,5,6-η:6,7,8-η)-Azulene]]heptacarbonyltriruthenium, 9CI

(μ$_3$-Azulene)-μ-carbonylhexacarbonyltriruthenium, 8CI

[11067-30-4]

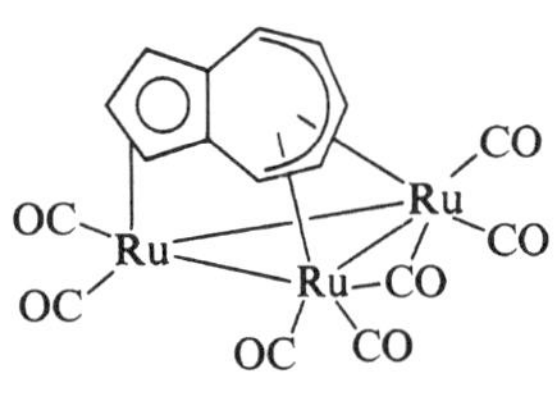

M 627.456

Thin red platelets.

Churchill, M.R. *et al*, *Inorg. Chem.*, 1973, **12**, 191 (*cryst struct*)

$C_{17}H_9F_{18}O_5PRu$ Ru-00452

Dicarbonyl[(1,2,3,4-η)-hexakis(trifluoromethyl)benzene](trimethyl phosphite-*P*)ruthenium

Dicarbonyl[hexakis(trifluoromethyl)benzene](phosphorous acid)ruthenium trimethyl ester, 8CI

[32823-70-4]

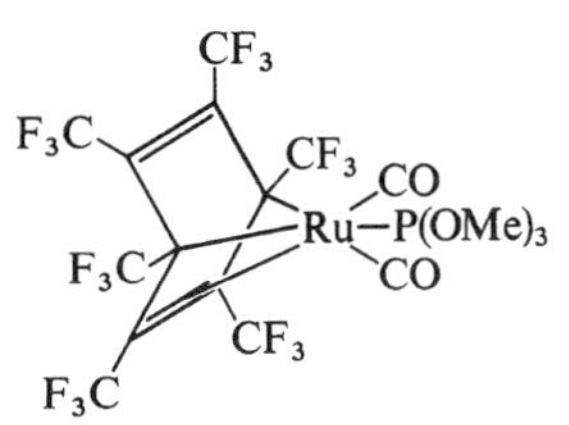

M 767.270

White cryst. (hexane). Mp 78-9°.

Burt, R. *et al*, *J. Chem. Soc., Dalton Trans.*, 1970, 2981 (*synth, ir, nmr, pmr, ms*)

$C_{17}H_{12}ORu$ Ru-00453

(η^5-2,4-Cyclopentadien-1-yl)[(1,2,3,3*a*,8*a*-η)-1,8-dihydro-8-oxocyclopenta[*a*]inden-1-yl]ruthenium, 9CI

π-Cyclopentadienyl(π-1,8-dihydro-8-oxocyclopent[a]indenyl)ruthenium, 8CI. 2,3-Ruthenocenoindenone

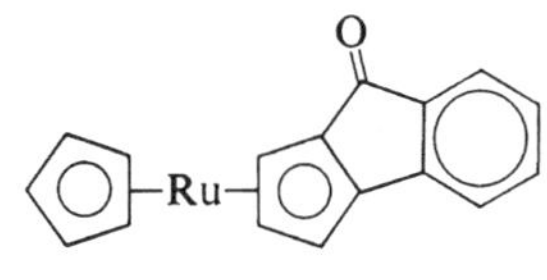

M 333.351

Orange needles (C_6H_6/hexane). Mp 134.2-135°.

Bublitz, D.E. *et al*, *J. Am. Chem. Soc.*, 1962, **84**, 1845 (*synth, uv, pmr*)

$C_{17}H_{14}ORu$ Ru-00454

Benzoylruthenocene

(Benzoyl-π-cyclopentadienyl)-π-cyclopentadienylruthenium, 8CI. Phenylruthenocenylmethanone. Phenyl ruthenocenyl ketone

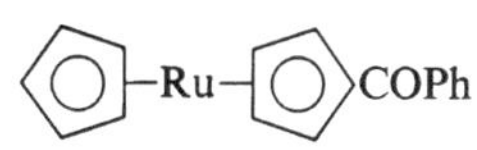

M 335.367

Yellow needles (pet. ether). Mp 124.5-125°.

Rausch, M.D. *et al*, *J. Am. Chem. Soc.*, 1960, **82**, 76 (*synth, ir*)
Rausch, M.D. *et al*, *J. Org. Chem.*, 1963, **28**, 3225 (*pmr*)

$C_{17}H_{15}Ru^{\oplus}$ Ru-00455

[(1,2,3,4,5,6-η)-1,1′-Biphenyl](η^5-2,4-cyclopentadien-1-yl)ruthenium(1+), 10CI

As (η^6-Chlorobenzene)(η^5-2,4-cyclopentadien-1-yl)ruthenium(1+), Ru-00183 with

X = Ph

M 320.376 (ion)

Hexafluorophosphate: [72804-59-2].
$C_{17}H_{15}F_6PRu$ M 465.340
Pale-yellow solid. Mp 129-31°.

Nesmeyanov, A.N. *et al*, *J. Organomet. Chem.*, 1979, **182**, C36 (*synth, pmr*)

$C_{17}H_{16}BRu^{\oplus}$ Ru-00456

(η^6-Benzene)[(1,2,3,4,5,6-η)-1-phenylboratabenzene]ruthenium(1+), 10CI

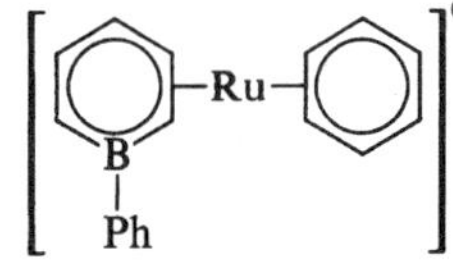

M 332.193 (ion)

Hexafluorophosphate: [69891-00-5].
$C_{17}H_{16}BF_6PRu$ M 477.158
Pearl-coloured plates. Mp 171-171.5°. Dec. >240°.

Herberich, G.E. *et al*, *Chem. Ber.*, 1979, **112**, 607 (*synth, pmr, nmr*)

$C_{17}H_{16}ORu$ Ru-00457

α-Phenylruthenocenemethanol

π-Cyclopentadienyl[(α-hydroxybenzyl)-π-cyclopentadienyl]ruthenium. (1-Hydroxy-1-phenylmethyl)ruthenocene. (α-Hydroxybenzyl)ruthenocene

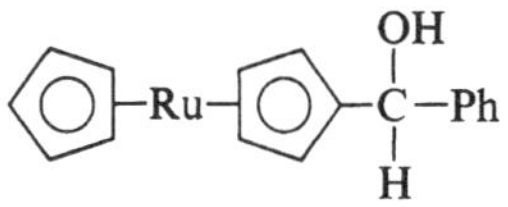

M 337.383

(±)-*form*

White needles (hexane). Mp 105.6-106.2°.

Bublitz, D.E. *et al*, *Chem. Ind. (London)*, 1960, 936 (*synth*)
Bublitz, D.E. *et al*, *J. Am. Chem. Soc.*, 1962, **84**, 1845 (*synth*)

$C_{17}H_{16}O_6Ru_2Si$ Ru-00458

Hexacarbonyl[μ-[[(2,3,6,7-η:3,4,5,6-η)-1,3,5,7-cyclooctatetraen-1-yl]trimethylsilane]]diruthenium, 10CI

[74238-32-7]

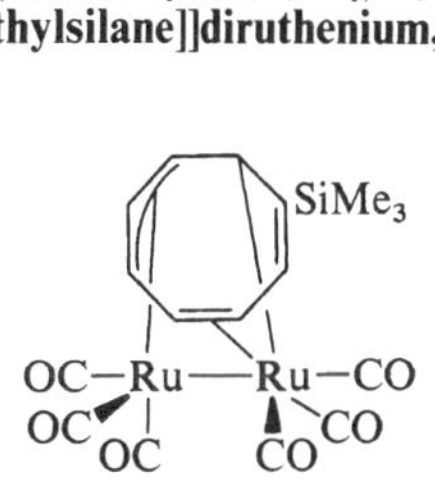

M 546.535

Fluxional. Yellow. Mp 58°.

Edwards, J.D. *et al*, *J. Chem. Soc., Dalton Trans.*, 1980, 552 (*synth, ir, pmr*)

$C_{17}H_{16}O_6Ru_3S$ Ru-00459

Hexacarbonyl[μ_3-[(1,2,3-η:4,5-η:6,7-η)-2,4,6-cyclohepta-trien-1-yl]][μ_3-(2-methyl-2-propanethiolato)]triruthenium, 9CI

[73107-21-8]

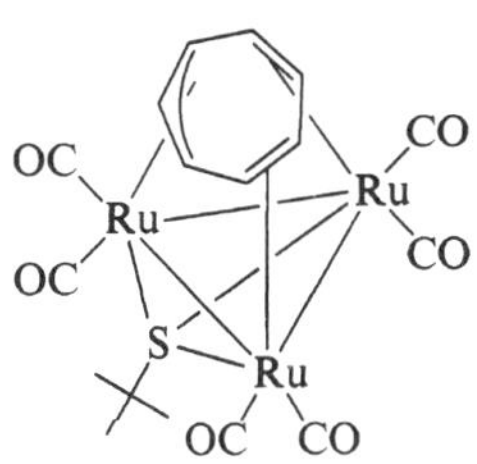

M 651.580

Yellow cryst. Mp 105-7°.

Knox, S.A.R. *et al*, *J. Chem. Soc., Dalton Trans.*, 1981, 2220, 2230 (*synth, ir, pmr, cryst struct*)

$C_{17}H_{16}Ru$ Ru-00460

Benzylruthenocene

(Phenylmethyl)ruthenocene

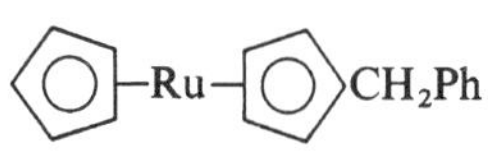

M 321.383

Mp 86-7°.

Mark, V. *et al*, *Inorg. Chem.*, 1964, **3**, 1067 (*synth, pmr*)

$C_{17}H_{20}O_2Ru_2$ Ru-00461

Dicarbonylbis(η^5-2,4-cyclopentadien-1-yl)-μ-ethylidene[μ-(1-methylethylidene)]diruthenium

[80303-44-2]

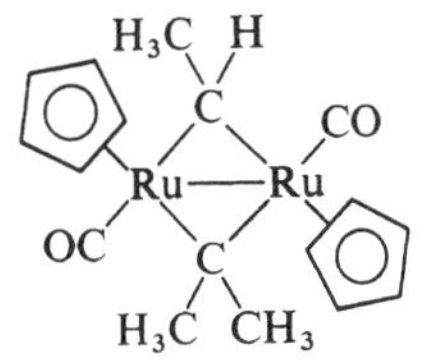

M 458.484

trans-form

Yellow cryst. Mp 161°.

Cooke, M. *et al*, *J. Chem. Soc., Chem. Commun.*, 1981, 862 (*synth, struct, ir, pmr, cmr*)

$C_{17}H_{22}ORu$ Ru-00462

Carbonyl[(1,2,5,6-η)-1,5-cyclooctadiene][(1,2,5,6-η-)-1,3,5-cyclooctatriene]ruthenium

[64331-16-4]

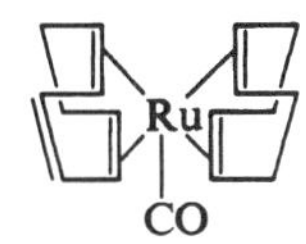

M 343.430

No phys. props. recorded.

Deganello, G. *et al*, *J. Organomet. Chem.*, 1977, **135**, 215 (*synth, ir, pmr*)

$C_{17}H_{23}NRu^{\oplus\oplus}$ Ru-00463

[(1,2,3,4,5,6-η)-*N,N*-Dimethylaniline][(1,2,3,4,5,6-η-)-1,3,5-trimethylbenzene]ruthenium(2+)

[(1,2,3,4,5,6-η)-N,N-Dimethylbenzenamine][(1,2,3,4,5,6-η)-1,3,5-trimethylbenzene]ruthenium(2+). (η^6-Dimethylaniline)(η^6-mesitylene)ruthenium(2+)

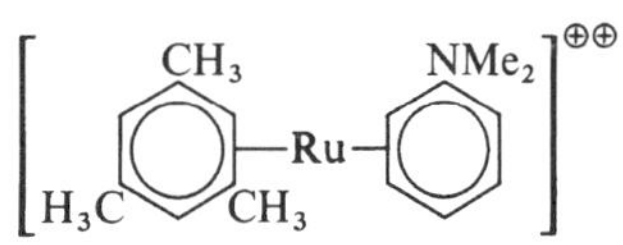

M 342.445 (ion)

Bis(tetrafluoroborate): [71836-19-6].
$C_{17}H_{23}B_2F_8NRu$ M 516.053
Colourless or pale-yellow air-stable cryst. (DMSO/$MeOH/Et_2O$).

Bennett, M.A. *et al*, *J. Organomet. Chem.*, 1979, **175**, 87 (*synth, pmr*)

$C_{17}H_{23}Ru^{\oplus}$ Ru-00464

(η^5-2,4-Cyclopentadien-1-yl)[(1,2,3,4,5,6-η)-hexamethylbenzene]ruthenium(1+), 10CI

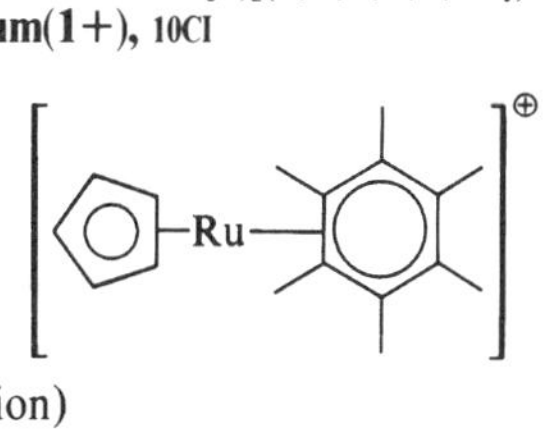

M 328.439 (ion)

Hexafluorophosphate: [72804-78-5].
$C_{17}H_{23}F_6PRu$ M 473.403
Cryst. (Me_2CO). Mp 250° dec.

Tetraphenylborate: [82466-21-5].

$C_{41}H_{43}BRu$ M 647.671
Grey-brown solid. Mp 252° dec.

Roman, E. *et al*, *Inorg. Chim. Acta*, 1979, **37**, L465 (*synth*, *pmr*, *ir*, *cmr*)
Robertson, I.W. *et al*, *J. Organomet. Chem.*, 1982, **228**, 171 (*synth*, *pmr*, *cmr*)

$C_{17}H_{30}OPRu^{\oplus}$ Ru-00465

Carbonyl[(1,2,3,4,5,6-η)-hexamethylbenzene]methyl(trimethylphosphine)ruthenium(1+), 10CI

As Carbonylchloro[(1,2,3,4,5,6-η)-hexamethylbenzene](trimethylphosphine)ruthenium(1+), Ru-00440 with

X = Me

M 382.467 (ion)
Hexafluorophosphate: [71685-08-0].
$C_{17}H_{30}F_6OP_2Ru$ M 527.431
No phys. props. reported.

Werner, H. *et al*, *J. Organomet. Chem.*, 1979, **174**, C67 (*synth*, *pmr*)

$C_{17}H_{40}O_9P_3Ru^{\oplus}$ Ru-00466

[(1,2,5,6-η)-1,5-Cyclooctadiene]hydrotris(trimethyl phosphite-*P*)ruthenium(1+), 9CI

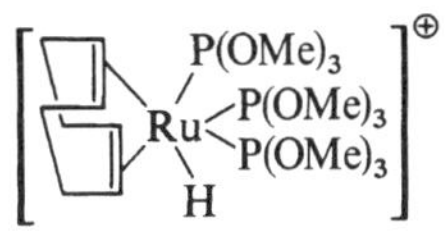

M 582.489 (ion)
Hexafluorophosphate: [61024-44-0].
$C_{17}H_{40}F_6O_9P_4Ru$ M 727.453
Colourless prisms (CH_2Cl_2/EtOH at 0°), stable in air-free solvs. Mp 142°.

Ashworth, T.V. *et al*, *J. Chem. Soc., Dalton Trans.*, 1978, 1032 (*synth*, *ir*, *pmr*)

$C_{17}O_{16}Ru_6^{\ominus\ominus}$ Ru-00467

[μ_6-(Carbon)]tri-μ-carbonyltridecacarbonylhexaruthenate(2−)

(μ_6-*Carbon*)*hexadecacarbonylhexaruthenate*(*2*−)

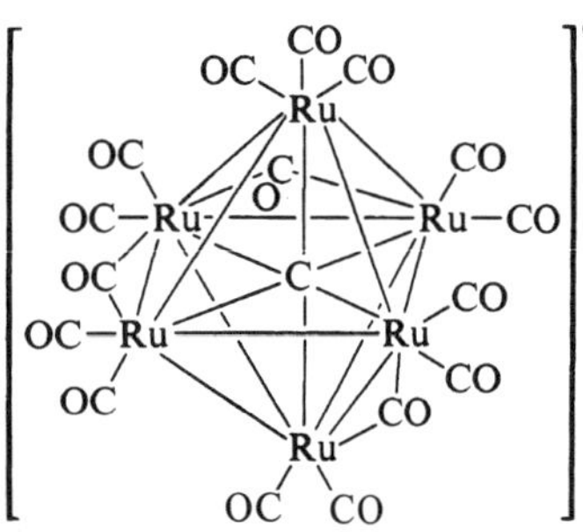

M 1066.597 (ion)
Isostructural with Fe analogue.
Bis(*tetraethylammonium*) *salt:* [73413-28-2].
$C_{33}H_{40}N_2O_{16}Ru_6$ M 1327.103
Bright-red cryst. (CH_2Cl_2).
Bis[*bis*(*triphenylphosphine*)*iminium*] *salt:* [73413-27-1].
$C_{89}H_{60}N_2O_{16}Ru_6P_4$ M 2143.772
No phys. props. reported.

Ansell, G.B. *et al*, *Acta Crystallogr., Sect. B*, 1980, **36**, 726 (*cryst struct*)
Bradley, J.S. *et al*, *J. Organomet. Chem.*, 1980, **184**, C33 (*synth*, *cmr*, *ir*, *struct*)
Johnson, B.F.G. *et al*, *J. Organomet. Chem.*, 1980, **191**, C3 (*synth*, *ir*, *struct*)
Bradley, J.S. *et al*, *Organometallics*, 1982, **1**, 748.

$C_{18}HO_{18}Ru_6^{\ominus}$ Ru-00468

Octadecacarbonyl-μ_6-hydrohexaruthenate(1−)

[62434-66-6]

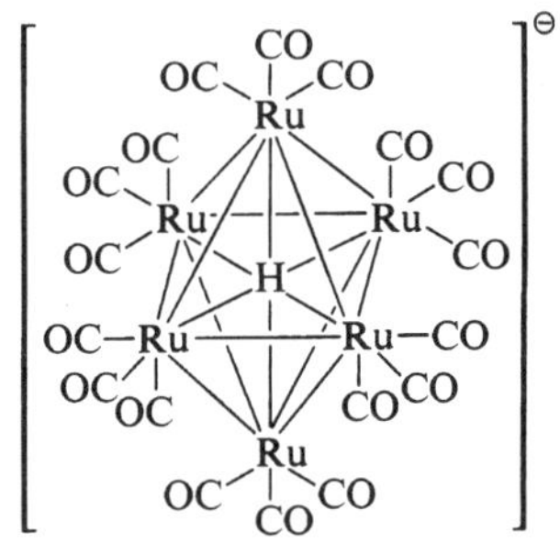

M 1111.615 (ion)
Contains interstitial hydride ligand. Salts exhibit polymorphism.
Tetraphenylarsonium salt: [72250-50-1].
$C_{42}H_{21}AsO_{18}Ru_6$ M 1494.959
Brown cryst.
Bis(*triphenylphosphine*)*iminium salt:* [62434-67-7].
$C_{54}H_{31}NO_{18}P_2Ru_6$ M 1650.202
Brown cryst. (CH_2Cl_2/MeOH).

Oxton, I.A. *et al*, *J. Chem. Soc., Chem. Commun.*, 1979, 687 (*ir*)
Jackson, P.F. *et al*, *J. Chem. Soc., Chem. Commun.*, 1980, 295; *J. Chem. Soc., Dalton Trans.*, 1980, 383 (*synth*, *ir*, *cryst struct*, *pmr*)

$C_{18}H_2O_{18}Ru_6$ Ru-00469

Octadecacarbonyldi-μ_3-hydrohexaruthenium, 8CI

[12568-51-3]

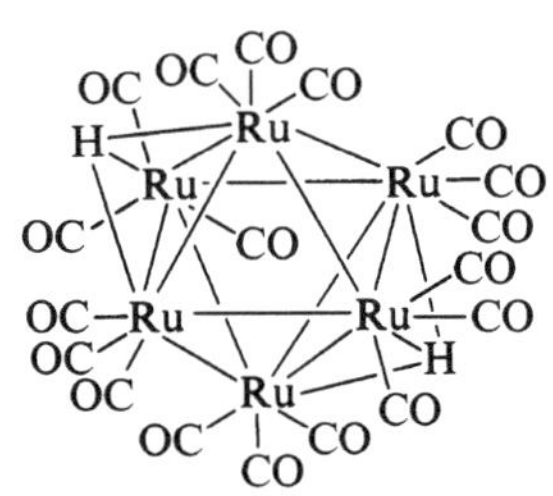

M 1112.623
Air-stable deep-violet cryst. (CH_2Cl_2 at −20°). Spar. sol. org. solvs., dec. by basic solvs.

Churchill, M.R. *et al*, *J. Chem. Soc., Chem. Commun.*, 1970, 458 (*synth*, *ms*, *ir*)
Churchill, M.R. *et al*, *J. Am. Chem. Soc.*, 1971, **93**, 5670 (*cryst struct*)
Knight, J. *et al*, *J. Chem. Soc., Dalton Trans.*, 1972, 1022 (*synth*)
Jackson, P.F. *et al*, *J. Chem. Soc., Chem. Commun.*, 1979, 735 (*synth*)
Andrews, J.A. *et al*, *Inorg. Chem.*, 1980, **19**, 3033 (*ir*)

$C_{18}H_4O_{16}Ru_6$ Ru-00470

Hexadecacarbonyl-μ-ethenylidenedi-μ-hydrohexaruthenium

[63280-32-0]

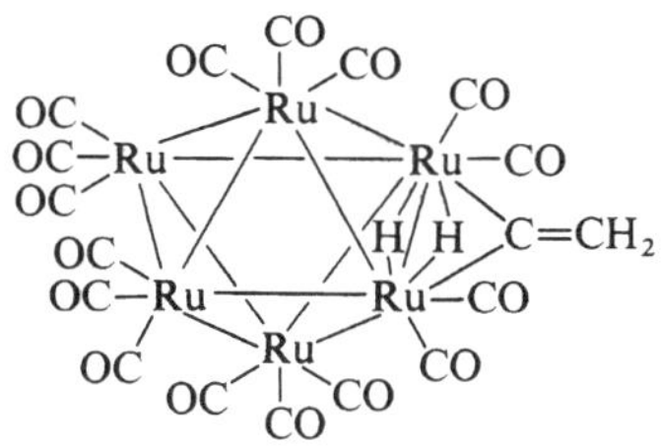

M 1082.640

No phys. description recorded. Isol. by tlc and characterised by ms.

Eady, C.R. *et al*, *J. Chem. Soc., Dalton Trans.*, 1977, 477 (*synth, ms*)

$C_{18}H_8O_{12}Ru_4$ Ru-00471

Dodecacarbonyl[μ₄-(1-η:1,2-η:1,2-η:2-η)-1-cyclohexene-1,2-diyl]tetraruthenium

[70755-72-5]

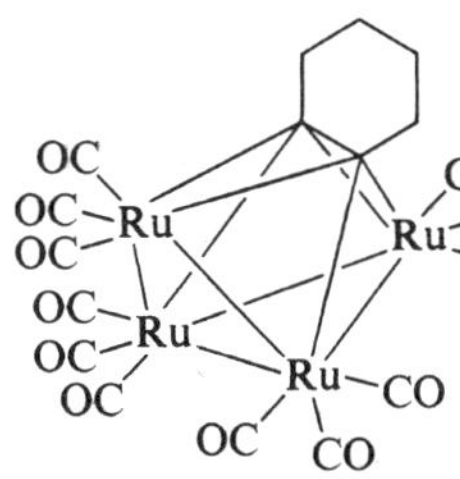

M 820.534

Purple powder. Sol. cyclohexane.

Aime, S. *et al*, *Inorg. Chim. Acta*, 1979, **34**, 49 (*synth, ir, pmr, ms*)

$C_{18}H_{10}N_2O_6Ru_2$ Ru-00472

Hexacarbonyl[μ-[N-phenyl-1,2-benzenediaminato(2−)]]diruthenium

Hexacarbonyl[μ-[N-phenyl-o-phenylenediaminato(2−)-]]diruthenium. (o-Semidine)hexacarbonyldiruthenium

[33846-32-1]

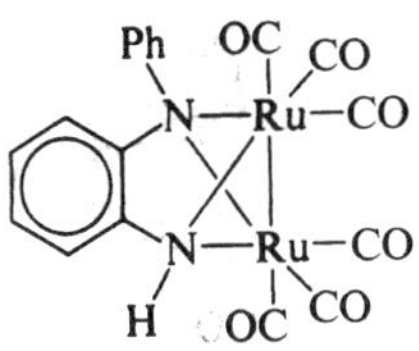

M 552.427

Rust-coloured cryst. (Et_2O/pet. ether). Mp 128-30° dec.

Bruce, M.I. *et al*, *J. Organomet. Chem.*, 1971, **31**, 275 (*synth, ir, pmr*)

$C_{18}H_{10}O_6Ru_2Se_2$ Ru-00473

Bis(μ-benzeneselenolato)hexacarbonyldiruthenium, 8CI

[33182-86-4]

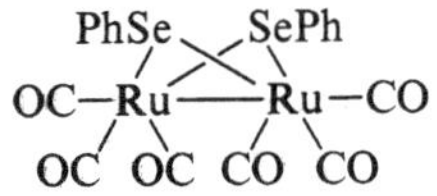

M 682.333

Yellow solid. Mp 97°.

Schermer, E.D. *et al*, *J. Organomet. Chem.*, 1971, **30**, 67 (*synth, ms, ir*)

$C_{18}H_{10}O_6Ru_2Te_2$ Ru-00474

Bis(μ-benzenetellurolato)hexacarbonyldiruthenium, 8CI

[33182-85-3]

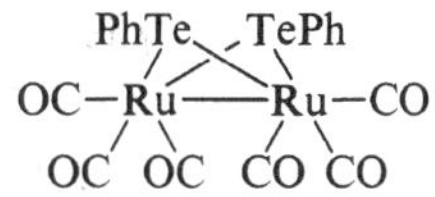

M 779.613

Yellow cryst. Mp 105-7° dec.

Schermer, E.D. *et al*, *J. Organomet. Chem.*, 1971, **30**, 67 (*synth, ms, ir*)

$C_{18}H_{10}O_8Ru_3$ Ru-00475

Octacarbonylbis(η⁵-2,4-cyclopentadien-1-yl)triruthenium

Tetracarbonylbis[dicarbonyl(η-cyclopentadienyl)ruthenio]ruthenium

[71817-46-4]

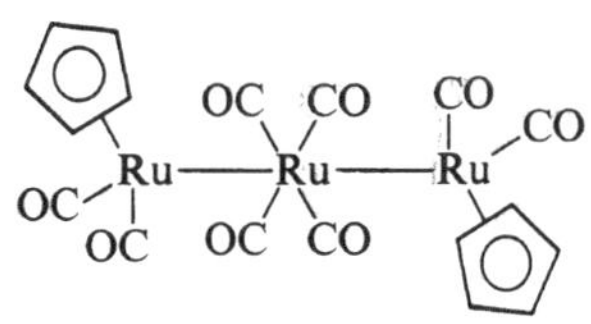

M 657.482

Yellow prisms (CH_2Cl_2/hexane).

Cook, N. *et al*, *J. Chem. Soc., Dalton Trans.*, 1979, 1032 (*synth, struct, ir, pmr*)

$C_{18}H_{12}O_6Ru_3$ Ru-00476

Di-μ-carbonyltetracarbonyl-[μ-[(1,2,3,4-η:1,5,6,7-η)-2,4,6-cycloheptatrien-1-yl]](η⁵-2,4-cyclopentadien-1-yl)triruthenium, 9CI

Hexacarbonyl(1,3,5-cycloheptatrienyl)(η⁵-cyclopentadienyl)triruthenium

[69210-23-7]

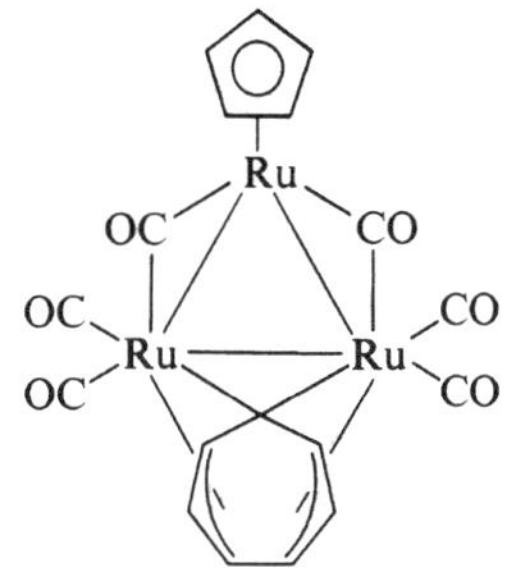

M 627.499

Fluxional. Orange cryst. (CH_2Cl_2/hexane). Mp 165-85° dec.

Humphries, A.P. *et al*, *J. Chem. Soc., Dalton Trans.*, 1978, 1523 (*synth, ir, pmr*)

$C_{18}H_{12}O_{10}Ru_3$ Ru-00477

Decacarbonyl(1,3-cyclooctadiene)triruthenium, 8CI

[12112-96-8]

$(C_8H_{12})Ru_3(CO)_{10}$

M 691.497
Not well characterised. Yellow prisms. Dec. at 115-20°.

Candlin, J.P. *et al*, *Chem. Ind.* (*London*), 1966, 1960

$C_{18}H_{14}CrO_6Ru$ Ru-00478

Pentacarbonyl(ethoxyruthenocenylmethylene)chromium, 10CI
[75280-69-2]

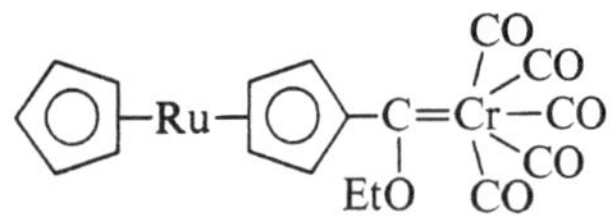

M 479.371
Orange-yellow cryst. (CH_2Cl_2/pentane). Mp 113°.

Fischer, E.O. *et al*, *J. Organomet. Chem.*, 1980, **191**, 261 (*synth, ir, ms, pmr, cmr*)

$C_{18}H_{14}MoO_6Ru$ Ru-00479

Pentacarbonyl(ethoxyruthenocenylmethylene)molybdenum, 10CI
[75280-68-1]

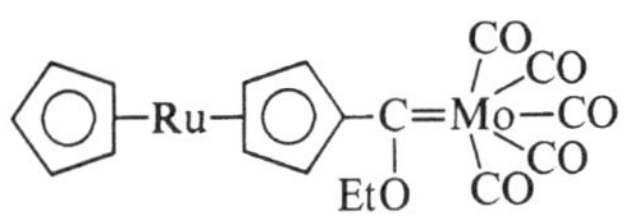

M 523.315
Orange-yellow cryst. (CH_2Cl_2/pentane). Mp 108°.

Fischer, E.O. *et al*, *J. Organomet. Chem.*, 1980, **191**, 261 (*synth, ir, ms, pmr, cmr*)

$C_{18}H_{14}N_2ORu$ Ru-00480

Carbonyl(η^5-2,4-cyclopentadien-1-yl)[(η^3-*C,N,N'*)2-(phenylazo)phenyl]ruthenium
(η^5-Cyclopentadienyl)[η^3-(2-phenylazo)phenyl]carbonylruthenium
[32825-30-2]

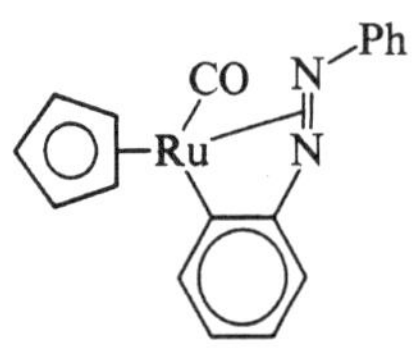

M 375.391
Red oil. $Bp_{0.1}$ 110-20°.

Bruce, M.I. *et al*, *J. Chem. Soc.* (*A*), 1970, 3204 (*synth, ir*)
Bruce, M.I. *et al*, *J. Chem. Soc., Dalton Trans.*, 1971, 2820 (*synth, ir, pmr*)

$C_{18}H_{14}O_6RuW$ Ru-00481

Pentacarbonyl(ethoxyruthenocenylmethylene)tungsten, 10CI
[75280-67-0]

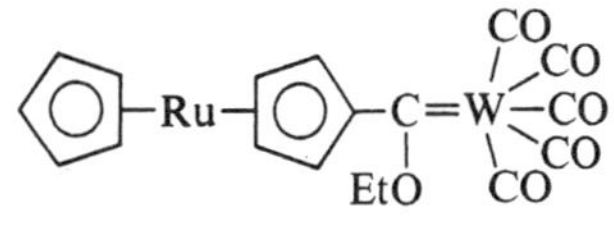

M 611.225
Orange-red cryst. (CH_2Cl_2/hexane). Mp 142°.

Fischer, E.O. *et al*, *J. Organomet. Chem.*, 1980, **191**, 261 (*synth, ir, ms, pmr, cmr, struct*)

$C_{18}H_{14}Ru$ Ru-00482

Bis[(1,2,3,3a,7a-η)-1*H*-inden-1-yl]ruthenium, 9CI
Di-π-indenylruthenium, 8CI. Bis(benzocyclopentadienyl)ruthenium

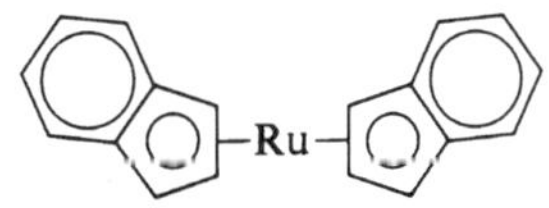

M 331.379
Orange cryst. Mp 200-1°.

Osiecki, J.H. *et al*, *J. Organomet. Chem.*, 1965, **3**, 107 (*synth*)
Webb, N.C. *et al*, *Acta Crystallogr.*, 1967, **22**, 382 (*struct*)

$C_{18}H_{22}Cl_2N_2Ru$ Ru-00483

Dichloro[(1,2,5,6-η)-1,5-cyclooctadiene]bis(pyridine)ruthenium, 9CI
[59992-85-7]

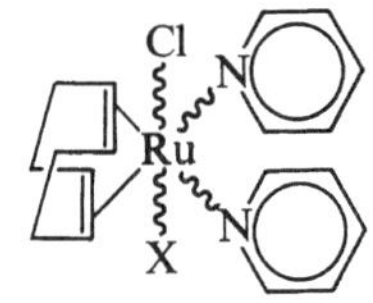

X = Cl

M 438.361
2 isomers detected. No phys. props. recorded.

Potvin, C. *et al*, *J. Organomet. Chem.*, 1976, **113**, 273 (*synth, pmr*)

$C_{18}H_{22}Ru$ Ru-00484

Bis[(1,2,3,3a,7a-η)-1*H*-4,5,6,7-tetrahydroinden-1-yl]ruthenium
Di(4,5,6,7-tetrahydro-π-indenyl)ruthenium. 1,2:1',2'-Bis(tetramethylene)ruthenocene

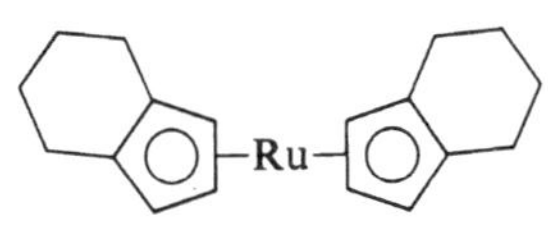

M 339.442
White cryst. Mp 66-7°.

Osiecki, J.H. *et al*, *J. Organomet. Chem.*, 1965, **3**, 107 (*synth*)

$C_{18}H_{23}ClN_2Ru$ Ru-00485

Chloro[(1,2,5,6-η)-1,5-cyclooctadiene]hydrobis(pyridine)ruthenium, 9CI
[53837-24-4]
As Dichloro[(1,2,5,6-η)-1,5-cyclooctadiene]bis(pyridine)ruthenium, Ru-00483 with

X = H

M 403.916
Large yellow prisms. Stable in solid state but dec. fairly rapidly in soln. Dec. >120° without melting.

Potvin, C. *et al*, *Bull. Soc. Chim. Fr.*, 1974, 783 (*synth*)
Ashworth, T.V. *et al*, *J. Chem. Soc., Dalton Trans.*, 1978, 1036 (*synth*)
Potvin, C. *et al*, *J. Organomet. Chem.*, 1981, **219**, 115 (*pmr, struct*)

$C_{18}H_{24}Cl_4O_2Ru_2$ **Ru-00486**

Dicarbonyldi-μ-chlorodichlorobis[(1,2,5,6-η)-1,5-cyclooctadiene]diruthenium, 9CI

Tetrachlorodicarbonylbis(1,5-cyclooctadiene)diruthenium

[12176-87-3]

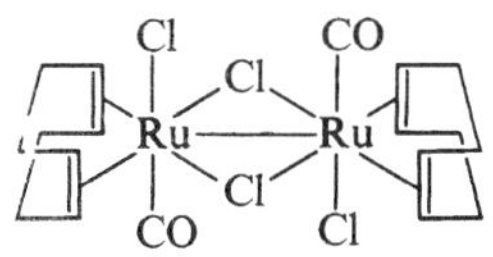

M 616.338

Bridged dimer. Brown cryst. Mp 233-45° dec.

Monomer: Carbonyldichloro(1,5-cyclooctadiene)-ruthenium.

$C_9H_{12}Cl_2ORu$ M 308.169

Unknown.

Robinson, S.D. *et al*, *J. Chem. Soc. (A)*, 1966, 300 (*synth*)
Tayim, H.A. *et al*, *J. Organomet. Chem.*, 1975, **92**, 107 (*synth*)

$C_{18}H_{24}Ge_2O_4Ru_2$ **Ru-00487**

Tetracarbonyl[μ-[(1,2,3,3a,6a-η:3a,4,5,6,6a-η)-pentalene-]]bis(trimethylgermyl)diruthenium, 9CI

[50648-32-3]

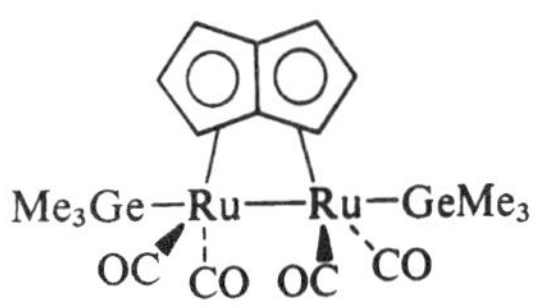

M 651.705

Pale-yellow cryst. Mp 235° dec.

Howard, J.A.K. *et al*, *J. Chem. Soc., Dalton Trans.*, 1978, 403, 412 (*synth, ir, pmr, cryst struct*)

$C_{18}H_{24}Ru^{\oplus\oplus}$ **Ru-00488**

(η^6-Benzene)[(1,2,3,4,5,6-η)-hexamethylbenzene]ruthenium(2+), 10CI

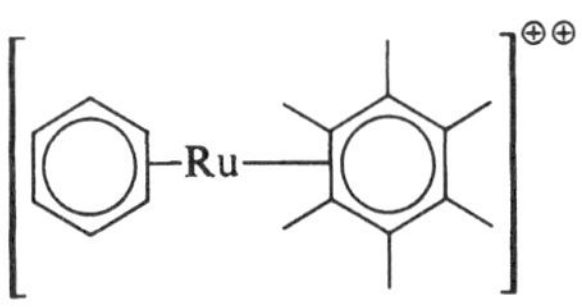

M 341.458 (ion)

Air-stable.

Bis(tetrafluoroborate): [71825-72-4].

$C_{18}H_{24}B_2F_8Ru$ M 515.065

No phys. props. reported.

Bennett, M.A. *et al*, *J. Organomet. Chem.*, 1979, **175**, 87 (*synth, pmr*)

$C_{18}H_{24}Ru^{\oplus\oplus}$ **Ru-00489**

Bis[(1,2,3,4,5,6-η)-1,3,5-trimethylbenzene]ruthenium(2+)

(Bis-π-mesitylene)ruthenium(2+)

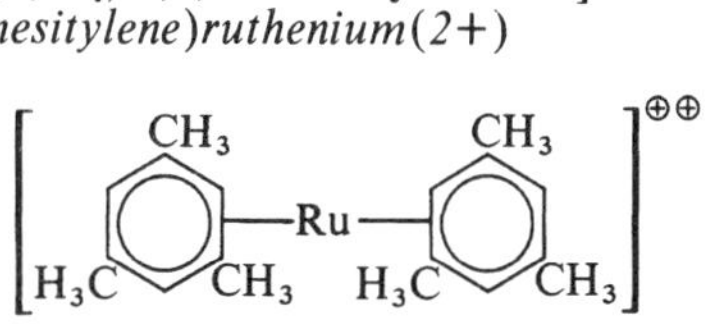

M 341.458 (ion)

Bis(tetraphenylborate):

$C_{66}H_{64}B_2Ru$ M 979.922

Pale-yellow solid (Me_2CO/Et_2O). Mp 198-200° dec.

Rybinskaya, M.I. *et al*, *J. Organomet. Chem.*, 1982, **235**, 215 (*synth*)

$C_{18}H_{26}O_4Ru$ **Ru-00490**

[(1,2,5,6-η)-1,5-Cyclooctadiene]bis(2,4-pentanedionato-O,O′)ruthenium, 9CI

Bis(acetylacetonato)(1,5-cyclooctadiene)ruthenium

[31742-01-5]

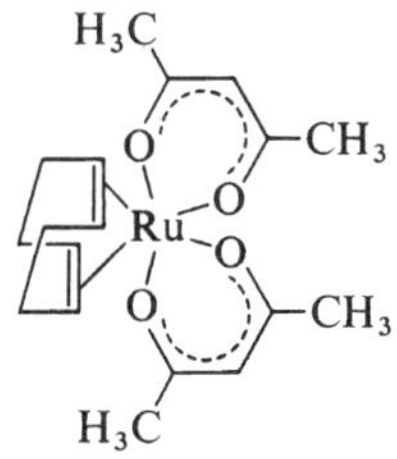

M 407.471

Yellow microcryst. (MeOH). Mp 101-2°.

Potvin, C. *et al*, *J. Less Common Metals*, 1970, **22**, 91; 1971, **24**, 334 (*synth, ir, pmr, cryst struct*)
Powell, P., *J. Organomet. Chem.*, 1974, **65**, 89 (*synth*)

$C_{18}H_{26}Ru$ **Ru-00491**

[(1,2,3,4-η)-1,3-Cyclohexadiene][(1,2,3,4,5,6-η)-hexamethylbenzene]ruthenium

[67421-01-6]

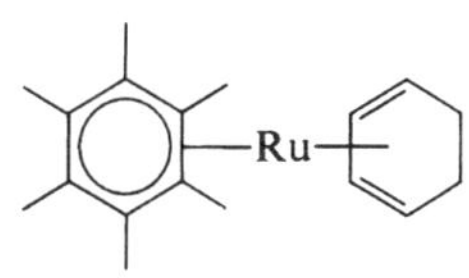

M 343.473

Yellow, air-sensitive cryst. (hexane).

Bennett, M.A. *et al*, *J. Organomet. Chem.*, 1978, **153**, C25 (*synth*)
Inorg. Synth., 1982, **21**, 77 (*synth, pmr*)

$C_{18}H_{26}RuS^{\oplus\oplus}$ **Ru-00492**

[η^6-(1-Isopropyl-4-methylbenzene)](η^5-tetramethylthiophene)ruthenium(2+)

[(1,2,3,4,5,6-η)-1-Methyl-4-(1-methylethyl)benzene][(1,2,3,4,5-η)tetramethylthiophene]ruthenium(2+), 10CI. (η^6-p-Cymene)(η^5-tetramethylthiophene)-ruthenium(2+)

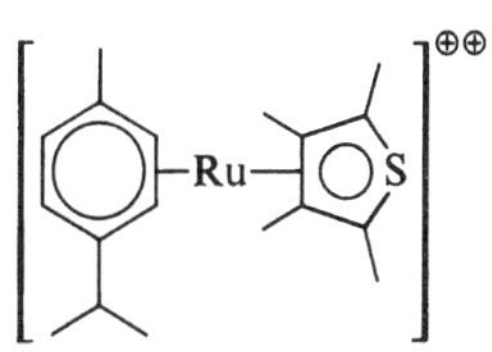

M 375.533 (ion)

Bis(hexafluorophosphate): [68246-45-7].

$C_{18}H_{26}F_{12}P_2RuS$ M 665.462

Brown oil. Not isol. in pure form. Dec. on recryst.

Russell, M.J.H. *et al*, *J. Chem. Soc., Dalton Trans.*, 1978, 857 (*synth, pmr, cmr*)

$C_{18}H_{27}O_{18}P_3Ru_3$ Ru-00493

Nonacarbonyltris(trimethyl phosphite-*P*)triruthenium, 9CI

[38686-18-9]

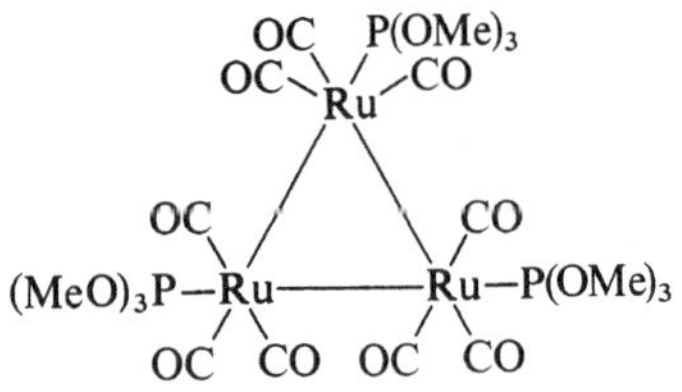

M 927.532

Red-orange cryst. Mp 111-4°.

Bruce, M.I. *et al*, *J. Chem. Soc., Dalton Trans.*, 1972, 2094 (*synth, ir, pmr*)
Alex, R.F. *et al*, *Organometallics*, 1982, **1**, 453 (*synth*)
Graaf, J.L. *et al*, *Organometallics*, 1982, **1**, 837 (*ir*)

$C_{18}H_{27}Ru^{\oplus}$ Ru-00494

[(1,2,3-η)-2-Cyclohexen-1-yl][(1,2,3,4,5,6-η)-hexamethylbenzene]ruthenium(1+), 10CI

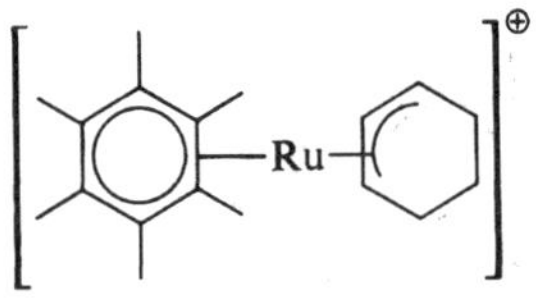

M 344.481 (ion)

Hexafluorophosphate: [67420-82-0].
$C_{18}H_{27}F_6PRu$ M 489.445
Characterised spectroscopically.

Bennett, M.A. *et al*, *J. Organomet. Chem.*, 1978, **153**, C25 (*synth, pmr*)

$C_{18}H_{31}O_{18}P_3Ru_4$ Ru-00495

Nonacarbonyltetra-μ-hydrotris(trimethyl phosphite-*P*)tetraruthenium

Nonacarbonyltetra-μ-hydrotris(phosphorous acid)tetraruthenium nonamethyl ester, 8CI

[34438-94-3]

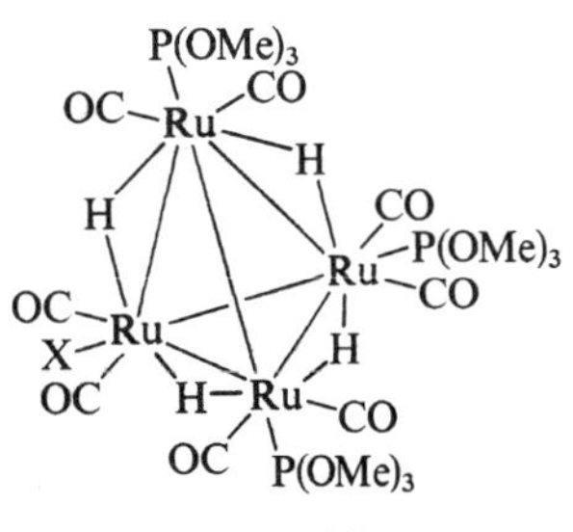

X = CO

M 1032.633

Air-stable orange cryst. Sol. hydrocarbons, halocarbons.

Kaesz, H.D. *et al*, *J. Am. Chem. Soc.*, 1971, **93**, 4594 (*synth, pmr*)

Having problems with locating a compound? Have you checked the indexes?

$C_{18}H_{34}Cl_2N_2Ru_2$ Ru-00496

μ-Chlorochlorobis[(1,2,5,6-η)-1,5-cyclooctadiene][μ-(1,1-dimethylhydrazine-*N*:*N'*)]-μ-hydrohydrodiruthenium

Dichlorobis(η⁴-1,5-cyclooctadiene)(1,1-dimethylhydrazine)dihydrodiruthenium

[66320-89-6]

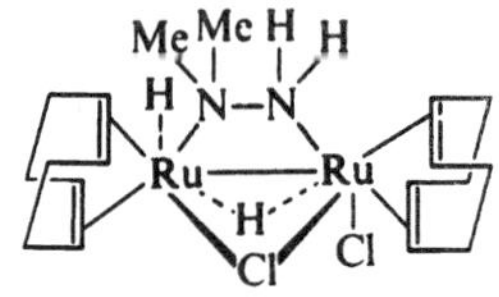

M 551.526

Dark-red needles (Me_2NNH_2 aq.). Mp 165°.

Ashworth, T.V. *et al*, *J. Chem. Soc., Dalton Trans.*, 1977, 1809; 1978, 1036, 1043 (*synth, ir, cryst struct*)

$C_{18}H_{42}P_4Ru^{\oplus\oplus}$ Ru-00497

Tris(trimethylphosphine)[(1,2,3,4,5-η)-6-(trimethylphosphonio)-2,4-cyclohexadien-1-yl]ruthenium(2+), 10CI

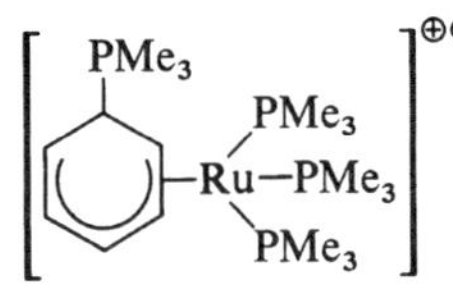

M 483.495 (ion)

Bis(hexafluorophosphate): [78343-01-8].
$C_{18}H_{42}F_{12}P_6Ru$ M 773.423
Colourless mod. air-stable cryst. V. sol. polar solvs.

Werner, R. *et al*, *J. Organomet. Chem.*, 1981, **210**, C11 (*synth, pmr, nmr*)

$C_{18}H_{54}O_{18}P_6Ru^{\oplus\oplus}$ Ru-00498

Hexakis(trimethyl phosphite-*P*)ruthenium(2+), 9CI

$Ru[P(OMe)_3]_6^{\oplus\oplus}$

M 845.526 (ion)

Bis(tetraphenylborate): [35083-08-0].
$C_{66}H_{94}B_2O_{18}P_6Ru$ M 1483.990
White cryst. ($MeOH/CH_2Cl_2$).

Hough, J. *et al*, *J. Chem. Soc., Chem. Commun.*, 1972, 371 (*synth*)
Couch, D.A. *et al*, *Inorg. Chim. Acta*, 1974, **9**, 39 (*synth, spectra*)

$C_{18}H_{58}P_6Ru_2$ Ru-00499

Di-μ-hydrodihydrohexakis(trimethylphosphine)diruthenium

Tetrahydrohexakis(trimethylphosphine)diruthenium

[76171-47-6]

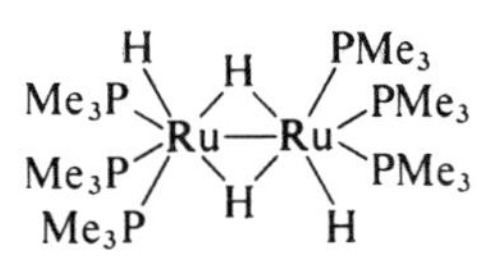

M 662.639

Bridged dimer. Red prisms.

Monomer: Dihydrotris(trimethylphosphine)ruthenium.
$C_9H_{29}P_3Ru$ M 331.319

Unknown.

Jones, R.A. *et al*, *J. Chem. Soc.*, *Dalton Trans.*, 1980, 2480 (*synth*, *ir*, *pmr*, *cmr*, *nmr*, *cryst struct*)

$C_{18}O_{17}Ru_6$ Ru-00500

μ_6-Carbon-μ-carbonylhexadecacarbonylhexaruthenium, 8CI

μ_6-Carbonheptadecacarbonylhexaruthenium

[12277-76-8]

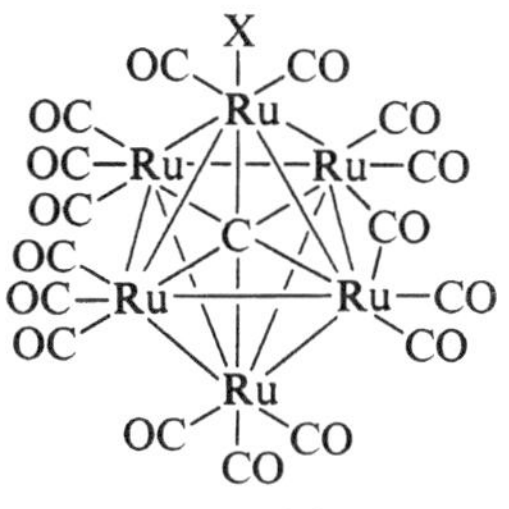

X = CO

M 1094.608

Deep red-violet cryst. (toluene), stable in air. Sol. most org. solvs. Involatile.

Johnson, B.F.G. *et al*, *J. Chem. Soc.* (*A*), 1968, 2865 (*synth*, *ir*, *ms*)
Sirigu, A. *et al*, *J. Chem. Soc.*, *Chem. Commun.*, 1969, 596 (*cryst struct*)
Eady, C.R. *et al*, *J. Organomet. Chem.*, 1973, **57**, C82.
Peake, B.M. *et al*, *J. Chem. Soc.*, *Chem. Commun.*, 1974, 945.
Johnson, B.F.G. *et al*, *J. Organomet. Chem.*, 1980, C3 (*synth*)

$C_{18}O_{18}Ru_6^{\ominus\ominus}$ Ru-00501

Di-μ-carbonyldi-μ_3-carbonyltetradecacarbonylhexaruthenate(2−), 10CI

Octadecacarbonylhexaruthenate(2−)

[62449-05-2]

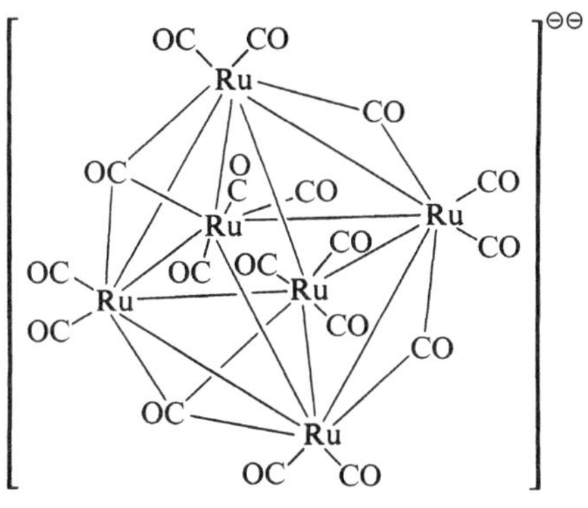

M 1110.607 (ion)

Bis(methyltriphenylphosphonium) salt: [72952-08-0]. $C_{56}H_{36}O_{18}P_2Ru_6$ M 1665.257
Cryst. ($MeOH/CH_2Cl_2$).

Bis[bis(triphenylphosphine)]iminium salt: [62501-17-1]. $C_{90}H_{60}N_2O_{18}P_4Ru_6$ M 2187.782
Dark-brown cryst. ($MeOH/CH_2Cl_2$).

Jackson, P.F. *et al*, *J. Chem. Soc.*, *Chem. Commun.*, 1979, 735 (*synth*, *cryst struct*)
Eady, C.R. *et al*, *J. Chem. Soc.*, *Dalton Trans.*, 1980, 383 (*synth*, *ir*, *cmr*)
Hayward, C.-M.T. *et al*, *Inorg. Chem.*, 1982, **21**, 3816 (*synth*, *struct*)

$C_{19}H_5O_{13}PRu_4$ Ru-00502

Tridecacarbonyl(μ_3-phenylphosphinidene)tetraruthenium

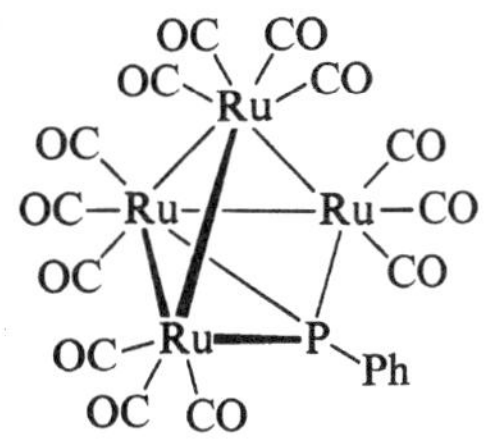

M 876.494

Obt. by pyrolysis of $Ru_3(CO)_9(PPh_2)(H)$. No phys. details reported.

Maclaughlin, S.A. *et al*, *Can. J. Chem.*, 1982, **60**, 87 (*synth*, *struct*, *ir*)

$C_{19}H_{10}O_{11}Ru_4$ Ru-00503

Undecacarbonyl[(1,2-η:5-η:5,6-η:5,6-η:6-η)-1-cycloocten-5-yne]tetraruthenium, 9CI

[37366-21-5]

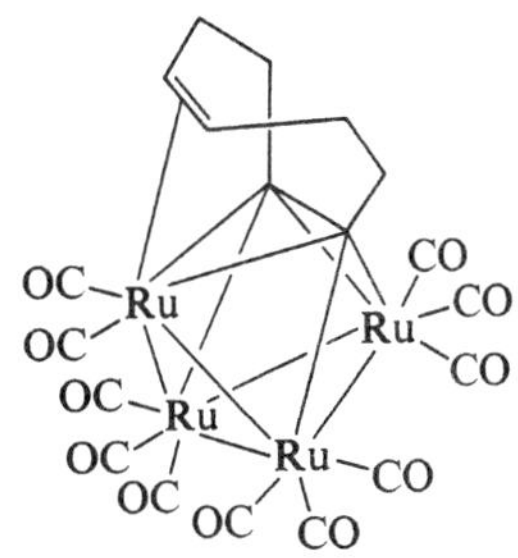

M 818.561

Black cryst. Sol. org. solvs.

Mason, R. *et al*, *J. Organomet. Chem.*, 1972, **43**, C39 (*cryst struct*)
Canty, A.J. *et al*, *J. Chem. Soc.*, *Dalton Trans.*, 1973, 2056 (*synth*, *ir*, *pmr*)

$C_{19}H_{14}N_2O_2Ru$ Ru-00504

Dicarbonyl(η^5-2,4-cyclopentadien-1-yl)[2-(phenylazo)phenyl]-ruthenium

π-Cyclopentadienyl-σ-[2-(phenylazo)phenyl]dicarbonylruthenium

[32629-41-7]

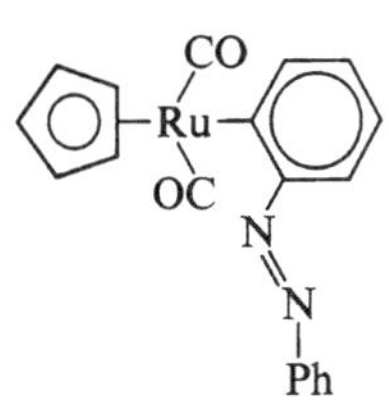

M 403.402

Orange-yellow needles (pet. ether). Mp 133-133.5°.

Bruce, M.I. *et al*, *J. Chem. Soc.*, *Dalton Trans.*, 1971, 2820 (*synth*, *ir*, *pmr*)

$C_{19}H_{14}O_3Ru$ Ru-00505

Tricarbonyl(η^4-1,4-diphenyl-1,3-butadiene)ruthenium

[1,1′-(η^4-1,3-Butadiene-1,4-diyl)bis[benzene]]tricarbonylruthenium, 9CI

[52649-43-1]

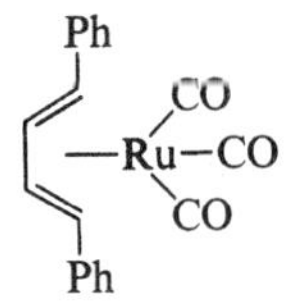

M 391.388

Yellow cryst. (hexane).

Gambino, O. *et al*, *Inorg. Chim. Acta*, 1974, **8**, 71 (*synth, ir, pmr, ms*)

$C_{19}H_{14}O_6Ru_3$ Ru-00506

Di-μ-carbonyltetracarbonyl[μ-(2,4,6-cyclooctatrien-1-yl)](η^5-2,4-cyclopentadien-1-yl)triruthenium, 9CI

Pentacarbonyl(1,3,5-cyclooctatrienyl)(η^5-cyclopentadienyl)triruthenium

[69553-71-5]

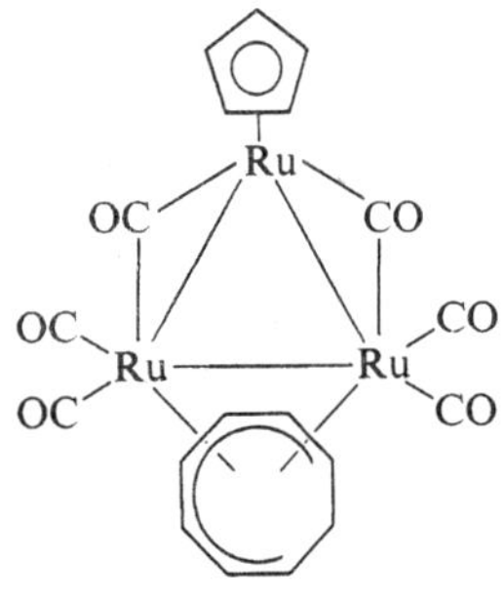

M 641.526

Fluxional. Purple cryst. Mp 195-210° dec.

Humphries, A.P. *et al*, *J. Chem. Soc., Dalton Trans.*, 1978, 1523 (*synth, ir, pmr*)

$C_{19}H_{15}O_2PRu$ Ru-00507

Dicarbonyl(η^5-2,4-cyclopentadien-1-yl)(diphenylphosphino)-ruthenium, 9CI

Dicarbonyl(η^5-cyclopentadienyl)ruthenium(diphenyl)-phosphine. Cyclopentadienyldicarbonylruthenium diphenylphosphine

[42802-19-7]

As Dicarbonyl(η^5-2,4-cyclopentadien-1-yl)-iodoruthenium, Ru-00051 with

$$X = -PPh_2$$

M 407.370

Prepd. and utilised *in situ*.

Haines, R.J. *et al*, *J. Organomet. Chem.*, 1973, **55**, 199 (*synth*)

$C_{19}H_{16}ORu$ Ru-00508

1-Phenyl-3-ruthenocenyl-2-propen-1-one

(3-Oxo-3-phenyl-1-propenyl)ruthenocene, 10CI

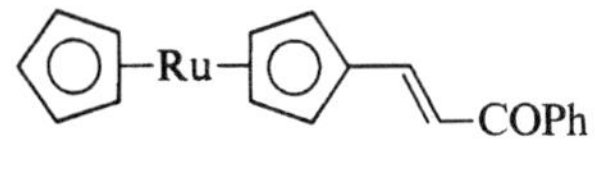

M 361.405

(*E*)-*form* [72277-78-2]

Yellow cryst. (C_6H_6/pet. ether). Mp 159-61°.

Toma, Š. *et al*, *Collect. Czech. Chem. Commun.*, 1981, **46**, 2531 (*synth, ir, pmr*)

$C_{19}H_{16}ORu$ Ru-00509

3-Phenyl-1-ruthenocenyl-2-propen-1-one

1-Oxo-3-phenyl-2-propenylruthenocene, 10CI. *Cinnamoylruthenocene*

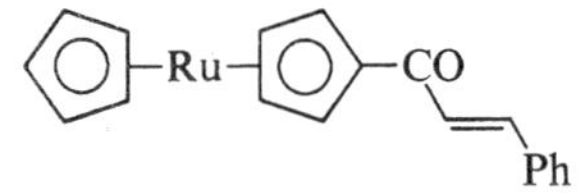

M 361.405

(*E*)-*form* [64884-14-6]

Yellow cryst. (C_6H_6/pet. ether). Mp 147-50°.

Toma, Š. *et al*, *Collect. Czech. Chem. Commun.*, 1981, **46**, 2531 (*synth, ir, pmr*)

$C_{19}H_{26}As_2Cl_2ORu$ Ru-00510

Carbonyldichlorobis(dimethylphenylarsine)(η^2-ethene)ruthenium

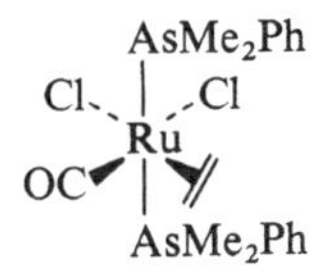

M 592.233

Cream cryst.

Stephenson, M. *et al*, *J. Chem. Soc., Dalton Trans.*, 1981, 2112 (*synth, ir, pmr, cmr*)

$C_{19}H_{26}Cl_2OP_2Ru$ Ru-00511

Carbonyldichlorobis(dimethylphenylphosphine)(η^2-ethene)-ruthenium, 9CI

[60595-73-5]

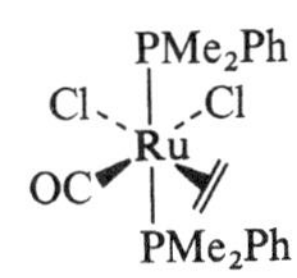

M 504.337

Cryst. (Me_2CO/pet. ether satd. with C_2H_4). Slowly loses ethylene in soln., but is stable over long periods in solid state.

Barnard, C.F.J. *et al*, *J. Chem. Soc., Dalton Trans.*, 1976, 953 (*synth, ir*)

Brown, L.D. *et al*, *Inorg. Chem.*, 1978, **17**, 2932 (*struct*)

Stephenson, M. *et al*, *J. Chem. Soc., Dalton Trans.*, 1981, 2112 (*pmr, cmr*)

$C_{19}H_{26}Ru$ Ru-00512

[(2,3,5,6-η)-Bicyclo[2.2.1]hepta-2,5-diene][(1,2,3,4,5,6-η)-hexamethylbenzene]ruthenium, 10CI

[(1,2,3,4,5,6-η)-Hexamethylbenzene][(2,3,5,6-η)-norbornadiene]ruthenium

[71933-82-9]

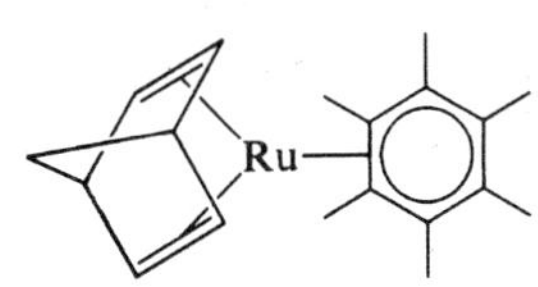

M 355.484

No phys. props. reported.

Bennett, M.A. *et al*, *J. Chem. Soc., Chem. Commun.*, 1979, 312 (*synth*)

$C_{19}H_{27}ORu^{\oplus}$ Ru-00513

Carbonyl[(1,2,3-η)-2-cyclohexen-1-yl][(1,2,3,4,5,6-η)-hexamethylbenzene]ruthenium(1+), 10CI

[67420-83-1]

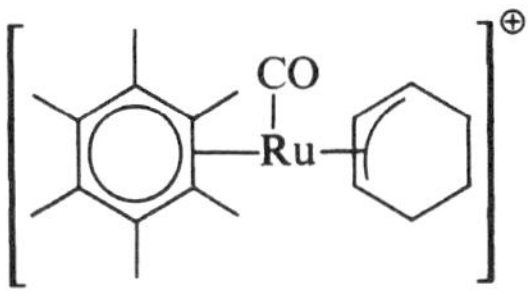

M 372.492 (ion)

Characterised spectroscopically.

Bennett, M.A. *et al*, *J. Organomet. Chem.*, 1978, **153**, C25 (*synth, pmr, ir*)

$C_{19}H_{27}Ru^{\oplus}$ Ru-00514

[(1,2,3,4,5-η)-2,4-Cycloheptadien-1-yl][(1,2,3,4,5,6-η)-hexamethylbenzene]ruthenium(1+), 10CI

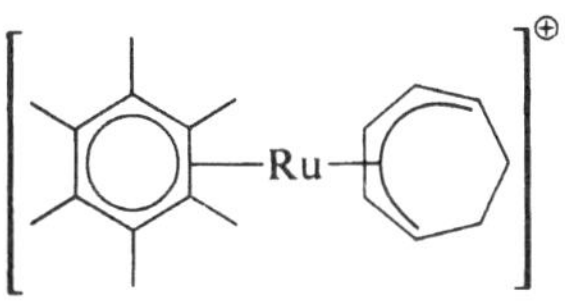

M 356.492 (ion)

Hexafluorophosphate: [67430-20-0].
$C_{19}H_{27}F_6PRu$ M 501.456
No phys. props. reported.

Bennett, M.A. *et al*, *J. Organomet. Chem.*, 1978, **153**, C25 (*synth, pmr*)

$C_{19}H_{34}N_2O_3Ru$ Ru-00515

Tricarbonyl[*N,N'*-1,2-ethanediylbis[2,4-dimethyl-3-pentanamine]-*N,N'*]ruthenium

Tricarbonyl[glyoxalbis(diisopropylmethylimine)]ruthenium

[78199-25-4]

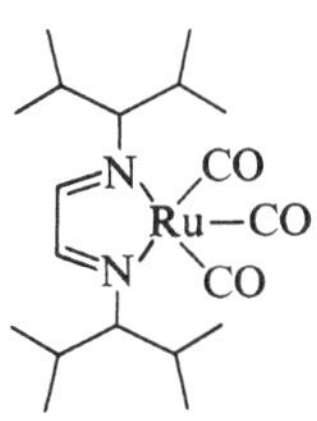

M 439.559

Black metallic cryst. (pentane at −70°).

Staal, L.H. *et al*, *Inorg. Chem.*, 1981, **20**, 3590 (*synth, ir, pmr*)

$C_{20}H_{10}O_6Ru_2$ Ru-00516

Hexacarbonyl[μ-[η²:η⁴-1,2-phenylene(2-phenyl-1,2-ethenediyl)]]diruthenium

Hexacarbonyl(diphenylacetylene)diruthenium, 8CI. Hexacarbonyl[2-phenyl-1-ruthenaindene]ruthenium

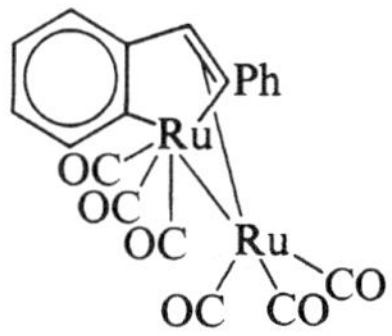

Possible structure

M 548.435

Red-brown powder. Mp 158-9°.

Cetini, G. *et al*, *J. Organomet. Chem.*, 1969, **17**, 437 (*synth, ir*)

$C_{20}H_{13}FeO_{12}PRu_3$ Ru-00517

Di-μ-carbonyloctacarbonyl(dicarbonyliron)di-μ-hydro(dimethylphenylphosphine)triruthenium

Dodecacarbonyldihydro(dimethylphenylphosphine)irontriruthenium

[74128-11-3]

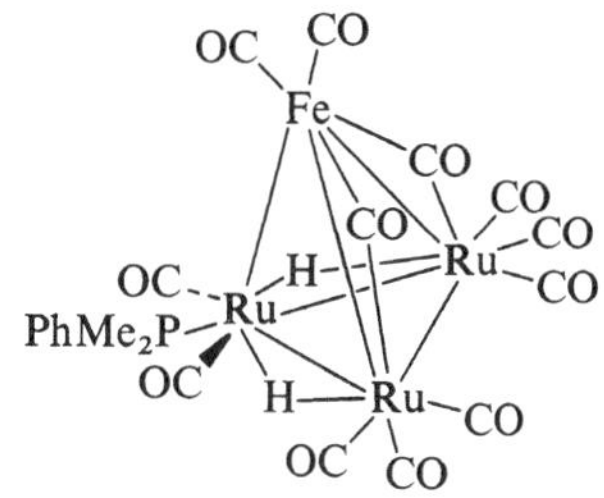

M 835.346

Fluxional molecule. Formed as a mixt. of isomers from which that shown crystallised preferentially. Red air-stable cryst. V. sol. CH_2Cl_2, spar. sol. hexane.

Fox, J.R. *et al*, *Inorg. Chem.*, 1981, **20**, 3214 (*synth, cryst struct, spectra*)

$C_{20}H_{15}I_3O_2PRu^{\ominus}$ Ru-00518

Dicarbonyltriiodo(triphenylphosphine)ruthenate(1−), 9CI

$Ph_3PRu(CO)_2I_3^{\ominus}$

M 800.095 (ion)

Stereochemistry not known.

Methyltriphenylphosphonium salt: [37820-98-7].
$C_{39}H_{33}I_3O_2P_2Ru$ M 1077.420
Yellow cryst.

Jeffery, J. *et al*, *J. Organomet. Chem.*, 1972, **40**, C42 (*synth, ir, pmr*)

$C_{20}H_{16}O_4Ru_3$ Ru-00519

Tetracarbonylbis[μ-(1,3,5,7-cyclooctatetraene)]triruthenium, 9CI

[12290-41-4]

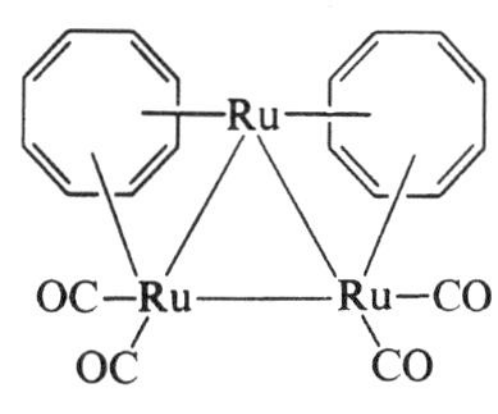

M 623.554

Air-stable red cryst. (C_6H_6). Mp 165-70° dec.

Cotton, F.A. *et al*, *J. Am. Chem. Soc.*, 1968, **90**, 6335; 1969, **91**, 6598 (*synth, pmr, cryst struct*)
Campbell, A.J. *et al*, *Inorg. Chem.*, 1976, **15**, 1321 (*pmr*)
Fyfe, C.A. *et al*, *J. Am. Chem. Soc.*, 1979, **101**, 1351 (*cmr*)

$C_{20}H_{16}O_6Ru_3$ **Ru-00520**

Di-μ-carbonyltetracarbonyl[(1,2,3,4,5-η)-2,4-cycloheptadien-1-yl][μ-[(1,2,3,4-η:1,5,6,7-η)-2,4,6-cycloheptatrien-1-yl]]triruthenium, 9CI

Hexacarbonyl(1,3-cycloheptadienyl)(cycloheptatrienyl)-triruthenium

[51404-01-4]

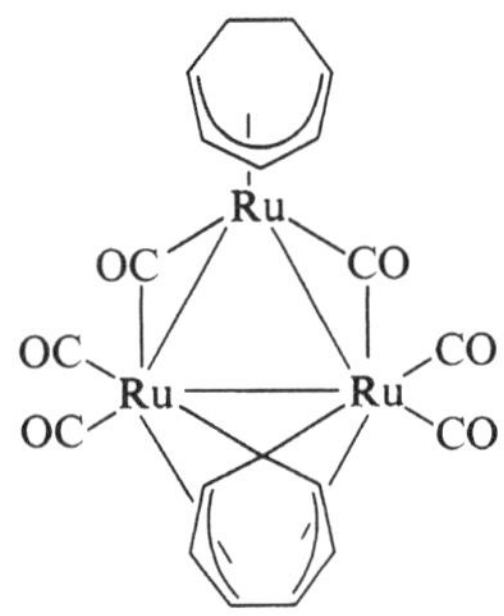

M 655.553

Fluxional C_7H_7 ring. Red cryst. Mp 155-60° dec.

Bau, R. *et al*, *J. Chem. Soc., Chem. Commun.*, 1973, 726 (*struct*)

Burt, J.C. *et al*, *J. Chem. Soc., Dalton Trans.*, 1975, 731 (*synth, ir, pmr*)

$C_{20}H_{16}Ru^{\oplus\oplus}$ **Ru-00521**

Bis[(1,2,3,4,4*a*,8*a*-η)naphthalene]ruthenium(2+)

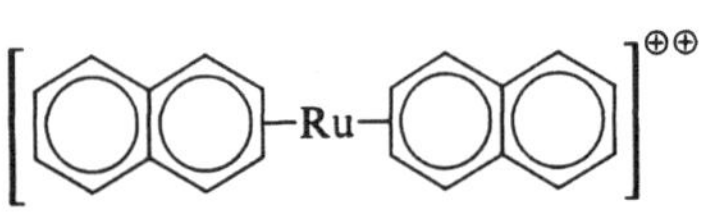

M 357.416 (ion)

Bis(hexafluorophosphate): [12118-06-8].

$C_{20}H_{16}F_{12}P_2Ru$ M 647.345

Yellow cryst.

Fischer, E.O. *et al*, *J. Organomet. Chem.*, 1967, **7**, 481 (*synth, ir, pmr*)

$C_{20}H_{18}HgRu_2$ **Ru-00522**

Diruthenocenylmercury, 9CI

[37298-86-5]

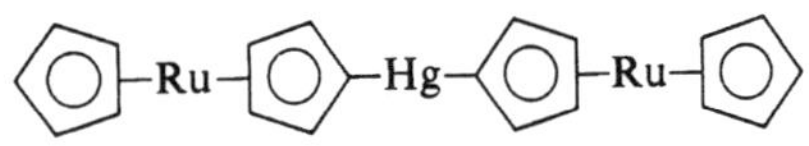

M 661.092

Cryst. (xylene). Dec. >190°.

Nesmeyanov, A.N. *et al*, *Izv. Akad. Nauk SSSR, Ser. Khim.*, 1972, 1823 (*synth*)

$C_{20}H_{18}Ru_2$ **Ru-00523**

1,1″-Biruthenocene, 10CI

[67126-04-9]

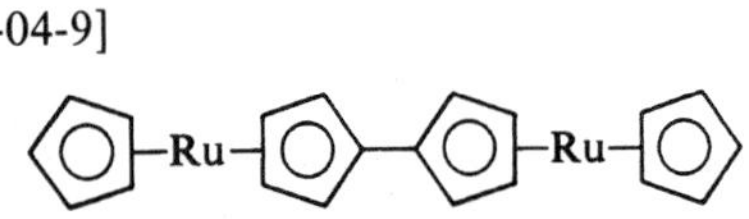

M 460.502

Colourless cryst. (heptane). Mp 238-9°.

Bednarik, L.A. *et al*, *J. Org. Chem.*, 1980, **45**, 2032 (*synth, uv*)

$C_{20}H_{24}O_2Ru_3S_3$ **Ru-00524**

μ-Carbonylcarbonyltris(η^5-2,4-cyclopentadien-1-yl-)tris[μ-(methanethiolato)]triruthenium, 9CI

Dicarbonyltricyclopentadienyltris[μ-(methanethiolato)-]triruthenium

[53537-19-2]

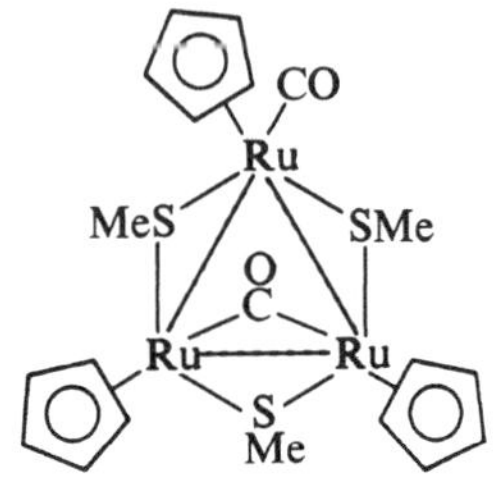

M 695.798

Red cryst. (CH_2Cl_2/pet. ether). Mp 175° dec.

Knox, G.R. *et al*, *J. Organomet. Chem.*, 1974, **74**, 105 (*synth, ir, pmr, ms*)

Killops, S.D. *et al*, *J. Chem. Soc., Dalton Trans.*, 1978, 1260.

$C_{20}H_{26}Ru$ **Ru-00525**

[(1,2,3,4-η)-1,3,5,7-Cyclooctatetraene][(1,2,3,4,5,6-η)-hexamethylbenzene]ruthenium

[62422-36-0]

As (η^6-Benzene)[(1,2,3,4-η)-1,3,5,7-cyclooctatetraene]ruthenium, Ru-00348 with

R = CH_3

M 367.495

Fluxional C_8H_8 ring. Air-sensitive orange-red cryst. (hexane).

Bennett, M.A. *et al*, *Inorg. Chem.*, 1980, **19**, 1014 (*synth, pmr, cmr, ir*)

$C_{20}H_{29}NRu^{\oplus\oplus}$ **Ru-00526**

(η^6-Dimethylaniline)(η^6-hexamethylbenzene)ruthenium(2+)

[(1,2,3,4,5,6-η)-N,N-Dimethylbenzenamine][(1,2,3,4,5,6-η)-hexamethylbenzene]ruthenium(2+), 10CI

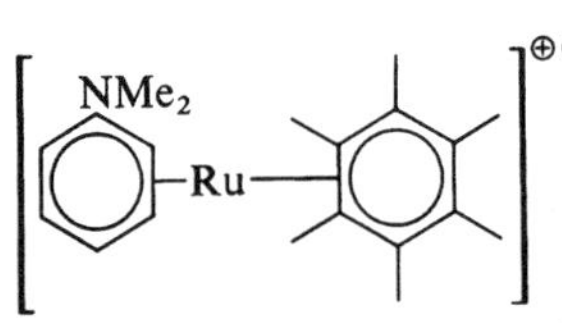

M 384.526 (ion)

Air-stable.

Bis(tetrafluoroborate): [71851-93-9].

$C_{20}H_{29}B_2F_8NRu$ M 558.133

No phys. props. reported.

Bennett, M.A. *et al*, *J. Organomet. Chem.*, 1979, **175**, 87 (*synth, pmr*)

$C_{20}H_{29}Ru^{\oplus}$ **Ru-00527**

[(1,2,3,4,5-η)-2,4,6-Cyclooctatrien-1-yl][(1,2,3,4,5,6-η)-hexamethylbenzene]ruthenium(1+), 10CI

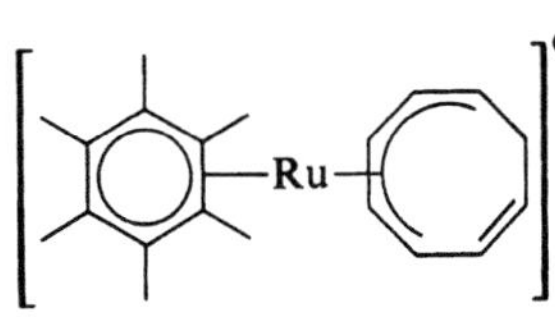

M 370.519 (ion)

Hexafluorophosphate: [77400-18-1].
$C_{20}H_{29}F_6PRu$ M 515.483
Pale-yellow cryst.

Bennett, M.A. *et al, Inorg. Chem.*, 1981, **20**, 2353 (*synth, pmr, cmr*)

$C_{20}H_{29}Ru^{\oplus}$ Ru-00528

[(1,2,3,6,7-η)-2,4,6-Cyclooctatrien-1-yl][(1,2,3,4,5,6-η)-hexamethylbenzene]ruthenium(1+), 10CI

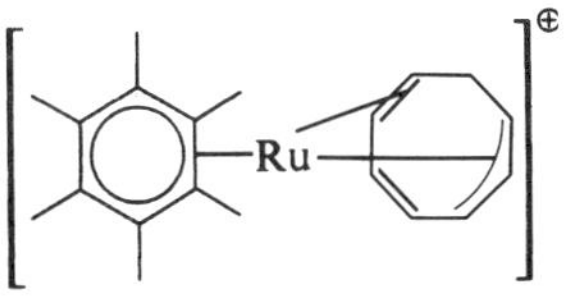

M 370.519 (ion)
Hexafluorophosphate: [77357-74-5].
$C_{20}H_{29}F_6PRu$ M 515.483
Yellow cryst. (Me_2CO/Et_2O or CH_2Cl_2/Et_2O).

Bennett, M.A. *et al, Inorg. Chem.*, 1981, **20**, 2353 (*synth, pmr, cmr*)

$C_{20}H_{32}As_4ClN_2Ru^{\oplus}$ Ru-00529

Chloro(dinitrogen)bis[1,2-phenylenebis[dimethylarsine]-*As,As'*]ruthenium(1+), 9CI

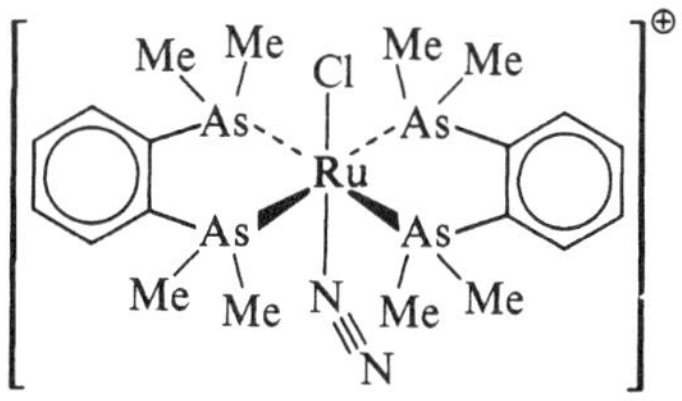

M 736.696 (ion)

trans-form

Hexafluorophosphate: [28909-62-8].
$C_{20}H_{32}As_4ClF_6N_2PRu$ M 881.660
White cryst. (CH_2Cl_2).
Hexafluoroantimonate:
$C_{20}H_{32}As_4ClF_6N_2RuSb$ M 972.436
White cryst. (Me_2CO aq.).

Douglas, P.G. *et al, J. Am. Chem. Soc.*, 1972, **94**, 5254 (*synth, pmr, ir*)
Feltham, R.D. *et al, Inorg. Chem.*, 1972, **11**, 2468 (*synth, ir*)

$C_{20}H_{32}Cl_4Ru_2$ Ru-00530

Di-μ-chlorodichlorobis[(1,2,3,6,7,8-η)-2,7-dimethyl-2,6-octadiene-1,8-diyl]diruthenium

Tetrachlorobis(2,7-dimethyl-2,6-octadiene-1,8-diyl)-diruthenium
[34801-97-3]

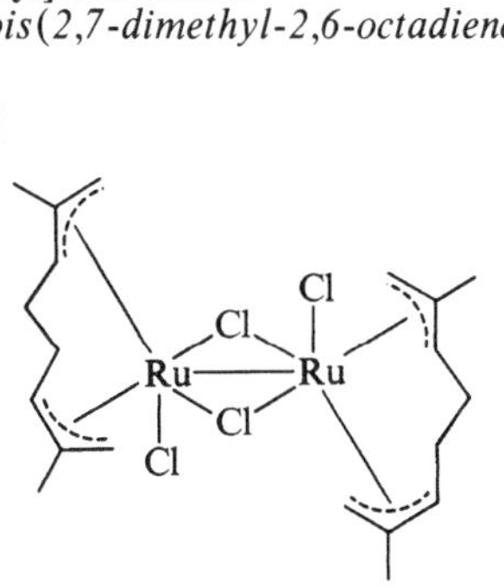

M 616.425
Bridged dimer. Red-brown.
Monomer: Dichloro[(1,2,3,6,7,8-η)-2,7-dimethyl-2,6-octadiene-1,8-diyl]ruthenium.
$C_{10}H_{16}Cl_2Ru$ M 308.212
Unknown.

Porri, L. *et al, Tetrahedron Lett.*, 1965, 4187 (*synth, struct*)
Colombo, A. *et al, Acta Crystallogr., Sect. B*, 1971, **27**, 1653 (*cryst struct*)

$C_{20}H_{32}N_4O_4Ru_2$ Ru-00531

[μ-[*N,N'*-Bis(1-methylethyl)-1,4-bis[(1-methylethyl)imino]-2,3-butanediaminato(2−)-N^1,N^2,N^3:N^2,N^3,N^4]]tetracarbonyldiruthenium

$Ru_2(CO)_4(IAE)$
[72918-62-8]

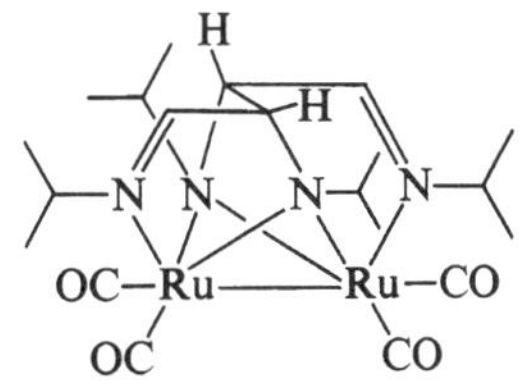

M 594.637
Yellow cryst. (toluene).

Staal, L.H. *et al, Inorg. Chim. Acta*, 1979, **37**, L485 (*synth, ir, pmr*)
Staal, L.H. *et al, Inorg. Chem.*, 1980, **19**, 3343.

$C_{20}H_{32}N_4O_4Ru_2$ Ru-00532

Tetracarbonylbis[μ-(η²-*N,N'*-1,2-ethanediylidenebis[2-propanamine]-*N,N'*]]diruthenium, 10CI

Tetracarbonylbis[glyoxalbis(isopropylimine)]diruthenium
[74552-72-0]

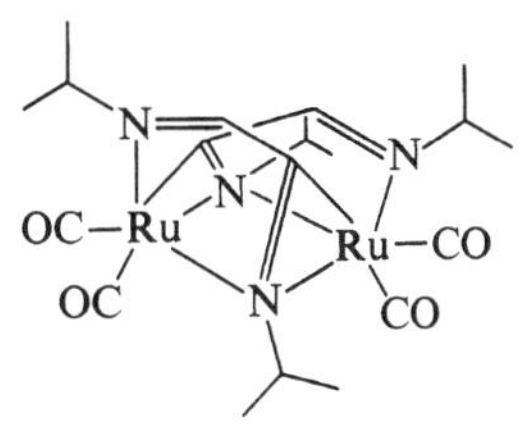

M 594.637
Yellow cryst.

Staal, L.H. *et al, Inorg. Chem.*, 1980, **19**, 3343 (*synth, pmr, cryst struct*)
Staal, L.H. *et al, Inorg. Chim. Acta*, 1981, **50**, 205 (*ms*)
Staal, L.H. *et al, J. Organomet. Chem.*, 1981, **204**, 101 (*cmr*)

$C_{20}H_{40}O_{20}P_4Ru_4$ Ru-00533

Octacarbonyltetra-μ-hydrotetrakis(trimethyl phosphite-*P*)-tetraruthenium

Octacarbonyltetra-μ-hydrotetrakis(phosphorous acid)-tetraruthenium dodecamethyl ester, 8CI
[34482-43-4]

As Nonacarbonyltetra-μ-hydrotris(trimethyl phosphite-*P*)tetraruthenium, Ru-00495 with

$$X = P(OMe)_3$$

M 1128.699
Air-stable orange cryst. Sol. hydrocarbons, halocarbons.

Kaesz, H.D. *et al, J. Am. Chem. Soc.*, 1971, **93**, 4594 (*synth, pmr*)

$C_{20}H_{58}P_6Ru_2^{\oplus\oplus}$ Ru-00534

Di-μ-methylenehexakis(trimethylphosphine)diruthenium(2+)

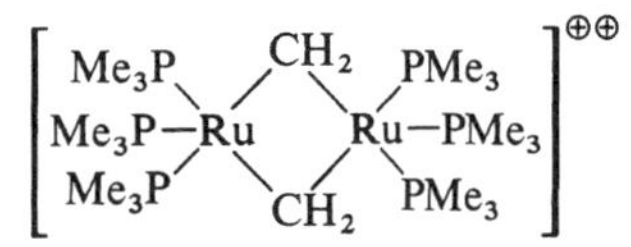

M 686.661 (ion)

Bis(tetrafluoroborate): [71296-09-8].
$C_{20}H_{58}B_2F_8P_6Ru_2$ M 860.268
Air-stable orange cryst. (MeOH). Sol. $MeNO_2$, MeCN, Me_2CO, MeOH. Insol. hydrocarbons, THF, Et_2O, CH_2Cl_2. Mp 240-5°. Diamagnetic. Dec. in $CHCl_3$ after ca. 30 min.

Bis(trifluoromethylsulfonate):
$C_{22}H_{58}F_6O_6P_6Ru_2S_2$ M 984.790
Air-stable orange cryst. (MeOH). Mp 203-7° dec. Diamagnetic.

Hursthouse, M.B. *et al*, *J. Am. Chem. Soc.*, 1979, **101**, 4128 (*synth, ir, pmr, struct*)

$C_{21}H_5O_{15}PRu_5$ Ru-00535

Pentadecacarbonyl[μ4-(phenylphosphoranetetrayl)]pentaruthenium

[78330-36-6]

As Pentadecacarbonyl[μ4-(methylphosphoranetetrayl)]-pentaruthenium, Ru-00415 with

R = Ph

M 1033.585

Green cryst. (CH_2Cl_2/pentane). Stable for weeks in air in solid state. Solns. stable under N_2. Mp 185-90° dec.

Natarajan, K. *et al*, *J. Organomet. Chem.*, 1981, **209**, 85 (*synth, ms, ir, pmr, cryst struct*)

$C_{21}H_6O_{14}Ru_6$ Ru-00536

(Benzene)[μ6-(carbon)]tetradecacarbonylhexaruthenium, 9CI

(η⁶-Benzene)(μ6-carbon)-μ-carbonyltridecacarbonylhexaruthenium

[37190-51-5]

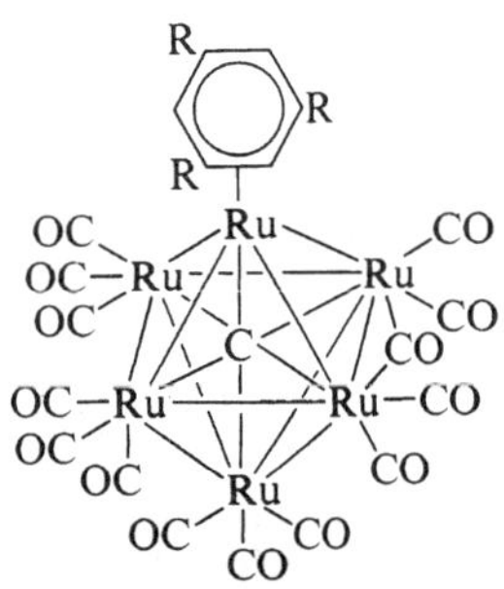

R = H

M 1088.690

Deep-purple air-stable cryst. (heptane/C_6H_6).

Eady, C.R. *et al*, *J. Organomet. Chem.*, 1972, **37**, C39 (*synth*)
Eady, C.R. *et al*, *J. Chem. Soc., Dalton Trans.*, 1975, 2606 (*synth, ir, pmr, ms*)

$C_{21}H_{10}N_2O_9Ru_3$ Ru-00537

Bis[μ3-[benzeneaminato(2−)]]nonacarbonyltriruthenium, 9CI

Bis(anilinato)nonacarbonyltriruthenium

[51233-18-2]

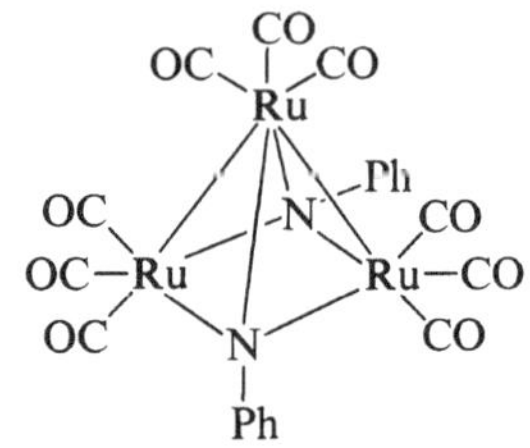

M 737.528

Orange cryst. Mp 125-30° dec.

Sappa, E. *et al*, *J. Organomet. Chem.*, 1973, **61**, 383 (*synth, ir, pmr*)

$C_{21}H_{10}N_2O_9Ru_3$ Ru-00538

Nonacarbonyl[μ3-[N-phenyl-o-phenylenediaminato(2−)]triruthenium, 8CI

Nonacarbonyl(μ³-o-semidine)triruthenium

[33846-31-0]

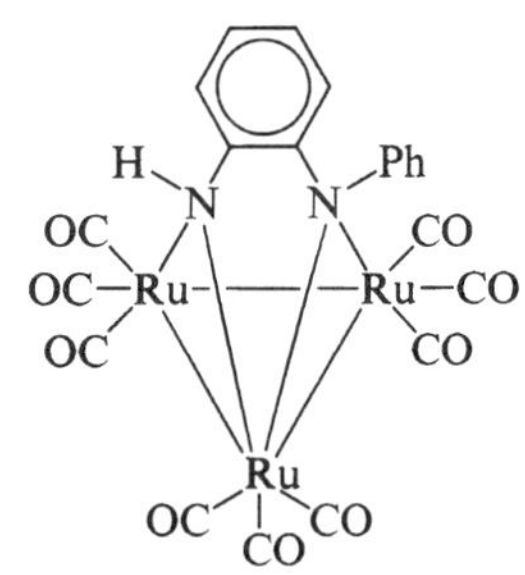

M 737.528

Deep-red needles (pet. ether). Mp 190° dec.

Bruce, M.I. *et al*, *J. Organomet. Chem.*, 1971, **31**, 275 (*synth, ir*)

$C_{21}H_{11}O_9PRu_3$ Ru-00539

Nonacarbonyl[μ3-(diphenylphosphino)]-μ-hydrotriruthenium, 9CI

[82647-79-8]

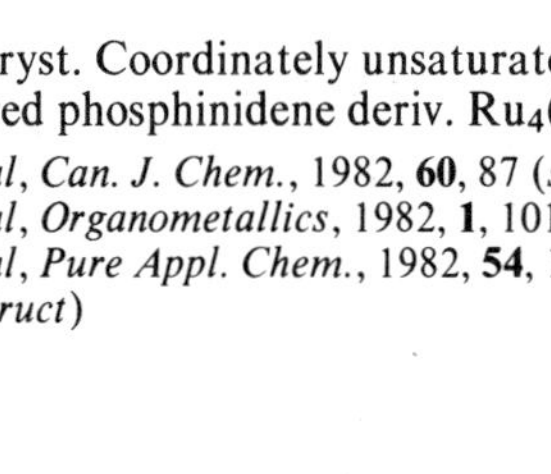

M 741.496

Orange-red cryst. Coordinately unsaturated. Thermolysis yields the red phosphinidene deriv. $Ru_4(CO)_{13}(PPh)$.

Carty, A.J. *et al*, *Can. J. Chem.*, 1982, **60**, 87 (*synth, ir, pmr*)
Carty, A.J. *et al*, *Organometallics*, 1982, **1**, 1013.
Carty, A.J. *et al*, *Pure Appl. Chem.*, 1982, **54**, 113 (*synth, ir, nmr, cryst struct*)

$C_{21}H_{14}O_9Ru_4$ Ru-00540

(η^6-Benzene)nonacarbonyl[μ_4-(1-η:1,2-η:1,2-η:2-η)-1-cyclohexene-1,2-diyl]tetraruthenium, 10CI

[70755-71-4]

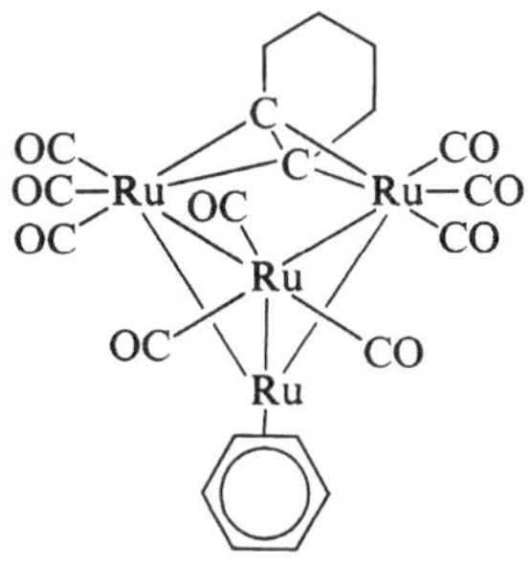

M 814.616

Violet needles. Sol. cyclohexane.

Aime, S. *et al*, *Inorg. Chim. Acta*, 1979, **34**, 49 (*synth*, *ir*, *cryst struct*, *pmr*, *ms*)

$C_{21}H_{15}Br_2O_3PRu$ Ru-00541

Dibromotricarbonyl(triphenylphosphine)ruthenium, 8CI

[22949-66-2]

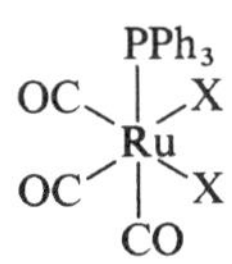

X = Br

M 607.199

Yellow-green solid.

Johnson, B.F.G. *et al*, *J. Chem. Soc. (A)*, 1969, 792 (*synth*, *ir*)
Kramer, G. *et al*, *Inorg. Chem.*, 1981, **20**, 1362 (*synth*)

$C_{21}H_{15}Cl_2O_3PRu$ Ru-00542

Tricarbonyldichloro(triphenylphosphine)ruthenium, 8CI

[22863-29-2]

As Dibromotricarbonyl(triphenylphosphine)ruthenium, Ru-00541 with

X = Cl

M 518.297

Pale-yellow cryst. (C_6H_6/pet. ether). Turns red at 160°. Dec. at 209-15°.

Bruce, M.I. *et al*, *Z. Naturforsch., B*, 1968, **23**, 1543 (*synth*)
Johnson, B.F.G. *et al*, *J. Chem. Soc. (A)*, 1969, 792 (*synth*, *ir*)

$C_{21}H_{15}I_2O_3PRu$ Ru-00543

Tricarbonyldiiodo(triphenylphosphine)ruthenium, 8CI

[15091-95-9]

As Dibromotricarbonyl(triphenylphosphine)ruthenium, Ru-00541 with

X = I

M 701.200

Orange prisms. Mp 126-7°.

Candlin, J.P. *et al*, *Chem. Ind. (London)*, 1966, 1960 (*synth*)
Johnson, B.F.G. *et al*, *J. Chem. Soc. (A)*, 1969, 792 (*synth*, *ir*)
Kramer, G. *et al*, *Inorg. Chem.*, 1981, **20**, 1362 (*synth*)

$C_{21}H_{16}O_9Ru_3$ Ru-00544

Nonacarbonyl[μ_3-[(1-η:1,2,3-η:3-η)-1,6,9-cyclododecatrien-1-yl-3-ylidene]]-μ-hydrotriruthenium, 9CI

[36655-50-2]

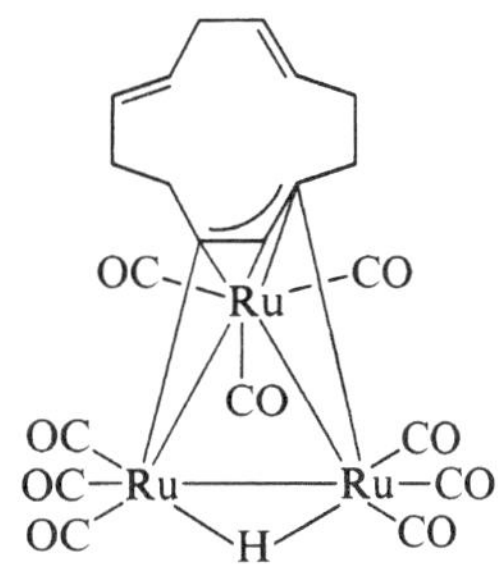

M 715.562

Yellow cryst. (pet. ether). Mp 138°.

Bruce, M.I. *et al*, *J. Chem. Soc., Chem. Commun.*, 1970, 735 (*synth*, *ir*)
Cox, A. *et al*, *J. Chem. Soc. (A)*, 1971, 3599 (*cryst struct*)

$C_{21}H_{17}O_3PRu$ Ru-00545

Tricarbonyldihydro(triphenylphosphine)ruthenium, 9CI

[42781-57-7]

$H_2Ru(CO)_3(PPh_3)$

M 449.407

Unstable. Stabilized at high H_2 pressure.

Whyman, R., *J. Organomet. Chem.*, 1973, **56**, 339 (*synth*, *ir*)

$C_{21}H_{18}FeORu$ Ru-00546

Ferrocenylruthenocenylmethanone

π-Cyclopentadienyl[(ferrocenylcarbonyl)-π-cyclopentadienyl]ruthenium, 8CI. Ferrocenyl ruthenocenyl ketone. Ruthenocenylcarbonylferrocene. Carbonylferrocenylruthenocene

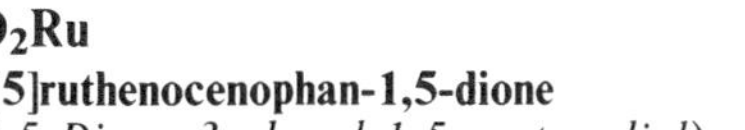

M 443.290

Orange-red needles (propanol). Mp 216°.

Rausch, M.D. *et al*, *J. Am. Chem. Soc.*, 1960, **82**, 76 (*synth*, *ir*)
Gubin, S.P. *et al*, *J. Organomet. Chem.*, 1970, **22**, 183 (*synth*, *pmr*)

$C_{21}H_{18}O_2Ru$ Ru-00547

3-Phenyl[5]ruthenocenophan-1,5-dione

1,1'-(1,5-Dioxo-3-phenyl-1,5-pentanediyl)ruthenocene, 10CI

[61182-41-0]

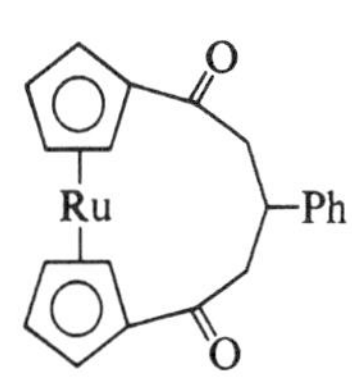

M 403.442

Cryst. with yellow tint. Highly insol. Mp 295°.

Nesmeyanov, A.N. *et al*, *Dokl. Akad. Nauk SSSR, Ser. Sci. Khim.*, 1976, **229**, 881 (*synth, ir, ms, uv*)

$C_{21}H_{18}O_3RuSi$ Ru-00548

Tricarbonyl[(2,3,4,5-η)-1,1-dimethyl-2,5-diphenylsilacyclopenta-2,4-diene]ruthenium, 10CI

[57583-96-7]

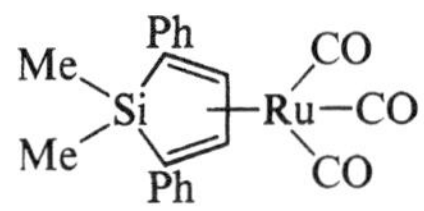

M 447.527

Yellow cryst. Mp 122°.

Abel, E.W. *et al*, *J. Chem. Soc., Dalton Trans.*, 1976, 2483 (*synth, ir, pmr, ms*)

Sakurai, H. *et al*, *J. Organomet. Chem.*, 1980, **201**, C14 (*nmr*)

$C_{21}H_{20}FeRu$ Ru-00549

(Ferrocenylmethyl)ruthenocene

π-Cyclopentadienyl[(ferrocenylmethyl)-π-cyclopentadienyl]ruthenium, 8CI. *Ferrocenyl(ruthenocenyl)methane. (Ruthenocenylmethyl)ferrocene*

[11105-99-0]

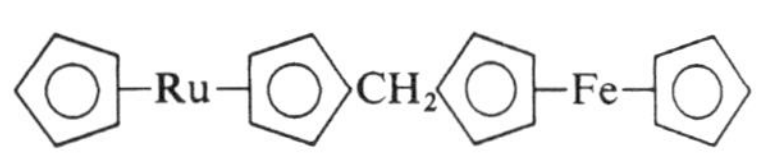

M 429.306

Yellow cryst. (hexane). Mp 139-40°.

Gubin, S.P. *et al*, *J. Organomet. Chem.*, 1970, **22**, 183 (*synth, pmr, polarog, ir*)

$C_{21}H_{22}Ru$ Ru-00550

3-Phenyl[5]ruthenocenophane

1,1'-(3-Phenyl-1,5-pentanediyl)ruthenocene, 10CI

[61182-42-1]

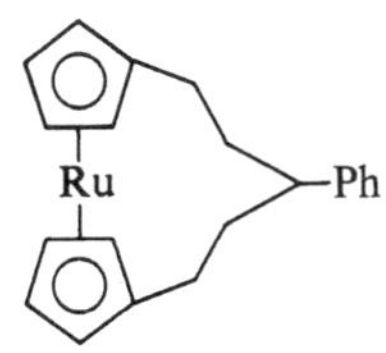

M 375.475

Cryst. with yellow tint. Mp 112-3°.

Nesmeyanov, A.N. *et al*, *Dokl. Akad. Nauk SSSR, Ser. Sci. Khim.*, 1976, **229**, 881 (*synth, ms, uv, pmr, cmr*)

$C_{21}H_{32}N_4O_5Ru_2$ Ru-00551

[μ-[N,N'-Bis(1-methylethyl)-1,4-bis[(1-methylethyl)imino]-2,3-butanediaminato(2−)-N^1,N^2,N^3:N^2,N^3,N^4]]-μ-carbonyltetracarbonyldiruthenium

$Ru_2(CO)_5(IAE)$

[72918-60-6]

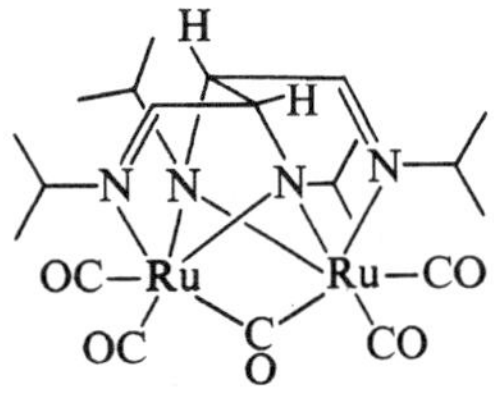

M 622.648

Bright-yellow cubic cryst. (toluene).

Staal, L.H. *et al*, *Inorg. Chim. Acta*, 1979, **37**, L485 (*synth, ir, pmr*)

Staal, L.H. *et al*, *Inorg. Chem.*, 1980, **19**, 3343 (*synth, uv*)

$C_{21}H_{36}O_3Ru^{\oplus\oplus}$ Ru-00552

[(1,2,3,4,5,6-η)-Hexamethylbenzene]tris(2-propanone)ruthenium(2+), 10CI

Tris(acetone)(η^6-hexamethylbenzene)ruthenium(2+)

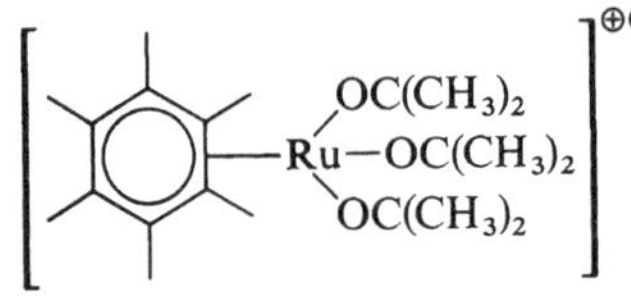

M 437.584 (ion)

Bis(hexafluorophosphate): [71303-84-9].

$C_{21}H_{36}F_{12}O_3P_2Ru$ M 727.512

No phys. props. reported.

Bennett, M.A. *et al*, *J. Chem. Soc., Chem. Commun.*, 1979, 32 (*synth, pmr*)

$C_{21}H_{43}O_9P_3Ru$ Ru-00553

[η^4-1,2,3,4-Tetramethyl-5,6-bis(methylene)-1,3-cyclohexadiene]tris(trimethyl phosphite-*P*)ruthenium

[81408-64-2]

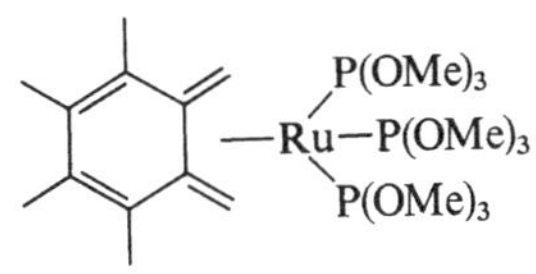

M 633.557

Yellow air-sensitive solid.

Bennett, M.A. *et al*, *Angew. Chem., Int. Ed. Engl.*, 1982, **21**, 378; *Angew. Chem., Int. Ed. Engl. Suppl*, 853 (*synth, pmr, cmr, nmr*)

$C_{21}H_{60}P_6Ru_2$ Ru-00554

Tri-μ-methylenehexakis(trimethylphosphine)diruthenium, 10CI

[66321-02-6]

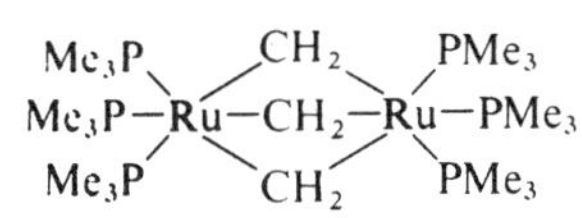

M 700.688

Orange-red prisms; moderately air-sensitive. Mp 186-8°.

Hursthouse, M.B. *et al*, *J. Am. Chem. Soc.*, 1979, **101**, 4128 (*synth, ir, pmr, cryst struct*)

$C_{21}H_{61}P_6Ru_2^{\oplus}$ Ru-00555

μ-Methyldi-μ-methylenehexakis(trimethylphosphine)diruthenium(1+), 10CI

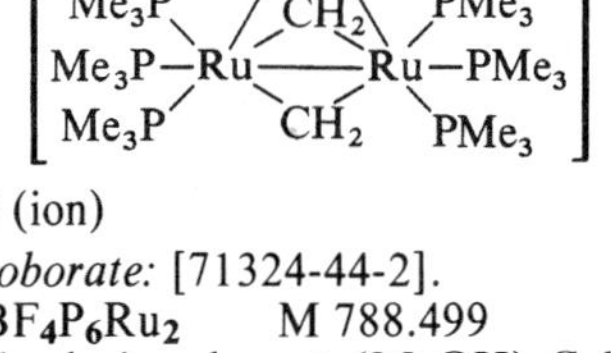

M 701.695 (ion)

Tetrafluoroborate: [71324-44-2].

$C_{21}H_{61}BF_4P_6Ru_2$ M 788.499

Air-stable, dark-red cryst. (MeOH). Sol. $MeNO_2$, MeCN, Me_2CO, MeOH, insol. hydrocarbons, THF, Et_2O. Mp 300-320° dec. Diamagnetic. Stable in CH_2Cl_2 soln. for long periods, dec. fairly rapidly in $CHCl_3$.

Hursthouse, M.B. *et al*, *J. Am. Chem. Soc.*, 1979, **101**, 4128 (*synth, cryst struct, ir, pmr*)

$C_{22}H_{10}O_9Ru_3$ Ru-00556

Nonacarbonyl[μ^3-[η:η:η^3-3,5-cyclohexadiene-1,2-diylidene-(phenylmethylidyne)]]-μ-hydrotriruthenium

[μ-(*Benzylidene*-o-*phenylene*)]*nonacarbonyl*-μ-*hydrotriruthenium*, 8CI

[32216-10-7]

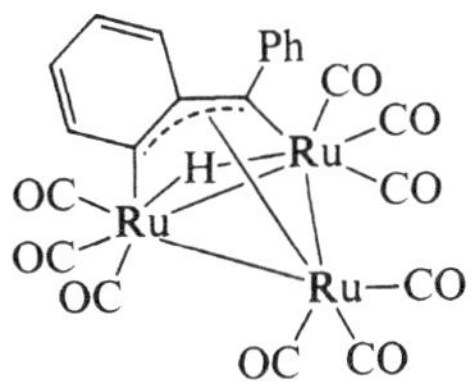

M 721.526

Red cryst. (pentane). Mp 139-42°.

Parkins, A.W. *et al*, *Angew. Chem., Int. Ed. Engl.*, 1970, **9**, 633 (*synth, ir, pmr, struct*)

$C_{22}H_{10}O_{15}Ru_6$ Ru-00557

[μ_6-(Carbon)]-μ-carbonyltetradecacarbonyl[μ-[(2,3-η:4,5-η)-2,4-hexadiene]]hexaruthenium

[μ_6-(*Carbon*)]*pentadecacarbonyl*(*2,4-hexadiene*)*hexaruthenium*

[76934-04-8]

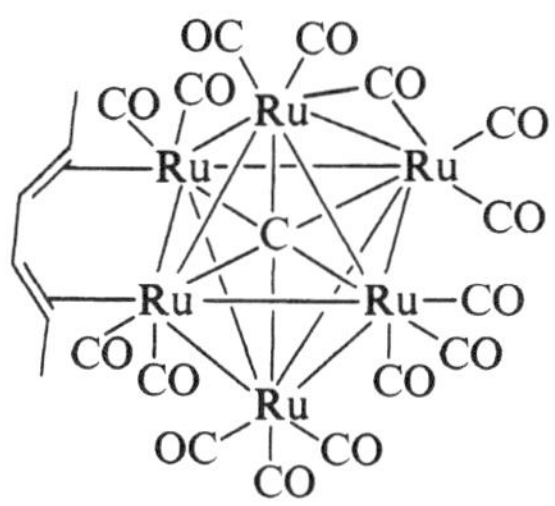

M 1120.732

No phys. description recorded. Isol. by tlc and characterised by ir, pmr and ms.

Jackson, P.F. *et al*, *J. Chem. Soc., Chem. Commun.*, 1980, 1190 (*synth, ms, ir, pmr*)

$C_{22}H_{14}O_4Ru_2$ Ru-00558

Di-μ-carbonyldicarbonylbis[(1,2,3,3*a*,7*a*-η)-1*H*-inden-1-yl-]diruthenium, 9CI

Tetracarbonyldi-π-indenyldiruthenium. Bis(indenyldicarbonylruthenium)

[60687-59-4]

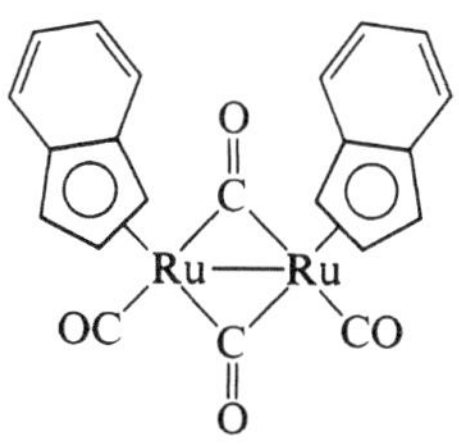

M 544.490

Exists in soln. as mixt. of bridged and non-bridged forms. Yellow cryst. (C_6H_6/hexane). Mp 246-7° dec.

McArdle, P. *et al*, *J. Chem. Soc.* (*A*), 1970, 2128 (*synth, ir*)

Gansow, O.A. *et al*, *J. Am. Chem. Soc.*, 1976, **98**, 5817 (*synth, pmr, cmr*)

Eisenstadt, A. *et al*, *J. Chem. Soc., Chem. Commun.*, 1982, 642 (*synth*)

$C_{22}H_{14}O_9Ru_4$ Ru-00559

Nonacarbonyl[μ_3-(4,6,8-trimethylazulene)]tetraruthenium, 8CI

4,6,8-Trimethylazulenetetraruthenium nonacarbonyl

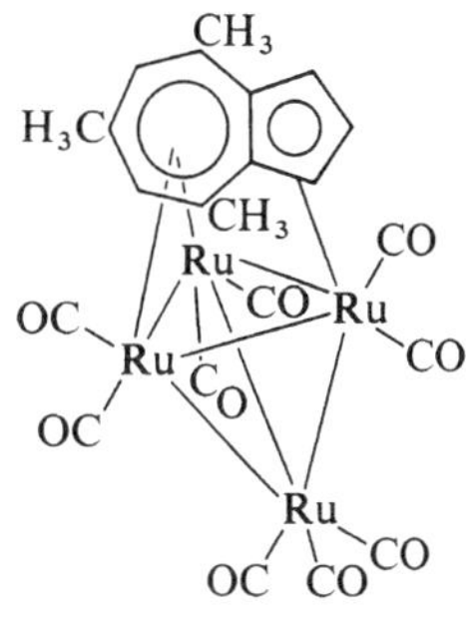

M 826.627

C_s symmetry. The azulene-Fe_3 bonding is delocalized. Bright-red cryst. (CH_2Cl_2/cyclohexane). Two cryst. forms, monoclinic and triclinic. Air-stable in solid state. Solns. dec. slowly on exposure to air.

Churchill, M.R. *et al*, *J. Am. Chem. Soc.*, 1968, **90**, 800 (*synth*)

Churchill, M.R. *et al*, *Inorg. Chem.*, 1969, **8**, 1956 (*synth, cryst struct*)

$C_{22}H_{15}O_4PRu$ Ru-00560

Tetracarbonyl(triphenylphosphine)ruthenium

[21192-23-4]

$Ph_3PRu(CO)_4$

M 475.402

Pale-yellow cryst. (C_6H_6/pentane). Mp 130° dec.

Johnson, B.F.G. *et al*, *J. Chem. Soc., Dalton Trans.*, 1975, 1876 (*synth, bibl*)

Yarrow, P. *et al*, *J. Organomet. Chem.*, 1981, **214**, 115 (*synth, pmr*)

Graaf, J.L. *et al*, *Organometallics*, 1982, **1**, 837 (*ir*)

$C_{22}H_{15}O_4RuSb$ Ru-00561

Tetracarbonyl(triphenylstibine)ruthenium, 10CI

[71356-99-5]

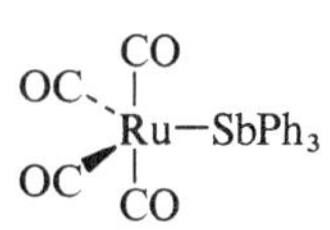

M 566.178

Small yellow needles (hexane). Mp 91-2°.

Forbes, E.J. *et al*, *J. Chem. Soc., Dalton Trans.*, 1979, 879 (*synth, ir, struct, ms*)

$C_{22}H_{16}O_{10}Ru_4$ Ru-00562

Decacarbonyl-2,8-cyclododecadiene-3,7-diyl-1-ylidenetetraruthenium, 9CI
Decacarbonyl(μ_4-π_2-cyclododecatrienylene)tetraruthenium, 8CI
[37279-20-2]

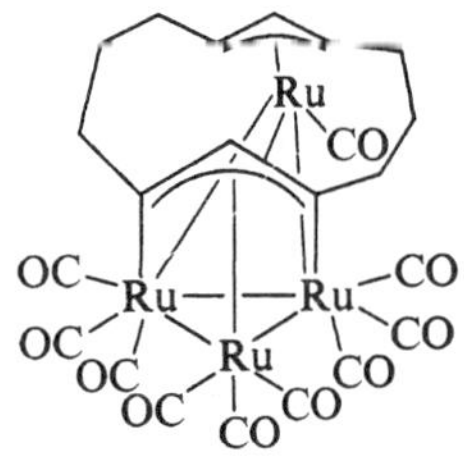

M 844.642
Dark-red plate-like cryst. (pentane/Et_2O). Mp 143° dec. Sublimes with some decomposition.

Bruce, M.I. *et al*, *J. Chem. Soc., Dalton Trans.*, 1972, 1293 (*synth, ir, pmr, ms*)
Belford, R. *et al*, *J. Chem. Soc., Dalton Trans.*, 1972, 2425 (*cryst struct*)

$C_{22}H_{16}O_{15}Ru_6S_3$ Ru-00563

[μ_6-(Carbon)]pentadecacarbonyltris[μ-(ethanethiolato)]-μ-hydrohexaruthenium
[74504-98-6]

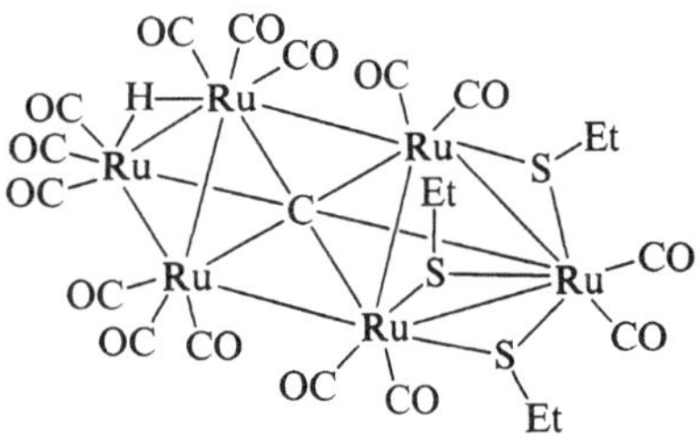

M 1222.959
Orange-red monoclinic cryst.

Johnson, B.F.G. *et al*, *J. Organomet. Chem.*, 1980, **185**, C17 (*synth, cryst struct*)

$C_{22}H_{18}FeRu$ Ru-00564

(Ruthenocenylethynyl)ferrocene
π-Cyclopentadienyl[(ferrocenylethynyl)-π-cyclopentadienyl]ruthenium, 8CI. *(Ferrocenylethynyl)ruthenocene. Ferrocenylruthenocenylacetylene*
[12337-26-7]

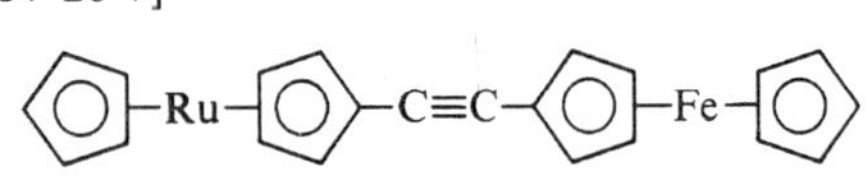

M 439.301
Orange cryst. by subl. Mp 259.5-261°. Bp_1 180° subl.

Rausch, M.D. *et al*, *J. Org. Chem.*, 1969, **34**, 1974 (*synth, ms, pmr*)

$C_{22}H_{18}O_6Ru_3$ Ru-00565

Di-μ-carbonyltetracarbonyl-[μ-[(1,2,6,7-η:2,3,4,5,6-η)-2,4,6-cyclooctatrien-1-yl]][(1,2,3,3a,6a-η)-1,4,5,6-tetrahydro-1-pentalenyl]triruthenium, 9CI
Pentacarbonyl(1,3,5-cyclooctatrienyl)(tetrahydropentalenyl)triruthenium
[54469-35-1]

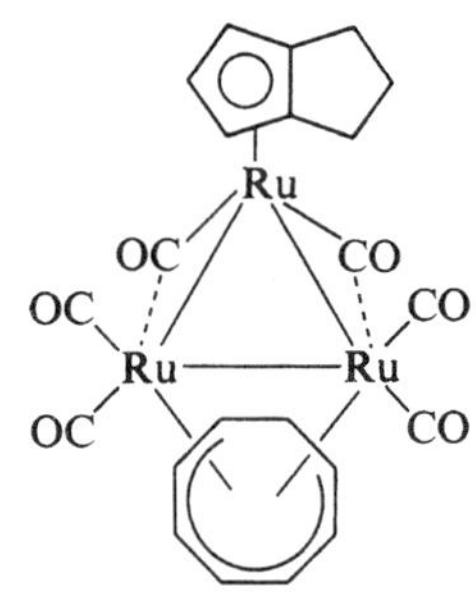

M 681.591
Red cryst. Mp 194°.

Bau, R. *et al*, *J. Organomet. Chem.*, 1974, **82**, C43 (*synth, ir, pmr, cryst struct*)

$C_{22}H_{20}B_2Ru$ Ru-00566

Bis[(1,2,3,4,5,6-η)-1-phenylboratabenzene]ruthenium, 9CI
Bis(1-phenylborinato)ruthenium
[60552-36-5]
As Bis[(1,2,3,4,5,6-η)-1-methylboratabenzene]ruthenium, Ru-00266 with

R = Ph

M 407.090
Air stable, colourless needles (pentane). Mp 136-136.5°. $Bp_{0.01}$ 140-60° subl.

Herberich, G.E. *et al*, *Chem. Ber.*, 1976, **109**, 2382 (*synth, pmr, nmr, ms*)

$C_{22}H_{20}NO_2PRu$ Ru-00567

Carbonylnitrosyl(η^3-2-propenyl)(triphenylphosphine)ruthenium
η^3-Allylcarbonylnitrosyl(triphenylphosphine)ruthenium
[65230-39-9]

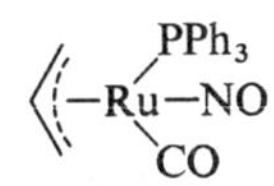

M 462.449
Orange soln. in THF, C_6H_6 or CH_2Cl_2.

Schoonover, M.W. *et al*, *J. Am. Chem. Soc.*, 1977, **99**, 8371 (*synth*)

$C_{22}H_{22}O_4Ru_2$ Ru-00568

Di-μ-carbonyldicarbonylbis[(1,2,3,3a,7a-η)-4,5,6,7-tetrahydro-1H-inden-1-yl]diruthenium, 9CI

Tetracarbonylbis(η⁵-tetrahydroindenyl)diruthenium

[60674-61-5]

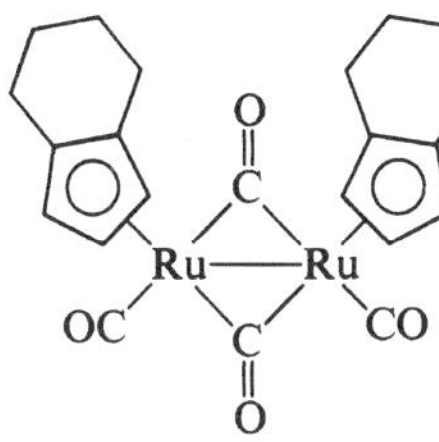

M 552.553

Yellow cryst. (CH_2Cl_2/pentane). Mp 123-4°.

Gansow, O.A. *et al*, *J. Am. Chem. Soc.*, 1976, **98**, 5817 (*synth, ir, pmr, cmr*)

$C_{22}H_{22}Ru^{\oplus\oplus}$ Ru-00569

(η⁶-Benzene)[[2.2]paracyclophane]ruthenium(2+)

(η⁶-Benzene)[(1,10,11,12,13,14-η)-tricyclo-[8.2.2.2^{4,7}]hexadeca-4,6,10,12,13,15-hexaene]ruthenium(2+), 10CI

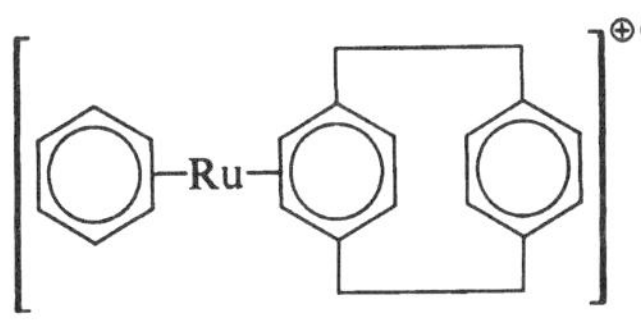

M 387.486 (ion)

Air-stable.

Bis(tetrafluoroborate): [77089-76-0].
$C_{22}H_{22}B_2F_8Ru$ M 561.093
White.

Laganis, E.D. *et al*, *Organometallics*, 1982, **1**, 1415 (*synth, pmr*)

$C_{22}H_{24}B_7O_2PRu$ Ru-00570

Dicarbonyl[η⁵-nonahydrodicarbanonaborato(4−)](triphenylphosphine)ruthenium, 10CI

closo-*6,6-Dicarbonyl-6-triphenylphosphine-6-ruthenanonahydro-2,3-dicarbanonaborane*

[76601-70-2]

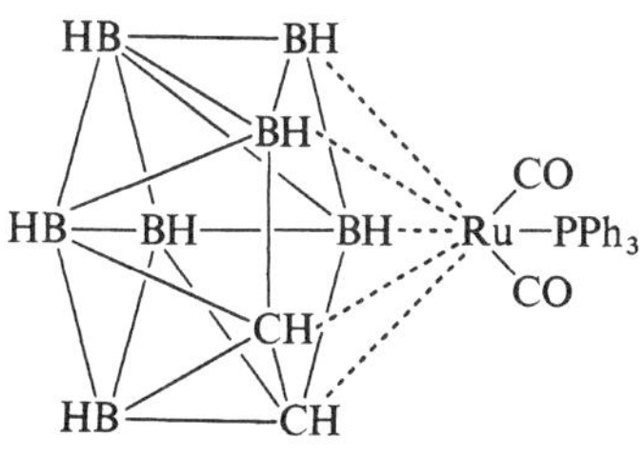

M 528.144

Yellow cryst. (C_6H_6/pentane).

Jung, C.W. *et al*, *J. Am. Chem. Soc.*, 1981, **103**, 811 (*synth, ir, pmr, nmr*)

$C_{22}H_{26}O_8Ru_2Si_4$ Ru-00571

[μ-[1,2,3,4-Benzenetetrayltetrakis[dimethylsilylene]]]octacarbonyldiruthenium, 9CI

[59432-90-5]

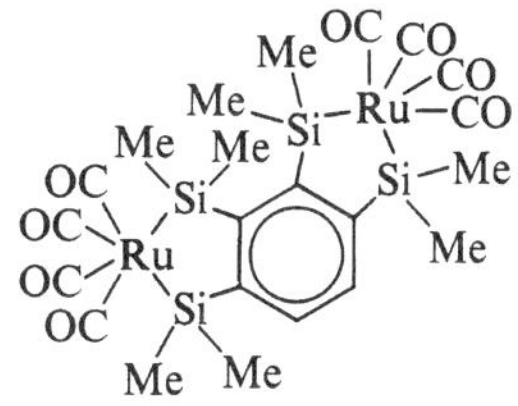

M 732.925

Colourless cryst. (hexane). Mp 220°.

Fink, W., *Helv. Chim. Acta*, 1976, **59**, 606 (*synth, pmr, ms*)

$C_{22}H_{26}O_8Ru_2Si_4$ Ru-00572

[μ-[1,2,4,5-Benzenetetrayltetrakis[dimethylsilylene]]]octacarbonyldiruthenium, 9CI

[59369-25-4]

M 732.925

Colourless cryst. (hexane). Mp 260-5°.

Fink, W., *Helv. Chim. Acta*, 1976, **59**, 606 (*synth, pmr, ms*)

$C_{22}H_{26}Ru^{\oplus\oplus}$ Ru-00573

[(1,2,3,4,5,6-η)-Hexamethylbenzene][(1,2,3,4,4a,8a-η)-naphthalene]ruthenium(2+), 10CI

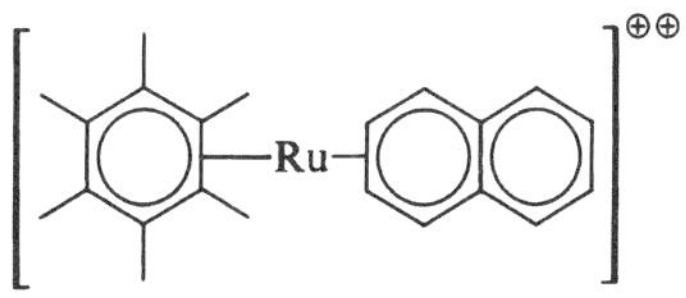

M 391.517 (ion)

Air-stable.

Bis(tetrafluoroborate): [71825-80-4].
$C_{22}H_{26}B_2F_8Ru$ M 565.125
No phys. props. reported.

Bennett, M.A. *et al*, *J. Organomet. Chem.*, 1979, **175**, 87 (*synth, pmr*)

$C_{22}H_{34}Cl_2Ru_2$ Ru-00574

Di-μ-chlorobis[(1,2,5,6-η)-1,5-cyclooctadiene]bis(η³-propenyl)diruthenium

Di-π-allyldi-μ-chlorodi-π-1,5-cyclooctadienyldiruthenium, 8CI

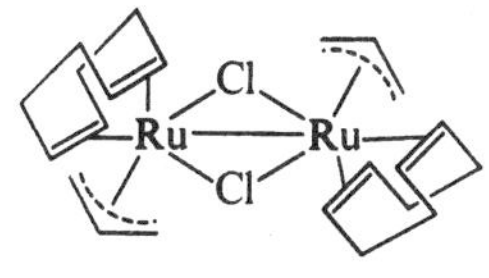

M 571.557

Golden-yellow prisms. Dec. at 135-40°.

Monomer: Chloro(1,5-cyclooctadiene)(η³-propenyl)ruthenium. Allylchloro(1,5-cyclooctadiene)ruthenium.
$C_{11}H_{17}ClRu$ M 285.778
Unknown.

Powell, J. *et al*, *J. Chem. Soc. (A)*, 1968, 159 (*synth, ir*)

$C_{23}H_{10}F_{10}ORu_2S_2$ Ru-00575

μ-Carbonylbis(η⁵-2,4-cyclopentadien-1-yl)bis[μ-(pentafluorobenzenethiolato)]diruthenium, 9CI

μ-Carbonyldi-π-cyclopentadienylbis(μ-pentafluorobenzenethiolato)diruthenium

[53612-32-1]

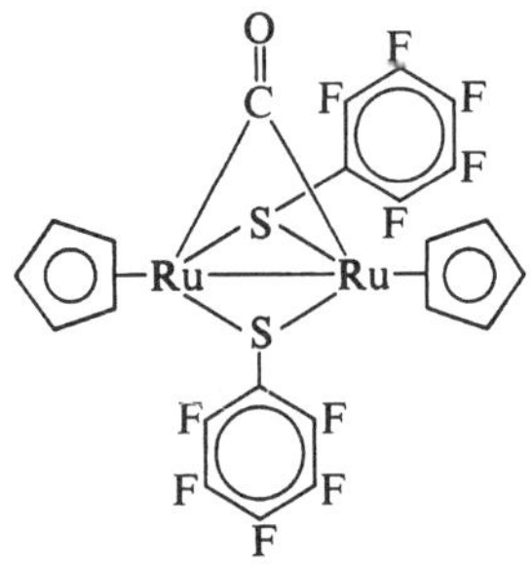

M 758.575

Deep-red cryst. (CH_2Cl_2/pet. ether). Mp 206°.

Knox, G.R. *et al*, *J. Organomet. Chem.*, 1974, **74**, 105 (*synth, ir, pmr*)

$C_{23}H_{14}O_5Ru_2$ Ru-00576

Pentacarbonyl[μ-[(η³:η⁵)-3,5-cyclohexadien-1-yl-2-ylidene-(phenylmethylidyne)-2,4-cyclopentadien-1-ylidene]]diruthenium, 9CI

Pentacarbonyl[μ-[η³:η⁵-2,4-cyclopentadien-1-ylidene-phenyl)methylbenzene]diruthenium, 9CI. Diphenylfulvenepentacarbonyldiruthenium. (6,6-Diphenylpentafulvene)pentacarbonyldiruthenium

[57177-16-9]

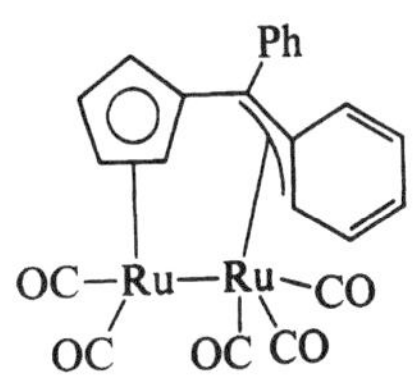

M 572.501

Orange cryst. Mp 159° dec.

Behrens, U. *et al*, *J. Organomet. Chem.*, 1975, **96**, 435 (*synth, ir, pmr, cryst struct*)

$C_{23}H_{20}BrO_2PRu$ Ru-00577

Bromodicarbonyl(η³-2-propenyl)(triphenylphosphine)ruthenium, 9CI

(η³-Allyl)bromodicarbonyl(triphenylphosphine)ruthenium

[56781-02-3]

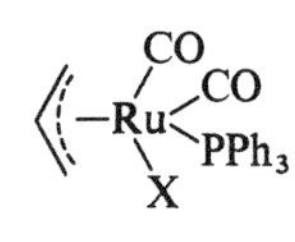

X = Br

M 540.358

Yellow cryst. (2,2,4-trimethylpentane). Mp 134-5°.

Sbrana, G. *et al*, *J. Chem. Soc., Dalton Trans.*, 1975, 754 (*synth*)

$C_{23}H_{20}ClO_2PRu$ Ru-00578

Dicarbonylchloro(η³-2-propenyl)(triphenylphosphine)ruthenium

(η³-Allyl)dicarbonylchloro(triphenylphosphine)ruthenium

[76736-33-9]

As Bromodicarbonyl(η³-2-propenyl)(triphenylphosphine)-ruthenium, Ru-00577 with

X = Cl

M 495.907

Cream-coloured cryst. (pet. ether). Mp 132-3°.

Barnard, C.F.J. *et al*, *J. Chem. Soc., Dalton Trans.*, 1980, 2418 (*synth*)

$C_{23}H_{20}FeORu$ Ru-00579

1-Ferrocenyl-3-ruthenocenyl-2-propen-1-one

(3-Ferrocenyl-3-oxo-1-propenyl)ruthenocene, 10CI

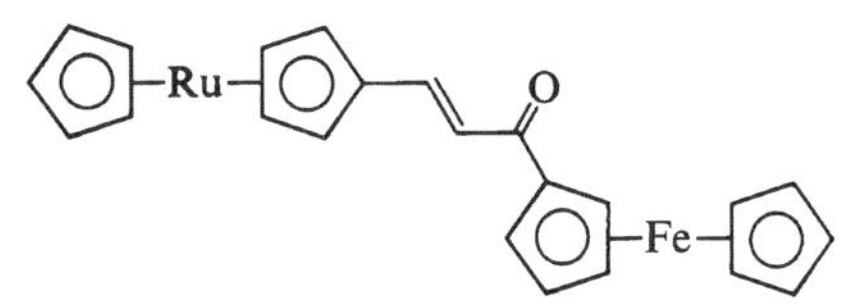

M 469.327

(*E*)-*form* [79643-37-1]

Cryst. (hexane/$CHCl_3$). Mp 257-9°.

Kamiyama, S. *et al*, *Bull. Chem. Soc. Jpn.*, 1981, **54**, 2079 (*synth, ir, pmr, uv*)

$C_{23}H_{20}FeORu$ Ru-00580

3-Ferrocenyl-1-ruthenocenyl-2-propen-1-one

(3-Ferrocenyl-1-oxo-2-propenyl)ruthenocene, 10CI

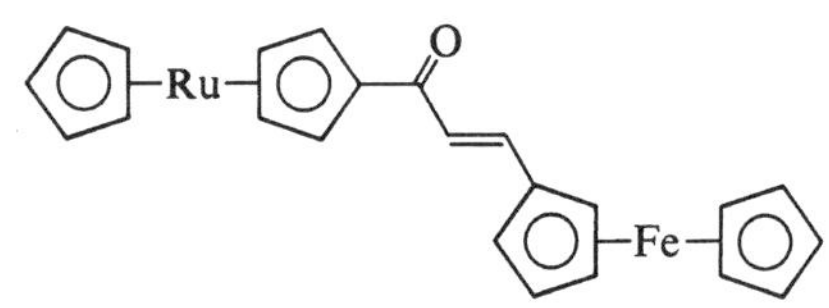

M 469.327

(*E*)-*form* [79354-76-0]

Cryst. (hexane/$CHCl_3$). Mp 218-20°.

Kamiyama, S. *et al*, *Bull. Chem. Soc. Jpn.*, 1981, **54**, 2079 (*synth, ir, pmr, uv*)

$C_{23}H_{20}ORu_2$ Ru-00581

1,3-Diruthenocenyl-2-propen-1-one

1,1″-(3-Oxo-1-propene-1,3-diyl)bisruthenocene, 10CI

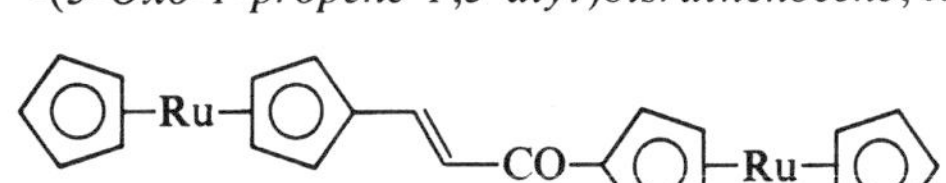

M 514.550

(*E*)-*form* [79354-75-9]

Cryst. (hexane/$CHCl_3$). Mp 255-8°.

Kamiyama, S. *et al*, *Bull. Chem. Soc. Jpn.*, 1981, **54**, 2079 (*synth, ir, pmr, ms, uv*)

$C_{23}H_{22}Ru^{\oplus\oplus}$ Ru-00582

[(1,2,3,4,4a,9a-η)-Anthracene][(1,2,3,4,5,6-η)-1,3,5-trimethylbenzene]ruthenium(2+), 10CI

(η⁶-Anthracene)(η⁶-mesitylene)ruthenium(2+)

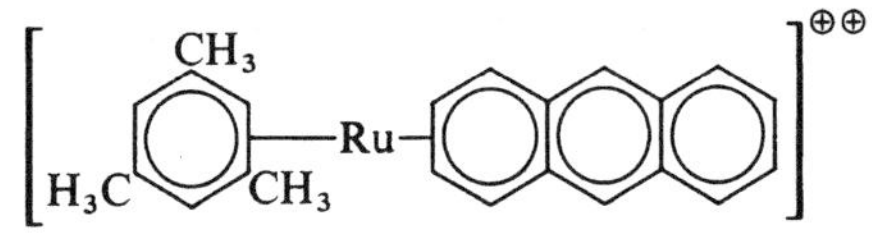

M 399.497 (ion)

Air-stable.

Bis(tetrafluoroborate): [71825-82-6].
$C_{23}H_{22}B_2F_8Ru$ M 573.104
No phys. props. reported.

Bennett, M.A. *et al*, *J. Organomet. Chem.*, 1979, **175**, 87 (*synth*, *pmr*)

$C_{24}H_{10}F_{10}O_2Ru_2S_2$ Ru-00583

μ-Carbonylcarbonylbis[η⁵-2,4-cyclopentadien-1-yl][μ-(pentafluorobenzenethiolato)](pentafluorobenzenethiolato)diruthenium

μ-Carbonylcarbonyldi-π-cyclopentadienyl[μ-(pentafluorobenzenethiolato)](pentafluorobenzenethiolato)diruthenium

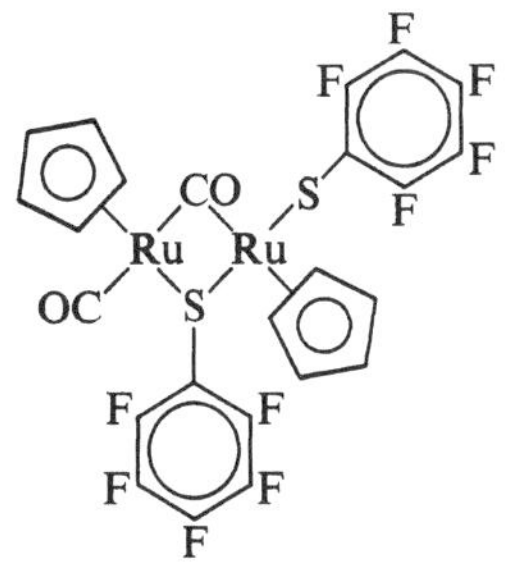

M 786.586

Red cryst. (CH_2Cl_2/pet. ether). Mp 221°.

Knox, G.R. *et al*, *J. Organomet. Chem.*, 1974, **74**, 105 (*synth*, *ir*, *pmr*)

$C_{24}H_{10}F_{10}O_2Ru_2S_2$ Ru-00584

Dicarbonylbis(η⁵-2,4-cyclopentadien-1-yl)bis[μ-(pentafluorobenzenethiolato)]diruthenium, 9CI

Dicarbonyldi-π-cyclopentadienylbis(μ-pentafluorobenzenethiolato)diruthenium

[53556-40-4]

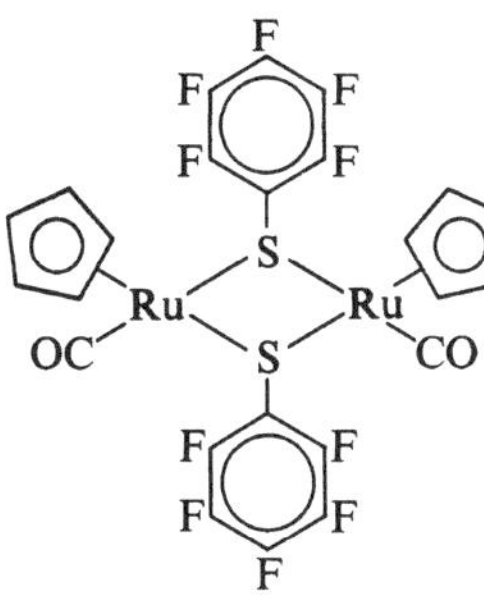

M 786.586

Obt. as an inseparable mixt. of isomers. Orange cryst. Mp 153°. Thermally isomerises to μ-Carbonylcarbonylbis[η⁵-2,4-cyclopentadien-1-yl][μ--(pentafluorobenzenethiolato)](pentafluorobenzenethiolato)diruthenium, Ru-00583 .

Knox, G.R. *et al*, *J. Organomet. Chem.*, 1974, **74**, 105 (*synth*, *ir*, *pmr*, *nmr*)

$C_{24}H_{12}O_{14}Ru_6$ Ru-00585

μ_6-Carbon-μ-carbonyltridecacarbonyl(1,3,5-trimethylbenzene)hexaruthenium

μ_6-Carbon-μ-carbonyltridecacarbonyl(mesitylene)hexaruthenium, 8CI.

μ_6-Carbontetradecacarbonyl(mesitylene)hexaruthenium

[12278-41-0]

As (Benzene)[μ_6-(carbon)]-tetradecacarbonylhexaruthenium, Ru-00536 with

R = CH_3

M 1130.770

Deep-purple cryst. (toluene), stable in air. Sol. most org. solvs. Involatile.

Johnson, B.F.G. *et al*, *J. Chem. Soc. (A)*, 1968, 2865 (*synth*, *ir*, *pmr*, *ms*)
Lewis, J. *et al*, *Acc. Chem. Res.*, 1968, **1**, 245 (*ms*)
Mason, R. *et al*, *J. Chem. Soc., Chem. Commun.*, 1968, 468 (*cryst struct*)

$C_{24}H_{18}O_2Ru$ Ru-00586

1,1′-Dibenzoylruthenocene

Bis(benzoylcyclopentadienyl)ruthenium

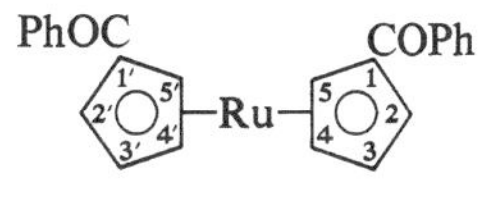

M 439.475

Golden-yellow leaflets (pet. ether). Mp 126-126.5°.

Rausch, M.D. *et al*, *J. Am. Chem. Soc.*, 1960, **82**, 76 (*synth*, *ir*)
Rausch, M.D. *et al*, *J. Org. Chem.*, 1963, **28**, 3225 (*pmr*)

$C_{24}H_{18}O_2Ru$ Ru-00587

1,2-Dibenzoylruthenocene

Cyclopentadienyl(dibenzoylcyclopentadienyl)ruthenium

M 439.475

Yellow needles (C_6H_6/hexane). Mp 141.8-142.4°.

Bublitz, D.E. *et al*, *J. Am. Chem. Soc.*, 1962, **84**, 1845 (*synth*)
Rausch, M.D. *et al*, *J. Org. Chem.*, 1963, **28**, 3225 (*pmr*)

$C_{24}H_{18}Ru_2$ Ru-00588

1,4-Diruthenocenyl-1,3-butadiyne

[μ-(Butadiynylenedi-π-cyclopentadienylene)]di-π-cyclopentadienyldiruthenium, 8CI. *1,3-Butynediylbisruthenocene*

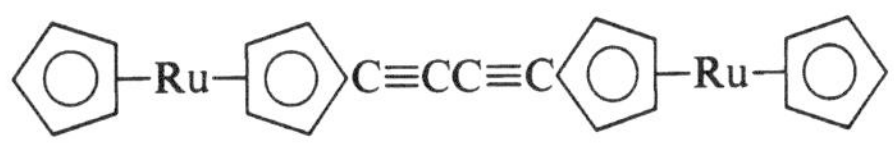

M 508.546

Cryst. (pet. ether). Mp 230-2°.

Hofer, O. *et al*, *J. Organomet. Chem.*, 1968, **13**, 443 (*synth*, *ir*)

$C_{24}H_{20}BrOPRu$ Ru-00589

Bromocarbonyl(η⁵-2,4-cyclopentadien-1-yl)(triphenylphosphine)ruthenium

Cyclopentadienylcarbonyl(triphenylphosphine)ruthenium bromide. (η⁵-Cyclopentadienyl)(bromo)carbonyl(triphenylphosphine)ruthenium

[32611-27-1]

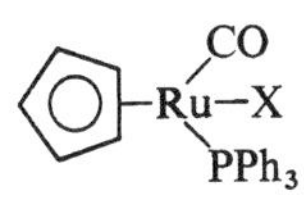

X = Br

M 536.369

Orange cryst. (CH_2Cl_2/hexane). Mp 236-8°.

Brown, D.A. *et al*, *Inorg. Chim. Acta*, 1970, **4**, 621 (*synth*, *ir*)

$C_{24}H_{20}BrO_4PRu$ **Ru-00590**

Bromocarbonyl(η^5-2,4-cyclopentadien-1-yl)(triphenyl phosphite-*P*)ruthenium

Bromocarbonyl-π-cyclopentadienyl(phosphorous acid)-ruthenium triphenyl ester, 8CI. *(η^5-Cyclopentadienyl)-bromocarbonyl(triphenyl phosphite)ruthenium. Cyclopentadienylcarbonyl(triphenyl phosphite)ruthenium bromide*

[32613-42-6]

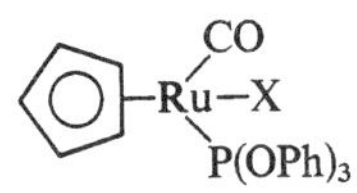

X = Br

M 584.367

Yellow-orange cryst. (CH_2Cl_2/hexane). Mp 125-7° dec.

Brown, D.A. *et al*, *Inorg. Chim. Acta*, 1970, **4**, 621 (*synth, ir*)

$C_{24}H_{20}ClOPRu$ **Ru-00591**

Carbonylchloro(η^5-2,4-cyclopentadien-1-yl)(triphenylphosphine)ruthenium

Cyclopentadienylcarbonyl(triphenylphosphine)ruthenium chloride. (η^5-Cyclopentadienyl)carbonyl(chloro)-(triphenylphosphine)ruthenium

[32613-25-5]

As Bromocarbonyl(η^5-2,4-cyclopentadien-1-yl)-(triphenylphosphine)ruthenium, Ru-00589 with

X = Cl

M 491.918

Orange cryst. (CH_2Cl_2/hexane). Mp 135-7°, 249-52° dec.

Brown, D.A. *et al*, *Inorg. Chim. Acta*, 1970, **4**, 621 (*synth, ir*)
Blackmore, T. *et al*, *J. Chem. Soc.* (*A*), 1971, 2376 (*synth, ir, pmr*)

$C_{24}H_{20}ClO_4PRu$ **Ru-00592**

Carbonylchloro(η^5-2,4-cyclopentadien-1-yl)(triphenyl phosphite-*P*)ruthenium

Carbonylchloro-π-cyclopentadienyl(phosphorous acid)-ruthenium triphenyl ester, 8CI. *(η^5-Cyclopentadienyl)-carbonylchloro(triphenyl phosphite)ruthenium. Cyclopentadienylcarbonyl(triphenyl phosphite)ruthenium chloride*

[32611-29-3]

As Bromocarbonyl(η^5-2,4-cyclopentadien-1-yl)(triphenyl phosphite-*P*)ruthenium, Ru-00590 with

X = Cl

M 539.916

Yellow-orange cryst. (CH_2Cl_2/hexane). Mp 147-9°.

Brown, D.A. *et al*, *Inorg. Chim. Acta*, 1970, **4**, 621 (*synth, ir*)

$C_{24}H_{20}IOPRu$ **Ru-00593**

Carbonyl(η^5-2,4-cyclopentadien-1-yl)iodo(triphenylphosphine)ruthenium

Cyclopentadienylcarbonyl(triphenylphosphine)ruthenium iodide. (η^5-Cyclopentadienyl)carbonyl(iodo)(triphenylphosphine)ruthenium

[32611-25-9]

As Bromocarbonyl(η^5-2,4-cyclopentadien-1-yl)-(triphenylphosphine)ruthenium, Ru-00589 with

X = I

M 583.370

Orange cryst. (CH_2Cl_2/hexane). Mp 190-4° dec.

Brown, D.A. *et al*, *Inorg. Chim. Acta*, 1970, **4**, 621 (*synth, ir*)

$C_{24}H_{20}IO_4PRu$ **Ru-00594**

Carbonyl(η^5-2,4-cyclopentadien-1-yl)iodo(triphenyl phosphite-*P*)ruthenium

Carbonyl-π-cyclopentadienyliodo(phosphorous acid)-ruthenium triphenyl ester, 8CI. *(η^5-Cyclopentadienyl)-carbonyl(iodo)(triphenyl phosphite)ruthenium. Cyclopentadienylcarbonyl(triphenyl phosphite)ruthenium iodide*

[31854-77-0]

As Bromocarbonyl(η^5-2,4-cyclopentadien-1-yl)(triphenyl phosphite-*P*)ruthenium, Ru-00590 with

X = I

M 631.368

Yellow-orange cryst. (CH_2Cl_2/hexane). Mp 170-2°.

Brown, D.A. *et al*, *Inorg. Chim. Acta*, 1970, **4**, 428, 621 (*synth, ir*)

$C_{24}H_{20}O_2Ru_2S_2$ **Ru-00595**

Bis[μ-(benzenethiolato)]dicarbonylbis(η^5-2,4-cyclopentadien-1-yl)diruthenium

Bis(μ-benzenethiolato)dicarbonyldi-π-cyclopentadienyldiruthenium

As Dicarbonylbis(η^5-2,4-cyclopentadien-1-yl)bis[μ-(methanethiolato)]diruthenium, Ru-00353 with

R = Ph

M 606.681

Two isomers known, *cis*- and *trans*- w.r.t. carbonyl/Cp groups. Unlike Bis[μ-(benzeneselenolato)] dicarbonylbis(η^5-2,4-cyclopentadien-1-yl)diruthenium, Ru-00596 no further stereoisomerism due to R-group configs. characterised even by nmr at −100°.

***cis*-form** [69047-78-5]

Orange cryst. (CH_2Cl_2/hexane). Mp 193-5°. Exhibits one ν_{CO} ir band in soln.

***trans*-form** [69088-23-9]

Red cryst. (CH_2Cl_2/hexane). Mp 89-90°. Thermodynamically most stable form; undergoes ready ring inversion in soln. at −100°; converted to *cis*-form by uv irradiation. Exhibits two ν_{CO} ir bands in soln.

Killops, S.D. *et al*, *J. Chem. Soc., Dalton Trans.*, 1978, 1260 (*synth, ir, pmr*)

$C_{24}H_{20}O_2Ru_2Se_2$ Ru-00596

Bis[μ-(benzeneselenolato)]dicarbonylbis(η^5-2,4-cyclopentadien-1-yl)diruthenium

Bis(μ-benzeneselenolato)dicarbonyldi-π-cyclopentadienyldiruthenium

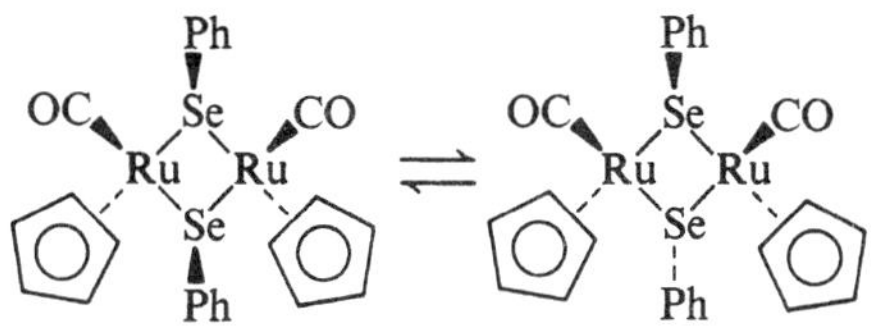

cis-form

M 700.481

cis-form [69047-79-6]

Orange cryst. (CH_2Cl_2/hexane). Mp 195-6°. Inseparable mixt. of isomers showing *cis-trans* Ph group isomerism, characterised by nmr. Less thermodynamically stable than *trans*-form. Exhibits one ν_{CO} band in soln.

trans-form [69088-24-0]

Red cryst. (CH_2Cl_2/hexane). Inseparable mixt. of *cis*- and *trans*- Ph-group isomers. Thermodynamically more stable form. Exhibits stereochemical non-rigidity in soln. Two ν_{CO} ir bands in soln.

Killops, S.D. *et al, J. Chem. Soc., Dalton Trans.*, 1978, 1260 (*synth, ir, pmr*)

$C_{24}H_{20}O_4Ru_4$ Ru-00597

Tetra-μ_3-carbonyltetrakis(η^5-2,4-cyclopentadien-1-yl)tetraruthenium

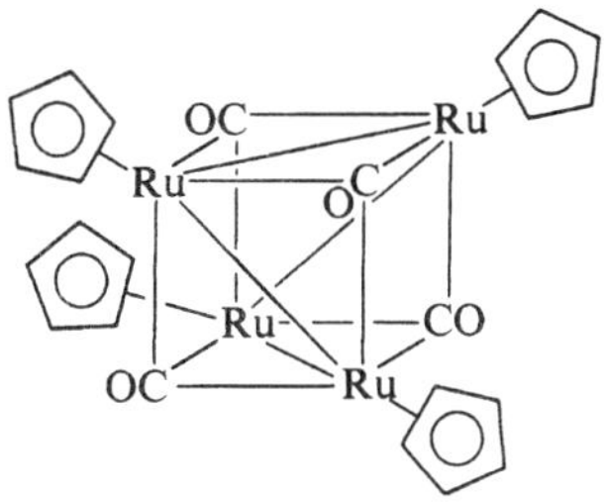

M 776.700

Deep-purple cryst. Mp >350°.

Blackmore, T. *et al, J. Chem. Soc. (A)*, 1968, 2931 (*synth, ir, ms*)

$C_{24}H_{21}Br_2PRu$ Ru-00598

(η^6-Benzene)dibromo(triphenylphosphine)ruthenium, 9CI

[39732-37-1]

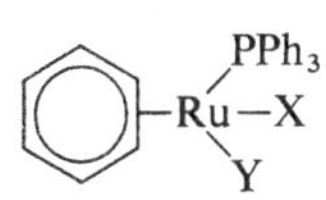

X = Y = Br

M 601.282

Orange-red cryst. (MeCN/Et_2O). Mp 202° dec.

Zelonka, R.A. *et al, J. Organomet. Chem.*, 1972, **44**, 383 (*nmr*)
Zelonka, R.A. *et al, Can. J. Chem.*, 1972, **50**, 3063 (*synth, pmr*)
Games, D.E. *et al, J. Organomet. Chem.*, 1980, **193**, 229 (*ms*)

$C_{24}H_{21}Cl_2PRu$ Ru-00599

(η^6-Benzene)dichloro(triphenylphosphine)ruthenium, 10CI, 9CI

[39732-57-5]

As (η^6-Benzene)dibromo(triphenylphosphine)ruthenium, Ru-00598 with

X = Y = Cl

M 512.380

Red cryst. ($CHCl_3$/heptane). Mp 184-6°.

Zelonka, R.A. *et al, Can. J. Chem.*, 1972, **50**, 3063 (*synth, pmr*)
Zelonka, R.A. *et al, J. Organomet. Chem.*, 1972, **44**, 383 (*nmr*)
Bennett, M.A. *et al, J. Chem. Soc., Dalton Trans.*, 1974, 233 (*synth, pmr, ir*)
Games, D.E. *et al, J. Organomet. Chem.*, 1980, **193**, 229 (*ms*)

$C_{24}H_{21}I_2PRu$ Ru-00600

(η^6-Benzene)diiodo(triphenylphosphine)ruthenium, 9CI

[39732-36-0]

As (η^6-Benzene)dibromo(triphenylphosphine)ruthenium, Ru-00598 with

X = Y = I

M 695.283

Orange-red cryst. (MeCN/Et_2O). Mp 196° dec.

Zelonka, R.A. *et al, J. Organomet. Chem.*, 1972, **44**, 383 (*nmr*)
Zelonka, R.A. *et al, Can J. Chem.*, 1972, **50**, 3063 (*synth, pmr*)

$C_{24}H_{21}OPRu$ Ru-00601

Carbonyl(η^5-2,4-cyclopentadien-1-yl)hydro(triphenylphosphine)ruthenium, 9CI

(η^5-Cyclopentadienyl)carbonyl(triphenylphosphine)ruthenium hydride. (η^5-Cyclopentadienyl)carbonyl(hydrido)(triphenylphosphine)ruthenium

[57327-06-7]

As Bromocarbonyl(η^5-2,4-cyclopentadien-1-yl)(triphenylphosphine)ruthenium, Ru-00589 with

X = H

M 457.473

Pale-yellow cryst. (CH_2Cl_2/hexane). Mp 125-7°, 133-6°.

Humphries, A.P. *et al, J. Chem. Soc., Dalton Trans.*, 1975, 1710 (*synth, ir, pmr*)
Bruce, M.I. *et al, J. Chem. Soc., Dalton Trans.*, 1979, 906 (*synth*)

$C_{24}H_{22}Ru$ Ru-00602

Bis[1-phenyl-[(2,3,4,5,6-η)-cyclohexa-2,4-dien-6-yl]]ruthenium

Bis(1-phenylcyclohexadienyl)ruthenium, 8CI

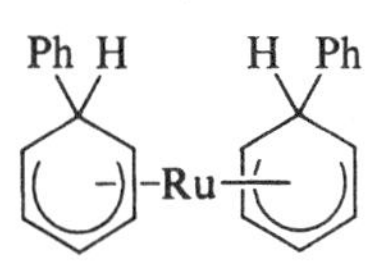

M 411.508

Bis-endo-phenyl-form

Yellow cryst. Mp 158°.

Jones, D. *et al, J. Chem. Soc.*, 1962, 4458 (*synth, ir, pmr*)

$C_{24}H_{24}GeIO_3PRu$ Ru-00603

Tricarbonyliodo(trimethylgermyl)(triphenylphosphine)ruthenium, 8CI

[30846-64-1]

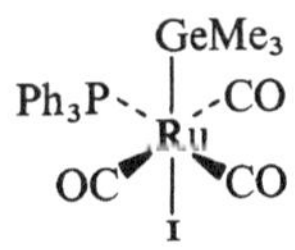

M 691.990

Orange cryst. (hexane). Mp 83-5°.

Stone, F.G.A. *et al*, *J. Chem. Soc. (A)*, 1971, 458 (*synth*, *ir*, *pmr*)

$C_{24}H_{24}IO_3PRuSi$ Ru-00604

Tricarbonyliodo(trimethylsilyl)(triphenylphosphine)ruthenium, 8CI

[30848-58-9]

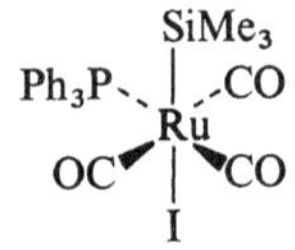

M 647.486

Yellow cryst. (CH_2Cl_2). Mp 89-91°.

Stone, F.G.A. *et al*, *J. Chem. Soc. (A)*, 1971, 458 (*synth*, *ir*, *pmr*)

$C_{24}H_{24}IO_3PRuSn$ Ru-00605

Tricarbonyliodo(trimethylstannyl)(triphenylphosphine)ruthenium, 8CI

[30846-65-2]

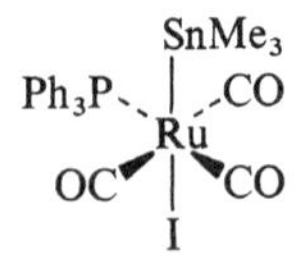

M 738.090

Yellow cryst. Mp 79-81°.

Stone, F.G.A. *et al*, *J. Chem. Soc. (A)*, 1971, 458 (*synth*, *ir*, *pmr*)

$C_{24}H_{25}O_3PRuSi$ Ru-00606

Tricarbonylhydro(trimethylsilyl)(triphenylphosphine)ruthenium, 8CI

[30848-61-4]

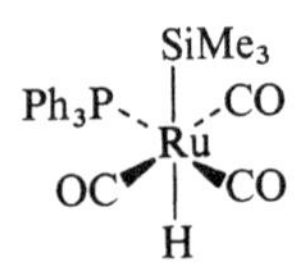

M 521.589

Pale-yellow cryst. (hexane). Mp 90-2°.

Stone, F.G.A. *et al*, *J. Chem. Soc. (A)*, 1971, 458 (*synth*, *ir*, *pmr*)

$C_{24}H_{25}O_3PRuSn$ Ru-00607

Tricarbonylhydro(trimethylstannyl)(triphenylphosphine)ruthenium, 8CI

[30846-67-4]

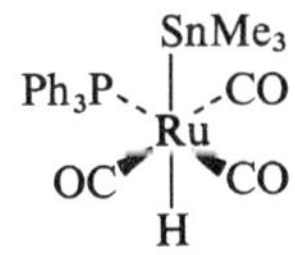

M 612.193

Pale-yellow. Characterised in soln. and not obt. pure in solid state.

Stone, F.G.A. *et al*, *J. Chem. Soc. (A)*, 1971, 458 (*synth*, *ir*, *pmr*)

$C_{24}H_{27}O_{23}P_3Ru_6$ Ru-00608

[μ_6-(Carbon)]-μ-carbonyltridecacarbonyltris(trimethyl phosphite-*P*)hexaruthenium

[μ_6-(Carbon)]-tetradecacarbonyltris(trimethylphosphite)hexaruthenium

[74535-09-4]

$Ru_6(CO)_{13}(\mu\text{-}CO)(\mu_6\text{-}C)[P(OMe)_3]_3$

M 1382.805

No phys. description recorded. Derived from μ_6-Carbon-μ-carbonylhexadecacarbonylhexaruthenium, Ru-00500 .

Johnson, B.F.G. *et al*, *J. Organomet. Chem.*, 1980, **185**, C17 (*synth*)

$C_{24}H_{30}O_4Ru_2$ Ru-00609

Di-μ-carbonyldicarbonylbis[(1,2,3,4,5-η)-1,2,3,4,5-pentamethyl-2,4-cyclopentadien-1-yl]diruthenium, 10CI

[70669-56-6]

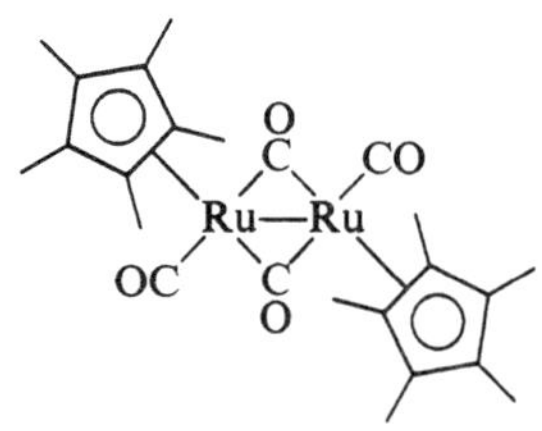

M 584.639

Orange-yellow cryst. (CH_2Cl_2/hexane). Mp 297°.

King, R.B. *et al*, *J. Organomet. Chem.*, 1979, **171**, 53 (*synth*, *ir*, *pmr*)

$C_{24}H_{32}N_4O_8Ru_4$ Ru-00610

Octacarbonylbis[μ-(η^4-*N,N'*-1,2-ethanediylidenebis[2-propanamine]-*N,N'*]]tetraruthenium

Octacarbonylbis[glyoxalbis(isopropylamine)]tetraruthenium

[78199-36-7]

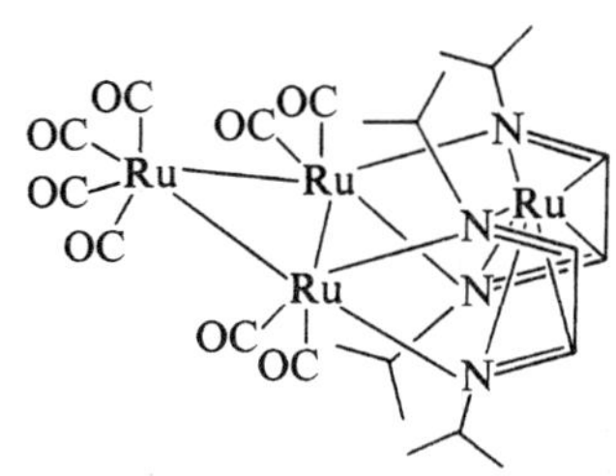

M 908.819

Dark-green cryst. (Et_2O or toluene).

Staal, L.H. *et al*, *J. Organomet. Chem.*, 1980, **199**, C13 (*cryst struct*)
Staal, L.H. *et al*, *Inorg. Chem.*, 1981, **20**, 3590 (*synth, ir, pmr, cryst struct*)

$C_{24}H_{34}Ru$ Ru-00611

[(1,2,3,4,5,6-η)-Hexamethylbenzene][(1,2,3,4-η)-1,2,3,4-tetramethyl-5,6-bis(methylene)-1,3-cyclohexadiene]-ruthenium

[83005-40-7]

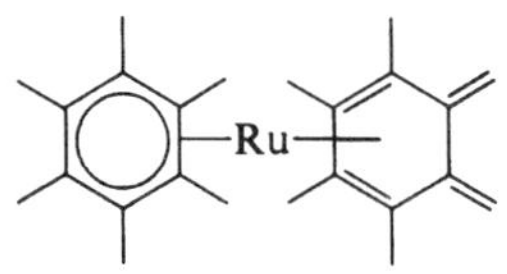

M 423.603
Yellow cryst. (toluene).

Hull, J.W. *et al*, *Organometallics*, 1982, **1**, 1716 (*synth, pmr, cmr, ms, struct*)

$C_{24}H_{36}Cl_4Ru_2$ Ru-00612

Di-μ-chlorodichlorobis[(1,2,3,4,5,6-η)-hexamethylbenzene]diruthenium

Tetrachlorobis(η⁶-hexamethylbenzene)diruthenium

[67421-02-7]

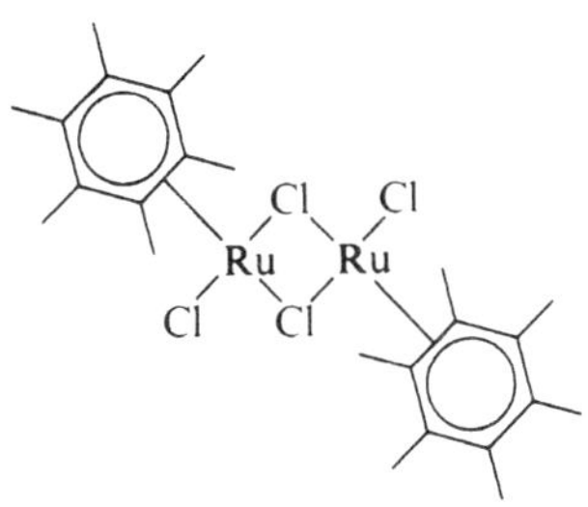

M 668.500
Bridged dimer. Hydrogenation catalyst precursor. Red-brown cryst. ($CHCl_3$ or CH_2Cl_2/hexane). Mp 270°.

Monomer: Dichloro(hexamethylbenzene)ruthenium.
$C_{12}H_{18}Cl_2Ru$ M 334.250
Unknown.

Bennett, M.A. *et al*, *Inorg. Chem.*, 1980, **19**, 1014 (*synth, ir, pmr*)
Grundy, S.L. *et al*, *J. Chem. Soc., Chem. Commun.*, 1982, 379 (*synth*)
Inorg. Synth., 1982, **21**, 75 (*synth, ir, pmr*)

$C_{24}H_{36}Ru^{\oplus\oplus}$ Ru-00613

Bis[(1,2,3,4,5,6-η)-hexamethylbenzene]ruthenium(2+), 10CI

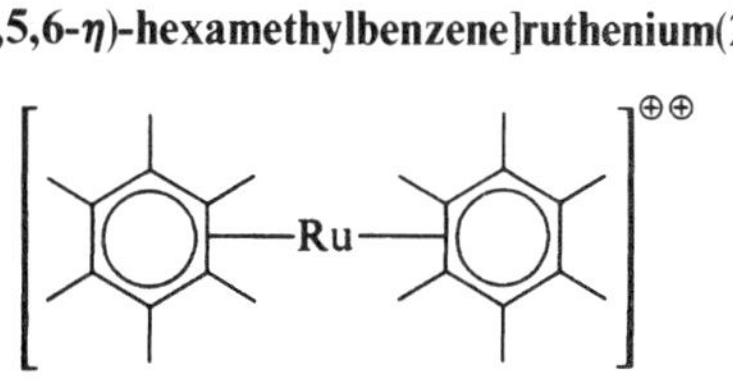

M 425.618 (ion)
Air-stable.

Bis(tetrafluoroborate): [71861-31-9].
$C_{24}H_{36}B_2F_8Ru$ M 599.226
White cryst. (DMSO/MeOH).

Fischer, E.O. *et al*, *Chem. Ber.*, 1970, **103**, 162 (*synth, pmr*)

Bennett, M.A. *et al*, *J. Organomet. Chem.*, 1979, **175**, 87 (*synth, pmr*)

$C_{24}H_{36}Ru$ Ru-00614

[(1,2,3,4-η)-Hexamethylbenzene][(1,2,3,4,5,6-η)-hexamethylbenzene]ruthenium, 9CI

Bis(hexamethylbenzene)ruthenium

[32732-05-1]

As [(1,2,3,4-η)-Benzene](η⁶-benzene)ruthenium, Ru-00250 with

$$R = CH_3$$

M 425.618
Fluxional molecule. Homogeneous hydrogenation catalyst. Air-sensitive orange needles (hexane). Mp 160° dec. Bp 90° subl. *in vacuo*.

Fischer, E.O. *et al*, *Chem. Ber.*, 1970, **103**, 162 (*synth, pmr*)
Muetterties, E.L. *et al*, *Acc. Chem. Res.*, 1970, **12**, 324 (*rev*)
Huttner, G. *et al*, *Acta Crystallogr., Sect. B*, 1972, **28**, 2049 (*struct*)
Johnson, J.W. *et al*, *J. Am. Chem. Soc.*, 1977, **99**, 7395 (*use*)
Darensbourg, M.Y. *et al*, *J. Am. Chem. Soc.*, 1978, **100**, 7425 (*pmr*)

$C_{24}H_{38}ClRu_2^{\oplus}$ Ru-00615

μ-Chlorobis[(1,2,3,4,5,6-η)-hexamethylbenzene]di-μ-hydro-diruthenium(1+), 10CI

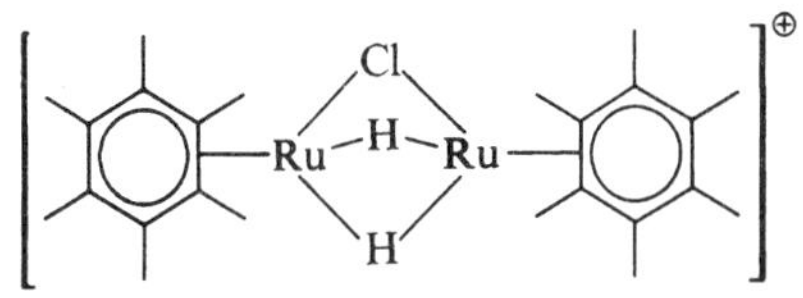

M 564.157 (ion)
Catalyst for hydrogenation of aromatics.

Chloride: [71933-81-8].
$C_{24}H_{38}Cl_2Ru_2$ M 599.610
Purple. Air-stable.

Bennett, M.A. *et al*, *J. Chem. Soc., Chem. Commun.*, 1979, 312 (*synth, use*)
Bennett, M.A., *CHEMTECH*, 1980, **10**, 444 (*synth, use*)

$C_{24}H_{62}O_{12}P_4Ru$ Ru-00616

Dihydrotetrakis(triethyl phosphite-*P*)ruthenium, 9CI

[53495-34-4]

$$H_2Ru[P(OEt)_3]_4$$

M 767.712
Mod. air-sensitive white needles (EtOH).

Gerlach, D.H. *et al*, *J. Am. Chem. Soc.*, 1972, **94**, 4545 (*synth, ir*)
Meakin, P. *et al*, *J. Am. Chem. Soc.*, 1973, **95**, 75 (*pmr, nmr*)
Inorg. Synth., 1974, **15**, 38 (*synth, ir, bibl*)

$C_{25}H_{20}O_2PRu^{\oplus}$ Ru-00617

Dicarbonyl(η^5-2,4-cyclopentadien-1-yl)(triphenylphosphine)-ruthenium(1+), 9CI

(η^5-Cyclopentadienyl)dicarbonyl(triphenylphosphine)-ruthenium(1+)

[57327-08-9]

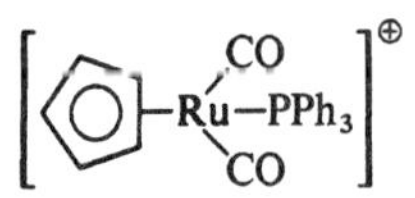

M 484.476 (ion)

Tetrafluoroborate: [57327-09-0].
$C_{25}H_{20}BF_4O_2PRu$ M 571.279
Yellow or white cryst. (Me_2CO). Mp 248-50° dec., 285°.

Humphries, A.P. *et al*, *J. Chem. Soc., Dalton Trans.*, 1975, 1710 (*synth, ir, pmr*)
Jungbauer, A. *et al*, *Z. Naturforsch.*, *B*, 1978, **33**, 1083 (*synth, ir, pmr*)

$C_{25}H_{22}NO_2PRu$ Ru-00618

(Aminocarbonyl)carbonyl(η^5-2,4-cyclopentadien-1-yl)(triphenylphosphine)ruthenium, 10CI

[72763-52-1]

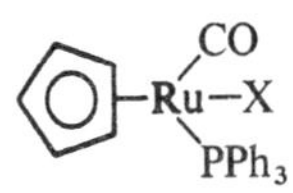

X = $CONH_2$

M 500.498
Colourless cryst. Mp 127° dec.

Behrens, H. *et al*, *Z. Naturforsch.*, *B*, 1979, **34**, 1477 (*synth, ir, pmr*)

$C_{25}H_{22}ORu_2$ Ru-00619

1,5-Diruthenocenyl-1,4-pentadien-3-one

1,1''-(3-Oxo-1,4-pentadien-1,5-diyl)bisruthenocene, 10CI

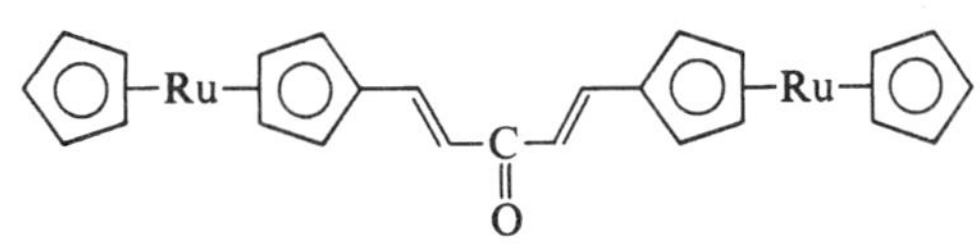

M 540.588

(E,E)-form [79354-77-1]
Cryst. ($CHCl_3/Et_2O$). Mp 131-4°.

Kamiyama, S. *et al*, *Bull. Chem. Soc. Jpn.*, 1981, **54**, 2079 (*synth, ir, pmr, ms, uv*)

$C_{25}H_{23}Cl_2PRu$ Ru-00620

Dichloro[(1,2,3,4,5,6-η)-1,3,5-cycloheptatriene](triphenylphosphine)ruthenium

[63630-82-0]

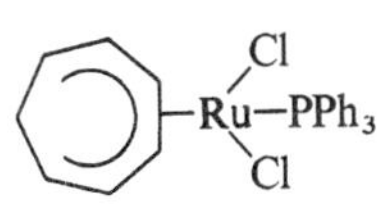

M 526.406
Orange cryst. (Me_2CO). Sol. $CHCl_3$.

Johnson, B.F.G. *et al*, *J. Chem. Soc., Dalton Trans.*, 1977, 719 (*synth, pmr, nmr*)

$C_{25}H_{23}OPRu$ Ru-00621

Carbonyl(η^5-2,4-cyclopentadien-1-yl)methyl(triphenylphosphine)ruthenium, 10CI

[71438-17-0]

As (Aminocarbonyl)carbonyl(η^5-2,4-cyclopentadien-1-yl)(triphenylphosphine)ruthenium, Ru-00618 with

X = Me

M 471.500
Pale-yellow cryst. (Et_2O/pet. ether). Mp 206-9°.

Bruce, M.I. *et al*, *J. Chem. Soc., Dalton Trans.*, 1979, 906 (*synth*)
Howell, J.A.S. *et al*, *J. Chem. Soc., Dalton Trans.*, 1980, 1845 (*synth, ir, pmr, cmr*)

$C_{25}H_{24}AuGeO_4PRu$ Ru-00622

Tetracarbonyl(trimethylgermyl)[(triphenylphosphine)aurio]-ruthenium, 8CI

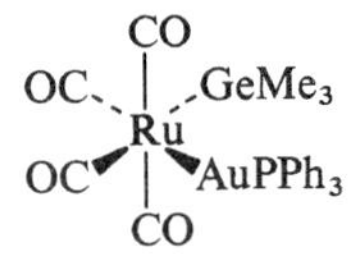

M 790.062

cis-form [34439-98-0]
Cream cryst. (CH_2Cl_2/hexane). Mp 95-105° dec.

Knox, S.A.R. *et al*, *J. Chem. Soc. (A)*, 1971, 2874 (*synth, ir, pmr*)

$C_{25}H_{24}AuO_4PRuSi$ Ru-00623

Tetracarbonyl(trimethylsilyl)[(triphenylphosphine)aurio]ruthenium, 8CI

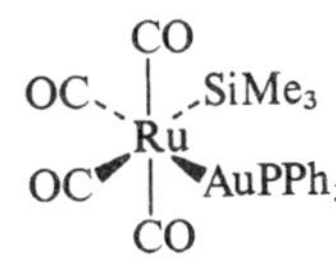

M 745.558

cis-form [26024-45-3]
Bronze cryst. Mp 106-7°.

Knox, S.A.R. *et al*, *J. Chem. Soc. (A)*, 1969, 2559 (*synth, ir*)

$C_{25}H_{24}ClPRu$ Ru-00624

(η^6-Benzene)chloromethyl(triphenylphosphine)ruthenium, 9CI

[40695-22-5]

As (η^6-Benzene)dibromo(triphenylphosphine)ruthenium, Ru-00598 with

X = Cl, Y = Me

M 491.961
Unstable yellow or orange cryst. ($MeCN/Et_2O$). Mp 139° dec.

Zelonka, R.A. *et al*, *J. Organomet. Chem.*, 1972, **44**, 383 (*synth, pmr, nmr*)
Bennett, M.A. *et al*, *J. Chem. Soc., Dalton Trans.*, 1974, 233 (*synth, pmr*)

$C_{25}H_{27}N_2OPRuS_4$ Ru-00625

Carbonylbis(dimethylcarbamodithioato-*S,S'*)(triphenylphosphine)ruthenium, 9CI

[56870-60-1]

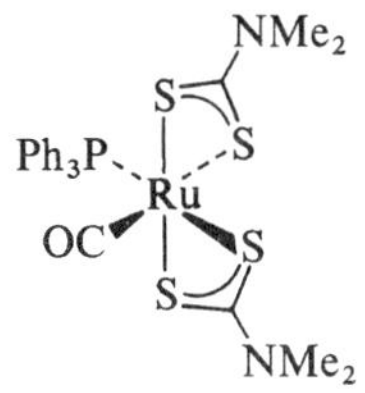

M 631.785

Dark-yellow cryst. Mp 248-50°.

Critchlow, P.B. *et al, J. Chem. Soc., Dalton Trans.*, 1975, 1367; *Inorg. Chem.*, 1978, **17**, 1902 (*synth, ir, pmr*)

$C_{25}H_{30}O_8Ru_3Si_3$ Ru-00626

Octacarbonyl[μ_3-[η^2:η^2:η^5-1,3,5-tris(trimethylsilyl)pentalene]]triruthenium

Octacarbonyl[μ_3-[η^2:η^2:η^5-*1,3,5-pentalenetriyltris*[*trimethylsilane*]]]*triruthenium*, 9CI

[56116-49-5]

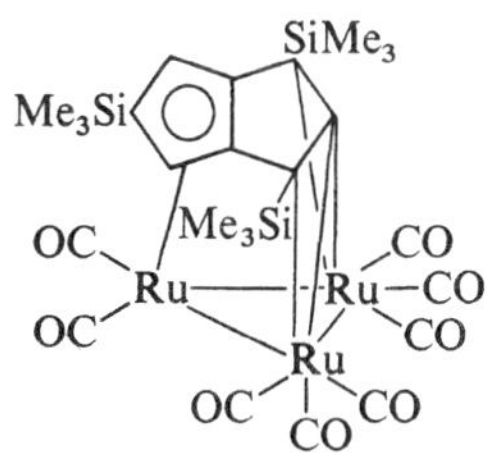

M 845.974

Face-bonded isomer; in soln. gives equilibrium mixture with Octacarbonyl[(1,2,3,3*a*,6*a*-η:3*a*,4,5,6,6*a*-η)-1,3,5-tris(trimethylsilyl)]pentalenetriruthenium, Ru-00627. Orange-red prisms. Sl. darker than Ru-00627. Mp 135°.

Knox, S.A.R. *et al, J. Chem. Soc., Dalton Trans.*, 1979, 1801 (*synth, ir, pmr, ms*)

Howard, J.A. *et al, J. Chem. Soc., Dalton Trans.*, 1979, 1812 (*cryst struct*)

$C_{25}H_{30}O_8Ru_3Si_3$ Ru-00627

Octacarbonyl[(1,2,3,3*a*,6*a*-η:3*a*,4,5,6,6*a*-η)-1,3,5-tris(trimethylsilyl)]pentalenetriruthenium

Octacarbonyl[μ-[[(*1,2,3,3*a*,6*a*-η:3*a*,4,5,6,6*a*-η-)-1,3,5-pentalenetriyl*]*tris*[*trimethylsilane*]]]*triruthenium*, 9CI

[72693-30-2]

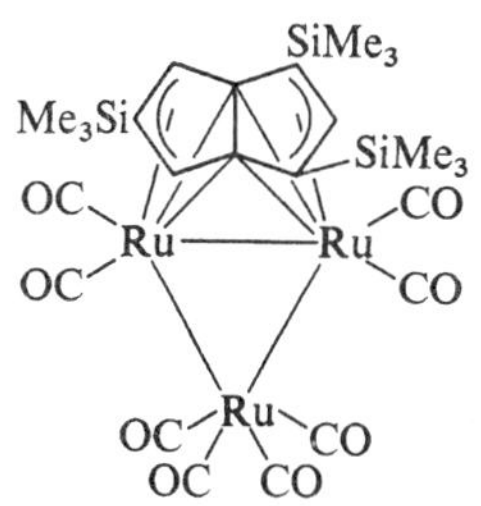

M 845.974

Edge bonded isomer; in soln. gives equilibrium mixture with Octacarbonyl[μ_3-[η^2:η^2:η^5-1,3,5-tris(trimethylsilyl)pentalene]]triruthenium, Ru-00626 . Yellow prisms (hexane). Mp 135°. Fluxional.

Knox, S.A.R. *et al, J. Chem. Soc., Dalton Trans.*, 1979, 1801 (*synth, ir, pmr, ms*)

Howard, J.A. *et al, J. Chem. Soc., Dalton Trans.*, 1979, 1812 (*cryst struct*)

$C_{26}H_{10}FeO_{12}Ru_3$ Ru-00628

Nonacarbonyl[μ_4-[(1-η:1,2-η:1,2-η:2-η)-1,2-diphenyl-1,2-ethenediyl]](tricarbonyliron)triruthenium

Dodecacarbonyl(*diphenylacetylene*)*irontriruthenium*

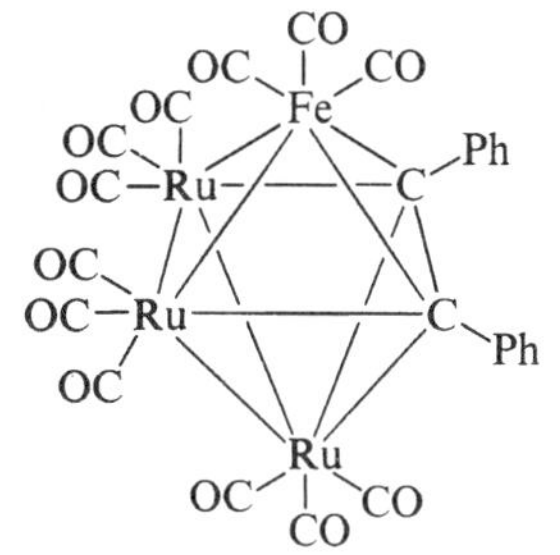

axial-form

M 873.415

axial-form [78248-93-8]

Black prismatic cryst. (hexane). Brown in soln.

equatorial-form [78248-94-9]

Black prismatic cryst. (hexane). Red-brown in soln. This isomer has the iron atom in the equatorial position.

Fox, J.R. *et al, Inorg. Chem.*, 1981, **20**, 3230 (*synth, cryst struct, ms, ir, pmr*)

$C_{26}H_{10}O_{12}Ru_4$ Ru-00629

Dodecacarbonyl[μ_4-[(1-η:1,2-η:1,2-η:2-η)-1,2-diphenyl-1,2-ethenediyl]]tetraruthenium, 9CI

μ-Diphenylacetylene-closo-*tetrakis*(*tricarbonylruthenium*)

[56802-15-4]

As Dodecacarbonyl[μ_4-(1-η:η^2:η^2:2-η-1,2-ethenediyl)]tetraruthenium, Ru-00325 with

R = Ph

M 918.638

Brown cryst.

Johnson, B.F.G. *et al, J. Organomet. Chem.*, 1975, **91**, C13 (*synth, ir, ms*)

Johnson, B.F.G. *et al, J. Chem. Soc., Dalton Trans.*, 1977, 1417 (*struct, cmr*)

$C_{26}H_{20}O_2Ru_2$ Ru-00630

1,1'':1',1'''-Bis(3-oxo-1-propene-1,3-diyl)bisruthenocene, 10CI

[*3.3*](*1,1'*)*Ruthenocenophane-2,14-diene-1,16-dione*

[79361-63-0]

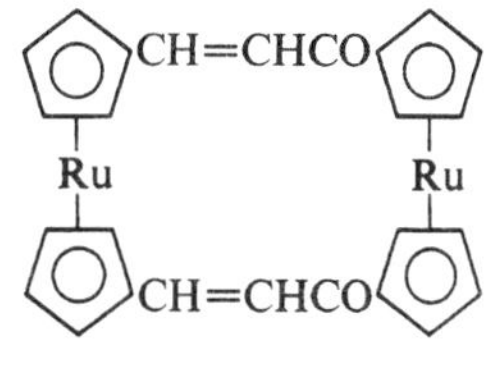

M 566.583

Cryst. ($CHCl_3$). Mp >300°.

Kamiyama, S. *et al*, *Bull. Chem. Soc. Jpn.*, 1981, **54**, 2079 (*synth, ir, pmr, ms, uv*)

$C_{26}H_{23}O_2PRu$ Ru-00631

Acetylcarbonyl(η^5-2,4-cyclopentadien-1-yl)(triphenylphosphine)ruthenium, 9CI

(η^5-*Cyclopentadienyl*)*acetylcarbonyl*(*triphenylphosphine*)*ruthenium*

[32610-57-4]

As Bromocarbonyl(η^5-2,4-cyclopentadien-1-yl)(triphenylphosphine)ruthenium, Ru-00589 with

X = Ac

M 499.510

Light-yellow cryst. (CH_2Cl_2/pet. ether).

Green, M.L.H. *et al*, *J. Chem. Soc. (A)*, 1971, 794 (*synth, ir, pmr*)
Van Doorn, J.A. *et al*, *J. Organomet. Chem.*, 1976, **105**, 245 (*pmr*)
Howell, J.A.S. *et al*, *J. Chem. Soc., Dalton Trans.*, 1980, 1845 (*synth, ir, pmr, cmr*)

$C_{26}H_{25}OPRu$ Ru-00632

Carbonyl(η^5-2,4-cyclopentadien-1-yl)ethyl(triphenylphosphine)ruthenium, 9CI

(η^5-*Cyclopentadienyl*)*carbonylethyl*(*triphenylphosphine*)*ruthenium.*
σ-*Ethyl*-π-*cyclopentadienylcarbonyl*(*triphenylphosphine*)*ruthenium*

[59140-13-5]

As Bromocarbonyl(η^5-2,4-cyclopentadien-1-yl)(triphenylphosphine)ruthenium, Ru-00589 with

X = Et

M 485.527

No phys. props. recorded.

Van Doorn, J.A. *et al*, *J. Organomet. Chem.*, 1976, **105**, 245 (*synth, pmr*)

$C_{26}H_{25}OPRuS_2$ Ru-00633

(η^5-2,4-Cyclopentadien-1-yl)(*O*-ethylcarbonodithioato-*S,S'*-)(triphenylphosphine)ruthenium, 10CI

[79408-47-2]

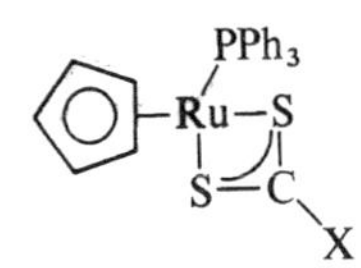

X = OEt

M 549.647

Yellow-orange solid. Mp 150-3°.

Wilczewski, T. *et al*, *J. Organomet. Chem.*, 1981, **215**, 87 (*synth, pmr, ir*)

$C_{26}H_{25}O_2PRu$ Ru-00634

Dicarbonyl[(1,2,3,4-η)-2,3-dimethyl-1,3-butadiene](triphenylphosphine)ruthenium

[75826-70-9]

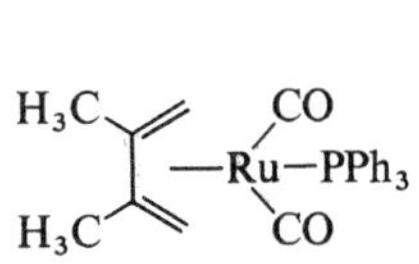

M 501.526

No phys. details recorded.

Amos, K. *et al*, *J. Organomet. Chem.*, 1980, **194**, C57 (*props*)

$C_{26}H_{25}PRu$ Ru-00635

(η^5-2,4-Cyclopentadien-1-yl)(η^3-2-propenyl)(triphenylphosphine)ruthenium, 10CI

π *Allyl* π *cyclopentadienyl*(*triphenylphosphine*)*ruthenium*

[74573-22-1]

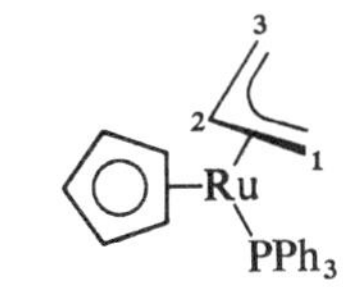

trans-form

M 469.527

Formed as a mixt. of isomers having C-2 either *cis*- or *trans*- with respect to PPh_3. Yellow powder (toluene/pentane).

Benn, R. *et al*, *Org. Magn. Reson.*, 1980, **14**, 435 (*pmr*)
Lehmkuhl, H. *et al*, *Justus Liebigs Ann. Chem.*, 1980, 754 (*synth, pmr, ms*)

$C_{26}H_{27}PRu$ Ru-00636

(η^6-Benzene)dimethyl(triphenylphosphine)ruthenium, 9CI

[52462-67-6]

As (η^6-Benzene)dibromo(triphenylphosphine)ruthenium, Ru-00598 with

X = Y = Me

M 471.543

Unstable brown oil. Corresponding Me_2PPh compd. is more stable.

Bennett, M.A. *et al*, *J. Chem. Soc., Dalton Trans.*, 1974, 233 (*synth, pmr*)

$C_{26}H_{27}PRu$ Ru-00637

Bis(η^4-1,3-butadiene)(triphenylphosphine)ruthenium

[62927-94-0]

M 471.543

Yellow cryst. (pet. ether). Mp 80-5° dec.

Cole-Hamilton, D.J. *et al*, *Nouv. J. Chim.*, 1977, **1**, 141 (*synth*)
Al-Ohaly, A.-R. *et al*, *J. Organomet. Chem.*, 1980, **202**, 297 (*synth, pmr, nmr*)

$C_{26}H_{30}O_8Ru_3$ Ru-00638

Di-μ-carbonylhexacarbonyl[μ_3-[(1,2,3,4-η:1,2,3,4-η:1,4-η)-1-(3,3-dimethyl-1-butenyl)-2,4-bis(1,1-dimethylethyl)-1,3-butadiene-1,4-diyl]]triruthenium, 9CI

*Octacarbonyl[μ_3-[2,4-di-*tert*-butyl-1-(3,3-dimethyl-1-butenyl)-1,3-butadiene-1,4-diyl]]triruthenium*

[75864-71-0]

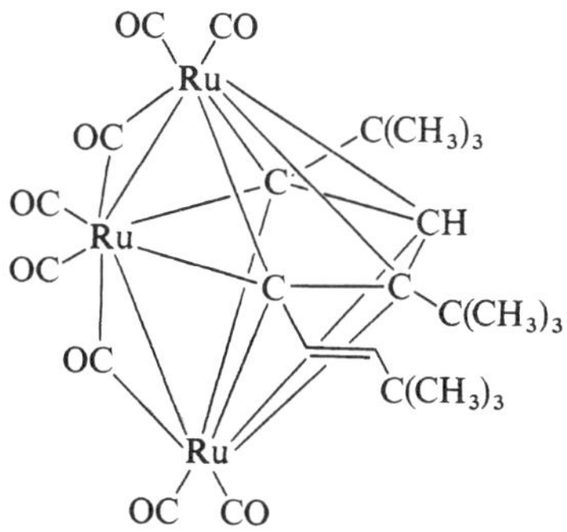

M 773.728

Yellow cryst. (heptane at 0°).

Sappa, E. *et al*, *Inorg. Chim. Acta*, 1980, **42**, 255 (*synth, ir, cryst struct*)

$C_{26}H_{36}O_{25}P_4Ru_6$ Ru-00639

[μ_6-(Carbon)]-μ-carbonyldodecacarbonyltetrakis(trimethyl-phosphite-*P*)hexaruthenium

[μ_6-(Carbon)]tridecacarbonyltetratris(trimethylphosphite)hexaruthenium

[74521-00-9]

$Ru_6(CO)_{12}(\mu\text{-}CO)(\mu_6\text{-}C)[P(OMe)_3]_4$

M 1478.870

No phys. descr. recorded. Derived from μ_6-Carbon-μ-carbonylhexadecacarbonylhexaruthenium, Ru-00500 .

Johnson, B.F.G. *et al*, *J. Organomet. Chem.*, 1980, **185**, C17 (*synth*)

$C_{27}H_{24}O_2PRu^{\oplus}$ Ru-00640

Dicarbonyl[(1,2,3,4,5-η)-2,4-cycloheptadien-1-yl](triphenylphosphine)ruthenium(1+), 9CI

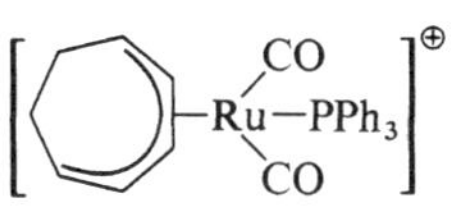

M 512.529 (ion)

Tetrafluoroborate: [57659-30-0].

$C_{27}H_{24}BF_4O_2PRu$ M 599.333

White cryst. (CH_2Cl_2/hexane). Mp 120° dec.

Sosinsky, B.A. *et al*, *J. Chem. Soc., Dalton Trans.*, 1975, 1633 (*synth, ir, pmr*)

$C_{27}H_{25}O_2PRu$ Ru-00641

Dicarbonyl[(1,2,3,4-η)-1,3-cycloheptadiene](triphenylphosphine)ruthenium, 9CI

[57556-81-7]

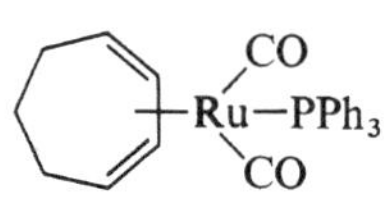

M 513.537

Yellow cryst. (CH_2Cl_2/hexane). Mp 121-3°.

Sosinsky, B.A. *et al*, *J. Chem. Soc., Dalton Trans.*, 1975, 1633 (*synth, ir, pmr*)

$C_{27}H_{25}O_2PRu$ Ru-00642

Dicarbonyl[(1,3,4,5-η)-4-cycloheptene-1,3-diyl](triphenylphosphine)ruthenium, 9CI

[57603-33-5]

As Tricarbonyl[(1,3,4,5-η)-4-cycloheptene-1,3-diyl]ruthenium, Ru-00145 with

$L = PPh_3$

M 513.537

Pale-yellow cryst. (CH_2Cl_2/hexane). Mp 141-5°.

Sosinsky, B.A. *et al*, *J. Chem. Soc., Dalton Trans.*, 1975, 1633 (*synth, ir, pmr*)

$C_{27}H_{27}PRu$ Ru-00643

(η^5-2,4-Cyclopentadien-1-yl)[(1,2,3-η)-2-methyl-2-propenyl-)](triphenylphosphine)ruthenium

π-Cyclopentadienyl-π-2-methylallyl(triphenylphosphine)ruthenium

[74573-23-2]

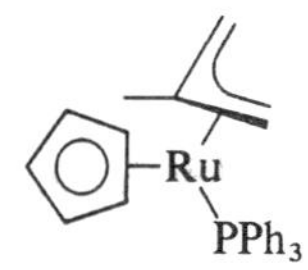

M 483.554

Exists only as *trans*-rotamer, in contrast to (η^5-2,4-Cyclopentadien-1-yl)(η^3-2-propenyl)(triphenylphosphine)ruthenium, Ru-00635 . Yellow powder (toluene/pentane).

Benn, R. *et al*, *Org. Magn. Reson.*, 1980, **14**, 435 (*pmr*)
Lehmkuhl, H. *et al*, *Justus Liebigs Ann. Chem.*, 1980, 754 (*synth, pmr, ms, cryst struct*)

$C_{28}H_{22}O_2Ru$ Ru-00644

1,1′-Bis(1-oxo-3-phenyl-2-propenyl)ruthenocene, 10CI

1,1′-Dicinnamoylruthenocene

[61182-40-9]

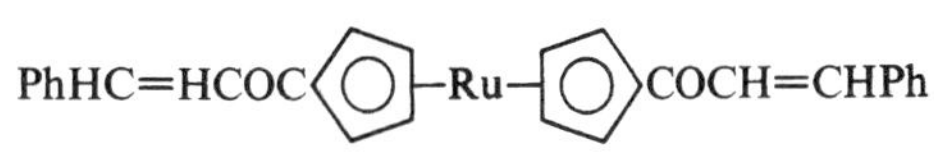

M 491.551

Bright-yellow cryst. Mp 250°.

Nesmeyanov, A.N. *et al*, *Dokl. Akad. Nauk SSSR, Ser. Sci. Khim.*, 1976, **229**, 881 (*synth, uv*)

$C_{28}H_{24}Br_2O_2P_2Ru$ Ru-00645

Dibromodicarbonyl[1,2-ethanediylbis(diphenylphosphine)-*P,P′*]ruthenium, 9CI

Dibromodicarbonyl[ethylenebis(diphenylphosphine)]-ruthenium, 8CI. *[1,2-Bis(diphenylphosphino)ethane]-dibromodicarbonylruthenium*

[29136-57-0]

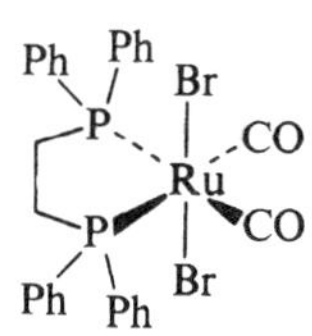

M 715.322

John, P., *Chem. Ber.*, 1970, **103**, 2178 (*struct, ir, pmr*)
Sandhu, S.S. *et al*, *Indian J. Chem.*, 1974, **12**, 834 (*synth, ir*)

$C_{28}H_{27}O_2PRu$ Ru-00646

Dicarbonyl[(1,4,5,6-η)-5-cyclooctene-1,4-diyl](triphenylphosphine)ruthenium, 9CI

[55763 66 1]

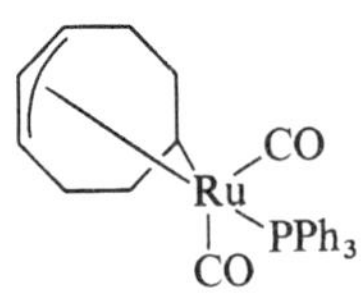

M 527.564

Yellow cryst. (pentane/Et_2O). Mp 126°.

Deeming, A.J. *et al*, *J. Chem. Soc., Dalton Trans.*, 1974, 2093 (*synth, ir, pmr*)

$C_{28}H_{30}NPRuS_2$ Ru-00647

(η⁵-2,4-Cyclopentadien-1-yl)(*N,N*-diethylcarbamodithioato-*S,S'*)(triphenylphosphine)ruthenium, 10CI

[79408-50-7]

As (η⁵-2,4-Cyclopentadien-1-yl)(*O*-ethylcarbonodithioato-*S,S'*)(triphenylphosphine)ruthenium, Ru-00633 with

X = NEt_2

M 576.715

Yellow-orange solid. Mp 173-7°.

Wilczewski, T. *et al*, *J. Organomet. Chem.*, 1981, **215**, 87 (*synth, ir, pmr*)

$C_{28}H_{32}N_4Ru^{\oplus\oplus}$ Ru-00648

[(1,2,5,6-η)-1,5-Cyclooctadiene]tetrakis(pyridine)ruthenium(2+), 9CI

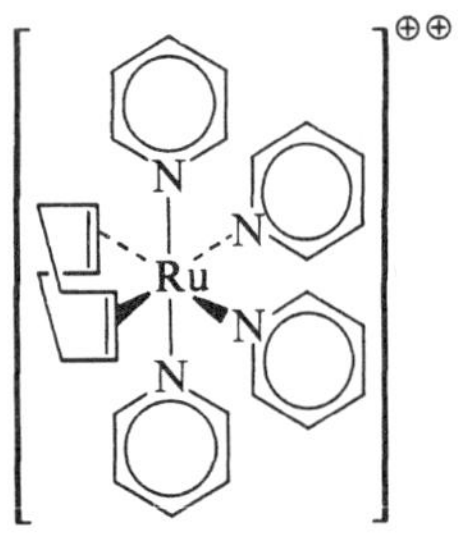

M 525.658 (ion)

Bis(tetraphenylborate): [37684-75-6].

$C_{76}H_{72}B_2N_4Ru$ M 1164.122

Efficient catalyst in reduction of nitrobenzenes with H_2O and CO. Yellow prisms. Mp 175° dec.

Ashworth, T.V. *et al*, *J. Chem. Soc., Dalton Trans.*, 1977, 1809 (*synth, ir, pmr*)
Okano, T. *et al*, *Chem. Lett.*, 1981, 1083 (*use*)

$C_{28}H_{40}As_4Cl_2Fe_2Ru$ Ru-00649

Dichlorobis[1,1'-ferrocenediylbis[dimethylarsine]]ruthenium, 8CI

[11058-58-5]

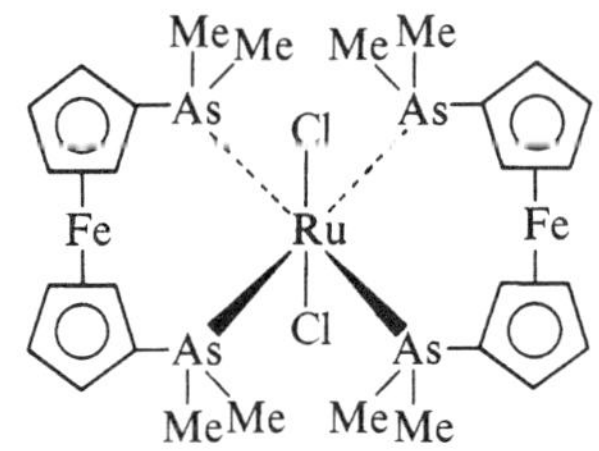

M 959.980

Orange cryst. (C_6H_6/heptane). Mp 280-281.5°.

Davison, A. *et al*, *Inorg. Chem.*, 1971, **10**, 832 (*synth, ir, pmr, uv*)

$C_{28}H_{40}P_3Ru^{\oplus}$ Ru-00650

(η⁴-1,3-Butadiene)tris(dimethylphenylphosphine)hydroruthenium(1+), 9CI

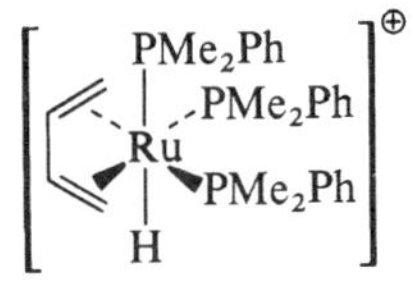

M 570.615 (ion)

Hexafluorophosphate: [61003-55-2].

$C_{28}H_{40}F_6P_4Ru$ M 715.579

Yellow prisms (CH_2Cl_2/MeOH). Mp 150°.

Ashworth, T.V. *et al*, *J. Chem. Soc., Dalton Trans.*, 1978, 1032 (*synth, pmr, struct*)

$C_{28}H_{80}P_8Ru_3^{\oplus\oplus}$ Ru-00651

Tetra-μ-methyleneoctakis(trimethylphosphine)triruthenium(2+), 10CI

Bis[bis(μ-methylene)tetrakis(trimethylphosphine)ruthenium]ruthenium(2+)

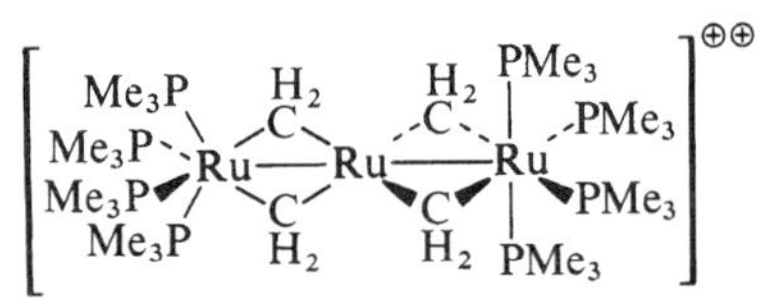

M 967.940 (ion)

Bis(tetrafluoroborate): [76210-26-9].

$C_{28}H_{80}B_2F_8P_8Ru_3$ M 1141.547

Dark-red cryst. (MeOH). Mp 260-80° dec.

Jones, R.A. *et al*, *J. Chem. Soc., Dalton Trans.*, 1980, 1771 (*synth, cryst struct, pmr, nmr*)

$C_{29}H_{15}Fe_2O_{11}PRu$ Ru-00652

Tricarbonyl(di-μ-carbonylhexacarbonyldiiron)(triphenylphosphine)ruthenium, 10CI

Di-μ-carbonylnonacarbonyl(triphenylphosphine)diironruthenium. Undecacarbonyl(triphenylphosphine)-diironruthenium

[77308-83-9]

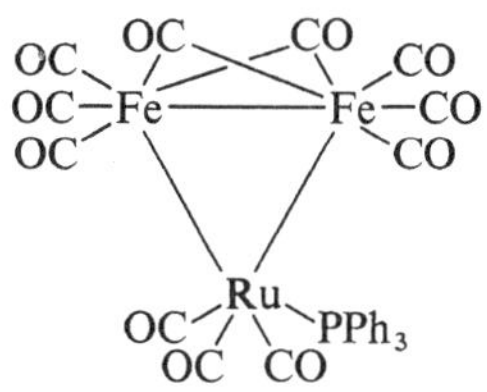

M 783.169

Violet cryst. Mp 136-40°.

Jones, D.F. *et al, Inorg. Chem.*, 1981, **20**, 3247 (*synth, ir*)

$C_{29}H_{15}O_{11}PRu_3$ Ru-00653

Undecacarbonyl(triphenylphosphine)triruthenium, 9CI

[38686-52-1]

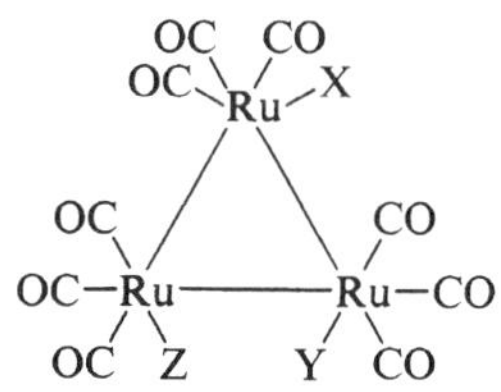

X = PPh_3, Y = Z = CO

M 873.615

Dark-red needles (Et_2O/hexane). Mp 131-3°.

Bruce, M.I. *et al, J. Chem. Soc., Dalton Trans.*, 1972, 2094 (*synth, ir*)
Malik, S.K. *et al, Inorg. Chem.*, 1978, **17**, 1484 (*synth, ir*)
Forbes, E.J., *J. Organomet. Chem.*, 1979, **182**, 143 (*synth, ir, cryst struct*)
Szostak, R. *et al, J. Organomet. Chem.*, 1980, **191**, 243 (*ir*)
Bruce, M.I. *et al, Aust. J. Chem.*, 1982, **35**, 935 (*synth*)
Bruce, M.I. *et al, J. Chem. Soc., Chem. Commun.*, 1982, 442 (*synth, ir*)

$C_{29}H_{19}O_{11}PRu_4$ Ru-00654

Undecacarbonyltetra-μ-hydro(triphenylphosphine)tetraruthenium, 9CI

[34742-79-5]

As Dodecacarbonyltetra-μ-hydrotetraruthenium, Ru-00224 with

X = PPh_3, Y = CO

M 978.716

Yellow-brown cryst. (cyclohexane/pentane). Thermally stable, even in air. Sol. most org. solvs. Reacts with halogenated solvs. at r.t.

Piacenti, F. *et al, Inorg. Chem.*, 1971, **10**, 2759 (*synth, ir, pmr*)
Vaglio, G.A. *et al, Inorg. Chim. Acta*, 1978, **30**, 161 (*synth, spectra*)

$C_{29}H_{25}BRu$ Ru-00655

(η^5-2,4-Cyclopentadien-1-yl)[(η^6-phenyl)triphenylborato(1–)]-ruthenium, 9CI

[33680-33-0]

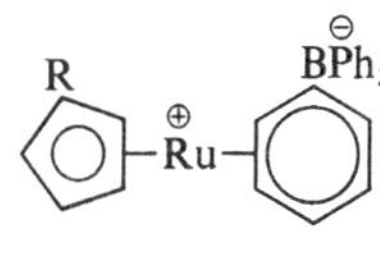

R = H

M 485.397

Zwitterionic complex. Brown cubes (MeOH), prob. air-stable. Sl. sol. most org. solvs. Λ 1.2 Ω^{-1} cm^2 mol^{-1} (Me_2CO).

Kruger, G.J. *et al, J. Chem. Soc., Dalton Trans.*, 1974, 1302 (*synth, pmr, ir, cryst struct*)
Haines, R.J. *et al, J. Organomet. Chem.*, 1975, **84**, 357 (*synth, ir, pmr*)

$C_{30}As_2F_{20}O_6Ru_2$ Ru-00656

Bis[μ-[bis(pentafluorophenyl)arsino]]hexacarbonyldiruthenium, 8CI

[20492-47-1]

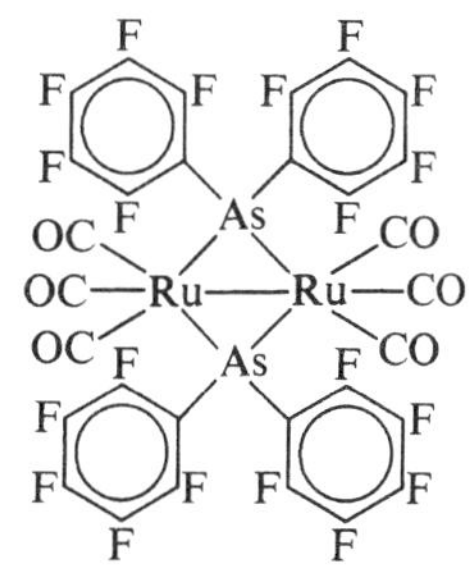

M 1188.278

Solid. Mp 231°.

Cooke, M. *et al, J. Chem. Soc. (A)*, 1968, 1507 (*synth, ir, nmr, ms*)

$C_{30}F_{20}O_6P_2Ru_2$ Ru-00657

Bis[μ-[bis(pentafluorophenyl)phosphino]]hexacarbonyldiruthenium, 8CI

[20909-98-2]

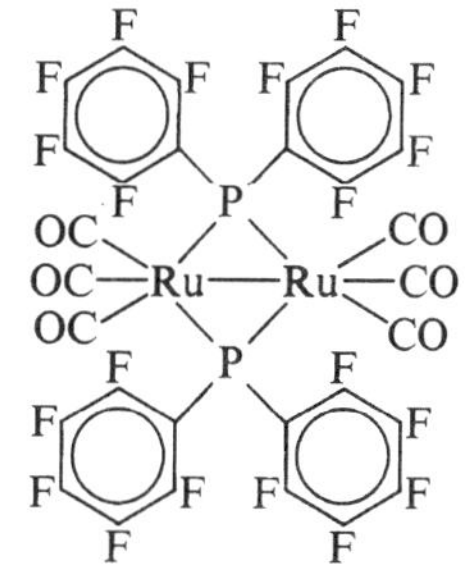

M 1100.382

Solid. Mp 227°.

Cooke, M. *et al, J. Chem. Soc. (A)*, 1968, 1507 (*synth, ir, nmr, ms*)

$C_{30}H_{20}O_6P_2Ru_2$ Ru-00658

Hexacarbonylbis[μ-(diphenylphosphino)]diruthenium, 9CI

[38685-25-5]

As Hexacarbonylbis[μ-(dimethylphosphino)]diruthenium, Ru-00156 with

R = Ph

M 740.572
White cryst.

Bruce, M.I. *et al*, *J. Chem. Soc., Dalton Trans.*, 1972, 2094 (*synth, ir, ms, pmr*)

$C_{30}H_{24}O_2Ru_2$ **Ru-00659**

1,1'':1',1'''-Bis(3-oxo-1,4-pentadiene-1,5-diyl)bisruthenocene, 10CI

[5.5](1,1')Ruthenocenophane-2,14,17,29-tetraene-1,16-dione

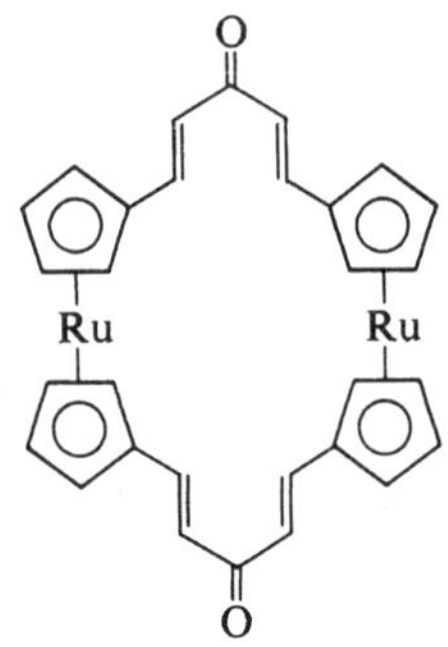

M 618.658

(all-E)-form [79354-72-6]

Cryst. ($CHCl_3$). Mp 287° dec.

Kamiyama, S. *et al*, *Bull. Chem. Soc. Jpn.*, 1981, **54**, 2079 (*synth, ir, pmr, ms, uv*)

$C_{30}H_{26}ClPRu$ **Ru-00660**

(η^6-Benzene)chlorophenyl(triphenylphosphine)ruthenium, 9CI

[41424-31-1]

As (η^6-Benzene)dibromo(triphenylphosphine)ruthenium, Ru-00598 with

X = Cl, Y = Ph

M 554.032

Yellow-brown cryst. ($MeCN/Et_2O$). Mp 150° dec.

Zelonka, R.A. *et al*, *J. Organomet. Chem.*, 1972, **44**, 383 (*synth, pmr, nmr*)

$C_{30}H_{26}Ru_3$ **Ru-00661**

1,1'':1''',1''''-Terruthenocene, 10CI

1,1'-Terruthenocene

[73178-90-2]

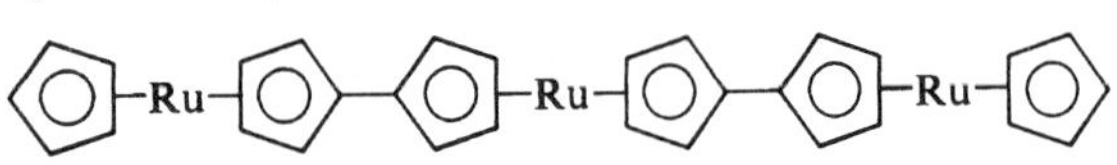

M 689.745

Colourless cryst. (heptane/C_6H_6). Mp 276°.

Bednarik, L.A. *et al*, *J. Org. Chem.*, 1980, **45**, 2032 (*synth, uv*)

$C_{30}H_{27}BRu$ **Ru-00662**

[(1,2,3,4,5-η)-1-Methyl-2,4-cyclopentadien-1-yl][(η^6-phenyl-)triphenylborato(1−)]ruthenium, 9CI

[55326-91-5]

As (η^5-2,4-Cyclopentadien-1-yl)[(η^6-phenyl)triphenylborato(1−)]ruthenium, Ru-00655 with

$R = CH_3$

M 499.423

Zwitterionic complex. Brown cryst. (MeOH), prob. air-stable.

Haines, R.J. *et al*, *J. Organomet. Chem.*, 1975, **84**, 357 (*synth, ir, pmr*)

$C_{30}H_{27}PRu^{\oplus\oplus}$ **Ru-00663**

(η^6-Benzene)[(1,2,3,4,5-η)-6-(triphenylphosphino)-2,4-cyclohexadien-1-yl]ruthenium(2+), 10CI

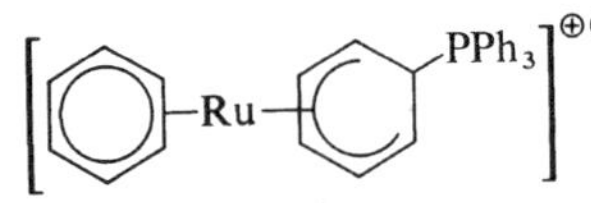

M 519.587 (ion)

Bis(hexafluorophosphate): [76932-90-6].

$C_{30}H_{27}F_{12}P_3Ru$ M 809.515

No phys. props. reported.

Domaille, P.J. *et al*, *J. Organomet. Chem.*, 1980, **202**, 191 (*synth, pmr, nmr*)

$C_{30}H_{34}ClPRu$ **Ru-00664**

Chloro[(1,2,3,4,5,6-η)-hexamethylbenzene]hydro(triphenylphosphine)ruthenium

[75182-14-8]

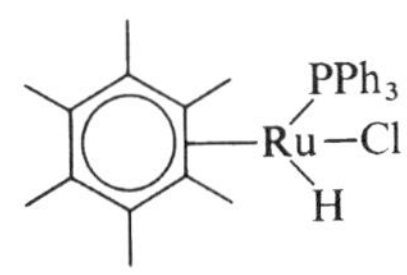

M 562.095

Homogeneous hydrogenation catalyst for aromatics. Air-sensitive yellow microcryst.

Bennett, M.A. *et al*, *J. Chem. Soc., Dalton Trans.*, 1974, 233.

Bennett, M.A. *et al*, *J. Chem. Soc., Chem. Commun.*, 1978, 582 (*synth, use, pmr, ir*)

$C_{30}H_{75}BrO_{15}P_5Ru^{\oplus}$ **Ru-00665**

Bromopentakis(triethyl phosphite-*P*)ruthenium(1+), 9CI

$BrRu[P(OEt)_3]_5^{\oplus}$

M 1011.756 (ion)

Tetraphenylborate: [35084-37-8].

$C_{54}H_{95}BBrO_{15}P_5Ru$ M 1330.988

Colourless cryst.

Couch, D.A. *et al*, *J. Chem. Soc., Chem. Commun.*, 1971, 1508 (*synth*)

Couch, D.A. *et al*, *Inorg. Chim. Acta*, 1974, **9**, 39 (*synth*)

$C_{30}H_{75}ClO_{15}P_5Ru^{\oplus}$ **Ru-00666**

Chloropentakis(triethyl phosphite-*P*)ruthenium(1+), 9CI

$ClRu[P(OEt)_3]_5^{\oplus}$

M 967.305 (ion)

Tetraphenylborate: [35084-36-7].

$C_{54}H_{95}BClO_{15}P_5Ru$ M 1286.537

Colourless cryst.

Couch, D.A. *et al*, *Inorg. Chim. Acta*, 1974, **9**, 39 (*synth, pmr*)

$C_{30}H_{76}O_{15}P_5Ru^{\oplus}$ **Ru-00667**

Hydropentakis(triethyl phosphite-*P*)ruthenium(1+), 9CI

$HRu[P(OEt)_3]_5^{\oplus}$

M 932.860 (ion)

Tetraphenylborate: [35083-13-7].
$C_{54}H_{96}BO_{15}P_5Ru$ M 1252.092
Colourless cryst.

Couch, D.A. *et al*, *Inorg. Chim. Acta*, 1974, **9**, 39 (*synth*, *ir*, *pmr*)

$C_{31}H_{20}O_3Ru$ Ru-00668

Tricarbonyl(1,2,3,4-tetraphenyl-1,3-cyclobutadiene)ruthenium, 8CI

Tricarbonyl[1,1′,1″,1‴-(η^4-1,3-cyclobutadiene-1,2,3,4-tetrayl)tetrakis[benzene]]ruthenium, 10CI

[12321-05-0]

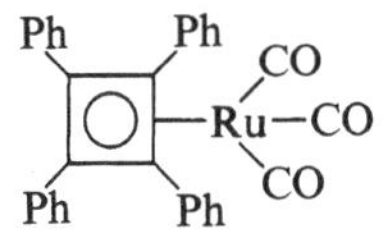

M 541.567

Yellow cryst. Mp 217-9°.

Sears, C.T. *et al*, *J. Organomet. Chem.*, 1968, **11**, 644 (*synth*, *ir*, *ms*)

$C_{31}H_{21}F_{12}PRu$ Ru-00669

(η^5-2,4-Cyclopentadien-1-yl)[(1,3,4-η)-5,5,5-trifluoro-1,2,3-tris(trifluoromethyl)-1,3-pentadienyl](triphenylphosphine)ruthenium, 9CI

[34802-04-5]

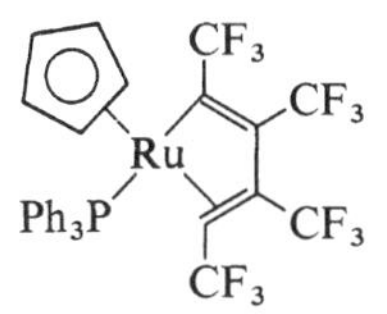

M 753.531

Pale-yellow cryst. (Et_2O/hexane). Mp 194-6°.

Blackmore, T. *et al*, *J. Chem. Soc., Chem. Commun.*, 1971, 852 (*synth*, *struct*)
Blackmore, T. *et al*, *J. Chem. Soc., Dalton Trans.*, 1974, 106 (*synth*, *pmr*, *nmr*, *ir*)

$C_{31}H_{27}OPRu$ Ru-00670

Benzylcarbonyl(η^5-2,4-cyclopentadien-1-yl)(triphenylphosphine)ruthenium

Carbonyl(η^5-2,4-cyclopentadien-1-yl)(phenylmethyl)-(triphenylphosphine)ruthenium, 10CI

[70586-96-8]

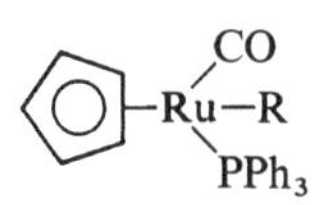

R = CH_2Ph

M 547.597

Yellow solid (CH_2Cl_2/MeOH). Mp 140-1°.

Bruce, M.I. *et al*, *J. Organomet. Chem.*, 1979, **171**, C5; *Aust. J. Chem.*, 1980, **33**, 1471 (*synth*, *ir*, *pmr*)

$C_{31}H_{29}ClP_2Ru$ Ru-00671

Chloro(η^5-2,4-cyclopentadien-1-yl)[1,2-ethanediylbis[diphenylphosphine]-*P*,*P′*]ruthenium, 10CI

[1,2-Bis(diphenylphosphino)ethane]chloro(η^5-cyclopentadienyl)ruthenium

[71397-10-9]

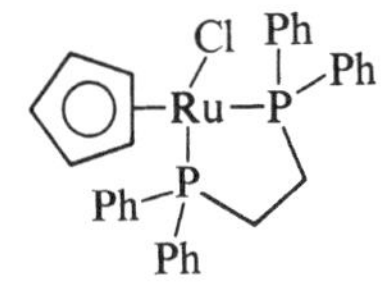

M 600.041

Orange cryst. (C_6H_6/hexane). Mp 232-4° (224-8°).

Ashby, G.S. *et al*, *Aust. J. Chem.*, 1979, **32**, 1003 (*synth*, *pmr*)
Treichel, P.M. *et al*, *Synth. React. Inorg. Metal-Org. Chem.*, 1980, **10**, 205 (*synth*, *pmr*)

$C_{32}H_{20}O_4Ru$ Ru-00672

Tricarbonyl[(2,3,4,5-η)-2,3,4,5-tetraphenyl-2,4-cyclopentadien-1-one]ruthenium, 9CI

Tetracarbonylbis(diphenylacetylene)ruthenium, 8CI. *Tetracycloneruthenium tricarbonyl*

[12321-08-3]

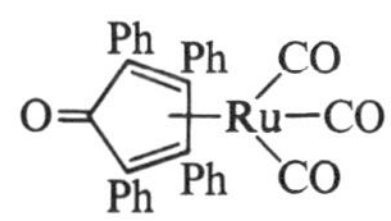

M 569.578

Pale-yellow cryst. (pentane). Mp 180-3° dec.

Bruce, M.I. *et al*, *J. Organomet. Chem.*, 1968, **12**, 411 (*synth*, *ir*, *ms*)
Kruczynski, L. *et al*, *Inorg. Chem.*, 1976, **15**, 3140 (*pmr*, *cmr*)

$C_{32}H_{24}ClN_2O_2PRu$ Ru-00673

Dicarbonylchloro[2-(phenylazo)phenyl](triphenylphosphine)-ruthenium, 8CI

*(Azobenzene-*C,N′*)dicarbonylchloro(triphenylphosphine)ruthenium*

[30178-79-1]

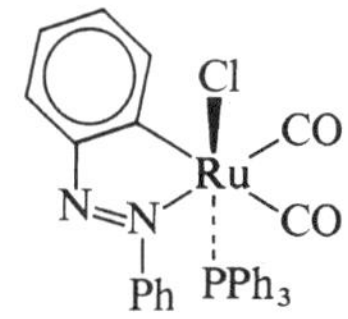

M 636.051

Yellow cryst. (CH_2Cl_2/hexane). Mp 181-2°.

Bruce, M.I. *et al*, *J. Chem. Soc.* (*A*), 1971, 2820 (*synth*, *ir*, *pmr*)

$C_{32}H_{24}O_4RuSi$ Ru-00674

Tricarbonyl[(2,3,4,5-η)-1-methyl-2,3,4,5-tetraphenylsilacyclopenta-2,4-dien-1-ol]ruthenium, 10CI

[62795-57-7]

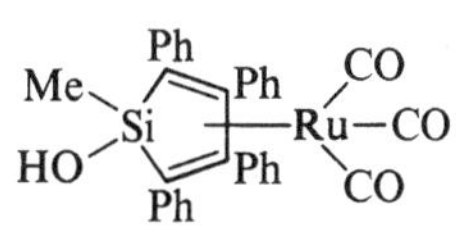

M 601.695

Solid. Mp 163-8° dec. Methyl group orientation not known.

Abel, E.W. *et al*, *J. Chem. Soc., Dalton Trans.*, 1976, 2484 (*synth*, *ms*)

$C_{32}H_{25}OPRu$ Ru-00675

Carbonyl(η^5-2,4-cyclopentadien-1-yl)(phenylethynyl)(triphenylphosphine)ruthenium, 10CI

[75592-69-7]

As Benzylcarbonyl(η^5-2,4-cyclopentadien-1-yl)(triphenylphosphine)ruthenium, Ru-00670 with

R = —C≡CPh

M 557.593

Yellow microcryst. (CH_2Cl_2/MeOH). Mp 218-20°.

Bruce, M.I. *et al*, *Aust. J. Chem.*, 1980, **33**, 1471 (*synth*, *ir*, *pmr*, *cmr*)

$C_{32}H_{29}OP_2Ru^{\oplus}$ Ru-00676

Carbonyl(η^5-2,4-cyclopentadien-1-yl)[1,2-ethanediylbis[diphenylphosphine]-*P*,*P'*]ruthenium(1+)

[1,2-Bis(diphenylphosphino)ethane]carbonyl(cyclopentadienyl)ruthenium(1+)

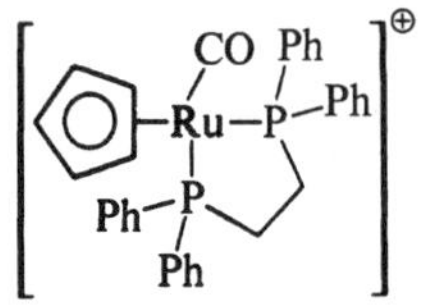

M 592.598 (ion)

Chloride: [68663-45-6].

$C_{32}H_{29}ClOP_2Ru$ M 628.051

Cryst. (Me_2CO/Et_2O). Mp 146°.

Jungbauer, A. *et al*, *Z. Naturforsch.*, *B*, 1978, **33**, 1083 (*synth*, *pmr*, *ir*)

Davies, S.G. *et al*, *J. Organomet. Chem.*, 1980, **188**, C41 (*synth*)

$C_{32}H_{31}BRu$ Ru-00677

[(1,2,3,5,6-η)-2,5-Cyclooctadien-1-yl][(η^6-phenyl)triphenylborato(1−)]ruthenium

[65297-47-4]

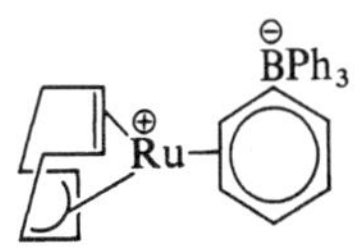

M 527.477

Pale-yellow cryst. Sl. sol. Py, extremely insol. other org. solvs. Mp 200°. Struct. and spectra also determined for the more sol. $PhBF_3$ analogue.

Ashworth, T.V. *et al*, *J. Chem. Soc., Chem. Commun.*, 1977, 937 (*synth*, *struct*)

Ashworth, T.V. *et al*, *J. Chem. Soc., Dalton Trans.*, 1977, 1809 (*synth*, *ir*)

$C_{32}H_{32}P_2Ru$ Ru-00678

(η^5-2,4-Cyclopentadien-1-yl)[1,2-ethanediylbis[diphenylphosphine]-*P*,*P'*]methylruthenium, 10CI

[1,2-Bis(diphenylphosphino)ethane](η^5-cyclopentadienyl)methylruthenium

[74481-94-0]

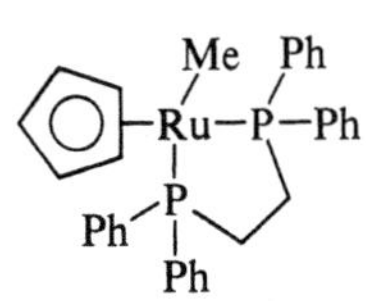

M 579.622

No props. reported.

Davies, S.G. *et al*, *J. Organomet. Chem.*, 1980, **188**, C41 (*synth*)

$C_{33}H_{15}O_{13}PRu_5$ Ru-00679

Tridecacarbonyl[μ-(diphenylphosphino)][μ_4-[(1-η:1-η:1,2-η-:1,2-η)phenylethynyl]]pentaruthenium, 10CI

[78179-75-6]

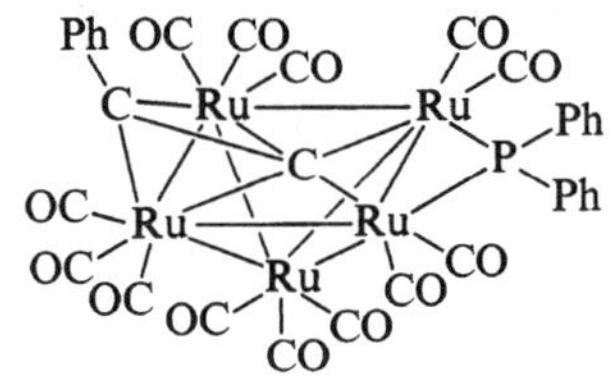

M 1155.797

Blue-black cryst. (heptane/C_6H_6). Mp 214-6°.

Carty, A.J. *et al*, *J. Am. Chem. Soc.*, 1981, **103**, 2456 (*synth*, *struct*, *nmr*, *ir*)

$C_{33}H_{15}O_{14}PRu_5$ Ru-00680

[μ_5-(Carbon)]tetradecacarbonyl(triphenylphosphine)pentaruthenium

[80050-10-8]

As (Carbon)pentadecacarbonylpentaruthenium, Ru-00447 with

X = PPh_3

M 1171.797

No phys. description recorded.

Farrar, D.H. *et al*, *J. Chem. Soc., Chem. Commun.*, 1981, 415 (*synth*, *cryst struct*)

$C_{34}H_{20}O_6Ru_2$ Ru-00681

Hexacarbonyl[μ-[(1,2,3,4-η:1,4-η)-1,2,3,4-tetraphenyl-1,3-butadiene-1,4-diyl]]diruthenium

Hexacarbonyl[μ-(1,2,3,4-tetraphenyl)-1,3-butadienylene)]diruthenium, 8CI. Hexacarbonylbis(diphenylacetylene)diruthenium, 8CI

[33310-08-6]

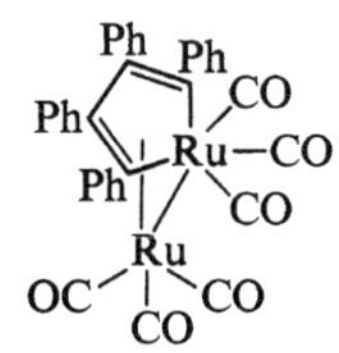

M 726.668

Yellow cryst. Mp 192°.

Sears, C.T. *et al*, *J. Organomet. Chem.*, 1968, **11**, 644 (*synth*, *ir*)

Cetini, G. *et al*, *J. Organomet. Chem.*, 1969, **17**, 437 (*synth*, *ir*)

Vaglio, G.A. *et al*, *Org. Mass Spectrom.*, 1971, **5**, 493 (*ms*)

$C_{34}H_{57}O_9P_3Ru_3Si$ Ru-00682

Tri-μ-carbonylhexacarbonyl[μ₃-[(methylsilylidyne)tris[dibutylphosphine]-*P*:*P*′:*P*″]]triruthenium

Nonacarbonyl[tris(dibutylphosphino)methylsilane]triruthenium

[69700-33-0]

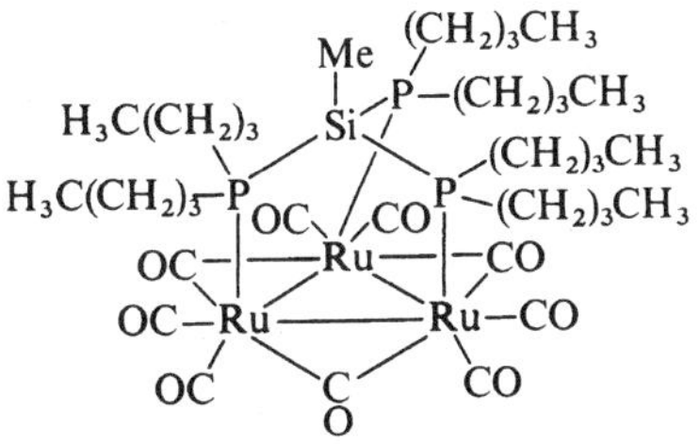

M 1034.036

Air-stable yellow cryst. (MeOH/Me_2CO).

De Boer, J.J. *et al*, *J. Chem. Soc., Chem. Commun.*, 1978, 1005 (*synth, cryst struct*)

$C_{34}H_{68}N_6RuTi_2$ Ru-00683

Hexakis(diethylaminato)-μ-1,1′-ruthenocenediyldititanium

Hexakis(N-ethylethanaminato)-μ-1,1′-ruthenocenediyldititanium, 9CI. *1,1′-Bis[tris(diethylamino)titanium]ruthenocene*

[60898-12-6]

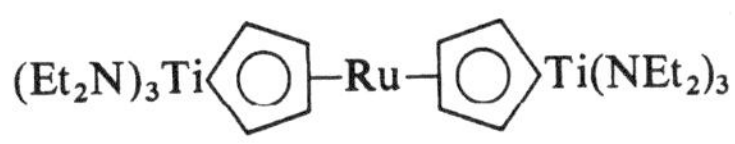

M 757.781

Yellow cryst. (Et_2O). Mp 90° dec.

Bürger, H. *et al*, *Z. Anorg. Allg. Chem.*, 1976, **423**, 112 (*synth, pmr, cmr*)

$C_{35}H_{15}O_{16}PRu_6$ Ru-00684

[μ₆-(Carbon)]-μ-carbonylpentadecacarbonyl(triphenylphosphine)hexaruthenium

(μ₆-Carbon)hexadecacarbonyl(triphenylphosphine)hexaruthenium

[12577-43-4]

As

μ₆-Carbon-μ-carbonylhexadecacarbonylhexaruthenium, Ru-00500 with

X = PPh_3

M 1328.888

Deep-red air-stable needles (hexane). V. sol. org. solvs.

Johnson, B.F.G. *et al*, *J. Chem. Soc. (A)*, 1970, 901 (*synth, ir*)
Johnson, B.F.G. *et al*, *J. Organomet. Chem.*, 1980, **185**, C17 (*struct*)

$C_{35}H_{20}O_7Ru_2$ Ru-00685

Hexacarbonyl[μ-[(1,2,5-η:1,4,5-η)-3-oxo-1,2,4,5-tetraphenyl-1,4-pentadiene-1,5-diyl]]diruthenium

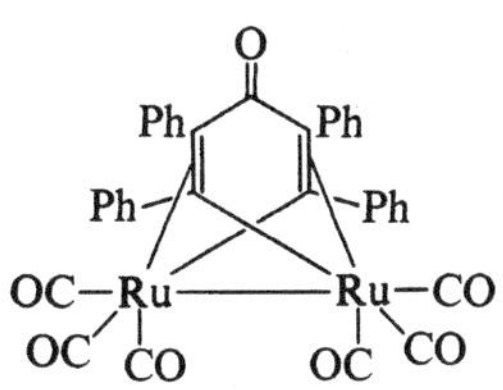

Possible structure

M 754.679

Yellow powder. Mp 108-10°.

Cetini, G. *et al*, *J. Organomet. Chem.*, 1969, **17**, 437 (*synth, ir*)

$C_{35}H_{29}N_2PRu$ Ru-00686

(η^5-2,4-Cyclopentadien-1-yl)[2-(phenylazo)phenyl](triphenylphosphine)ruthenium, 9CI

[34807-86-8]

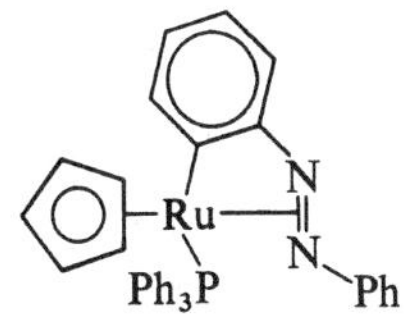

M 609.671

Deep green-black cryst. (hexane). Mp 199-199.5° dec.

Bruce, M.I. *et al*, *J. Chem. Soc. (A)*, 1971, 2820 (*synth, ir*)

$C_{35}H_{38}ClO_7P_2Ru^{\oplus}$ Ru-00687

Carbonylchloro(η^4-tetraphenylcyclobutadiene)bis(trimethyl phosphite-*P*)ruthenium(1+)

Carbonylchloro[1,1′,1″,1‴-(η^4-1,3-cyclobutadiene-1,2,3,4-tetrayl)tetrakis[benzene]]bis(trimethyl phosphite-P)ruthenium(1+), 10CI

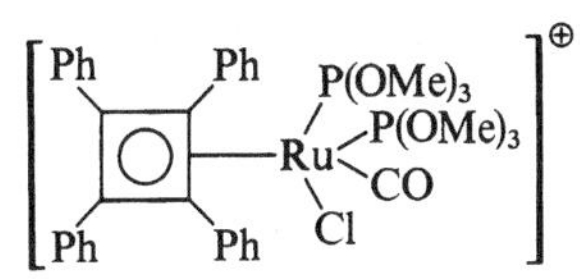

M 769.152 (ion)

Trichloride: [77347-77-4].

$C_{35}H_{38}Cl_4O_7P_2Ru$ M 875.511

No phys. props. recorded. Unstable in soln.

Connelly, N.G. *et al*, *J. Chem. Soc., Dalton Trans.*, 1981, 34 (*synth, ir, props*)

$C_{35}H_{38}O_7P_2Ru$ Ru-00688

Carbonyl(η^4-tetraphenylcyclobutadiene)bis(trimethylphosphite-*P*)ruthenium

Carbonyl[1,1′,1″,1‴-(η^4-1,3-cyclobutadiene-1,2,3,4-tetrayl)tetrakis[benzene]]bis(trimethyl phosphite-P)ruthenium

[77370-83-3]

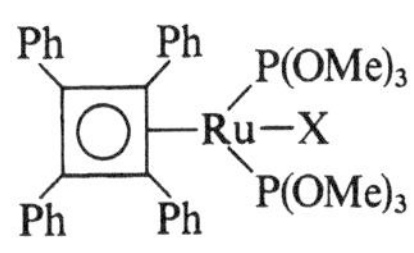

X = CO

M 733.699

Stable yellow cryst. (hexane). Solns. mod. air-stable. Sol. polar solvs. Mp 164-6°.

Connelly, N.G. *et al*, *J. Chem. Soc., Dalton Trans.*, 1981, 34 (*synth, pmr*)

$C_{36}H_{20}O_8Ru_3$ Ru-00689

Di-μ-carbonylhexacarbonyl(1,2,3,4-tetraphenyl-1,3-butadiene-1,4-diyl)triruthenium

Di-μ-carbonylhexacarbonyl[μ_3-(1,2,3,4-tetraphenyl-1,3-butadienylene)]triruthenium. Octacarbonyl(1,2,3,4-tetraphenyl-1,3-butadiene-1,4-diyl)triruthenium

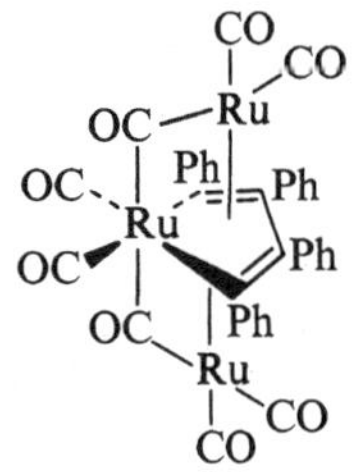

M 883.759

Orange-yellow isomer of Octacarbonylbis[1,1′-(1,2-ethynediyl)bis[benzene]]triruthenium, Ru-00690 . Bright-orange powder. Mp 170° dec.

Cetini, G. *et al*, *J. Organomet. Chem.*, 1969, **17**, 437 (*synth, ir*)

$C_{36}H_{20}O_8Ru_3$ Ru-00690

Octacarbonylbis[1,1′-(1,2-ethynediyl)bis[benzene]]triruthenium, 9CI

Octacarbonylbis[μ_3-(diphenylvinylene)]triruthenium. Octacarbonylbis(diphenylacetylene)triruthenium

[12577-61-6]

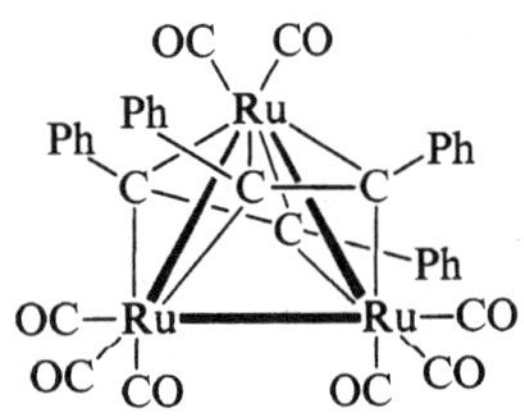

M 883.759

Violet isomer of Di-μ-carbonylhexacarbonyl(1,2,3,4-tetraphenyl-1,3-butadiene-1,4-diyl)triruthenium, Ru-00689 . Red-violet powder. Mp 125-7° dec. Rearranges to Di-μ-carbonylhexacarbonyl(1,2,3,4-tetraphenyl-1,3-butadiene-1,4-diyl)triruthenium, Ru-00689 at 60°.

Cetini, G. *et al*, *J. Organomet. Chem.*, 1969, **17**, 437 (*synth, ir*)
Sappa, E. *et al*, *CA*, 1976, **85**, 86145 (*ms*)

$C_{36}H_{28}O_{10}P_2Ru_4$ Ru-00691

Decacarbonyl[μ-[1,2-ethanediylbis[diphenylphosphine]-*P*:*P′*]]tetra-μ-hydrotetraruthenium, 10CI

[1,2-Bis(diphenylphosphine)ethane]decacarbonyltetrahydrotetraruthenium

[66322-95-0]

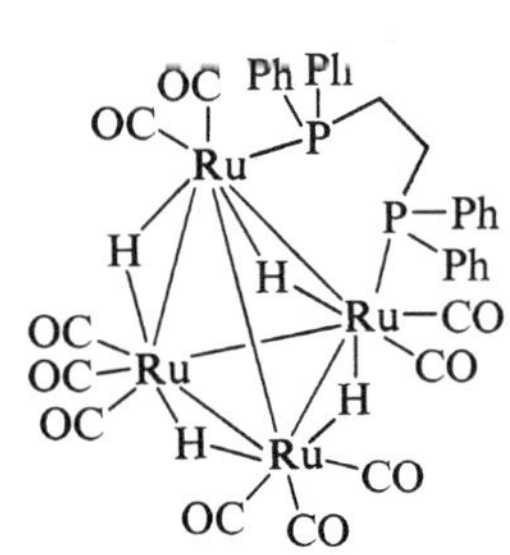

vicinal-form

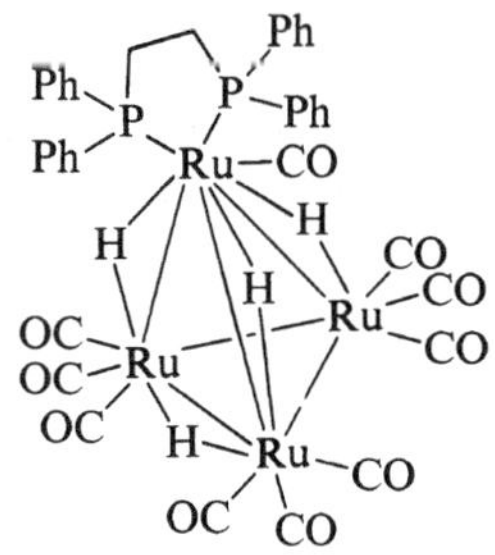

geminal-form

M 1086.839

Two structural isomers known.

geminal-form

Red-orange cryst.

vicinal-form

Red cryst. (CH_2Cl_2/MeOH).

Shapley, J.R. *et al*, *J. Am. Chem. Soc.*, 1977, **99**, 7384 (*cryst struct*)
Churchill, M.R. *et al*, *Inorg. Chem.*, 1978, **17**, 1950 (*synth, ir, cryst struct*)
Churchill, M.R. *et al*, *Inorg. Chem.*, 1980, **19**, 1277 (*synth, ms, ir, nmr, pmr, isom*)

$C_{36}H_{30}As_2Cl_3NRu$ Ru-00692

Trichloronitridobis(triphenylarsine)ruthenium

[55449-95-1]

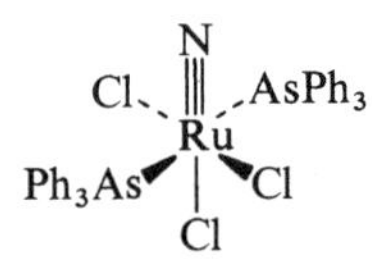

M 833.912

Yellow cryst. Mp 169°.

Pawson, D. *et al*, *J. Chem. Soc., Dalton Trans.*, 1975, 417 (*synth, ir*)

$C_{36}H_{30}Br_3NOP_2Ru$ Ru-00693

Tribromonitrosylbis(triphenylphosphine)ruthenium, 9CI, 8CI

[62961-03-9]

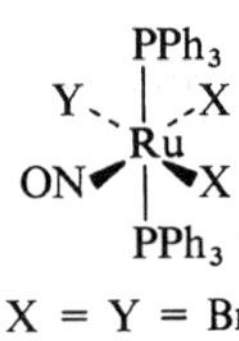

X = Y = Br

M 895.369

trans-form [29826-82-2]

Brown or orange cryst. Mp 228-30°, 274-6°.

Laing, K.R. *et al*, *J. Chem. Soc. (A)*, 1970, 2149 (*synth, ir, nmr*)
Robinson, S.D. *et al*, *J. Chem. Soc., Dalton Trans.*, 1972, 1 (*synth, spectra*)
Jain, K.C. *et al*, *Indian J. Chem., Sect. A*, 1981, **20**, 1124 (*synth*)

$C_{36}H_{30}ClI_2NOP_2Ru$ Ru-00694

Chlorodiiodonitrosylbis(triphenylphosphine)ruthenium, 9CI, 8CI

[29826-81-1]

As Tribromonitrosylbis(triphenylphosphine)ruthenium, Ru-00693 with

X = I, Y = Cl

M 944.919

Probably obt. in 2 isomeric forms. Orange-brown cryst. (CH_2Cl_2/EtOH), brown platelets (MeOH/CH_2Cl_2). Mp 210-2°, 285-8°.

Laing, K.R. *et al*, *J. Chem. Soc.* (*A*), 1970, 2149 (*synth*, *ir*)
Robinson, S.D. *et al*, *J. Chem. Soc.*, *Dalton Trans.*, 1972, 1 (*synth*, *spectra*)

$C_{36}H_{30}ClNOP_2Ru$ Ru-00695

Chloronitrosylbis(triphenylphosphine)ruthenium, 9CI

[38856-98-3]

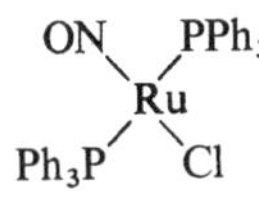

M 691.110

Coordinatively unsaturated. Hydrogenation catalyst. Green cryst., extremely reactive to O_2.

Stiddard, M.H.B. *et al*, *J. Chem. Soc.*, *Chem. Commun.*, 1969, 1372 (*synth*, *ir*, *props*)
Gaughan, A.P. *et al*, *Inorg. Chem.*, 1974, **13**, 786 (*synth*, *ir*)
Gandolfi, O. *et al*, *J. Organomet. Chem.*, 1976, **104**, C41 (*props*, *ir*)

$C_{36}H_{30}ClNO_3P_2Ru$ Ru-00696

Chloro(dioxygen)nitrosylbis(triphenylphosphine)ruthenium, 9CI, 8CI

[22804-23-5]

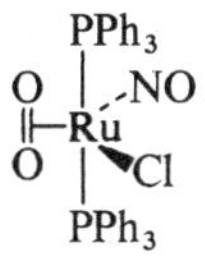

M 723.108

Oxidation catalyst. Dark-brown cryst. (MeOH).

Laing, K.R. *et al*, *J. Chem. Soc.*, *Chem. Commun.*, 1968, 1556 (*ir*)
Graham, B.W. *et al*, *J. Chem. Soc.*, *Dalton Trans.*, 1972, 1237 (*synth*, *struct*)
Ungermann, C.B. *et al*, *J. Am. Chem. Soc.*, 1976, **98**, 3862 (*synth*)

$C_{36}H_{30}ClN_2O_2P_2Ru^{\oplus}$ Ru-00697

Chlorodinitrosylbis(triphenylphosphine)ruthenium(1+), 8CI

[53093-70-2]

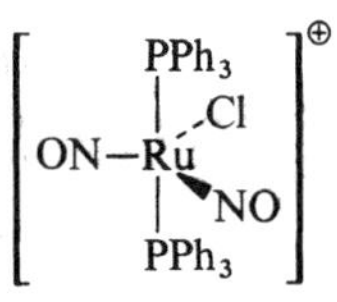

M 721.116 (ion)

Hexafluorophosphate: [36237-03-3].
$C_{36}H_{30}ClF_6N_2O_2P_3Ru$ M 866.080
Orange cryst. + $1C_6H_6$.

Tetrafluoroborate: [54890-53-8].
$C_{36}H_{30}BClF_4N_2O_2P_2Ru$ M 807.919
Orange cryst. Sol. MeOH, CH_2Cl_2, insol. Et_2O, hexane. Mp 196°. Mod. air-stable.

Pierpont, C.G. *et al*, *J. Am. Chem. Soc.*, 1970, **92**, 4760 (*struct*)
Inorg. Synth., 1976, **16**, 21 (*synth*, *ir*)
Reed, J., *J. Inorg. Nucl. Chem.*, 1976, **38**, 2239 (*synth*)

$C_{36}H_{30}Cl_2P_2Ru$ Ru-00698

Dichlorobis(triphenylphosphine)ruthenium, 9CI, 8CI

[34076-51-2]

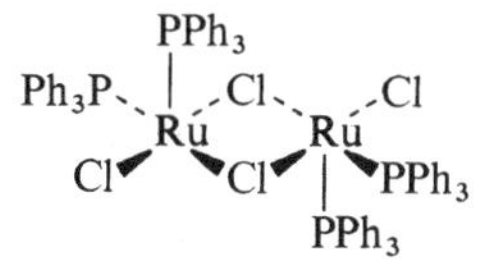

Dimer

M 696.557

Dimeric or polymeric species formed by dissociation of $(Ph_3P)_2RuCl_2$ in soln. Olefin isomerisation catalyst.

Dimer: Di-μ-chlorodichlorobis(triphenylphosphine)diruthenium. Tetrachlorobis(triphenylphosphine)diruthenium.
$C_{72}H_{60}Cl_4P_4Ru_2$ M 1393.113
No phys. props. reported.

Polymer: [32698-53-6]. Green solid.

Attridge, C.J. *et al*, *J. Chem. Soc.* (*C*), 1971, 2999 (*use*)
James, B.R. *et al*, *J. Chem. Soc.*, *Chem. Commun.*, 1974, 439 (*synth*)
Hoffman, P.R. *et al*, *J. Am. Chem. Soc.*, 1975, **97**, 4221 (*synth*, *nmr*)

$C_{36}H_{30}Cl_3NOP_2Ru$ Ru-00699

Trichloronitrosylbis(triphenylphosphine)ruthenium

[15349-78-7]

As Tribromonitrosylbis(triphenylphosphine)ruthenium, Ru-00693 with

X = Y = Cl

M 762.016

Selective hydrogenation catalyst.

trans-form [29826-79-7]
Yellow-orange cryst. (CH_2Cl_2/EtOH). Mp 215-30°, 250°, 290° (sealed tube in N_2).

Laing, K.R. *et al*, *J. Chem. Soc.* (*A*), 1970, 2149 (*synth*, *ir*)
Baird, M.C., *Inorg. Chim. Acta*, 1971, **5**, 46 (*synth*)
Harris, R.O. *et al*, *J. Organomet. Chem.*, 1973, **54**, 259 (*synth*)
Inorg. Synth., 1974, **15**, 45 (*synth*, *ir*, *bibl*)
Haymore, B.L. *et al*, *Inorg. Chem.*, 1975, **14**, 3060 (*synth*, *cryst struct*)
Holderegger, R. *et al*, *Helv. Chim. Acta*, 1979, **62**, 2159 (*synth*)
Jain, K.C. *et al*, *Indian J. Chem.*, *Sect. A*, 1980, **19**, 1089; 1981, **20**, 1124 (*synth*, *spectra*)

$C_{36}H_{30}Cl_3NP_2Ru$ **Ru-00700**

Trichloronitridobis(triphenylphosphine)ruthenium, 9CI

[55449-96-2]

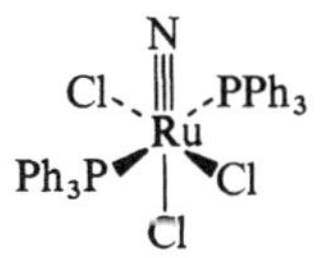

M 746.016

Not obt. pure.

Pawson, D. *et al*, *J. Chem. Soc., Dalton Trans.*, 1975, 417 (*synth, ir*)

$C_{36}H_{30}Cl_3NRuSb_2$ **Ru-00701**

Trichloronitridobis(triphenylstibine)ruthenium

[55411-27-3]

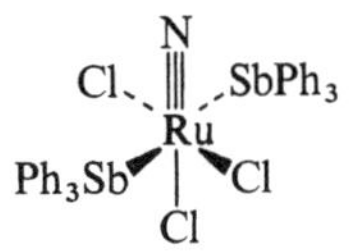

M 927.569

Orange cryst. Mp 165°.

Pawson, D. *et al*, *J. Chem. Soc., Dalton Trans.*, 1975, 417 (*synth, ir*)

$C_{36}H_{30}Cl_4P_2Ru^{\ominus}$ **Ru-00702**

Tetrachlorobis(triphenylphosphine)ruthenate(1–), 9CI, 8CI

[47841-10-1]

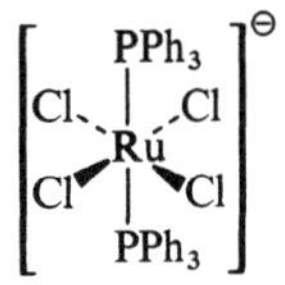

M 767.463 (ion)

trans-form

Tetramethylammonium salt: [22863-33-8].

$C_{40}H_{42}Cl_4NP_2Ru$ M 841.608

Orange cryst. Insol. CH_2Cl_2. Mp 220° dec.

Stephenson, T.A., *J. Chem. Soc. (A)*, 1970, 889 (*synth, spectra*)
Evans, R.S. *et al*, *Inorg. Chem.*, 1976, **15**, 1139 (*mcd*)
Contreras, R. *et al*, *J. Organomet. Chem.*, 1979, **179**, C55.

$C_{36}H_{30}N_2O_2P_2Ru$ **Ru-00703**

Dinitrosylbis(triphenylphosphine)ruthenium, 9CI

[30352-63-7]

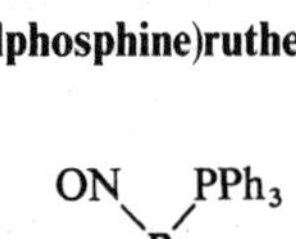

M 685.663

Orange-red needles ($MeOH/CH_2Cl_2$). Mp 185-6° (sealed tube in N_2), 144-5°.

Benzene adduct (2:1):

$C_{78}H_{66}N_4O_4P_4Ru_2$ M 1449.439

Deep-red cryst. (C_6H_6/hexane). Mp 150°.

Inorg. Synth., 1974, **15**, 52 (*synth, ir, bibl*)
Gaughan, A.P. *et al*, *Inorg. Chem.*, 1974, **13**, 786 (*synth, ir, cryst struct*)
Bhaduri, S. *et al*, *Acta Crystallogr., Sect. B*, 1975, **31**, 897 (*cryst struct*)
Cenini, S. *et al*, *Gazz. Chim. Ital.*, 1975, **105**, 255 (*synth*)
Feltham, R.D. *et al*, *J. Am. Chem. Soc.*, 1982, **104**, 641 (*pe*)

$C_{36}H_{66}Cl_2P_2Ru_2^{\oplus\oplus}$ **Ru-00704**

Bis(η^6-benzene)di-μ-chlorobis(tributylphosphine)diruthenium(2+)

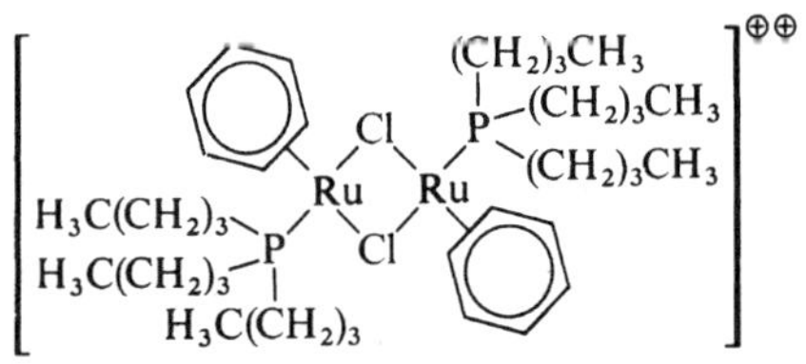

M 833.911 (ion)

Dichloride: [12119-52-7].

$C_{36}H_{66}Cl_4P_2Ru_2$ M 904.817

Red cryst. (CH_2Cl_2/pet. ether).

Monomer: (η^6-Benzene)chloro(tributylphosphine)-ruthenium(1+).

$C_{18}H_{33}ClPRu$ M 416.955

Unknown.

Winkhaus, G. *et al*, *J. Organomet. Chem.*, 1967, **7**, 487 (*synth*)

$C_{36}H_{90}O_{18}P_6Ru^{\oplus\oplus}$ **Ru-00705**

Hexakis(triethyl phosphite-*P*)ruthenium(2+), 9CI

$Ru[P(OEt)_3]_6^{\oplus\oplus}$

M 1098.009 (ion)

Bis(tetraphenylborate): [35083-09-1].

$C_{84}H_{130}B_2O_{18}P_6Ru$ M 1736.473

White cryst. ($MeOH/CH_2Cl_2$).

Couch, D.A. *et al*, *J. Chem. Soc., Chem. Commun.*, 1971, 1508 (*synth*)
Couch, D.A. *et al*, *Inorg. Chim. Acta*, 1974, **9**, 39.

$C_{37}H_{20}FeO_{13}P_2Ru_3$ **Ru-00706**

Nonacarbonylbis[μ-(diphenylphosphino)](tetracarbonyliron)-triruthenium, 10CI

Tridecacarbonylbis[μ(diphenylphosphino)]irontriruthenium

[79043-00-8]

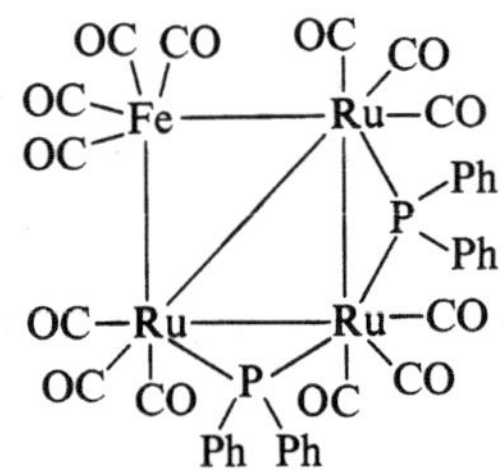

M 1093.562

Dark-red cryst. (hexane/Et_2O).

Churchill, M.R. *et al*, *J. Organomet. Chem.*, 1981, **213**, 139 (*synth, ir, struct*)

$C_{37}H_{30}As_2Cl_3ORu^{\ominus}$ Ru-00707

Carbonyltrichlorobis(triphenylarsine)ruthenate(1–), 9CI

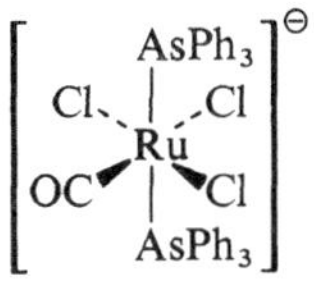

M 847.916 (ion)

Benzyltriphenylphosphonium salt: [54163-49-4].
$C_{62}H_{52}As_2Cl_3OPRu$ M 1201.338
Orange cryst. (CH_2Cl_2/pentane). Mp 215-7° dec.

Stephenson, T.A. *et al*, *J. Chem. Soc., Dalton Trans.*, 1974, 1640 (*synth, ir*)

$C_{37}H_{30}BrNO_2P_2Ru$ Ru-00708

Bromocarbonylnitrosylbis(triphenylphosphine)ruthenium, 8CI

[22804-20-2]

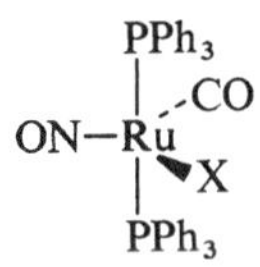

X = Br

M 763.571
Yellow-brown cryst. Mp 172-4° dec.

Laing, K.R. *et al*, *J. Chem. Soc. (A)*, 1970, 2149 (*synth, ir*)
McCleverty, J.A. *et al*, *J. Chem. Soc., Chem. Commun.*, 1976, 1061 (*synth*)

$C_{37}H_{30}Br_3RuS$ Ru-00709

Tribromocarbonothioylbis(triphenylphosphine)ruthenium, 9CI

[50938-34-6]

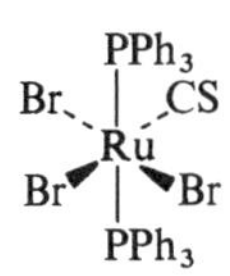

M 847.486
Dark-purple cryst. Mp 185° dec.

Ruiz-Ramirez, L. *et al*, *J. Chem. Soc., Dalton Trans.*, 1973, 1770 (*synth, ir*)

$C_{37}H_{30}ClNO_2P_2Ru$ Ru-00710

Carbonylchloronitrosylbis(triphenylphosphine)ruthenium, 8CI

[22863-37-2]

As Bromocarbonylnitrosylbis(triphenylphosphine)ruthenium, Ru-00708 with

X = Cl

M 719.120
Yellow-brown cryst. Mp 164-6°.

Laing, K.R. *et al*, *J. Chem. Soc. (A)*, 1970, 2149 (*synth, ir*)
Doyle, G., *J. Organomet. Chem.*, 1982, **224**, 355 (*synth, ir*)

$C_{37}H_{30}Cl_3P_2RuS^{\ominus}$ Ru-00711

(Carbonothioyl)trichlorobis(triphenylphosphine)ruthenate(1–), 9CI

Trichloro(thiocarbonyl)bis(triphenylphosphine)ruthenate(1–), 8CI

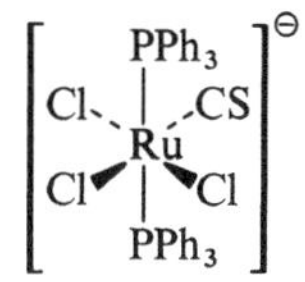

M 776.081 (ion)

Tetraphenylarsonium salt: [34214-62-5].
$C_{61}H_{50}AsCl_3P_2RuS$ M 1159.424
Red cryst. Mp 180-3°.

Stephenson, T.A. *et al*, *J. Chem. Soc., Dalton Trans.*, 1974, 1134 (*synth, ir*)

$C_{37}H_{30}INO_2P_2Ru$ Ru-00712

Carbonyliodonitrosylbis(triphenylphosphine)ruthenium, 9CI, 8CI

[29826-66-2]

As Bromocarbonylnitrosylbis(triphenylphosphine)ruthenium, Ru-00708 with

X = I

M 810.572
Brown cryst. Mp 136-8° dec.

Laing, K.R. *et al*, *J. Chem. Soc. (A)*, 1970, 2149 (*synth, ir*)
Hall, D. *et al*, *Cryst. Struct. Commun.*, 1974, **3**, 327 (*cryst struct*)

$C_{37}H_{47}IO_9P_3Ru^{\oplus}$ Ru-00713

Iodo(η^4-tetraphenylcyclobutadiene)tris(trimethyl phosphite-*P*)ruthenium(1+)

*[1,1',1'',1'''-(η^4-1,3-Cyclobutadiene-1,2,3,4-tetrayl)tetrakis[benzene]]iodotris(trimethyl phosphite-*P*)ruthenium(1+), 10CI*

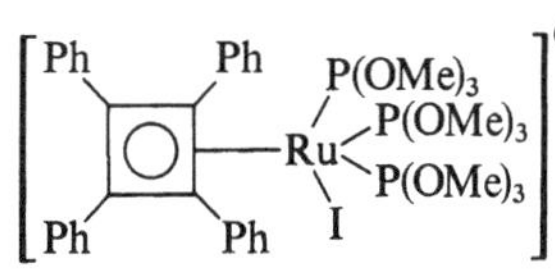

M 956.669 (ion)

Triiodide: [77347-86-5].
$C_{37}H_{47}I_4O_9P_3Ru$ M 1337.382
Air-stable yellow-orange cryst. (CH_2Cl_2/hexane). Sol. polar solvs. Mp 104° dec. Mod. air-stable in soln.

Connelly, N.G. *et al*, *J. Chem. Soc., Dalton Trans.*, 1981, 34 (*synth*)

$C_{37}H_{47}O_9P_3Ru$ Ru-00714

(η^4-Tetraphenylcyclobutadiene)tris(trimethyl phosphite-*P*)ruthenium

*[1,1',1'',1'''-(η^4-1,3-Cyclobutadiene-1,2,3,4-tetrayl)tetrakis[benzene]]tris(trimethyl phosphite-*P*)ruthenium, 10CI*

[77347-82-1]

As Carbonyl(η^4-tetraphenylcyclobutadiene)bis(trimethylphosphite-*P*)ruthenium, Ru-00688 with

X = $P(OMe)_3$

M 829.764

Yellow cryst. (hexane). Stable in air. Solns. moderately air-stable. Sol. polar solvs. Mp 188-92° dec.

Connelly, N.G. *et al*, *J. Chem. Soc., Dalton Trans.*, 1981, 34 (*synth, pmr*)

$C_{38}H_{30}Br_2OP_2RuSe$ **Ru-00715**

Dibromo(carbonoselenoyl)carbonylbis(triphenylphosphine)-ruthenium, 9CI

[56260-11-8]

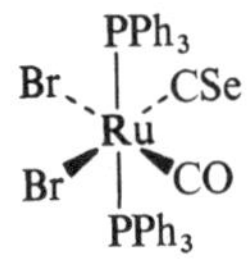

M 904.440

Colourless, stable to air and moisture.

Clark, G.R. *et al*, *J. Organomet. Chem.*, 1975, **90**, C37 (*synth, ir*)

$C_{38}H_{30}Br_2O_2P_2Ru$ **Ru-00716**

Dibromodicarbonylbis(triphenylphosphine)ruthenium

[14564-32-0]

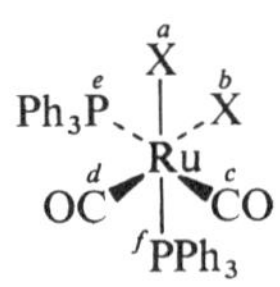

ab-dibromo-*cd*-dicarbonyl-*ef*-diphosphine-*form*

X = Br

M 841.479

ab-dibromo-cd-dicarbonyl-ef-diphosphine-form [29099-84-1]

cis-*form*

White cryst. (CH_2Cl_2/MeOH). Mp 270°, 335°.

af-dibromo-bd-dicarbonyl-ce-diphosphine-form

trans-*form*

Fawn.

Stephenson, T.A. *et al*, *J. Inorg. Nucl. Chem.*, 1966, **28**, 945 (*synth, ir*)
Cleare, M.J. *et al*, *J. Chem. Soc.* (*A*), 1969, 372 (*synth, ir*)
Hieber, W. *et al*, *Chem. Ber.*, 1970, **103**, 2161 (*synth, ir*)
John, P., *Chem. Ber.*, 1970, **103**, 2178 (*struct*)
James, B.R. *et al*, *J. Chem. Soc., Dalton Trans.*, 1973, 2247 (*synth*)
Barnard, C.F.J. *et al*, *J. Chem. Soc., Dalton Trans.*, 1976, 953 (*synth, ir*)
Kuhnen, H.B. *et al*, *J. Organomet. Chem.*, 1976, **105**, 357 (*synth*)

$C_{38}H_{30}ClF_4NOP_2Ru$ **Ru-00717**

Chloronitrosyl(η^2-tetrafluoroethene)bis(triphenylphosphine)-ruthenium, 9CI

[40999-72-2]

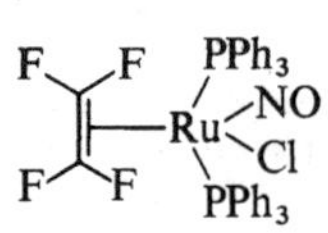

M 791.125

Orange cryst. Dissociates readily in soln.

Clemens, J. *et al*, *J. Chem. Soc., Dalton Trans.*, 1973, 375 (*synth, ir*)

$C_{38}H_{30}Cl_2OP_2RuS$ **Ru-00718**

(Carbonothioyl)carbonyldichlorobis(triphenylphosphine)ruthenium, 9CI

[56298-19-2]

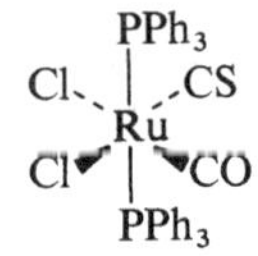

M 768.638

Light-red cryst. (toluene). Mp 157-8°.

Gilbert, J.D. *et al*, *J. Chem. Soc.* (*A*), 1968, 2198 (*synth, ir*)
Grundy, K.R. *et al*, *J. Organomet. Chem.*, 1975, **90**, C34 (*synth*)

$C_{38}H_{30}Cl_2OP_2RuSe$ **Ru-00719**

(Carbonylselenoyl)carbonyldichlorobis(triphenylphosphine)-ruthenium

[56260-19-6]

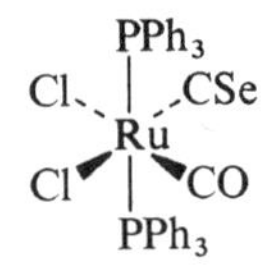

M 815.538

Colourless, air- and moisture-stable cryst.

Clark, G.R. *et al*, *J. Organomet. Chem.*, 1975, **90**, C37; 1977, **134**, 229 (*struct, synth, ir*)

$C_{38}H_{30}Cl_2O_2P_2Ru$ **Ru-00720**

Dicarbonyldichlorobis(triphenylphosphine)ruthenium

[14564-35-3]

As Dibromodicarbonylbis(triphenylphosphine)ruthenium, Ru-00716 with

X = Cl

M 752.577

ab-dicarbonyl-df-dichloro-ce-diphosphine-form [15709-57-6]

cis,cis,trans-*form*

White solid. Mp 309-12°. Exhibits two ν_{CO} bands in ir.

af-dicarbonyl-bd-dichloro-ce-diphosphine-form [15709-73-6]

all-trans-*form*

Pale-yellow. Exhibits one ν_{CO} band in ir.

ab-dicarbonyl-cd-dichloro-ef-diphosphine-form

cis-*form*

Solid. Mp 230-5°.

Collman, J.P. *et al*, *J. Am. Chem. Soc.*, 1965, **87**, 4008 (*synth, ir*)
Stephenson, T.A. *et al*, *J. Inorg. Nucl. Chem.*, 1966, **28**, 945 (*synth, ir, isom*)
Lupin, M.S. *et al*, *J. Chem. Soc.* (*A*), 1968, 741 (*ir*)
Cleare, M.J. *et al*, *J. Chem. Soc.* (*A*), 1969, 372 (*synth, ir*)
Fahey, D.R. *et al*, *J. Org. Chem.*, 1973, **38**, 80 (*synth, ir, nmr*)
James, B.R. *et al*, *J. Chem. Soc., Dalton Trans.*, 1973, 2247 (*ir*)
Barnard, C.F.J. *et al*, *J. Chem. Soc., Dalton Trans.*, 1976, 953 (*synth, spectra*)
Kuhlmann, E.J. *et al*, *J. Organomet. Chem.*, 1979, **174**, 81 (*synth*)
Rauchfuss, T.B., *J. Am. Chem. Soc.*, 1979, **101**, 1045 (*synth*)

$C_{38}H_{30}Cl_2O_8P_2Ru$ Ru-00721

Dicarbonyldichlorobis(triphenyl phosphite-*P*)ruthenium

Dicarbonyldichlorobis(phosphorous acid)ruthenium hexaphenyl ester, 8CI

[28577-82-4]

$Ru(CO)_2Cl_2[P(OPh)_3]_2$

M 848.574

White cryst. ($MeOH/CH_2Cl_2$). Mp 158-60°.

Levison, J.J. *et al*, *J. Chem. Soc. (A)*, 1970, 639 (*synth, ir*)

$C_{38}H_{30}Cl_2P_2RuS_2$ Ru-00722

Bis(carbonothioyl)dichlorobis(triphenylphosphine)ruthenium, 9CI

[51405-78-8]

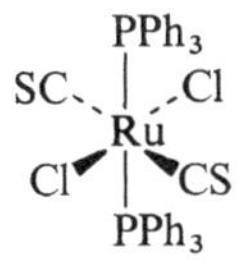

M 784.699

Dark-brown solid. Mp 160°. Exhibits single ν_{CS} ir frequency.

Poddar, R.K. *et al*, *J. Inorg. Nucl. Chem.*, 1973, **35**, 3769 (*synth, ir, uv*)

$C_{38}H_{30}I_2O_2P_2Ru$ Ru-00723

Dicarbonyldiodobis(triphenylphosphine)ruthenium

[14564-39-7]

As Dibromodicarbonylbis(triphenylphosphine)ruthenium, Ru-00716 with

X = I

M 935.480

ab-dicarbonyl-cd-diiodo-ef-diphosphine-form [62´77-06-4]

cis-*form*

Yellow cryst. ($CHCl_3/C_6H_6$). Mp 320°. ir ν_{CO} 2055, 1994 cm^{-1} ($CHCl_3$).

af-dicarbonyl-bd-diiodo-ce-diphosphine-form [29091-98-3]

trans-*form*

Obt. by irradiation of *cis*-form. Buff cryst. ($CHCl_3$), red cryst. + THF. Mp 135° (red form). ir ν_{CO} 1998 cm^{-1} (nujol).

Cleare, M.J. *et al*, *J. Chem. Soc. (A)*, 1969, 372 (*synth, ir*)
John, P. *et al*, *Chem. Ber.*, 1970, **103**, 2161 (*synth, isom*)
Jeffery, J. *et al*, *J. Organomet. Chem.*, 1972, **40**, C42 (*isom*)
Kingston, J.V. *et al*, *J. Inorg. Nucl. Chem.*, 1972, **34**, 227 (*synth, ir*)
Barnard, C.F.J. *et al*, *J. Chem. Soc., Dalton Trans.*, 1976, 953 (*synth, ir, isom*)
Cenini, S. *et al*, *Inorg. Chim. Acta*, 1976, **20**, 119 (*synth*)
Kuhnhen, H.B., *J. Organomet. Chem.*, 1977, **129**, 215 (*synth, ir, nmr*)

$C_{38}H_{30}NO_3P_2Ru^{\oplus}$ Ru-00724

Dicarbonylnitrosylbis(triphenylphosphine)ruthenium(1+), 8CI

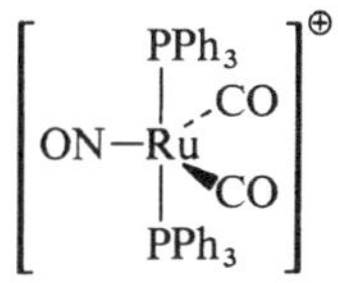

M 711.677 (ion)

Hexafluorophosphate: [33881-64-0].
$C_{38}H_{30}F_6NO_3P_3Ru$ M 856.642
Orange cryst. (CH_2Cl_2/Et_2O).

Tetraphenylborate: [33792-22-2].
$C_{62}H_{50}BNO_3P_2Ru$ M 1030.909
Cream cryst. (CH_2Cl_2/Et_2O).

Johnson, B.F.G. *et al*, *J. Chem. Soc., Dalton Trans.*, 1973, 478 (*synth, ir*)

$C_{38}H_{30}O_2P_2RuS_2$ Ru-00725

Dicarbonyl(dithioperoxy)bis(triphenylphosphine)ruthenium, 10CI

[64479-11-4]

As Dicarbonyl(dioxygen)bis(triphenylphosphine)ruthenium, Ru-00726 with

X = S

M 745.791

Clark, G.R. *et al*, *J. Organomet. Chem.*, 1977, **136**, C1 (*synth, ir*)
Farrar, D.H. *et al*, *J. Am. Chem. Soc.*, 1979, **101**, 6577.

$C_{38}H_{30}O_4P_2Ru$ Ru-00726

Dicarbonyl(dioxygen)bis(triphenylphosphine)ruthenium, 10CI

[35880-56-9]

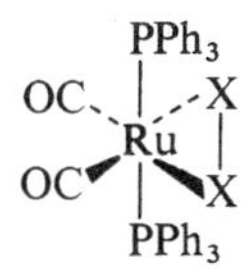

X = O

M 713.670

No phys. props. reported.

Cavit, B.E. *et al*, *J. Chem. Soc., Chem. Commun.*, 1972, 60 (*synth, ir*)
Porta, F. *et al*, *J. Organomet. Chem.*, 1978, **150**, 261 (*synth, ir*)

$C_{38}H_{30}O_4P_2RuS$ Ru-00727

Dicarbonyl[(*O,S*-η)-sulfur dioxide]bis(triphenylphosphine)-ruthenium, 9CI

[31200-91-6]

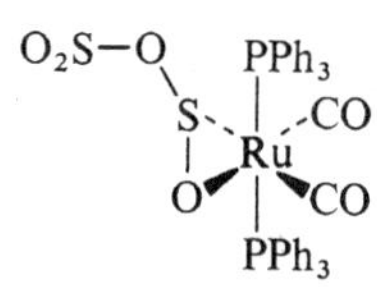

SO₂ complex

M 745.730

Mustard-yellow air-sensitive cryst.

SO_2 complex: [76612-05-0].
$C_{38}H_{30}O_6P_2RuS_2$ M 809.789
Cryst. (C_6H_6/heptane).

Collman, J.P. *et al*, *Inorg. Chem.*, 1971, **10**, 219 (*synth, ir*)

Moody, D.C. *et al*, *J. Chem. Soc., Chem. Commun.*, 1980, 1230 (*cryst struct*)

$C_{38}H_{30}O_6P_2RuS$ Ru-00728

Dicarbonyl[sulfata(2–)-*O,O'*]bis(triphenylphosphine)ruthenium, 9CI

Dicarbonylbis(triphenylphosphine)ruthenium sulfate

[31075-99-7]

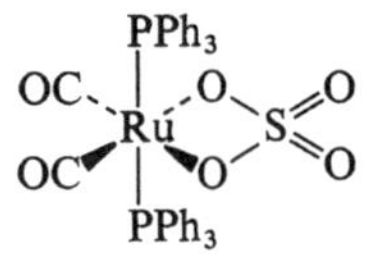

M 777.729

White powder.

Valentine, J. *et al*, *Inorg. Chem.*, 1971, **10**, 219 (*synth, ir*)
Moody, D.C. *et al*, *Cryst. Struct. Commun.*, 1976, **5**, 145 (*struct*)

$C_{38}H_{31}BrO_2P_2Ru$ Ru-00729

Bromodicarbonylhydrobis(triphenylphosphine)ruthenium, 9CI

[52538-83-7]

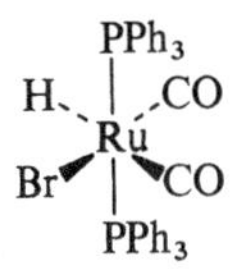

M 762.583

White cryst. ($MeOH/CH_2Cl_2$).

James, B.R. *et al*, *J. Chem. Soc., Dalton Trans.*, 1973, 2247 (*synth, ir*)

$C_{38}H_{31}ClO_2P_2Ru$ Ru-00730

Dicarbonylchlorohydrobis(triphenylphosphine)ruthenium, 8CI, 9CI

[56144-64-0]

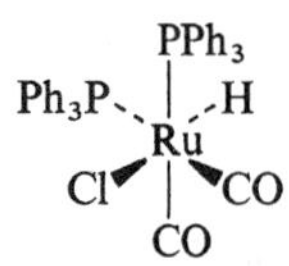

M 718.132

Colourless cryst. ($MeOH/CH_2Cl_2$). Mp 126-7°. Air stable.

James, B.R. *et al*, *J. Chem. Soc., Dalton Trans.*, 1973, 2247 (*synth*)
Cenini, S. *et al*, *Gazz. Chim. Ital.*, 1975, **105**, 255 (*synth*)
Young, R.J. *et al*, *J. Chem. Soc., Dalton Trans.*, 1976, 719.
Geffroy, G.L. *et al*, *Inorg. Chem.*, 1977, **16**, 744 (*nmr, ir, uv*)

$C_{38}H_{32}O_2P_2Ru$ Ru-00731

Dicarbonyldihydrobis(triphenylphosphine)ruthenium, 9CI

[21029-29-8]

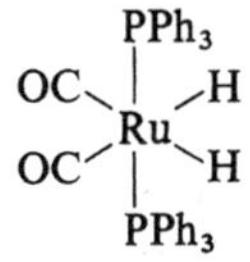

M 683.687

Isomerisation, hydrogenation and hydroformylation catalyst. Cream ppt. (THF/heptane). Mp 166-7°.

Cotton, J.D. *et al*, *J. Chem. Soc. (A)*, 1968, 2162 (*synth, ir*)
L'Eplattenier, F. *et al*, *Inorg. Chem.*, 1968, **7**, 1290 (*synth, ir, pmr*)
Porta, F. *et al*, *J. Organomet. Chem.*, 1968, **150**, 261 (*use*)
U.S.P., 3 597 461, (*1971*); *CA*, **75**, 110424 (*synth*)
Cenini, S. *et al*, *Inorg. Chim. Acta*, 1976, **20**, 119 (*synth*)
Sanchez-Delgado, R.A. *et al*, *J. Chem. Soc., Dalton Trans.*, 1976, 399 (*use*)

$C_{38}H_{33}Cl_2NO_2P_2Ru$ Ru-00732

Acetyldichloronitrosylbis(triphenylphosphine)ruthenium, 9CI

[58448-30-9]

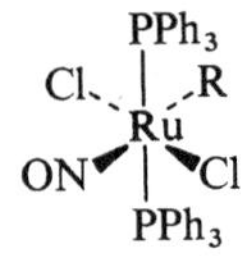

R = Ac

M 769.608

Yellow.

Giovannitti, B. *et al*, *J. Organomet. Chem.*, 1977, **129**, 207 (*synth, ir*)

$C_{38}H_{33}O_3P_2Ru^{\oplus}$ Ru-00733

Aquadicarbonylhydrobis(triphenylphosphine)ruthenium(1+), 9CI

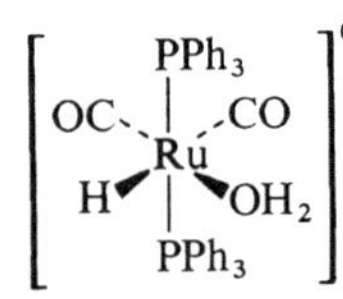

M 700.694 (ion)

Tetrafluoroborate: [60020-25-9].
$C_{38}H_{33}BF_4O_3P_2Ru$ M 787.498
Colourless needles + 1 EtOH or large colourless cryst. + 1 H_2O (CH_2Cl_2/EtOH aq.). Mp 122-5° (monohydrate).

Young, R.J. *et al*, *J. Chem. Soc., Dalton Trans.*, 1976, 719 (*synth*)
Boniface, S.M. *et al*, *J. Organomet. Chem.*, 1981, **206**, 109 (*synth, ir, cryst struct, pmr*)

$C_{38}H_{39}B_7P_2Ru$ Ru-00734

[(4,5,6,7,8,9-η)-Nonahydro-6,8-dicarbanonaborato(4–)]bis(triphenylphosphine)ruthenium, 10CI

Hyper-closo-*6,6-Bis(triphenylphosphine)-6-ruthenanonahydro-2,3-dicarbanonaborane*

[61081-79-6]

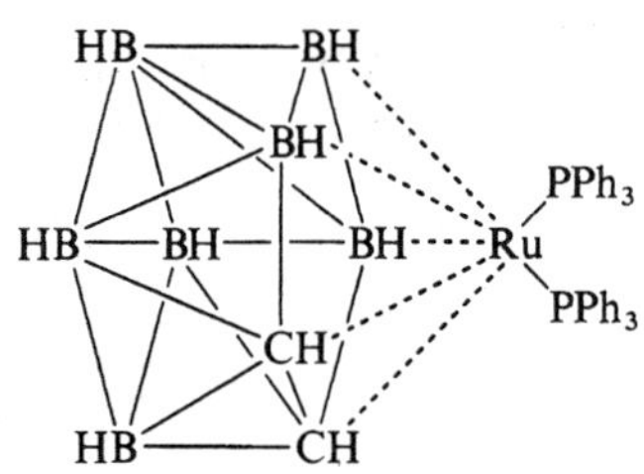

M 734.414

Catalyst for homogeneous hydrogenation of alkenes. Shiny black needles (CH_2Cl_2/EtOH). Mp 209-12°.

Jung, C.W. *et al*, *J. Am. Chem. Soc.*, 1981, **103**, 811 (*synth, pmr, nmr*)

$C_{38}H_{42}B_9ClP_2Ru$ **Ru-00735**

Chlorohydrobis(triphenylphosphine)[(7,8,9,10,11-η)undecahydro-7,8-dicarbaundecaborato(2−)]ruthenium

[67328-91-0]

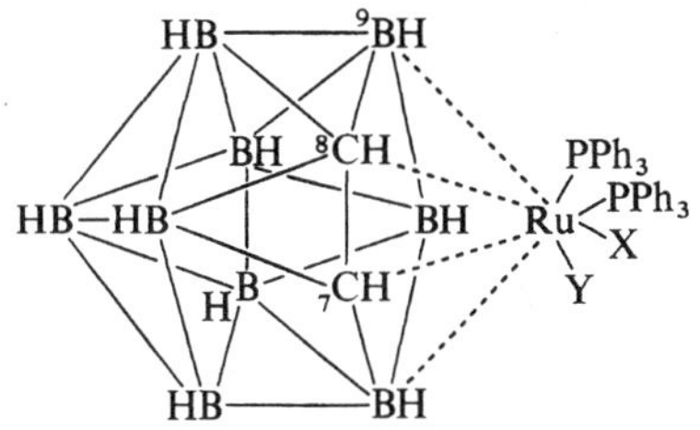

X = Cl, Y = H

M 794.510

Orange-red cryst.

Wong, E.H.S. *et al*, *Inorg. Chem.*, 1978, **17**, 2863 (*synth*, *ir*)

$C_{38}H_{42}B_9ClP_2Ru$ **Ru-00736**

Chlorohydrobis(triphenylphosphine)[(7,8,9,10,11-η)-undecahydro-7,9-dicarbaundecaborato(2−)]ruthenium

[67350-80-5]

M 794.510

Yellow cryst. Spar. sol. org. solvs.

Wong, E.H.S. *et al*, *Inorg. Chem.*, 1978, **17**, 2863 (*synth*, *ir*)

$C_{38}H_{43}B_9P_2Ru$ **Ru-00737**

Dihydrobis(triphenylphosphine)[(7,8,9,10,11-η)undecahydro-7,8-dicarbaundecaborato(2−)]ruthenium

[60172-21-6]

As Chlorohydrobis(triphenylphosphine)[(7,8,9,10,11-η)undecahydro-7,8-dicarbaundecaborato(2−)]ruthenium, Ru-00735 with

X = Y = H

M 760.065

Colourless cryst.

Wong, E.H.S. *et al*, *J. Chem. Soc., Chem. Commun.*, 1976, 257; *Inorg. Chem.*, 1978, **17**, 2863 (*synth*, *pmr*, *nmr*, *ir*)

$C_{38}H_{43}B_9P_2Ru$ **Ru-00738**

Dihydrobis(triphenylphosphine)[(7,8,9,10,11-η)-undecahydro-7,9-dicarbaundecaborato(2−)]ruthenium, 9CI

[*2,2-Bis(triphenylphosphine)-2,2-dihydro-2-ruthena]-undecahydro-1,7-dicarbaundecaborane*

[60172-17-0]

M 760.065

Stereochemically nonrigid. Light-blue cryst. (dioxan or toluene/EtOH), grey-white when pure. Reversibly forms a dark-blue dehydro complex on heating.

Wong, E.H.S. *et al*, *J. Chem. Soc., Chem. Commun.*, 1976, 257; *Inorg. Chem.*, 1978, **17**, 2863 (*synth*, *struct*, *pmr*, *nmr*, *ir*)

Marder, T.B. *et al*, *J. Am. Chem. Soc.*, 1981, **103**, 2988 (*pmr*, *nmr*)

$C_{39}H_{30}As_2O_3Ru$ **Ru-00739**

Tricarbonylbis(triphenylarsine)ruthenium, 9CI, 8CI

[14655-23-3]

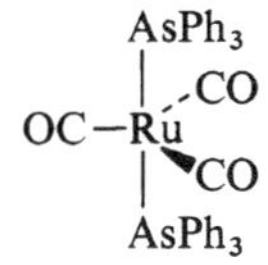

M 797.577

Pale-yellow cryst.

Collman, J.P. *et al*, *J. Am. Chem. Soc.*, 1965, **87**, 4008 (*synth*, *ir*)

Magoon, E.F. *et al*, *J. Organomet. Chem.*, 1975, **99**, 135 (*use*)

$C_{39}H_{30}ClO_3P_2Ru^{\oplus}$ **Ru-00740**

Tricarbonylchlorobis(triphenylphosphine)ruthenium(1+), 8CI

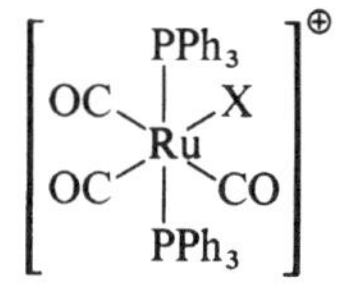

X = Cl

M 745.135 (ion)

Tetrachloroaluminate: [16594-36-8].

$C_{39}H_{30}AlCl_5O_3P_2Ru$ M 913.928

Yellow cryst. Dec. at 250-2°.

Perchlorate: [83111-40-4].

$C_{39}H_{30}Cl_2O_7P_2Ru$ M 844.585

No phys. props. reported.

Hieber, W. *et al*, *Chem. Ber.*, 1967, **100**, 1961 (*synth*, *ir*)

Jungbauer, A. *et al*, *Z. Naturforsch., B*, 1979, **34**, 1641 (*synth*)

Clark, G.R. *et al*, *J. Organomet. Chem.*, 1982, **234**, C9 (*synth*)

$C_{39}H_{30}IO_3P_2Ru^{\oplus}$ **Ru-00741**

Tricarbonyliodobis(triphenylphosphine)ruthenium(1+), 9CI

As Tricarbonylchlorobis(triphenylphosphine)ruthenium(1+), Ru-00740 with

X = I

M 836.586 (ion)

Iodide: [59169-04-9].

$C_{39}H_{30}I_2O_3P_2Ru$ M 963.491

No phys. props. reported.

Trichloroiodoaluminate: [16592-04-4].

$C_{39}H_{30}AlCl_3I_2O_3P_2Ru$ M 1096.831

Yellow cryst. Dec. at 276-8°.

Hieber, W. *et al*, *Chem. Ber.*, 1967, **100**, 1961 (*synth*, *ir*)

Kuhnhen, H.B., *J. Organomet. Chem.*, 1976, **105**, 357; 1977, **129**, 215 (*synth*, *ir*)

$C_{39}H_{30}O_3P_2Ru$ **Ru-00742**

Tricarbonylbis(triphenylphosphine)ruthenium, 9CI

[14741-36-7]

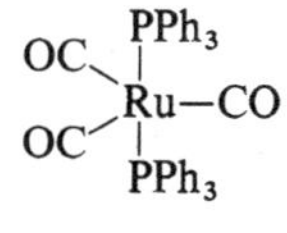

M 709.682

***trans*-form** [20332-49-4]

Hydroformylation catalyst. Pale-yellow cryst. (THF/heptane or C_6H_6/hexane). Sol. C_6H_6, CH_2Cl_2. Mp 262-6° (sealed under N_2), 170-3° (in air).

Collman, J.P. *et al*, *J. Am. Chem. Soc.*, 1965, **87**, 4008 (*synth*)
L'Eplattenier, F. *et al*, *Inorg. Chem.*, 1968, **7**, 1290 (*synth, ir*)
U.S.P., 3 597 461, (*1971*); *CA*, **75**, 110424 (*synth*)
Ahmad, N. *et al*, *J. Chem. Soc., Dalton Trans.*, 1972, 843 (*synth*)
Inorg. Synth., 1974, **15**, 50 (*synth, ir*)
Cenini, S. *et al*, *Gazz. Chim. Ital.*, 1975, **105**, 255 (*synth*)
Johnson, B.F.G. *et al*, *J. Chem. Soc., Dalton Trans.*, 1975, 1876 (*synth, ir*)
Klopsch, A. *et al*, *Ber. Bunsenges. Phys. Chem.*, 1976, **80**, 500 (*ir*)
Graaf, J.L. *et al*, *Organometallics*, 1982, **1**, 837 (*ir*)

$C_{39}H_{30}O_9P_2Ru$ Ru-00743

Tricarbonylbis(triphenyl phosphite-*P*)ruthenium

Tricarbonylbis(phosphorous acid)ruthenium hexaphenyl ester, 8CI

[31447-11-7]

$[(PhO)_3P]_2Ru(CO)_3$

M 805.678

No phys. props. reported.

Ger. Pat., 2 026 926, (*1971*); *CA*, **74**, 124822

$C_{39}H_{31}O_3P_2Ru^{\oplus}$ Ru-00744

Tricarbonylhydrobis(triphenylphosphine)ruthenium(1+), 9CI, 8CI

As Tricarbonylchlorobis(triphenylphosphine)ruthenium(1+), Ru-00740 with

X = H

M 710.690 (ion)

Hexafluorophosphate: [40530-71-0].
$C_{39}H_{31}F_6O_3P_3Ru$ M 855.654
White cryst.

Johnson, B.F.G. *et al*, *J. Chem. Soc., Dalton Trans.*, 1973, 478 (*synth, ir, pmr*)

$C_{39}H_{33}ClOP_2RuSe_2$ Ru-00745

Carbonylchloro[η^2-(methylseleno)selenoxomethyl]bis(triphenylphosphine)ruthenium, 9CI

[56260-08-3]

$(H_3CCSe_2)RuCl(CO)(PPh_3)_2$

M 874.080

No phys. props. reported. ir ν_{CO} 1945, ν_{CSe} 950 cm^{-1}.

Clark, G.R. *et al*, *J. Organomet. Chem.*, 1975, **90**, C37.

$C_{39}H_{33}ClO_2P_2Ru$ Ru-00746

Acetylcarbonylchlorobis(triphenylphosphine)ruthenium, 8CI

[34808-05-4]

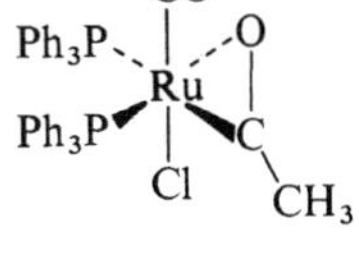

M 732.159

Characterised spectroscopically.

Sears, D. *et al*, *J. Chem. Soc., Chem. Commun.*, 1971, 777 (*synth, ir, pmr*)

$C_{39}H_{33}ClP_2RuS_2$ Ru-00747

Carbonylchloro[η^2-(methylthio)thioxomethyl]bis(triphenylphosphine)ruthenium, 9CI

[56260-16-3]

$(H_3CCS_2)RuCl(CO)(PPh_3)_2$

M 764.280

Bonding of dithioester group not certain. No phys. props. reported. ir ν_{CO} 1948, ν_{CS} 1085 cm^{-1}.

Grundy, K.R. *et al*, *J. Organomet. Chem.*, 1975, **90**, C34 (*synth, ir*)

$C_{39}H_{35}NOP_2Ru$ Ru-00748

Nitrosyl(η^3-2-propenyl)bis(triphenylphosphine)ruthenium

(η^3-Allyl)nitrosylbis(triphenylphosphine)ruthenium

[65230-38-8]

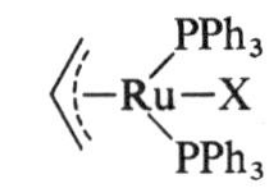

X = NO

M 696.729

Red cryst. (THF/Et_2O), stable in air for several months but v. air-sensitive in soln. Mod. sol. C_6H_6, THF, CH_2Cl_2, dec. rapidly $CHCl_3$, CCl_4.

Schoonover, M.W. *et al*, *J. Am. Chem. Soc.*, 1977, **99**, 8371 (*synth, ir, pmr, cryst struct*)
Schoonover, M.W. *et al*, *Inorg. Chem.*, 1978, **17**, 3050 (*cryst struct*)

$C_{39}H_{41}B_9OP_2Ru$ Ru-00749

Carbonylbis(triphenylphosphine)[(7,8,9,10,11-η)undecahydro-7,8-dicarbaundecaborato(2−)]ruthenium, 9CI

[56465-05-5]

As Chlorohydrobis(triphenylphosphine)[(7,8,9,10,11-η)undecahydro-7,8-dicarbaundecaborato(2−)]ruthenium, Ru-00735 with

X = Y = CO

M 786.060

Yellow cryst. (EtOH/CH_2Cl_2).

Wong, E.H.J. *et al*, *Inorg. Chem.*, 1978, **17**, 2863 (*synth, ir*)

$C_{39}H_{41}B_9OP_2Ru$ Ru-00750

Carbonylbis(triphenylphosphine)[(7,8,9,10,11-η)-undecahydro-7,9-dicarbaundecaborato(2−)]ruthenium, 9CI

[60172-19-2]

M 786.060

Yellow cryst. (CH_2Cl_2/EtOH).

Wong, E.H.S. *et al*, *Inorg. Chem.*, 1978, **17**, 2863 (*synth, ir*)

$C_{40}H_{30}ClF_6NOP_2Ru$ Ru-00751

Chloro[(2,3-η)-1,1,1,4,4,4-hexafluoro-2-butyne]nitrosylbis-(triphenylphosphine)ruthenium

[Bis(trifluoromethyl)acetylene]chloronitrosylbis(triphenylphosphine)ruthenium

[40999-71-1]

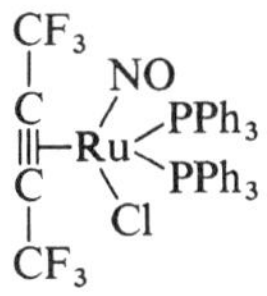

M 853.144

Red cryst. Sol. C_6H_6. Mp 209-12° dec.

Clemens, J. *et al*, *J. Chem. Soc., Dalton Trans.*, 1973, 375 (*synth, ir, nmr*)

$C_{40}H_{30}Cl_4O_4P_2Ru_2$ Ru-00752

Tetracarbonyldi-μ-chlorodichlorobis(triphenylphosphine)diruthenium, 9CI

Tetracarbonyltetrachlorobis(triphenylphosphine)diruthenium

[21216-91-1]

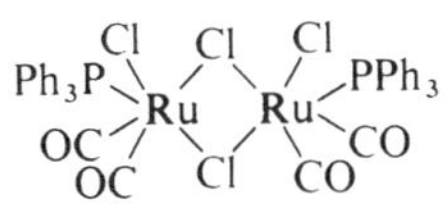

M 980.574

Bridged dimer. Hydrogenation catalyst. Yellow cryst. (CH_2Cl_2/pentane). Mp >250°.

Monomer: Dicarbonyldichloro(triphenylphosphine)-ruthenium.

$C_{20}H_{15}Cl_2O_2PRu$ M 490.287

Unknown.

Piacenti, F. *et al*, *Inorg. Chem.*, 1968, **7**, 1815 (*synth, ir*)
Johnson, B.F.G. *et al*, *J. Chem. Soc. (A)*, 1969, 792 (*synth, ir*)
U.S.P., 3 906 045, (*1975*); *CA*, **84**, 4637 (*use*)

$C_{40}H_{30}O_4RuSn_2$ Ru-00753

Tetracarbonylbis(triphenylstannyl)ruthenium

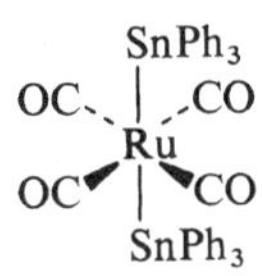

M 913.125

trans-form [17653-14-4]

Pale-yellow solid. Mp 180-2°.

Cotton, J.D. *et al*, *J. Chem. Soc. (A)*, 1968, 2162, 2758 (*synth, ir*)

$C_{40}H_{33}IO_4P_2Ru$ Ru-00754

Dicarbonyliodo(methoxycarbonyl)bis(triphenylphosphine)-ruthenium

[16594-61-9]

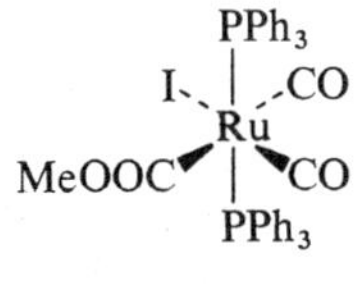

M 867.620

Characterised spectroscopically.

Hieber, W. *et al*, *Chem. Ber.*, 1967, **100**, 1961 (*synth, ir*)

$C_{40}H_{34}O_2P_2Ru$ Ru-00755

Dicarbonyl(η^2-ethene)bis(triphenylphosphine)ruthenium

Dicarbonyl(ethylene)bis(triphenylphosphine)ruthenium

[35795-47-2]

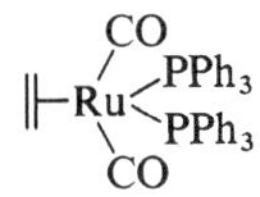

M 709.725

No phys. props. recorded.

Cavit, B.E. *et al*, *J. Chem. Soc., Chem. Commun.*, 1972, 60 (*synth, ir*)
Schramm, K.D. *et al*, *Inorg. Chem.*, 1980, **19**, 2441 (*synth*)

$C_{40}H_{34}Ru_4$ Ru-00756

1,1'':1''',1'''':1''''',1''''''-Quaterruthenocene, 10CI

1,1'-Quaterruthenocene

[73178-91-3]

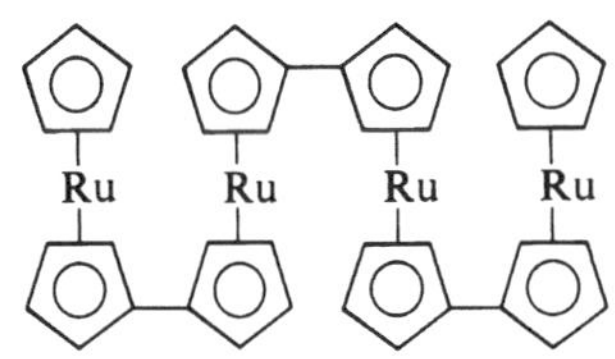

M 918.989

Bednarik, L. *et al*, *J. Org. Chem.*, 1980, **45**, 2032 (*synth, uv*)

$C_{40}H_{35}ClO_2P_2Ru$ Ru-00757

Carbonylchloro(1-oxopropyl)bis(triphenylphosphine)ruthenium, 9CI

Carbonylchloropropionylbis(triphenylphosphine)ruthenium, 8CI

[34808-04-3]

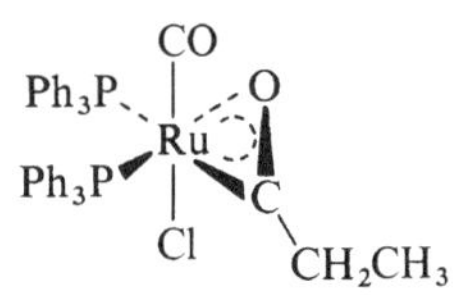

M 746.186

Forms 1:1 adducts with benzene and dichloromethane.

Sears, D. *et al*, *J. Chem. Soc., Chem. Commun.*, 1971, 777 (*synth, ir, pmr*)
Braca, G. *et al*, *Chem. Ind. (Milan)*, 1974, **56**, 110 (*synth*)
Cole-Hamilton, D.J. *et al*, *Nouv. J. Chim.*, 1977, **1**, 141.

$C_{40}H_{36}Cl_2N_2P_2Ru$ Ru-00758

Bis(acetonitrile)dichlorobis(triphenylphosphine)ruthenium, 9CI

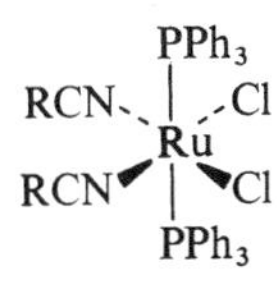

R = H_3C—

M 778.661

***cis,cis,trans*-form** [26407-33-0]
Yellow cryst. (Me_2CO). Dec. at 160°.

Gilbert, J.D. *et al*, *J. Chem. Soc. (A)*, 1969, 1749 (*synth*, *ir*)
Poddar, R.J. *et al*, *J. Inorg. Nucl. Chem.*, 1973, **35**, 3769 (*synth*, *ir*)
Cole-Hamilton, D.J. *et al*, *J. Chem. Soc., Dalton Trans.*, 1979, 1283 (*synth*, *ir*, *nmr*)

$C_{40}H_{36}O_2P_2RuS_4$ Ru-00759

Bis(*O*-methylcarbonodithioato-*S,S'*)bis(triphenylphosphine)-ruthenium, 9CI
Bis(O-methyldithiocarbonato)bis(triphenylphosphine)-ruthenium

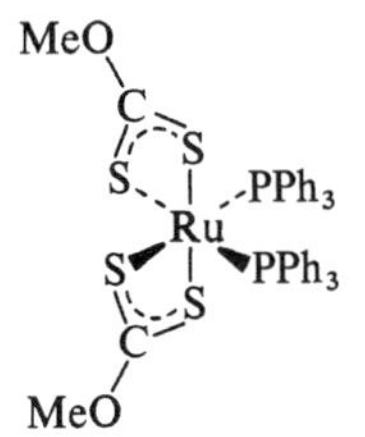

M 839.981

***cis*-form** [56846-19-6]
Orange cryst. (CH_2Cl_2/MeOH). Mp 170-1°.

O'Connor, C. *et al*, *J. Chem. Soc. (A)*, 1969, 84 (*synth*, *ir*)
Critchlow, P.B. *et al*, *J. Chem. Soc., Dalton Trans.*, 1975, 1367 (*synth*, *pmr*, *ir*)

$C_{41}H_{34}O_6P_2Ru$ Ru-00760

(η^5-2,4-Cyclopentadien-1-yl)[2-[(diphenoxyphosphino)oxy]-phenyl-*C,P*](triphenyl phosphite-*P*)ruthenium, 9CI
[37668-63-6]

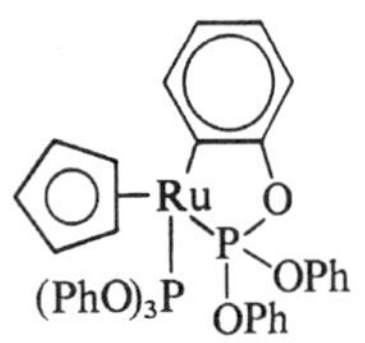

M 785.734
Pale-yellow cryst. Mp 116-8°.

Bruce, M.I. *et al*, *J. Chem. Soc., Dalton Trans.*, 1976, 81 (*synth*, *ir*, *pmr*)

$C_{41}H_{34}P_2Ru$ Ru-00761

(η^5-2,4-Cyclopentadien-1-yl)[2-(diphenylphosphino)phenyl-*C,P*](triphenylphosphine)ruthenium, 9CI
[37668-62-5]

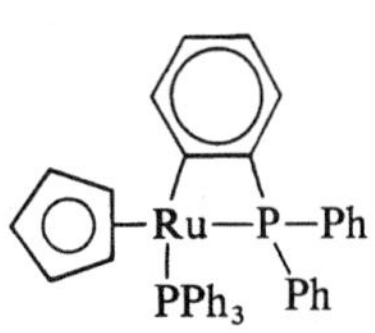

M 689.737
Pale-yellow cryst. Mp 223-5°.

Bruce, M.I. *et al*, *J. Chem. Soc., Dalton Trans.*, 1976, 81 (*synth*, *ir*, *pmr*)

$C_{41}H_{35}As_2ClRu$ Ru-00762

Chloro(η^5-2,4-cyclopentadien-1-yl)bis(triphenylarsine)ruthenium
(η^5-Cyclopentadienyl)chlorobis(triphenylarsine)ruthenium. Cyclopentadienylbis(triphenylarsine)ruthenium chloride
[34692-08-5]

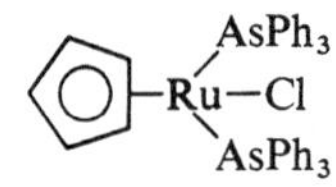

M 814.094
Orange cryst. (CH_2Cl_2/hexane). Mp 158-61° dec.

Blackmore, T. *et al*, *J. Chem. Soc. (A)*, 1971, 2376 (*synth*, *ir*, *pmr*)

$C_{41}H_{35}BrP_2Ru$ Ru-00763

Bromo(η^5-2,4-cyclopentadien-1-yl)bis(triphenylphosphine)-ruthenium
(η^5-Cyclopentadienyl)bromobis(triphenylphosphine)-ruthenium. Cyclopentadienylbis(triphenylphosphine)-ruthenium bromide
[32993-06-9]

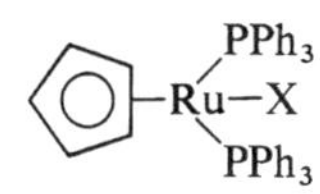

X = Br

M 770.649
Orange cryst. Mp 135-8°, 212-8°.

Gilbert, J.D. *et al*, *J. Chem. Soc. (A)*, 1969, 1749 (*synth*, *pmr*)
Bruce, M.I. *et al*, *Aust. J. Chem.*, 1977, **30**, 1601 (*synth*)
Wilczewski, T. *et al*, *J. Organomet. Chem.*, 1981, **215**, 87 (*synth*, *pmr*, *ms*)

$C_{41}H_{35}ClO_6P_2Ru$ Ru-00764

Chloro(η^5-2,4-cyclopentadien-1-yl)bis(triphenyl phosphite-*P*)ruthenium, 9CI
Chloro-π-cyclopentadienylbis(phosphorous acid)ruthenium hexaphenyl ester, 8CI. Cyclopentadienylbis(triphenyl phosphite)ruthenium chloride
[59557-52-7]

As Chloro(η^5-2,4-cyclopentadien-1-yl)bis(trimethyl phosphite-*P*)ruthenium, Ru-00206 with

R = Ph

M 822.194
Orange cryst. Mp 152-4°.

Bruce, M.I. *et al*, *J. Chem. Soc., Dalton Trans.*, 1976, 81 (*synth*, *ir*, *pmr*)

$C_{41}H_{35}ClP_2Ru$ Ru-00765

Chloro(η^5-2,4-cyclopentadien-1-yl)bis(triphenylphosphine)-ruthenium, 9CI
(η^5-Cyclopentadienyl)chlorobis(triphenylphosphine)-ruthenium. Cyclopentadienylbis(triphenylphosphine)-ruthenium chloride
[32993-05-8]

As Bromo(η^5-2,4-cyclopentadien-1-yl)-bis(triphenylphosphine)ruthenium, Ru-00763 with

X = Cl

M 726.198

Air-stable orange cryst. (CH_2Cl_2/hexane or MeOH). Sol. C_6H_6, CH_2Cl_2, insol. H_2O, pet. ether. Mp 131-7°, 236-48°. Polymorphic.

Gilbert, J.D. *et al*, *J. Chem. Soc. (A)*, 1969, 1749 (*synth, pmr, ir*)
Blackmore, T. *et al*, *J. Chem. Soc. (A)*, 1971, 2376 (*synth, ir, pmr*)
Bruce, M.I. *et al*, *J. Chem. Soc., Dalton Trans.*, 1981, 1398 (*struct*)
Wilczewski, T. *et al*, *J. Organomet. Chem.*, 1981, **215**, 87 (*synth, pmr, nmr, ms*)
Inorg. Synth., 1982, **21**, 79 (*synth, pmr*)

$C_{41}H_{35}FP_2Ru$ Ru-00766

(η^5-2,4-Cyclopentadien-1-yl)fluorobis(triphenylphosphine)-ruthenium
Cyclopentadienylbis(triphenylphosphine)ruthenium fluoride
[59557-58-3]
As Bromo(η^5-2,4-cyclopentadien-1-yl)-bis(triphenylphosphine)ruthenium, Ru-00763 with

X = F

M 709.743
Yellow cryst. (MeOH). Mp 115-20°. F not detected by nmr.

Bruce, M.I. *et al*, *J. Chem. Soc., Dalton Trans.*, 1976, 81 (*synth, pmr*)

$C_{41}H_{35}IP_2Ru$ Ru-00767

(η^5-2,4-Cyclopentadien-1-yl)iodobis(triphenylphosphine)ruthenium
Cyclopentadienylbis(triphenylphosphine)ruthenium iodide
[34692-10-9]
As Bromo(η^5-2,4-cyclopentadien-1-yl)-bis(triphenylphosphine)ruthenium, Ru-00763 with

X = I

M 817.650
Brick-red cryst. (C_6H_6/hexane). Mp 153-5°, 228-34°.

Blackmore, T. *et al*, *J. Chem. Soc. (A)*, 1971, 2376 (*synth, ir, pmr*)
Wilczewski, T. *et al*, *J. Organomet. Chem.*, 1981, **215**, 87 (*synth, pmr, ms, nmr*)

$C_{41}H_{36}P_2Ru$ Ru-00768

(η^5-2,4-Cyclopentadien-1-yl)hydrobis(triphenylphosphine)-ruthenium, 9CI
Cyclopentadienylbis(triphenylphosphine)ruthenium hydride
[34676-08-9]
As Bromo(η^5-2,4-cyclopentadien-1-yl)-bis(triphenylphosphine)ruthenium, Ru-00763 with

X = H

M 691.753
Yellow air-stable cryst. (Et_2O/hexane). Mp 135-40° dec., 181-4°. Corresponding deuteride also prepd.

Blackmore, T. *et al*, *J. Chem. Soc. (A)*, 1971, 2376 (*synth, ir, pmr*)
Lehmkuhl, H. *et al*, *J. Organomet. Chem.*, 1981, **217**, C5 (*synth, nmr*)
Wilczewski, T. *et al*, *J. Organomet. Chem.*, 1981, **215**, 87 (*synth, pmr, nmr, ir, ms*)

$C_{41}H_{38}NP_2Ru$ Ru-00769

(Acetonitrile)(η^3-2-propenyl)bis(triphenylphosphine)ruthenium, 10CI
Acetonitrile(η^3-allyl)bis(triphenylphosphine)ruthenium
[67507-27-1]
As Nitrosyl(η^3-2-propenyl)bis(triphenylphosphine)-ruthenium, Ru-00748 with

X = H_3CCN—

M 707.775
Pale-yellow solid. Spar. sol. org. solvs. Mp 193° dec.

Sherman, E.O. *et al*, *J. Chem. Soc., Chem. Commun.*, 1978, 223 (*synth, ir*)

$C_{41}H_{39}BP_2Ru$ Ru-00770

(η^5-2,4-Cyclopentadien-1-yl)[tetrahydroborato(1−)-*H*]bis-(triphenylphosphine)ruthenium, 10CI
[79408-42-7]
As Bromo(η^5-2,4-cyclopentadien-1-yl)bis(triphenylphosphine)ruthenium, Ru-00763 with

X = BH_4

M 705.587
Ash-grey or white cryst. (Et_2O). Mp 95-103° dec., 145-50°.

Blackmore, T. *et al*, *J. Chem. Soc. (A)*, 1971, 2376 (*synth, ir, pmr*)
Wilczewski, T. *et al*, *J. Organomet. Chem.*, 1981, **215**, 87 (*synth, pmr, ms*)

$C_{42}H_{30}F_8O_2P_2Ru$ Ru-00771

Dicarbonyl(1,1,2,2,3,3,4,4-octafluoro-1,4-butanediyl)bis[triphenylphosphine]ruthenium
[63374-19-6]

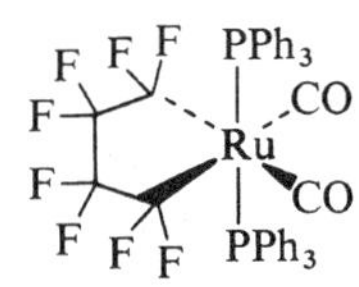

M 881.703
Obt. by reacn. of $Ru(CO)_2(PPh_3)_2$ with $F_2C{=}CF_2$. Off-white cryst.

Kuwae, R. *et al*, *Inorg. Chim. Acta*, 1977, **22**, 39 (*synth, pmr*)

$C_{42}H_{35}NOP_2Ru$ Ru-00772

(Cyanato-*N*)(η^5-2,4-cyclopentadien-1-yl)bis(triphenylphosphine)ruthenium, 10CI
(η^5-Cyclopentadienyl)isocyanatobis(triphenylphosphine)ruthenium
[79408-45-0]
As Bromo(η^5-2,4-cyclopentadien-1-yl)-bis(triphenylphosphine)ruthenium, Ru-00763 with

X = —NCO

M 732.762
Dark-yellow cryst. Mp 254-7°.

Wilczewski, T. *et al*, *J. Organomet. Chem.*, 1981, **215**, 87 (*synth, pmr, ir*)

$C_{42}H_{35}NP_2Ru$ **Ru-00773**

(Cyano-*C*)(η^5-2,4-cyclopentadien-1-yl)bis(triphenylphosphine)ruthenium, 9CI

(η^5-Cyclopentadienyl)cyanobis(triphenylphosphine)ruthenium. Cyclopentadienylbis(triphenylphosphine)ruthenium cyanide

[55272-32-7]

As Bromo(η^5-2,4-cyclopentadien-1-yl)-bis(triphenylphosphine)ruthenium, Ru-00763 with

X = CN

M 716.763

Yellow cryst. (CH_2Cl_2/pet. ether). Mp 268-71°.

CH_2Cl_2 solvate: Mp 253-8°.

Haines, R.J. *et al*, *J. Organomet. Chem.*, 1975, **84**, 357 (*synth, ir, pmr*)
Bruce, M.I. *et al*, *Aust. J. Chem.*, 1981, **34**, 209 (*synth*)
Wilczewski, T. *et al*, *J. Organomet. Chem.*, 1981, **215**, 87 (*synth, pmr, ir, ms*)

$C_{42}H_{35}NP_2RuS$ **Ru-00774**

(η^5-2,4-Cyclopentadien-1-yl)(thiocyanato-*N*)bis(triphenylphosphine)ruthenium, 10CI

(η^5-Cyclopentadienyl)isothiocyanatobis(triphenylphosphine)ruthenium

[79408-44-9]

As Bromo(η^5-2,4-cyclopentadien-1-yl)-bis(triphenylphosphine)ruthenium, Ru-00763 with

X = —NCS

M 748.823

Golden cryst. Mp 202-4°.

Wilczewski, T. *et al*, *J. Organomet. Chem.*, 1981, **215**, 87 (*synth, pmr, ir*)

$C_{42}H_{35}OP_2Ru^{\oplus}$ **Ru-00775**

Carbonyl(η^5-2,4-cyclopentadien-1-yl)bis(triphenylphosphine)-ruthenium(1+), 9CI

(η^5-Cyclopentadienyl)carbonylbis(triphenylphosphine)-ruthenium(1+)

[47859-32-5]

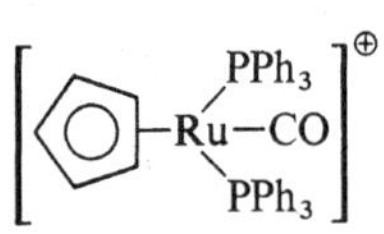

M 718.755 (ion)

Tetraphenylborate: [34676-02-3].
$C_{66}H_{55}BOP_2Ru$ M 1037.987
Yellow powder (CH_2Cl_2/pet. ether). Mp 226-7°.

Blackmore, T. *et al*, *J. Chem. Soc. (A)*, 1971, 2376 (*pmr, ir*)
Haines, R.J. *et al*, *J. Organomet. Chem.*, 1975, **84**, 357 (*synth, ir, pmr*)

$C_{42}H_{36}ClP_2Ru^{\oplus}$ **Ru-00776**

(η^6-Benzene)chlorobis(triphenylphosphine)ruthenium(1+), 10CI

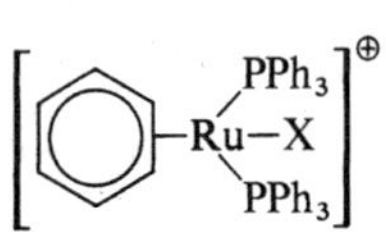

X = Cl

M 739.217 (ion)

Hexafluorophosphate: [69058-55-5].
$C_{42}H_{36}ClF_6P_3Ru$ M 884.181
Yellow cryst. Mp 172°.

Arthur, T. *et al*, *J. Organomet. Chem.*, 1978, **162**, 121 (*synth, pmr*)

$C_{42}H_{36}OP_2Ru$ **Ru-00777**

Hydro[(1,2,3,4,5-η)-6-oxo-2,4-cyclohexadien-1-yl]bis(triphenylphosphine)ruthenium

[61817-37-6]

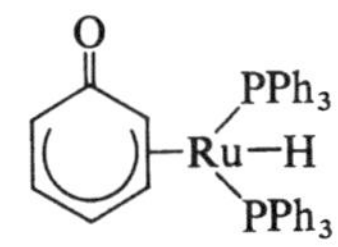

M 719.763

Yellow cryst. Dec. at ca. 160°. Diamagnetic and nonconducting in CH_2Cl_2.

Cole-Hamilton, D.J. *et al*, *J. Chem. Soc., Dalton Trans.*, 1976, 1995 (*synth, ir, pmr*)

$C_{42}H_{37}ClP_2Ru$ **Ru-00778**

Chloro[(1,2,3,4,5-η)-1-methyl-2,4-cyclopentadien-1-yl]bis-(triphenylphosphine)ruthenium, 9CI

(η^5-Methylcyclopentadienyl)chlorobis(triphenylphosphine)ruthenium. (Methylcyclopentadienyl)bis(triphenylphosphine)ruthenium chloride

[55272-36-1]

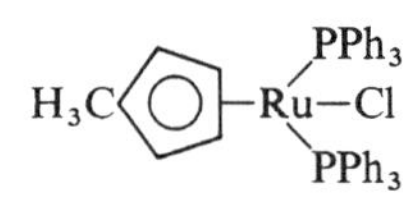

M 740.225

Yellow cryst. ($CHCl_3$/pet. ether).

Haines, R.J. *et al*, *J. Organomet. Chem.*, 1975, **84**, 357 (*synth, pmr*)

$C_{42}H_{37}P_2Ru^{\oplus}$ **Ru-00779**

(η^6-Benzene)hydrobis(triphenylphosphine)ruthenium(1+), 10CI

As (η^6-Benzene)chlorobis(triphenylphosphine)ruthenium(1+), Ru-00776 with

X = H

M 704.772 (ion)

Tetrafluoroborate: [62662-80-0].
$C_{42}H_{37}BF_4P_2Ru$ M 791.575
Yellow cryst.

Cole-Hamilton, D.J. *et al*, *J. Chem. Soc., Dalton Trans.*, 1976, 1995 (*synth, ir, pmr*)

$C_{42}H_{38}P_2Ru$ **Ru-00780**

[(1,2,3,4,5-η)-2,4-Cyclohexadien-1-yl]hydrobis(triphenylphosphine)ruthenium

[69450-47-1]

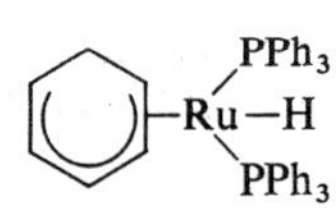

M 705.780

Yellow microcryst. Mp 90-5° dec.

Chaudret, B.N. *et al*, *Acta Chem. Scand., Ser. A*, 1978, **32**, 763 (*synth, pmr, nmr, ir*)

$C_{42}H_{38}P_2Ru$ **Ru-00781**

(η^5-2,4-Cyclopentadien-1-yl)methylbis(triphenylphosphine)-ruthenium, 9CI

σ-Methyl-π-cyclopentadienylbis(triphenylphosphine)ruthenium

[34692-09-6]

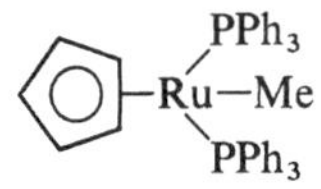

M 705.780

Orange-yellow cryst. (Et_2O/hexane). Mp 150-1° dec.

Blackmore, T. *et al*, *J. Chem. Soc. (A)*, 1971, 2376 (*synth, ir, pmr*)

Lehmkuhl, H. *et al*, *J. Organomet. Chem.*, 1981, **217**, C5 (*synth, nmr*)

$C_{42}H_{40}P_2Ru$ **Ru-00782**

Bis(η^3-2-propenyl)bis(triphenylphosphine)ruthenium

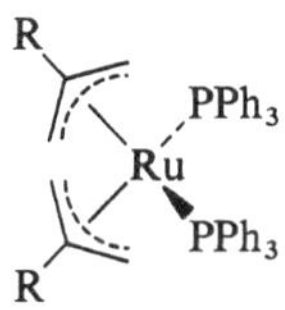

R = H

M 707.796

Pale-yellow solid + ½toluene (toluene). Spar. sol. aromatic solvs., sol. $CHCl_3$. Mp 90-100° dec.

Smith, A.E., *Inorg. Chem.*, 1972, **11**, 2306 (*cryst struct*)

Cole-Hamilton, D.J. *et al*, *J. Chem. Soc., Chem. Commun.*, 1977, 59 (*synth*)

Cole-Hamilton, D.J. *et al*, *Nouv. J. Chim.*, 1977, **1**, 141 (*synth*)

$C_{42}H_{42}N_2P_2RuS_4$ **Ru-00783**

Bis(dimethylcarbamodithioato-*S,S'*)bis(triphenylphosphine)-ruthenium, 9CI

[22391-74-8]

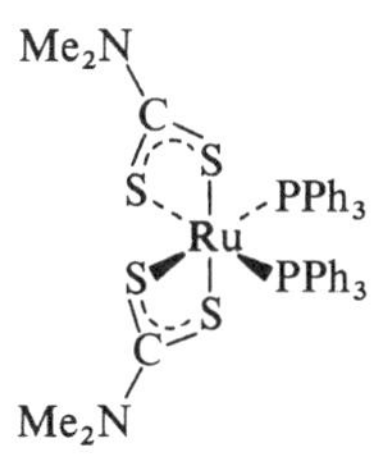

M 866.065

cis-form [52646-85-2]

Mustard-yellow cryst. (CH_2Cl_2/MeOH). Mp 235-6°. Forms an acetone solvate, Mp 153° dec.

O'Connor, C. *et al*, *J. Chem. Soc. (A)*, 1969, 84 (*synth, ir, pmr*)

Cole-Hamilton, D.J. *et al*, *J. Chem. Soc., Dalton Trans.*, 1974, 739 (*pmr*)

Critchlow, P.B. *et al*, *J. Chem. Soc., Dalton Trans.*, 1975, 1367 (*synth, ir, pmr*)

$C_{42}H_{45}P_3Ru$ **Ru-00784**

Hydrotris(methyldiphenylphosphine)(η^3-2-propenyl)ruthenium

(η^3-Allyl)hydridotris(methyldiphenylphosphine)ruthenium

[67507-29-3]

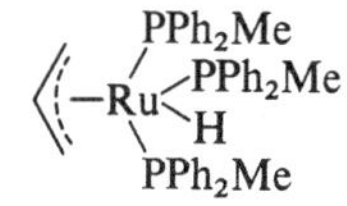

M 743.809

Off-white solid. Mp 164° dec.

Sherman, E.O. *et al*, *J. Chem. Soc., Chem. Commun.*, 1978, 223 (*synth, ir*)

$C_{43}H_{30}Cl_4N_2O_2P_2Ru$ **Ru-00785**

Dicarbonyl[1,2,3,4-tetrachloro-5-(η^2-diazo)-1,3-cyclopentadiene]bis(triphenylphosphine)ruthenium

[73531-62-1]

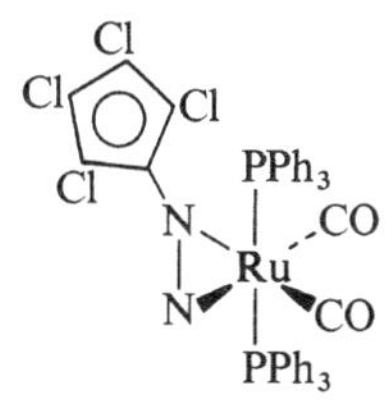

M 911.552

Yellow-orange microcryst. +1CH_2Cl_2 (CH_2Cl_2/hexane).

Schramm, K.D. *et al*, *Inorg. Chem.*, 1980, **19**, 2441 (*synth, cryst struct*)

$C_{43}H_{35}Cl_2NO_2P_2Ru$ **Ru-00786**

Benzoyldichloronitrosylbis(triphenylphosphine)ruthenium, 9CI

[58410-32-5]

As Acetyldichloronitrosylbis(triphenylphosphine)ruthenium, Ru-00732 with

R = COPh

M 831.679

Yellow.

Giovannitti, B. *et al*, *J. Organomet. Chem.*, 1977, **129**, 207 (*synth, ir*)

$C_{43}H_{36}N_2OP_2Ru$ **Ru-00787**

Carbonylhydro(phenylazo)bis(triphenylphosphine)ruthenium, 9CI

[56335-53-6]

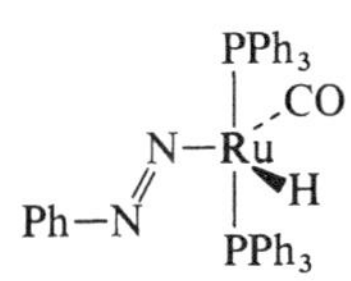

M 759.788

Brown cryst. (CH_2Cl_2).

Haymore, B.L. *et al*, *Inorg. Chem.*, 1975, **14**, 2784 (*synth, ir*)

$C_{43}H_{37}Cl_2NOP_2Ru$ **Ru-00788**

Benzyldichloronitrosylbis(triphenylphosphine)ruthenium

Dichloronitrosyl(phenylmethyl)bis(triphenylphosphine)-ruthenium, 9CI

[58410-34-7]

As Acetyldichloronitrosylbis(triphenylphosphine)ruthenium, Ru-00732 with

R = $-CH_2Ph$

M 817.695

Yellow cryst.

Giovannitti, B. *et al*, *J. Organomet. Chem.*, 1977, **129**, 207 (*synth*, *ir*)

$C_{43}H_{37}Cl_3N_2P_2Ru$ Ru-00789

Trichloro(4-methylphenylazo)bis(triphenylphosphine)ruthenium

[39556-32-6]

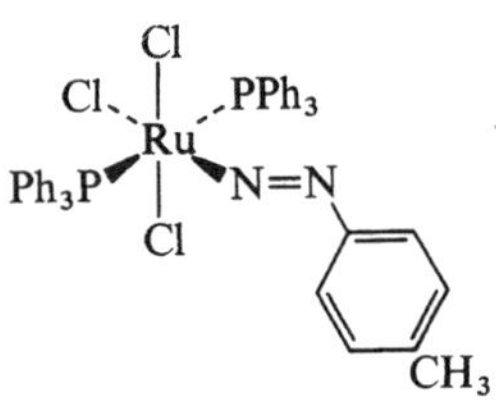

M 851.155

Brown cryst. (Me_2CO/hexane). Forms solvates with CH_2Cl_2 and with Me_2CO.

Laing, K.R. *et al*, *J. Chem. Soc., Dalton Trans.*, 1973, 2731 (*synth*, *ir*)

McArdle, J.V. *et al*, *Inorg. Chem.*, 1973, **12**, 1676 (*struct*)

Haymore, B.L. *et al*, *Inorg. Chem.*, 1975, **14**, 2784, 3060 (*ir*, *struct*)

$C_{43}H_{38}Cl_2P_2Ru$ Ru-00790

[(2,3,5,6-η)-Bicyclo[2.2.1]hepta-2,5-diene]dichlorobis(triphenylphosphine)ruthenium, 9CI

Dichloro(2,5-norbornadiene)bis(triphenylphosphine)-ruthenium

[12216-80-7]

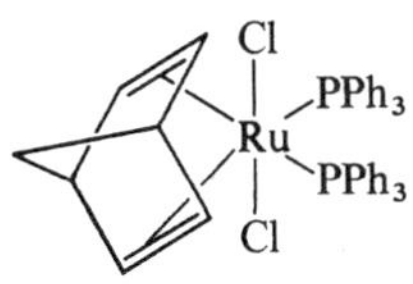

M 788.697

Orange-yellow cryst. Stable in solid state but dec. rapidly in soln. Sol. Me_2CO, v. insol. other common solvs. Mp 220°.

Robinson, S.D. *et al*, *J. Chem. Soc. (A)*, 1966, 300 (*synth*)

Ruiz-Ramirez, L. *et al*, *J. Chem. Soc., Dalton Trans.*, 1973, 1770 (*synth*, *ir*)

Bergbreiter, D.E. *et al*, *J. Organomet. Chem.*, 1981, **205**, 407 (*cryst struct*)

$C_{43}H_{38}NP_2Ru^{\oplus}$ Ru-00791

Acetonitrile(η^5-2,4-cyclopentadien-1-yl)bis(triphenylphosphine)ruthenium(1+)

(η^5-Cyclopentadienyl)acetonitrilebis(triphenylphosphine)ruthenium(1+)

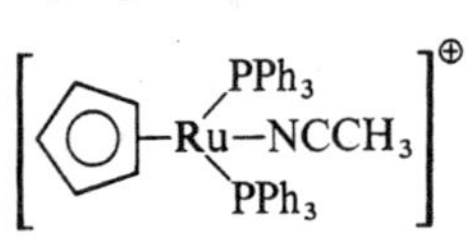

M 731.797 (ion)

Hexafluorophosphate: [34676-03-4].
$C_{43}H_{38}F_6NP_3Ru$ M 876.762
Yellow cryst. (CH_2Cl_2/Et_2O). Mp 173-6°.

Tetraphenylborate: [34738-27-7].
$C_{67}H_{58}BNP_2Ru$ M 1051.029
Yellow cryst. ($MeCN/Et_2O$). Mp 157-60°.

Hexafluoroarsenate: [34676-04-5].
$C_{43}H_{38}AsF_6NP_2Ru$ M 920.709
Yellow cryst. (CH_2Cl_2/Et_2O). Mp 168-70°.

Blackmore, T. *et al*, *J. Chem. Soc. (A)*, 1971, 2376 (*synth*, *ir*, *pmr*)

$C_{43}H_{38}OP_2RuS_2$ Ru-00792

(η^5-2,4-Cyclopentadien-1-yl)(*O*-methylcarbonodithioato-*S*)-bis(triphenylphosphine)ruthenium

π-Cyclopentadienylbis(triphenylphosphine)ruthenium methylxanthate

As Bromo(η^5-2,4-cyclopentadien-1-yl)-bis(triphenylphosphine)ruthenium, Ru-00763 with

X = $-SC(S)OMe$

M 797.910

Yellow-orange solid. Mp 119-20°.

Wilczewski, T. *et al*, *J. Organomet. Chem.*, 1981, **215**, 87 (*synth*, *ir*, *pmr*)

$C_{43}H_{40}P_2Ru$ Ru-00793

(η^5-2,4-Cyclopentadien-1-yl)ethylbis(triphenylphosphine)ruthenium, 10CI

[79594-92-6]

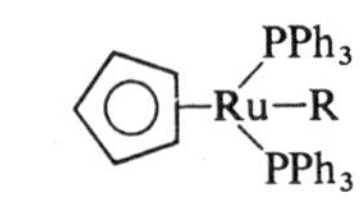

R = Et

M 719.807

Lehmkuhl, H. *et al*, *J. Organomet. Chem.*, 1981, **217**, C5 (*synth*, *nmr*)

$C_{44}H_{30}Cl_4O_4P_2Ru$ Ru-00794

Dicarbonyl(3,4,5,6-tetrachloro-1,2-benzenediolato-*O,O'*)bis-(triphenylphosphine)ruthenium, 9CI

Dicarbonyl[tetrachloropyrocatecholato(2−)]bis(triphenylphosphine)ruthenium, 8CI

[41326-30-1]

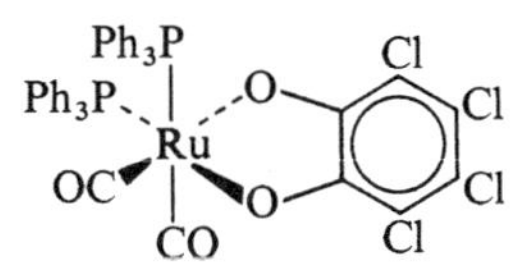

M 927.548

Orange cryst. (CH_2Cl_2/heptane). Mp 265-75° dec. Diamagnetic. Undergoes stepwise one-electron oxidn.

Balch, A.L. *et al*, *J. Organomet. Chem.*, 1971, **30**, C31 (*synth*, *ir*)

Balch, A.L., *J. Am. Chem. Soc.*, 1973, **95**, 2723 (*props*)

Girgis, A.Y. *et al*, *Inorg. Chem.*, 1975, **14**, 2327 (*synth*, *uv*, *ir*)

$C_{44}H_{30}F_{12}P_2RuS_4$ Ru-00795

Bis(1,1,1,4,4,4-hexafluoro-2,3-butanedithionato-*S,S'*)bis(triphenylphosphine)ruthenium

Bis(1,1,1,4,4,4-hexafluoro-2,3-butanedithione)bis(triphenylphosphine)ruthenium, 8CI

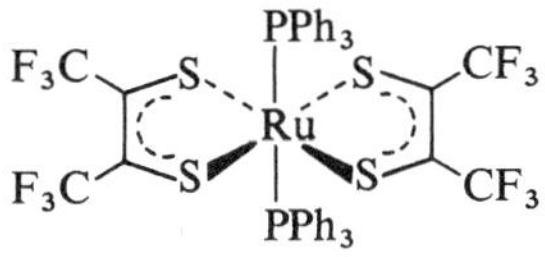

M 1077.959

trans-form [33573-19-2]

White cryst. (CH_2Cl_2/hexane). Mp 196-7°.

Balch, A.L. *et al*, Inorg. Chem., 1971, **10**, 1410 (*synth, uv*)

$C_{44}H_{35}BrN_2O_2P_2Ru$ Ru-00796

Bromodicarbonyl(phenylazo)bis(triphenylphosphine)ruthenium, 9CI

Bromodicarbonyl(phenyldiazo)bis(triphenylphosphine)-ruthenium

[56335-47-8]

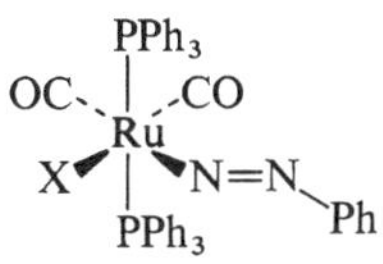

X = Br

M 866.694

Chocolate-brown solid + $1Me_2CO$. ir ν_{CO} 2043, 1984, ν_{NN} 1455 cm^{-1}. Reacts with $H^⊕$ forming diazene complexes.

Haymore, B.L. *et al*, Inorg. Chem., 1975, **14**, 2784 (*synth, ir*)

$C_{44}H_{35}IN_2O_2P_2Ru$ Ru-00797

Dicarbonyliodo(phenylazo)bis(triphenylphosphine)ruthenium, 9CI

Iododicarbonyl(phenyldiazo)bis(triphenylphosphine)-ruthenium

[56335-49-0]

As Bromodicarbonyl(phenylazo)bis(triphenylphosphine)-ruthenium, Ru-00796 with

X = I

M 913.695

Brown solid + $\frac{1}{2}Me_2CO$. ir ν_{CO} 2043, 1970, ν_{NN} 1458 cm^{-1}.

Haymore, B.L. *et al*, Inorg. Chem., 1975, **14**, 2784 (*synth, ir*)

$C_{44}H_{35}N_2O_2P_2Ru^⊕$ Ru-00798

Dicarbonyl(phenylazo)bis(triphenylphosphine)ruthenium(1+), 9CI

Dicarbonyl(phenyldiazo)bis(triphenylphosphine)ruthenium(1+)

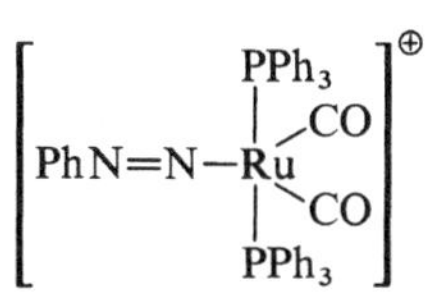

M 786.790 (ion)

Tetrafluoroborate: [54733-59-4].

$C_{44}H_{35}BF_4N_2O_2P_2Ru$ M 873.594

Yellow cryst. + $1C_6H_6$ (C_6H_6). Mp 193-5°.

Hexafluorophosphate: [56389-71-0].

$C_{44}H_{35}F_6N_2O_2P_3Ru$ M 931.754

Lemon-yellow cryst. + $1C_6H_6$ (C_6H_6).

Haymore, B.L. *et al*, Inorg. Chem., 1975, **14**, 2784 (*synth, ir*)
Cenini, S. *et al*, Inorg. Chim. Acta, 1976, **20**, 119 (*synth, ir, pmr*)

$C_{44}H_{36}ClN_2O_2P_2Ru^⊕$ Ru-00799

Dicarbonylchloro(phenyldiazene-*N*²)bis(triphenylphosphine)-ruthenium(1+), 9CI

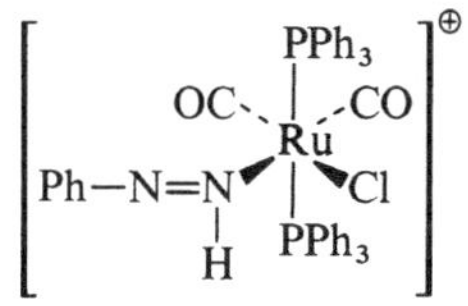

M 823.251 (ion)

Perchlorate: [57307-34-3].

$C_{44}H_{36}Cl_2N_2O_6P_2Ru$ M 922.702

Yellow cryst. (Me_2CO), yellow cryst. + $1CH_2Cl_2$ (CH_2Cl_2).

Haymore, B.L. *et al*, J. Am. Chem. Soc., 1975, **97**, 5369 (*synth, ir, pmr, raman, struct*)

$C_{44}H_{36}N_2O_2P_2Ru$ Ru-00800

Dicarbonylhydro(phenylazo)bis(triphenylphosphine)ruthenium, 9CI

Hydridodicarbonyl(phenyldiazo)bis(triphenylphosphine)ruthenium

[56335-50-3]

As Bromodicarbonyl(phenylazo)bis(triphenylphosphine)-ruthenium, Ru-00796 with

X = H

M 787.798

Lime-green cryst. which slowly convert to more stable brown form in solid state. ir ν_{CO} 2026, 1973, ν_{RuH} 1915, ν_{NN} 1455 cm^{-1} (perfluorokerosene).

Haymore, B.L. *et al*, Inorg. Chem., 1975, **14**, 2784 (*synth, ir*)

$C_{44}H_{36}N_2O_2P_2Ru^{⊕⊕}$ Ru-00801

Dicarbonyl(phenyldiazene-*N*²)bis(triphenylphosphine)ruthenium(2+), 9CI

Dicarbonyl(phenyldiimine)bis(triphenylphosphine)ruthenium(2+)

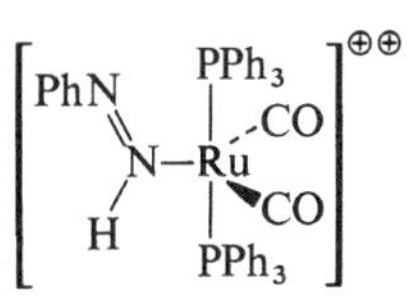

M 787.798 (ion)

Bis(tetrafluoroborate): [54734-25-7].

$C_{44}H_{36}B_2F_8N_2O_2P_2Ru$ M 961.405

White cryst. + $1H_2O$. Mp 165°.

Cenini, S. *et al*, Inorg. Chim. Acta, 1976, **20**, 119 (*synth, ir, pmr*)

$C_{44}H_{38}O_6P_2Ru$ Ru-00802

Dicarbonyl[(2,3-η)-dimethyl-2-butenedioate]bis(triphenylphosphine)ruthenium

[63374-18-5]

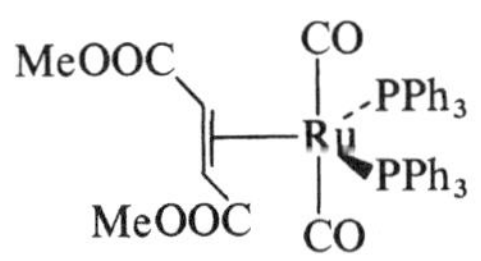

M 825.798

(E)-form

Air-stable cryst. Prob. does not undergo ligand scrambling in soln. like related complexes.

Kuwae, R. *et al*, *Inorg. Chim. Acta*, 1977, **22**, 39 (*synth, pmr*)

$C_{44}H_{38}P_2Ru$ Ru-00803

[(1,2,3,4,5,6-η)-1,3,5,7-Cyclooctatetraene]bis(triphenylphosphine)ruthenium, 9CI

[42516-75-6]

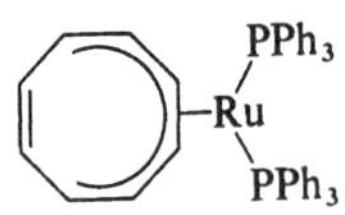

M 729.802

Yellow cryst.

Schrock, R.R. *et al*, *J. Am. Chem. Soc.*, 1973, **95**, 4102 (*synth*)

$C_{44}H_{38}P_2Ru$ Ru-00804

(η^5-2,4-Cyclopentadien-1-yl)-1-propynylbis(triphenylphosphine)ruthenium, 10CI

[69175-65-1]

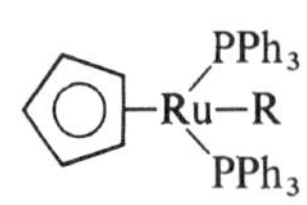

R = $-C{\equiv}CCH_3$

M 729.802

Yellow cryst. Mp 209-15°.

Bruce, M.I. *et al*, *Aust. J. Chem.*, 1979, **32**, 1471 (*synth, ir, pmr, cmr*)

$C_{44}H_{40}P_2Ru$ Ru-00805

(η^5-2,4-Cyclopentadien-1-yl)-2-propenylbis(triphenylphosphine)ruthenium, 10CI

σ-Allyl-π-cyclopentadienylbis(triphenylphosphine)ruthenium

[74573-16-3]

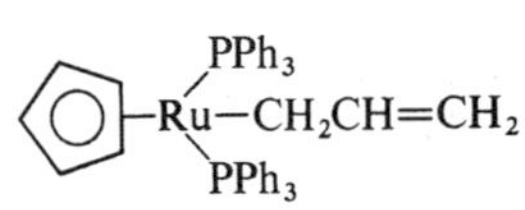

M 731.818

Yellow cryst. (toluene/pentane). At 80° transformed into (η^5-2,4-Cyclopentadien-1-yl)(η^3-2-propenyl)(triphenylphosphine)ruthenium, Ru-00635 .

Lehmkuhl, H. *et al*, *Justus Liebigs Ann. Chem.*, 1980, 754 (*synth, pmr*)

$C_{44}H_{42}P_2Ru$ Ru-00806

(η^5-2,4-Cyclopentadien-1-yl)propylbis(triphenylphosphine)ruthenium, 10CI

[79619-57-1]

As (η^5-2,4-Cyclopentadien-1-yl)ethylbis(triphenylphosphine)ruthenium, Ru-00793 with

R = $-CH_2CH_2CH_3$

M 733.833

Lehmkuhl, H. *et al*, *J. Organomet. Chem.*, 1981, **217**, C5 (*synth, pmr*)

$C_{44}H_{44}P_2Ru$ Ru-00807

Bis[(1,2,3-η)-2-methyl-2-propenyl]bis(triphenylphosphine)ruthenium

Bis(2-methyl-π-allyl)bis(triphenylphosphine)ruthenium, 8CI

As Bis(η^3-2-propenyl)bis(triphenylphosphine)ruthenium, Ru-00782 with

R = CH_3

M 735.849

Pale-yellow prisms. Mp 115-23° dec. Complex too insoluble for nmr studies.

Powell, J. *et al*, *J. Chem. Soc. (A)*, 1968, 159 (*synth*)

$C_{44}H_{48}Ge_2O_2P_2Ru$ Ru-00808

Dicarbonylbis(trimethylgermyl)bis(triphenylphosphine)ruthenium, 8CI

[34397-62-1]

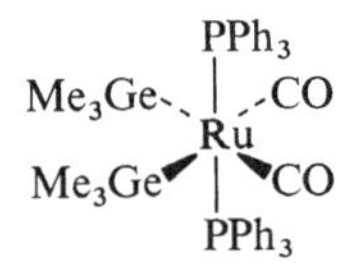

M 917.060

White cryst. Mp 112-4°.

Knox, S.A.R. *et al*, *J. Chem. Soc. (A)*, 1971, 2874 (*synth, ir*)

$C_{44}H_{48}O_2P_2RuSi_2$ Ru-00809

Dicarbonylbis(trimethylsilyl)bis(triphenylphosphine)ruthenium, 8CI

[26024-47-5]

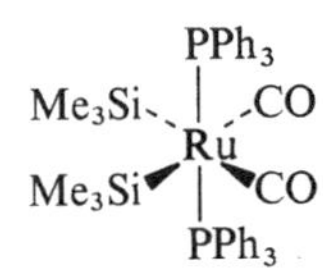

M 828.051

Yellow cryst. (CH_2Cl_2/hexane). Mp 118-20°.

Knox, S.A.R. *et al*, *J. Chem. Soc. (A)*, 1969, 2559 (*synth, ir*)

$C_{45}H_{36}F_6P_2Ru$ Ru-00810

(η^5-2,4-Cyclopentadien-1-yl)[3,3,3-trifluoro-1-(trifluoromethyl)-1-propenyl]bis(triphenylphosphine)ruthenium, 9CI

[52462-32-5]

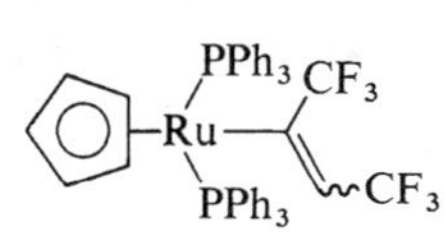

M 853.787

Yellow cryst. (Et_2O/hexane). Mp 127-8°.

Blackmore, T. *et al*, *J. Chem. Soc., Dalton Trans.*, 1974, 106 (*synth, pmr, ir, nmr*)

$C_{45}H_{44}P_2Ru$ Ru-00811

Butyl(η^5-2,4-cyclopentadien-1-yl)bis(triphenylphosphine)ruthenium, 10CI

[79594-93-7]

As (η^5-2,4-Cyclopentadien-1-yl)ethylbis(triphenylphosphine)ruthenium, Ru-00793 with

$$R = -(CH_2)_3CH_3$$

M 747.860

Lehmkuhl, H. *et al*, *J. Organomet. Chem.*, 1981, **217**, C5 (*synth, nmr*)

$C_{46}H_{30}O_{10}P_2Ru_3$ Ru-00812

Decacarbonylbis(triphenylphosphine)triruthenium

[66172-60-9]

As Undecacarbonyl(triphenylphosphine)triruthenium, Ru-00653 with

$$X = Y = PPh_3, Z = CO$$

M 1107.895

Red cryst.

Bruce, M.I. *et al*, *J. Chem. Soc., Dalton Trans.*, 1972, 2094.
Jones, D.F. *et al*, *J. Organomet. Chem.*, 1981, **204**, C1 (*synth*)
Bruce, M.I. *et al*, *J. Chem. Soc., Chem. Commun.*, 1982, 442 (*synth, ir*)

$C_{46}H_{34}O_{10}P_2Ru_4$ Ru-00813

Decacarbonyltetra-μ-hydrobis(triphenylphosphine)tetraruthenium, 9CI

[34742-80-8]

As Dodecacarbonyltetra-μ-hydrotetraruthenium, Ru-00224 with

$$X = Y = PPh_3$$

M 1212.996

Brown-red plates (CH_2Cl_2/pentane). Thermally stable even in air. Sol. most org. solvs., reacts with halogenated solvs.

Piacenti, F. *et al*, *Inorg. Chem.*, 1971, **10**, 2759 (*synth, ir, pmr*)
Banditelli, P. *et al*, *Thermochim. Acta*, 1975, **12**, 277.
Wilson, R.D. *et al*, *Inorg. Chem.*, 1978, **17**, 1271 (*synth, cryst struct*)
Marzocchi, M.P. *et al*, *Indian J. Pure Appl. Phys.*, 1978, **16**, 304 (*raman*)
Sasvari, K. *et al*, *Acta Crystallogr., Sect. B*, 1979, **35**, 87 (*cryst struct*)

$C_{46}H_{36}NO_4P_2Ru^{\oplus}$ Ru-00814

[(Benzoylamino)carbonyl]dicarbonylbis(triphenylphosphine)ruthenium(1+), 9CI

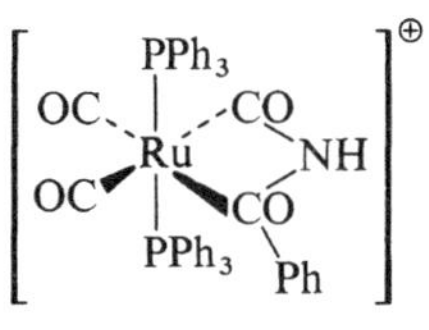

M 829.812 (ion)

Tetrafluoroborate: [56259-36-0].
$C_{46}H_{36}BF_4NO_4P_2Ru$ M 916.616
White cryst. Mp 235°.

Cenini, S. *et al*, *J. Organomet. Chem.*, 1975, **88**, 237 (*synth, ir, pmr*)

$C_{47}H_{35}F_5P_2Ru$ Ru-00815

(η^5-2,4-Cyclopentadien-1-yl)(pentafluorophenyl)bis(triphenylphosphine)ruthenium

σ-Pentafluorophenyl-π-cyclopentadienylbis(triphenylphosphine)ruthenium

[34692-11-0]

As Bromo(η^5-2,4-cyclopentadien-1-yl)-bis(triphenylphosphine)ruthenium, Ru-00763 with

$$X = C_6F_5$$

M 857.803

Yellow cryst. (CH_2Cl_2/hexane). Mp 130-3°.

Blackmore, T. *et al*, *J. Chem. Soc. (A)*, 1971, 2376 (*synth, ir, pmr*)

$C_{48}H_{40}Cl_4N_4O_4P_4Ru_4$ Ru-00816

Tetra-μ-chlorotetrakis[μ-(diphenylphosphino)]tetranitrosyltetraruthenium, 9CI

[38386-79-7]

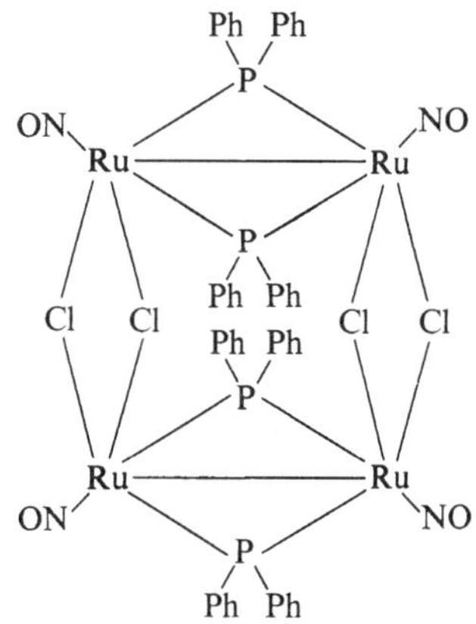

M 1406.855

Cryst. (C_6H_6/MeOH).

Eisenberg, R. *et al*, *J. Am. Chem. Soc.*, 1972, **94**, 6240 (*synth, struct*)

$C_{48}H_{42}N_2P_2S_2Ru$ Ru-00817

Bis(2-aminobenzenethiolato-*N,S*)bis(triphenylphosphine)ruthenium, 9CI

[26455-77-6]

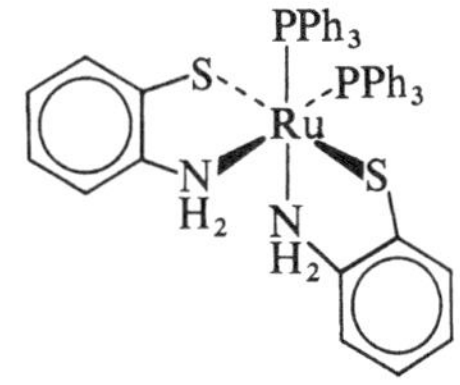

M 874.011

Yellow cryst. Mp ca. 130° dec.

Gilbert, J.D. *et al*, *J. Chem. Soc. (A)*, 1969, 1749 (*synth*)
Rosete, R.O. *et al*, *J. Chem. Soc., Dalton Trans.*, 1979, 1618 (*synth, ir*)

$C_{48}H_{42}P_2Ru$ Ru-00818

Benzyl(η^5-2,4-cyclopentadien-1-yl)bis(triphenylphosphine)ruthenium

(η^5-2,4-Cyclopentadien-1-yl)(phenylmethyl)bis(triphenylphosphine)ruthenium, 10CI, 9CI

[59568-94-4]

As (η^5-2,4-Cyclopentadien-1-yl)-1-propynylbis(triphenylphosphine)ruthenium, Ru-00804 with

$$R = —CH_2Ph$$

M 781.877
Yellow cryst. (Et_2O). Mp 127-32° dec.

Bruce, M.I. *et al*, *J. Chem. Soc., Dalton Trans.*, 1977, 621 (*synth*)

$C_{48}H_{44}P_2Ru$ Ru-00819

(η^4-1,3-Butadiene)[(η^2-ethenyl)benzene]bis(triphenylphosphine)ruthenium, 10CI

(η^4-1,3-Butadiene)styrenebis(triphenylphosphine)ruthenium

[69450-61-9]

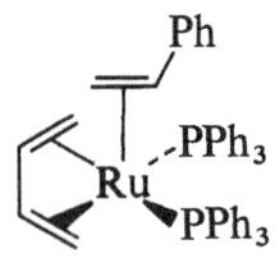

cis-form

M 783.893
Exists in soln. as a mixture of *cis*- and *trans*-isomers. Pale-yellow cryst. + 1 toluene (toluene). Mp 147°.

Chaudret, B. *et al*, *J. Chem. Soc., Dalton Trans.*, 1978, 1739 (*synth, ir, nmr*)

$C_{48}H_{48}P_2Ru$ Ru-00820

[(1,2,3-η)-2,4-Cyclohexadien-1-yl][(1,2-η)-cyclohexene]hydrobis(triphenylphosphine)ruthenium

[69450-46-0]

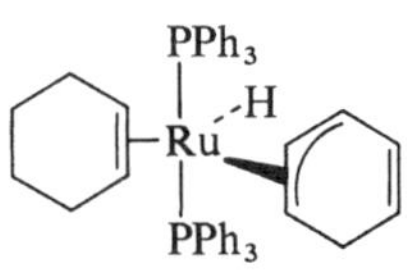

diapical-diphosphine-form

M 787.925

diapical-diphosphine-form
Pale-yellow microcryst. Mp 85-9° dec. Cannot be recryst. without dec.

apial-equatorial-diphosphine-form
Yellow solid. Mp 115° dec. Cannot be recryst. without dec.

Chaudret, B.N. *et al*, *Acta Chem. Scand., Ser. A*, 1978, **32**, 763 (*synth, pmr, nmr, ir*)

$C_{49}H_{40}P_2Ru$ Ru-00821

(η^5-2,4-Cyclopentadien-1-yl)(phenylethynyl)bis(triphenylphosphine)ruthenium, 9CI

[58355-23-0]

As (η^5-2,4-Cyclopentadien-1-yl)-1-propynylbis(triphenylphosphine)ruthenium, Ru-00804 with

R = —C≡CPh

M 791.873
Yellow cryst. (CH_2Cl_2/MeOH). Sol. CH_2Cl_2, THF, insol. pet. ether, Et_2O, alcohols. Mp 202-5° dec.

CuCl adduct:
$C_{49}H_{40}ClCuP_2Ru$ M 890.872
Mp 185-8° dec.

Abu Salah, O.M. *et al*, *J. Organomet. Chem.*, 1974, **64**, C48 (*struct, deriv*)
Abu Salah, O.M. *et al*, *J. Chem. Soc., Dalton Trans.*, 1975, 2351 (*synth, ir, pmr*)
Bruce, M.I. *et al*, *Aust. J. Chem.*, 1979, **32**, 1471 (*synth, ir, pmr, cmr*)
Inorg. Synth., 1982, **21**, 82 (*synth, ir, pmr*)

$C_{50}H_{40}Cl_2N_2P_2Ru$ Ru-00822

Bis(benzonitrile)dichlorobis(triphenylphosphine)ruthenium, 8CI

[26407-38-5]

As Bis(acetonitrile)dichlorobis(triphenylphosphine)ruthenium, Ru-00758 with

R = Ph

M 902.803
Yellow cryst. Mp 168-70° dec.

Gilbert, J.D. *et al*, *J. Chem. Soc. (A)*, 1969, 1749 (*synth, ir*)

$C_{50}H_{46}N_2O_2P_4Ru_2$ Ru-00823

Bis[μ-(diphenylphosphino)]dinitrosylbis(methyldiphenylphosphine)diruthenium

[39045-39-1]

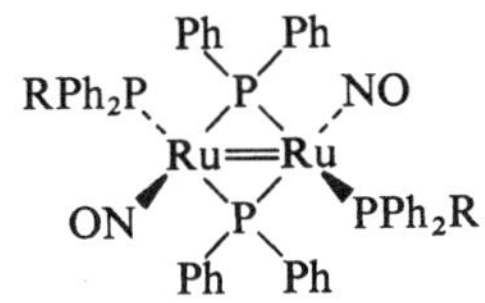

R = Me

M 1032.961
Dark-purple cryst. (THF/MeOH or THF/Et_2O). ir ν_{NO} 1625 cm^{-1}.

Reed, J. *et al*, *Inorg. Chem.*, 1973, **12**, 2949 (*synth, struct*)

$C_{50}H_{48}P_2Ru$ Ru-00824

[(η^2-Ethenyl)benzene][(1,2,3,4-η)-1,3-hexadiene]bis(triphenylphosphine)ruthenium, 10CI

(η^4-1,3-Hexadiene)(η^2-styrene)bis(triphenylphosphine)ruthenium

[69450-63-1]

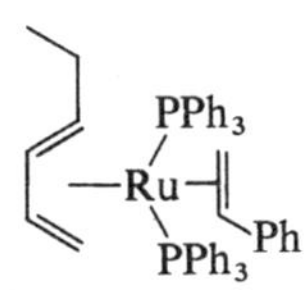

M 811.947
Yellow powder. Mp 122°.

Chaudret, B. *et al*, *J. Chem. Soc., Dalton Trans.*, 1978, 1739 (*synth, ir, nmr*)

$C_{50}H_{50}P_2Ru$ Ru-00825

[η^2-(Ethenylbenzene)][(1,2,3-η)-2-hexenyl]hydrobis(triphenylphosphine)ruthenium

(η^3-2-Hexenyl)hydro(η^2-styrene)bis(triphenylphosphine)ruthenium

[69500-43-2]

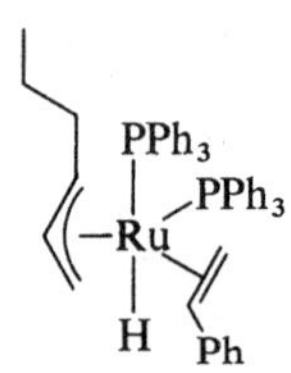

M 813.963
^{31}P nmr spectrum suggests equimolar mixt. of *syn*- and *anti*-isomers. Yellow cryst. or yellow cryst. + $\frac{1}{2}$toluene (toluene/pet. ether). Mp 119°.

Chaudret, B. *et al*, *J. Chem. Soc., Dalton Trans.*, 1978, 1739 (*synth, ir, pmr, nmr*)

$C_{51}H_{45}N_3OP_2Ru$ Ru-00826

[1,3-Bis(4-methylphenyl)-1-triazenato-N^1,N^3]carbonylhydrobis(triphenylphosphine)ruthenium, 9CI

[54206-37-0]

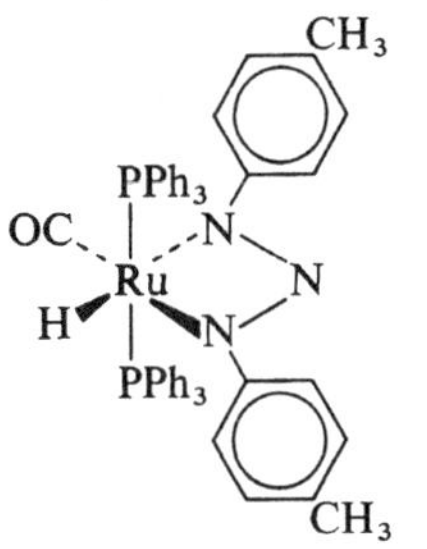

M 878.954

Orange cryst. (EtOH/C_6H_6). Mp 217-8°.

Laing, K.R. *et al, J. Chem. Soc., Dalton Trans.*, 1974, 1205 (*synth, ir, pmr*)
Brown, L.D. *et al, Inorg. Chem.*, 1976, **15**, 2788 (*cryst struct*)
Cresswell, C.J. *et al, Inorg. Chim. Acta*, 1982, **60**, 157 (*synth, pmr*)

$C_{52}H_{30}O_{15}P_2Ru_6$ Ru-00827

[μ_6-(Carbon)]-μ-carbonyltetradecacarbonylbis(triphenylphosphine)hexaruthenium

μ_6-Carbonpentadecacarbonylbis(triphenylphosphine)hexaruthenium

[74904-04-4]

$Ru_6(CO)_{14}(\mu\text{-}CO)(\mu_6\text{-}C)(PPh_3)_2$

M 1563.168

No phys. description recorded. Derived from μ_6-Carbon-μ-carbonylhexadecacarbonylhexaruthenium, Ru-00500 .

Johnson, B.F.G. *et al, J. Organomet. Chem.*, 1980, **185**, C17 (*synth*)

$C_{52}H_{44}ClNOP_2Ru$ Ru-00828

Carbonylchloro[η^2-(4-methylphenyl)[(4-methylphenyl)imino]-methyl]bis(triphenylphosphine)ruthenium

Carbonylchloro[η^2-(p-tolyl)[(p-tolylimino)methyl]bis-(triphenylphosphine)ruthenium

[68124-76-5]

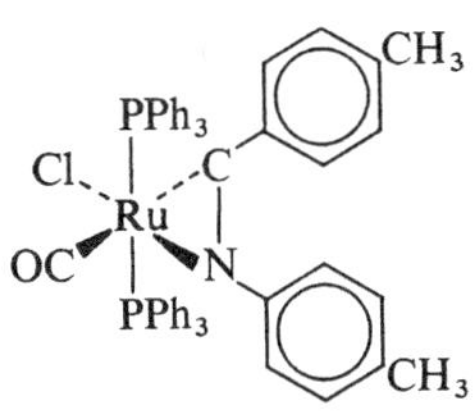

M 897.396

Orange cryst. Readily protonated.

Roper, W.R. *et al, J. Organomet. Chem.*, 1978, **157**, C27 (*synth, ir, cryst struct*)

$C_{52}H_{46}P_2Ru$ Ru-00829

Bis[(η^2-ethenyl)benzene]bis(triphenylphosphine)ruthenium, 10CI

Bis(styrene)bis(triphenylphosphine)ruthenium. Bis-(phenylethylene)bis(triphenylphosphine)ruthenium

[68088-93-7]

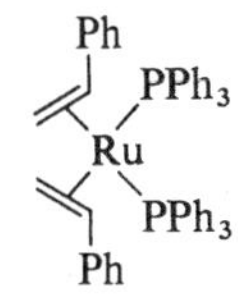

M 833.953

Red-purple cryst. (styrene/pet. ether), stable in dry air for several minutes. Unstable in common org. solvs.

Komiya, S. *et al, Bull. Chem. Soc. Jpn.*, 1976, **49**, 2553 (*synth*)
Carrondo, M.A. *et al, J. Chem. Soc., Chem. Commun.*, 1978, 463 (*nmr, struct*)
Chaudret, B.N. *et al, J. Chem. Soc., Dalton Trans.*, 1978, 1739 (*struct*)

$C_{52}H_{50}P_4Ru$ Ru-00830

Bis[1,2-ethanediylbis[diphenylphosphine]-P,P']dihydroruthenium, 9CI

Bis[(1,2-diphenylphosphino)ethane]dihydroruthenium

[41753-60-0]

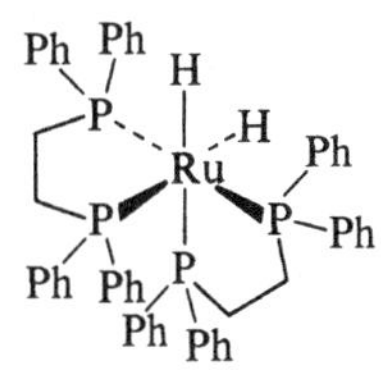

M 899.932

Colourless cryst. (C_6H_6/pentane).

Meakin, P. *et al, J. Am. Chem. Soc.*, 1973, **95**, 75 (*pmr*)
Pertici, P. *et al, Inorg. Chim. Acta*, 1979, **37**, L521 (*synth, struct*)

$C_{53}H_{44}N_2O_3P_2Ru$ Ru-00831

[N,N'-Bis(4-methylphenyl)ureato(2–)-N,N']dicarbonylbis-(triphenylphosphine)ruthenium, 10CI

Dicarbonyl(p-tolylurylene)bis(triphenylphosphine)ruthenium

[69403-53-8]

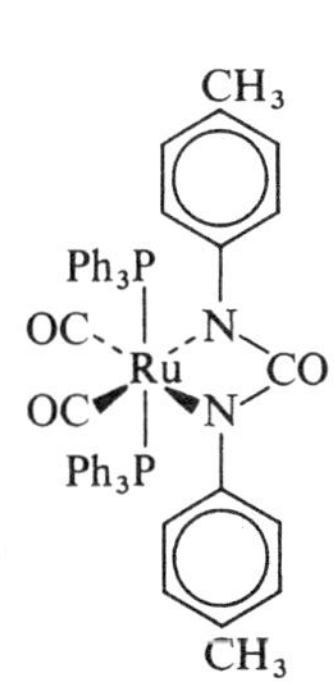

M 919.960

Orange air-stable cryst. (CH_2Cl_2/MeOH). Mp 194-7°. ir ν_{CO} 2020, 1950 cm^{-1}.

Sahajpal, A. *et al, Inorg. Chem.*, 1979, **18**, 3572 (*synth, ir, pmr*)

$C_{54}H_{45}Br_2P_3Ru$ Ru-00832

Dibromotris(triphenylphosphine)ruthenium, 9CI

[15709-75-8]

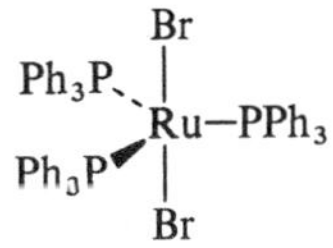

M 1047.749

Hydrogenation and oxidation catalyst. Dark-brown cryst. Mp 162°. Not homogeneous.

Stephenson, T.A. *et al*, *J. Inorg. Nucl. Chem.*, 1966, **28**, 945 (*synth, spectra*)
Armit, P.W. *et al*, *J. Chem. Soc., Dalton Trans.*, 1975, 1663 (*rev, nmr, ir*)
Hoffman, P.R. *et al*, *J. Am. Chem. Soc.*, 1975, **97**, 4221 (*nmr*)

$C_{54}H_{45}Cl_2P_3Ru$ Ru-00833

Dichlorotris(triphenylphosphine)ruthenium

Tris(triphenylphosphine)ruthenium chloride

[15529-49-4]

$(Ph_3P)_3RuCl_2$

M 958.847

Hydrosilation and homogeneous catalytic transfer-hydrogenation catalyst; catalyst for isomerisation of alkenes and for electrophilic addition of halogenated compds. to alkenes. Red-brown or shiny black cryst. Mp 132-4°. PPh_3 dissociation occurs in nonpolar media.

La Placa, S.J. *et al*, *Inorg. Chem.*, 1965, **4**, 778 (*cryst struct*)
Stephenson, T.A. *et al*, *J. Inorg. Nucl. Chem.*, 1966, **28**, 945 (*synth*)
Inorg. Synth., 1971, **12**, 238 (*synth*)
Sanchez-Delgado, R.A. *et al*, *J. Mol. Catal.*, 1976, **6**, 303 (*use*)
Masumoto, H. *et al*, *Chem. Lett.*, 1979, 1011 (*use*)
Pri-Bar, I. *et al*, *J. Org. Chem.*, 1980, **45**, 4418 (*use, nmr*)
Strohmier, W. *et al*, *Z. Naturforsch., B*, 1980, **35**, 648 (*use*)
Watanabe, H. *et al*, *J. Organomet. Chem.*, 1980, **195**, 363 (*use*)
Fieser, M. *et al*, *Reagents for Organic Synthesis*, Wiley, 1967-83, **8**, 159 (*use*)
Tsuji, J. *et al*, *Chem. Lett.*, 1981, 1169 (*use*)

$C_{54}H_{46}BrP_3Ru$ Ru-00834

Bromohydrotris(triphenylphosphine)ruthenium, 9CI, 8CI

[22391-69-1]

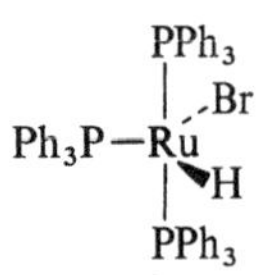

M 968.853

Dark-violet-red cryst. +1 C_6H_6. Mp 238-40°.

Hallman, P.S. *et al*, *J. Chem. Soc. (A)*, 1968, 3143 (*synth, spectra*)
Hoffman, P.R. *et al*, *J. Am. Chem. Soc.*, 1974, **97**, 4221 (*nmr*)
Kono, H. *et al*, *Chem. Lett.*, 1975, 955 (*synth*)

$C_{54}H_{46}ClP_3Ru$ Ru-00835

Chlorohydrotris(triphenylphosphine)ruthenium, 9CI, 8CI

[55102-19-7]

$HRu(PPh_3)_3Cl$

M 924.402

Stereochemically non-rigid. Hydrogenation catalyst. Purple cryst. (toluene), dec. slowly in air. Sol. CH_2Cl_2, insol. Et_2O, EtOH. Forms toluene and dimethylaniline solvates.

Skapski, A.C. *et al*, *J. Chem. Soc., Chem. Commun.*, 1968, 1230 (*cryst struct*)
Inorg. Synth., 1971, **13**, 131 (*synth, ir, pmr*)
Hoffmann, P.R. *et al*, *J. Am. Chem. Soc.*, 1975, **97**, 4221 (*pmr, nmr*)
Geoffroy, G.L. *et al*, *J. Chem. Soc., Chem. Commun.*, 1976, 20 (*synth*)
Spencer, A., *Inorg. Nucl. Chem. Lett.*, 1976, **12**, 661 (*spectra*)
Kono, H. *et al*, *J. Organomet. Chem.*, 1977, **132**, 53 (*synth*)
Diamond, S.E. *et al*, *J. Organomet. Chem.*, 1977, **142**, C55 (*synth*)
Fieser, M. *et al*, *Reagents for Organic Synthesis*, Wiley, 1967-83, **7**, 60.

$C_{54}H_{46}NOP_3Ru$ Ru-00836

Hydronitrosyltris(triphenylphosphine)ruthenium, 9CI

[33991-11-6]

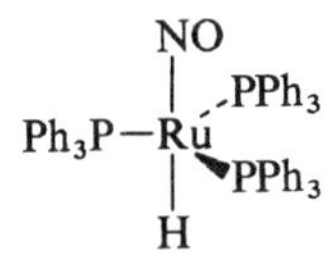

M 918.955

Red-brown air-stable cryst. Mp 197-200° dec.

Pierpont, C.G. *et al*, *Inorg. Chem.*, 1972, **11**, 1094 (*cryst struct*)
Reed, J. *et al*, *Inorg. Chem.*, 1973, **12**, 2949 (*synth*)
Inorg. Synth., 1977, **17**, 73 (*synth, ir, pmr*)

$C_{54}H_{46}P_3Ru^{\oplus}$ Ru-00837

Hydro[(η^6-phenyl)diphenylphosphine]bis(triphenylphosphine)-ruthenium(1+), 9CI

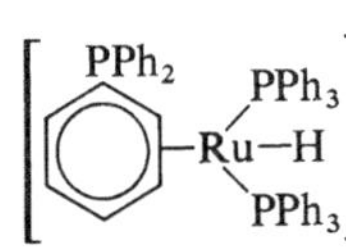

M 888.949 (ion)

Tetrafluoroborate: [41392-83-0].
$C_{54}H_{46}BF_4P_3Ru$ M 975.752
Pale lemon-yellow cryst.

McConway, J.C. *et al*, *J. Chem. Soc., Chem. Commun.*, 1974, 327 (*cryst struct*)
Young, R. *et al*, *J. Chem. Soc., Dalton Trans.*, 1976, 719, 1995 (*synth, nmr*)
Inorg. Synth., 1977, **17**, 77 (*synth*)

$C_{54}H_{47}N_2P_3Ru$ Ru-00838

(Dinitrogen)dihydrotris(triphenylphosphine)ruthenium, 9CI

[22337-84-4]

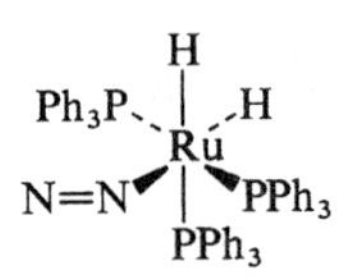

M 917.970

Colourless cryst. (C_6H_6/hexane), dec. in air.

Inorg. Synth. , 1974, **15**, 31 (*synth, ir*)
Pennella, F., *J. Organomet. Chem.*, 1974, **65**, C17 (*ir, pmr*)
Cenini, S. *et al*, *Gazz. Chim. Ital.*, 1975, **105**, 255 (*synth*)

$C_{54}H_{49}P_3Ru$ **Ru-00839**
Tetrahydrotris(triphenylphosphine)ruthenium, 9CI
[31275-06-6]

$H_4Ru(PPh_3)_3$

M 891.972
Colourless cryst. (toluene), dec. in air.

Harris, R.O. *et al*, *J. Organomet. Chem.*, 1973, **54**, 259 (*synth*)
Inorg. Synth., 1974, **15**, 31 (*synth, ir*)
Al-Ohaly, A.-R. *et al*, *J. Organomet. Chem.*, 1980, **202**, 297 (*synth*)

$C_{55}H_{45}Br_2O_{10}P_3Ru$ **Ru-00840**
Dibromocarbonyltris(triphenyl phosphite-*P*)ruthenium
Dibromocarbonyltris(phosphorous acid)ruthenium nonaphenyl ester, 8CI
[28577-77-7]

$[(PhO)_3P]_3Ru(CO)Br_2$

M 1219.754
Stereochem. not known. Cream prisms. Mp 141-5°.

Levison, J.J. *et al*, *J. Chem. Soc. (A)*, 1970, 639 (*synth, ir*)

$C_{55}H_{45}Cl_2OP_3Ru$ **Ru-00841**
Carbonyldichlorotris(triphenylphosphine)ruthenium, 8CI
[15390-63-3]

$(Ph_3P)_3Ru(CO)Cl_2$

M 986.857
Catalyst for carboxylation of alcohols. Orange cryst. Mp 70-2°.

Cleare, M.J., *Platinum Metals Rev.*, 1967, **11**, 148 (*synth*)
Ruiz-Ramirez, L. *et al*, *J. Chem. Soc., Dalton Trans.*, 1974, 1640 (*synth, ir*)
Svoboda, P. *et al*, *J. Organomet. Chem.*, 1974, **65**, C37 (*synth*)

$C_{55}H_{45}Cl_2O_{10}P_3Ru$ **Ru-00842**
Carbonyldichlorotris(triphenyl phosphite-*P*)ruthenium
Carbonyldichlorotris(phosphorous acid)ruthenium nonaphenyl ester, 8CI
[28577-76-6]

$[(PhO)_3P]_3Ru(CO)Cl_2$

M 1130.852
Stereochemistry not known. Yellow cryst. Mp 128-9°.

Levison, J.J. *et al*, *J. Chem. Soc. (A)*, 1970, 639 (*synth, ir*)

$C_{55}H_{45}I_2O_{10}P_3Ru$ **Ru-00843**
Carbonyldiiodotris(triphenyl phosphite-*P*)ruthenium
Carbonyldiiodotris(phosphorous acid)ruthenium nonaphenyl ester, 8CI
[28577-78-8]

$[(PhO)_3P]_3Ru(CO)I_2$

M 1313.755
Stereochemistry not known. Orange cryst. Mp 148-51°.

Levison, J.J. *et al*, *J. Chem. Soc. (A)*, 1970, 639 (*synth, ir*)

$C_{55}H_{45}NO_2P_3Ru^{\oplus}$ **Ru-00844**
Carbonylnitrosyltris(triphenylphosphine)ruthenium(1+), 9CI

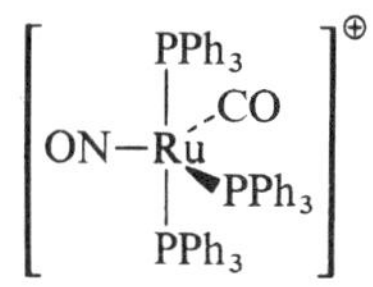

M 945.957 (ion)
Hexafluorophosphate: [40464-70-8].
$C_{55}H_{45}F_6NO_2P_4Ru$ M 1090.921
Buff-yellow cryst.

Johnson, B.F.G. *et al*, *J. Chem. Soc., Dalton Trans.*, 1973, 478 (*synth, ir*)

$C_{55}H_{46}BrOP_3Ru$ **Ru-00845**
Bromocarbonylhydrotris(triphenylphosphine)ruthenium, 9CI, 8CI
[16971-34-9]

$(Ph_3P)_3RuH(CO)Br$

M 996.863

Vaska, L., *J. Am. Chem. Soc.*, 1960, **82**, 1263; 1961, **83**, 1262; *CA*, 1967, **66**, 121641 (*synth, ir*)

$C_{55}H_{46}ClOP_3Ru$ **Ru-00846**
Carbonylchlorohydrotris(triphenylphosphine)ruthenium, 9CI
[16971-33-8]

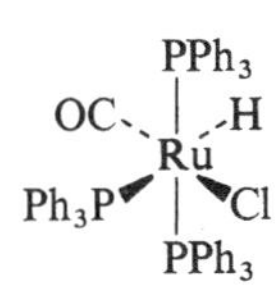

M 952.412
Catalyst for isomerisation of acetylenic silyl ethers to dienol silyl ethers. Creamy-white microcryst. Mp 235-7° (sealed under N_2), 209-10° (in air).

Vaska, L. *et al*, *J. Am. Chem. Soc.*, 1961, **83**, 1262 (*synth*)
Levison, J.J. *et al*, *J. Chem. Soc. (A)*, 1970, 2947 (*synth, ir*)
Inorg. Synth., 1974, **15**, 45 (*synth, ir*)

$C_{55}H_{47}OP_3Ru$ Ru-00847

Carbonyldihydrotris(triphenylphosphine)ruthenium, 9CI

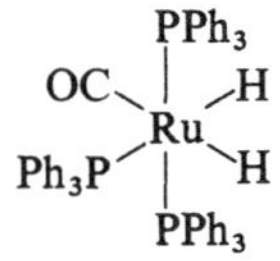

White isomer

M 917.967

White-form [25360-32-1]

Hydroformylation catalyst. Needles (C_6H_6 or C_6H_6/hexane), cryst. + $1CH_2Cl_2$ (CH_2Cl_2). Mod. sol. $CHCl_3$, CH_2Cl_2, C_6H_6. Mp 213-5° (sealed under N_2), 160-2°, 147-8° (in air).

Orange-form [22337-78-6]

Insol. common org. solvs. Mp 147° (darkens at 141°).

Hallman, P.S. *et al*, *J. Chem. Soc.* (*A*), 1968, 3143 (*synth*)
Levison, J.J. *et al*, *J. Chem. Soc.* (*A*), 1970, 2947 (*synth*, *ir*)
Ahmad, N. *et al*, *J. Chem. Soc., Dalton Trans.*, 1972, 843 (*synth*)
Harris, R.O. *et al*, *J. Organomet. Chem.*, 1973, **54**, 259 (*synth*, *ir*)
Inorg. Synth., 1974, **15**, 48 (*synth*, *ir*, *pmr*)
U.S.P., 3 804 914, (*1974*); *CA*, **80**, 145588 (*synth*)
Cenini, S. *et al*, *Gazz. Chim. Ital.*, 1975, **105**, 255 (*synth*)
Sanchez-Delgado, R.A. *et al*, *J. Chem. Soc., Dalton Trans.*, 1976, 399 (*use*)

$C_{56}H_{45}O_2P_3Ru$ Ru-00848

Dicarbonyltris(triphenylphosphine)ruthenium, 9CI

[61647-76-5]

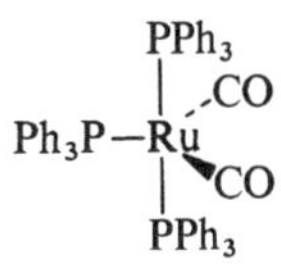

M 943.962

Ph_3P dissociation occurs in soln. Yellow cryst. Mp 130°.

Cavit, B.E., *J. Chem. Soc., Chem. Commun.*, 1972, 61 (*synth*, *ir*)
Cenini, S. *et al*, *Inorg. Chim. Acta*, 1976, **20**, 119 (*synth*)
Schramm, K.D. *et al*, *Inorg. Chem.*, 1980, **19**, 2441 (*synth*)
Grundy, K.R., *Inorg. Chim. Acta*, 1981, **53**, L225 (*synth*)
Gaffney, T.R. *et al*, *Inorg. Chem.*, 1982, **21**, 2851 (*synth*)
Roper, W.R. *et al*, *J. Organomet. Chem.*, 1982, **234**, C5 (*synth*)

$C_{56}H_{46}O_2P_3Ru^{\oplus}$ Ru-00849

Dicarbonylhydrotris(triphenylphosphine)ruthenium(1+), 9CI

[35798-10-8]

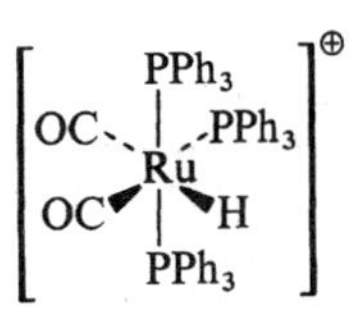

M 944.969 (ion)

Characterised spectroscopically.

Cavit, B.E. *et al*, *J. Chem. Soc., Chem. Commun.*, 1972, 60 (*synth*, *ir*, *pmr*)

$C_{56}H_{49}P_3Ru$ Ru-00850

(η^2-Ethene)tris(triphenylphosphine)ruthenium

(η^2-Ethylene)tris(triphenylphosphine)ruthenium

[38814-44-7]

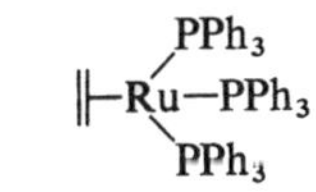

M 915.994

White cryst. Sol. C_6H_6. Mp 150-5° dec.

Komiya, S. *et al*, *Bull. Chem. Soc. Jpn.*, 1976, **49**, 2553 (*synth*, *pmr*)

$C_{57}H_{50}ClP_3Ru$ Ru-00851

Chloro(η^3-2-propenyl)tris(triphenylphosphine)ruthenium

π-Allylchlorotris(triphenylphosphine)ruthenium

[31725-25-4]

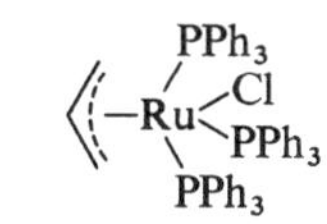

M 964.466

Yellow cryst. (C_6H_6/THF).

Maxfield, P.L., *Inorg. Nucl. Chem. Lett.*, 1970, **6**, 707 (*synth*)

$C_{58}H_{48}O_3P_3Ru^{\oplus}$ Ru-00852

Acetyldicarbonyltris(triphenylphosphine)ruthenium(1+), 8CI

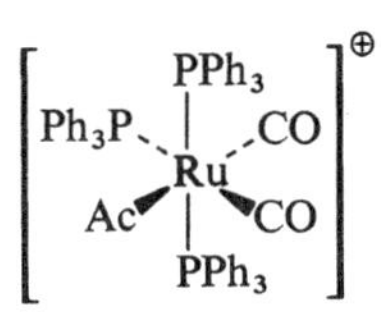

M 987.007 (ion)

Chloride: [34275-37-1].

$C_{58}H_{48}ClO_3P_3Ru$ M 1022.460

Characterised spectroscopically.

Sears, C.T. *et al*, *J. Chem. Soc., Chem. Commun.*, 1971, 777 (*synth*, *ir*)

$C_{58}H_{51}P_3Ru$ Ru-00853

(η^4-1,3-Butadiene)tris(triphenylphosphine)ruthenium, 10CI

[61226-84-4]

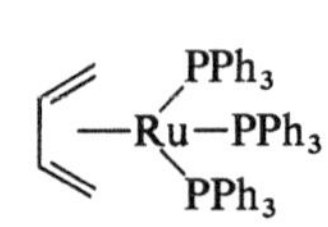

M 942.032

Greenish-yellow needles (C_6H_6 or toluene). Mp 185° dec.

Komiya, S. *et al*, *Bull. Chem. Soc. Jpn.*, 1976, **49**, 2553 (*synth*, *pmr*, *nmr*)

$C_{60}H_{50}N_2O_2P_4Ru_2$ Ru-00854

Bis[μ-(diphenylphosphino)]dinitrosylbis(triphenylphosphine)-diruthenium, 9CI

[41666-87-9]

As Bis[μ-(diphenylphosphino)]dinitrosylbis(methyldiphenylphosphine)diruthenium, Ru-00823 with

R = Ph

M 1157.102

Struct. by analogy with Bis[μ-(diphenylphosphino)]dinitrosylbis(methyldiphenylphosphine)diruthenium, Ru-00823 .

Eisenberg, R. *et al*, *J. Am. Chem. Soc.*, 1972, **94**, 6240 (*synth*)

$C_{63}H_{45}O_9P_3Ru_3$ Ru-00855

Nonacarbonyltris(triphenylphosphine)triruthenium, 9CI

[15453-53-9]

As Undecacarbonyl(triphenylphosphine)triruthenium, Ru-00653 with

X = Y = Z = PPh_3

M 1342.174

Purple-red cryst. Mp 178-81° dec.

Bruce, M.I. *et al*, *J. Chem. Soc., Dalton Trans.*, 1972, 2094 (*synth, pmr, bibl*)
Tyler, D.R. *et al*, *J. Am. Chem. Soc.*, 1978, **100**, 7888 (*uv*)
Angeloni, L. *et al*, *J. Mol. Struct.*, 1982, **79**, 93 (*raman*)
Bruce, M.I. *et al*, *Aust. J. Chem.*, 1982, **35**, 935 (*synth*)
Bruce, M.I. *et al*, *J. Chem. Soc., Chem. Commun.*, 1982, 442 (*synth, ir*)
Graaf, J.L. *et al*, *Organometallics*, 1982, **1**, 837 (*ir*)

$C_{63}H_{49}O_9P_3Ru_4$ Ru-00856

Nonacarbonyltetra-μ-hydrotris(triphenylphosphine)tetraruthenium, 9CI

[34742-81-9]

$Ru_4H_4(CO)_9(PPh_3)_3$

M 1447.276

Structure not determined. Catalyst for hydrogenation of cyclohexanone, pentyne, ethene. Deep-red cryst. Thermally stable even in air. Sol. most org. solvs. Reacts with halogenated solvs. at r.t.

Piacenti, F. *et al*, *Inorg. Chem.*, 1971, **10**, 2759 (*synth, ir, pmr*)
Banditelli, P. *et al*, *Thermochim. Acta*, 1975, **12**, 277.

$C_{72}H_{60}Br_2O_{12}P_4Ru$ Ru-00857

Dibromotetrakis(triphenyl phosphite-*P*)ruthenium

Dibromotetrakis(phosphorous acid)ruthenium dodecaphenyl ester, 8CI

[28577-70-0]

As Dibromotetrakis(triphenylphosphine)ruthenium, Ru-00858 with

R = OPh

M 1502.032

Cream prisms (CH_2Cl_2/hexane). Mp 163-6°.

Levison, J.J. *et al*, *J. Chem. Soc. (A)*, 1970, 639 (*synth*)

$C_{72}H_{60}Br_2P_4Ru$ Ru-00858

Dibromotetrakis(triphenylphosphine)ruthenium, 9CI

[15555-76-7]

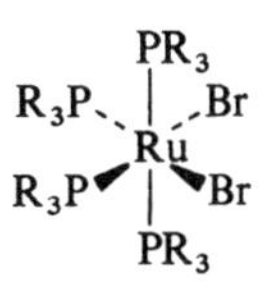

R = Ph

M 1310.039

Red-brown cryst. Mp 143-4°. Dissociates to Ph_3P in soln.

Stephenson, T.A. *et al*, *J. Inorg. Nucl. Chem.*, 1966, **28**, 945 (*synth*)
Armit, P.W. *et al*, *J. Organomet. Chem.*, 1973, **57**, C80.

$C_{72}H_{60}Cl_2P_4Ru$ Ru-00859

Dichlorotetrakis(triphenylphosphine)ruthenium

[15555-77-8]

$Ru(PPh_3)_4Cl_2$

M 1221.137

Catalyst for dimerization of acrylonitrile and electrophilic additions to olefins. Dark-brown cryst. Mp 130-2°. Ph_3P dissociates in soln.

Inorg. Synth., 1970, **12**, 237 (*synth*)
Armit, P.W. *et al*, *J. Chem. Soc., Dalton Trans.*, 1975, 1663 (*synth*)
Hoffman, P.R. *et al*, *J. Am. Chem. Soc.*, 1975, **97**, 4221 (*nmr*)

$C_{72}H_{61}ClO_{12}P_4Ru$ Ru-00860

Chlorohydrotetrakis(triphenyl phosphite-*P*)ruthenium

Chlorohydrotetrakis(phosphorous acid)ruthenium dodecaphenyl ester, 8CI

[24651-66-9]

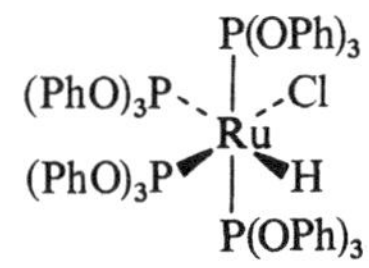

M 1378.685

White powder. Mp 166-9° (softens at 55°).

Parshall, G.W. *et al*, *J. Am. Chem. Soc.*, 1969, **91**, 4990 (*synth, pmr, ir*)

$C_{72}H_{62}P_4Ru$ Ru-00861

Dihydrotetrakis(triphenylphosphine)ruthenium

[19529-00-1]

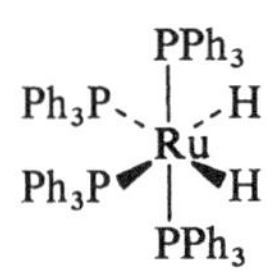

M 1152.247

Dissociates in soln. Transfer hydrogenation catalyst. Causes isomerisation of allyl to vinyl ethers. Bright-yellow cryst., mod. stable in air. Mp 150°.

Immirzi, A. *et al*, *Cryst. Struct. Commun.*, 1972, **1**, 317 (*cryst struct*)
Harris, R.O. *et al*, *J. Organomet. Chem.*, 1973, **54**, 259 (*synth, ir*)
Inorg. Synth., 1977, **17**, 75 (*synth, ir, bibl*)
Pertici, P. *et al*, *Inorg. Chim. Acta*, 1979, **37**, L521 (*synth*)
Suzuki, H. *et al*, *Tetrahedron Lett.*, 1979, 1415 (*use*)
Fieser, M. *et al*, *Reagents for Organic Synthesis*, Wiley, 1967-83, **7**, 109 (*use*)

$C_{73}H_{60}Cl_4P_4Ru_2S$ Ru-00862

(Carbonothioyl)tri-μ-chlorochlorotetrakis(triphenylphosphine)diruthenium, 9CI

Tetrachloro(thiocarbonyl)tetrakis(triphenylphosphine)-diruthenium

[54275-51-3]

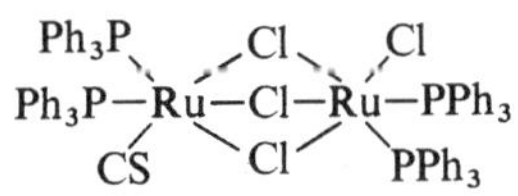

M 1437.184

Red cryst.

Me_2CO complex: [83242-24-4].
$C_{76}H_{66}Cl_4OP_4Ru_2S$ M 1495.264
Red-brown cryst. Mp 167-8°.

Stephenson, T.A. *et al*, *J. Chem. Soc., Dalton Trans.*, 1974, 1134 (*synth, nmr, ir*)
Fraser, A.J.F. *et al*, *J. Chem. Soc., Dalton Trans.*, 1974, 1139 (*cryst struct*)
Inorg. Synth., 1982, **21**, 28 (*synth, ir, pmr, bibl*)

$C_{74}H_{60}Cl_4O_2P_4Ru_2$ Ru-00863

Dicarbonyldi-μ-chlorodichlorotetrakis(triphenylphosphine)diruthenium, 9CI

Dicarbonyltetrachlorotetrakis(triphenylphosphine)diruthenium

[41060-95-1]

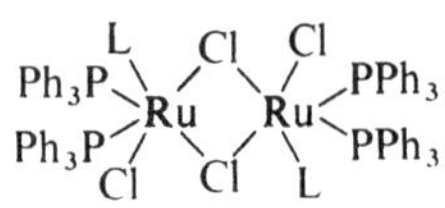

L = CO

M 1449.134

Bridged dimer. Orange cryst. (CH_2Cl_2/pentane). Mp 168-72° dec.

Monomer: Carbonyldichlorobis(triphenylphosphine)-ruthenium.
$C_{37}H_{30}Cl_2OP_2Ru$ M 724.567
Unknown.

Ruiz-Ramirez, L. *et al*, *J. Chem. Soc., Dalton Trans.*, 1974, 1640 (*synth, pmr, ir*)

$C_{74}H_{60}Cl_4P_4Ru_2S_2$ Ru-00864

Bis(carbonothioyl)di-μ-chlorodichlorotetrakis(triphenylphosphine)diruthenium, 9CI

Tetrachloro(thiocarbonyl)tetrakis(triphenylphosphine)-diruthenium

[54217-59-3]

As Dicarbonyldi-μ-chlorodichlorotetrakis(triphenylphosphine)diruthenium, Ru-00863 with

L = CS

M 1481.255

Bridged dimer. Purple-red cryst. Mp 165-6° dec.

Monomer: (Carbonothioyl)-dichlorobis(triphenylphosphine)ruthenium.
$C_{37}H_{30}Cl_2P_2RuS$ M 740.628
Unknown.

Armit, P.W. *et al*, *J. Chem. Soc., Dalton Trans.*, 1974, 1134 (*synth, ir, nmr*)

$C_{80}H_{64}O_8P_4Ru_4$ Ru-00865

Octacarbonyltetra-μ-hydrotetrakis(triphenylphosphine)tetraruthenium, 9CI

[34802-75-0]

$Ru_4H_4(CO)_8(PPh_3)_4$

M 1681.556

Structure not determined. Catalyst for hydrogenation of ethene and cyclohexanone. Red microcryst. (CH_2Cl_2/pentane). Thermally stable even in air. Spar. sol. aliphatic solvs., mod. sol. aromatic solvs.

Piacenti, F. *et al*, *Inorg. Chem.*, 1971, **10**, 2759 (*synth, ir, pmr*)
Banditelli, P. *et al*, *Thermochim. Acta*, 1975, **12**, 277.

$Cl_5NORu^{\ominus\ominus}$ Ru-00866

Pentachloronitrosylruthenate(2−), 9CI, 8CI

[19592-09-7]

$Cl_5Ru(NO)^{\ominus\ominus}$

M 308.341 (ion)

Di-NH_4 salt: [13820-58-1].
$Cl_5H_8N_3ORu$ M 344.418
Red cryst. Dec. at 300-75°.

Di-K salt: [14854-54-7].
Cl_5K_2NORu M 386.538
Red cryst. Dec. at 400-70°.

Cleare, M.J. *et al*, *J. Chem. Soc. (A)*, 1969, 372 (*synth, ir, raman*)
Veal, J.T. *et al*, *Acta Crystallogr., Sect. B*, 1972, **28**, 3525 (*cryst struct*)
Psalidas, V.S. *et al*, *CA*, 1973, **78**, 154390 (*synth*)
Sueur, S. *et al*, *Bull. Soc. Chim. Fr.*, 1976, 1051 (*synth*)
Cheney, R.P. *et al*, *Inorg. Chem.*, 1977, **16**, 3338 (*synth, uv*)
Suvorov, A.V. *et al*, *Zh. Obshch. Khim.*, 1978, **48**, 2169 (*nmr*)

F_5P_5Ru Ru-00867

Pentakis(phosphorous trifluoride)ruthenium, 9CI

[19702-30-8]

$Ru(PF_3)_5$

M 350.931

Stereochemically non-rigid. Colourless cryst. Mp 30°. $Bp_{0.001}$ 30° subl.

▷Dec. at 155°

Kruck, T. *et al*, *Z. Anorg. Allg. Chem.*, 1968, **356**, 118 (*synth, nmr*)
Meakin, P. *et al*, *J. Am. Chem. Soc.*, 1972, **94**, 5271 (*nmr*)
Udovich, C.A. *et al*, *J. Organomet. Chem.*, 1972, **36**, 355 (*synth, ir, ms*)
Head, R.A. *et al*, *J. Chem. Soc., Dalton Trans.*, 1975, 2054 (*uv*)

$F_{12}HP_4Ru^{\ominus}$ Ru-00868

Hydrotetrakis(phosphorous trifluoride)ruthenate(1−), 9CI

[32236-95-6]

$HRu(PF_3)_4^{\ominus}$

M 453.954 (ion)

Stereochemically nonrigid.

Tris(phenanthroline)iron(2+) salt:
$C_{36}H_{26}F_{24}FeP_8Ru_2$ M 1420.340
Air-stable bright-red solid.

Kruck, T., *Angew. Chem., Int. Ed. Engl.*, 1967, **6**, 53 (*rev*)
Kruck, T. *et al*, *Z. Anorg. Allg. Chem.*, 1969, **371**, 1 (*synth, ir, pmr, nmr*)
Meakin, P. *et al*, *J. Am. Chem. Soc.*, 1972, **94**, 5271 (*pmr, nmr*)

$F_{12}H_2P_4Ru$ **Ru-00869**

Dihydrotetrakis(phosphorous trifluoride)ruthenium, 9CI

[24899-53-4]

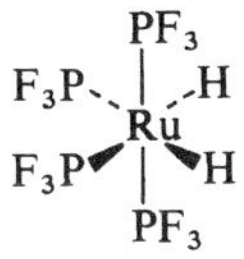

M 454.962

Mp −76°. Dec. at 290°.

Kruck, T. *et al*, *Z. Anorg. Allg. Chem.*, 1969, **371**, 1 (*synth, spectra*)

Meakin, P. *et al*, *J. Am. Chem. Soc.*, 1973, **95**, 75 (*pmr, nmr*)

$F_{12}P_4Ru^{\ominus\ominus}$ **Ru-00870**

Tetrakis(phosphorous trifluoride)ruthenate(2−)

$Ru(PF_3)_4^{\ominus\ominus}$

M 452.946 (ion)

Di-K salt: [26876-73-3].

$F_{12}K_2P_4Ru$ M 531.142

Bright-yellow air-sensitive solid (Me_2CO/pentane).

Mp 150° dec.

Kruck, T. *et al*, *Z. Anorg. Allg. Chem.*, 1969, **371**, 1 (*synth, ir, nmr*)

Name Index

Acetonitrile(η^3-allyl)bis(triphenylphosphine)ruthenium, *see* Ru-00769
(Acetonitrile)(η^6-benzene)bis(trimethylphosphine)ruthenium(2+), Ru-00371
(Acetonitrile)(η^6-benzene)dichloroosmium, *in* Os-00036
(Acetonitrile)carbonyldichloro[(1,2,5,6-η)-1,5-cyclooctadiene]ruthenium, Ru-00199
Acetonitrile(η^5-2,4-cyclopentadien-1-yl)-bis(triphenylphosphine)osmium(1+), Os-00419
Acetonitrile(η^5-2,4-cyclopentadien-1-yl)-bis(triphenylphosphine)ruthenium(1+), Ru-00791
[(Acetonitrile)decacarbonyl-μ-hydrotriosmium]pentacarbonylrhenium, Os-00290
(Acetonitrile)dicarbonyl(η^5-2,4-cyclopentadien-1-yl)osmium(1+), Os-00074
(Acetonitrile)dicarbonyl(η^5-2,4-cyclopentadien-1-yl)ruthenium(1+), Ru-00096
(Acetonitrile)pentadecacarbonyl-μ-hydrotriosmiumrhenium, *see* Os-00290
(Acetonitrile)(η^3-2-propenyl)bis(triphenylphosphine)ruthenium, Ru-00769
(Acetonitrile)undecacarbonyltriosmium, Os-00213
(η^6-Acetophenone)(η^6-benzene)ruthenium(2+), *see* Ru-00342
Acetoxyruthenocene, *in* Ru-00142
(η^6-Acetylbenzene)(η^6-benzene)ruthenium(2+), *see* Ru-00342
Acetylcarbonylchlorobis(triphenylphosphine)ruthenium, Ru-00746
Acetylcarbonyl(η^5-2,4-cyclopentadien-1-yl)-(triphenylphosphine)ruthenium, Ru-00631
[(1,2,3,4,5-η)-1-Acetyl-2,4-cyclopentadien-1-yl](η^6-benzene)ruthenium(1+), Ru-00308
[(1,2,3,4,5-η)-1-Acetyl-2,4-cyclopentadien-1-yl](η^6-chlorobenzene)ruthenium(1+), Ru-00303
(Acetyl-π-cyclopentadienyl)-π-cyclopentadienylruthenocene, *see* Ru-00240
(1-Acetyl-1*H*-1,2-diazepine)tricarbonylruthenium, *in* Ru-00073
Acetyldicarbonyl(η^5-2,4-cyclopentadien-1-yl)ruthenium, Ru-00100
Acetyldicarbonyltris(triphenylphosphine)ruthenium(1+), Ru-00852
Acetyldichloronitrosylbis(triphenylphosphine)ruthenium, Ru-00732
Acetylenecarbonylnitrosylbis(triphenylphosphine)osmium(1+), *see* Os-00398
η^2-Acetylene(pentaamine)ruthenium(2+), *see* Ru-00013
(1-Acetyl-2-methyl-π-cyclopentadienyl)-π-cyclopentadienylruthenium, *see* Ru-00309
1-Acetyl-2-methylruthenocene, Ru-00309
Acetylosmocene, Os-00177
(Acetylphenyldiazene-N^2,O^1)dichlorobis(triphenylphosphine)osmium, Os-00426
Acetylruthenocene, Ru-00240
(Acrylonitrile)tetracarbonylruthenium, *see* Ru-00044
(η^3-Allyl)(η^6-benzene)chloroosmium, *see* Os-00081
η^3-Allyl(η^6-benzene)chlororuthenium, *see* Ru-00109
(η^3-Allyl)bromodicarbonyl(triphenylphosphine)ruthenium, *see* Ru-00577
π-Allylbromotricarbonylruthenium, *see* Ru-00034
π-Allylcarbonyl-π-cyclopentadienylruthenium, *see* Ru-00105
(π-Allyl)carbonyl(π-indenyl)ruthenium, *see* Ru-00304
η^3-Allylcarbonylnitrosyl(triphenylphosphine)ruthenium, *see* Ru-00567
Allylchloro(1,5-cyclooctadiene)ruthenium, *in* Ru-00574
π-Allylchlorotris(triphenylphosphine)ruthenium, *see* Ru-00851
σ-Allyl-π-cyclopentadienylbis(triphenylphosphine)ruthenium, *see* Ru-00805
η^1-Allyl(η^5-cyclopentadienyl)dicarbonylruthenium, *see* Ru-00143
σ-Allyl-π-cyclopentadienylruthenium dicarbonyl, *see* Ru-00143
π-Allyl-π-cyclopentadienyl(triphenylphosphine)ruthenium, *see* Ru-00635
(η^3-Allyl)dicarbonylchloro(triphenylphosphine)ruthenium, *see* Ru-00578
Allyldicarbonyl-π-cyclopentadienylruthenium, *see* Ru-00143
(η^3-Allyl)hydridotris(methyldiphenylphosphine)ruthenium, *see* Ru-00784
(η^3-Allyl)nitrosylbis(triphenylphosphine)ruthenium, *see* Ru-00748
π-Allyltricarbonylchlororuthenium, *see* Ru-00035
π-Allyltricarbonyliodoruthenium, *see* Ru-00036
μ-Amidodecacarbonyl-μ-hydrotriosmium, Os-00097
(Aminocarbonyl)carbonyl(η^5-2,4-cyclopentadien-1-yl)-(triphenylphosphine)ruthenium, Ru-00618
(Aminocarbonyl)dicarbonyl(η^5-2,4-cyclopentadien-1-yl)osmium, Os-00060
(Aminocarbonyl)dicarbonyl(η^5-2,4-cyclopentadien-1-yl)ruthenium, Ru-00077
(Aminocarbonyl)ruthenocene, *in* Ru-00185
Ammine(η^6-benzene)(1,2-ethanediamine-*N*,*N*′)osmium(2+), Os-00066
Amminedicarbonyl(η^5-2,4-cyclopentadien-1-yl)ruthenium(1+), Ru-00060
μ_3-Anilinatodecacarbonyltriruthenium, *see* Ru-00417
(μ_3-Anilinato)nonacarbonyldi-μ-hydrotriosmium, Os-00253
μ_3-Anilinatononacarbonyldihydrotriruthenium, *see* Ru-00381
(η^6-Aniline)(η^5-cyclopentadienyl)ruthenium(1+), Ru-00190
(η^6-Anisole)(η^6-benzene)ruthenium(2+), *see* Ru-00310
(η^6-Anisole)(η^5-cyclopentadienyl)ruthenium(1+), *see* Ru-00253
(η^6-Anthracene)(η^6-mesitylene)ruthenium(2+), *see* Ru-00582
[(1,2,3,4,4*a*,9*a*-η)-Anthracene][(1,2,3,4,5,6-η)-1,3,5-trimethylbenzene]ruthenium(2+), Ru-00582
Aquacarbonyltetrachlororuthenate(2−), Ru-00006
Aquadicarbonylhydrobis(triphenylphosphine)ruthenium(1+), Ru-00733
[(2,3,4,5-η)-1*H*-Azepine-1-carboxylic acid]tricarbonylruthenium, Ru-00128
(Azobenzene-*C*,*N*′)dicarbonylchlororuthenium, *see* Ru-00330
(Azobenzene-*C*,*N*′)dicarbonylchloro(triphenylphosphine)ruthenium, *see* Ru-00673
[(Azobenzene)-*C*,*N*′]dicarbonyldichlororuthenate(1−), *see* Ru-00331
(μ_3-Azulene)-μ-carbonylhexacarbonyltriruthenium, *see* Ru-00451
[μ-[(1,2,3,3*a*,8*a*-η:4,5,6-η:6,7,8-η)-Azulene]]heptacarbonyltriruthenium, Ru-00451
[μ-[(1,2,3,3*a*,8*a*-η:4,5,6-η)-azulene]]pentacarbonyldiruthenium, Ru-00383
[μ_3-[Benzenaminato(2−)]]-μ_3-carbonylnonacarbonyltriruthenium, Ru-00417
[μ_3-[Benzenaminato(2−)]]nonacarbonyldi-μ-hydrotriosmium, *see* Os-00253
[μ_3-[Benzenaminato(2−)]]nonacarbonyldi-μ-hydrotriruthenium, Ru-00381
3,1,2-(η^6-Benzene)-*$RuC_2B_9H_{11}$*, *see* Ru-00087
[μ_2-[Benzeneaminato(1−)]]decacarbonyl-μ-hydrotriosmium, Os-00276
(η^6-Benzeneamine)(η^5-2,4-cyclopentadien-1-yl)ruthenium(1+), *see* Ru-00190
[(1,2,3,4-η)-Benzene](η^6-benzene)ruthenium, Ru-00250
(η^6-Benzene)[(1,2,3,4,5,6-η)-benzoic acid]ruthenium(2+), Ru-00306
(η^6-Benzene)bis(trimethylphosphine)ruthenium, Ru-00276
(η^6-Benzene)bis(trimethyl phosphite-*P*)osmium, Os-00195
(η^6-Benzene)bis(trimethyl phosphite-*P*)ruthenium, Ru-00275
(η^6-Benzene)bis(triphenylphosphine)osmium, Os-00413
(η^6-Benzene)(μ_6-carbon)-μ-carbonyltridecacarbonylhexaruthenium, *see* Ru-00536

(Benzene)[μ_6-(carbon)]tetradecacarbonylhexaruthenium, Ru-00536
(η^6-Benzene)carbonylbis(trichlorogermyl)ruthenium, Ru-00053
(η^6-Benzene)carbonylbis(trichlorosilyl)ruthenium, Ru-00054
(η^6-Benzene)carbonyldichloroosmium, Os-00044
(η^6-Benzene)(η^6-chlorobenzene)ruthenium(2+), Ru-00234
(η^6-Benzene)chlorobis(pyridine)ruthenium(1+), Ru-00427
(η^6-Benzene)chlorobis(triphenylphosphine)ruthenium(1+), Ru-00776
(η^6-Benzene)chloro[(1,2,3,4-η)-1,3-cyclohexadiene]ruthenium(1+), Ru-00255
(η^6-Benzene)chloro(1,2-ethanediamine-*N*,*N'*)osmium(1+), Os-00065
(η^6-Benzene)chloro(η^2-ethene)(trimethylphosphine)ruthenium(1+), Ru-00202
(η^6-Benzene)chloro(η^2-ethylene)(trimethylphosphine)ruthenium, *see* Ru-00202
(η^6-Benzene)chloromethyl(triphenylphosphine)ruthenium, Ru-00624
(η^6-Benzene)chlorophenyl(triphenylphosphine)ruthenium, Ru-00660
(η^6-Benzene)chloro(η^3-2-propenyl)osmium, Os-00081
(η^6-Benzene)chloro(η^3-2-propenyl)ruthenium, Ru-00109
(η^6-Benzene)chloro(tributylphosphine)ruthenium(1+), *in* Ru-00704
(η^6-Benzene)[(1,2,3,4-η)-1,3-Cyclohexadiene]osmium, Os-00183
(η^6-Benzene)[(1,2,3,4-η)-1,3-cyclohexadiene]ruthenium, Ru-00259
(η^6-Benzene)[(1,2,5,6-η)-1,5-cyclooctadiene]ruthenium, Ru-00365
(η^6-Benzene)[(1,2,3,4-η)-1,3,5,7-cyclooctatetraene]osmium, Os-00240
(η^6-Benzene)[(1,2,3,4-η)-1,3,5,7-cyclooctatetraene]ruthenium, Ru-00348
(η^6-Benzene)[(1,2,3,4,5-η)-2,4,6-cyclooctatrien-1-yl]ruthenium(1+), Ru-00358
(η^6-Benzene)(η^5-2,4-cyclopentadien-1-yl)osmium(1+), Os-00140
(η^6-Benzene)(η^5-2,4-cyclopentadien-1-yl)ruthenium(1+), Ru-00189
η^6-Benzenedibromoosmium, Os-00034
(η^6-Benzene)dibromoruthenium, *in* Ru-00237
(η^6-Benzene)dibromo(triphenylphosphine)ruthenium, Ru-00598
(Benzene)dichloro(methylsulfoxide)ruthenium, *in* Ru-00038
η^6-Benzenedichloroosmium, Os-00036
(η^6-Benzene)dichloro(pyridine)osmium, Os-00137
(η^6-Benzene)dichlororuthenium, Ru-00038
(η^6-Benzene)dichloro[sulfinylbis[methane]-*O*]ruthenium, *in* Ru-00038
(η^6-Benzene)dichloro(trimethylphosphine)ruthenium, Ru-00113
(η^6-Benzene)dichloro(triphenylphosphine)ruthenium, Ru-00599
(η^6-Benzene)diiodoosmium, Os-00037
(η^6-Benzene)diiodoruthenium, *in* Ru-00239
(η^6-Benzene)diiodo(trimethylphosphine)osmium, Os-00082
(η^6-Benzene)diiodo(triphenylphosphine)osmium, Os-00337
η^6-Benzenediiodo(triphenylphosphine)osmium, Os-00431
(η^6-Benzene)diiodo(triphenylphosphine)ruthenium, Ru-00600
(η^6-Benzene)[(1,2,3,4,5,6-η)-6-(dimethoxyphosphinyl)-2,4-cyclohexadien-1-yl]ruthenium(1+), Ru-00362
(η^6-Benzene)dimethyl(triphenylphosphine)ruthenium, Ru-00636
(η^6-Benzene)(η^2-ethene)bis(trimethylphosphine)ruthenium(2+), Ru-00372
(η^6-Benzene)(η^2-ethene)hydro(trimethylphosphine)osmium(1+), Os-00146
(η^6-Benzene)(η^2-ethene)hydro(trimethylphosphine)ruthenium(1+), Ru-00204
(η^6-Benzene)(η^2-ethene)iodo(trimethylphosphine)osmium(1+), Os-00144
(η^6-Benzene)(η^2-ethene)methyl(trimethylphosphine)osmium(1+), Os-00194
(η^6-Benzene)(η^2-ethene)methyl(trimethylphosphine)ruthenium(1+), Ru-00273
(η^6-Benzene)(η^2-ethene)(trimethylphosphine)osmium, Os-00145
(η^6-Benzene)(η^2-ethene)(trimethylphosphine)ruthenium, Ru-00203
(η^6-Benzene)[(1,2,3,4,5-η)-1-ethyl-2,4-cyclopentadien-1-yl]ruthenium(1+), Ru-00315
(η^6-Benzene)(η^2-ethylene)hydro(trimethylphosphine)ruthenium(1+), *see* Ru-00204
(η^6-Benzene)(η^2-ethylene)(trimethylphosphine)osmium, *see* Os-00145
(η^6-Benzene)ethylene(trimethylphosphine)ruthenium, *see* Ru-00203
[μ_3-[(1-η:1,2-η:2-η)-Benzene]]heptacarbonylbis[μ-(dimethylphosphino)]triosmium, Os-00298
(η^6-Benzene)[(1,2,3,4,5,6-η)-hexamethylbenzene]ruthenium(2+), Ru-00488
(η^6-Benzene)hydrobis(triphenylphosphine)osmium(1+), Os-00414
(η^6-Benzene)hydrobis(triphenylphosphine)ruthenium(1+), Ru-00779
(η^6-Benzene)[hydrotris(1*H*-pyrazolato-*N'*)borato(1−)-N^2,$N^{2'}$,$N^{2''}$]ruthenium(1+), Ru-00395
(η^6-Benzene)iodobis(triphenylphosphine)osmium(1+), Os-00412
(η^6-Benzene)iodomethyl(trimethylphosphine)osmium, Os-00111
π-Benzene-π-mesityleneruthenium(2+), *see* Ru-00400
(η^6-Benzene)[(1,2,3,4,5,6-η)-methoxybenzene]ruthenium(2+), Ru-00310
(η^6-Benzene)(η^6-methoxycarbonylbenzene)ruthenium(2+) bis(tetrafluoroborate), *in* Ru-00306
(η^6-Benzene)[(1,2,3,4,5,6-η)-methylbenzoate]ruthenium(2+) bis(tetrafluoroborate), *in* Ru-00306
(η^6-Benzene)methylbis(triphenylphosphine)osmium(1+), Os-00420
(η^6-Benzene)nonacarbonyl[μ_4-(1-η:1,2-η:1,2-η:2-η)-1-cyclohexene-1,2-diyl]tetraruthenium, Ru-00540
(Benzene)nonacarbonyl-μ-hydrotriosmium(1+), Os-00254
π-Benzeneosmium dibromide, *see* Os-00034
π-Benzeneosmium dichloride, *see* Os-00036
(π-Benzene)osmium diiodide, *see* Os-00037
(η^6-Benzene)[[2.2]paracyclophane]ruthenium(2+), Ru-00569
(η^6-Benzene)(η^6-phenol)ruthenium(2+), Ru-00241
(η^6-Benzene)[(1,2,3,4,5,6-η)-1-phenylboratabenzene]ruthenium(1+), Ru-00456
(η^6-Benzene)[1-(η^6-phenyl)ethanone]ruthenium(2+), Ru-00342
(Benzeneselenolato)dicarbonyl(η^5-2,4-cyclopentadien-1-yl)ruthenium, Ru-00301
[μ-[1,2,3,4-Benzenetetrayltetrakis[dimethylsilylene]]]octacarbonyldiruthenium, Ru-00571
[μ-[1,2,4,5-Benzenetetrayltetrakis[dimethylsilylene]]]octacarbonyldiruthenium, Ru-00572
[(μ-Benzenethiolato)]decacarbonyl-μ-hydrotriosmium, Os-00275
(Benzenethiolato)dicarbonyl(η^5-2,4-cyclopentadien-1-yl)ruthenium, Ru-00300
(η^6-Benzene)trichlororuthenate(1−), Ru-00039
(η^6-Benzene)[(1,10,11,12,13,14-η)-tricyclo[8.2.2.2^{4,7}]-hexadeca-4,6,10,12,13,15-hexaene]ruthenium(2+), *see* Ru-00569
(η^6-Benzene)[(1,2,3,4,5,6-η)-1,3,5-trimethylbenzene]ruthenium(2+), Ru-00400
(η^6-Benzene)[(1,2,3,4,5-η)-6-(triphenylphosphino)-2,4-cyclohexadien-1-yl]ruthenium(2+), Ru-00663
(η^6-Benzene)tris(trimethylphosphine)ruthenium(2+), Ru-00407
(η^6-Benzene)[(7,8,9,10,11-η)-undecahydro-7,8-dicarbaundecaborato(2−)]ruthenium, Ru-00087
[(1,2,3,4,5,6-η)-Benzoato][(1,2,3,4,5,6-η)-1,3,5-trimethylbenzene]ruthenium(1+) hydrogen tetrafluoroborate(1:1:2), *in* Ru-00429
(η^6-Benzoic acid)(η^6-mesitylene)ruthenium(2+), *see* Ru-00429
(η^6-Benzoic acid)(η^6-1,3,5-trimethylbenzene)ruthenium(2+), Ru-00429
(Benzonitrile)tricarbonyldichlororuthenium, Ru-00125
[μ-(Benzophenone hydrazonato)]decacarbonyl-μ-hydrotriosmium, *see* Os-00331

[μ_3-(2(3*H*)-Benzothiazolethionato-N^3:S^2:S^2)]nonacarbonyl-μ-hydrotriruthenium, Ru-00416
[(Benzoylamino)carbonyl]dicarbonylbis(triphenylphosphine)ruthenium(1+), Ru-00814
(Benzoyl-π-cyclopentadienyl)-π-cyclopentadienylruthenium, *see* Ru-00454
Benzoyldichloronitrosylbis(triphenylphosphine)ruthenium, Ru-00786
Benzoylosmocene, Os-00295
Benzoylruthenocene, Ru-00454
Benzylcarbonyl(η^5-2,4-cyclopentadien-1-yl)-(triphenylphosphine)ruthenium, Ru-00670
Benzyl(η^5-2,4-cyclopentadien-1-yl)bis(triphenylphosphine)ruthenium, Ru-00818
σ-Benzyl-π-cyclopentadienylruthenium dicarbonyl, *see* Ru-00337
Benzyldicarbonyl(η^5-2,4-cyclopentadien-1-yl)ruthenium, Ru-00337
Benzyldichloronitrosylbis(triphenylphosphine)ruthenium, Ru-00788
[μ-(Benzylidene-*o*-phenylene)]nonacarbonyl-μ-hydrotriruthenium, *see* Ru-00556
Benzylosmocene, Os-00299
Benzylruthenocene, Ru-00460
[(2,3,5,6-η)-Bicyclo[2.2.1]hepta-2,5-diene]bis(η^3-2-propenyl)ruthenium, Ru-00322
Bicyclo[2.2.1]hepta-2,5-dienecarbonyldichlororuthenium, *in* Ru-00428
[(2,3,5,6-η)-Bicyclo[2.2.1]hepta-2,5-diene]carbonyltrichlororuthenate(1−), Ru-00078
(Bicyclo[2.2.1]hepta-2,5-diene)(1,3,5-cycloheptatriene)ruthenium, Ru-00355
[(2,3,5,6-η)-Bicyclo[2.2.1]hepta-2,5-diene][(1,2,3,4,5,6-η)-1,3,5,7-cyclooctatetraene]ruthenium, Ru-00397
[(2,3,5,6-η)-Bicyclo[2.2.1]hepta-2,5-diene]-dichlorobis(triphenylphosphine)ruthenium, Ru-00790
[(2,3,5,6-η)-Bicyclo[2.2.1]hepta-2,5-diene]dichlororuthenium, Ru-00058
[(2,3,5,6-η)-Bicyclo[2.2.1]hepta-2,5-diene][(1,2,3,4,5,6-η)-hexamethylbenzene]ruthenium, Ru-00512
[(2,3,4,5-η)Bicyclo[4.2.0]octa-2,4-diene][(1,2,3,4,5,6-η)-1,3,5-cyclooctatriene]ruthenium, Ru-00432
[(2,3,4,5-η)-Bicyclo[4.2.0]octa-2,4-diene]tricarbonylruthenium, Ru-00186
[η^5-(Bicyclo[3.3.0]octa-2,4-dien-1-yl)]-dicarbonyl(trimethylgermyl)osmium, *see* Os-00221
[η^5-(Bicyclo[3.3.0]octa-2,4-dien-1-yl)]-dicarbonyl(trimethylsilyl)osmium, *see* Os-00222
π-Bicyclo[5.1.0]octadienyliumtricarbonylosmium(1+), *see* Os-00135
[(2,3,4,5,6-η)-Bicyclo[5.1.0]octa-3,5-dien-2-yl]tricarbonylosmium(1+), Os-00135
[(2,3,4,6,7-η)-Bicyclo[3.2.1]octa-3,6-dien-2-yl]tricarbonylruthenium(1+), Ru-00178
[(2,3,4,6-η)Bicyclo[3.2.1]oct-3-ene-2,6-diyl]tricarbonylruthenium, Ru-00187
[(1,2,3,4,5,6-η)-1,1′-Biphenyl](η^5-2,4-cyclopentadien-1-yl)ruthenium(1+), Ru-00455
1,1″-Biruthenocene, Ru-00523
Bis[μ-(acetato-*O*:*O*′)]hexacarbonyldiosmium, Os-00102
Bis(acetonitrile)(η^3-allyl)(1,5-cyclooctadi ne)osmium, *see* Os-00265
Bis(acetonitrile)(η^6-benzene)chlororuthenium(1+), Ru-00154
Bis(acetonitrile)chloro[(1,2,3,4,5,6-η)-1,3,5-cycloheptatriene]ruthenium(1+), Ru-00198
Bis(acetonitrile)[(1,2,5,6-η)-1,5-cyclooctadiene](η^3-2-propenyl)osmium(1+), Os-00265
Bis(acetonitrile)decacarbonyltriosmium, Os-00226
Bis(acetonitrile)dicarbonyldichlororuthenium, Ru-00037
Bis(acetonitrile)dichlorobis(triphenylphosphine)ruthenium, Ru-00758
Bis(acetonitrile)dodecacarbonyltri-μ-hydrotetraosmium(1+), Os-00279
Bis(acetylacetonato)(1,5-cyclooctadiene)ruthenium, *see* Ru-00490
Bis(acetyl-π-cyclopentadienyl)ruthenium, *see* Ru-00346
Bis(η^3-allyl)(1,5-cyclooctadiene)osmium, *see* Os-00244
Bis(π-allyl)(1,5-cyclooctadiene)ruthenium, *see* Ru-00369
Bis(2-aminobenzenethiolato-*N*,*S*)bis(triphenylphosphine)ruthenium, Ru-00817
Bis(*o*-aminobenzenethiolato)dicarbonylruthenium, *see* Ru-00336
Bis(2-aminothiophenol)dicarbonylruthenium, Ru-00336
Bis(anilinato)nonacarbonyltriruthenium, *see* Ru-00537
Bis(aniline)dibromodicarbonylruthenium, Ru-00340
Bis(aniline)dicarbonyldichlororuthenium, Ru-00341
Bis(benzenamine)dicarbonyldichlororuthenium, *see* Ru-00341
Bis[μ_3-[benzeneaminato(2−)]]nonacarbonyltriruthenium, Ru-00537
Bis(benzeneamine)dibromodicarbonylruthenium, *see* Ru-00340
Bis(η^6-benzene)di-μ-bromodibromodiruthenium, Ru-00237
Bis(η^6-benzene)di-μ-chlorobis(tributylphosphine)diruthenium(2+), Ru-00704
Bis(η^6-benzene)di-μ-chlorodichlorodiosmium, *in* Os-00036
Bis(η^6-benzene)di-μ-chlorodichlorodiruthenium, *in* Ru-00038
Bis(η^6-benzene)di-μ-iododiiododiruthenium, Ru-00239
Bis(η^6-benzene)osmium(2+), Os-00179
Bis(η^6-Benzene)ruthenium(2+), Ru-00251
[1,2-Bis(benzeneseleno)ethane-*Se*,*Se*′]dibromodicarbonylruthenium, Ru-00424
Bis[μ-(benzeneselenolato)]dicarbonylbis(η^5-2,4-cyclopentadien-1-yl)diruthenium, Ru-00596
Bis(μ-benzeneselenolato)dicarbonyldi-π-cyclopentadienyldiruthenium, *see* Ru-00596
Bis(μ-benzeneselenolato)hexacarbonyldiruthenium, Ru-00473
Bis(μ-benzenetellurolato)hexacarbonyldiruthenium, Ru-00474
Bis(η^6-benzene)tetrabromodiruthenium, *see* Ru-00237
Bis(η^6-benzene)tetrachlorodiosmium, *in* Os-00036
Bis(η^6-benzene)tetrachlorodiruthenium, *in* Ru-00038
Bis(benzene)tetraiodobis(triphenylphosphine)diosmium, *in* Os-00431
Bis(η^6-benzene)tetraiododiruthenium, *see* Ru-00239
Bis[μ-(benzenethiolato)]dicarbonylbis(η^5-2,4-cyclopentadien-1-yl)diruthenium, Ru-00595
Bis(μ-benzenethiolato)dicarbonyldi-π-cyclopentadienyldiruthenium, *see* Ru-00595
[1,2-Bis(benzenethiolato)ethane-*S*,*S*′]dibromodicarbonylruthenium, Ru-00423
Bis(η^6-benzene)tri-μ-chlorodiosmium(1+), Os-00176
Bis(η^6-benzene)tri-μ-chlorodiruthenium(1+), Ru-00238
Bis(benzocyclopentadienyl)ruthenium, *see* Ru-00482
Bis(benzonitrile)dicarbonyldichlororuthenium, Ru-00421
Bis(benzonitrile)dichlorobis(triphenylphosphine)ruthenium, Ru-00822
Bis(benzoylcyclopentadienyl)ruthenium, *see* Ru-00586
Bis(bicyclo[2.2.1]hepta-2,5-diene)dicarbonyltetrachlorodiruthenium, Ru-00428
Bis[1,2-bis(dimethylphosphino)ethane](η^2-ethylene)ruthenium, *see* Ru-00377
Bis[bis(μ-methylene)tetrakis(trimethylphosphine)ruthenium]ruthenium(2+), *see* Ru-00651
Bis[1,2-bis(methylthio)ethane-*S*,*S*′]dibromoruthenium, Ru-00088
Bis[μ-[bis(pentafluorophenyl)arsino]]hexacarbonyldiruthenium, Ru-00656
Bis[μ-[bis(pentafluorophenyl)phosphino]]hexacarbonyldiruthenium, Ru-00657
Bis(η^4-1,3-butadiene)(triphenylphosphine)ruthenium, Ru-00637
Bis(*tert*-butyl isocyanide)tricarbonylruthenium, *see* Ru-00320
Bis[μ-(carbonothioyl)]dicarbonylbis(η^5-2,4-cyclopentadien-1-yl)diruthenium, Ru-00333
Bis(carbonothioyl)dichlorobis(triphenylphosphine)ruthenium, Ru-00722
Bis(carbonothioyl)di-μ-chlorodichlorotetrakis(triphenylphosphine)diruthenium, Ru-00864
Bis(carboxycyclopentadienyl)ruthenium, *see* Ru-00232
1,1′-Bis(chloromercuri)ruthenocene, Ru-00130
Bis[(1,2,3,4,5-η)-2,4-cycloheptadien-1-yl]ruthenium, Ru-00366
Bis(η^5-cyclohexa-2,4-dien-1-yl)ruthenium, Ru-00260
Bis[(1,2,3,4,5-η)-2,4-cyclooctadien-1-yl]ruthenium, Ru-00434

Bis(cyclopentadienyldicarbonylruthenium), *see* Ru-00335
Bis(η^5-2,4-cyclopentadien-1-yl)iodoruthenium(1+), Ru-00141
Bis(cyclopentadienyl)osmium, *see* Os-00109
Bis(cyclopentadienylosmium dicarbonyl), *see* Os-00233
Bis(cyclopentadienyl)ruthenium, *see* Ru-00149
Bis(cyclopentadienyl)tetracarbonyldiruthenium, *see* Ru-00335
Bis(dimethylcarbamodithioato-*S,S'*)bis(triphenylphosphine)ruthenium, Ru-00783
[1,2-Bis(dimethylsilyl)ethane]tetracarbonylruthenium, *see* Ru-00160
[1,2-Bis(diphenylphosphine)ethane]decacarbonyltetrahydrotetraruthenium, *see* Ru-00691
Bis[μ-(diphenylphosphino)]-dinitrosylbis(methyldiphenylphosphine)diruthenium, Ru-00823
Bis[μ-(diphenylphosphino)]-dinitrosylbis(triphenylphosphine)diruthenium, Ru-00854
[1,2-Bis(diphenylphosphino)ethane]bromo(η^5-cyclopentadienyl)osmium, *see* Os-00355
[1,2-Bis(diphenylphosphino)ethane]-carbonyl(cyclopentadienyl)ruthenium(1+), *see* Ru-00676
[1,2-Bis(diphenylphosphino)ethane]chloro(η^5-cyclopentadienyl)ruthenium, *see* Ru-00671
[1,2-Bis(diphenylphosphino)ethane](η^5-cyclopentadienyl)methylruthenium, *see* Ru-00678
[1,2-Bis(diphenylphosphino)ethane]dibromodicarbonylruthenium, *see* Ru-00645
Bis[1,2-(diphenylphosphino)ethane]dicarbonyldiiodoosmium, *see* Os-00346
Bis[(1,2-diphenylphosphino)ethane]dihydroruthenium, *see* Ru-00830
Bis[1,2-ethanediylbis[dimethylphosphine]-*P,P'*](η^2-ethene)ruthenium, Ru-00377
Bis[1,2-ethanediylbis[diphenylphosphine]-*P,P'*]dihydroruthenium, Ru-00830
Bis(η^2-ethene)[(1,2,3,4,5,6-η)-hexamethylbenzene]ruthenium, Ru-00438
Bis[(η^2-ethenyl)benzene]bis(triphenylphosphine)ruthenium, Ru-00829
1,1'-Bis(ethenyl)ruthenocene, *see* Ru-00350
Bis(ethyl-π-cyclopentadienyl)ruthenium, *see* Ru-00367
Bis(ethylene)(hexamethylbenzene)ruthenium, *see* Ru-00438
[1,2-Bis(ethylthio)ethane-*S,S'*]dicarbonyldiiodoruthenium, Ru-00086
Bis(1,1,1,4,4,4-hexafluoro-2,3-butanedithionato-*S,S'*)-bis(triphenylphosphine)ruthenium, Ru-00795
Bis(1,1,1,4,4,4-hexafluoro-2,3-butanedithione)-bis(triphenylphosphine)ruthenium, *see* Ru-00795
Bis(hexamethylbenzene)ruthenium, *see* Ru-00614
Bis[(1,2,3,4,5,6-η)-hexamethylbenzene]ruthenium(2+), Ru-00613
1,1'-Bis(hydroxymethyl)ruthenocene, Ru-00258
Bis(indenyldicarbonylruthenium), *see* Ru-00558
Bis[(1,2,3,3*a*,7*a*-η)-1*H*-inden-1-yl]ruthenium, Ru-00482
Bis(2-isocyano-2-methylpropane)tricarbonylruthenium, Ru-00320
(Bis-π-mesitylene)ruthenium(2+), *see* Ru-00489
Bis(2-methyl-π-allyl)bis(phosphorous acid)ruthenium hexamethyl ester, *see* Ru-00374
Bis(2-methyl-π-allyl)bis(triphenylphosphine)ruthenium, *see* Ru-00807
Bis[(1,2,3,4,5,6-η)-1-methylboratabenzene]ruthenium, Ru-00266
Bis(1-methylborinato)ruthenium, *see* Ru-00266
Bis(*O*-methylcarbonodithioato-*S,S'*)bis(triphenylphosphine)ruthenium, Ru-00759
Bis(methylcyclopentadienyldicarbonylruthenium), *see* Ru-00425
Bis(methyl-π-cyclopentadienyl)ruthenium, *see* Ru-00261
Bis(*O*-methyldithiocarbonato)bis(triphenylphosphine)ruthenium, *see* Ru-00759
[μ-[*N,N'*-Bis(1-methylethyl)-1,4-bis[(1-methylethyl)imino]-2,3-butanediaminato(2−)-N^1,N^2,N^3:N^2,N^3,N^4]]-μ-carbonyl-tetracarbonyldiruthenium, Ru-00551
[μ-[*N,N'*-Bis(1-methylethyl)-1,4-bis[(1-methylethyl)imino]-2,3-butanediaminato(2−)-N^1,N^2,N^3:N^2,N^3,N^4]]tetracarbonyldiruthenium, Ru-00531
[1,3-Bis(4-methylphenyl)-1-triazenato-N^1,N^3]-carbonylhydrobis(triphenylphosphine)ruthenium, Ru-00826
[*N,N'*-Bis(4-methylphenyl)ureato(2−)-*N,N'*]-dicarbonylbis(triphenylphosphine)ruthenium, Ru-00831
Bis[2-methyl-2-propanaminoato(2−)]dioxoosmium, Os-00067
Bis[(1,2,3-η)-2-methyl-2-propenyl]bis(trimethyl phosphite-*P*)ruthenium, Ru-00374
Bis[(1,2,3-η)-2-methyl-2-propenyl]bis(triphenylphosphine)ruthenium, Ru-00807
Bis[(1,2,3,4,4*a*,8*a*-η)naphthalene]ruthenium(2+), Ru-00521
Bis[μ-[η^6-nonahydrodicarbanonaborato(2−)-*B,H*]]-tetrakis(triethylphosphine)diruthenium, *in* Ru-00379
1,1'':1',1'''-Bis(3-oxo-1,4-pentadiene-1,5-diyl)bisruthenocene, Ru-00659
1,1'-Bis(1-oxo-3-phenyl-2-propenyl)ruthenocene, Ru-00644
1,1'':1',1'''-Bis(3-oxo-1-propene-1,3-diyl)bisruthenocene, Ru-00630
Bis[(1,2,3,4,5,6-η)-1-phenylboratabenzene]osmium, Os-00327
Bis[(1,2,3,4,5,6-η)-1-phenylboratabenzene]ruthenium, Ru-00566
Bis(1-phenylborinato)osmium, *see* Os-00327
Bis(1-phenylborinato)ruthenium, *see* Ru-00566
Bis(1-phenylcyclohexadienyl)ruthenium, *see* Ru-00602
Bis[1-phenyl-[(2,3,4,5,6-η)-cyclohexa-2,4-dien-6-yl]]ruthenium, Ru-00602
Bis(phenylethylene)bis(triphenylphosphine)ruthenium, *see* Ru-00829
[1,2-Bis(phenylseleno)ethane]dibromodicarbonylruthenium, *see* Ru-00424
[1,2-Bis(phenylthio)ethane]dibromodicarbonylruthenium, *see* Ru-00423
Bis(η^3-2-propenyl)bis(triphenylphosphine)ruthenium, Ru-00782
Bis(styrene)bis(triphenylphosphine)ruthenium, *see* Ru-00829
Bis[(1,2,3,3*a*,7*a*-η)-1*H*-4,5,6,7-tetrahydroinden-1-yl]ruthenium, Ru-00484
1,2:1',2'-Bis(tetramethylene)ruthenocene, *see* Ru-00484
Bis[(thiocarboxy)-π-cyclopentadienyl]ruthenium, *see* Ru-00231
[Bis(trifluoromethyl)acetylene]-chloronitrosylbis(triphenylphosphine)ruthenium, *see* Ru-00751
Bis[(1,2,3,4,5,6-η)-1,3,5-trimethylbenzene]ruthenium(2+), Ru-00489
[2,2-Bis(triphenylphosphine)-2,2-dihydro-2-ruthena]-undecahydro-1,7-dicarbaundecaborane, *see* Ru-00738
Hyper-closo-6,6-Bis(triphenylphosphine)-6-ruthenanonahydro-2,3-dicarbanonaborane, *see* Ru-00734
Bis(triphenylphosphine)tetracarbonylosmium(2+), *see* Os-00403
1,1'-Bis[tris(diethylamino)titanium]ruthenocene, *see* Ru-00683
Bis(vinyl-π-cyclopentadienyl)ruthenium, *see* Ru-00350
Bononoruthenocene, *see* Ru-00151
[μ_4-(Boryl-*B*:*B*:*H*:*H'*)]dodecacarbonyl-μ-hydrotetraruthenium, Ru-00219
Bromocarbonyl-π-cyclopentadienyl(phosphorous acid)ruthenium trimethyl ester, *see* Ru-00110
Bromocarbonyl-π-cyclopentadienyl(phosphorous acid)ruthenium triphenyl ester, *see* Ru-00590
Bromocarbonyl(η^5-2,4-cyclopentadien-1-yl)-(trimethylphosphite-*P*)ruthenium, Ru-00110
Bromocarbonyl(η^5-2,4-cyclopentadien-1-yl)-(triphenylphosphine)osmium, Os-00336
Bromocarbonyl(η^5-2,4-cyclopentadien-1-yl)-(triphenylphosphine)ruthenium, Ru-00589
Bromocarbonyl(η^5-2,4-cyclopentadien-1-yl)(triphenyl phosphite-*P*)ruthenium, Ru-00590
Bromocarbonylhydrobis(tricyclohexylphosphine)osmium, Os-00373
Bromocarbonylhydrotris(triphenylphosphine)osmium, Os-00440
Bromocarbonylhydrotris(triphenylphosphine)ruthenium, Ru-00845
Bromocarbonylnitrosylbis(triphenylphosphine)ruthenium, Ru-00708
Bromo(η^5-2,4-cyclopentadien-1-yl)bis(trimethylphosphine)ruthenium, Ru-00205

Bromo(η^5-2,4-cyclopentadien-1-yl)bis(trimethyl phosphite-*P*)osmium, Os-00147
Bromo(η^5-2,4-cyclopentadien-1-yl)bis(triphenylphosphine)osmium, Os-00410
Bromo(η^5-2,4-cyclopentadien-1-yl)bis(triphenylphosphine)ruthenium, Ru-00763
Bromo(η^5-2,4-cyclopentadien-1-yl)bis(triphenyl phosphite-*P*)osmium, Os-00409
Bromo(η^5-2,4-cyclopentadien-1-yl)[1,2-ethanediylbis[diphenylphosphine]-*P*,*P'*]osmium, Os-00355
Bromodecacarbonyl[(2,3-η)-1,1,1,4,4,4-hexafluoro-2-butene]triosmate(1−), Os-00224
μ-Bromodecacarbonyl-μ-hydrotriosmium, Os-00092
Bromodicarbonyl[(1,2,3,4,5-η)-2,4-cycloheptadien-1-yl]osmium, Os-00077
Bromodicarbonyl[(1,2,3,4,5-η)-2,4-cycloheptadien-1-yl]ruthenium, Ru-00102
Bromodicarbonyl(η^5-2,4-cyclopentadien-1-yl)osmium, Os-00041
Bromodicarbonyl(η^5-2,4-cyclopentadien-1-yl)ruthenium, Ru-00046
Bromodicarbonylhydrobis(tricyclohexylphosphine)osmium, Os-00393
Bromodicarbonylhydrobis(triphenylphosphine)ruthenium, Ru-00729
Bromodicarbonyl[hydrotris(pyrazolato)borato(1−)]ruthenium, Ru-00180
Bromodicarbonyl[(1,2,3,4,5-η)-1,2,3,4,5-pentamethyl-2,4-cyclopentadien-1-yl]osmium, Os-00186
Bromodicarbonyl(phenydiazo)bis(triphenylphosphine)osmium, *see* Os-00421
Bromodicarbonyl(phenylazo)bis(triphenylphosphine)osmium, Os-00421
Bromodicarbonyl(phenylazo)bis(triphenylphosphine)ruthenium, Ru-00796
Bromodicarbonyl(phenyldiazo)bis(triphenylphosphine)ruthenium, *see* Ru-00796
Bromodicarbonyl(η^3-2-propenyl)(triphenylphosphine)ruthenium, Ru-00577
Bromododecacarbonylmethyltriosmium, Os-00211
Bromohydrotris(triphenylphosphine)ruthenium, Ru-00834
[μ_3-(Bromomethylidyne)]nonacarbonyltri-μ-hydrotriosmium, Os-00096
[μ_3-(Bromomethylidyne)]nonacarbonyltri-μ-hydrotriruthenium, Ru-00122
Bromooctacarbonylmethyldiosmium, Os-00071
Bromopentadecacarbonyl-μ-hydro-μ_5-methanetetraylpentaruthenium, Ru-00414
Bromopentakis(triethyl phosphite-*P*)osmium(1+), Os-00350
Bromopentakis(triethyl phosphite-*P*)ruthenium(1+), Ru-00665
Bromopentakis(trimethyl phosphite-*P*)osmium(1+), Os-00266
Bromopentakis(trimethyl phosphite-*P*)ruthenium(1+), Ru-00409
Bromoruthenocene, Ru-00134
Bromotetracarbonylethylosmium, Os-00033
Bromotetracarbonylmethylosmium, Os-00024
Bromotetracarbonyl(ruthenocenylmethylidyne)tungsten, Ru-00385
Bromotetracarbonyl(trimethylgermyl)osmium, Os-00047
Bromotetracarbonyl(trimethylsilyl)osmium, Os-00048
Bromotetracarbonyl(trimethylsilyl)ruthenium, Ru-00061
Bromotricarbonyl(η^3-2-propenyl)ruthenium, Ru-00034
[μ-[(1,2-η:3,4-η)-1,3-Butadiene]]decacarbonyltriosmium, Os-00229
[μ-[(1,2-η:3,4-η)-1,3-Butadiene]]decacarbonyltriosmium, Os-00228
[1,1'-(η^4-1,3-Butadiene-1,4-diyl)bis[benzene]]tricarbonylruthenium, *see* Ru-00505
[μ-[(1,2,3,4-η:1,4-η)-1,3-Butadiene-1,4-diyl]]hexacarbonyldiosmium, Os-00098
[μ-[(1,2,3,4-η:1,4-η)-1,3-Butadiene-1,4-diyl]]nonacarbonyltriosmium, Os-00216
(η^4-1,3-Butadiene)[(η^2-ethenyl)benzene]-bis(triphenylphosphine)ruthenium, Ru-00819
(η^4-1,3-Butadiene)styrenebis(triphenylphosphine)ruthenium, *see* Ru-00819
(η^4-1,3-Butadiene)tricarbonylosmium, Os-00046
(η^4-Butadiene)tricarbonylruthenium, Ru-00056
(η^4-1,3-Butadiene)tris(dimethylphenylphosphine)hydroruthenium(1+), Ru-00650
(η^4-1,3-Butadiene)tris(triphenylphosphine)ruthenium, Ru-00853
[μ-(Butadiynylenedi-π-cyclopentadienylene)]di-π-cyclopentadienyldiruthenium, *see* Ru-00588
1,1'-(1,4-Butanediyl)ruthenocene, *see* Ru-00357
[(2,3-η)-2-Butenedioic acid]tetracarbonylruthenium, Ru-00066
[(3,4-η)-3-Buten-2-one]tetracarbonylruthenium, Ru-00075
1,2-(1-Butenylene)ruthenocene, *see* Ru-00349
[μ-(*tert*-Butylcarbonimidoyl)]decacarbonyl-μ-hydrotriosmate(1−), Os-00257
Butyl(η^5-2,4-cyclopentadien-1-yl)bis(triphenylphosphine)ruthenium, Ru-00811
(*tert*-Butylethylene)nonacarbonyldihydrotriruthenium(2+), *see* Ru-00392
(*tert*-Butyl isocyanide)decacarbonyldihydrotriosmium, *see* Os-00259
(*tert*-Butylisocyanide)nonacarbonyldi-μ-hydrotriosmium, *see* Os-00234
N-*tert*-Butyl osmiamate, *see* Os-00020
1,3-Butynediylbisruthenocene, *see* Ru-00588
[μ_6-(Carbon)]-μ-carbonyldodecacarbonyltetrakis(trimethylphosphite-*P*)hexaruthenium, Ru-00639
μ_6-Carbon-μ-carbonylhexadecacarbonylhexaruthenium, Ru-00500
[μ_6-(Carbon)]-μ-carbonylpentadecacarbonyl(triphenylphosphine)hexaruthenium, Ru-00684
[μ_6-(Carbon)]-μ-carbonyltetradecacarbonylbis(triphenylphosphine)hexaruthenium, Ru-00827
[μ_6-(Carbon)]-μ-carbonyltetradecacarbonyl[μ-[(2,3-η:4,5-η)-2,4-hexadiene]]hexaruthenium, Ru-00557
μ_6-Carbon-μ-carbonyltridecacarbonyl(mesitylene)hexaruthenium, *see* Ru-00585
μ_6-Carbon-μ-carbonyltridecacarbonyl(1,3,5-trimethylbenzene)hexaruthenium, Ru-00585
[μ_6-(Carbon)]-μ-carbonyltridecacarbonyltris(trimethyl phosphite-*P*)hexaruthenium, Ru-00608
(Carbon)heneicosacarbonyloctaosmium, Os-00329
μ_6-Carbonheptadecacarbonylhexaruthenium, *see* Ru-00500
[μ_6-(Carbon)]hexadecacarbonyldihydrohexaruthenium, Ru-00448
(μ_6-Carbon)hexadecacarbonylhexaruthenate(2−), *see* Ru-00467
(μ_6-Carbon)hexadecacarbonyl(triphenylphosphine)hexaruthenium, *see* Ru-00684
(Carbon)nonadecacarbonyldihydroheptaosmium, Os-00310
(Carbonoselenoyl)carbonyldichlorobis(triphenylphosphine)osmium, Os-00380
(Carbonotelluroyl)carbonyldichlorobis(triphenylphosphine)osmium, Os-00381
(Carbonothioyl)carbonyl(η^5-2,4-cyclopentadien-1-yl)iodoruthenium, Ru-00050
[μ-(Carbonothioyl)]-μ-carbonyldicarbonylbis(η^5-2,4-cyclopentadien-1-yl)diruthenium, Ru-00334
(Carbonothioyl)carbonyldichlorobis(triphenylphosphine)ruthenium, Ru-00718
(Carbonothioyl)carbonyldichlorobis(triphenylphosphine)osmium, Os-00379
(Carbonothioyl)carbonyl(η^2-ethene)bis(triphenylphosphine)osmium, Os-00406
(Carbonothioyl)dicarbonyl(η^5-2,4-cyclopentadien-1-yl)ruthenium(1+), Ru-00071
(Carbonothioyl)dichlorobis(triphenylphosphine)ruthenium, *in* Ru-00864
(Carbonothioyl)heptacarbonyldi-μ-hydrodi-μ_3-thioxotriosmium, Os-00055
(Carbonothioyl)octacarbonyldi-μ_3-thioxotriosmium, Os-00085
μ-(Carbonothioyl)tricarbonyl(η^5-cyclopentadienyl)diruthenium, *see* Ru-00334
(Carbonothioyl)trichlorobis(triphenylphosphine)ruthenate(1−), Ru-00711
(Carbonothioyl)tri-μ-chlorochlorotetrakis(triphenylphosphine)diruthenium, Ru-00862

μ_6-Carbonpentadecacarbonylbis(triphenylphosphine)hexaruthenium, *see* Ru-00827
[μ_5-(Carbon)]pentadecacarbonylchloropentaruthenate(1–), Ru-00413
[μ_6-(Carbon)]pentadecacarbonyl(2,4-hexadiene)hexaruthenium, *see* Ru-00557
[μ_5-(Carbon)]pentadecacarbonyliodopentaosmate(1–), Os-00285
(Carbon)pentadecacarbonylpentaosmium, Os-00286
(Carbon)pentadecacarbonylpentaruthenium, Ru-00447
[μ_6-(Carbon)]pentadecacarbonyltris[μ-(ethanethiolato)]-μ-hydrohexaruthenium, Ru-00563
(Carbon)tetracosacarbonylbis-μ_2-iododecaosmium, Os-00341
[μ_6-(Carbon)]tetracosacarbonyldecaosmate(2–), Os-00342
[μ_6-(Carbon)]tetracosacarbonyl-μ-iododecaosmate(1–), Os-00340
[μ_5-(Carbon)]tetradecacarbonyl[μ-(dimethyl phosphito-O'':P)]-μ-hydropentaosmium, Os-00291
μ_6-Carbontetradecacarbonyl(mesitylene)hexaruthenium, *see* Ru-00585
[μ_5-(Carbon)]tetradecacarbonyl(triphenylphosphine)pentaruthenium, Ru-00680
[μ_6-(Carbon)]-tetradecacarbonyltris(trimethylphosphite)hexaruthenium, *see* Ru-00608
[μ_6-(Carbon)]tri-μ-carbonyltridecacarbonylhexaruthenate(2–), Ru-00467
[μ_6-(Carbon)]tridecacarbonyltetratris(trimethylphosphite)hexaruthenium, *see* Ru-00639
μ-Carbonylbis(η^5-2,4-cyclopentadien-1-yl)bis[μ-(pentafluorobenzenethiolato)]diruthenium, Ru-00575
Carbonylbis(dimethylcarbamodithioato-S,S')-(triphenylphosphine)ruthenium, Ru-00625
Carbonylbis(triphenylphosphine)[(7,8,9,10,11-η)-undecahydro-7,8-dicarbaundecaborato(2–)]ruthenium, Ru-00749
Carbonylbis(triphenylphosphine)[(7,8,9,10,11-η)-undecahydro-7,9-dicarbaundecaborato(2–)]ruthenium, Ru-00750
μ-Carbonylcarbonylbis[η^5-2,4-cyclopentadien-1-yl][μ-(pentafluorobenzenethiolato)]-(pentafluorobenzenethiolato)diruthenium, Ru-00583
μ-Carbonylcarbonyldi-π-cyclopentadienyl[μ-(pentafluorobenzenethiolato)]-(pentafluorobenzenethiolato)diruthenium, *see* Ru-00583
μ-Carbonylcarbonyltris(η^5-2,4-cyclopentadien-1-yl)tris[μ-(methanethiolato)]triruthenium, Ru-00524
Carbonylchloro[1,1′,1″,1‴-(η^4-1,3-cyclobutadiene-1,2,3,4-tetrayl)tetrakis[benzene]]bis(trimethyl phosphite-P)ruthenium(1+), *see* Ru-00687
Carbonylchloro-π-cyclopentadienyl(phosphorous acid)ruthenium trimethyl ester, *see* Ru-00111
Carbonylchloro-π-cyclopentadienyl(phosphorous acid)ruthenium triphenyl ester, *see* Ru-00592
Carbonylchloro(η^5-2,4-cyclopentadien-1-yl)-(trimethylphosphite-P)ruthenium, Ru-00111
Carbonylchloro(η^5-2,4-cyclopentadien-1-yl)-(triphenylphosphine)ruthenium, Ru-00591
Carbonylchloro(η^5-2,4-cyclopentadien-1-yl)(triphenyl phosphite-P)ruthenium, Ru-00592
Carbonylchloro[(1,2,3,4,5,6-η)-hexamethylbenzene]-(trimethylphosphine)ruthenium(1+), Ru-00440
Carbonylchlorohydrobis(tricyclohexylphosphine)osmium, Os-00374
Carbonylchlorohydrotris(triphenylarsine)osmium, Os-00439
Carbonylchlorohydrotris(triphenylphosphine)osmium, Os-00441
Carbonylchlorohydrotris(triphenylphosphine)ruthenium, Ru-00846
Carbonylchloro[(4-methylphenyl)methylidyne]-bis(triphenylphosphine)osmium, Os-00427
Carbonylchloro[η^2-(4-methylphenyl)[(4-methylphenyl)imino]-methyl]bis(triphenylphosphine)ruthenium, Ru-00828
Carbonylchloro[η^2-(methylseleno)selenoxomethyl]-bis(triphenylphosphine)ruthenium, Ru-00745
Carbonylchloro[η^2-(methylthio)thioxomethyl]-bis(triphenylphosphine)ruthenium, Ru-00747
Carbonylchloronitrosylbis(triphenylphosphine)osmium, Os-00370
Carbonylchloronitrosylbis(triphenylphosphine)ruthenium, Ru-00710
Carbonylchloro(1-oxopropyl)bis(triphenylphosphine)ruthenium, Ru-00757
Carbonylchloropropionylbis(triphenylphosphine)ruthenium, *see* Ru-00757
Carbonylchloro(η^4-tetraphenylcyclobutadiene)bis(trimethyl phosphite-P)ruthenium(1+), Ru-00687
Carbonylchloro[η^2-(p-tolyl)[(p-tolylimino)methyl]-bis(triphenylphosphine)ruthenium, *see* Ru-00828
Carbonyl[1,1′,1″,1‴-(η^4-1,3-cyclobutadiene-1,2,3,4-tetrayl)tetrakis[benzene]]bis(trimethyl phosphite-P)ruthenium, *see* Ru-00688
Carbonyl[(1,2,3,4,5-η)-2,4-cycloheptadien-1-yl]-bis(trimethyl phosphite-P)osmium(1+), Os-00245
Carbonyl[(1,2,3-η)-2-cyclohexen-1-yl][(1,2,3,4,5,6-η)-hexamethylbenzene]ruthenium(1+), Ru-00513
Carbonyl[(1,2,5,6-η)-1,5-cyclooctadiene][(1,2,5,6-η)-1,3,5-cyclooctatriene]ruthenium, Ru-00462
Carbonyl(η^5-2,4-cyclopentadien-1-yl)-bis(trimethylphosphine)ruthenium(1+), Ru-00274
Carbonyl(η^5-2,4-cyclopentadien-1-yl)-bis(triphenylphosphine)ruthenium(1+), Ru-00775
Carbonyl(η^5-2,4-cyclopentadien-1-yl)[1,2-ethanediylbis[diphenylphosphine]-P,P']ruthenium(1+), Ru-00676
Carbonyl(η^5-2,4-cyclopentadien-1-yl)-ethyl(triphenylphosphine)ruthenium, Ru-00632
Carbonyl(η^5-2,4-cyclopentadien-1-yl)-hydro(triphenylphosphine)ruthenium, Ru-00601
Carbonyl-π-cyclopentadienyliodo(phosphorous acid)ruthenium trimethyl ester, *see* Ru-00112
Carbonyl-π-cyclopentadienyliodo(phosphorous acid)ruthenium triphenyl ester, *see* Ru-00594
Carbonyl(η^5-2,4-cyclopentadien-1-yl)iodo(trimethyl phosphite-P)ruthenium, Ru-00112
Carbonyl(η^5-2,4-cyclopentadien-1-yl)-iodo(triphenylphosphine)ruthenium, Ru-00593
Carbonyl(η^5-2,4-cyclopentadien-1-yl)iodo(triphenyl phosphite-P)ruthenium, Ru-00594
Carbonyl(η^5-2,4-cyclopentadien-1-yl)[(1,2,3-η)-2-methyl-2-propenyl]ruthenium, Ru-00155
Carbonyl(η^5-2,4-cyclopentadien-1-yl)-methyl(triphenylphosphine)ruthenium, Ru-00621
Carbonyl(η^5-2,4-cyclopentadien-1-yl)[(η^3-C,N,N')2-(phenylazo)phenyl]ruthenium, Ru-00480
Carbonyl(η^5-2,4-cyclopentadien-1-yl)(phenylethynyl)-(triphenylphosphine)ruthenium, Ru-00675
Carbonyl(η^5-2,4-cyclopentadien-1-yl)(phenylmethyl)-(triphenylphosphine)ruthenium, *see* Ru-00670
Carbonyl(η^5-2,4-cyclopentadien-1-yl)(η^3-2-propenyl)ruthenium, Ru-00105
μ-Carbonyldecacarbonyl-μ-hydrotriosmate(1–), Os-00121
μ-Carbonyldecacarbonyl-μ-hydrotriruthenate(1–), Ru-00168
μ-Carbonyldecacarbonyl-μ-methylenetriosmium, Os-00157
μ-Carbonyldecacarbonyl[μ-(trimethylsilyl)methylene]triosmium, Os-00258
μ-Carbonyldecacarbonyltriosmate(2–), Os-00148
μ-Carbonyldecacarbonyltriruthenate(2–), Ru-00210
μ-Carbonyldicarbonylbis(η^5-2,4-cyclopentadien-1-yl)-μ-ethenylidenediruthenium, Ru-00390
μ-Carbonyldicarbonylbis(η^5-2,4-cyclopentadien-1-yl)(μ-ethylidiyne)diruthenium(1+), Ru-00393
μ-Carbonyldicarbonylbis(η^5-2,4-cyclopentadien-1-yl)[μ-(thiocarbonyl)]diruthenium, *see* Ru-00334
Carbonyldichlorobis(dimethylphenylarsine)(η^2-ethene)ruthenium, Ru-00510
Carbonyldichlorobis(dimethylphenylphosphine)(η^2-ethene)ruthenium, Ru-00511
Carbonyldichlorobis(triphenylphosphine)ruthenium, *in* Ru-00863
Carbonyldichloro(1,5-cyclooctadiene)ruthenium, *in* Ru-00486
Carbonyldichloro(dichloromethylene)bis(triphenylphosphine)osmium, Os-00384

Carbonyldichloro[(1,2,3,4,5,6-η)-hexamethylbenzene]ruthenium, Ru-00318
Carbonyldichloro(thiocarbonyl)bis(triphenylphosphine)osmium, *see* Os-00379
Carbonyldichlorotris(phosphorous acid)ruthenium nonaphenyl ester, *see* Ru-00842
Carbonyldichlorotris(triphenylphosphine)ruthenium, Ru-00841
Carbonyldichlorotris(triphenyl phosphite-*P*)ruthenium, Ru-00842
μ-Carbonyldi-π-cyclopentadienylbis(μ-pentafluorobenzenethiolato)diruthenium, *see* Ru-00575
μ-Carbonyldi-μ-hydrobis[(1,2,3,4,5-η)-1,2,3,4,5-pentamethyl-2,4-cyclopentadien-1-yl]diosmium, Os-00324
Carbonyldihydronitrosylbis(triphenylphosphine)osmium(1+), Os-00372
Carbonyldihydrotris(triphenylphosphine)osmium, Os-00442
Carbonyldihydrotris(triphenylphosphine)ruthenium, Ru-00847
Carbonyldiiodotris(phosphorous acid)ruthenium nonaphenyl ester, *see* Ru-00843
Carbonyldiiodotris(triphenyl phosphite-*P*)ruthenium, Ru-00843
μ-Carbonyldodecacarbonyl-μ-chlorotetraruthenate(1–), Ru-00281
μ_3-Carbonyldodecacarbonyldi-μ-hydrotetraosmium, Os-00210
μ-Carbonyldodecacarbonyl-μ-hydrotetraosmate(1–), Os-00205
μ_3-Carbonyldodecacarbonylhydrotetraruthenate(1–), Ru-00284
μ_3-Carbonyldodecacarbonyltetraruthenate(2–), Ru-00324
Carbonyl(ethylene)bis(triphenylphosphine)(thiocarbonyl)osmium, *see* Os-00406
Carbonyl(η^2-ethyne)nitrosylbis(triphenylphosphine)osmium(1+), Os-00398
Carbonylferrocenylruthenocene, *see* Ru-00546
μ-Carbonylhexacarbonyl(μ-carbonylpentacarbonyldiiron)di-μ-hydrodiruthenium, Ru-00289
Carbonyl[(1,2,3,4,5,6-η)-hexamethylbenzene]-hydro(trimethylphosphine)ruthenium(1+), Ru-00443
Carbonyl[(1,2,3,4,5,6-η)-hexamethylbenzene]-methyl(trimethylphosphine)ruthenium(1+), Ru-00465
Carbonyl[(1,2,3,4,5,6-η)-hexamethylbenzene]-(trimethylphosphine)ruthenium, Ru-00441
Carbonylhydridochlorotris(triphenylphosphine)osmium, *see* Os-00441
Carbonylhydronitrosylbis(triphenylphosphine)osmium, Os-00371
Carbonylhydro(phenylazo)bis(triphenylphosphine)osmium, Os-00418
Carbonylhydro(phenylazo)bis(triphenylphosphine)ruthenium, Ru-00787
Carbonyl[(1,2,3,3*a*,7*a*-η)-1*H*-inden-1-yl](η^3-2-propenyl)ruthenium, Ru-00304
Carbonyliodonitrosylbis(triphenylphosphine)ruthenium, Ru-00712
Carbonyl(methoxycarbonyl)nitrosylbis(triphenylphosphine)osmium, Os-00401
(μ_3-Carbonylmethylidyne)nonacarbonyldi(μ-hydrido)triosmium, *see* Os-00124
Carbonylnitrosyl(η^3-2-propenyl)(triphenylphosphine)ruthenium, Ru-00567
Carbonylnitrosyltris(triphenylphosphine)osmium(1+), Os-00438
Carbonylnitrosyltris(triphenylphosphine)ruthenium(1+), Ru-00844
μ-Carbonylnonacarbonyl[μ_3-[(1-η:1,2-η:2-η)-1,2-ethenediyl]]-triosmium, Os-00156
μ-Carbonylnonacarbonyl[μ_3-[(1-η:1,2-η:2-η)ethyne]]triosmium, *see* Os-00156
μ_3-Carbonylnonacarbonyl[μ_3-[1,1,1-trimethylsilanaminato(2–)]]triruthenium, Ru-00298
Carbonylnorbornadienyltrichlororuthenate(1–), *see* Ru-00078
μ-Carbonyloctacarbonyldiosmium, Os-00086
μ-Carbonyloctacarbonyldiruthenium, Ru-00116
Carbonylpentachloroosmate(2–), Os-00003
Carbonylpentachloroosmate(3–), Os-00004
Carbonylpentachlororuthenate(2–), Ru-00004
(Carbonylselenoyl)carbonyldichlorobis(triphenylphosphine)ruthenium, Ru-00719
μ-Carbonyltetracarbonyl[μ-[(1,2,3,4,5,6,7,8-η)-1,3,5,7-cyclooctatetraene]]diruthenium, Ru-00296
Carbonyltetrakis(phosphorous trifluoride)ruthenium, Ru-00005
Carbonyl(η^4-tetraphenylcyclobutadiene)-bis(trimethylphosphite-*P*)ruthenium, Ru-00688
Carbonyltrichlorobis(triphenylarsine)ruthenate(1–), Ru-00707
Carbonyltrichlororuthenium, Ru-00003
Carbonyltrihydro[(1,2,3,4,5-η)-1,2,3,4,5-pentamethyl-2,4-cyclopentadien-1-yl]osmium, Os-00143
Carbonyltriiodoruthenium, Ru-00007
(Carboxy-π-cyclopentadienyl)-π-cyclopentadienylruthenium, *see* Ru-00185
(Carboxy-π-cyclopentadienyl)(methyl-π-cyclopentadienyl)ruthenium, *see* Ru-00245
(2-Carboxyethenyl)ruthenocene, *see* Ru-00307
(2-Carboxyethyl)ruthenocene, *see* Ru-00312
(1-Carboxy-2-methyl-π-cyclopentadienyl)π-cyclopentadienylruthenium, *see* Ru-00246
(1-Carboxy-3-methyl-π-cyclopentadienyl)-π-cyclopentadienylruthenium, *see* Ru-00247
1-Carboxy-1′-methylruthenocene, *see* Ru-00245
1-Carboxy-2-methylruthenocene, *see* Ru-00246
1-Carboxy-3-methylruthenocene, *see* Ru-00247
Carboxyosmocene, *see* Os-00136
[(3-Carboxypropionyl)-π-cyclopentadienyl]-π-cyclopentadienylruthenium, *see* Ru-00347
[(3-Carboxypropyl)-π-cyclopentadienyl]-π-cyclopentadienylruthenium, *see* Ru-00352
Carboxyruthenocene, *see* Ru-00185
(η^6-Chlorobenzene)(η^5-2,4-cyclopentadien-1-yl)ruthenium(1+), Ru-00183
μ-Chlorobis[(1,2,3,4,5,6-η)-hexamethylbenzene]di-μ-hydrodiruthenium(1+), Ru-00615
μ-Chlorochlorobis[(1,2,5,6-η)-1,5-cyclooctadiene][μ-(1,1-dimethylhydrazine-*N*:*N*′)]-μ-hydrohydrodiruthenium, Ru-00496
Chloro[(1,2,5,6-η)-1,5-cyclooctadiene]hydrobis(pyridine)ruthenium, Ru-00485
Chloro(1,5-cyclooctadiene)(η^3-propenyl)ruthenium, *in* Ru-00574
Chloro-π-cyclopentadienylbis(*tert*-butylisonitrile)ruthenium, *see* Ru-00406
Chloro(η^5-2,4-cyclopentadien-1-yl)bis(2-isocyano-2-methylpropane)ruthenium, Ru-00406
Chloro-π-cyclopentadienylbis(phosphorous acid)ruthenium hexamethyl ester, *see* Ru-00206
Chloro-π-cyclopentadienylbis(phosphorous acid)ruthenium hexaphenyl ester, *see* Ru-00764
Chloro(η^5-2,4-cyclopentadien-1-yl)bis(trimethylphosphine)ruthenium, Ru-00207
Chloro(η^5-2,4-cyclopentadien-1-yl)bis(trimethyl phosphite-*P*)ruthenium, Ru-00206
Chloro(η^5-2,4-cyclopentadien-1-yl)bis(triphenylarsine)ruthenium, Ru-00762
Chloro(η^5-2,4-cyclopentadien-1-yl)bis(triphenylphosphine)ruthenium, Ru-00765
Chloro(η^5-2,4-cyclopentadien-1-yl)bis(triphenyl phosphite-*P*)ruthenium, Ru-00764
(Chloro-π-cyclopentadienyl)-π-cyclopentadienylruthenium, *see* Ru-00135
(η^5-1-Chloro-2,4-cyclopentadien-1-yl)(η^5-2,4-cyclopentadien-1-yl)ruthenium, *see* Ru-00135
Chloro(η^5-2,4-cyclopentadien-1-yl)[1,2-ethanediylbis[diphenylphosphine]-*P*,*P*′]ruthenium, Ru-00671
Chlorodicarbonyl(phenyldiazo)bis(triphenylphosphine)osmium, *see* Os-00422
Chlorodiiodonitrosylbis(triphenylphosphine)ruthenium, Ru-00694
Chloro(dinitrogen)bis[1,2-phenylenebis[dimethylarsine]-*As*,*As*′]ruthenium(1+), Ru-00529
Chlorodinitrosylbis(triphenylphosphine)ruthenium(1+), Ru-00697
Chloro(dioxygen)nitrosylbis(triphenylphosphine)ruthenium, Ru-00696
(1-Chloroethenyl)ruthenocene, Ru-00235
[(1-Chloro-2-formylvinyl)-π-cyclopentadienyl]-π-cyclopentadienylruthenium, *see* Ru-00302
Chloro[(2,3-η)-1,1,1,4,4,4-hexafluoro-2-butyne]-nitrosylbis(triphenylphosphine)ruthenium, Ru-00751

Chloro[(1,2,3,4,5,6-η)-hexamethylbenzene]-hydro(triphenylphosphine)ruthenium, Ru-00664
Chlorohydridodicarbonylbis(triphenylphosphine)osmium, *see* Os-00391
Chlorohydrobis(triphenylphosphine)[(7,8,9,10,11-η)-undecahydro-7,8-dicarbaundecaborato(2−)]ruthenium, Ru-00735
Chlorohydrobis(triphenylphosphine)[(7,8,9,10,11-η)-undecahydro-7,9-dicarbaundecaborato(2−)]ruthenium, Ru-00736
Chlorohydrotetrakis(phosphorous acid)ruthenium dodecaphenyl ester, *see* Ru-00860
Chlorohydrotetrakis(triphenyl phosphite-*P*)ruthenium, Ru-00860
Chlorohydrotris(triphenylphosphine)ruthenium, Ru-00835
Chloro[(1,2,3,4,5-η)-1-methyl-2,4-cyclopentadien-1-yl]-bis(triphenylphosphine)ruthenium, Ru-00778
Chloronitrosylbis(triphenylphosphine)ruthenium, Ru-00695
Chloronitrosyl(η^2-tetrafluoroethene)-bis(triphenylphosphine)ruthenium, Ru-00717
Chloropentakis(triethyl phosphite-*P*)osmium(1+), Os-00351
Chloropentakis(triethyl phosphite-*P*)ruthenium(1+), Ru-00666
Chloropentakis(trimethyl phosphite-*P*)osmium(1+), Os-00267
Chloropentakis(trimethyl phosphite-*P*)ruthenium(1+), Ru-00410
Chloro(η^3-2-propenyl)tris(triphenylphosphine)ruthenium, Ru-00851
Chlororuthenocene, Ru-00135
1-Chloro-1-ruthenocenylethylene, *see* Ru-00235
3-Chloro-3-ruthenocenyl-2-propen-1-al, Ru-00302
(Chlorosulfonyl)ruthenocene, *in* Ru-00146
[(1-Chlorovinyl)-π-cyclopentadienyl]-π-cyclopentadienylruthenium, *see* Ru-00235
(1-Chlorovinyl)ruthenocene, *see* Ru-00235
Cinnamoylruthenocene, *see* Ru-00509
hyper-closo-6,6-Bis(triethylphosphine)-6,2,3-$RuC_2B_7H_9$, *see* Ru-00379
(Cyanato-*N*)(η^5-2,4-cyclopentadien-1-yl)-bis(triphenylphosphine)ruthenium, Ru-00772
[(1,2,3,4,5-η)-6-Cyano-2,4-cycloheptadien-1-yl]-[(1,2,3,4,5-η)-2,4-cycloheptadien-1-yl]ruthenium, Ru-00398
[(η-1,2,3,6)-5-Cyanocyclooctadienyl]ruthenium tricarbonyl, *see* Ru-00236
(Cyano-*C*)(η^5-2,4-cyclopentadien-1-yl)-bis(triphenylphosphine)ruthenium, Ru-00773
(Cyano-π-cyclopentadienyl)-π-cyclopentadienylruthenium, *in* Ru-00185
Cyano-η^5-cyclopentadienyldicarbonylruthenium, *see* Ru-00067
(Cyano-π-cyclopentadienyl)(methyl-π-cyclopentadienyl)ruthenium, *in* Ru-00245
(1-Cyano-2-methyl-π-cyclopentadienyl)(π-cyclopentadienyl)ruthenium, *in* Ru-00246
(1-Cyano-3-methyl-π-cyclopentadienyl)(π-cyclopentadienyl)ruthenium, *in* Ru-00247
1-Cyano-1′-methylruthenocene, *in* Ru-00245
1-Cyano-2-methylruthenocene, *in* Ru-00246
1-Cyano-3-methylruthenocene, *in* Ru-00247
Cyanoruthenocene, *in* Ru-00185
[1,1′,1″,1‴-(η^4-1,3-Cyclobutadiene-1,2,3,4-tetrayl)-tetrakis[benzene]]iodotris(trimethyl phosphite-*P*)ruthenium(1+), *see* Ru-00713
[1,1′,1″,1‴-(η^4-1,3-Cyclobutadiene-1,2,3,4-tetrayl)-tetrakis[benzene]]tris(trimethyl phosphite-*P*)ruthenium, *see* Ru-00714
[(1,2,3,4,5-η)-2,4-Cycloheptadien-1-yl][(1,2,3,4,5,6-η)-1,3,5-cycloheptatriene]ruthenium(1+), Ru-00359
[(1,2,3,4,5-η)-2,4-Cycloheptadien-1-yl][(1,2,3,4,5-η)-2,4,6-cycloheptatrien-1-yl]ruthenium, Ru-00356
[(1,2,3,4,5-η)-2,4-Cycloheptadien-1-yl][(1,2,3,4,5,6-η)-hexamethylbenzene]ruthenium(1+), Ru-00514
[(1,2,3,4,5,6-η)-1,3,5-Cycloheptatriene][(1,2,5,6-η)-1,5-cyclooctadiene]ruthenium, Ru-00404
(1,3,5-Cycloheptatriene)(2,5-norbornadiene)ruthenium, *see* Ru-00355
(η^1-Cycloheptatrienyl)(η^5-cyclopentadienyl)dicarbonylosmium, *see* Os-00235
(η^7-Cycloheptatrienylium)[(1,2,5,6-η)-1,5-cyclooctadiene]ruthenium(1+), Ru-00401
μ_3-(3,5-Cyclohexadien-1,2-diylidene)-di-μ-hydridotris(tricarbonylosmium), *see* Os-00252
[(1,2,3,4-η)-1,3-Cyclohexadiene][(1,2,3,4,5,6-η)-hexamethylbenzene]ruthenium, Ru-00491
(η^4-1,3-Cyclohexadiene)(η^6-mesitylene)ruthenium, *see* Ru-00405
(η^4-1,3-Cyclohexadiene)[(1,2,3,4,5,6-η)-1,3,5-trimethylbenzene]ruthenium, Ru-00405
[(1,2,3,4-η)-1,3-Cyclohexadiene]tris(phosphorous trifluoride)ruthenium, Ru-00040
(η^4-1,3-Cyclohexadiene)tris(trifluorophosphine)ruthenium, *see* Ru-00040
[(1,2,3,4-η)-1,3-Cyclohexadiene]tris(trimethyl phosphite-*P*)ruthenium, Ru-00408
[(1,2,3-η)-2,4-Cyclohexadien-1-yl][(1,2-η)-cyclohexene]-hydrobis(triphenylphosphine)ruthenium, Ru-00820
[(1,2,3,4,5-η)-2,4-Cyclohexadien-1-yl](η^5-2,4-cyclopentadien-1-yl)ruthenium, Ru-00195
[(1,2,3,4,5-η)-2,4-Cyclohexadien-1-yl]-hydrobis(triphenylphosphine)ruthenium, Ru-00780
(η^5-2,4-Cyclohexadienyl)(η^6-mesitylene)ruthenium(1+), *see* Ru-00402
(η^5-2,4-Cyclohexadien-1-yl)[(1,2,3,4,5,6-η)-1,3,5-trimethylbenzene]ruthenium(1+), Ru-00402
[(1,2,3-η)-2-Cyclohexen-1-yl][(1,2,3,4,5,6-η)-hexamethylbenzene]ruthenium(1+), Ru-00494
[(1,2,5,6-η)-1,5-Cyclooctadiene]bis[(1,2,3-η)-2-methyl-2-propenyl]ruthenium, Ru-00439
[(1,2,5,6-η)-1,5-Cyclooctadiene]bis(2,4-pentanedionato-*O,O′*)ruthenium, Ru-00490
[(1,2,5,6-η)-1,5-Cyclooctadiene]bis(η^3-2-propenyl)osmium, Os-00244
[(1,2,5,6-η)-1,5-Cyclooctadiene]bis(η^3-2-propenyl)ruthenium, Ru-00369
[(1,2,5,6-η)-1,5-Cyclooctadiene][(1,2,3,4,5,6-η)-1,3,5-cyclooctatriene]osmium, Os-00281
[(1,2,5,6-η)-1,5-Cyclooctadiene][(1,2,3,4,5,6-η)-1,3,5-cyclooctatriene]ruthenium, Ru-00435
[(1,2,5,6-η)-1,5-Cyclooctadiene]hydrotris(trimethyl phosphite-*P*)ruthenium(1+), Ru-00466
[(1,2,5,6-η)-1,5-Cyclooctadiene]tetrakis(pyridine)ruthenium(2+), Ru-00648
(1,5-Cyclooctadiene)tris(*tert*-butylisocyanide)osmium, *see* Os-00333
[(1,2,5,6-η)-1,5-Cyclooctadiene]tris(1,1-dimethylhydrazine-N^2)hydroruthenium(1+), Ru-00378
[(1,2,5,6-η)-1,5-Cyclooctadiene]tris(2-isocyano-2-methylpropane)osmium, Os-00333
π-1,5-Cyclooctadienylbis(2-methyl-π-allyl)ruthenium, *see* Ru-00439
[(1,2,3,5,6-η)-2,5-Cyclooctadien-1-yl][(η^6-phenyl)-triphenylborato(1−)]ruthenium, Ru-00677
[(1,2,3,4,5,6-η)-1,3,5,7-Cyclooctatetraene]-bis(triphenylphosphine)ruthenium, Ru-00803
Cyclooctatetraenehexacarbonyldiosmium, *see* Os-00231
[(1,2,3,4-η)-1,3,5,7-Cyclooctatetraene][(1,2,3,4,5,6-η)-hexamethylbenzene]ruthenium, Ru-00525
(η^4-Cyclooctatetraene)(η^6-mesitylene)osmium, *see* Os-00300
(Cyclooctatetraene)pentacarbonyldiruthenium, *see* Ru-00296
Cyclooctatetraenetricarbonylruthenium, *see* Ru-00176
[(1,2,3,4-η)-1,3,5,7-cyclooctatetraene][(1,2,3,4,5,6-η)-1,3,5-trimethylbenzene]osmium, Os-00300
[(1,2,3,4,5-η)-2,4,6-Cyclooctatrien-1-yl][(1,2,3,4,5,6-η)-hexamethylbenzene]ruthenium(1+), Ru-00527
[(1,2,3,6,7-η)-2,4,6-Cyclooctatrien-1-yl][(1,2,3,4,5,6-η)-hexamethylbenzene]ruthenium(1+), Ru-00528
(η^5-Cyclopentadienyl)acetonitrilebis(triphenylphosphine)ruthenium(1+), *see* Ru-00791
(η^5-Cyclopentadienyl)acetylcarbonyl(triphenylphosphine)ruthenium, *see* Ru-00631
Cyclopentadienylbis(trimethyl phosphite)ruthenium chloride, *see* Ru-00206
Cyclopentadienylbis(triphenylarsine)ruthenium chloride, *see* Ru-00762
Cyclopentadienylbis(triphenylphosphine)osmium bromide, *see* Os-00410

Cyclopentadienylbis(triphenylphosphine)ruthenium bromide, *see* Ru-00763
Cyclopentadienylbis(triphenylphosphine)ruthenium chloride, *see* Ru-00765
Cyclopentadienylbis(triphenylphosphine)ruthenium cyanide, *see* Ru-00773
Cyclopentadienylbis(triphenylphosphine)ruthenium fluoride, *see* Ru-00766
Cyclopentadienylbis(triphenylphosphine)ruthenium hydride, *see* Ru-00768
Cyclopentadienylbis(triphenylphosphine)ruthenium iodide, *see* Ru-00767
π-Cyclopentadienylbis(triphenylphosphine)ruthenium methylxanthate, *see* Ru-00792
Cyclopentadienylbis(triphenyl phosphite)ruthenium chloride, *see* Ru-00764
(η^5-Cyclopentadienyl)bromobis(triphenylphosphine)osmium, *see* Os-00410
(η^5-Cyclopentadienyl)bromobis(triphenylphosphine)ruthenium, *see* Ru-00763
(η^5-Cyclopentadienyl)bromocarbonyl(triphenylphosphine)osmium, *see* Os-00336
(η^5-Cyclopentadienyl)(bromo)carbonyl(triphenylphosphine)ruthenium, *see* Ru-00589
(η^5-Cyclopentadienyl)bromocarbonyl(triphenyl phosphite)ruthenium, *see* Ru-00590
η^5-Cyclopentadienylbromodicarbonylosmium, *see* Os-00041
η^5-Cyclopentadienylbromodicarbonylruthenium, *see* Ru-00046
(η^5-Cyclopentadienyl)carbonylbis(triphenylphosphine)ruthenium(1+), *see* Ru-00775
(η^5-Cyclopentadienyl)carbonyl(chloro)(triphenylphosphine)ruthenium, *see* Ru-00591
(η^5-Cyclopentadienyl)carbonylchloro(triphenyl phosphite)ruthenium, *see* Ru-00592
(η^5-Cyclopentadienyl)carbonylethyl(triphenylphosphine)ruthenium, *see* Ru-00632
(η^5-Cyclopentadienyl)carbonyl(hydrido)(triphenylphosphine)ruthenium, *see* Ru-00601
(η^5-Cyclopentadienyl)carbonyl(iodo)(triphenylphosphine)ruthenium, *see* Ru-00593
(η^5-Cyclopentadienyl)carbonyl(iodo)(triphenyl phosphite)ruthenium, *see* Ru-00594
Cyclopentadienylcarbonyl(triphenylphosphine)osmium bromide, *see* Os-00336
Cyclopentadienylcarbonyl(triphenylphosphine)ruthenium iodide, *see* Ru-00593
Cyclopentadienylcarbonyl(triphenylphosphine)ruthenium bromide, *see* Ru-00589
Cyclopentadienylcarbonyl(triphenylphosphine)ruthenium chloride, *see* Ru-00591
(η^5-Cyclopentadienyl)carbonyl(triphenylphosphine)-ruthenium hydride, *see* Ru-00601
Cyclopentadienylcarbonyl(triphenyl phosphite)ruthenium iodide, *see* Ru-00594
Cyclopentadienylcarbonyl(triphenyl phosphite)ruthenium bromide, *see* Ru-00590
Cyclopentadienylcarbonyl(triphenyl phosphite)ruthenium chloride, *see* Ru-00592
(η^5-Cyclopentadienyl)chlorobis(triphenylarsine)ruthenium, *see* Ru-00762
(η^5-Cyclopentadienyl)chlorobis(triphenylphosphine)ruthenium, *see* Ru-00765
(η^5-Cyclopentadienyl)cyanobis(triphenylphosphine)ruthenium, *see* Ru-00773
σ-Cyclopentadienyl-π-cyclopentadienyldicarbonylruthenium, *see* Ru-00229
Cyclopentadienyl(dibenzoylcyclopentadienyl)ruthenium, *see* Ru-00587
η^5-Cyclopentadienyldicarbonylacetylruthenium, *see* Ru-00100
η^5-Cyclopentadienyldicarbonylbenzylruthenium, *see* Ru-00337
Cyclopentadienyldicarbonylchloroosmium, *see* Os-00042
Cyclopentadienyldicarbonylchlororuthenium, *see* Ru-00047
η^5-Cyclopentadienyldicarbonylethylruthenium, *see* Ru-00106
(η^5-Cyclopentadienyl)dicarbonyliodoosmium, *see* Os-00043
η^5-Cyclopentadienyldicarbonyliodoruthenium, *see* Ru-00051
η^5-Cyclopentadienyldicarbonyl(isocyanato)ruthenium, *see* Ru-00070
(η^5-Cyclopentadienyldicarbonyl(isothiocyanato)ruthenium, *see* Ru-00068
η^5-Cyclopentadienyldicarbonylmethylruthenium, *see* Ru-00079
Cyclopentadienyldicarbonylosmium bromide, *see* Os-00041
η^5-Cyclopentadienyldicarbonylosmium chloride, *see* Os-00042
Cyclopentadienyldicarbonylosmium hydride, *see* Os-00045
η^5-Cyclopentadienyldicarbonyl(pentafluorophenyl)ruthenium, *see* Ru-00292
η^5-Cyclopentadienyldicarbonylphenylruthenium, *see* Ru-00299
(η^5-Cyclopentadienyl)dicarbonylruthenium(1−), *see* Ru-00052
Cyclopentadienyldicarbonylruthenium bromide, *see* Ru-00046
η^5-Cyclopentadienyldicarbonylruthenium chloride, *see* Ru-00047
Cyclopentadienyldicarbonylruthenium cyanide, *see* Ru-00067
Cyclopentadienyldicarbonylruthenium diphenylphosphine, *see* Ru-00507
Cyclopentadienyldicarbonylruthenium hydride, *see* Ru-00055
Cyclopentadienyldicarbonylruthenium iodide, *see* Ru-00051
Cyclopentadienyldicarbonylruthenium isocyanate, *see* Ru-00070
Cyclopentadienyldicarbonylruthenium isothiocyanate, *see* Ru-00068
Cyclopentadienyldicarbonylruthenium methylmercaptide, *see* Ru-00080
(η^5-Cyclopentadienyl)dicarbonylrutheniumpentacarbonylrhenium, *see* Ru-00225
Cyclopentadienyldicarbonylruthenium pentafluorophenylmercaptide, *see* Ru-00293
Cyclopentadienyldicarbonylruthenium phenylmercaptide, *see* Ru-00300
Cyclopentadienyldicarbonylruthenium phenylselenide, *see* Ru-00301
(η^5-Cyclopentadienyl)dicarbonylrutheniumtetracarbonylcobalt, *see* Ru-00172
Cyclopentadienyldicarbonylruthenium thiocyanate, *see* Ru-00069
(Cyclopentadienyldicarbonylruthenium)tin trichloride, *see* Ru-00049
(Cyclopentadienyldicarbonylruthenium)trimethylsilane, *see* Ru-00159
η^5-Cyclopentadienyldicarbonyl(thiocyanato)ruthenium, *see* Ru-00069
η^5-Cyclopentadienyldicarbonyl(trifluoroacetyl)ruthenium, *see* Ru-00093
η^5-Cyclopentadienyldicarbonyl(trimethylsilyl)ruthenium, *see* Ru-00159
(η^5-Cyclopentadienyl)dicarbonyl(triphenylphosphine)ruthenium(1+), *see* Ru-00617
(η^5-2,4-Cyclopentadien-1-yl)(*N*,*N*-diethylcarbamodithioato-*S*,*S'*)(triphenylphosphine)ruthenium, Ru-00647
π-Cyclopentadienyl(4,5-dihydro-π-indenyl)ruthenium, Ru-00349
(η^5-2,4-Cyclopentadien-1-yl)[(1,2,3,3*a*,8*a*-η)-1,8-dihydro-8-oxocyclopenta[*a*]inden-1-yl]ruthenium, Ru-00453
π-Cyclopentadienyl(π-1,8-dihydro-8-oxocyclopent[*a*]indenyl)ruthenium, *see* Ru-00453
π-Cyclopentadienyl[[(dimethylamino)methyl]-π-cyclopentadienyl]ruthenium, *see* Ru-00317
(π-Cyclopentadienyl)(1,2-dimethyl-π-cyclopentadienyl)ruthenium, *see* Ru-00262
(π-Cyclopentadienyl)(1,3-dimethyl-π-cyclopentadienyl)ruthenium, *see* Ru-00263
(η^5-2,4-Cyclopentadien-1-yl)[2-[(diphenoxyphosphino)oxy]-phenyl-*C*,*P*](triphenyl phosphite-*P*)ruthenium, Ru-00760
(η^5-2,4-Cyclopentadien-1-yl)[2-(diphenylphosphino)phenyl-*C*,*P*](triphenylphosphine)ruthenium, Ru-00761
(η^5-2,4-Cyclopentadien-1-yl)[1,2-ethanediylbis[diphenylphosphine]-*P*,*P'*]methylruthenium, Ru-00678
(η^5-Cyclopentadienyl)(η^2-ethene)dicarbonylruthenium(1+), *see* Ru-00104
(η^5-2,4-Cyclopentadien-1-yl)ethylbis(triphenylphosphine)ruthenium, Ru-00793
(η^5-2,4-Cyclopentadien-1-yl)(*O*-ethylcarbonodithioato-*S*,*S'*)-(triphenylphosphine)ruthenium, Ru-00633
π-Cyclopentadienyl(ethyl-π-cyclopentadienyl)ruthenium, *see* Ru-00264
(π-Cyclopentadienyl)(ethynyl-π-cyclopentadienyl)ruthenium, *see* Ru-00233

π-Cyclopentadienyl[(ferrocenylcarbonyl)-π-cyclopentadienyl]ruthenium, *see* Ru-00546
π-Cyclopentadienyl[(ferrocenylethynyl)-π-cyclopentadienyl]ruthenium, *see* Ru-00564
π-Cyclopentadienyl[(ferrocenylmethyl)-π-cyclopentadienyl]osmium, *see* Os-00323
π-Cyclopentadienyl[(ferrocenylmethyl)-π-cyclopentadienyl]ruthenium, *see* Ru-00549
(η^5-2,4-Cyclopentadien-1-yl)fluorobis(triphenylphosphine)ruthenium, Ru-00766
π-Cyclopentadienyl(formyl-π-cyclopentadienyl)ruthenium, *see* Ru-00184
π-Cyclopentadienyl(1-formyl-2-methyl-π-cyclopentadienyl)ruthenium, *see* Ru-00243
π-Cyclopentadienyl(1-formyl-3-methyl-π-cyclopentadienyl)ruthenium, *see* Ru-00244
(η^5-2,4-Cyclopentadien-1-yl)[(1,2,3,4,5,6-η)-hexamethylbenzene]ruthenium(1+), Ru-00464
(η^5-2,4-Cyclopentadien-1-yl)hydrobis(triphenylphosphine)ruthenium, Ru-00768
π-Cyclopentadienyl[(α-hydroxybenzyl)-π-cyclopentadienyl]ruthenium, *see* Ru-00457
(π-Cyclopentadienyl)(π-hydroxycyclopentadienyl)ruthenium, *see* Ru-00142
π-Cyclopentadienyl[(hydroxymethyl)-π-cyclopentadienyl]ruthenium, *see* Ru-00191
π-Cyclopentadienyl[1-(hydroxymethyl)-2-methyl-π-cyclopentadienyl]ruthenium, *see* Ru-00257
(η^5-2,4-Cyclopentadien-1-yl)[(1,2,3,4,5-η)-1-(1-hydroxy-2-methylpropyl)-2,4-cyclopentadien-1-yl]osmium, *see* Os-00242
(η^5-2,4-Cyclopentadien-1-yl)iodobis(trimethylphosphine)ruthenium, Ru-00208
(η^5-2,4-Cyclopentadien-1-yl)iodobis(triphenylphosphine)ruthenium, Ru-00767
(η^5-Cyclopentadienyl)isocyanatobis(triphenylphosphine)ruthenium, *see* Ru-00772
(η^5-Cyclopentadienyl)isothiocyanatobis(triphenylphosphine)ruthenium, *see* Ru-00774
(η^5-2,4-Cyclopentadien-1-yl)[(1,2,3,4,5,6-η)-methoxybenzene]ruthenium(1+), Ru-00253
π-Cyclopentadienyl-π-(2-methylallyl)carbonylruthenium, *see* Ru-00155
π-Cyclopentadienyl-π-2-methylallyl(triphenylphosphine)ruthenium, *see* Ru-00643
(η^5-2,4-Cyclopentadien-1-yl)[(1,2,3,4,5,6-η)-methylbenzene]ruthenium(1+), Ru-00254
(η^5-2,4-Cyclopentadien-1-yl)methylbis(triphenylphosphine)ruthenium, Ru-00781
(η^5-2,4-Cyclopentadien-1-yl)(*O*-methylcarbonodithioato-*S*)-bis(triphenylphosphine)ruthenium, Ru-00792
π-Cyclopentadienyl(methyl-π-cyclopentadienyl)ruthenium, *see* Ru-00196
(η^5-2,4-Cyclopentadien-1-yl)[(1,2,3-η)-2-methyl-2-propenyl)](triphenylphosphine)ruthenium, Ru-00643
(η^5-2,4-Cyclopentadien-1-yl)-nitrosylbis(trimethylphosphine)ruthenium(2+), Ru-00209
Cyclopentadienylosmiumdicarbonyl iodide, *see* Os-00043
(η^5-2,4-Cyclopentadien-1-yl)(pentafluorophenyl)-bis(triphenylphosphine)ruthenium, Ru-00815
(η^5-Cyclopentadienyl)[η^3-(2-phenylazo)phenyl]carbonylruthenium, *see* Ru-00480
π-Cyclopentadienyl-σ-[2-(phenylazo)phenyl]dicarbonylruthenium, *see* Ru-00504
(η^5-2,4-Cyclopentadien-1-yl)[2-(phenylazo)phenyl]-(triphenylphosphine)ruthenium, Ru-00686
(η^5-2,4-Cyclopentadien-1-yl)(phenylethynyl)-bis(triphenylphosphine)osmium, Os-00433
(η^5-2,4-Cyclopentadien-1-yl)(phenylethynyl)-bis(triphenylphosphine)ruthenium, Ru-00821
(η^5-2,4-Cyclopentadien-1-yl)(phenylmethyl)-bis(triphenylphosphine)ruthenium, *see* Ru-00818
(η^5-2,4-Cyclopentadien-1-yl)[(η^6-phenyl)-triphenylborato(1−)]ruthenium, Ru-00655
(η^5-2,4-Cyclopentadien-1-yl)[(1,2-η)-1-propene]-bis(trimethylphosphine)ruthenium(1+), Ru-00373
(η^5-2,4-Cyclopentadien-1-yl)-2-propenylbis(triphenylphosphine)ruthenium, Ru-00805
(η^5-2,4-Cyclopentadien-1-yl)(η^3-2-propenyl)-(triphenylphosphine)ruthenium, Ru-00635
π-Cyclopentadienyl(propionyl-π-cyclopentadienyl)ruthenium, *see* Ru-00311
(η^5-2,4-Cyclopentadien-1-yl)propylbis(triphenylphosphine)ruthenium, Ru-00806
(η^5-2,4-Cyclopentadien-1-yl)-1-propynylbis(triphenylphosphine)ruthenium, Ru-00804
Cyclopentadienylrutheniumdicarbonylpentacarbonylrhenium, *see* Ru-00225
Cyclopentadienylrutheniumdicarbonyltetracarbonylcobalt, *see* Ru-00172
(η^5-2,4-Cyclopentadien-1-yl)[tetrahydroborato(1−)-*H*]-bis(triphenylphosphine)ruthenium, Ru-00770
π-Cyclopentadienyl(4,5,6,7-tetrahydro-4-hydroxy-π-indenyl)ruthenium, Ru-00351
π-Cyclopentadienyl(4,5,6,7-tetrahydro-4-oxo-π-indenyl)ruthenium, Ru-00343
(η^5-2,4-Cyclopentadien-1-yl)(thiocyanato-*N*)-bis(triphenylphosphine)ruthenium, Ru-00774
(η^5-Cyclopentadienyl)(η^6-toluene)ruthenium(1+), *see* Ru-00254
(η^5-Cyclopentadienyl)tricarbonylruthenium(1+), *see* Ru-00072
(η^5-2,4-Cyclopentadien-1-yl)[3,3,3-trifluoro-1-(trifluoromethyl)-1-propenyl]bis(triphenylphosphine)ruthenium, Ru-00810
(η^5-2,4-Cyclopentadien-1-yl)[(1,3,4-η)-5,5,5-trifluoro-1,2,3-tris(trifluoromethyl)-1,3-pentadienyl]-(triphenylphosphine)ruthenium, Ru-00669
π-Cyclopentadienyl-(trimethylsilyl-π-cyclopentadienyl)ruthenium, *see* Ru-00323
(η^5-2,4-Cyclopentadien-1-yl)tris(trimethylphosphine)ruthenium(1+), Ru-00376
(η^5-2,4-Cyclopentadien-1-yl)tris(trimethyl phosphite-*P*)ruthenium(1+), Ru-00375
π-Cyclopentadienyl(vinyl-π-cyclopentadienyl)osmium, *see* Os-00180
π-Cyclopentadienyl(vinyl-π-cyclopentadienyl)ruthenium, *see* Ru-00252
(η^6-*p*-Cymene)(η^5-tetramethylthiophene)ruthenium(2+), *see* Ru-00492
Decacarbonylbis(trimethyl phosphite-*P*)triruthenium, Ru-00430
Decacarbonylbis(triphenylphosphine)triosmium, Os-00428
Decacarbonylbis(triphenylphosphine)triruthenium, Ru-00812
Decacarbonyl-μ-chloro-μ-hydrotriosmium, Os-00093
Decacarbonyl-2,8-cyclododecadiene-3,7-diyl-1-ylidenetetraruthenium, Ru-00562
Decacarbonyl(μ_4-π_2-cyclododecatrienylene)tetraruthenium, *see* Ru-00562
Decacarbonyl[(1,2,3,4-η)-1,3-cyclohexadiene]triosmium, Os-00278
Decacarbonyl[(1,2,3,4,5-η)-2,4-cyclohexadien-1-yl]triosmium(1+), Os-00277
Decacarbonyl(1,3-cyclooctadiene)triruthenium, Ru-00477
Decacarbonyl(decacarbonyldi-μ-hydrotriosmium)diruthenium, Os-00311
Decacarbonyldi-μ-chlorotriosmium, Os-00091
Decacarbonyldi-μ-hydro-μ-methylenetriosmium, *see* Os-00127
Decacarbonyldi-μ-hydro-μ-methylenetriruthenium, Ru-00171
Decacarbonyldi-μ-hydrotriosmium, Os-00095
1,1,1,1,2,2,2,3,3,3-Decacarbonyl-2,3-dihydrotriosmium, *see* Os-00095
Decacarbonyldihydro(triphenylphosphine)triosmium, *see* Os-00345
Decacarbonyldi-μ-iodotriosmium, Os-00118
Decacarbonyl[μ-[(dimethylamino)carbonyl-*C*:*O*]]-μ-hydrotriruthenium, Ru-00295
Decacarbonyl[μ-[(1,1-dimethylethyl)carbonimidoyl]]-μ-hydrotriosmate(1−), *see* Os-00257
Decacarbonyl[μ-[(dimethyliminio)methylene]]-μ-hydrotriruthenium, Ru-00294
Decacarbonyldi-μ-nitrosyltriosmium, Os-00120
Decacarbonyldi-μ-nitrosyltriruthenium, Ru-00167

Decacarbonyldinitrosyltriruthenium, *see* Ru-00167
Decacarbonyl[μ-(diphenylmethanone hydrazonato)]-μ-hydrotriosmium, Os-00331
Decacarbonyl[μ-[1,2-ethanediylbis[diphenylphosphine]-*P:P'*]]-tetra-μ-hydrotetraruthenium, Ru-00691
Decacarbonyl(μ-ethanethiolato)di-μ-hydrotriruthenium(1+), Ru-00228
Decacarbonyl[μ-(ethanethiolato)]-μ-hydrotriosmium, Os-00169
Decacarbonyl[μ-(ethanethiolato)]-μ-hydrotriruthenium, Ru-00226
Decacarbonyl[μ-[(1-η:1,2-η)ethenyl]]-μ-hydrotriosmium, Os-00164
Decacarbonyl-μ-ethenylidenedi-μ-hydrotriruthenium, Ru-00222
Decacarbonyl[μ_3-[1,1'-[(1-η:1,2-η:2-η)-1,2-ethynediyl]-bis[benzene]]]triosmium, Os-00335
Decacarbonyl-μ-hydro[μ-(acetyl-*C:O*)]triosmium, Os-00165
Decacarbonyl-μ-hydrohydro(2-isocyano-2-methylpropane)triosmium, Os-00259
Decacarbonyl-μ-hydrohydro(triphenylphosphine)triosmium, Os-00345
Decacarbonyl-μ-hydro[μ-(methanaminato)]triosmium, Os-00128
Decacarbonyl-μ-hydro[μ-(methoxymethylidyne)]triosmium, Os-00166
Decacarbonyl-μ-hydro[μ-(methoxymethylidyne)]triruthenium, Ru-00223
Decacarbonyl-μ-hydro-μ_3-methylidynetriosmium, Os-00123
Decacarbonyl-μ-hydro[μ-[(methylimino)phenylmethyl-*C:N*]]triosmium, Os-00303
Decacarbonyl-μ-hydro[μ-[(4-methylphenyl)azo]]triosmium, Os-00292
Decacarbonyl-μ-hydro-μ-methyltriosmium, Os-00127
Decacarbonyl-μ-hydro-μ-nitrosyltriosmium, Os-00094
Decacarbonyl-μ-hydro-μ-nitrosyltriruthenium, Ru-00121
Decacarbonyl-μ-hydro[μ-(phenylphosphino)]triruthenium, Ru-00420
Decacarbonyl-μ-hydro[μ-(2-pyridinyl-*C:N*)]triosmium, Os-00251
Decacarbonyl-μ-nitrosyltriosmate(1−), Os-00119
Decacarbonyl-μ-nitrosyltriruthenate(1−), Ru-00166
Decacarbonyl[μ-(pyridazine-N^1,N^2)]triruthenium, *see* Ru-00326
Decacarbonyltetra-μ-hydrobis(phosphorous acid)tetraruthenium hexamethyl ester, *see* Ru-00433
Decacarbonyltetra-μ-hydrobis(trimethyl phosphite-*P*)tetraruthenium, Ru-00433
Decacarbonyltetra-μ-hydrobis(triphenylphosphine)tetraruthenium, Ru-00813
Decacarbonyl(μ_3-trimethylsilylaminato)triruthenium, *see* Ru-00298
Decachlororuthenocene, Ru-00120
Diacetyldibromodicarbonylosmate(2−), Os-00035
1,1'-Diacetylosmocene, Os-00239
1,1'-Diacetylruthenocene, Ru-00346
Diallyl(1,5-cyclooctadiene)osmium, *see* Os-00244
Di-π-allyldicarbonylruthenium, *see* Ru-00083
Di-π-allyldi-μ-chlorodi-π-1,5-cyclooctadienyldiruthenium, *see* Ru-00574
Di-π-allyl-π-2,5-norbornadienylruthenium, *see* Ru-00322
Dibenzeneruthenium, *see* Ru-00250
1,1'-Dibenzoylruthenocene, Ru-00586
1,2-Dibenzoylruthenocene, Ru-00587
Dibromo(carbonoselenoyl)carbonylbis(triphenylphosphine)ruthenium, Ru-00715
Dibromocarbonyltris(phosphorous acid)ruthenium nonaphenyl ester, *see* Ru-00840
Dibromocarbonyltris(triphenyl phosphite-*P*)ruthenium, Ru-00840
Di-μ-bromodecacarbonyltriosmium, Os-00090
Di-μ-bromodibromohexacarbonyldiosmium, Os-00029
Di-μ-bromodibromohexacarbonyldiruthenium, Ru-00032
Dibromodicarbonylbis(triphenylarsine)osmium, Os-00375
Dibromodicarbonylbis(triphenylphosphine)osmium, Os-00377
Dibromodicarbonylbis(triphenylphosphine)ruthenium, Ru-00716
Dibromodicarbonylbis(triphenylstibine)osmium, Os-00378
Dibromodicarbonyl[1,2-ethanediylbis(diphenylphosphine)-*P,P'*]ruthenium, Ru-00645
Dibromodicarbonyl[ethylenebis(diphenylphosphine)]ruthenium, *see* Ru-00645
Dibromodicarbonylruthenium, Ru-00008
Dibromododecacarbonyltriosmium, Os-00149
Di-μ-bromohexacarbonyldiosmium, Os-00028
Dibromooctacarbonyldiosmium, Os-00053
1,1'-Dibromoruthenocene, Ru-00129
Dibromotetracarbonylosmium, Os-00015
Dibromotetracarbonylruthenium, Ru-00022
Dibromotetrakis(isocyanomethane)osmium, Os-00063
Dibromotetrakis(methyl isocyanide)osmium, *see* Os-00063
Dibromotetrakis(phosphorous acid)ruthenium dodecaphenyl ester, *see* Ru-00857
Dibromotetrakis(triphenylphosphine)osmium, Os-00445
Dibromotetrakis(triphenylphosphine)ruthenium, Ru-00858
Dibromotetrakis(triphenyl phosphite-*P*)ruthenium, Ru-00857
Dibromotricarbonylosmium, *in* Os-00029
Dibromotricarbonylruthenium, *in* Ru-00032
Dibromotricarbonyl(triphenylphosphine)osmium, Os-00317
Dibromotricarbonyl(triphenylphosphine)ruthenium, Ru-00541
Dibromotris(triphenylphosphine)osmium, Os-00434
Dibromotris(triphenylphosphine)ruthenium, Ru-00832
Dicarbonylbis(2-aminothiophenol)ruthenium, *see* Ru-00336
Dicarbonylbis(η^5-2,4-cyclopentadien-1-yl)bis[μ-(methanethiolato)]diruthenium, Ru-00353
Dicarbonylbis(η^5-2,4-cyclopentadien-1-yl)bis[μ-(pentafluorobenzenethiolato)]diruthenium, Ru-00584
Dicarbonylbis(η^5-2,4-cyclopentadien-1-yl)bis[μ-(thiocarbonyl)]diruthenium, *see* Ru-00333
Dicarbonylbis(η^5-2,4-cyclopentadien-1-yl)-μ-ethylidene[μ-(1-methylethylidene)]diruthenium, Ru-00461
Dicarbonylbis(cyclopentadienyl)ruthenium, *see* Ru-00229
Dicarbonylbis[μ-(dimethylcarbamodithioato-*S:S,S'*)]-bis(dimethylcarbamodithioato-*S,S'*)-diruthenium, Ru-00370
Dicarbonylbis(dimethylcarbamodithioato-*S,S'*)ruthenium, Ru-00085
Dicarbonylbis(dimethyldithiocarbamato)ruthenium, *see* Ru-00085
Dicarbonylbis(2-methyl-π-allyl)ruthenium, *see* Ru-00158
Dicarbonylbis[(1,2,3-η)-2-methyl-2-propenyl]ruthenium, Ru-00158
Dicarbonylbis(phosphorous acid)(tetrafluoroethylene)-osmium hexamethyl ester, *see* Os-00113
Dicarbonylbis(phosphorous acid)(tetrafluoroethylene)-ruthenium hexamethyl ester, *see* Ru-00161
Dicarbonylbis(λ^3-2-propenyl)ruthenium, Ru-00083
Dicarbonyl[bis(trifluoromethyl)vinylene]bis(phosphorous acid)osmium hexamethyl ester, *see* Os-00189
Dicarbonylbis(trimethylgermyl)bis(triphenylphosphine)ruthenium, Ru-00808
Dicarbonylbis(trimethylsilyl)bis(triphenylphosphine)ruthenium, Ru-00809
Dicarbonylbis(triphenylarsine)dibromoosmium, *see* Os-00375
Dicarbonylbis(triphenylarsine)dichloroosmium, *see* Os-00376
Dicarbonylbis(triphenylphosphine)dibromoosmium, *see* Os-00377
Dicarbonylbis(triphenylphosphine)dichloroosmium, *see* Os-00382
Dicarbonylbis(triphenylphosphine)diiodoosmium, *see* Os-00385
Dicarbonylbis(triphenylphosphine)ruthenium sulfate, *see* Ru-00728
Dicarbonylbis(triphenylstibine)dibromoosmium, *see* Os-00378
Dicarbonylbis(triphenylstibine)dichloroosmium, *see* Os-00383
Dicarbonylbis(triphenylstibine)diiodoosmium, *see* Os-00386
Dicarbonylchloro(η^5-2,4-cyclopentadien-1-yl)osmium, Os-00042
Dicarbonylchloro(η^5-2,4-cyclopentadien-1-yl)ruthenium, Ru-00047
Dicarbonylchlorohydrobis(triphenylphosphine)osmium, Os-00391
Dicarbonylchlorohydrobis(triphenylphosphine)ruthenium, Ru-00730
Dicarbonylchloro[hydrotris(pyrazolato)borato(1−)]ruthenium, Ru-00181

Dicarbonylchloro(methoxycarbonyl)bis(triphenylphosphine)osmium, Os-00405
Dicarbonylchloro[4-methyl(isocyanobenzene)]-bis(triphenylphosphine)osmium(1+), Os-00429
Dicarbonylchloro(phenylazo)bis(triphenylphosphine)osmium, Os-00422
Dicarbonylchloro[2-(phenylazo)phenyl]ruthenium, Ru-00330
Dicarbonylchloro[2-(phenylazo)phenyl](triphenylphosphine)ruthenium, Ru-00673
Dicarbonylchloro(phenyldiazene-N^2)bis(triphenylphosphine)ruthenium(1+), Ru-00799
Dicarbonylchloro(η^3-2-propenyl)(triphenylphosphine)ruthenium, Ru-00578
Dicarbonylchloro(thioxomethyl)bis(triphenylphosphine)osmium, Os-00396
Dicarbonyl(cyanato-*N*)(η^5-2,4-cyclopentadien-1-yl)ruthenium, Ru-00070
Dicarbonyl(cyano-*C*)(η^5-2,4-cyclopentadien-1-yl)ruthenium, Ru-00067
Dicarbonyl(cyanomethane)(cyclopentadienyl)osmium(1+), *see* Os-00074
Dicarbonyl[(1,2,3,4-η)-1,3-cycloheptadiene]-(triphenylphosphine)ruthenium, Ru-00641
Dicarbonyl[(1,2,3,4,5-η)-2,4-cycloheptadien-1-yl]iodoosmium, Os-00078
Dicarbonyl[(1,2,3,4,5-η)-2,4-cycloheptadien-1-yl]iodoruthenium, Ru-00103
Dicarbonyl[(1,2,3,4,5-η)-2,4-cycloheptadien-1-yl]-(trimethylgermyl)osmium, Os-00190
Dicarbonyl[(1,2,3,4,5-η)-2,4-cycloheptadien-1-yl]-(trimethylgermyl)ruthenium, Ru-00268
Dicarbonyl[(1,2,3,4,5-η)-2,4-cycloheptadien-1-yl]-(trimethylsilyl)ruthenium, Ru-00271
Dicarbonyl[(1,2,3,4,5-η)-2,4-cycloheptadien-1-yl]-(triphenylphosphine)ruthenium(1+), Ru-00640
Dicarbonyl[(1,2-η)-1,3,5-cycloheptatriene](η^5-2,4-cyclopentadien-1-yl)osmium(1+), Os-00237
Dicarbonyl[(1,2,3,4,5,6-η)-1,3,5-cycloheptatriene]ruthenium, Ru-00098
Dicarbonyl(1,3,5-cycloheptatrien-1-yl)(η^5-2,4-cyclopentadien-1-yl)osmium, Os-00235
Dicarbonyl[(1,3,4,5-η)-4-cycloheptene-1,3-diyl]-(triphenylphosphine)ruthenium, Ru-00642
Dicarbonyl[(1,2,3,4,5-η)-2,4-cyclohexadien-1-yl]-(methoxycarbonyl)osmium, Os-00106
Dicarbonyl[(1,2,3,4,5-η)-2,4-cyclohexadien-1-yl]-(methoxycarbonyl)ruthenium, Ru-00147
Dicarbonyl[(1,2,3,4,5-η)-2,4-cyclohexadien-1-yl]-(trimethylgermyl)ruthenium, Ru-00200
Dicarbonyl[(1,2,3,4,5-η)-2,4-cyclohexadien-1-yl]-(trimethylsilyl)ruthenium, Ru-00201
Dicarbonyl[(1,2-η)-cyclohexene](η^5-2,4-cyclopentadien-1-yl)ruthenium(1+), Ru-00314
Dicarbonyl[(1,4,5,6-η)-5-cyclooctene-1,4-diyl]-(triphenylphosphine)ruthenium, Ru-00646
Dicarbonyl-2,4-cyclopentadien-1-yl(η^5-2,4-cyclopentadien-1-yl)ruthenium, Ru-00229
Dicarbonyl(η^5-2,4-cyclopentadien-1-yl)(diphenylphosphino)ruthenium, Ru-00507
Dicarbonyl(η^5-2,4-cyclopentadien-1-yl)(η^2-ethene)ruthenium(1+), Ru-00104
Dicarbonyl-π-cyclopentadienyl(ethylene)ruthenium(1+), *see* Ru-00104
Dicarbonyl(η^5-2,4-cyclopentadien-1-yl)ethylruthenium, Ru-00106
Dicarbonyl-π-cyclopentadienylhydridoosmium, *see* Os-00045
Dicarbonyl-π-cyclopentadienylhydridoruthenium, *see* Ru-00055
Dicarbonyl(η^5-2,4-cyclopentadien-1-yl)hydroosmium, Os-00045
Dicarbonyl(η^5-2,4-cyclopentadien-1-yl)hydroruthenium, Ru-00055
Dicarbonyl(η^5-2,4-cyclopentadien-1-yl)iodoosmium, Os-00043
Dicarbonyl(η^5-2,4-cyclopentadien-1-yl)iodoruthenium, Ru-00051
Dicarbonyl-π-cyclopentadienyl(isocyanato)ruthenium, *see* Ru-00070
Dicarbonyl(η^5-2,4-cyclopentadien-1-yl)(isocyanomethane)ruthenium(1+), Ru-00097
Dicarbonyl-π-cyclopentadienyl(isothiocyanato)ruthenium, *see* Ru-00068
Dicarbonyl(η^5-2,4-cyclopentadien-1-yl)(methanethiolato)ruthenium, Ru-00080
Dicarbonyl(η^5-2,4-cyclopentadienyl)(methyl isocyanide)ruthenium(1+), *see* Ru-00097
Dicarbonyl(η^5-2,4-cyclopentadien-1-yl)methylruthenium, Ru-00079
Dicarbonyl-η^5-cyclopentadienyl(methyl-*S*-sulfinato)ruthenium, *see* Ru-00082
Dicarbonyl(η^5-2,4-cyclopentadien-1-yl)(methylsulfonyl)ruthenium, Ru-00082
Dicarbonyl-π-cyclopentadienyl(methylthio)ruthenium, *see* Ru-00080
Dicarbonyl(η^5-2,4-cyclopentadien-1-yl)[(methylthio)-thioxomethyl]ruthenium, Ru-00099
Dicarbonyl(η^5-2,4-cyclopentadien-1-yl)-(pentacarbonylrhenio)ruthenium, Ru-00225
Dicarbonyl(η^5-2,4-cyclopentadien-1-yl)-(pentafluorobenzenethiolato)ruthenium, Ru-00293
Dicarbonyl(η^5-2,4-cyclopentadien-1-yl)(pentafluorophenyl)ruthenium, Ru-00292
Dicarbonyl-π-cyclopentadienyl(pentafluorophenylthio)ruthenium, *see* Ru-00293
Dicarbonyl(η^5-2,4-cyclopentadien-1-yl)(1,2,3,3,3-pentafluoro-1-propenyl)ruthenium, Ru-00126
Dicarbonyl(η^5-2,4-cyclopentadien-1-yl)[2-(phenylazo)-phenyl]ruthenium, Ru-00504
Dicarbonyl(η^5-2,4-cyclopentadien-1-yl)phenylmethylruthenium, *see* Ru-00337
Dicarbonyl(η^5-2,4-cyclopentadien-1-yl)phenylruthenium, Ru-00299
Dicarbonyl-π-cyclopentadienyl(phenylthio)ruthenium, *see* Ru-00300
Dicarbonyl(η^5-2,4-cyclopentadien-1-yl)[(1,2-η)-1-propene]ruthenium(1+), Ru-00152
Dicarbonyl(η^5-2,4-cyclopentadien-1-yl)-2-propenylruthenium, Ru-00143
Dicarbonyl(η^5-2,4-cyclopentadien-1-yl)ruthenate(1−), Ru-00052
Dicarbonyl(η^5-cyclopentadienyl)ruthenium(diphenyl)phosphine, *see* Ru-00507
(Dicarbonyl-π-cyclopentadienylruthenium)trichlorosilane, *see* Ru-00048
(Dicarbonyl-π-cyclopentadienylruthenium)trichlorostannane, *see* Ru-00049
Dicarbonyl(η^5-2,4-cyclopentadien-1-yl)-(tetracarbonylcobalt)ruthenium, Ru-00172
Dicarbonyl(η^5-2,4-cyclopentadien-1-yl)(thiocyanato-*N*)ruthenium, Ru-00068
Dicarbonyl(η^5-2,4-cyclopentadien-1-yl)(thiocyanato-*S*)ruthenium, Ru-00069
Dicarbonyl(η^5-2,4-cyclopentadien-1-yl)(trichlorosilyl)ruthenium, Ru-00048
Dicarbonyl(η^5-2,4-cyclopentadien-1-yl)(trichlorostannyl)ruthenium, Ru-00049
Dicarbonyl(η^5-2,4-cyclopentadien-1-yl)(trifluoroacetyl)ruthenium, Ru-00093
Dicarbonyl(η^5-2,4-cyclopentadien-1-yl)(trimethylphosphine)ruthenium(1+), Ru-00157
Dicarbonyl[(1,2,3,4,5-η)-2,4-cyclopentadien-1-yl]-(trimethylsilyl)osmium, Os-00193
Dicarbonyl(η^5-2,4-cyclopentadien-1-yl)(trimethylsilyl)ruthenium, Ru-00159
Dicarbonyl(η^5-2,4-cyclopentadien-1-yl)(triphenylphosphine)ruthenium(1+), Ru-00617
Di-μ-carbonyldecacarbonyldiironosmium, *see* Os-00152
Di-μ-carbonyldicarbonylbis(η^5-2,4-cyclopentadien-1-yl)diruthenium, Ru-00335
Di-μ-carbonyldicarbonylbis[(1,2,3,3*a*,7*a*-η)-1*H*-inden-1-yl]diruthenium, Ru-00558
Di-μ-carbonyldicarbonylbis[(1,2,3,4,5-η)-1-methyl-2,4-cyclopentadien-1-yl]diruthenium, Ru-00425
Di-μ-carbonyldicarbonylbis[(1,2,3,4,5-η)-1,2,3,4,5-pentamethyl-2,4-cyclopentadien-1-yl]diruthenium, Ru-00609

Di-μ-carbonyldicarbonylbis[(1,2,3,3*a*,7*a*-η)-4,5,6,7-tetrahydro-1*H*-inden-1-yl]diruthenium, Ru-00568
Di-μ-carbonyldicarbonyldi-μ-hydrotris[μ_3-(tricarbonylruthenio)]iron, Ru-00287
Di-μ-carbonyldi-μ_3-carbonyltetradecacarbonylhexaruthenate(2−), Ru-00501
Dicarbonyldichlorobis(cyanobenzene)ruthenium, *see* Ru-00421
Dicarbonyldichlorobis(phosphorous acid)ruthenium hexaphenyl ester, *see* Ru-00721
Dicarbonyldichlorobis(triphenylarsine)osmium, Os-00376
Dicarbonyldichlorobis(triphenylphosphine)osmium, Os-00382
Dicarbonyldichlorobis(triphenylphosphine)ruthenium, Ru-00720
Dicarbonyldichlorobis(triphenyl phosphite-*P*)ruthenium, Ru-00721
Dicarbonyldichlorobis(triphenylstibine)osmium, Os-00383
Dicarbonyldi-μ-chlorodichlorobis[(1,2,5,6-η)-1,5-cyclooctadiene]diruthenium, Ru-00486
Dicarbonyldi-μ-chlorodichlorotetrakis(triphenylphosphine)diruthenium, Ru-00863
Dicarbonyldichloro[2-(phenylazo)phenyl]ruthenate(1−), Ru-00331
Dicarbonyldichlororuthenium, Ru-00010
Dicarbonyldichloro(triphenylphosphine)ruthenium, *in* Ru-00752
Dicarbonyldi-π-cyclopentadienylbis(methanethiolato)diruthenium, *see* Ru-00353
Dicarbonyldi-π-cyclopentadienylbis(μ-pentafluorobenzenethiolato)diruthenium, *see* Ru-00584
Dicarbonyldiethylbis(triphenylphosphine)osmium, Os-00415
Dicarbonyldi-μ-hydrobis[(1,2,3,4,5-η)-1,2,3,4,5-pentamethyl-2,4-cyclopentadien-1-yl]diosmium, Os-00328
Dicarbonyldihydrobis(triphenylphosphine)osmium, Os-00392
Dicarbonyldihydrobis(triphenylphosphine)ruthenium, Ru-00731
Dicarbonyldiiodobis(triphenylphosphine)osmium, Os-00385
Dicarbonyldiiodobis(triphenylstibine)osmium, Os-00386
Dicarbonyldiiodoosmium, Os-00008
Dicarbonyldiiodoruthenium, Ru-00015
Dicarbonyl[(1,2,3,4-η)-2,3-dimethyl-1,3-butadiene]-(triphenylphosphine)ruthenium, Ru-00634
Dicarbonyl[(2,3-η)-dimethyl-2-butenedioate]-bis(triphenylphosphine)ruthenium, Ru-00802
Dicarbonyldinitrosylosmium, Os-00011
Dicarbonyldiodobis(triphenylphosphine)ruthenium, Ru-00723
Dicarbonyl(dioxygen)bis(triphenylphosphine)osmium, Os-00390
Dicarbonyl(dioxygen)bis(triphenylphosphine)ruthenium, Ru-00726
Dicarbonyl(diseleno)bis(triphenylphosphine)osmium, Os-00389
Dicarbonyl(dithioperoxy)bis(triphenylphosphine)osmium, Os-00388
Dicarbonyl(dithioperoxy)bis(triphenylphosphine)ruthenium, Ru-00725
Dicarbonyl[1,2-ethanediylbis[diphenylphosphine]-*P*,*P'*]diiodoosmium, Os-00346
Dicarbonyl(η^2-ethene)bis(triphenylphosphine)osmium, Os-00407
Dicarbonyl(η^2-ethene)bis(triphenylphosphine)ruthenium, Ru-00755
Dicarbonyl[ethylenebis(diphenylphosphine)]diiodoosmium, *see* Os-00346
Dicarbonyl(η^2-ethylene)bis(triphenylphosphine)osmium, *see* Os-00407
Dicarbonyl(ethylene)bis(triphenyphosphine)ruthenium, *see* Ru-00755
Dicarbonyl(η^2-formaldehyde)bis(triphenylphosphine)osmium, Os-00400
Di-μ-carbonylhexacarbonyl(dicarbonyliron)di-μ-hydro(tricarbonylruthenium)diosmium, Os-00207
Di-μ-carbonylhexacarbonyl[μ_3-[(1,2,3,4-η:1,2,3,4-η:1,4-η)-1-(3,3-dimethyl-1-butenyl)-2,4-bis(1,1-dimethylethyl)-1,3-butadiene-1,4-diyl]]triruthenium, Ru-00638
Di-μ-carbonylhexacarbonyl[μ-(tetracarbonylruthenio)]diiron, *see* Ru-00215
Di-μ-carbonylhexacarbonyl(1,2,3,4-tetraphenyl-1,3-butadiene-1,4-diyl)triruthenium, Ru-00689
Di-μ-carbonylhexacarbonyl[μ_3-(1,2,3,4-tetraphenyl-1,3-butadienylene)]triruthenium, *see* Ru-00689
Dicarbonyl[hexakis(trifluoromethyl)benzene](phosphorous acid)ruthenium trimethyl ester, *see* Ru-00452
Dicarbonyl[(1,2,3,4-η)-hexakis(trifluoromethyl)benzene]-(trimethyl phosphite-*P*)ruthenium, Ru-00452
Dicarbonylhydro[(methylthio)thioxomethyl]-bis(triphenylphosphine)osmium, Os-00408
Dicarbonylhydro[(1,2,3,4,5-η)-1,2,3,4,5-pentamethyl-2,4-cyclopentadien-1-yl]osmium, Os-00188
Dicarbonylhydro(phenylazo)bis(triphenylphosphine)osmium, Os-00425
Dicarbonylhydro(phenylazo)bis(triphenylphosphine)ruthenium, Ru-00800
Dicarbonyl[hydrotris(pyrazolato)borato(1−)]iodoruthenium, Ru-00182
Dicarbonylhydrotris(triphenylphosphine)ruthenium(1+), Ru-00849
Di-μ-carbonylicosacarbonyloctaosmate(2−), Os-00330
Dicarbonyl[(1,2,3,3*a*,7*a*-η)-1-*H*-inden-1-yl]ruthenate(1−), Ru-00175
Dicarbonyliodo(methoxycarbonyl)bis(triphenylphosphine)ruthenium, Ru-00754
Dicarbonyliodo(phenylazo)bis(triphenylphosphine)osmium, Os-00423
Dicarbonyliodo(phenylazo)bis(triphenylphosphine)ruthenium, Ru-00797
Dicarbonyliodo[(1,2,3,3*a*,6*a*-η)-1,4,5,6-tetrahydro-1-pentalenyl]ruthenium, Ru-00136
Dicarbonyl(η^2-methanethial)bis(triphenylphosphine)osmium, Os-00399
Dicarbonyl[(4-methyl-2,5-cyclohexadien-1-ylidene)-methylene]bis(triphenylphosphine)osmium, Os-00430
Dicarbonylnitrosylbis(triphenylphosphine)osmium(1+), Os-00387
Dicarbonylnitrosylbis(triphenylphosphine)ruthenium(1+), Ru-00724
Di-μ-carbonylnonacarbonyl(dicarbonylferrate)-μ-hydrotriruthenate(1−), Ru-00282
Di-μ-carbonylnonacarbonyl(dicarbonyliron)di-μ-hydrotriosmium, Os-00208
Di-μ-carbonylnonacarbonyl(dicarbonylruthenium)di-μ-hydrotriosmium, Os-00209
Di-μ-carbonylnonacarbonyl(triphenylphosphine)diironruthenium, *see* Ru-00652
Dicarbonyl[η^5-nonahydrodicarbanonaborato(4−)]-(triphenylphosphine)ruthenium, Ru-00570
Di-μ-carbonyloctacarbonyl(dicarbonyliron)di-μ-hydro(dimethylphenylphosphine)triruthenium, Ru-00517
Dicarbonyl(1,1,2,2,3,3,4,4-octafluoro-1,4-butanediyl)-bis[triphenylphosphine]ruthenium, Ru-00771
Dicarbonyl[(1,2,3,4,5-η)-1,2,3,4,5-pentamethyl-2,4-cyclopentadien-1-yl]iodoosmium, Os-00187
Dicarbonyl[(1,2,3,4,5-η)-1,2,3,4,5-pentamethyl-2,4-cyclopentadien-1-yl](tetrahydrofuran)osmium(1+), Os-00282
Dicarbonyl(phenylazo)bis(triphenylphosphine)osmium(1+), Os-00424
Dicarbonyl(phenylazo)bis(triphenylphosphine)ruthenium(1+), Ru-00798
Dicarbonyl[2-*C*,*N'*-(phenylazo)phenyl]chlororuthenium, *see* Ru-00330
Dicarbonyl(phenyldiazene-N^2)bis(triphenylphosphine)ruthenium(2+), Ru-00801
Dicarbonyl(phenyldiazo)bis(triphenylphosphine)osmium(1+), *see* Os-00424
Dicarbonyl(phenyldiazo)bis(triphenylphosphine)ruthenium(1+), *see* Ru-00798
Dicarbonyl(phenyldiimine)bis(triphenylphosphine)ruthenium(2+), *see* Ru-00801
Dicarbonyl[sulfata(2−)-*O*,*O'*]bis(triphenylphosphine)ruthenium, Ru-00728
Dicarbonyl[(*O*,*S*-η)-sulfur dioxide]bis(triphenylphosphine)ruthenium, Ru-00727
Di-μ-carbonyltetracarbonyl[(1,2,3,4,5-η)-2,4-cycloheptadien-1-yl][μ-[(1,2,3,4-η:1,5,6,7-η)-2,4,6-cycloheptatrien-1-yl]]triruthenium, Ru-00520

Di-μ-carbonyltetracarbonyl[μ-[(1,2,3,4-η:1,5,6,7-η)-2,4,6-cycloheptatrien-1-yl]](η^5-2,4-cyclopentadien-1-yl)triruthenium, Ru-00476
Di-μ-carbonyltetracarbonyl[μ-(2,4,6-cyclooctatrien-1-yl)]-(η^5-2,4-cyclopentadien-1-yl)triruthenium, Ru-00506
Di-μ-carbonyltetracarbonyl[μ-[(1,2,6,7-η:2,3,4,5,6-η)-2,4,6-cyclooctatrien-1-yl]][(1,2,3,3*a*,6*a*-η)-1,4,5,6-tetrahydro-1-pentalenyl]triruthenium, Ru-00565
Dicarbonyl(3,4,5,6-tetrachloro-1,2-benzenediolato-*O*,*O′*)-bis(triphenylphosphine)ruthenium, Ru-00794
Dicarbonyl[1,2,3,4-tetrachloro-5-(η^2-diazo)-1,3-cyclopentadiene]bis(triphenylphosphine)ruthenium, Ru-00785
Dicarbonyltetrachloroosmate(2−), Os-00007
Dicarbonyl[tetrachloropyrocatecholato(2−)]-bis(triphenylphosphine)ruthenium, *see* Ru-00794
Dicarbonyltetrachlororuthenate(2−), Ru-00011
Dicarbonyltetrachlorotetrakis(triphenylphosphine)diruthenium, *see* Ru-00863
Dicarbonyl(η^2-tetrafluoroethene)bis(trimethyl phosphite-*P*)osmium, Os-00113
Dicarbonyl(tetrafluoroethene)bis(trimethyl phosphite-*P*)ruthenium, Ru-00161
Dicarbonyl[(1,2,3,3*a*,6*a*-η)-1,4,5,6-tetrahydro-1-pentalenyl](trimethylgermyl)osmium, Os-00221
Dicarbonyl[(1,2,3,3*a*,6*a*-η)-1,4,5,6-tetrahydro-1-pentalenyl](trimethylgermyl)ruthenium, Ru-00319
Dicarbonyl[(1,2,3,3*a*,6*a*-η)-1,4,5,6-tetrahydro-1-pentalenyl](trimethylsilyl)osmium, Os-00222
Dicarbonyl[(1,2,3,3*a*,6*a*-η)-1,4,5,6-tetrahydro-1-pentalenyl](trimethylsilyl)ruthenium, Ru-00321
Dicarbonyltetraiodoosmate(1−), Os-00009
Dicarbonyltetraiodoosmate(2−), Os-00010
Dicarbonyltetraiodoruthenate(2−), Ru-00016
Dicarbonyl(*p*-tolylurylene)bis(triphenylphosphine)ruthenium, *see* Ru-00831
Di-μ-carbonyltricarbonyl(dicarbonyliron)di-μ-hydrobis(tricarbonylruthenium)osmium, *see* Os-00206
Di-μ-carbonyltricarbonyl(dicarbonyliron)(hexacarbonyl-μ-hydrodiruthenium)-μ-hydroosmium, Os-00206
Dicarbonyltricyclopentadienyltris[μ-(methanethiolato)]triruthenium, *see* Ru-00524
Dicarbonyltriiodo(triphenylphosphine)ruthenate(1−), Ru-00518
Dicarbonyl(trimethylsilyl)[(1,2,3,6,7-η)-8-(trimethylsilyl)-2,4,6-cyclooctatrien-1-yl]ruthenium, Ru-00437
closo-6,6-Dicarbonyl-6-triphenylphosphine-6-ruthenanonahydro-2,3-dicarbanonaborane, *see* Ru-00570
Dicarbonyltris(phosphorous trifluoride)ruthenium, Ru-00012
Dicarbonyltris(triphenylphosphine)osmium, Os-00443
Dicarbonyltris(triphenylphosphine)ruthenium, Ru-00848
Di-μ-carbonylundecacarbonyldi-μ-hydrotetraruthenium, Ru-00291
1,1′-Dicarboxyosmocene, *see* Os-00174
1,1′-Dicarboxyruthenocene, *see* Ru-00232
Di-μ-chlorobis[(1,2,5,6-η)-1,5-cyclooctadiene]bis(η^3-propenyl)diruthenium, Ru-00574
Dichlorobis(η^4-1,5-cyclooctadiene)(1,1-dimethylhydrazine)dihydrodiruthenium, *see* Ru-00496
Dichlorobis[1,1′-ferrocenediylbis[dimethylarsine]]ruthenium, Ru-00649
Dichlorobis(triphenylphosphine)ruthenium, Ru-00698
Dichloro(chloromercurio)nitrosylbis(triphenylphosphine)osmium, Os-00361
Dichloro(cyanobenzene)tricarbonylruthenium, *see* Ru-00125
Dichloro[(1,2,3,4-η)-1,3-cycloheptadiene]ruthenium, Ru-00065
Dichloro[(1,2,3,4-η)-1,3,5-cycloheptatriene]ruthenium, Ru-00059
Dichloro[(1,2,3,4,5,6-η)-1,3,5-cycloheptatriene]-(triphenylphosphine)ruthenium, Ru-00620
Dichloro[(1,2,5,6-η)-1,5-cyclooctadiene]bis(pyridine)ruthenium, Ru-00483
Dichloro[(1,2,5,6-η)-1,5-cyclooctadiene]osmium, Os-00064
Dichloro[(1,2,5,6-η)-1,5-cyclooctadiene]ruthenium, Ru-00084
Di-μ-chlorodichlorobis(1,3,5-cycloheptatriene)diruthenium, *in* Ru-00059
Di-μ-chlorodichlorobis[(1,2,3,6,7,8-η)-2,7-dimethyl-2,6-octadiene-1,8-diyl]diruthenium, Ru-00530
Di-μ-chlorodichlorobis[(1,2,3,4,5,6-η)-hexamethylbenzene]diruthenium, Ru-00612
Di-μ-chlorodichlorobis(triphenylphosphine)diruthenium, *in* Ru-00698
Dichloro[(1,2,3,6,7,8-η)-2,7-dimethyl-2,6-octadiene-1,8-diyl]ruthenium, *in* Ru-00530
Dichloro(dinitrogen)tris(triethylphosphine)osmium, Os-00306
Dichloro[(1,2,3,6,7,10,11,12-η)-2,6,10-dodecatriene-1,12-diyl]ruthenium, Ru-00267
Dichloro-π_2-2,6,10-dodecatrienyleneruthenium, *see* Ru-00267
Dichloro(hexamethylbenzene)ruthenium, *in* Ru-00612
Dichloro(3-hexenylenedi-π-allyl)ruthenium, *see* Ru-00267
Dichloronitrosyl(phenylmethyl)bis(triphenylphosphine)ruthenium, *see* Ru-00788
Dichloro(2,5-norbornadiene)bis(triphenylphosphine)ruthenium, *see* Ru-00790
Dichloro(norbornadiene)ruthenium, *see* Ru-00058
Dichlorotetracarbonylosmium, *see* Os-00017
Dichlorotetrakis(triphenylphosphine)ruthenium, Ru-00859
Dichlorotricarbonyl(phenyl cyanide)ruthenium, *see* Ru-00125
Dichlorotris(triphenylphosphine)osmium, Os-00435
Dichlorotris(triphenylphosphine)ruthenium, Ru-00833
1,1′-Dicinnamoylruthenocene, *see* Ru-00644
Dicyclohexadienylruthenium, *see* Ru-00260
Di-π-cyclopentadienylruthenium, *see* Ru-00149
1,1′-Diethenylruthenocene, *see* Ru-00350
1,1′-Diethylruthenocene, Ru-00367
1,1′-Diformylruthenocene, *see* Ru-00230
Di-μ-hydrido-μ_3-(phenylphosphiniden)-*triangulo*-tris(tricarbonylruthenium), *see* Ru-00382
Dihydrobis(triphenylphosphine)[(7,8,9,10,11-η)undecahydro-7,8-dicarbaundecaborato(2−)]ruthenium, Ru-00737
Dihydrobis(triphenylphosphine)[(7,8,9,10,11-η)-undecahydro-7,9-dicarbaundecaborato(2−)]ruthenium, Ru-00738
Di-μ-hydrodihydrohexakis(trimethylphosphine)diruthenium, Ru-00499
Dihydrotetrakis(phosphorous trifluoride)osmium, Os-00447
Dihydrotetrakis(phosphorous trifluoride)ruthenium, Ru-00869
Dihydrotetrakis(triethyl phosphite-*P*)ruthenium, Ru-00616
Dihydrotetrakis(trimethylphosphine)osmium, Os-00196
Dihydrotetrakis(trimethylphosphine)ruthenium, Ru-00278
Dihydrotetrakis(trimethyl phosphite-*P*)ruthenium, Ru-00277
Dihydrotetrakis(triphenylphosphine)ruthenium, Ru-00861
Dihydrotris(trimethylphosphine)ruthenium, *in* Ru-00499
Di-π-indenylruthenium, *see* Ru-00482
1,1′-Diiodoruthenocene, Ru-00131
1,1′-Dilithioruthenocene, Ru-00132
Dimanganese osmium tetradecacarbonyl, *see* Os-00246
1-[(Dimethylamino)ethyl]ruthenocene, Ru-00368
(*N*,*N*-Dimethylaminomethyl)ruthenocene, Ru-00317
(η^6-Dimethylaniline)(η^6-hexamethylbenzene)ruthenium(2+), Ru-00526
(η^6-Dimethylaniline)(η^6-mesitylene)ruthenium(2+), *see* Ru-00463
[(1,2,3,4,5,6-η)-*N*,*N*-Dimethylaniline][(1,2,3,4,5,6-η)-1,3,5-trimethylbenzene]ruthenium(2+), Ru-00463
[(1,2,3,4,5,6-η)-*N*,*N*-Dimethylbenzenamine][(1,2,3,4,5,6-η)-hexamethylbenzene]ruthenium(2+), *see* Ru-00526
[(1,2,3,4,5,6-η)-*N*,*N*-Dimethylbenzenamine][(1,2,3,4,5,6-η)-1,3,5-trimethylbenzene]ruthenium(2+), *see* Ru-00463
Di-μ-methylenehexakis(trimethylphosphine)diruthenium(2+), Ru-00534
(Dimethyl fumarate)tetracarbonylruthenium, *in* Ru-00066
(Dimethyl maleate)tetracarbonylruthenium, *in* Ru-00066
1,1′-Dimethylosmocene, Os-00184
1,1′-Dimethylruthenocene, Ru-00261
1,2-Dimethylruthenocene, Ru-00262
1,3-Dimethylruthenocene, Ru-00263
N,*N*-Dimethylruthenocenemethylamine, *see* Ru-00317
N,*N*-Dimethyl-1-ruthenocenylethylamine, *see* Ru-00368

(Dinitrogen)dihydrotris(triphenylphosphine)ruthenium, Ru-00838
(Dinitrogen)tris(ethyldiphenylphosphine)dihydroosmium, Os-00416
Dinitrosylbis(triphenylphosphine)osmium, Os-00366
Dinitrosylbis(triphenylphosphine)ruthenium, Ru-00703
Diosmium enneacarbonyl, *see* Os-00086
Diosmium nonacarbonyl, *see* Os-00086
Dioxobis(*tert*-butylimido)osmium, *see* Os-00067
1,1′-(1,5-Dioxo-3-phenyl-1,5-pentanediyl)ruthenocene, *see* Ru-00547
μ-Diphenylacetylene-*closo*-tetrakis(tricarbonylruthenium), *see* Ru-00629
Diphenylfulvenepentacarbonyldiruthenium, *see* Ru-00576
(6,6-Diphenylpentafulvene)pentacarbonyldiruthenium, *see* Ru-00576
Diruthenium enneacarbonyl, *see* Ru-00116
Diruthenium nonacarbonyl, *see* Ru-00116
1,4-Diruthenocenyl-1,3-butadiyne, Ru-00588
Diruthenocenylmercury, Ru-00522
1,5-Diruthenocenyl-1,4-pentadien-3-one, Ru-00619
1,3-Diruthenocenyl-2-propen-1-one, Ru-00581
Di(4,5,6,7-tetrahydro-π-indenyl)ruthenium, *see* Ru-00484
1,1′-Di(thiocarboxy)ruthenocene, *see* Ru-00231
1,1′-Divinylruthenocene, Ru-00350
Docosacarbonyloctaosmate(2−), *see* Os-00330
Dodecacarbonydichlorotriosmium, Os-00150
Dodecacarbonyl[μ_4-(1-η:1,2-η:1,2-η:2-η)-1-cyclohexene-1,2-diyl]tetraruthenium, Ru-00471
Dodecacarbonyldihydro(dimethylphenylphosphine)irontriruthenium, *see* Ru-00517
Dodecacarbonyldi-μ-hydrodi-μ_3-selenoxotetraosmium, Os-00161
Dodecacarbonyldi-μ-hydrodi-μ_3-thioxotetraosmium, Os-00160
Dodecacarbonyldi-μ-hydro[μ-(2-phenylethylidene)]tetraosmium, Os-00313
Dodecacarbonyldi-μ-hydrotetraosmate(2−), Os-00159
Dodecacarbonyldihydrotetraruthenate(2−), *see* Ru-00218
Dodecacarbonyldihydrotriosmium, Os-00158
Dodecacarbonyldiiodotriosmium, Os-00197
Dodecacarbonyldiironosmium, *see* Os-00152
Dodecacarbonyldiironruthenium, *see* Ru-00215
Dodecacarbonyldimethyltriosmium, Os-00230
Dodecacarbonyldiosmiumruthenium, *see* Os-00200
Dodecacarbonyl(diphenylacetylene)irontriruthenium, *see* Ru-00628
Dodecacarbonyl[μ_4-[(1-η:1,2-η:1,2-η:2-η)-1,2-diphenyl-1,2-ethenediyl]]tetraruthenium, Ru-00629
Dodecacarbonyldi-μ_3-thioxotetraosmium, Os-00203
Dodecacarbonyl[μ_4-[(1-η:1,2-η:1,2-η:2-η)-1,2-ethenediyl]]tetraosmium, Os-00225
Dodecacarbonyl[μ_4-(1-η:η^2:η^2:2-η-1,2-ethenediyl)]tetraruthenium, Ru-00325
Dodecacarbonylhexachlorotriruthenium, Ru-00211
Dodecacarbonylhydromanganesediosmium, Os-00153
Dodecacarbonylhydronitridoirontriruthenium, *see* Ru-00216
Dodecacarbonyl-μ-hydrotriosmium(1+), Os-00155
Dodecacarbonyl-μ-hydrotriruthenium(1+), Ru-00217
Dodecacarbonylirondiosmium, Os-00151
Dodecacarbonylirondiruthenium, *see* Ru-00214
Dodecacarbonylmanganatediosmate(1−), Os-00198
Dodecacarbonyl[μ_3-[methanaminato(2−)]]tetraosmium, Os-00214
Dodecacarbonyl(nitrosylferrate)triruthenate(1−), *see* Ru-00213
Dodecacarbonyloctafluorotetraruthenium, *see* Ru-00212
Dodecacarbonylosmiumdiruthenium, *see* Os-00181
Dodecacarbonylpenta-μ-hydrotriosmiumrhenium, *see* Os-00168
Dodecacarbonylrhenatediosmate(1−), Os-00199
Dodecacarbonyltetra-μ-fluorotetrafluorotetraruthenium, Ru-00212
Dodecacarbonyltetrahydro(iron)triruthenium, Ru-00221
Dodecacarbonyltetra-μ-hydrotetraosmium, Os-00167
Dodecacarbonyltetra-μ-hydrotetraruthenium, Ru-00224
Dodecacarbonyltetra-μ_3-oxotetraosmium, Os-00204
Dodecacarbonyl-μ_3-thioxotetraosmium, Os-00202
Dodecacarbonyltri-μ-hydro-μ-iodotetraosmium, Os-00162
Dodecacarbonyltri-μ-hydrotetraosmate(1−), Os-00163
Dodecacarbonyltrihydrotetraruthenate(1−), Ru-00220
Dodecacarbonyltriosmium, Os-00201
Dodecacarbonyltriruthenium, Ru-00280
Eicosacarbonyldi-μ-hydrotriosmiumdiruthenium, *see* Os-00311
Eicosacarbonylheptaosmate(2−), Os-00315
Eicosacarbonylhexaosmium, *see* Os-00314
(η^2-Ethene)tris(triphenylphosphine)ruthenium, Ru-00850
[(η^2-Ethenyl)benzene][(1,2,3,4-η)-1,3-hexadiene]-bis(triphenylphosphine)ruthenium, Ru-00824
[η^2-(Ethenylbenzene)][(1,2,3-η)-2-hexenyl]-hydrobis(triphenylphosphine)ruthenium, Ru-00825
Ethenylosmocene, *see* Os-00180
Ethenylruthenocene, *see* Ru-00252
σ-Ethyl-π-cyclopentadienylcarbonyl(triphenylphosphine)ruthenium, *see* Ru-00632
σ-Ethyl-π-cyclopentadienylruthenium dicarbonyl, *see* Ru-00106
(η^2-Ethylene)tris(triphenylphosphine)ruthenium, *see* Ru-00850
Ethylosmocene, Os-00185
Ethylruthenocene, Ru-00264
Ethyl ruthenocenyl ketone, *see* Ru-00311
Ethynylruthenocene, Ru-00233
(Ferrocenylcarbonyl)osmocene, *see* Os-00322
(Ferrocenylethynyl)ruthenocene, *see* Ru-00564
Ferrocenylmethylosmocene, Os-00323
(Ferrocenylmethyl)ruthenocene, Ru-00549
Ferrocenyl osmocenyl ketone, *see* Os-00322
Ferrocenylosmocenylmethanone, *see* Os-00322
(3-Ferrocenyl-1-oxo-2-propenyl)ruthenocene, *see* Ru-00580
(3-Ferrocenyl-3-oxo-1-propenyl)ruthenocene, *see* Ru-00579
Ferrocenylruthenocenylacetylene, *see* Ru-00564
Ferrocenyl ruthenocenyl ketone, *see* Ru-00546
Ferrocenyl(ruthenocenyl)methane, *see* Ru-00549
Ferrocenylruthenocenylmethanone, Ru-00546
1-Ferrocenyl-3-ruthenocenyl-2-propen-1-one, Ru-00579
3-Ferrocenyl-1-ruthenocenyl-2-propen-1-one, Ru-00580
(Formyl-π-cyclopentadienyl)(methyl-π-cyclopentadienyl)ruthenium, *see* Ru-00242
1-Formyl-1′-methylruthenocene, *see* Ru-00242
1-Formyl-2-methylruthenocene, *see* Ru-00243
1-Formyl-3-methylruthenocene, *see* Ru-00244
Formylruthenocene, *see* Ru-00184
(Fumaric acid)tetracarbonylruthenium, *in* Ru-00066
Heneicosacarbonylheptaosmium, Os-00325
Heptacarbonylbis[μ-(diphenylphosphino)][μ_3-[(1-η:1,2-η:2-η)-1,3-cyclohexadien-5-yne]]triosmium, Os-00369
Heptacarbonyl[μ_3-(5-η:5,6-η:6-η)-1,3-cyclohexadien-5-yne]-bis[μ-(dimethylphosphine)]triosmium, *see* Os-00298
Heptacarbonyl[μ_3-[(5-η:5,6-η:6-η)-1,3-cyclohexadien-5-yne]]-[μ-(diphenylphosphino)]hydro(triphenylphosphine)triosmium, Os-00417
Heptacarbonyl(η^5-cyclopentadienyl)rheniumruthenium, *see* Ru-00225
Heptacarbonyldihydro(triphenylphosphine)diosmium, Os-00339
Heptacarbonyl[μ_3-[1,1′-(η^1:η^1:η^2-1,2-ethynediyl)-bis[benzene]]][μ-(1,2,3,4-η:1,4-η)-1,2,3,4-tetraphenyl-1,3-butadiene-1,4-diyl]triosmium, Os-00432
Heptacarbonyl-μ-hydro-μ-nitrosyltris(trimethyl phosphite-*P*)triruthenium, Ru-00442
Heptacosacarbonyl[μ_3-(carboxylato-*C*:*O*:*O*′)]-μ-hydrononaosmate(1−), Os-00344
Heptadecacarbonyltetrakis(trimethylphosphite-*P*)hexaosmium, Os-00349
Heptaosmium heneicosacarbonyl, *see* Os-00325
Hexacarbonylbis[μ-(dimethylarsino)]diruthenium, Ru-00153
Hexacarbonylbis[μ-(dimethylgermylene)]bis(trimethylgermyl)diruthenium, Ru-00444
Hexacarbonylbis[μ-(dimethylgermylene)]bis(trimethylgermyl)diosmium, Os-00283
Hexacarbonylbis[μ-(dimethylphosphino)]diruthenium, Ru-00156
Hexacarbonylbis[μ-(dimethylsilylene)]bis(trimethylsilyl)diruthenium, Ru-00445

Hexacarbonylbis[μ-(dimethylsilylene)]bis(trimethylsilyl)diosmium, Os-00284
Hexacarbonylbis[μ-(dimethylstannylene)]-bis(trimethylstannyl)diruthenium, Ru-00446
Hexacarbonylbis(diphenylacetylene)diruthenium, *see* Ru-00681
Hexacarbonylbis[μ-(diphenylphosphino)]diruthenium, Ru-00658
Hexacarbonyl(μ-carbonylhexacarbonyldiferrate)-μ-hydrodiruthenate(1−), Ru-00283
Hexacarbonyl(1,3-cycloheptadienyl)(cycloheptatrienyl)triruthenium, *see* Ru-00520
Hexacarbonyl[μ-[(1,2,3-η:4,5,6-η)-1,3,5-cycloheptatriene]]diruthenium, Ru-00297
Hexacarbonyl(1,3,5-cycloheptatrienyl)(η^5-cyclopentadienyl)triruthenium, *see* Ru-00476
Hexacarbonyl[μ_3-[(1,2,3-η:4,5-η:6,7-η)-2,4,6-cycloheptatrien-1-yl]][μ_3-(2-methyl-2-propanethiolato)]triruthenium, Ru-00459
Hexacarbonyl[μ-[(1,2,3,4,5,8-η)-1,3,5,7-cyclooctatetraene]]-diosmium, Os-00231
Hexacarbonyl(μ-1,3,5,7-cyclooctatetraene)diruthenium, *see* Ru-00328
Hexacarbonyl[μ-[(1,2,3,4-η:1,4,5,8-η)-1,3,5,7-cyclooctatetraene]]diruthenium, Ru-00328
Hexacarbonyl[μ_3-[(1,2,3,4-η:5,6-η:7,8-η)-1,3,5,7-cyclooctatetraene]]-μ_3-thioxotriruthenium, Ru-00329
Hexacarbonyl[μ-[[(2,3,6,7-η:3,4,5,6-η)-1,3,5,7-cyclooctatetraen-1-yl]trimethylsilane]]diruthenium, Ru-00458
Hexacarbonyldi-μ-chlorodichlorodiosmium, Os-00031
Hexacarbonyldi-μ-chlorodichlorodiruthenium, Ru-00033
Hexacarbonyldi-μ-chlorodiosmium, Os-00030
Hexacarbonyldihydrobis(triphenylphosphine)diosmium, Os-00411
Hexacarbonyldi-μ-iododiiododiosmium, Os-00040
Hexacarbonyldi-μ-iododiiododiruthenium, Ru-00043
Hexacarbonyldi-μ-iododiosmium, Os-00039
Hexacarbonyl(diphenylacetylene)diruthenium, *see* Ru-00516
Hexacarbonyl[μ-[η^2-*N*,*N′*-1,2-ethanediylidenebis[2-methyl-2-propanamine]-*N*,*N′*]]diosmium, Os-00280
Hexacarbonyl[μ-[η^2-*N*,*N′*-1,2-ethanediylidenebis[2-methyl-2-propanamine]-*N*,*N′*]]diruthenium, Ru-00431
Hexacarbonyl[μ-(η^2:η^4-1,2-ethenediyl-1,2-phenylene)]diosmium, Os-00227
Hexacarbonyl[μ-(η^2:η^4-1,2-ethenediyl-1,2-phenylene)]diruthenium, Ru-00327
Hexacarbonyl[glyoxalbis(*tert*-butylimine)]diosmium, *see* Os-00280
Hexacarbonyl[glyoxalbis(*tert*-butylimine)]diruthenium, *see* Ru-00431
Hexacarbonyl-μ-hydro[μ-[(*O*,1-η:1,2-η)-3-oxo-1-phenyl-1-butenyl]]diruthenium, Ru-00422
Hexacarbonyl(osmacyclopentadiene)osmium, *see* Os-00098
Hexacarbonyl[μ-[(1,2,5-η:1,4,5-η)-3-oxo-1,2,4,5-tetraphenyl-1,4-pentadiene-1,5-diyl]]diruthenium, Ru-00685
Hexacarbonyl[μ-[*N*-phenyl-1,2-benzenediaminato(2−)]]diruthenium, Ru-00472
Hexacarbonyl[μ-[η^2:η^4-1,2-phenylene(2-phenyl-1,2-ethenediyl)]]diruthenium, Ru-00516
Hexacarbonyl[μ-[*N*-phenyl-*o*-phenylenediaminato(2−)]]diruthenium, *see* Ru-00472
Hexacarbonyl[2-phenyl-1-ruthenaindene]ruthenium, *see* Ru-00516
Hexacarbonyltetrachlorodiosmium, *see* Os-00031
Hexacarbonyltetrachlorodiruthenium, *see* Ru-00033
Hexacarbonyltetraiododiruthenium, *see* Ru-00043
Hexacarbonyl[μ-[(1,2,3,4-η:1,4-η)-1,2,3,4-tetramethyl-1,3-butadiene-1,4-diyl]]diosmium, Os-00236
Hexacarbonyl[μ-[(1,2,3,4-η:1,4-η)-1,2,3,4-tetramethyl-1,3-butadiene-1,4-diyl]]diruthenium, Ru-00338
Hexacarbonyl[μ-(1,2,3,4-tetramethyl-1,3-butadienylene)]diosmium, *see* Os-00236
Hexacarbonyl(2,2,6,6-tetramethyl-3,6-diaza-3,5-octadiene)diosmium, *see* Os-00280
Hexacarbonyl(2,2,6,6-tetramethyl-3,6-diaza-3,5-octadiene)diruthenium, *see* Ru-00431
Hexacarbonyl[2,3,4,5-tetramethyl(osmacyclopentadiene)]osmium, *see* Os-00236
Hexacarbonyl[μ-[(1,2,4,5-η:1,5-η)-1,2,4,5-tetramethyl-3-oxo-1,4-pentadiene-1,5-diyl]]diosmium, Os-00260
Hexacarbonyl[μ-[(1,2,3,4-η:1,4-η)-1,2,3,4-tetraphenyl-1,3-butadiene-1,4-diyl]]diosmium, Os-00356
Hexacarbonyl[μ-[(1,2,3,4-η:1,4-η)-1,2,3,4-tetraphenyl-1,3-butadiene-1,4-diyl]]diruthenium, Ru-00681
Hexacarbonyl[μ-(1,2,3,4-tetraphenyl-1,3-butadienylene)]diosmium, *see* Os-00356
Hexacarbonyl[μ-(1,2,3,4-tetraphenyl)-1,3-butadienylene)]diruthenium, *see* Ru-00681
Hexacarbonyltris[μ-(dimethylgermylene)]diosmium, Os-00191
Hexacarbonyltris[μ-(dimethylgermylene)]diruthenium, Ru-00269
Hexacarbonyltris[μ-(dimethylsilyl)]diruthenium, Ru-00272
Hexacarbonyltris[μ-(dimethylsilylene)]diruthenium, *see* Ru-00272
Hexadecacarbonyldi-μ-hydropentaosmium, Os-00273
Hexadecacarbonyl[μ_3-[(1-η:1,2-η:2-η)-1,2-dimethyl-1,2-ethenediyl]](μ_5-methanetetrayl)hexaosmium, Os-00316
Hexadecacarbonyl-μ-ethenylidenedi-μ-hydrohexaruthenium, Ru-00470
Hexadecacarbonyl-μ_3-ethylidyne-μ_4-ethylidynehexaosmium, Os-00312
Hexadecacarbonylhexaoxohexaosmium, Os-00289
Hexadecacarbonylhydromanganesetriosmium, Os-00271
Hexadecacarbonylhydrorheniumtriosmium, Os-00272
Hexadecacarbonylpentaosmium, Os-00287
Hexadecacarbonyl-μ_3-thioxo-μ_4-thioxohexaosmium, Os-00288
(η^4-1,3-Hexadiene)(η^2-styrene)bis(triphenylphosphine)ruthenium, *see* Ru-00824
(Hexafluoro-2-butyne)dicarbonylbis(trimethyl phosphite-*P*)osmium, Os-00189
Hexakis(diethylaminato)-μ-1,1′-ruthenocenediyldititanium, Ru-00683
Hexakis(*N*-ethylethanaminato)-μ-1,1′-ruthenocenediyldititanium, *see* Ru-00683
Hexakis(isocyanomethane)osmium(2+), Os-00192
Hexakis(isocyanomethane)ruthenium(2+), Ru-00270
Hexakis(methyl isocyanide)osmium(2+), *see* Os-00192
Hexakis(methyl isocyanide)ruthenium(2+), *see* Ru-00270
Hexakis(triethyl phosphite-*P*)osmium(2+), Os-00367
Hexakis(triethyl phosphite-*P*)ruthenium(2+), Ru-00705
Hexakis(trimethyl phosphite-*P*)ruthenium(2+), Ru-00498
[(1,2,3,4-η)-Hexamethylbenzene][(1,2,3,4,5,6-η)-hexamethylbenzene]ruthenium, Ru-00614
[(1,2,3,4,5,6-η)-Hexamethylbenzene][(1,2,3,4,4*a*,8*a*-η)-naphthalene]ruthenium(2+), Ru-00573
[(1,2,3,4,5,6-η)-Hexamethylbenzene][(2,3,5,6-η)-norbornadiene]ruthenium, *see* Ru-00512
[(1,2,3,4,5,6-η)-Hexamethylbenzene][1,2,3,4-η)-1,2,3,4-tetramethyl-5,6-bis(methylene)-1,3-cyclohexadiene]ruthenium, Ru-00611
[(1,2,3,4,5,6-η)-Hexamethylbenzene]tris(2-propanone)ruthenium(2+), Ru-00552
Hexaosmium icosacarbonyl, *see* Os-00314
Hexaosmium octadecacarbonyl, *see* Os-00308
(η^3-2-Hexenyl)hydro(η^2-styrene)bis(triphenylphosphine)ruthenium, *see* Ru-00825
μ-Hydrido-μ-(benzylideneacetonato)hexacarbonyldiruthenium, *see* Ru-00422
Hydridobromocarbonyltris(triphenylphosphine)osmium, *see* Os-00440
Hydridocarbonyl(phenyldiazo)bis(triphenylphosphine)osmium, *see* Os-00418
Hydridodicarbonyl(phenyldiazo)bis(triphenylphosphine)osmium, *see* Os-00425
Hydridodicarbonyl(phenyldiazo)bis(triphenylphosphine)ruthenium, *see* Ru-00800
Hydrobromocarbonyltris(triphenylphosphine)osmium, *see* Os-00440
Hydrochlorocarbonyltris(triphenylphosphine)osmium, *see* Os-00441

Hydrodicarbonyl-π-cyclopentadienylosmium, *see* Os-00045
Hydrodicarbonyl-π-cyclopentadienylruthenium, *see* Ru-00055
Hydronitrosyltris(triphenylphosphine)osmium, Os-00436
Hydronitrosyltris(triphenylphosphine)ruthenium, Ru-00836
Hydro[(1,2,3,4,5-η)-6-oxo-2,4-cyclohexadien-1-yl]-bis(triphenylphosphine)ruthenium, Ru-00777
Hydropentakis(triethyl phosphite-*P*)ruthenium(1+), Ru-00667
Hydropentakis(trimethyl phosphite-*P*)ruthenium(1+), Ru-00412
Hydro[(η^6-phenyl)diphenylphosphine]bis(triphenylphosphine)ruthenium(1+), Ru-00837
Hydrotetrakis(phosphorous trifluoride)osmate(1−), Os-00446
Hydrotetrakis(phosphorous trifluoride)ruthenate(1−), Ru-00868
Hydrotris(methyldiphenylphosphine)(η^3-2-propenyl)ruthenium, Ru-00784
(α-Hydroxybenzyl)osmocene, *see* Os-00297
(α-Hydroxybenzyl)ruthenocene, *see* Ru-00457
(1-Hydroxyethyl)osmocene, Os-00182
(1-Hydroxyethyl)ruthenocene, Ru-00256
(1-Hydroxy-1-methylethyl)ruthenocene, Ru-00316
1-(Hydroxymethyl)-2-methylruthenocene, Ru-00257
(1-Hydroxy-2-methylpropyl)osmocene, Os-00242
Hydroxymethylruthenocene, *see* Ru-00191
(1-Hydroxy-1-phenylmethyl)osmocene, Os-00297
(1-Hydroxy-1-phenylmethyl)ruthenocene, *see* Ru-00457
Hydroxyruthenocene, Ru-00142
α-Hydroxy-1,2-tetramethyleneruthenocene, *see* Ru-00351
Icosacarbonyldi-μ-hydrotriosmiumdiruthenium, *see* Os-00311
Icosacarbonylheptaosmate(2−), *see* Os-00315
Icosacarbonylhexaosmium, Os-00314
(η^5-Indenyl)dicarbonylruthenium(1−), *see* Ru-00175
Iododicarbonyl(phenyldiazo)bis(triphenylphosphine)osmium, *see* Os-00423
Iododicarbonyl(phenyldiazo)bis(triphenylphosphine)ruthenium, *see* Ru-00797
Iodoruthenocene, Ru-00137
Iodo(η^4-tetraphenylcyclobutadiene)tris(trimethyl phosphite-*P*)ruthenium(1+), Ru-00713
[η^6-(1-Isopropyl-4-methylbenzene)](η^5-tetramethylthiophene)ruthenium(2+), Ru-00492
Lithioruthenocene, Ru-00138
(Maleic acid)tetracarbonylruthenium, *in* Ru-00066
(Maleic anhydride)tetracarbonylruthenium, *in* Ru-00066
Methoxycarbonylosmocene, *in* Os-00136
Methoxycarbonylruthenocene, *in* Ru-00185
Methoxyruthenocene, *in* Ru-00142
(Methylcyclopentadienyl)bis(triphenylphosphine)ruthenium chloride, *see* Ru-00778
σ-Methyl-π-cyclopentadienylbis(triphenylphosphine)ruthenium, *see* Ru-00781
(η^5-Methylcyclopentadienyl)chlorobis(triphenylphosphine)ruthenium, *see* Ru-00778
[(1,2,3,4,5-η)-1-Methyl-2,4-cyclopentadien-1-yl][(η^6-phenyl)triphenylborato(1−)]ruthenium, Ru-00662
σ-Methyl-π-cyclopentadienylruthenium dicarbonyl, *see* Ru-00079
μ-Methyldi-μ-methylenehexakis(trimethylphosphine)diruthenium(1+), Ru-00555
[(1,2,3,4,5,6-η)-1-Methyl-4-(1-methylethyl)benzene]-[(1,2,3,4,5-η)tetramethylthiophene]ruthenium(2+), *see* Ru-00492
α-Methylosmocenemethanol, *see* Os-00182
Methylosmocenylcarbinol, *see* Os-00182
Methyl osmocenyl ketone, *see* Os-00177
(2-Methyl-1-oxopropyl)osmocene, Os-00241
[2-Methyl-2-propanaminato(2-)]trioxoosmium, *see* Os-00020
(η^3-2-Methyl-2-propenyl)(η^5-cyclopentadienyl)carbonylruthenium, *see* Ru-00155
Methylruthenocene, Ru-00196
1′-Methylruthenocenecarboxaldehyde, Ru-00242
2-Methylruthenocenecarboxaldehyde, Ru-00243
3-Methylruthenocenecarboxaldehyde, Ru-00244
1′-Methylruthenocenecarboxylic acid, Ru-00245
2-Methylruthenocenecarboxylic acid, Ru-00246
3-Methylruthenocenecarboxylic acid, Ru-00247
α-Methylruthenoceneethanol, *see* Ru-00316
2-Methylruthenocenemethanol, *see* Ru-00257
α-Methylruthenocenemethanol, *see* Ru-00256
Methyl ruthenocenyl ether, *in* Ru-00142
Methyl ruthenocenyl ketone, *see* Ru-00240
Nitrosyl(η^3-2-propenyl)bis(triphenylphosphine)ruthenium, Ru-00748
Nonacarbonylbis[μ-(diphenylphosphino)](tetracarbonyliron)triruthenium, Ru-00706
Nonacarbonyl(*tert*-butylacetylene)hydrotriruthenium, *see* Ru-00388
Nonacarbonyl(*tert*-butylacetylene)triruthenate(1−), *see* Ru-00387
Nonacarbonyl[μ_3-(chloromethylidyne)]tri-μ-hydrotriruthenium, Ru-00123
Nonacarbonyl[μ_3-[(1-η:1,2,3-η:3-η)-1,6,9-cyclododecatrien-1-yl-3-ylidene]]-μ-hydrotriruthenium, Ru-00544
Nonacarbonyl[μ_3-[(1-η:1,2-η:2-η)-1,3-cyclohexadiene-1,2-diyl]]di-μ-hydrotriruthenium, Ru-00384
Nonacarbonyl[μ^3-[η:η:η^3-3,5-cyclohexadiene-1,2-diylidene(phenylmethylidyne)]]-μ-hydrotriruthenium, Ru-00556
Nonacarbonyl[μ_3-[(1,2-η:3-η:4,5-η)-2,4-cyclohexadien-1-yl]]-μ-hydrotriosmium, Os-00256
Nonacarbonyl[μ_3-[(5-η:5,6-η:6-η)-1,3-cyclohexadien-5-yne]]-di-μ-hydrotriosmium, Os-00252
Nonacarbonyl[μ_3-[(1-η:1,2-η:2-η)cyclooctyne]]di-μ-hydrotriosmium, Os-00296
1,1,1,2,2,2,3,3,3-Nonacarbonyl-1,2;1,3-di-μ-hydrido-μ_3-sulfido-*triangulo*-triosmium, *see* Os-00070
Nonacarbonyldi-μ-hydro(2-isocyano-2-methylpropane)triosmium, Os-00234
Nonacarbonyldi-μ-hydro[μ_3-[methanaminato(2−)]]triosmium, Os-00101
Nonacarbonyldi-μ-hydro[μ_3-(oxoethenylidene)]triosmium, Os-00124
Nonacarbonyldi-μ-hydro[μ_3-[(1-η:1-η:1,2-η)-2-phenylethenylidene]]triosmium, Os-00293
Nonacarbonyldi-μ-hydro[μ_3-(phenylphosphinidene)]triosmium, Os-00255
Nonacarbonyldi-μ-hydro[μ_3-(phenylphosphinidene)]triruthenium, Ru-00382
Nonacarbonyldi-μ-hydro-μ_3-selenoxotriruthenium, Ru-00090
Nonacarbonyldi-μ-hydro-μ_3-telluroxotriruthenium, Ru-00091
Nonacarbonyldi-μ-hydro-μ_3-thioxotriosmium, Os-00070
Nonacarbonyldi-μ-hydro-μ_3-thioxotriruthenium, Ru-00089
Nonacarbonyldihydro(triphenylphosphine)triosmium, Os-00343
Nonacarbonyl(3,3-dimethyl-1-butene)dihydrotriruthenium(2+), *see* Ru-00392
Nonacarbonyl[μ_3-[(1-η:1-η:1,2-η)-3,3-dimethyl-1-butenylidene]]di-μ-hydrotriruthenium, Ru-00391
Nonacarbonyl[μ_3-(3,3-dimethylbutylidyne)]tri-μ-hydrotriruthenium, Ru-00394
Nonacarbonyl[μ_3-[(1-η:1,2-η:1,2-η)-3,3-dimethyl-1-butynyl]]-μ-hydrotriruthenium, Ru-00388
Nonacarbonyl[μ_3-[(1-η:1,2-η:1,2-η)-3,3-dimethyl-1-butynyl]]-triruthenate(1−), Ru-00387
Nonacarbonyl[μ_3-[(1-η:1,2-η:2-η)-1-(1,1-dimethylethyl)-1,2-ethenediyl]]di-μ-hydrotriruthenium(2+), Ru-00392
Nonacarbonyl(*N,N*-dimethylmethanamine)di-μ-nitrosyltriosmium, *see* Os-00173
Nonacarbonyldi-μ-nitrosyl(trimethylamine)triosmium, Os-00173
Nonacarbonyldi-μ-nitrosyl(trimethyl phosphite-*P*)triosmium, Os-00172
Nonacarbonyldinitrosyltriosmium, Os-00084
Nonacarbonyldiosmium, *see* Os-00086
Nonacarbonyl[μ_4-[(1-η:1,2-η:1,2-η:2-η)-1,2-diphenyl-1,2-ethenediyl]](tricarbonyliron)triruthenium, Ru-00628
Nonacarbonyl[μ_3-(diphenylphosphino)]-μ-hydrotriruthenium, Ru-00539
Nonacarbonyldiruthenium, *see* Ru-00116
Nonacarbonyldi-μ_3-selenoxotriosmium, Os-00088
Nonacarbonyldi-μ_3-selenoxotriruthenium, Ru-00118
Nonacarbonyldi-μ_3-telluroxotriosmium, Os-00089
Nonacarbonyldi-μ_3-thioxotriosmium, Os-00087
Nonacarbonyldi-μ_3-thioxotriruthenium, Ru-00117
Nonacarbonyl(η^2-ethene)-μ-hydro[μ-(methanethiolato)]triosmium, Os-00171

Nonacarbonyl[μ_3-[(1-η:1-η:1,2-η)ethenylidene]]di-μ-hydrotriosmium, Os-00126
Nonacarbonyl[μ-[(1-η:1-η:1,2-η)-ethenylidene]]di-μ-hydrotriruthenium, Ru-00170
Nonacarbonyl-μ_3-ethylidynetri-μ-hydrotriosmium, Os-00129
Nonacarbonyl-μ_3-ethylidynetri-μ-hydrotriruthenium, Ru-00173
Nonacarbonyl[μ_3-[(1-η:1,2-η:1,2-η)ethynyl]]-μ-hydrotriosmium, Os-00122
Nonacarbonyl-μ-hydro[μ_3-(methanethiolato)]triosmium, Os-00100
Nonacarbonyl-μ-hydro-μ_4-nitrido(tricarbonyliron)triruthenium, Ru-00216
Nonacarbonyl-μ-hydro-μ_3-thioxotriosmate(1−), Os-00069
Nonacarbonyl[μ_3-[*N*-phenyl-*o*-phenylenediaminato(2−)]triruthenium, Ru-00538
Nonacarbonyl(μ^3-*o*-semidine)triruthenium, *see* Ru-00538
Nonacarbonyltetra-μ-hydrotris(phosphorous acid)tetraruthenium nonamethyl ester, *see* Ru-00495
Nonacarbonyltetra-μ-hydrotris(trimethyl phosphite-*P*)tetraruthenium, Ru-00495
Nonacarbonyltetra-μ-hydrotris(triphenylphosphine)tetraruthenium, Ru-00856
Nonacarbonyl[μ-[(1,2,3,4-η:1,4-η)-1,2,3,4-tetraphenyl-1,3-butadiene-1,4-diyl]]triosmium, Os-00368
Nonacarbonyl[μ-(1,2,3,4-tetraphenyl-1,3-butadienylene)]triosmium, *see* Os-00368
Nonacarbonyltri-μ-hydro[μ_3-(methoxymethylidyne)]triosmium, Os-00130
Nonacarbonyltri-μ-hydro[μ_3-(methoxymethylidyne)]triruthenium, Ru-00174
Nonacarbonyltri-μ-hydro[μ_3-(2-methoxy-2-oxoethylidyne)]triosmium, Os-00170
Nonacarbonyltri-μ-hydro[μ_3-(2-methoxy-2-oxoethylidyne)]triruthenium, Ru-00227
Nonacarbonyltri-μ-hydro-μ_3-methylidynetriosmium, Os-00099
Nonacarbonyltri-μ-hydro-μ_3-methylidynetriruthenium, Ru-00124
Nonacarbonyltri-μ-hydro-μ_3-thioxotriruthenium(1+), Ru-00092
Nonacarbonyl[μ_3-(4,6,8-trimethylazulene)]tetraruthenium, Ru-00559
Nonacarbonyl[tris(dibutylphosphino)methylsilane]triruthenium, *see* Ru-00682
Nonacarbonyltris[μ-(dimethylgermylene)]triosmium, Os-00263
Nonacarbonyltris[μ-(dimethylgermylene)]triruthenium, Ru-00399
Nonacarbonyltris[μ-(dimethylsilylene)]triosmium, Os-00264
Nonacarbonyltris(μ-hydrido)(μ-chloromethylidyne)triruthenium, *see* Ru-00123
Nonacarbonyltris(μ-hydrido)(μ-methylidyne)triruthenium, *see* Ru-00124
Nonacarbonyltris(trimethyl phosphite-*P*)triruthenium, Ru-00493
Nonacarbonyltris(triphenylphosphine)triosmium, Os-00444
Nonacarbonyltris(triphenylphosphine)triruthenium, Ru-00855
Nonadecacarbonylpentaosmium, Os-00309
[η^6-Nonahydrodicarbanonaborato(2−)-*B*,*H*]-bis(triethylphosphine)ruthenium, Ru-00379
[(4,5,6,7,8,9-η)-Nonahydro-6,8-dicarbanonaborato(4−)]-bis(triphenylphosphine)ruthenium, Ru-00734
Norbornadienerutheniumcarbonyl dichloride, *in* Ru-00428
Octacarbonylbis(η^5-2,4-cyclopentadien-1-yl)triruthenium, Ru-00475
Octacarbonylbis[μ-(dimethylstannylene)]diruthenium, Ru-00248
Octacarbonylbis(diphenylacetylene)triruthenium, *see* Ru-00690
Octacarbonylbis[μ_3-(diphenylvinylene)]triruthenium, *see* Ru-00690
Octacarbonylbis[μ-(η^4-*N*,*N*′-1,2-ethanediylidenebis[2-propanamine]-*N*,*N*′]]tetraruthenium, Ru-00610
Octacarbonylbis[1,1′-(1,2-ethynediyl)bis[benzene]]triruthenium, Ru-00690
Octacarbonylbis[glyoxalbis(isopropylamine)]tetraruthenium, *see* Ru-00610
Octacarbonylbis(trimethylgermyl)diruthenium, Ru-00360
Octacarbonylbis(trimethylsilyl)diosmium, Os-00243
Octacarbonylbis(trimethylsilyl)diruthenium, Ru-00363
Octacarbonylbis(trimethylstannyl)diruthenium, Ru-00364
Octacarbonyl[μ_3-[2,4-di-*tert*-butyl-1-(3,3-dimethyl-1-butenyl)-1,3-butadiene-1,4-diyl]]triruthenium, *see* Ru-00638
Octacarbonyldichlorodiosmium, Os-00054
Octacarbonyldihydrodiosmium, Os-00056
Octacarbonyldi-μ-hydrohydro-μ_3-thioxotriosmate(1−), Os-00057
Octacarbonyldiiododiosmium, Os-00068
Octacarbonyldinitrosyl(trimethyl phosphite)triosmium, Os-00133
Octacarbonyl[(1,2,3,3*a*,6*a*-η;3*a*,4,5,6,6*a*-η)pentalene]triruthenium, Ru-00418
Octacarbonyl[μ_3-[(1,2-η:2,3-η:3*a*,4,5,6,6*a*-η)-pentalene]]triruthenium, Ru-00419
Octacarbonyl[μ_3-[η^2:η^2:η^5-1,3,5-pentalenetriyltris[trimethylsilane]]]triruthenium, *see* Ru-00626
Octacarbonyl[μ-[[(1,2,3,3*a*,6*a*-η:3*a*,4,5,6,6*a*-η)-1,3,5-pentalenetriyl]tris[trimethylsilane]]]triruthenium, *see* Ru-00627
Octacarbonyl[μ-(tetracarbonylferrio)]diruthenium, *see* Ru-00214
Octacarbonyl(tetracarbonyliron)diruthenium, Ru-00214
Octacarbonyl[μ-(tetracarbonylruthenio)]diosmium, *see* Os-00200
Octacarbonyl(tetracarbonylruthenium)diosmium, Os-00200
Octacarbonyltetra-μ-hydrotetrakis(phosphorous acid)tetraruthenium dodecamethyl ester, *see* Ru-00533
Octacarbonyltetra-μ-hydrotetrakis(trimethyl phosphite-*P*)tetraruthenium, Ru-00533
Octacarbonyltetra-μ-hydrotetrakis(triphenylphosphine)tetraruthenium, Ru-00865
Octacarbonyl(1,2,3,4-tetraphenyl-1,3-butadiene-1,4-diyl)triruthenium, *see* Ru-00689
Octacarbonyl[μ-[(1,2,3,4-η:1,4-η)-1,2,3,4-tetraphenyl-2-butene-1,4-diylidene]]triosmium, Os-00357
Octacarbonyltrihydro-μ_3-thioxotrisomate(1−), *see* Os-00057
Octacarbonyl[μ_3-[η^2:η^2:η^5-1,3,5-tris(trimethylsilyl)-pentalene]]triruthenium, Ru-00626
Octacarbonyl[(1,2,3,3*a*,6*a*-η:3*a*,4,5,6,6*a*-η)-1,3,5-tris(trimethylsilyl)]pentalenetriruthenium, Ru-00627
1,1′,2,2′,3,3′,4,4′-Octachloro-5,5′-diiodoruthenocene, Ru-00119
Octadecacarbonyldihydrohexaosmium, Os-00302
Octadecacarbonyldi-μ_3-hydrohexaruthenium, Ru-00469
Octadecacarbonylhexaosmate(2−), Os-00307
Octadecacarbonylhexaosmium, Os-00308
Octadecacarbonylhexaruthenate(2−), *see* Ru-00501
Octadecacarbonyl-μ_3-hydrohexaosmate(1−), Os-00301
Octadecacarbonyl-μ_6-hydrohexaruthenate(1−), Ru-00468
Octadecacarbonyl-μ-hydro-μ_3-hydrohexaosmium, *see* Os-00302
Octaosmium tricosacarbonyl, *see* Os-00334
Osmium carbonyl, *see* Os-00027
Osmium carbonyl[$Os_5(CO)_{16}$], *see* Os-00287
Osmium carbonyl[$Os_5(CO)_{19}$], *see* Os-00309
Osmium carbonyl[$Os_6(CO)_{18}$], *see* Os-00308
Osmium carbonyl $Os_7(CO)_{21}$, *see* Os-00325
Osmium carbonyl [$Os_8(CO)_{23}$], *see* Os-00334
Osmium dicarbonyl diiodide, *see* Os-00008
Osmium dirhenium tetradecacarbonyl, *see* Os-00247
Osmium pentacarbonyl, *see* Os-00027
Osmocene, Os-00109
Osmocenecarboxylic acid, Os-00136
1,1′-Osmocenedicarboxylic acid, Os-00174
Osmocenium, Os-00110
(Osmocenylcarbonyl)ferrocene, Os-00322
α-Osmocenylethanol, *see* Os-00182
Osmocenylmethylferrocene, *see* Os-00323
Osmocenyl phenyl ketone, *see* Os-00295
Osmocenylphenylmethanone, *see* Os-00295
1,1′-(1-Oxo-1,4-butanediyl)ruthenocene, *see* Ru-00344
1,1′-(2-Oxo-1,4-butanediyl)ruthenocene, *see* Ru-00345
1,1″-(3-Oxo-1,4-pentadien-1,5-diyl)bisruthenocene, *see* Ru-00619

1-Oxo-3-phenyl-2-propenylruthenocene, *see* Ru-00509
(3-Oxo-3-phenyl-1-propenyl)ruthenocene, *see* Ru-00508
1,1′-(1-Oxo-1,3-propanediyl)ruthenocene, *see* Ru-00305
1,1″-(3-Oxo-1-propene-1,3-diyl)bisruthenocene, *see* Ru-00581
γ-Oxoruthenocenebutanoic acid, Ru-00347
1,2-(α-Oxotetramethylene)ruthenocene, *see* Ru-00343
Pentaamine(η^2-ethene)ruthenium(2+), Ru-00014
Pentaammine(η^2-ethyne)ruthenium(2+), Ru-00013
Pentabromocarbonylosmate(2−), Os-00001
Pentabromocarbonylosmate(3−), Os-00002
Pentabromocarbonylruthenate(2−), Ru-00002
Pentacarbonyl-μ-2,4,6-cycloheptatrien-1-yl(trimethylgermyl)diosmium, Os-00261
Pentacarbonyl-μ-2,4,6-cycloheptatrien-1-yl(trimethylsilyl)diosmium, Os-00262
Pentacarbonyl[μ-[(η^3:η^5)-3,5-cyclohexadien-1-yl-2-ylidene(phenylmethylidyne)-2,4-cyclopentadien-1-ylidene]]-diruthenium, Ru-00576
Pentacarbonyl(1,3,5-cyclooctatrienyl)(η^5-cyclopentadienyl)triruthenium, *see* Ru-00506
Pentacarbonyl(1,3,5-cyclooctatrienyl)-(tetrahydropentalenyl)triruthenium, *see* Ru-00565
Pentacarbonyl[μ-[η^3:η^5-2,4-cyclopentadien-1-ylidenephenyl)-methylbenzene]diruthenium, *see* Ru-00576
Pentacarbonyl(ethoxyruthenocenylmethylene)chromium, Ru-00478
Pentacarbonyl(ethoxyruthenocenylmethylene)molybdenum, Ru-00479
Pentacarbonyl(ethoxyruthenocenylmethylene)tungsten, Ru-00481
Pentacarbonylhydroosmium(1+), Os-00023
Pentacarbonylosmium, Os-00027
Pentacarbonylruthenium, Ru-00031
1,2,3,4,5-Pentacarboxyruthenocene, *see* Ru-00389
Pentachlorocarbonylosmate(2−), *see* Os-00003
Pentachloronitrosylruthenate(2−), Ru-00866
Pentachloro(triphenylphosphine)osmate(1−), Os-00305
Pentadecacarbonyldi-μ-hydroiodopentaosmate(1−), Os-00249
Pentadecacarbonyldi-μ-hydropentaosmium, Os-00250
Pentadecacarbonyldi-μ-hydro(trimethyl phosphite-*P*)pentaosmium, Os-00304
Pentadecacarbonyl-μ-ethenylidenedi-μ-hydropentaruthenium, Ru-00449
Pentadecacarbonyl[μ_4-(ethylphosphoranetetrayl)]pentaruthenium, Ru-00450
Pentadecacarbonyl-μ-hydropentaosmate(1−), Os-00248
Pentadecacarbonyliodopentaosmate(1−), Os-00269
Pentadecacarbonyl[μ_4-[methyl phosphinito(2−)-*P*:*P*:*P*:*P*]]pentaosmium, Os-00274
Pentadecacarbonyl[μ_4-(methylphosphoranetetrayl)]pentaruthenium, Ru-00415
Pentadecacarbonylpentaosmate(2−), Os-00270
Pentadecacarbonyl[μ_4-(phenylphosphoranetetrayl)]pentaruthenium, Ru-00535
σ-Pentafluorophenyl-π-cyclopentadienylbis(triphenylphosphine)ruthenium, *see* Ru-00815
σ-(Pentafluorophenyl)-π-cyclopentadienylruthenium dicarbonyl, *see* Ru-00292
σ-(Pentafluoro-1-propenyl)-π-cyclopentadienylruthenium dicarbonyl, *see* Ru-00126
1,2,3,4,5-Pentakis(methoxycarbonyl)ruthenocene, *in* Ru-00389
Pentakis(phosphorous trifluoride)osmium, Os-00449
▷Pentakis(phosphorous trifluoride)ruthenium, Ru-00867
Pentakis(trimethyl phosphite-*P*)osmium, Os-00268
Pentakis(trimethyl phosphite-*P*)ruthenium, Ru-00411
Pentammine(ethylene)ruthenium(2+), *see* Ru-00014
Pentaosmium hexadecacarbonyl, *see* Os-00287
Pentaosmium nonadecacarbonyl, *see* Os-00309
Perchlororuthenocene, *see* Ru-00120
σ-Phenyl-π-cyclopentadienylruthenium dicarbonyl, *see* Ru-00299
Phenylmethylosmocene, *see* Os-00299
(Phenylmethyl)ruthenocene, *see* Ru-00460
α-Phenylosmocenemethanol, *see* Os-00297
1,1′-(3-Phenyl-1,5-pentanediyl)ruthenocene, *see* Ru-00550
Phenylruthenocene, Ru-00426
α-Phenylruthenocenemethanol, Ru-00457
3-Phenyl[5]ruthenocenophan-1,5-dione, Ru-00547
3-Phenyl[5]ruthenocenophane, Ru-00550
Phenyl ruthenocenyl ketone, *see* Ru-00454
Phenylruthenocenylmethanone, *see* Ru-00454
1-Phenyl-3-ruthenocenyl-2-propen-1-one, Ru-00508
3-Phenyl-1-ruthenocenyl-2-propen-1-one, Ru-00509
1,1′-(1,3-Propanediyl)ruthenocene, *see* Ru-00313
Propanoylruthenocene, Ru-00311
(η^3-2-Propenyl)(η^5-cyclopentadienyl)carbonylruthenium, *see* Ru-00105
1,1′-Quaterruthenocene, *see* Ru-00756
1,1″:1‴,1⁗:1‴″,1‴‴-Quaterruthenocene, Ru-00756
$Ru_2(CO)_4$(IAE), *see* Ru-00531
$Ru_2(CO)_5$(IAE), *see* Ru-00551
Ruthenium carbonyl [$Ru_3(CO)_{12}$], *see* Ru-00280
Ruthenium pentacarbonyl, *see* Ru-00031
Ruthenocene, Ru-00149
1,1′-Ruthenocenebis(thioic acid), Ru-00231
Ruthenoceneboronic acid, Ru-00151
Ruthenocenebutanoic acid, Ru-00352
Ruthenocenecarboxaldehyde, Ru-00184
Ruthenocenecarboxamide, *in* Ru-00185
Ruthenocenecarboxylic acid, Ru-00185
1,1′-Ruthenocenedicarboxaldehyde, Ru-00230
1,1′-Ruthenocenedicarboxylic acid, Ru-00232
1,1′-Ruthenocenedimethanol, *see* Ru-00258
μ-1,1′-Ruthenocenediyldilithium, *see* Ru-00132
Ruthenocenemethanol, Ru-00191
Ruthenocenenitrile, *in* Ru-00185
1,2,3,4,5-Ruthenocenepentacarboxylic acid, Ru-00389
Ruthenocenesulfonic acid, Ru-00146
Ruthenocenesulfonyl chloride, *in* Ru-00146
Ruthenocenium, Ru-00150
2,3-Ruthenocenoindenone, *see* Ru-00453
Ruthenocenol, *see* Ru-00142
[3]Ruthenocenophane, Ru-00313
[4]Ruthenocenophane, Ru-00357
[3.3](1,1′)Ruthenocenophane-2,14-diene-1,16-dione, *see* Ru-00630
[5.5](1,1′)Ruthenocenophane-2,14,17,29-tetraene-1,16-dione, *see* Ru-00659
[3]Ruthenocenophan-1-one, Ru-00305
[4]-Ruthenocenophan-1-one, Ru-00344
[4]Ruthenocenophan-2-one, Ru-00345
Ruthenocenylacetylene, *see* Ru-00233
3-Ruthenocenylacrylic acid, *see* Ru-00307
Ruthenocenylcarbonylferrocene, *see* Ru-00546
β-Ruthenocenyl-β-chloroacrolein, *see* Ru-00302
1-Ruthenocenylethanol, *see* Ru-00256
1-Ruthenocenylethanone, *see* Ru-00240
(Ruthenocenylethynyl)ferrocene, Ru-00564
Ruthenocenyllithium, *see* Ru-00138
(Ruthenocenylmethyl)ferrocene, *see* Ru-00549
3-Ruthenocenylpropanoic acid, Ru-00312
2-Ruthenocenyl-2-propanol, *see* Ru-00316
1-Ruthenocenylpropanone, *see* Ru-00311
3-Ruthenocenyl-2-propenoic acid, Ru-00307
(*o*-Semidine)hexacarbonyldiruthenium, *see* Ru-00472
Sulforuthenocene, *see* Ru-00146
1,1′-Terruthenocene, *see* Ru-00661
1,1″:1‴,1⁗-Terruthenocene, Ru-00661
Tetrabromodicarbonylosmate(1−), Os-00005
Tetrabromodicarbonylosmate(2−), Os-00006
Tetrabromodicarbonylruthenate(2−), Ru-00009
Tetrabromohexacarbonyldiosmium, *see* Os-00029
Tetrabromohexacarbonyldiruthenium, *see* Ru-00032
Tetracarbonyl(μ^2-acetylene)[glyoxalbis(isopropylimine)]diruthenium, *see* Ru-00361
Tetracarbonylbis(chloromercurio)osmium, Os-00016
Tetracarbonylbis(chloromercurio)ruthenium, Ru-00023
Tetracarbonylbis[μ-(1,3,5,7-cyclooctatetraene)]triruthenium, Ru-00519

Tetracarbonylbis(η^5-2,4-cyclopentadien-1-yl)diosmium, Os-00233
Tetracarbonylbis(η^5-2,4-cyclopentadien-1-yl)(mercury)diruthenium, Ru-00332
Tetracarbonylbis[dicarbonyl(η-cyclopentadienyl)ruthenio]ruthenium, *see* Ru-00475
Tetracarbonylbis(diphenylacetylene)ruthenium, *see* Ru-00672
Tetracarbonylbis[μ-(η^2-*N,N'*-1,2-ethanediylidenebis[2-propanamine]-*N,N'*]]diruthenium, Ru-00532
Tetracarbonylbis[glyoxalbis(isopropylimine)]diruthenium, *see* Ru-00532
Tetracarbonylbis(methyl-π-cyclopentadienyl)diruthenium, *see* Ru-00425
Tetracarbonylbis(pentacarbonylmanganio)osmium, Os-00246
Tetracarbonylbis(pentacarbonylmanganio)ruthenium, Ru-00380
Tetracarbonylbis(pentacarbonylrhenio)osmium, Os-00247
Tetracarbonylbis(η^5-tetrahydroindenyl)diruthenium, *see* Ru-00568
Tetracarbonylbis(trichlorogermyl)ruthenium, Ru-00025
Tetracarbonylbis(trimethylgermyl)osmium, Os-00114
Tetracarbonylbis(trimethylgermyl)ruthenium, Ru-00162
Tetracarbonylbis(trimethylplumbyl)osmium, Os-00115
Tetracarbonylbis(trimethylplumbyl)ruthenium, Ru-00163
Tetracarbonylbis(trimethylsilyl)osmium, Os-00116
Tetracarbonylbis(trimethylsilyl)ruthenium, Ru-00164
Tetracarbonylbis(trimethylstannyl)osmium, Os-00117
Tetracarbonylbis(trimethylstannyl)ruthenium, Ru-00165
Tetracarbonylbis[(triphenylphosphine)gold]osmium, Os-00402
Tetracarbonylbis(triphenylphosphine)osmium(2+), Os-00403
Tetracarbonylbis(triphenylstannyl)osmium, Os-00404
Tetracarbonylbis(triphenylstannyl)ruthenium, Ru-00753
Tetracarbonylchloro(ruthenocenomethylidyne)tungsten, Ru-00386
Tetracarbonyl(cyanoethylene)ruthenium, *see* Ru-00044
Tetracarbonyl[μ-(1,2,3,4-η:1,5,6,7-η)-2,4,6-cycloheptatrien-1-yl][μ-(2-methyl-2-propanethiolato)]diruthenium, Ru-00396
Tetracarbonyl(di-μ-carbonylhexacarbonyldiferrio)ruthenium, *see* Ru-00215
Tetracarbonyl(di-μ-carbonylhexacarbonyldiiron)osmium, Os-00152
Tetracarbonyl(di-μ-carbonylhexacarbonyldiiron)ruthenium, Ru-00215
Tetracarbonyldi-μ-chlorobis[2-(phenylazo)phenyl-*C,N'*]diruthenium, *in* Ru-00330
Tetracarbonyldi-μ-chlorodichlorobis(triphenylphosphine)diruthenium, Ru-00752
Tetracarbonyldichloroosmium, Os-00017
Tetracarbonyldichlororuthenium, Ru-00024
Tetracarbonyldi-π-cyclopentadienyldiosmium, *see* Os-00233
Tetracarbonyldi-π-cyclopentadienyl-μ-mercuriodiruthenium, *see* Ru-00332
Tetracarbonyldiethylosmium, Os-00062
Tetracarbonyldihydroosmium, Os-00019
Tetracarbonyldihydroruthenium, Ru-00027
Tetracarbonyldi-π-indenyldiruthenium, *see* Ru-00558
Tetracarbonyldiiodoosmium, Os-00021
Tetracarbonyldiiodoruthenium, Ru-00028
Tetracarbonyldimethylosmium, Os-00038
Tetracarbonyl[3,4-dimethyl-2(3*H*)-thiazolylidene]osmium, Os-00072
Tetracarbonyldipropionylosmium, Os-00108
Tetracarbonyl[1,2-ethanediylbis[dimethylsilylene]]osmium, Os-00112
Tetracarbonyl[1,2-ethanediylbis[dimethylsilylene]]ruthenium, Ru-00160
Tetracarbonyl[μ-[η^4-*N,N'*-1,2-ethanediylidenebis[2-propanamine]-*N,N'*]]-μ-1,2-ethenediyldiruthenium, Ru-00361
Tetracarbonyl(η^2-ethene)osmium, Os-00032
Tetracarbonyl(ethylene)osmium, *see* Os-00032
Tetracarbonyl[(3,4-η)-2,5-furandione]ruthenium, *in* Ru-00066
Tetracarbonylhydromethylosmium, Os-00026
Tetracarbonylhydroosmate(1−), Os-00018
Tetracarbonylhydro(trimethylgermyl)osmium, Os-00051
Tetracarbonylhydro(trimethylsilyl)osmium, Os-00052
Tetracarbonyliodomethylosmium, Os-00025
Tetracarbonyliodo(trimethylgermyl)osmium, Os-00049
Tetracarbonyliodo(trimethylgermyl)ruthenium, Ru-00062
Tetracarbonyliodo(trimethylsilyl)osmium, Os-00050
Tetracarbonyliodo(trimethylsilyl)ruthenium, Ru-00063
Tetracarbonyliodo(trimethylstannyl)ruthenium, Ru-00064
Tetracarbonyl(methyl acrylate)ruthenium, *see* Ru-00076
Tetracarbonyl[(2,3-η)-methyl 2-propenoate]ruthenium, Ru-00076
Tetracarbonyl(octacarbonyldiruthenio)iron, *see* Ru-00214
Tetracarbonyl(octacarbonyldiruthenio)osmium, *see* Os-00181
Tetracarbonyl(octacarbonyldiruthenium)osmium, Os-00181
Tetracarbonylosmate(2−), Os-00022
Tetracarbonyl[μ-[(1,2,3,3*a*,6*a*-η:3*a*,4,5,6,6*a*-η)-pentalene]]-bis(trimethylgermyl)diruthenium, Ru-00487
Tetracarbonyl[1,2-phenylenebis[dimethylsilylene]]ruthenium, Ru-00354
Tetracarbonyl(phosphorous trifluoride)ruthenium, Ru-00026
Tetracarbonyl[(2,3-η)-2-propenenitrile]ruthenium, Ru-00044
Tetracarbonylruthenate(2−), Ru-00029
Tetracarbonyltetrachlorobis(triphenylphosphine)diruthenium, *see* Ru-00752
Tetra-μ_3-carbonyltetrakis(η^5-2,4-cyclopentadien-1-yl)tetraruthenium, Ru-00597
Tetracarbonyl(trimethylgermyl)[(triphenylphosphine)aurio]ruth enium, Ru-00622
Tetracarbonyl(trimethylsilyl)osmium, *in* Os-00243
Tetracarbonyl(trimethylsilyl)[(triphenylphosphine)aurio]ruthenium, Ru-00623
Tetracarbonyl(triphenylphosphine)osmium, Os-00326
Tetracarbonyl(triphenylphosphine)ruthenium, Ru-00560
Tetracarbonyl(triphenylstibine)ruthenium, Ru-00561
Tetrachlorobis(1,3,5-cycloheptatriene)diruthenium, *in* Ru-00059
Tetrachlorobis(2,7-dimethyl-2,6-octadiene-1,8-diyl)diruthenium, *see* Ru-00530
Tetrachlorobis(η^6-hexamethylbenzene)diruthenium, *see* Ru-00612
Tetrachlorobis(triphenylphosphine)diruthenium, *in* Ru-00698
Tetrachlorobis(triphenylphosphine)osmate(1−), Os-00364
Tetrachlorobis(triphenylphosphine)osmium, Os-00365
Tetrachlorobis(triphenylphosphine)ruthenate(1−), Ru-00702
Tetrachlorodicarbonylbis(1,5-cyclooctadiene)diruthenium, *see* Ru-00486
Tetrachlorodicarbonylbis(2,5-norbornadiene)diruthenium, *see* Ru-00428
Tetra-μ-chlorotetrakis[μ-(diphenylphosphino)]tetranitrosyltetraruthenium, Ru-00816
Tetrachloro(thiocarbonyl)tetrakis(triphenylphosphine)diruthenium, *see* Ru-00862
Tetrachloro(thiocarbonyl)tetrakis(triphenylphosphine)diruthenium, *see* Ru-00864
Tetracycloneruthenium tricarbonyl, *see* Ru-00672
Tetradecacarbonyldimanganeseosmium, *see* Os-00246
Tetradecacarbonyldimanganeseruthenium, *see* Ru-00380
Tetradecacarbonylosmiumdirhenium, *see* Os-00247
Tetrahydrohexakis(trimethylphosphine)diruthenium, *see* Ru-00499
Tetrahydro-π-pentalenyldicarbonyl(trimethylsilyl)ruthenium, *see* Ru-00321
Tetrahydrotris(triphenylphosphine)osmium, Os-00437
Tetrahydrotris(triphenylphosphine)ruthenium, Ru-00839
Tetrakis(acetonitrile)[(2,3,5,6-η)bicyclo[2.2.1]hepta-2,5-diene]ruthenium(2+), Ru-00403
Tetrakis(acetonitrile)[(1,2,5,6-η)-1,5-cyclooctadiene]ruthenium(2+), Ru-00436
Tetrakis(acetonitrile)norbornadieneruthenium, *see* Ru-00403
Tetrakis(phosphorous trifluoride)osmate(2−), Os-00448
Tetrakis(phosphorous trifluoride)ruthenate(2−), Ru-00870
[η^4-1,2,3,4-Tetramethyl-5,6-bis(methylene)-1,3-cyclohexadiene]tris(trimethyl phosphite-*P*)ruthenium, Ru-00553
Tetra-μ-methyleneoctakis(trimethylphosphine)triruthenium(2+), Ru-00651
1,1,3,3-Tetramethyl-2,2,2,2-tetracarbonyl-1,3-dihydrobenzo[*d*][2,1,3]ruthenadisilole, *see* Ru-00354
(η^4-Tetraphenylcyclobutadiene)tris(trimethyl phosphite-*P*)ruthenium, Ru-00714

Molecular Formula Index

CBr_3RuO
Tribromocarbonylruthenium, Ru-00001
CBr_5Cs_2ORu
Pentabromocarbonylruthenate(2−); Di-Cs salt, *in* Ru-00002
CBr_5Cs_3OOs
Pentabromocarbonylosmate(3−); Tri-Cs salt, *in* Os-00002
CBr_5K_2OOs
Pentabromocarbonylosmate(2−); Di-K salt, *in* Os-00001
$CBr_5OOs^{\ominus\ominus}$
Pentabromocarbonylosmate(2−), Os-00001
Pentabromocarbonylosmate(3−), Os-00002
$CBr_5ORu^{\ominus\ominus}$
Pentabromocarbonylruthenate(2−), Ru-00002
CCl_3ORu
Carbonyltrichlororuthenium, Ru-00003
CCl_5Cs_2OOs
Carbonylpentachloroosmate(2−); Di-Cs salt, *in* Os-00003
CCl_5Cs_2ORu
Carbonylpentachlororuthenate(2−); Di-Cs salt, *in* Ru-00004
CCl_5Cs_3OOs
Carbonylpentachloroosmate(3−); Tri-Cs salt, *in* Os-00004
CCl_5K_2ORu
Carbonylpentachlororuthenate(2−); Di-K salt, *in* Ru-00004
$CCl_5OOs^{\ominus\ominus}$
Carbonylpentachloroosmate(2−), Os-00003
Carbonylpentachloroosmate(3−), Os-00004
$CCl_5ORu^{\ominus\ominus}$
Carbonylpentachlororuthenate(2−), Ru-00004
$CF_{12}OP_4Ru$
Carbonyltetrakis(phosphorous trifluoride)ruthenium, Ru-00005
$CH_2Cl_4Cs_2O_2Ru$
Aquacarbonyltetrachlororuthenate(2−); Di-Cs salt, *in* Ru-00006
$CH_2Cl_4O_2Ru^{\ominus\ominus}$
Aquacarbonyltetrachlororuthenate(2−), Ru-00006
$CH_{10}Cl_4N_2O_2Ru$
Aquacarbonyltetrachlororuthenate(2−); Di-NH_4 salt, *in* Ru-00006
CI_3ORu
Carbonyltriiodoruthenium, Ru-00007
$C_2Br_2O_2Ru$
Dibromodicarbonylruthenium, Ru-00008
$C_2Br_4Cs_2O_2OS$
Tetrabromodicarbonylosmate(2−); Di-Cs salt, *in* Os-00006
$C_2Br_4Cs_2O_2Ru$
Tetrabromodicarbonylruthenate(2−); Di-Cs salt, *in* Ru-00009
$C_2Br_4O_2Os^{\ominus}$
Tetrabromodicarbonylosmate(1−), Os-00005
Tetrabromodicarbonylosmate(2−), Os-00006
$C_2Br_4O_2Ru^{\ominus\ominus}$
Tetrabromodicarbonylruthenate(2−), Ru-00009
$C_2Cl_2O_2Ru$
Dicarbonyldichlororuthenium, Ru-00010
$C_2Cl_4Cs_2O_2Os$
Dicarbonyltetrachloroosmate(2−); Di-Cs salt, *in* Os-00007
$C_2Cl_4Cs_2O_2Ru$
Dicarbonyltetrachlororuthenate(2−); Di-Cs salt, *in* Ru-00011
$C_2Cl_4O_2Os^{\ominus\ominus}$
Dicarbonyltetrachloroosmate(2−), Os-00007
$C_2Cl_4O_2Ru^{\ominus\ominus}$
Dicarbonyltetrachlororuthenate(2−), Ru-00011
$C_2Cs_2I_4O_2Os$
Dicarbonyltetraiodoosmate(2−); Di-Cs salt, *in* Os-00010
$C_2Cs_2I_4O_2Ru$
Dicarbonyltetraiodoruthenate(2−); Di-Cs salt, *in* Ru-00016
$C_2F_9O_2P_3Ru$
Dicarbonyltris(phosphorous trifluoride)ruthenium, Ru-00012
$C_2H_{17}N_5Ru^{\oplus\oplus}$
Pentaammine(η^2-ethyne)ruthenium(2+), Ru-00013
$C_2H_{19}N_5Ru^{\oplus\oplus}$
Pentaamine(η^2-ethene)ruthenium(2+), Ru-00014
$C_2I_2O_2Os$
Dicarbonyldiiodoosmium, Os-00008
$C_2I_2O_2Ru$
Dicarbonyldiiodoruthenium, Ru-00015
$C_2I_4O_2Os^{\ominus}$
Dicarbonyltetraiodoosmate(1−), Os-00009
Dicarbonyltetraiodoosmate(2−), Os-00010
$C_2I_4O_2Ru^{\ominus\ominus}$
Dicarbonyltetraiodoruthenate(2−), Ru-00016
$C_2N_2O_4Os$
Dicarbonyldinitrosylosmium, Os-00011
$C_3Br_2O_3Os$
Dibromotricarbonylosmium, *in* Os-00029
$C_3Br_2O_3Ru$
Dibromotricarbonylruthenium, *in* Ru-00032
$C_3Br_3CsO_3Os$
Tribromotricarbonylosmate(1−); Cs salt, *in* Os-00012
$C_3Br_3CsO_3Ru$
Tribromotricarbonylruthenate(1−); Cs salt, *in* Ru-00017
$C_3Br_3O_3Os^{\ominus}$
Tribromotricarbonylosmate(1−), Os-00012
$C_3Br_3O_3Ru^{\ominus}$
Tribromotricarbonylruthenate(1−), Ru-00017
$C_3Cl_2O_3Os$
Tricarbonydichloroosmium, *in* Os-00031
$C_3Cl_2O_3Ru$
Tricarbonyldichlororuthenium, *in* Ru-00033
$C_3Cl_3CsO_3Ru$
Tricarbonyltrichlororuthenate(1−); Cs salt, *in* Ru-00018
$C_3Cl_3O_3Os^{\ominus}$
Tricarbonyltrichloroosmate(1−), Os-00013
$C_3Cl_3O_3Ru^{\ominus}$
Tricarbonyltrichlororuthenate(1−), Ru-00018
$C_3CsI_3O_3Ru$
Tricarbonyltriiodoruthenate(1−); Cs salt, *in* Ru-00021
$C_3F_2O_3Ru$
Tricarbonyldifluororuthenium, *in* Ru-00212
$C_3F_3O_3Ru$
Tricarbonyltrifluororuthenium, Ru-00019
$C_3F_6O_3P_2Ru$
Tricarbonylbis(phosphorous trifluoride)ruthenium, Ru-00020
$C_3I_2O_3Ru$
Tricarbonyldiiodoruthenium, *in* Ru-00043
$C_3I_3O_3Os^{\ominus}$
Tricarbonyltriiodoosmate(1−), Os-00014
$C_3I_3O_3Ru^{\ominus}$
Tricarbonyltriiodoruthenate(1−), Ru-00021

$C_4Br_2O_4Os$
Dibromotetracarbonylosmium, Os-00015
$C_4Br_2O_4Ru$
Dibromotetracarbonylruthenium, Ru-00022
$C_4Cl_2Hg_2O_4Os$
Tetracarbonylbis(chloromercurio)osmium, Os-00016
$C_4Cl_2Hg_2O_4Ru$
Tetracarbonylbis(chloromercurio)ruthenium, Ru-00023
$C_4Cl_2O_4Os$
Tetracarbonyldichloroosmium, Os-00017
$C_4Cl_2O_4Ru$
Tetracarbonyldichlororuthenium, Ru-00024
$C_4Cl_6Ge_2O_4Ru$
Tetracarbonylbis(trichlorogermyl)ruthenium, Ru-00025
$C_4F_3O_4PRu$
Tetracarbonyl(phosphorous trifluoride)ruthenium, Ru-00026
C_4HNaO_4Os
Tetracarbonylhydroosmate(1−); Na salt, *in* Os-00018
$C_4HO_4Os^{\ominus}$
Tetracarbonylhydroosmate(1−), Os-00018
$C_4H_2O_4Os$
Tetracarbonyldihydroosmium, Os-00019
$C_4H_2O_4Ru$
Tetracarbonyldihydroruthenium, Ru-00027
$C_4H_4Na_2Os$
Tetracarbonylosmate(2−); Di-Na salt, *in* Os-00022
$C_4H_9NO_3Os$
Trioxo(*tert*-butylnitrido)osmium, Os-00020
$C_4I_2O_4Os$
Tetracarbonyldiiodoosmium, Os-00021
$C_4I_2O_4Ru$
Tetracarbonyldiiodoruthenium, Ru-00028
$C_4Na_2O_4Ru$
Tetracarbonylruthenate(2−); Di-Na salt, *in* Ru-00029
$C_4O_4Os^{\ominus\ominus}$
Tetracarbonylosmate(2−), Os-00022
$C_4O_4Ru^{\ominus\ominus}$
Tetracarbonylruthenate(2−), Ru-00029
$C_5HF_6O_5OsP$
Pentacarbonylhydroosmium(1+); Hexafluorophosphate, *in* Os-00023
$C_5HO_5Os^{\oplus}$
Pentacarbonylhydroosmium(1+), Os-00023
$C_5H_3BrO_4Os$
Bromotetracarbonylmethylosmium, Os-00024
$C_5H_3IO_4Os$
Tetracarbonyliodomethylosmium, Os-00025
$C_5H_4O_4Os$
Tetracarbonylhydromethylosmium, Os-00026
$C_5H_{11}B_9O_3Ru$
Tricarbonyl[(7,8,9,10,11-η)undecahydro-7,8-dicarbaundecaborato(2−)]ruthenium, Ru-00030
C_5O_5Os
Pentacarbonylosmium, Os-00027
C_5O_5Ru
Pentacarbonylruthenium, Ru-00031
$C_6Br_2O_6Os_2$
Di-μ-bromohexacarbonyldiosmium, Os-00028
$C_6Br_4O_6Os_2$
Di-μ-bromodibromohexacarbonyldiosmium, Os-00029
$C_6Br_4O_6Ru_2$
Di-μ-bromodibromohexacarbonyldiruthenium, Ru-00032
$C_6Cl_2O_6Os_2$
Hexacarbonyldi-μ-chlorodiosmium, Os-00030
$C_6Cl_4O_6Os_2$
Hexacarbonyldi-μ-chlorodichlorodiosmium, Os-00031
$C_6Cl_4O_6Ru_2$
Hexacarbonyldi-μ-chlorodichlorodiruthenium, Ru-00033
$C_6Cl_7O_6Ru_2Sb$
Hexacarbonyldi-μ-chlorodichlorodiruthenium; Trichlorostibine adduct, *in* Ru-00033
$C_6H_4O_4Os$
Tetracarbonyl(η^2-ethene)osmium, Os-00032
$C_6H_5BrO_3Ru$
Bromotricarbonyl(η^3-2-propenyl)ruthenium, Ru-00034
$C_6H_5BrO_4Os$
Bromotetracarbonylethylosmium, Os-00033
$C_6H_5ClO_3Ru$
Tricarbonylchloro(η^3-2-propenyl)ruthenium, Ru-00035
$C_6H_5IO_3Ru$
Tricarbonyliodo(η^3-2-propenyl)ruthenium, Ru-00036
$C_6H_6Br_2Os$
η^6-Benzenedibromoosmium, Os-00034
$C_6H_6Br_2OsO_4^{\ominus\ominus}$
Diacetyldibromodicarbonylosmate(2−), Os-00035
$C_6H_6Br_2Ru$
(η^6-Benzene)dibromoruthenium, *in* Ru-00237
$C_6H_6Br_4Mg_2OsO_4$
Diacetyldibromodicarbonylosmate(2−); Bis(bromomagnesium) salt, *in* Os-00035
$C_6H_6Cl_2N_2O_2Ru$
Bis(acetonitrile)dicarbonyldichlororuthenium, Ru-00037
$C_6H_6Cl_2Os$
η^6-Benzenedichloroosmium, Os-00036
$C_6H_6Cl_2Ru$
(η^6-Benzene)dichlororuthenium, Ru-00038
$C_6H_6Cl_3CsRu$
(η^6-Benzene)trichlororuthenate(1−); Cs salt, *in* Ru-00039
$C_6H_6Cl_3Ru^{\ominus}$
(η^6-Benzene)trichlororuthenate(1−), Ru-00039
$C_6H_6I_2Os$
(η^6-Benzene)diiodoosmium, Os-00037
$C_6H_6I_2Ru$
(η^6-Benzene)diiodoruthenium, *in* Ru-00239
$C_6H_6O_4Os$
Tetracarbonyldimethylosmium, Os-00038
$C_6H_8F_9P_3Ru$
[(1,2,3,4-η)-1,3-Cyclohexadiene]tris(phosphorous trifluoride)ruthenium, Ru-00040
$C_6H_{12}O_3Ru^{\oplus\oplus}$
Triaqua(η^6-benzene)ruthenium(2+), Ru-00041
$C_6H_{15}F_{12}N_3P_2Ru$
Triammine(η^6-benzene)ruthenium(2+); Bis(hexafluorophosphate), *in* Ru-00042
$C_6H_{15}N_3Ru^{\oplus\oplus}$
Triammine(η^6-benzene)ruthenium(2+), Ru-00042
$C_6I_2O_6Os_2$
Hexacarbonyldi-μ-iododiosmium, Os-00039
$C_6I_4O_6Os_2$
Hexacarbonyldi-μ-iododiiododiosmium, Os-00040
$C_6I_4O_6Ru_2$
Hexacarbonyldi-μ-iododiiododiruthenium, Ru-00043
$C_7H_3NO_4Ru$
Tetracarbonyl[(2,3-η)-2-propenenitrile]ruthenium, Ru-00044
$C_7H_4O_3Ru$
Tricarbonyl(η^4-1,3-cyclobutadiene)ruthenium, Ru-00045
$C_7H_5BrO_2Os$
Bromodicarbonyl(η^5-2,4-cyclopentadien-1-yl)osmium, Os-00041

$C_7H_5BrO_2Ru$
Bromodicarbonyl(η^5-2,4-cyclopentadien-1-yl)ruthenium, Ru-00046

$C_7H_5ClO_2Os$
Dicarbonylchloro(η^5-2,4-cyclopentadien-1-yl)osmium, Os-00042

$C_7H_5ClO_2Ru$
Dicarbonylchloro(η^5-2,4-cyclopentadien-1-yl)ruthenium, Ru-00047

$C_7H_5Cl_3O_2RuSi$
Dicarbonyl(η^5-2,4-cyclopentadien-1-yl)(trichlorosilyl)-ruthenium, Ru-00048

$C_7H_5Cl_3O_2RuSn$
Dicarbonyl(η^5-2,4-cyclopentadien-1-yl)-(trichlorostannyl)ruthenium, Ru-00049

C_7H_5IORuS
(Carbonothioyl)carbonyl(η^5-2,4-cyclopentadien-1-yl)-iodoruthenium, Ru-00050

$C_7H_5IO_2Os$
Dicarbonyl(η^5-2,4-cyclopentadien-1-yl)iodoosmium, Os-00043

$C_7H_5IO_2Ru$
Dicarbonyl(η^5-2,4-cyclopentadien-1-yl)iodoruthenium, Ru-00051

$C_7H_5NaO_2Ru$
Dicarbonyl(η^5-2,4-cyclopentadien-1-yl)ruthenate(1−); Na salt, *in* Ru-00052

$C_7H_5O_2Ru^{\ominus}$
Dicarbonyl(η^5-2,4-cyclopentadien-1-yl)ruthenate(1−), Ru-00052

$C_7H_6Cl_2OOs$
(η^6-Benzene)carbonyldichloroosmium, Os-00044

$C_7H_6Cl_6Ge_2ORu$
(η^6-Benzene)carbonylbis(trichlorogermyl)ruthenium, Ru-00053

$C_7H_6Cl_6ORuSi_2$
(η^6-Benzene)carbonylbis(trichlorosilyl)ruthenium, Ru-00054

$C_7H_6O_2Os$
Dicarbonyl(η^5-2,4-cyclopentadien-1-yl)hydroosmium, Os-00045

$C_7H_6O_2Ru$
Dicarbonyl(η^5-2,4-cyclopentadien-1-yl)hydroruthenium, Ru-00055

$C_7H_6O_3Os$
(η^4-1,3-Butadiene)tricarbonylosmium, Os-00046

$C_7H_6O_3Ru$
(η^4-Butadiene)tricarbonylruthenium, Ru-00056

$C_7H_7ClO_3Ru$
Tricarbonylchloro[(1,2,3-η)-2-methyl-2-propenyl]-ruthenium, Ru-00057

$C_7H_8Cl_2Ru$
[(2,3,5,6-η)-Bicyclo[2.2.1]hepta-2,5-diene]-dichlororuthenium, Ru-00058
Dichloro[(1,2,3,4-η)-1,3,5-cycloheptatriene]ruthenium, Ru-00059

$C_7H_8F_6NO_2PRu$
Amminedicarbonyl(η^5-2,4-cyclopentadien-1-yl)-ruthenium(1+); Hexafluorophosphate, *in* Ru-00060

$C_7H_8NO_2Ru^{\oplus}$
Amminedicarbonyl(η^5-2,4-cyclopentadien-1-yl)-ruthenium(1+), Ru-00060

$C_7H_9BrGeO_4Os$
Bromotetracarbonyl(trimethylgermyl)osmium, Os-00047

$C_7H_9BrO_4OsSi$
Bromotetracarbonyl(trimethylsilyl)osmium, Os-00048

$C_7H_9BrO_4RuSi$
Bromotetracarbonyl(trimethylsilyl)ruthenium, Ru-00061

$C_7H_9GeIO_4Os$
Tetracarbonyliodo(trimethylgermyl)osmium, Os-00049

$C_7H_9GeIO_4Ru$
Tetracarbonyliodo(trimethylgermyl)ruthenium, Ru-00062

$C_7H_9IO_4OsSi$
Tetracarbonyliodo(trimethylsilyl)osmium, Os-00050

$C_7H_9IO_4RuSi$
Tetracarbonyliodo(trimethylsilyl)ruthenium, Ru-00063

$C_7H_9IO_4RuSn$
Tetracarbonyliodo(trimethylstannyl)ruthenium, Ru-00064

$C_7H_9O_4OsSi$
Tetracarbonyl(trimethylsilyl)osmium, *in* Os-00243

$C_7H_{10}Cl_2Ru$
Dichloro[(1,2,3,4-η)-1,3-cycloheptadiene]ruthenium, Ru-00065

$C_7H_{10}GeO_4Os$
Tetracarbonylhydro(trimethylgermyl)osmium, Os-00051

$C_7H_{10}O_4OsSi$
Tetracarbonylhydro(trimethylsilyl)osmium, Os-00052

$C_8Br_2O_8Os_2$
Dibromooctacarbonyldiosmium, Os-00053

$C_8Cl_2O_8Os_2$
Octacarbonyldichlorodiosmium, Os-00054

$C_8H_2O_7Os_3S_3$
(Carbonothioyl)heptacarbonyldi-μ-hydrodi-μ_3-thioxotriosmium, Os-00055

$C_8H_2O_7Ru$
Tetracarbonyl[(3,4-η)-2,5-furandione]ruthenium, *in* Ru-00066

$C_8H_2O_8Os_2$
Octacarbonyldihydrodiosmium, Os-00056

$C_8H_3O_8Os_3S^{\ominus}$
Octacarbonyldi-μ-hydrohydro-μ_3-thioxotriosmate(1−), Os-00057

$C_8H_4O_8Ru$
[(2,3-η)-2-Butenedioic acid]tetracarbonylruthenium, Ru-00066

$C_8H_5BF_4O_3Os$
Tricarbonyl(η^5-2,4-cyclopentadien-1-yl)osmium(1+); Tetrafluoroborate, *in* Os-00058

$C_8H_5BF_4O_3Ru$
Tricarbonyl(η^5-2,4-cyclopentadien-1-yl)ruthenium(1+); Tetrafluoroborate, *in* Ru-00072

$C_8H_5F_6O_3PRu$
Tricarbonyl(η^5-2,4-cyclopentadien-1-yl)ruthenium(1+); Hexafluorophosphate, *in* Ru-00072

$C_8H_5NO_2Ru$
Dicarbonyl(cyano-*C*)(η^5-2,4-cyclopentadien-1-yl)-ruthenium, Ru-00067

$C_8H_5NO_2RuS$
Dicarbonyl(η^5-2,4-cyclopentadien-1-yl)(thiocyanato-*N*)-ruthenium, Ru-00068
Dicarbonyl(η^5-2,4-cyclopentadien-1-yl)(thiocyanato-*S*)-ruthenium, Ru-00069

$C_8H_5NO_3Ru$
Dicarbonyl(cyanato-*N*)(η^5-2,4-cyclopentadien-1-yl)-ruthenium, Ru-00070

$C_8H_5O_2RuS^{\oplus}$
(Carbonothioyl)dicarbonyl(η^5-2,4-cyclopentadien-1-yl)-ruthenium(1+), Ru-00071

$C_8H_5O_3Os^{\oplus}$
Tricarbonyl(η^5-2,4-cyclopentadien-1-yl)osmium(1+), Os-00058

$C_8H_5O_3Ru^{\oplus}$
Tricarbonyl(η^5-2,4-cyclopentadien-1-yl)ruthenium(1+), Ru-00072

$C_8H_6N_2O_3Ru$
Tricarbonyl[(4,5,6,7-η)-1*H*-1,2-diazepine]ruthenium, Ru-00073

$C_8H_6O_3Os$
Tricarbonyl[(1,2,3,4-η)-1,3-cyclopentadiene]osmium, Os-00059

$C_8H_6O_3Ru$
Tricarbonyl[(1,2,3,4-η)-1,3-cyclopentadiene]ruthenium, Ru-00074

$C_8H_6O_5Ru$
[(3,4-η)-3-Buten-2-one]tetracarbonylruthenium, Ru-00075

$C_8H_6O_6Ru$
Tetracarbonyl[(2,3-η)-methyl 2-propenoate]ruthenium, Ru-00076

$C_8H_7NO_3Os$
(Aminocarbonyl)dicarbonyl(η^5-2,4-cyclopentadien-1-yl)-osmium, Os-00060

$C_8H_7NO_3Ru$
(Aminocarbonyl)dicarbonyl(η^5-2,4-cyclopentadien-1-yl)-ruthenium, Ru-00077

$C_8H_8Cl_2ORu$
Bicyclo[2.2.1]hepta-2,5-dienecarbonyldichlororuthenium, *in* Ru-00428

$C_8H_8Cl_3CsORu$
[(2,3,5,6-η)-Bicyclo[2.2.1]hepta-2,5-diene]-carbonyltrichlororuthenate(1−); Cs salt, *in* Ru-00078

$C_8H_8Cl_3ORu^{\ominus}$
[(2,3,5,6-η)-Bicyclo[2.2.1]hepta-2,5-diene]-carbonyltrichlororuthenate(1−), Ru-00078

$C_8H_8O_2Ru$
Dicarbonyl(η^5-2,4-cyclopentadien-1-yl)methylruthenium, Ru-00079

$C_8H_8O_2RuS$
Dicarbonyl(η^5-2,4-cyclopentadien-1-yl)(methanethiolato)-ruthenium, Ru-00080

$C_8H_8O_3Os$
Tricarbonyl[(1,2,3,4-η)-1,3-pentadiene]osmium, Os-00061

$C_8H_8O_3Ru$
Tricarbonyl[(1,2,3,4-η)-1,3-pentadiene]ruthenium, Ru-00081

$C_8H_8O_4RuS$
Dicarbonyl(η^5-2,4-cyclopentadien-1-yl)(methylsulfonyl)-ruthenium, Ru-00082

$C_8H_9Cl_2NO_5$
(Acetonitrile)(η^6-benzene)dichloroosmium, *in* Os-00036

$C_8H_{10}O_2Ru$
Dicarbonylbis(λ^3-2-propenyl)ruthenium, Ru-00083

$C_8H_{10}O_4Os$
Tetracarbonyldiethylosmium, Os-00062

$C_8H_{12}Br_2N_4Os$
Dibromotetrakis(isocyanomethane)osmium, Os-00063

$C_8H_{12}Cl_2ORuS$
(η^6-Benzene)dichloro[sulfinylbis[methane]-*O*]ruthenium, *in* Ru-00038

$C_8H_{12}Cl_2Os$
Dichloro[(1,2,5,6-η)-1,5-cyclooctadiene]osmium, Os-00064

$C_8H_{12}Cl_2Ru$
Dichloro[(1,2,5,6-η)-1,5-cyclooctadiene]ruthenium, Ru-00084

$C_8H_{12}N_2O_2RuS_4$
Dicarbonylbis(dimethylcarbamodithioato-*S,S'*)ruthenium, Ru-00085

$C_8H_{14}ClF_6N_2OsP$
(η^6-Benzene)chloro(1,2-ethanediamine-*N,N'*)osmium(1+); Hexafluorophosphate, *in* Os-00065

$C_8H_{14}ClN_2Os^{\oplus}$
(η^6-Benzene)chloro(1,2-ethanediamine-*N,N'*)osmium(1+), Os-00065

$C_8H_{14}I_2O_2RuS_2$
[1,2-Bis(ethylthio)ethane-*S,S'*]-dicarbonyldiiodoruthenium, Ru-00086

$C_8H_{17}B_9Ru$
(η^6-Benzene)[(7,8,9,10,11-η)-undecahydro-7,8-dicarbaundecaborato(2−)]ruthenium, Ru-00087

$C_8H_{17}F_{12}N_3OsP_2$
Ammine(η^6-benzene)(1,2-ethanediamine-*N,N'*)osmium(2+); Bis(hexafluorophosphate), *in* Os-00066

$C_8H_{17}N_3Os^{\oplus\oplus}$
Ammine(η^6-benzene)(1,2-ethanediamine-*N,N'*)osmium(2+), Os-00066

$C_8H_{18}N_2O_2Os$
Bis[2-methyl-2-propanaminoato(2−)]dioxoosmium, Os-00067

$C_8H_{20}Br_2RuS_4$
Bis[1,2-bis(methylthio)ethane-*S,S'*]dibromoruthenium, Ru-00088

$C_8I_2O_8Os_2$
Octacarbonyldiiododiosmium, Os-00068

$C_9HO_9Os_3S^{\ominus}$
Nonacarbonyl-μ-hydro-μ_3-thioxotriosmate(1−), Os-00069

$C_9H_2O_9Os_3S$
Nonacarbonyldi-μ-hydro-μ_3-thioxotriosmium, Os-00070

$C_9H_2O_9Ru_3S$
Nonacarbonyldi-μ-hydro-μ_3-thioxotriruthenium, Ru-00089

$C_9H_2O_9Ru_3Se$
Nonacarbonyldi-μ-hydro-μ_3-selenoxotriruthenium, Ru-00090

$C_9H_2O_9Ru_3Te$
Nonacarbonyldi-μ-hydro-μ_3-telluroxotriruthenium, Ru-00091

$C_9H_3BrO_8Os_2$
Bromooctacarbonylmethyldiosmium, Os-00071

$C_9H_3O_9Ru_3S^{\oplus}$
Nonacarbonyltri-μ-hydro-μ_3-thioxotriruthenium(1+), Ru-00092

$C_9H_5F_3O_3Ru$
Dicarbonyl(η^5-2,4-cyclopentadien-1-yl)(trifluoroacetyl)-ruthenium, Ru-00093

$C_9H_5F_3O_5RuS_2$
(Carbonothioyl)dicarbonyl(η^5-2,4-cyclopentadien-1-yl)-ruthenium(1+); Trifluoromethanesulfonate, *in* Ru-00071

$C_9H_6O_4Ru$
Tricarbonyl[(2,3,4,5-η)-2,4-cyclohexadien-1-one]-ruthenium, Ru-00094

$C_9H_7BF_4O_3Os$
Tricarbonyl[(1,2,3,4,5-η)-2,4-cyclohexadien-1-yl]-osmium(1+); Tetrafluoroborate, *in* Os-00073

$C_9H_7BF_4O_3Ru$
Tricarbonyl[(1,2,3,4,5-η)-2,4-cyclohexadien-1-yl]-ruthenium(1+); Tetrafluoroborate, *in* Ru-00095

$C_9H_7NO_4OsS$
Tetracarbonyl[3,4-dimethyl-2(3*H*)-thiazolylidene]osmium, Os-00072

$C_9H_7O_3Os^{\oplus}$
Tricarbonyl[(1,2,3,4,5-η)-2,4-cyclohexadien-1-yl]-osmium(1+), Os-00073

$C_9H_7O_3Ru^{\oplus}$
Tricarbonyl[(1,2,3,4,5-η)-2,4-cyclohexadien-1-yl]-ruthenium(1+), Ru-00095

$C_9H_8BF_4NO_2Os$
(Acetonitrile)dicarbonyl(η^5-2,4-cyclopentadien-1-yl)-osmium(1+); Tetrafluoroborate, *in* Os-00074

$C_9H_8F_6NO_2PRu$
(Acetonitrile)dicarbonyl(η^5-2,4-cyclopentadien-1-yl)-ruthenium(1+); Hexafluorophosphate, *in* Ru-00096
Dicarbonyl(η^5-2,4-cyclopentadien-1-yl)(isocyanomethane)-ruthenium(1+); Hexafluorophosphate, *in* Ru-00097

$C_9H_8NO_2Os^{\oplus}$
(Acetonitrile)dicarbonyl(η^5-2,4-cyclopentadien-1-yl)-osmium(1+), Os-00074

$C_9H_8NO_2Ru^{\oplus}$
(Acetonitrile)dicarbonyl(η^5-2,4-cyclopentadien-1-yl)-ruthenium(1+), Ru-00096
Dicarbonyl(η^5-2,4-cyclopentadien-1-yl)(isocyanomethane)-ruthenium(1+), Ru-00097

$C_9H_8O_2Ru$
Dicarbonyl[(1,2,3,4,5,6-η)-1,3,5-cycloheptatriene]-ruthenium, Ru-00098

$C_9H_8O_2S_2Ru$
Dicarbonyl(η^5-2,4-cyclopentadien-1-yl)[(methylthio)-thioxomethyl]ruthenium, Ru-00099

$C_9H_8O_3Os$
Tricarbonyl[(1,2,3,4-η)-1,3-cyclohexadiene]osmium, Os-00075
Tricarbonyl[(1,3,4,5-η)-4-cyclohexene-1,3-diyl]osmium, Os-00076

$C_9H_8O_3Ru$
Acetyldicarbonyl(η^5-2,4-cyclopentadien-1-yl)ruthenium, Ru-00100
Tricarbonyl[(1,2,3,4-η)-1,3-cyclohexadiene]ruthenium, Ru-00101

$C_9H_9BF_4O_2Ru$
Dicarbonyl(η^5-2,4-cyclopentadien-1-yl)(η^2-ethene)-ruthenium(1+); Tetrafluoroborate, *in* Ru-00104

$C_9H_9BrO_2Os$
Bromodicarbonyl[(1,2,3,4,5-η)-2,4-cycloheptadien-1-yl]-osmium, Os-00077

$C_9H_9BrO_2Ru$
Bromodicarbonyl[(1,2,3,4,5-η)-2,4-cycloheptadien-1-yl]-ruthenium, Ru-00102

$C_9H_9F_6O_2Ru$
Dicarbonyl(η^5-2,4-cyclopentadien-1-yl)(η^2-ethene)-ruthenium(1+); Hexafluorophosphate, *in* Ru-00104

$C_9H_9IO_2Os$
Dicarbonyl[(1,2,3,4,5-η)-2,4-cycloheptadien-1-yl]-iodoosmium, Os-00078

$C_9H_9IO_2Ru$
Dicarbonyl[(1,2,3,4,5-η)-2,4-cycloheptadien-1-yl]-iodoruthenium, Ru-00103

$C_9H_9O_2Ru^{\oplus}$
Dicarbonyl(η^5-2,4-cyclopentadien-1-yl)(η^2-ethene)-ruthenium(1+), Ru-00104

$C_9H_{10}ORu$
Carbonyl(η^5-2,4-cyclopentadien-1-yl)(η^3-2-propenyl)-ruthenium, Ru-00105

$C_9H_{10}O_2Ru$
Dicarbonyl(η^5-2,4-cyclopentadien-1-yl)ethylruthenium, Ru-00106

$C_9H_{10}O_3Os$
Tricarbonyl[(1,2,3,4-η)-2,3-dimethyl-1,3-butadiene]-osmium, Os-00079
Tricarbonyl[(2,3,4,5-η)-2,4-hexadiene]osmium, Os-00080

$C_9H_{10}O_3Ru$
Tricarbonyl[(1,2,3,4-η)-2,3-dimethyl-1,3-butadiene]-ruthenium, Ru-00107
Tricarbonyl(2,4-hexadiene)ruthenium, Ru-00108

$C_9H_{11}ClOs$
(η^6-Benzene)chloro(η^3-2-propenyl)osmium, Os-00081

$C_9H_{11}ClRu$
(η^6-Benzene)chloro(η^3-2-propenyl)ruthenium, Ru-00109

$C_9H_{12}Cl_2ORu$
Carbonyldichloro(1,5-cyclooctadiene)ruthenium, *in* Ru-00486

$C_9H_{14}BrO_4PRu$
Bromocarbonyl(η^5-2,4-cyclopentadien-1-yl)-(trimethylphosphite-*P*)ruthenium, Ru-00110

$C_9H_{14}ClO_4PRu$
Carbonylchloro(η^5-2,4-cyclopentadien-1-yl)-(trimethylphosphite-*P*)ruthenium, Ru-00111

$C_9H_{14}IO_4PRu$
Carbonyl(η^5-2,4-cyclopentadien-1-yl)iodo(trimethyl phosphite-*P*)ruthenium, Ru-00112

$C_9H_{15}Cl_2PRu$
(η^6-Benzene)dichloro(trimethylphosphine)ruthenium, Ru-00113

$C_9H_{15}I_2OsP$
(η^6-Benzene)diiodo(trimethylphosphine)osmium, Os-00082

$C_9H_{18}O_3P_2Ru$
Tricarbonylbis(trimethylphosphine)ruthenium, Ru-00114

$C_9H_{18}O_9OsP_2$
Tricarbonylbis(trimethyl phosphite-*P*)osmium, Os-00083

$C_9H_{18}O_9P_2Ru$
Tricarbonylbis(trimethyl phosphite-*P*)ruthenium, Ru-00115

$C_9H_{24}Br_5N_2OOs$
Pentabromocarbonylosmate(2−); Bis(tetramethylammonium) salt, *in* Os-00001

$C_9H_{29}P_3Ru$
Dihydrotris(trimethylphosphine)ruthenium, *in* Ru-00499

$C_9N_2O_{11}Os_3$
Nonacarbonyldinitrosyltriosmium, Os-00084

$C_9O_8Os_3S_3$
(Carbonothioyl)octacarbonyldi-μ_3-thioxotriosmium, Os-00085

$C_9O_9Os_2$
μ-Carbonyloctacarbonyldiosmium, Os-00086

$C_9O_9Os_3S_2$
Nonacarbonyldi-μ_3-thioxotriosmium, Os-00087

$C_9O_9Os_3Se_2$
Nonacarbonyldi-μ_3-selenoxotriosmium, Os-00088

$C_9O_9Os_3Te_2$
Nonacarbonyldi-μ_3-telluroxotriosmium, Os-00089

$C_9O_9Ru_2$
μ-Carbonyloctacarbonyldiruthenium, Ru-00116

$C_9O_9Ru_3S_2$
Nonacarbonyldi-μ_3-thioxotriruthenium, Ru-00117

$C_9O_9Ru_3Se_2$
Nonacarbonyldi-μ_3-selenoxotriruthenium, Ru-00118

$C_{10}Br_2O_{10}Os_3$
Di-μ-bromodecacarbonyltriosmium, Os-00090

$C_{10}Cl_2O_{10}Os_3$
Decacarbonyldi-μ-chlorotriosmium, Os-00091

$C_{10}Cl_8I_2Ru$
1,1′,2,2′,3,3′,4,4′-Octachloro-5,5′-diiodoruthenocene, Ru-00119

$C_{10}Cl_{10}Ru$
Decachlororuthenocene, Ru-00120

$C_{10}HBrO_{10}Os_3$
μ-Bromodecacarbonyl-μ-hydrotriosmium, Os-00092

$C_{10}HClO_{10}Os_3$
Decacarbonyl-μ-chloro-μ-hydrotriosmium, Os-00093

$C_{10}HNO_{11}Os_3$
Decacarbonyl-μ-hydro-μ-nitrosyltriosmium, Os-00094

$C_{10}HNO_{11}Ru_3$
Decacarbonyl-μ-hydro-μ-nitrosyltriruthenium, Ru-00121

$C_{10}H_2O_{10}Os_3$
Decacarbonyldi-μ-hydrotriosmium, Os-00095

$C_{10}H_3BrO_9Os_3$
[μ_3-(Bromomethylidyne)]nonacarbonyltri-μ-hydrotriosmium, Os-00096

$C_{10}H_3BrO_9Ru_3$
[μ_3-(Bromomethylidyne)]nonacarbonyltri-μ-hydrotriruthenium, Ru-00122

$C_{10}H_3ClO_9Ru_3$
Nonacarbonyl[μ_3-(chloromethylidyne)]tri-μ-hydrotriruthenium, Ru-00123

$C_{10}H_3NO_{10}Os_3$
μ-Amidodecacarbonyl-μ-hydrotriosmium, Os-00097

$C_{10}H_4O_6Os_2$
[μ-[(1,2,3,4-η:1,4-η)-1,3-Butadiene-1,4-diyl]]-hexacarbonyldiosmium, Os-00098

$C_{10}H_4O_9Os_3$
Nonacarbonyltri-μ-hydro-μ_3-methylidynetriosmium, Os-00099

$C_{10}H_4O_9Os_3S$
Nonacarbonyl-μ-hydro[μ_3-(methanethiolato)]triosmium, Os-00100

$C_{10}H_4O_9Ru_3$
Nonacarbonyltri-μ-hydro-μ_3-methylidynetriruthenium, Ru-00124

$C_{10}H_5Cl_2NO_3Ru$
(Benzonitrile)tricarbonyldichlororuthenium, Ru-00125

$C_{10}H_5F_5O_2Ru$
Dicarbonyl(η^5-2,4-cyclopentadien-1-yl)(1,2,3,3,3-pentafluoro-1-propenyl)ruthenium, Ru-00126

$C_{10}H_5NO_9Os_3$
Nonacarbonyldi-μ-hydro[μ_3-[methanaminato(2−)]]-triosmium, Os-00101

$C_{10}H_6O_4Ru$
Tricarbonyl[(2,3,4,5-η)-2,4,6-cycloheptatrien-1-one]-ruthenium, Ru-00127

$C_{10}H_6O_{10}Os_2$
Bis[μ-(acetato-*O*:*O*′)]hexacarbonyldiosmium, Os-00102

$C_{10}H_7NO_3Os$
Tricarbonyl[(1,3,4,5-η)-2-cyano-4-cyclohexene-1,3-diyl]-osmium, Os-00103
Tricarbonyl[(2,3,4,5-η)-2,4-cyclohexadiene-1-carbonitrile]osmium, Os-00104

$C_{10}H_7NO_5Ru$
[(2,3,4,5-η)-1*H*-Azepine-1-carboxylic acid]tricarbonylruthenium, Ru-00128

$C_{10}H_8Br_2Ru$
1,1′-Dibromoruthenocene, Ru-00129

$C_{10}H_8Cl_2Hg_2Ru$
1,1′-Bis(chloromercuri)ruthenocene, Ru-00130

$C_{10}H_8I_2Ru$
1,1′-Diiodoruthenocene, Ru-00131

$C_{10}H_8Li_2Ru$
1,1′-Dilithioruthenocene, Ru-00132

$C_{10}H_8N_2O_4Ru$
(1-Acetyl-1*H*-1,2-diazepine)tricarbonylruthenium, *in* Ru-00073

$C_{10}H_8O_3Ru$
Tricarbonyl[(1,2,3,4-η)-1,3,5-cycloheptatriene]-ruthenium, Ru-00133

$C_{10}H_8O_8Ru$
(Dimethyl fumarate)tetracarbonylruthenium, *in* Ru-00066

$C_{10}H_9BF_4O_3Ru$
Tricarbonyl[(1,2,3,4,5-η)-2,4-cycloheptadien-1-yl]-ruthenium(1+); Tetrafluoroborate, *in* Ru-00139

$C_{10}H_9BF_4O_4Ru$
Tricarbonyl[(2,3,4,5,6-η)-1-methoxy-2,4-cyclohexadien-1-yl]ruthenium(1+); Tetrafluoroborate, *in* Ru-00140

$C_{10}H_9BrRu$
Bromoruthenocene, Ru-00134

$C_{10}H_9ClO_2RuS$
Ruthenocenesulfonyl chloride, *in* Ru-00146

$C_{10}H_9ClRu$
Chlororuthenocene, Ru-00135

$C_{10}H_9F_6O_3PRu$
Tricarbonyl[(1,2,3,4,5-η)-2,4-cycloheptadien-1-yl]-ruthenium(1+); Hexafluorophosphate, *in* Ru-00139

$C_{10}H_9IO_2Ru$
Dicarbonyliodo[(1,2,3,3*a*,6*a*-η)-1,4,5,6-tetrahydro-1-pentalenyl]ruthenium, Ru-00136

$C_{10}H_9IRu$
Iodoruthenocene, Ru-00137

$C_{10}H_9LiRu$
Lithioruthenocene, Ru-00138

$C_{10}H_9O_3Ru^{\oplus}$
Tricarbonyl[(1,2,3,4,5-η)-2,4-cycloheptadien-1-yl]-ruthenium(1+), Ru-00139

$C_{10}H_9O_4Ru^{\oplus}$
Tricarbonyl[(2,3,4,5,6-η)-1-methoxy-2,4-cyclohexadien-1-yl]ruthenium(1+), Ru-00140

$C_{10}H_{10}BF_4Os$
Osmocenium; Tetrafluoroborate, *in* Os-00110

$C_{10}H_{10}IRu^{\oplus}$
Bis(η^5-2,4-cyclopentadien-1-yl)iodoruthenium(1+), Ru-00141

$C_{10}H_{10}I_4Ru$
Bis(η^5-2,4-cyclopentadien-1-yl)iodoruthenium(1+); Triiodide, *in* Ru-00141

$C_{10}H_{10}ORu$
Hydroxyruthenocene, Ru-00142

$C_{10}H_{10}O_2Ru$
Dicarbonyl(η^5-2,4-cyclopentadien-1-yl)-2-propenylruthenium, Ru-00143

$C_{10}H_{10}O_3Os$
Tricarbonyl[(1,2,3,4-η)-1,3-cycloheptadiene]osmium, Os-00105

$C_{10}H_{10}O_3Ru$
Tricarbonyl[(1,2,3,4-η)-1,3-cycloheptadiene]ruthenium, Ru-00144
Tricarbonyl[(1,3,4,5-η)-4-cycloheptene-1,3-diyl]-ruthenium, Ru-00145

$C_{10}H_{10}O_3RuS$
Ruthenocenesulfonic acid, Ru-00146

$C_{10}H_{10}O_4Os$
Dicarbonyl[(1,2,3,4,5-η)-2,4-cyclohexadien-1-yl]-(methoxycarbonyl)osmium, Os-00106
Tricarbonyl[(1,2,3,4-η)-5-methoxy-1,3-cyclohexadiene]-osmium, Os-00107

$C_{10}H_{10}O_4Ru$
Dicarbonyl[(1,2,3,4,5-η)-2,4-cyclohexadien-1-yl]-(methoxycarbonyl)ruthenium, Ru-00147
Tricarbonyl[(1,2,3,4-η)-1-methoxy-1,3-cyclohexadiene]-ruthenium, Ru-00148

$C_{10}H_{10}O_6Os$
Tetracarbonyldipropionylosmium, Os-00108

$C_{10}H_{10}Os$
Osmocene, Os-00109
Osmocenium, Os-00110

$C_{10}H_{10}Ru$
Ruthenocene, Ru-00149
Ruthenocenium, Ru-00150

$C_{10}H_{11}BO_2Ru$
Ruthenoceneboronic acid, Ru-00151

$C_{10}H_{11}F_6O_2PRu$
Dicarbonyl(η^5-2,4-cyclopentadien-1-yl)[(1,2-η)-1-propene]ruthenium(1+); Hexafluorophosphate, *in* Ru-00152

$C_{10}H_{11}O_2Ru^{\oplus}$
Dicarbonyl(η^5-2,4-cyclopentadien-1-yl)[(1,2-η)-1-propene]ruthenium(1+), Ru-00152

$C_{10}H_{12}As_2O_6Ru_2$
Hexacarbonylbis[μ-(dimethylarsino)]diruthenium, Ru-00153

$C_{10}H_{12}ClN_2Ru^{\oplus}$
Bis(acetonitrile)(η^6-benzene)chlororuthenium(1+), Ru-00154

$C_{10}H_{12}Cl_4HgN_2Ru$
Bis(acetonitrile)(η^6-benzene)chlororuthenium(1+); Trichloromercurate, *in* Ru-00154

$C_{10}H_{12}ORu$
Carbonyl(η^5-2,4-cyclopentadien-1-yl)[(1,2,3-η)-2-methyl-2-propenyl]ruthenium, Ru-00155

$C_{10}H_{12}O_6P_2Ru_2$
Hexacarbonylbis[μ-(dimethylphosphino)]diruthenium, Ru-00156

$C_{10}H_{14}F_6O_2P_2Ru$
Dicarbonyl(η^5-2,4-cyclopentadien-1-yl)-(trimethylphosphine)ruthenium(1+); Hexafluorophosphate, *in* Ru-00157

$C_{10}H_{14}O_2PRu^{\oplus}$
Dicarbonyl(η^5-2,4-cyclopentadien-1-yl)-(trimethylphosphine)ruthenium(1+), Ru-00157

$C_{10}H_{14}O_2Ru$
Dicarbonylbis[(1,2,3-η)-2-methyl-2-propenyl]ruthenium, Ru-00158

$C_{10}H_{14}O_2RuSi$
Dicarbonyl(η^5-2,4-cyclopentadien-1-yl)(trimethylsilyl)-ruthenium, Ru-00159

$C_{10}H_{15}IOsP$
(η^6-Benzene)iodomethyl(trimethylphosphine)osmium, Os-00111

$C_{10}H_{16}Cl_2Ru$
Dichloro[(1,2,3,6,7,8-η)-2,7-dimethyl-2,6-octadiene-1,8-diyl]ruthenium, *in* Ru-00530

$C_{10}H_{16}O_4OsSi_2$
Tetracarbonyl[1,2-ethanediylbis[dimethylsilylene]]-osmium, Os-00112

$C_{10}H_{16}O_4RuSi_2$
Tetracarbonyl[1,2-ethanediylbis[dimethylsilylene]]-ruthenium, Ru-00160

$C_{10}H_{18}F_4O_8OsP_2$
Dicarbonyl(η^2-tetrafluoroethene)bis(trimethyl phosphite-*P*)osmium, Os-00113

$C_{10}H_{18}F_4O_8P_2Ru$
Dicarbonyl(tetrafluoroethene)bis(trimethyl phosphite-*P*)ruthenium, Ru-00161

$C_{10}H_{18}Ge_2O_4Os$
Tetracarbonylbis(trimethylgermyl)osmium, Os-00114

$C_{10}H_{18}Ge_2O_4Ru$
Tetracarbonylbis(trimethylgermyl)ruthenium, Ru-00162

$C_{10}H_{18}O_4OsPb_2$
Tetracarbonylbis(trimethylplumbyl)osmium, Os-00115

$C_{10}H_{18}O_4OsSi_2$
Tetracarbonylbis(trimethylsilyl)osmium, Os-00116

$C_{10}H_{18}O_4OsSn_2$
Tetracarbonylbis(trimethylstannyl)osmium, Os-00117

$C_{10}H_{18}O_4Pb_2Ru$
Tetracarbonylbis(trimethylplumbyl)ruthenium, Ru-00163

$C_{10}H_{18}O_4RuSi_2$
Tetracarbonylbis(trimethylsilyl)ruthenium, Ru-00164

$C_{10}H_{18}O_4RuSn_2$
Tetracarbonylbis(trimethylstannyl)ruthenium, Ru-00165

$C_{10}I_2O_{10}Os_3$
Decacarbonyldi-μ-iodotriosmium, Os-00118

$C_{10}NO_{11}Os_3^{\ominus}$
Decacarbonyl-μ-nitrosyltriosmate(1−), Os-00119

$C_{10}NO_{11}Ru_3^{\ominus}$
Decacarbonyl-μ-nitrosyltriruthenate(1−), Ru-00166

$C_{10}N_2O_{12}Os_3$
Decacarbonyldi-μ-nitrosyltriosmium, Os-00120

$C_{10}N_2O_{12}Ru_3$
Decacarbonyldi-μ-nitrosyltriruthenium, Ru-00167

$C_{11}HO_{11}Os_3^{\ominus}$
μ-Carbonyldecacarbonyl-μ-hydrotriosmate(1−), Os-00121

$C_{11}HO_{11}Ru_3^{\ominus}$
μ-Carbonyldecacarbonyl-μ-hydrotriruthenate(1−), Ru-00168

$C_{11}H_2O_9Os_3$
Nonacarbonyl[μ_3-[(1-η:1,2-η:1,2-η)ethynyl]]-μ-hydrotriosmium, Os-00122

$C_{11}H_2O_{10}Os_3$
Decacarbonyl-μ-hydro-μ_3-methylidynetriosmium, Os-00123
Nonacarbonyldi-μ-hydro[μ_3-(oxoethenylidene)]triosmium, Os-00124

$C_{11}H_2O_{11}Os_3$
Undecacarbonyl-μ-hydrohydrotriosmium, Os-00125

$C_{11}H_2O_{11}Ru_3$
Undecacarbonyl-μ-hydrohydrotriruthenium, Ru-00169

$C_{11}H_4O_9Os_3$
Nonacarbonyl[μ_3-[(1-η:1-η:1,2-η)ethenylidene]]di-μ-hydrotriosmium, Os-00126

$C_{11}H_4O_9Ru_3$
Nonacarbonyl[μ-[(1-η:1-η:1,2-η)-ethenylidene]]di-μ-hydrotriruthenium, Ru-00170

$C_{11}H_4O_{10}Os_3$
Decacarbonyl-μ-hydro-μ-methyltriosmium, Os-00127

$C_{11}H_4O_{10}Ru_3$
Decacarbonyldi-μ-hydro-μ-methylenetriruthenium, Ru-00171

$C_{11}H_5CoO_6Ru$
Dicarbonyl(η^5-2,4-cyclopentadien-1-yl)-(tetracarbonylcobalt)ruthenium, Ru-00172

$C_{11}H_5NO_{10}Os_3$
Decacarbonyl-μ-hydro[μ-(methanaminato)]triosmium, Os-00128

$C_{11}H_6O_9Os_3$
Nonacarbonyl-μ_3-ethylidynetri-μ-hydrotriosmium, Os-00129

$C_{11}H_6O_9Ru_3$
Nonacarbonyl-μ_3-ethylidynetri-μ-hydrotriruthenium, Ru-00173

$C_{11}H_6O_{10}Os_3$
Nonacarbonyltri-μ-hydro[μ_3-(methoxymethylidyne)]-triosmium, Os-00130

$C_{11}H_6O_{10}Ru_3$
Nonacarbonyltri-μ-hydro[μ_3-(methoxymethylidyne)]-triruthenium, Ru-00174

$C_{11}H_7O_2Ru^{\ominus}$
Dicarbonyl[(1,2,3,3*a*,7*a*-η)-1-*H*-inden-1-yl]ruthenate(1−), Ru-00175

$C_{11}H_8O_3Os$
Tricarbonyl[(1,2,3,4-η)-1,3,5,7-cyclooctatetraene]-osmium, Os-00131
Tricarbonyl[(1,2,3,6-η)-1,3,5,7-cyclooctatetraene]-osmium, Os-00132

$C_{11}H_8O_3Ru$
Tricarbonyl[(1,2,3,4-η)-1,3,5,7-cyclooctatetraene]-ruthenium, Ru-00176

$C_{11}H_9BF_4O_3Ru$
[(2,3,4,6,7-η)-Bicyclo[3.2.1]octa-3,6-dien-2-yl]-tricarbonylruthenium(1+); Tetrafluoroborate, *in* Ru-00178
Tricarbonyl[(1,2,3,6,7-η)-2,4,6-cyclooctatrien-1-yl]-ruthenium(1+); Tetrafluoroborate, *in* Ru-00179

$C_{11}H_9ClO_3Ru$
Tricarbonylchloro[(1,2,3-η)-2,4,7-cyclooctatrien-1-yl]-ruthenium, Ru-00177

$C_{11}H_9F_6O_3PRu$
Tricarbonyl[(1,2,3,6,7-η)-2,4,6-cyclooctatrien-1-yl]-ruthenium(1+); Hexafluorophosphate, *in* Ru-00179

$C_{11}H_9NO_5Ru$
Tricarbonyl[(2,3,4,5-η)methyl-1*H*-azepine-1-carboxylate]-ruthenium, *in* Ru-00128

$C_{11}H_9NRu$
(Cyano-π-cyclopentadienyl)-π-cyclopentadienylruthenium, *in* Ru-00185

$C_{11}H_9N_2O_{13}Os_3P$
Octacarbonyldinitrosyl(trimethyl phosphite)triosmium, Os-00133

$C_{11}H_9O_3Os^{\oplus}$
Tricarbonyl[(1,2,5,6,7-η)-1,3,5-cyclooctatrien-1-yl]-osmium(1+), Os-00134

$C_{11}H_9O_3Os_3^{\oplus}$
[(2,3,4,5,6-η)-Bicyclo[5.1.0]octa-3,5-dien-2-yl]-tricarbonylosmium(1+), Os-00135

$C_{11}H_9O_3Ru^{\oplus}$
[(2,3,4,6,7-η)-Bicyclo[3.2.1]octa-3,6-dien-2-yl]-tricarbonylruthenium(1+), Ru-00178
Tricarbonyl[(1,2,3,6,7-η)-2,4,6-cyclooctatrien-1-yl]-ruthenium(1+), Ru-00179

$C_{11}H_{10}BBrN_6O_2Ru$
Bromodicarbonyl[hydrotris(pyrazolato)borato(1–)]-ruthenium, Ru-00180

$C_{11}H_{10}BClF_4Ru$
(η^6-Chlorobenzene)(η^5-2,4-cyclopentadien-1-yl)-ruthenium(1+); Tetrafluoroborate, *in* Ru-00183

$C_{11}H_{10}BClN_6O_2Ru$
Dicarbonylchloro[hydrotris(pyrazolato)borato(1–)]-ruthenium, Ru-00181

$C_{11}H_{10}BIN_6O_2Ru$
Dicarbonyl[hydrotris(pyrazolato)borato(1–)]-iodoruthenium, Ru-00182

$C_{11}H_{10}ClF_6PRu$
(η^6-Chlorobenzene)(η^5-2,4-cyclopentadien-1-yl)-ruthenium(1+); Hexafluorophosphate, *in* Ru-00183

$C_{11}H_{10}ClRu^{\oplus}$
(η^6-Chlorobenzene)(η^5-2,4-cyclopentadien-1-yl)-ruthenium(1+), Ru-00183

$C_{11}H_{10}ORu$
Ruthenocenecarboxaldehyde, Ru-00184

$C_{11}H_{10}O_2Os$
Osmocenecarboxylic acid, Os-00136

$C_{11}H_{10}O_2Ru$
Ruthenocenecarboxylic acid, Ru-00185

$C_{11}H_{10}O_3Ru$
[(2,3,4,5-η)-Bicyclo[4.2.0]octa-2,4-diene]-tricarbonylruthenium, Ru-00186
[(2,3,4,6-η)Bicyclo[3.2.1]oct-3-ene-2,6-diyl]-tricarbonylruthenium, Ru-00187

$C_{11}H_{11}BF_4O_3Os$
Tricarbonyl[(1,2,3,4,5-η)-2,4-cyclooctadien-1-yl]-osmium(1+); Tetrafluoroborate, *in* Os-00138

$C_{11}H_{11}BF_4O_3Ru$
Tricarbonyl[(1,2,3,5,6-η)-2,5-cyclooctadien-1-yl]-ruthenium(1+); Tetrafluoroborate, *in* Ru-00188

$C_{11}H_{11}ClRu$
(η^6-Benzene)(η^5-2,4-cyclopentadien-1-yl)ruthenium(1+); Chloride, *in* Ru-00189

$C_{11}H_{11}Cl_2NOs$
(η^6-Benzene)dichloro(pyridine)osmium, Os-00137

$C_{11}H_{11}NORu$
(Aminocarbonyl)ruthenocene, *in* Ru-00185

$C_{11}H_{11}O_3Os^{\oplus}$
Tricarbonyl[(1,2,3,4,5-η)-2,4-cyclooctadien-1-yl]-osmium(1+), Os-00138
Tricarbonyl[(1,2,3,5,6-η)-2,5-cyclooctadien-1-yl]-osmium(1+), Os-00139

$C_{11}H_{11}O_3Ru^{\oplus}$
Tricarbonyl[(1,2,3,5,6-η)-2,5-cyclooctadien-1-yl]-ruthenium(1+), Ru-00188

$C_{11}H_{11}Os^{\oplus}$
(η^6-Benzene)(η^5-2,4-cyclopentadien-1-yl)osmium(1+), Os-00140

$C_{11}H_{11}Ru^{\oplus}$
(η^6-Benzene)(η^5-2,4-cyclopentadien-1-yl)ruthenium(1+), Ru-00189

$C_{11}H_{12}BF_4NRu$
(η^6-Aniline)(η^5-cyclopentadienyl)ruthenium(1+); Tetrafluoroborate, *in* Ru-00190

$C_{11}H_{12}NRu^{\oplus}$
(η^6-Aniline)(η^5-cyclopentadienyl)ruthenium(1+), Ru-00190

$C_{11}H_{12}ORu$
Methoxyruthenocene, *in* Ru-00142
Ruthenocenemethanol, Ru-00191

$C_{11}H_{12}O_3Os$
Tricarbonyl[(1,2,3,4-η)-1,3-cyclooctadiene]osmium, Os-00141
Tricarbonyl[(1,3,4,5-η)-4-cyclooctene-1,3-diyl]osmium, Os-00142

$C_{11}H_{12}O_3Ru$
Tricarbonyl[(1,2,5,6-η)-1,5-cyclooctadiene]ruthenium, Ru-00192
Tricarbonyl[(1,4,5,6-η)-5-cyclooctene-1,4-diyl]-ruthenium, Ru-00193

$C_{11}H_{12}O_7Ru$
Tricarbonylbis[(2,3-η)-2-methyl-2-propenoate]ruthenium, Ru-00194

$C_{11}H_{12}Ru$
[(1,2,3,4,5-η)-2,4-Cyclohexadien-1-yl](η^5-2,4-cyclopentadien-1-yl)ruthenium, Ru-00195
Methylruthenocene, Ru-00196

$C_{11}H_{13}ClO_3Ru$
Tricarbonylchloro[(1,2,3-η)-2,4,7-cyclooctatrien-1-yl]-ruthenium, Ru-00197

$C_{11}H_{14}BClF_4N_2Ru$
Bis(acetonitrile)chloro[(1,2,3,4,5,6-η)-1,3,5-cycloheptatriene]ruthenium(1+); Tetrafluoroborate, *in* Ru-00198

$C_{11}H_{14}ClN_2Ru^{\oplus}$
Bis(acetonitrile)chloro[(1,2,3,4,5,6-η)-1,3,5-cycloheptatriene]ruthenium(1+), Ru-00198

$C_{11}H_{15}Cl_2NORu$
(Acetonitrile)carbonyldichloro[(1,2,5,6-η)-1,5-cyclooctadiene]ruthenium, Ru-00199

$C_{11}H_{16}GeO_2Ru$
Dicarbonyl[(1,2,3,4,5-η)-2,4-cyclohexadien-1-yl]-(trimethylgermyl)ruthenium, Ru-00200

$C_{11}H_{16}O_2RuSi$
Dicarbonyl[(1,2,3,4,5-η)-2,4-cyclohexadien-1-yl]-(trimethylsilyl)ruthenium, Ru-00201

$C_{11}H_{17}ClRu$
Chloro(1,5-cyclooctadiene)(η^3-propenyl)ruthenium, *in* Ru-00574

$C_{11}H_{17}NaO_2Ru$
Dicarbonyl[(1,2,3,3*a*,7*a*-η)-1-*H*-inden-1-yl]ruthenate(1–); Na salt, *in* Ru-00175

$C_{11}H_{18}OOs$
Carbonyltrihydro[(1,2,3,4,5-η)-1,2,3,4,5-pentamethyl-2,4-cyclopentadien-1-yl]osmium, Os-00143

$C_{11}H_{19}ClF_6P_2Ru$
(η^6-Benzene)chloro(η^2-ethene)(trimethylphosphine)-ruthenium(1+); Hexafluorophosphate, *in* Ru-00202

$C_{11}H_{19}ClPRu^{\oplus}$
(η^6-Benzene)chloro(η^2-ethene)(trimethylphosphine)-ruthenium(1+), Ru-00202

$C_{11}H_{19}F_6IOsP_2$
(η^6-Benzene)(η^2-ethene)iodo(trimethylphosphine)-osmium(1+); Hexafluorophosphate, *in* Os-00144

$C_{11}H_{19}IOsP^{\oplus}$
(η^6-Benzene)(η^2-ethene)iodo(trimethylphosphine)-osmium(1+), Os-00144

$C_{11}H_{19}OsP$
(η^6-Benzene)(η^2-ethene)(trimethylphosphine)osmium, Os-00145

$C_{11}H_{19}PRu$
(η^6-Benzene)(η^2-ethene)(trimethylphosphine)ruthenium, Ru-00203

$C_{11}H_{20}F_6OsP_2$
(η^6-Benzene)(η^2-ethene)hydro(trimethylphosphine)-osmium(1+); Hexafluorophosphate, *in* Os-00146

$C_{11}H_{20}F_6P_2Ru$
(η^6-Benzene)(η^2-ethene)hydro(trimethylphosphine)-ruthenium(1+); Hexafluorophosphate, *in* Ru-00204

$C_{11}H_{20}I_3NO_3Ru$
Tricarbonyltriiodoruthenate(1−); Tetraethylammonium salt, *in* Ru-00021

$C_{11}H_{20}OsP^{\oplus}$
(η^6-Benzene)(η^2-ethene)hydro(trimethylphosphine)-osmium(1+), Os-00146

$C_{11}H_{20}PRu^{\oplus}$
(η^6-Benzene)(η^2-ethene)hydro(trimethylphosphine)-ruthenium(1+), Ru-00204

$C_{11}H_{23}BrO_6OsP_2$
Bromo(η^5-2,4-cyclopentadien-1-yl)bis(trimethyl phosphite-*P*)osmium, Os-00147

$C_{11}H_{23}BrP_2Ru$
Bromo(η^5-2,4-cyclopentadien-1-yl)-bis(trimethylphosphine)ruthenium, Ru-00205

$C_{11}H_{23}ClO_6P_2Ru$
Chloro(η^5-2,4-cyclopentadien-1-yl)bis(trimethyl phosphite-*P*)ruthenium, Ru-00206

$C_{11}H_{23}ClP_2Ru$
Chloro(η^5-2,4-cyclopentadien-1-yl)-bis(trimethylphosphine)ruthenium, Ru-00207

$C_{11}H_{23}F_{12}NOP_4Ru$
(η^5-2,4-Cyclopentadien-1-yl)-nitrosylbis(trimethylphosphine)ruthenium(2+); Bis(hexafluorophosphate), *in* Ru-00209

$C_{11}H_{23}IP_2Ru$
(η^5-2,4-Cyclopentadien-1-yl)iodobis(trimethylphosphine)-ruthenium, Ru-00208

$C_{11}H_{23}NOP_2Ru^{\oplus\oplus}$
(η^5-2,4-Cyclopentadien-1-yl)-nitrosylbis(trimethylphosphine)ruthenium(2+), Ru-00209

$C_{11}K_2O_{11}Os_3$
μ-Carbonyldecacarbonyltriosmate(2−); Di-K salt, *in* Os-00148

$C_{11}O_{11}CaRu$
μ-Carbonyldecacarbonyltriruthenate(2−); Ca salt, *in* Ru-00210

$C_{11}O_{11}Os_3^{\ominus\ominus}$
μ-Carbonyldecacarbonyltriosmate(2−), Os-00148

$C_{11}O_{11}Ru_3^{\ominus\ominus}$
μ-Carbonyldecacarbonyltriruthenate(2−), Ru-00210

$C_{12}Br_2O_{12}Os_3$
Dibromododecacarbonyltriosmium, Os-00149

$C_{12}Cl_2O_{12}Os_3$
Dodecacarbonydichlorotriosmium, Os-00150

$C_{12}Cl_6O_{12}Ru_3$
Dodecacarbonylhexachlorotriruthenium, Ru-00211

$C_{12}F_8O_{12}Ru_4$
Dodecacarbonyltetra-μ-fluorotetrafluorotetraruthenium, Ru-00212

$C_{12}FeNO_{13}Ru_3^{\ominus}$
Tri-μ-carbonylnonacarbonyl(nitrosylferrate)-triruthenate(1−), Ru-00213

$C_{12}FeO_{12}Os_2$
Dodecacarbonylirondiosmium, Os-00151

$C_{12}FeO_{12}Ru_2$
Octacarbonyl(tetracarbonyliron)diruthenium, Ru-00214

$C_{12}Fe_2O_{12}Os$
Tetracarbonyl(di-μ-carbonylhexacarbonyldiiron)osmium, Os-00152

$C_{12}Fe_2O_{12}Ru$
Tetracarbonyl(di-μ-carbonylhexacarbonyldiiron)-ruthenium, Ru-00215

$C_{12}HF_6O_{12}Os_3P$
Dodecacarbonyl-μ-hydrotriosmium(1+); Hexafluorophosphate, *in* Os-00155

$C_{12}HF_6O_{12}PRu_3$
Dodecacarbonyl-μ-hydrotriruthenium(1+); Hexafluorophosphate, *in* Ru-00217

$C_{12}HFeNO_{12}Ru_3$
Nonacarbonyl-μ-hydro-μ_4-nitrido(tricarbonyliron)-triruthenium, Ru-00216

$C_{12}HMnO_{12}Os_3$
Dodecacarbonylhydromanganesediosmium, Os-00153

$C_{12}HO_{12}Os_3^{\oplus}$
Dodecacarbonyl-μ-hydrotriosmium(1+), Os-00155

$C_{12}HO_{12}Ru_3^{\oplus}$
Dodecacarbonyl-μ-hydrotriruthenium(1+), Ru-00217

$C_{12}H_2K_2O_{12}Ru_4$
Tri-μ-carbonylnonacarbonyldi-μ-hydrotetraruthenate(2−); Di-K salt, *in* Ru-00218

$C_{12}H_2O_{10}Os_3$
μ-Carbonylnonacarbonyl[μ_3-[(1-η:1,2-η:2-η)-1,2-ethenediyl]]triosmium, Os-00156

$C_{12}H_2O_{11}Os_3$
μ-Carbonyldecacarbonyl-μ-methylenetriosmium, Os-00157

$C_{12}H_2O_{12}Os_3$
Dodecacarbonyldihydrotriosmium, Os-00158

$C_{12}H_2O_{12}Os_4^{\ominus\ominus}$
Dodecacarbonyldi-μ-hydrotetraosmate(2−), Os-00159

$C_{12}H_2O_{12}Os_4S_2$
Dodecacarbonyldi-μ-hydrodi-μ_3-thioxotetraosmium, Os-00160

$C_{12}H_2O_{12}Os_4Se_2$
Dodecacarbonyldi-μ-hydrodi-μ_3-selenoxotetraosmium, Os-00161

$C_{12}H_2O_{12}Ru_4^{\ominus\ominus}$
Tri-μ-carbonylnonacarbonyldi-μ-hydrotetraruthenate(2−), Ru-00218

$C_{12}H_3BO_{12}Ru_4$
[μ_4-(Boryl-*B*:*B*:*H*:*H′*)]dodecacarbonyl-μ-hydrotetraruthenium, Ru-00219

$C_{12}H_3IO_{12}Os_4$
Dodecacarbonyltri-μ-hydro-μ-iodotetraosmium, Os-00162

$C_{12}H_3O_{12}Os_4^{\ominus}$
Dodecacarbonyltri-μ-hydrotetraosmate(1−), Os-00163

$C_{12}H_3O_{12}Ru_4^{\ominus}$
Dodecacarbonyltrihydrotetraruthenate(1−), Ru-00220

$C_{12}H_4FeO_{12}Ru_3$
Dodecacarbonyltetrahydro(iron)triruthenium, Ru-00221

$C_{12}H_4O_{10}Os_3$
Decacarbonyl[μ-[(1-η:1,2-η)ethenyl]]-μ-hydrotriosmium, Os-00164

$C_{12}H_4O_{10}Ru_3$
Decacarbonyl-μ-ethenylidenedi-μ-hydrotriruthenium, Ru-00222

$C_{12}H_4O_{11}Os_3$
Decacarbonyl-μ-hydro[μ-(acetyl-*C*:*O*)]triosmium, Os-00165
Decacarbonyl-μ-hydro[μ-(methoxymethylidyne)]triosmium, Os-00166

$C_{12}H_4O_{11}Ru_3$
Decacarbonyl-μ-hydro[μ-(methoxymethylidyne)]-triruthenium, Ru-00223

$C_{12}H_4O_{12}Os_4$
Dodecacarbonyltetra-μ-hydrotetraosmium, Os-00167

$C_{12}H_4O_{12}Ru_4$
Dodecacarbonyltetra-μ-hydrotetraruthenium, Ru-00224

$C_{12}H_5O_7ReRu$
Dicarbonyl(η^5-2,4-cyclopentadien-1-yl)-(pentacarbonylrhenio)ruthenium, Ru-00225

$C_{12}H_5O_{12}Os_3Re$
Tricarbonyltri-μ-hydro(nonacarbonyldi-μ-hydrotriosmium)-rhenium, Os-00168

$C_{12}H_6O_{10}Os_3S$
Decacarbonyl[μ-(ethanethiolato)]-μ-hydrotriosmium, Os-00169

$C_{12}H_6O_{10}Ru_3S$
Decacarbonyl[μ-(ethanethiolato)]-μ-hydrotriruthenium, Ru-00226

$C_{12}H_6O_{11}Os_3$
Nonacarbonyltri-μ-hydro[μ_3-(2-methoxy-2-oxoethylidyne)]-triosmium, Os-00170

$C_{12}H_6O_{11}Ru_3$
Nonacarbonyltri-μ-hydro[μ_3-(2-methoxy-2-oxoethylidyne)]-triruthenium, Ru-00227

$C_{12}H_7O_{10}Ru_3S^{\oplus}$
Decacarbonyl(μ-ethanethiolato)di-μ-hydrotriruthenium(1+), Ru-00228

$C_{12}H_8O_9Os_3S$
Nonacarbonyl(η^2-ethene)-μ-hydro[μ-(methanethiolato)]-triosmium, Os-00171

$C_{12}H_9N_2O_{14}Os_3$
Nonacarbonyldi-μ-nitrosyl(trimethyl phosphite-*P*)triosmium, Os-00172

$C_{12}H_9N_3O_{11}Os_3$
Nonacarbonyldi-μ-nitrosyl(trimethylamine)triosmium, Os-00173

$C_{12}H_{10}O_2Ru$
Dicarbonyl-2,4-cyclopentadien-1-yl(η^5-2,4-cyclopentadien-1-yl)ruthenium, Ru-00229
1,1′-Ruthenocenedicarboxaldehyde, Ru-00230

$C_{12}H_{10}O_2RuS_2$
1,1′-Ruthenocenebis(thioic acid), Ru-00231

$C_{12}H_{10}O_4Os$
1,1′-Osmocenedicarboxylic acid, Os-00174

$C_{12}H_{10}O_4Ru$
1,1′-Ruthenocenedicarboxylic acid, Ru-00232

$C_{12}H_{10}Ru$
Ethynylruthenocene, Ru-00233

$C_{12}H_{11}B_2ClF_8Ru$
(η^6-Benzene)(η^6-chlorobenzene)ruthenium(2+); Bis(tetrafluoroborate), *in* Ru-00234

$C_{12}H_{11}ClRu^{\oplus\oplus}$
(η^6-Benzene)(η^6-chlorobenzene)ruthenium(2+), Ru-00234
(1-Chloroethenyl)ruthenocene, Ru-00235

$C_{12}H_{11}NO_3Os$
Tricarbonyl[(1,2,3,5-η)-4-cyano-2-cyclooctene-1,4-diyl]-osmium, Os-00175

$C_{12}H_{11}NO_3Ru$
Tricarbonyl[(1,4,5,6-η)-2-cyano-5-cyclooctene-1,4-diyl]-ruthenium, Ru-00236

$C_{12}H_{11}NRu$
(Cyano-π-cyclopentadienyl)(methyl-π-cyclopentadienyl)-ruthenium, *in* Ru-00245
(1-Cyano-2-methyl-π-cyclopentadienyl)(π-cyclopentadienyl)ruthenium, *in* Ru-00246
(1-Cyano-3-methyl-π-cyclopentadienyl)(π-cyclopentadienyl)ruthenium, *in* Ru-00247

$C_{12}H_{12}BCl_3F_4Os_2$
Bis(η^6-benzene)tri-μ-chlorodiosmium(1+); Tetrafluoroborate, *in* Os-00176

$C_{12}H_{12}BCl_3F_4Ru_2$
Bis(η^6-benzene)tri-μ-chlorodiruthenium(1+); Tetrafluoroborate, *in* Ru-00238

$C_{12}H_{12}B_2F_8ORu$
(η^6-Benzene)(η^6-phenol)ruthenium(2+); Bis(tetrafluoroborate), *in* Ru-00241

$C_{12}H_{12}Br_4Ru_2$
Bis(η^6-benzene)di-μ-bromodibromodiruthenium, Ru-00237

$C_{12}H_{12}Cl_2O_8Ru$
▷Bis(η^6-Benzene)ruthenium(2+); Diperchlorate, *in* Ru-00251

$C_{12}H_{12}Cl_3F_6Ru_2$
Bis(η^6-benzene)tri-μ-chlorodiruthenium(1+); Hexafluorophosphate, *in* Ru-00238

$C_{12}H_{12}Cl_3Os_2^{\oplus}$
Bis(η^6-benzene)tri-μ-chlorodiosmium(1+), Os-00176

$C_{12}H_{12}Cl_3Ru_2^{\oplus}$
Bis(η^6-benzene)tri-μ-chlorodiruthenium(1+), Ru-00238

$C_{12}H_{12}Cl_4Os_2$
Bis(η^6-benzene)di-μ-chlorodichlorodiosmium, *in* Os-00036

$C_{12}H_{12}Cl_4Ru_2$
Bis(η^6-benzene)di-μ-chlorodichlorodiruthenium, *in* Ru-00038

$C_{12}H_{12}F_{12}OsP_2$
Bis(η^6-benzene)osmium(2+); Bis(hexafluorophosphate), *in* Os-00179

$C_{12}H_{12}F_{12}P_2Ru$
Bis(η^6-Benzene)ruthenium(2+); Bis(hexafluorophosphate), *in* Ru-00251

$C_{12}H_{12}I_4Ru_2$
Bis(η^6-benzene)di-μ-iododiiododiruthenium, Ru-00239

$C_{12}H_{12}OOs$
Acetylosmocene, Os-00177

$C_{12}H_{12}ORu$
Acetylruthenocene, Ru-00240
(η^6-Benzene)(η^6-phenol)ruthenium(2+), Ru-00241
1′-Methylruthenocenecarboxaldehyde, Ru-00242
2-Methylruthenocenecarboxaldehyde, Ru-00243
3-Methylruthenocenecarboxaldehyde, Ru-00244

$C_{12}H_{12}O_2Os$
Methoxycarbonylosmocene, *in* Os-00136

$C_{12}H_{12}O_2Ru$
Acetoxyruthenocene, *in* Ru-00142
Methoxycarbonylruthenocene, *in* Ru-00185
1′-Methylruthenocenecarboxylic acid, Ru-00245
2-Methylruthenocenecarboxylic acid, Ru-00246
3-Methylruthenocenecarboxylic acid, Ru-00247

$C_{12}H_{12}O_4Os$
Tricarbonyl[(2,3,4,5-η)-2,3,4,5-tetramethyl-2,4-cyclopentadien-1-one]osmium, Os-00178

$C_{12}H_{12}O_8Ru_2Sn_2$
Octacarbonylbis[μ-(dimethylstannylene)]diruthenium, Ru-00248

$C_{12}H_{12}Os^{\oplus\oplus}$
Bis(η^6-benzene)osmium(2+), Os-00179
Vinylosmocene, Os-00180

$C_{12}H_{12}OsRu_2$
Tetracarbonyl(octacarbonyldiruthenium)osmium, Os-00181

$C_{12}H_{12}Ru$
[(1,2,3,4-η)-Benzene](η^6-benzene)ruthenium, Ru-00250
Bis(η^6-Benzene)ruthenium(2+), Ru-00251
Vinylruthenocene, Ru-00252

$C_{12}H_{13}BF_4ORu$
(η^5-2,4-Cyclopentadien-1-yl)[(1,2,3,4,5,6-η)-methoxybenzene]ruthenium(1+); Tetrafluoroborate, *in* Ru-00253

$C_{12}H_{13}F_6PRu$
(η^5-2,4-Cyclopentadien-1-yl)[(1,2,3,4,5,6-η)-methylbenzene]ruthenium(1+); Hexafluorophosphate, *in* Ru-00254

$C_{12}H_{13}ORu^{\oplus}$
(η^5-2,4-Cyclopentadien-1-yl)[(1,2,3,4,5,6-η)-methoxybenzene]ruthenium(1+), Ru-00253

$C_{12}H_{13}Ru^{\oplus}$
(η^5-2,4-Cyclopentadien-1-yl)[(1,2,3,4,5,6-η)-methylbenzene]ruthenium(1+), Ru-00254

$C_{12}H_{14}ClRu^{\oplus}$
(η^6-Benzene)chloro[(1,2,3,4-η)-1,3-cyclohexadiene]-ruthenium(1+), Ru-00255

$C_{12}H_{14}OOs$
(1-Hydroxyethyl)osmocene, Os-00182

$C_{12}H_{14}ORu$
(1-Hydroxyethyl)ruthenocene, Ru-00256
1-(Hydroxymethyl)-2-methylruthenocene, Ru-00257

$C_{12}H_{14}O_2Ru$
1,1′-Bis(hydroxymethyl)ruthenocene, Ru-00258

$C_{12}H_{14}O_3RuS$
Ruthenocenesulfonic acid; Et ester, *in* Ru-00146

$C_{12}H_{14}Os$
(η^6-Benzene)[(1,2,3,4-η)-1,3-Cyclohexadiene]osmium, Os-00183
1,1′-Dimethylosmocene, Os-00184
Ethylosmocene, Os-00185

$C_{12}H_{14}Ru$
(η^6-Benzene)[(1,2,3,4-η)-1,3-cyclohexadiene]ruthenium, Ru-00259
Bis(η^5-cyclohexa-2,4-dien-1-yl)ruthenium, Ru-00260
1,1′-Dimethylruthenocene, Ru-00261
1,2-Dimethylruthenocene, Ru-00262
1,3-Dimethylruthenocene, Ru-00263
Ethylruthenocene, Ru-00264

$C_{12}H_{15}B_2F_8N_3Ru$
Tris(acetonitrile)(η^6-benzene)ruthenium(2+); Bis(tetrafluoroborate), *in* Ru-00265

$C_{12}H_{15}BrO_2Os$
Bromodicarbonyl[(1,2,3,4,5-η)-1,2,3,4,5-pentamethyl-2,4-cyclopentadien-1-yl]osmium, Os-00186

$C_{12}H_{15}IO_2Os$
Dicarbonyl[(1,2,3,4,5-η)-1,2,3,4,5-pentamethyl-2,4-cyclopentadien-1-yl]iodoosmium, Os-00187

$C_{12}H_{15}N_3Ru^{\oplus\oplus}$
Tris(acetonitrile)(η^6-benzene)ruthenium(2+), Ru-00265

$C_{12}H_{16}B_2Ru$
Bis[(1,2,3,4,5,6-η)-1-methylboratabenzene]ruthenium, Ru-00266

$C_{12}H_{16}O_2Os$
Dicarbonylhydro[(1,2,3,4,5-η)-1,2,3,4,5-pentamethyl-2,4-cyclopentadien-1-yl]osmium, Os-00188

$C_{12}H_{18}B_2F_8N_6Ru$
Hexakis(isocyanomethane)ruthenium(2+); Bis(tetrafluoroborate), *in* Ru-00270

$C_{12}H_{18}Cl_2N_6O_8Os$
Hexakis(isocyanomethane)osmium(2+); Diperchlorate, *in* Os-00192

$C_{12}H_{18}Cl_2Ru$
Dichloro[(1,2,3,6,7,10,11,12-η)-2,6,10-dodecatriene-1,12-diyl]ruthenium, Ru-00267
Dichloro(hexamethylbenzene)ruthenium, *in* Ru-00612

$C_{12}H_{18}F_6O_8OsP_2$
(Hexafluoro-2-butyne)dicarbonylbis(trimethyl phosphite-*P*)osmium, Os-00189

$C_{12}H_{18}F_{12}N_6P_2Ru$
Hexakis(isocyanomethane)ruthenium(2+); Bis(hexafluorophosphate), *in* Ru-00270

$C_{12}H_{18}GeO_2Os$
Dicarbonyl[(1,2,3,4,5-η)-2,4-cycloheptadien-1-yl]-(trimethylgermyl)osmium, Os-00190

$C_{12}H_{18}GeO_2Ru$
Dicarbonyl[(1,2,3,4,5-η)-2,4-cycloheptadien-1-yl]-(trimethylgermyl)ruthenium, Ru-00268

$C_{12}H_{18}Ge_3O_6Os_2$
Hexacarbonyltris[μ-(dimethylgermylene)]diosmium, Os-00191

$C_{12}H_{18}Ge_3O_6Ru_2$
Hexacarbonyltris[μ-(dimethylgermylene)]diruthenium, Ru-00269

$C_{12}H_{18}N_6Os^{\oplus\oplus}$
Hexakis(isocyanomethane)osmium(2+), Os-00192

$C_{12}H_{18}N_6Ru^{\oplus\oplus}$
Hexakis(isocyanomethane)ruthenium(2+), Ru-00270

$C_{12}H_{18}O_2OsSi$
Dicarbonyl[(1,2,3,4,5-η)-2,4-cyclopentadien-1-yl]-(trimethylsilyl)osmium, Os-00193

$C_{12}H_{18}O_2RuSi$
Dicarbonyl[(1,2,3,4,5-η)-2,4-cycloheptadien-1-yl]-(trimethylsilyl)ruthenium, Ru-00271

$C_{12}H_{18}O_6Ru_2Si_3$
Hexacarbonyltris[μ-(dimethylsilyl)]diruthenium, Ru-00272

$C_{12}H_{22}F_6OsP_2$
(η^6-Benzene)(η^2-ethene)methyl(trimethylphosphine)-osmium(1+); Hexafluorophosphate, *in* Os-00194

$C_{12}H_{22}F_6P_2Ru$
(η^6-Benzene)(η^2-ethene)methyl(trimethylphosphine)-ruthenium(1+); Hexafluorophosphate, *in* Ru-00273

$C_{12}H_{22}OsP^{\oplus}$
(η^6-Benzene)(η^2-ethene)methyl(trimethylphosphine)-osmium(1+), Os-00194

$C_{12}H_{22}PRu^{\oplus}$
(η^6-Benzene)(η^2-ethene)methyl(trimethylphosphine)-ruthenium(1+), Ru-00273

$C_{12}H_{23}F_6OP_3Ru$
Carbonyl(η^5-2,4-cyclopentadien-1-yl)-bis(trimethylphosphine)ruthenium(1+); Hexafluorophosphate, *in* Ru-00274

$C_{12}H_{23}OP_2Ru^{\oplus}$
Carbonyl(η^5-2,4-cyclopentadien-1-yl)-bis(trimethylphosphine)ruthenium(1+), Ru-00274

$C_{12}H_{24}O_6OsP_2$
(η^6-Benzene)bis(trimethyl phosphite-*P*)osmium, Os-00195

$C_{12}H_{24}O_6P_2Ru$
(η^6-Benzene)bis(trimethyl phosphite-*P*)ruthenium, Ru-00275

$C_{12}H_{24}P_2Ru$
(η^6-Benzene)bis(trimethylphosphine)ruthenium, Ru-00276

$C_{12}H_{38}O_{12}P_4Ru$
Dihydrotetrakis(trimethyl phosphite-*P*)ruthenium, Ru-00277

$C_{12}H_{38}OsP_4$
Dihydrotetrakis(trimethylphosphine)osmium, Os-00196

$C_{12}H_{38}P_4Ru$
Dihydrotetrakis(trimethylphosphine)ruthenium, Ru-00278

$C_{12}I_2O_{12}Os_3$
Dodecacarbonyldiiodotriosmium, Os-00197

$C_{12}MnO_{12}Os_2^{\ominus}$
Dodecacarbonylmanganatediosmate(1−), Os-00198

$C_{12}O_{12}Os_2Re^{\ominus}$
Dodecacarbonylrhenatediosmate(1−), Os-00199

$C_{12}O_{12}Os_2Ru$
Octacarbonyl(tetracarbonylruthenium)diosmium, Os-00200

$C_{12}O_{12}Os_3$
Dodecacarbonyltriosmium, Os-00201

$C_{12}O_{12}Os_4S$
Dodecacarbonyl-μ_3-thioxotetraosmium, Os-00202

$C_{12}O_{12}Os_4S_2$
Dodecacarbonyldi-μ3-thioxotetraosmium, Os-00203

$C_{12}O_{12}Ru_3$
Dodecacarbonyltriruthenium, Ru-00280

$C_{12}O_{16}Os_4$
Dodecacarbonyltetra-μ3-oxotetraosmium, Os-00204

$C_{13}ClO_{13}Ru_4^{\ominus}$
μ-Carbonyldodecacarbonyl-μ-chlorotetraruthenate(1−), Ru-00281

$C_{13}HFeO_{13}Ru_3^{\ominus}$
Di-μ-carbonylnonacarbonyl(dicarbonylferrate)-μ-hydrotriruthenate(1−), Ru-00282

$C_{13}HFe_2O_{13}Ru_2^{\ominus}$
Hexacarbonyl(μ-carbonylhexacarbonyldiferrate)-μ-hydrodiruthenate(1−), Ru-00283

$C_{13}HKO_{13}Ru_4$
μ3-Carbonyldodecacarbonylhydrotetraruthenate(1−); K salt, *in* Ru-00284

$C_{13}HO_{13}Os_4^{\ominus}$
μ-Carbonyldodecacarbonyl-μ-hydrotetraosmate(1−), Os-00205

$C_{13}HO_{13}Ru_4^{\ominus}$
μ3-Carbonyldodecacarbonylhydrotetraruthenate(1−), Ru-00284

$C_{13}H_2FeO_{13}OsRu_2$
Di-μ-carbonyltricarbonyl(dicarbonyliron)(hexacarbonyl-μ-hydrodiruthenium)-μ-hydroosmium, Os-00206

$C_{13}H_2FeO_{13}Os_2Ru$
Di-μ-carbonylhexacarbonyl(dicarbonyliron)di-μ-hydro(tricarbonylruthenium)diosmium, Os-00207

$C_{13}H_2FeO_{13}Os_3$
Di-μ-carbonylnonacarbonyl(dicarbonyliron)di-μ-hydrotriosmium, Os-00208

$C_{13}H_2FeO_{13}Ru_3$
Di-μ-carbonyldicarbonyldi-μ-hydrotris[μ3-(tricarbonylruthenio)]iron, Ru-00287
Tridecacarbonyldihydroirontriruthenium, Ru-00288

$C_{13}H_2Fe_2O_{13}Ru_2$
μ-Carbonylhexacarbonyl(μ-carbonylpentacarbonyldiiron)-di-μ-hydrodiruthenium, Ru-00289

$C_{13}H_2O_{13}Os_3Ru$
Di-μ-carbonylnonacarbonyl(dicarbonylruthenium)di-μ-hydrotriosmium, Os-00209

$C_{13}H_2O_{13}Os_4$
μ3-Carbonyldodecacarbonyldi-μ-hydrotetraosmium, Os-00210

$C_{13}H_2O_{13}Ru_4$
Di-μ-carbonylundecacarbonyldi-μ-hydrotetraruthenium, Ru-00291

$C_{13}H_3BrO_{12}Os_3$
Bromododecacarbonylmethyltriosmium, Os-00211

$C_{13}H_3MnO_{13}Os_3$
Tridecacarbonyltrihydromanganesetriosmium, Os-00212

$C_{13}H_3NO_{11}Os_3$
(Acetonitrile)undecacarbonyltriosmium, Os-00213

$C_{13}H_3NO_{12}Os_4$
Dodecacarbonyl[μ3-[methanaminato(2−)]]tetraosmium, Os-00214

$C_{13}H_3O_{13}Os_3Re$
Tridecacarbonyltrihydrorheniumtriosmium, Os-00215

$C_{13}H_4O_9Os_3$
[μ-[(1,2,3,4-η:1,4-η)-1,3-Butadiene-1,4-diyl]]-nonacarbonyltriosmium, Os-00216

$C_{13}H_4O_{11}Os_4$
Undecacarbonyl[μ3-[(1-η:1,2-η:2-η)-1,2-ethenediyl]]di-μ-hydrotetraosmium, Os-00217

$C_{13}H_5F_5O_2Ru$
Dicarbonyl(η5-2,4-cyclopentadien-1-yl)-(pentafluorophenyl)ruthenium, Ru-00292

$C_{13}H_5F_5O_2RuS$
Dicarbonyl(η5-2,4-cyclopentadien-1-yl)-(pentafluorobenzenethiolato)ruthenium, Ru-00293

$C_{13}H_6O_{11}Os_4$
Undecacarbonyl[μ-(η:η2-ethenyl)]tri-μ-hydrotetraosmium, Os-00218

$C_{13}H_7NO_{10}Ru_3$
Decacarbonyl[μ-[(dimethyliminio)methylene]]-μ-hydrotriruthenium, Ru-00294

$C_{13}H_7NO_{11}Ru_3$
Decacarbonyl[μ-[(dimethylamino)carbonyl-*C*:*O*]]-μ-hydrotriruthenium, Ru-00295

$C_{13}H_7O_{13}Os_4^{\ominus}$
Undecacarbonyltetra-μ-hydro(methoxycarbonyl)-tetraosmate(1−), Os-00219

$C_{13}H_8O_5Ru_2$
μ-Carbonyltetracarbonyl[μ-[(1,2,3,4,5,6,7,8-η)-1,3,5,7-cyclooctatetraene]]diruthenium, Ru-00296

$C_{13}H_8O_6Ru_2$
Hexacarbonyl[μ-[(1,2,3-η:4,5,6-η)-1,3,5-cycloheptatriene]]diruthenium, Ru-00297

$C_{13}H_9NO_{10}Ru_3Si$
μ3-Carbonylnonacarbonyl[μ3-[1,1,1-trimethylsilanaminato(2−)]]triruthenium, Ru-00298

$C_{13}H_{10}O_2Ru$
Dicarbonyl(η5-2,4-cyclopentadien-1-yl)phenylruthenium, Ru-00299

$C_{13}H_{10}O_2RuS$
(Benzenethiolato)dicarbonyl(η5-2,4-cyclopentadien-1-yl)-ruthenium, Ru-00300

$C_{13}H_{10}O_2RuSe$
(Benzeneselenolato)dicarbonyl(η5-2,4-cyclopentadien-1-yl)ruthenium, Ru-00301

$C_{13}H_{11}ClORu$
3-Chloro-3-ruthenocenyl-2-propen-1-al, Ru-00302

$C_{13}H_{12}ClF_6OPRu$
[(1,2,3,4,5-η)-1-Acetyl-2,4-cyclopentadien-1-yl](η6-chlorobenzene)ruthenium(1+); Hexafluorophosphate, *in* Ru-00303

$C_{13}H_{12}ClORu^{\oplus}$
[(1,2,3,4,5-η)-1-Acetyl-2,4-cyclopentadien-1-yl](η6-chlorobenzene)ruthenium(1+), Ru-00303

$C_{13}H_{12}ORu$
Carbonyl[(1,2,3,3*a*,7*a*-η)-1*H*-inden-1-yl](η3-2-propenyl)-ruthenium, Ru-00304
[3]Ruthenocenophan-1-one, Ru-00305

$C_{13}H_{12}O_2Ru^{\oplus\oplus}$
(η6-Benzene)[(1,2,3,4,5,6-η)-benzoic acid]ruthenium(2+), Ru-00306
3-Ruthenocenyl-2-propenoic acid, Ru-00307

$C_{13}H_{13}F_6OPRu$
[(1,2,3,4,5-η)-1-Acetyl-2,4-cyclopentadien-1-yl](η6-benzene)ruthenium(1+); Hexafluorophosphate, *in* Ru-00308

$C_{13}H_{13}ORu^{\oplus}$
[(1,2,3,4,5-η)-1-Acetyl-2,4-cyclopentadien-1-yl](η6-benzene)ruthenium(1+), Ru-00308

$C_{13}H_{14}B_2F_8ORu$
(η6-Benzene)[(1,2,3,4,5,6-η)-methoxybenzene]-ruthenium(2+); Bis(tetrafluoroborate), *in* Ru-00310

$C_{13}H_{14}ORu$
1-Acetyl-2-methylruthenocene, Ru-00309
(η6-Benzene)[(1,2,3,4,5,6-η)-methoxybenzene]-ruthenium(2+), Ru-00310
Propanoylruthenocene, Ru-00311

$C_{13}H_{14}O_2Ru$
1′-Methylruthenocenecarboxylic acid; Me ester, *in* Ru-00245
2-Methylruthenocenecarboxylic acid; Me ester, *in* Ru-00246

3-Methylruthenocenecarboxylic acid; Me ester, *in* Ru-00247
3-Ruthenocenylpropanoic acid, Ru-00312

$C_{13}H_{14}Ru$
[3]Ruthenocenophane, Ru-00313

$C_{13}H_{15}BF_4O_3Os$
Tricarbonyl[(1,2,3,4,5-η)-1,2,3,4,5-pentamethyl-2,4-cyclopentadien-1-yl]osmium(1+); Tetrafluoroborate, *in* Os-00220

$C_{13}H_{15}F_6O_2PRu$
Dicarbonyl[(1,2-η)-cyclohexene](η^5-2,4-cyclopentadien-1-yl)ruthenium(1+); Hexafluorophosphate, *in* Ru-00314

$C_{13}H_{15}F_6PRu$
(η^6-Benzene)[(1,2,3,4,5-η)-1-ethyl-2,4-cyclopentadien-1-yl]ruthenium(1+); Hexafluorophosphate, *in* Ru-00315

$C_{13}H_{15}O_2Ru^{\oplus}$
Dicarbonyl[(1,2-η)-cyclohexene](η^5-2,4-cyclopentadien-1-yl)ruthenium(1+), Ru-00314

$C_{13}H_{15}O_3Os^{\oplus}$
Tricarbonyl[(1,2,3,4,5-η)-1,2,3,4,5-pentamethyl-2,4-cyclopentadien-1-yl]osmium(1+), Os-00220

$C_{13}H_{15}Ru^{\oplus}$
(η^6-Benzene)[(1,2,3,4,5-η)-1-ethyl-2,4-cyclopentadien-1-yl]ruthenium(1+), Ru-00315

$C_{13}H_{16}ORu$
(1-Hydroxy-1-methylethyl)ruthenocene, Ru-00316

$C_{13}H_{17}NRu$
(*N*,*N*-Dimethylaminomethyl)ruthenocene, Ru-00317

$C_{13}H_{18}Cl_2ORu$
Carbonyldichloro[(1,2,3,4,5,6-η)-hexamethylbenzene]ruthenium, Ru-00318

$C_{13}H_{18}GeO_2Os$
Dicarbonyl[(1,2,3,3*a*,6*a*-η)-1,4,5,6-tetrahydro-1-pentalenyl](trimethylgermyl)osmium, Os-00221

$C_{13}H_{18}GeO_2Ru$
Dicarbonyl[(1,2,3,3*a*,6*a*-η)-1,4,5,6-tetrahydro-1-pentalenyl](trimethylgermyl)ruthenium, Ru-00319

$C_{13}H_{18}N_2O_3Ru$
Bis(2-isocyano-2-methylpropane)tricarbonylruthenium, Ru-00320

$C_{13}H_{18}O_2OsSi$
Dicarbonyl[(1,2,3,3*a*,6*a*-η)-1,4,5,6-tetrahydro-1-pentalenyl](trimethylsilyl)osmium, Os-00222

$C_{13}H_{18}O_2RuSi$
Dicarbonyl[(1,2,3,3*a*,6*a*-η)-1,4,5,6-tetrahydro-1-pentalenyl](trimethylsilyl)ruthenium, Ru-00321

$C_{13}H_{18}OsSi$
Trimethylsilylosmocene, Os-00223

$C_{13}H_{18}Ru$
[(2,3,5,6-η)-Bicyclo[2.2.1]hepta-2,5-diene]bis(η^3-2-propenyl)ruthenium, Ru-00322

$C_{13}H_{18}RuSi$
Trimethylsilylruthenocene, Ru-00323

$C_{13}K_2O_{13}Ru_4$
μ_3-Carbonyldodecacarbonyltetraruthenate(2−); Di-K salt, *in* Ru-00324

$C_{13}O_{13}Ru_4^{\ominus\ominus}$
μ_3-Carbonyldodecacarbonyltetraruthenate(2−), Ru-00324

$C_{14}H_2BrF_6O_{10}Os_3^{\ominus}$
Bromodecacarbonyl[(2,3-η)-1,1,1,4,4,4-hexafluoro-2-butene]triosmate(1−), Os-00224

$C_{14}H_2O_{12}Os_4$
Dodecacarbonyl[μ_4-[(1-η:1,2-η:1,2-η:2-η)-1,2-ethenediyl]]tetraosmium, Os-00225

$C_{14}H_2O_{12}Ru_4$
Dodecacarbonyl[μ_4-(1-η:η^2:η^2:2-η-1,2-ethenediyl)]tetraruthenium, Ru-00325

$C_{14}H_4N_2O_{10}Ru_3$
Tri-μ-carbonylheptacarbonyl[μ-(pyridazine-N^1:N^2)]triruthenium, Ru-00326

$C_{14}H_6N_2O_{10}Os_3$
Bis(acetonitrile)decacarbonyltriosmium, Os-00226

$C_{14}H_6O_6Os_2$
Hexacarbonyl[μ-(η^2:η^4-1,2-ethenediyl-1,2-phenylene)]diosmium, Os-00227

$C_{14}H_6O_6Ru_2$
Hexacarbonyl[μ-(η^2:η^4-1,2-ethenediyl-1,2-phenylene)]diruthenium, Ru-00327

$C_{14}H_6O_{10}Os_3$
[(1,2,3,4-η)-1,3-Butadiene]decacarbonyltriosmium, Os-00229
Os-00229
[μ-[(1,2-η:3,4-η)-1,3-Butadiene]]decacarbonyltriosmium, Os-00228

$C_{14}H_6O_{12}Os_3$
Dodecacarbonyldimethyltriosmium, Os-00230

$C_{14}H_8O_6Os_2$
Hexacarbonyl[μ-[(1,2,3,4,5,8-η)-1,3,5,7-cyclooctatetraene]]diosmium, Os-00231

$C_{14}H_8O_6Ru_2$
Hexacarbonyl[μ-[(1,2,3,4-η:1,4,5,8-η)-1,3,5,7-cyclooctatetraene]]diruthenium, Ru-00328

$C_{14}H_8O_6Ru_3S$
Hexacarbonyl[μ_3-[(1,2,3,4-η:5,6-η:7,8-η)-1,3,5,7-cyclooctatetraene]]-μ_3-thioxotriruthenium, Ru-00329

$C_{14}H_9ClN_2O_2Ru$
Dicarbonylchloro[2-(phenylazo)phenyl]ruthenium, Ru-00330

$C_{14}H_9Cl_2N_2O_2Ru^{\ominus}$
Dicarbonyldichloro[2-(phenylazo)phenyl]ruthenate(1−), Ru-00331

$C_{14}H_9O_{14}Os_3P$
Undecacarbonyl(trimethyl phosphite-*P*)triosmium, Os-00232

$C_{14}H_{10}HgO_4Ru_2$
Tetracarbonylbis(η^5-2,4-cyclopentadien-1-yl)(mercury)diruthenium, Ru-00332

$C_{14}H_{10}O_2Ru_2S_2$
Bis[μ-(carbonothioyl)]dicarbonylbis(η^5-2,4-cyclopentadien-1-yl)diruthenium, Ru-00333

$C_{14}H_{10}O_3Ru_2S$
[μ-(Carbonothioyl)]-μ-carbonyldicarbonylbis(η^5-2,4-cyclopentadien-1-yl)diruthenium, Ru-00334

$C_{14}H_{10}O_4Os_2$
Tetracarbonylbis(η^5-2,4-cyclopentadien-1-yl)diosmium, Os-00233

$C_{14}H_{10}O_4Ru_2$
Di-μ-carbonyldicarbonylbis(η^5-2,4-cyclopentadien-1-yl)diruthenium, Ru-00335

$C_{14}H_{11}NO_9Os_3$
Nonacarbonyldi-μ-hydro(2-isocyano-2-methylpropane)triosmium, Os-00234

$C_{14}H_{12}N_2O_2RuS_2$
Bis(2-aminothiophenol)dicarbonylruthenium, Ru-00336

$C_{14}H_{12}O_2Os$
Dicarbonyl(1,3,5-cycloheptatrien-1-yl)(η^5-2,4-cyclopentadien-1-yl)osmium, Os-00235

$C_{14}H_{12}O_2Ru$
Benzyldicarbonyl(η^5-2,4-cyclopentadien-1-yl)ruthenium, Ru-00337

$C_{14}H_{12}O_6Os_2$
Hexacarbonyl[μ-[(1,2,3,4-η:1,4-η)-1,2,3,4-tetramethyl-1,3-butadiene-1,4-diyl]]diosmium, Os-00236

$C_{14}H_{12}O_6Ru_2$
Hexacarbonyl[μ-[(1,2,3,4-η:1,4-η)-1,2,3,4-tetramethyl-1,3-butadiene-1,4-diyl]]diruthenium, Ru-00338

$C_{14}H_{13}BF_4O_2Os$
Dicarbonyl[(1,2-η)-1,3,5-cycloheptatriene](η⁵-2,4-cyclopentadien-1-yl)osmium(1+); Tetrafluoroborate, *in* Os-00237

$C_{14}H_{13}O_2Os^{\oplus}$
Dicarbonyl[(1,2-η)-1,3,5-cycloheptatriene](η⁵-2,4-cyclopentadien-1-yl)osmium(1+), Os-00237

$C_{14}H_{13}O_{14}Os_4P$
Undecacarbonyltetra-μ-hydro(trimethyl phosphite-*P*)tetraosmium, Os-00238

$C_{14}H_{13}O_{14}PRu_4$
Undecacarbonyltetra-μ-hydro(trimethyl phosphite-*P*)tetraruthenium, Ru-00339

$C_{14}H_{14}B_2F_8ORu$
(η⁶-Benzene)[1-(η⁶-phenyl)ethanone]ruthenium(2+); Bis-(tetrafluoroborate), *in* Ru-00342

$C_{14}H_{14}B_2F_8O_2Ru$
(η⁶-Benzene)[(1,2,3,4,5,6-η)-methylbenzoate]-ruthenium(2+) bis(tetrafluoroborate), *in* Ru-00306

$C_{14}H_{14}Br_2N_2O_2Ru$
Bis(aniline)dibromodicarbonylruthenium, Ru-00340

$C_{14}H_{14}Cl_2N_2O_2Ru$
Bis(aniline)dicarbonyldichlororuthenium, Ru-00341

$C_{14}H_{14}ORu^{\oplus\oplus}$
(η⁶-Benzene)[1-(η⁶-phenyl)ethanone]ruthenium(2+), Ru-00342
π-Cyclopentadienyl(4,5,6,7-tetrahydro-4-oxo-π-indenyl)-ruthenium, Ru-00343
[4]-Ruthenocenophan-1-one, Ru-00344
[4]Ruthenocenophan-2-one, Ru-00345

$C_{14}H_{14}O_2Os$
1,1′-Diacetylosmocene, Os-00239

$C_{14}H_{14}O_2Ru$
1,1′-Diacetylruthenocene, Ru-00346

$C_{14}H_{14}O_3Ru$
γ-Oxoruthenocenebutanoic acid, Ru-00347

$C_{14}H_{14}Os$
(η⁶-Benzene)[(1,2,3,4-η)-1,3,5,7-cyclooctatetraene]-osmium, Os-00240

$C_{14}H_{14}Ru$
(η⁶-Benzene)[(1,2,3,4-η)-1,3,5,7-cyclooctatetraene]-ruthenium, Ru-00348
π-Cyclopentadienyl(4,5-dihydro-π-indenyl)ruthenium, Ru-00349
1,1′-Divinylruthenocene, Ru-00350

$C_{14}H_{16}Cl_4Ru_2$
Di-μ-chlorodichlorobis(1,3,5-cycloheptatriene)-diruthenium, *in* Ru-00059

$C_{14}H_{16}OOs$
(2-Methyl-1-oxopropyl)osmocene, Os-00241

$C_{14}H_{16}ORu$
π-Cyclopentadienyl(4,5,6,7-tetrahydro-4-hydroxy-π-indenyl)ruthenium, Ru-00351

$C_{14}H_{16}O_2Os$
(1-Hydroxyethyl)osmocene; Ac, *in* Os-00182

$C_{14}H_{16}O_2Ru$
(1-Hydroxyethyl)ruthenocene; Ac, *in* Ru-00256
Ruthenocenebutanoic acid, Ru-00352

$C_{14}H_{16}O_2Ru_2S_2$
Dicarbonylbis(η⁵-2,4-cyclopentadien-1-yl)bis[μ-(methanethiolato)]diruthenium, Ru-00353

$C_{14}H_{16}O_4RuSi_2$
Tetracarbonyl[1,2-phenylenebis[dimethylsilylene]]-ruthenium, Ru-00354

$C_{14}H_{16}Ru$
(Bicyclo[2.2.1]hepta-2,5-diene)(1,3,5-cycloheptatriene)-ruthenium, Ru-00355
[(1,2,3,4,5-η)-2,4-Cycloheptadien-1-yl][(1,2,3,4,5-η)-2,4,6-cycloheptatrien-1-yl]ruthenium, Ru-00356
[4]Ruthenocenophane, Ru-00357

$C_{14}H_{17}BF_4Ru$
[(1,2,3,4,5-η)-2,4-Cycloheptadien-1-yl][(1,2,3,4,5,6-η)-1,3,5-cycloheptatriene]ruthenium(1+); Tetrafluoroborate, *in* Ru-00359

$C_{14}H_{17}F_6PRu$
(η⁶-Benzene)[(1,2,3,4,5-η)-2,4,6-cyclooctatrien-1-yl]-ruthenium(1+); Hexafluorophosphate, *in* Ru-00358
[(1,2,3,4,5-η)-2,4-Cycloheptadien-1-yl][(1,2,3,4,5,6-η)-1,3,5-cycloheptatriene]ruthenium(1+); Hexafluorophosphate, *in* Ru-00359

$C_{14}H_{17}Ru^{\oplus}$
(η⁶-Benzene)[(1,2,3,4,5-η)-2,4,6-cyclooctatrien-1-yl]-ruthenium(1+), Ru-00358
[(1,2,3,4,5-η)-2,4-Cycloheptadien-1-yl][(1,2,3,4,5,6-η)-1,3,5-cycloheptatriene]ruthenium(1+), Ru-00359

$C_{14}H_{18}F_6O_3P_2Ru$
(η⁶-Benzene)[(1,2,3,4,5,6-η)-6-(dimethoxyphosphinyl)-2,4-cyclohexadien-1-yl]ruthenium(1+); Hexafluorophosphate, *in* Ru-00362

$C_{14}H_{18}Ge_2O_8Ru_2$
Octacarbonylbis(trimethylgermyl)diruthenium, Ru-00360

$C_{14}H_{18}N_2O_4Ru_2$
Tetracarbonyl[μ-[η⁴-*N*,*N*′-1,2-ethanediylidenebis[2-propanamine]-*N*,*N*′]]-μ-1,2-ethenediyldiruthenium, Ru-00361

$C_{14}H_{18}OOs$
(1-Hydroxy-2-methylpropyl)osmocene, Os-00242

$C_{14}H_{18}O_3PRu^{\oplus}$
(η⁶-Benzene)[(1,2,3,4,5,6-η)-6-(dimethoxyphosphinyl)-2,4-cyclohexadien-1-yl]ruthenium(1+), Ru-00362

$C_{14}H_{18}O_8Os_2Si_2$
Octacarbonylbis(trimethylsilyl)diosmium, Os-00243

$C_{14}H_{18}O_8Ru_2Si_2$
Octacarbonylbis(trimethylsilyl)diruthenium, Ru-00363

$C_{14}H_{18}O_8Ru_2Sn_2$
Octacarbonylbis(trimethylstannyl)diruthenium, Ru-00364

$C_{14}H_{18}Ru$
(η⁶-Benzene)[(1,2,5,6-η)-1,5-cyclooctadiene]ruthenium, Ru-00365
Bis[(1,2,3,4,5-η)-2,4-cycloheptadien-1-yl]ruthenium, Ru-00366
1,1′-Diethylruthenocene, Ru-00367

$C_{14}H_{19}NRu$
1-[(Dimethylamino)ethyl]ruthenocene, Ru-00368

$C_{14}H_{20}INRu$
(*N*,*N*-Dimethylaminomethyl)ruthenocene; B,MeI, *in* Ru-00317

$C_{14}H_{22}Os$
[(1,2,5,6-η)-1,5-Cyclooctadiene]bis(η³-2-propenyl)-osmium, Os-00244

$C_{14}H_{22}Ru$
[(1,2,5,6-η)-1,5-Cyclooctadiene]bis(η³-2-propenyl)-ruthenium, Ru-00369

$C_{14}H_{24}N_4O_2Ru_2S_8$
Dicarbonylbis[μ-(dimethylcarbamodithioato-*S*:*S*,*S*′)]-bis(dimethylcarbamodithioato-*S*,*S*′)-diruthenium, Ru-00370

$C_{14}H_{27}F_{12}NP_4Ru$
(Acetonitrile)(η⁶-benzene)bis(trimethylphosphine)-ruthenium(2+); Bis(hexafluorophosphate), *in* Ru-00371

$C_{14}H_{27}IO_7OsP_2$
Carbonyl[(1,2,3,4,5-η)-2,4-cycloheptadien-1-yl]-bis(trimethyl phosphite-*P*)osmium(1+); Iodide, *in* Os-00245

$C_{14}H_{27}NP_2Ru^{\oplus\oplus}$
(Acetonitrile)(η⁶-benzene)bis(trimethylphosphine)-ruthenium(2+), Ru-00371

$C_{14}H_{27}O_7OsP_2^{\oplus}$
Carbonyl[(1,2,3,4,5-η)-2,4-cycloheptadien-1-yl]-bis(trimethyl phosphite-*P*)osmium(1+), Os-00245

$C_{14}H_{28}F_{12}P_4Ru$
(η^6-Benzene)(η^2-ethene)bis(trimethylphosphine)ruthenium(2+); Bis(hexafluorophosphate), *in* Ru-00372
$C_{14}H_{28}P_2Ru^{\oplus\oplus}$
(η^6-Benzene)(η^2-ethene)bis(trimethylphosphine)ruthenium(2+), Ru-00372
$C_{14}H_{29}F_6P_3Ru$
(η^5-2,4-Cyclopentadien-1-yl)[(1,2-η)-1-propene]bis(trimethylphosphine)ruthenium(1+); Hexafluorophosphate, *in* Ru-00373
$C_{14}H_{29}P_2Ru^{\oplus}$
(η^5-2,4-Cyclopentadien-1-yl)[(1,2-η)-1-propene]bis(trimethylphosphine)ruthenium(1+), Ru-00373
$C_{14}H_{32}F_6O_9P_4Ru$
(η^5-2,4-Cyclopentadien-1-yl)tris(trimethyl phosphite-*P*)ruthenium(1+); Hexafluorophosphate, *in* Ru-00375
$C_{14}H_{32}F_6P_4Ru$
(η^5-2,4-Cyclopentadien-1-yl)tris(trimethylphosphine)ruthenium(1+); Hexafluorophosphate, *in* Ru-00376
$C_{14}H_{32}O_6P_2Ru$
Bis[(1,2,3-η)-2-methyl-2-propenyl]bis(trimethyl phosphite-*P*)ruthenium, Ru-00374
$C_{14}H_{32}O_9P_3Ru^{\oplus}$
(η^5-2,4-Cyclopentadien-1-yl)tris(trimethyl phosphite-*P*)ruthenium(1+), Ru-00375
$C_{14}H_{32}P_3Ru^{\oplus}$
(η^5-2,4-Cyclopentadien-1-yl)tris(trimethylphosphine)ruthenium(1+), Ru-00376
$C_{14}H_{36}P_4Ru$
Bis[1,2-ethanediylbis[dimethylphosphine]-*P,P'*](η^2-ethene)ruthenium, Ru-00377
$C_{14}H_{37}F_6N_6PRu$
[(1,2,5,6-η)-1,5-Cyclooctadiene]tris(1,1-dimethylhydrazine-N^2)hydroruthenium(1+); Hexafluorophosphate, *in* Ru-00378
$C_{14}H_{37}N_6Ru^{\oplus}$
[(1,2,5,6-η)-1,5-Cyclooctadiene]tris(1,1-dimethylhydrazine-N^2)hydroruthenium(1+), Ru-00378
$C_{14}H_{39}B_7P_2Ru$
[η^6-Nonahydrodicarbanonaborato(2−)-*B,H*]bis(triethylphosphine)ruthenium, Ru-00379
$C_{14}Mn_2O_{14}Os$
Tetracarbonylbis(pentacarbonylmanganio)osmium, Os-00246
$C_{14}Mn_2O_{14}Ru$
Tetracarbonylbis(pentacarbonylmanganio)ruthenium, Ru-00380
$C_{14}O_{14}OsRe_2$
Tetracarbonylbis(pentacarbonylrhenio)osmium, Os-00247
$C_{15}HO_{15}Os_5^{\ominus}$
Pentadecacarbonyl-μ-hydropentaosmate(1−), Os-00248
$C_{15}H_2IO_{15}Os_5^{\ominus}$
Pentadecacarbonyldi-μ-hydroiodopentaosmate(1−), Os-00249
$C_{15}H_2O_{15}Os_5$
Pentadecacarbonyldi-μ-hydropentaosmium, Os-00250
$C_{15}H_5NO_{10}Os_3$
Decacarbonyl-μ-hydro[μ-(2-pyridinyl-*C:N*)]triosmium, Os-00251
$C_{15}H_6O_9Os_3$
Nonacarbonyl[μ_3-[(5-η:5,6-η:6-η)-1,3-cyclohexadien-5-yne]]di-μ-hydrotriosmium, Os-00252
$C_{15}H_7BF_4O_9Os_3$
(Benzene)nonacarbonyl-μ-hydrotriosmium(1+); Tetrafluoroborate, *in* Os-00254
$C_{15}H_7NO_9Os_3$
(μ_3-Anilinato)nonacarbonyldi-μ-hydrotriosmium, Os-00253
$C_{15}H_7NO_9Ru_3$
[μ_3-[Benzenaminato(2−)]]nonacarbonyldi-μ-hydrotriruthenium, Ru-00381
$C_{15}H_7O_9Os_3^{\oplus}$
(Benzene)nonacarbonyl-μ-hydrotriosmium(1+), Os-00254
$C_{15}H_7O_9Os_3P$
Nonacarbonyldi-μ-hydro[μ_3-(phenylphosphinidene)]triosmium, Os-00255
$C_{15}H_7O_9PRu_3$
Nonacarbonyldi-μ-hydro[μ_3-(phenylphosphinidene)]triruthenium, Ru-00382
$C_{15}H_8O_5Ru_2$
[μ-[(1,2,3,3*a*,8*a*-η:4,5,6-η)-azulene]]pentacarbonyldiruthenium, Ru-00383
$C_{15}H_8O_9Os_3$
Nonacarbonyl[μ_3-[(1,2-η:3-η:4,5-η)-2,4-cyclohexadien-1-yl]]-μ-hydrotriosmium, Os-00256
$C_{15}H_8O_9Ru_3$
Nonacarbonyl[μ_3-[(1-η:1,2-η:2-η)-1,3-cyclohexadiene-1,2-diyl]]di-μ-hydrotriruthenium, Ru-00384
$C_{15}H_9BrO_4RuW$
Bromotetracarbonyl(ruthenocenylmethylidyne)tungsten, Ru-00385
$C_{15}H_9ClO_4RuW$
Tetracarbonylchloro(ruthenocenomethylidyne)tungsten, Ru-00386
$C_{15}H_9O_9Ru_3^{\ominus}$
Nonacarbonyl[μ_3-[(1-η:1,2-η:1,2-η)-3,3-dimethyl-1-butynyl]]triruthenate(1−), Ru-00387
$C_{15}H_{10}NO_{10}Os_3^{\ominus}$
[μ-(*tert*-Butylcarbonimidoyl)]decacarbonyl-μ-hydrotriosmate(1−), Os-00257
$C_{15}H_{10}O_9Ru_3$
Nonacarbonyl[μ_3-[(1-η:1,2-η:1,2-η)-3,3-dimethyl-1-butynyl]]-μ-hydrotriruthenium, Ru-00388
$C_{15}H_{10}O_{10}Ru$
1,2,3,4,5-Ruthenocenepentacarboxylic acid, Ru-00389
$C_{15}H_{10}O_{11}Os_3Si$
μ-Carbonyldecacarbonyl[μ-(trimethylsilyl)methylene]triosmium, Os-00258
$C_{15}H_{11}NO_{10}Os_3$
[μ-(*tert*-Butylcarbonimidoyl)]decacarbonyl-μ-hydrotriosmate(1−); Hydrogen compd., *in* Os-00257
Decacarbonyl-μ-hydrohydro(2-isocyano-2-methylpropane)triosmium, Os-00259
$C_{15}H_{12}O_3Ru_2$
μ-Carbonyldicarbonylbis(η^5-2,4-cyclopentadien-1-yl)-μ-ethenylidenediruthenium, Ru-00390
$C_{15}H_{12}O_7Os_2$
Hexacarbonyl[μ-[(1,2,4,5-η:1,5-η)-1,2,4,5-tetramethyl-3-oxo-1,4-pentadiene-1,5-diyl]]diosmium, Os-00260
$C_{15}H_{12}O_9Ru_3$
Nonacarbonyl[μ_3-[(1-η:1-η:1,2-η)-3,3-dimethyl-1-butenylidene]]di-μ-hydrotriruthenium, Ru-00391
Nonacarbonyl[μ_3-[(1-η:1,2-η:2-η)-1-(1,1-dimethylethyl)-1,2-ethenediyl]]di-μ-hydrotriruthenium(2+), Ru-00392
$C_{15}H_{13}BF_4O_3Ru_2$
μ-Carbonyldicarbonylbis(η^5-2,4-cyclopentadien-1-yl)(μ-ethylidiyne)diruthenium(1+); Tetrafluoroborate, *in* Ru-00393
$C_{15}H_{13}O_3Ru_2^{\oplus}$
μ-Carbonyldicarbonylbis(η^5-2,4-cyclopentadien-1-yl)(μ-ethylidiyne)diruthenium(1+), Ru-00393
$C_{15}H_{14}O_9Ru_3$
Nonacarbonyl[μ_3-(3,3-dimethylbutylidyne)]tri-μ-hydrotriruthenium, Ru-00394
$C_{15}H_{16}BF_6N_6PRu$
(η^6-Benzene)[hydrotris(1*H*-pyrazolato-*N'*)borato(1−)-$N^2,N^{2'},N^{2''}$]ruthenium(1+); Hexafluorophosphate, *in* Ru-00395

$C_{15}H_{16}BN_6Ru^{\oplus}$
(η^6-Benzene)[hydrotris(1*H*-pyrazolato-*N*′)borato(1−)-$N^2,N^{2\prime},N^{2\prime\prime}$]ruthenium(1+), Ru-00395

$C_{15}H_{16}GeO_5Os_2$
Pentacarbonyl-*μ*-2,4,6-cycloheptatrien-1-yl(trimethylgermyl)diosmium, Os-00261

$C_{15}H_{16}NO_{11}Os_3$
μ-Carbonyldecacarbonyl-*μ*-hydrotriosmate(1−); Tetramethylammonium salt, *in* Os-00121

$C_{15}H_{16}O_4Ru_2S$
Tetracarbonyl[*μ*-(1,2,3,4-*η*:1,5,6,7-*η*)-2,4,6-cycloheptatrien-1-yl][*μ*-(2-methyl-2-propanethiolato)]diruthenium, Ru-00396

$C_{15}H_{16}O_5Os_2Si$
Pentacarbonyl-*μ*-2,4,6-cycloheptatrien-1-yl(trimethylsilyl)diosmium, Os-00262

$C_{15}H_{16}Ru$
[(2,3,5,6-*η*)-Bicyclo[2.2.1]hepta-2,5-diene]-[(1,2,3,4,5,6-*η*)-1,3,5,7-cyclooctatetraene]-ruthenium, Ru-00397

$C_{15}H_{17}NRu$
[(1,2,3,4,5-*η*)-6-Cyano-2,4-cycloheptadien-1-yl]-[(1,2,3,4,5-*η*)-2,4-cycloheptadien-1-yl]ruthenium, Ru-00398

$C_{15}H_{18}Ge_3O_9Os_3$
Nonacarbonyltris[*μ*-(dimethylgermylene)]triosmium, Os-00263

$C_{15}H_{18}Ge_3O_9Ru_3$
Nonacarbonyltris[*μ*-(dimethylgermylene)]triruthenium, Ru-00399

$C_{15}H_{18}O_9Os_3Si_3$
Nonacarbonyltris[*μ*-(dimethylsilylene)]triosmium, Os-00264

$C_{15}H_{18}Ru^{\oplus\oplus}$
(η^6-Benzene)[(1,2,3,4,5,6-*η*)-1,3,5-trimethylbenzene]-ruthenium(2+), Ru-00400

$C_{15}H_{19}BF_4Ru$
(η^7-Cycloheptatrienylium)[(1,2,5,6-*η*)-1,5-cyclooctadiene]ruthenium(1+); Tetrafluoroborate, *in* Ru-00401

$C_{15}H_{19}F_6PRu$
(η^5-2,4-Cyclohexadien-1-yl)[(1,2,3,4,5,6-*η*)-1,3,5-trimethylbenzene]ruthenium(1+); Hexafluorophosphate, *in* Ru-00402

$C_{15}H_{19}Ru^{\oplus}$
(η^7-Cycloheptatrienylium)[(1,2,5,6-*η*)-1,5-cyclooctadiene]ruthenium(1+), Ru-00401
(η^5-2,4-Cyclohexadien-1-yl)[(1,2,3,4,5,6-*η*)-1,3,5-trimethylbenzene]ruthenium(1+), Ru-00402

$C_{15}H_{20}Cl_2N_4O_8Ru$
Tetrakis(acetonitrile)[(2,3,5,6-*η*)bicyclo[2.2.1]hepta-2,5-diene]ruthenium(2+); Diperchlorate, *in* Ru-00403

$C_{15}H_{20}N_4Ru^{\oplus\oplus}$
Tetrakis(acetonitrile)[(2,3,5,6-*η*)bicyclo[2.2.1]hepta-2,5-diene]ruthenium(2+), Ru-00403

$C_{15}H_{20}Ru$
[(1,2,3,4,5,6-*η*)-1,3,5-Cycloheptatriene][(1,2,5,6-*η*)-1,5-cyclooctadiene]ruthenium, Ru-00404
(η^4-1,3-Cyclohexadiene)[(1,2,3,4,5,6-*η*)-1,3,5-trimethylbenzene]ruthenium, Ru-00405

$C_{15}H_{23}BF_4N_2Os$
Bis(acetonitrile)[(1,2,5,6-*η*)-1,5-cyclooctadiene](η^3-2-propenyl)osmium(1+); Tetrafluoroborate, *in* Os-00265

$C_{15}H_{23}ClN_2Ru$
Chloro(η^5-2,4-cyclopentadien-1-yl)bis(2-isocyano-2-methylpropane)ruthenium, Ru-00406

$C_{15}H_{23}N_2Os^{\oplus}$
Bis(acetonitrile)[(1,2,5,6-*η*)-1,5-cyclooctadiene](η^3-2-propenyl)osmium(1+), Os-00265

$C_{15}H_{33}F_{12}P_5Ru$
(η^6-Benzene)tris(trimethylphosphine)ruthenium(2+); Bis(hexafluorophosphate), *in* Ru-00407

$C_{15}H_{33}P_3Ru^{\oplus\oplus}$
(η^6-Benzene)tris(trimethylphosphine)ruthenium(2+), Ru-00407

$C_{15}H_{35}O_9P_3Ru$
[(1,2,3,4-*η*)-1,3-Cyclohexadiene]tris(trimethyl phosphite-*P*)ruthenium, Ru-00408

$C_{15}H_{45}BrO_{15}OsP_5^{\oplus}$
Bromopentakis(trimethyl phosphite-*P*)osmium(1+), Os-00266

$C_{15}H_{45}BrO_{15}P_5Ru^{\oplus}$
Bromopentakis(trimethyl phosphite-*P*)ruthenium(1+), Ru-00409

$C_{15}H_{45}ClO_{15}OsP_5^{\oplus}$
Chloropentakis(trimethyl phosphite-*P*)osmium(1+), Os-00267

$C_{15}H_{45}ClO_{15}P_5Ru^{\oplus}$
Chloropentakis(trimethyl phosphite-*P*)ruthenium(1+), Ru-00410

$C_{15}H_{45}O_{15}OsP_5$
Pentakis(trimethyl phosphite-*P*)osmium, Os-00268

$C_{15}H_{45}O_{15}P_5Ru$
Pentakis(trimethyl phosphite-*P*)ruthenium, Ru-00411

$C_{15}H_{46}O_{15}P_5Ru^{\oplus}$
Hydropentakis(trimethyl phosphite-*P*)ruthenium(1+), Ru-00412

$C_{15}IO_{15}Os_5^{\ominus}$
Pentadecacarbonyliodopentaosmate(1−), Os-00269

$C_{15}O_{15}Os_5^{\ominus\ominus}$
Pentadecacarbonylpentaosmate(2−), Os-00270

$C_{16}ClO_{15}Ru_5^{\ominus}$
[μ_5-(Carbon)]pentadecacarbonylchloropentaruthenate(1−), Ru-00413

$C_{16}HBrO_{15}Ru_5$
Bromopentadecacarbonyl-*μ*-hydro-μ_5-methanetetraylpentaruthenium, Ru-00414

$C_{16}HMnO_{16}Os_3$
Hexadecacarbonylhydromanganesetriosmium, Os-00271

$C_{16}HO_{16}Os_3Re$
Hexadecacarbonylhydrorheniumtriosmium, Os-00272

$C_{16}H_2O_{16}Os_5$
Hexadecacarbonyldi-*μ*-hydropentaosmium, Os-00273

$C_{16}H_3O_{15}PRu_5$
Pentadecacarbonyl[μ_4-(methylphosphoranetetrayl)]-pentaruthenium, Ru-00415

$C_{16}H_3O_{16}Os_5P$
Pentadecacarbonyl[μ_4-[methyl phosphinito(2−)-*P*:*P*:*P*:*P*]]-pentaosmium, Os-00274

$C_{16}H_5NO_9Ru_3S_2$
[μ_3-(2(3*H*)-Benzothiazolethionato-N^3:S^2:S^2)]-nonacarbonyl-*μ*-hydrotriruthenium, Ru-00416

$C_{16}H_5NO_{10}Ru_3$
[μ_3-[Benzenaminato(2−)]]-μ_3-carbonylnonacarbonyltriruthenium, Ru-00417

$C_{16}H_6O_8Ru_3$
Octacarbonyl[(1,2,3,3*a*,6*a*-*η*;3*a*,4,5,6,6*a*-*η*)pentalene]-triruthenium, Ru-00418
Octacarbonyl[μ_3-[(1,2-*η*:2,3-*η*:3*a*,4,5,6,6*a*-*η*)-pentalene]]-triruthenium, Ru-00419

$C_{16}H_6O_{10}Os_3S$
[(*μ*-Benzenethiolato)]decacarbonyl-*μ*-hydrotriosmium, Os-00275

$C_{16}H_7BF_4O_{10}Os_3$
Decacarbonyl[(1,2,3,4,5-*η*)-2,4-cyclohexadien-1-yl]-triosmium(1+); Tetrafluoroborate, *in* Os-00277

$C_{16}H_7NO_{10}Os_3$
[μ_2-[Benzeneaminato(1−)]]decacarbonyl-*μ*-hydrotriosmium, Os-00276

$C_{16}H_7O_{10}Os_3^{\oplus}$
Decacarbonyl[(1,2,3,4,5-*η*)-2,4-cyclohexadien-1-yl]-triosmium(1+), Os-00277

$C_{16}H_7O_{10}PRu_3$
Decacarbonyl-μ-hydro[μ-(phenylphosphino)]triruthenium, Ru-00420

$C_{16}H_8O_{10}Os_3$
Decacarbonyl[(1,2,3,4-η)-1,3-cyclohexadiene]triosmium, Os-00278

$C_{16}H_9BF_4N_2O_{12}Os_4$
Bis(acetonitrile)dodecacarbonyltri-μ-hydrotetraosmium(1+); Tetrafluoroborate, *in* Os-00279

$C_{16}H_9N_2O_{12}Os_4^{\oplus}$
Bis(acetonitrile)dodecacarbonyltri-μ-hydrotetraosmium(1+), Os-00279

$C_{16}H_{10}Cl_2N_2O_2Ru$
Bis(benzonitrile)dicarbonyldichlororuthenium, Ru-00421

$C_{16}H_{10}O_7Ru_2$
Hexacarbonyl-μ-hydro[μ-[(*O*,1-η:1,2-η)-3-oxo-1-phenyl-1-butenyl]]diruthenium, Ru-00422

$C_{16}H_{12}MnNO_{12}Os_2$
Dodecacarbonylmanganatediosmate(1−); Tetramethylammonium salt, *in* Os-00198

$C_{16}H_{12}NO_{12}Os_2Re$
Dodecacarbonylrhenatediosmate(1−); Tetramethylammonium salt, *in* Os-00199

$C_{16}H_{14}Br_2O_2RuS_2$
[1,2-Bis(benzenethiolato)ethane-*S*,*S*′]-dibromodicarbonylruthenium, Ru-00423

$C_{16}H_{14}Br_2O_2RuSe_2$
[1,2-Bis(benzeneseleno)ethane-*Se*,*Se*′]-dibromodicarbonylruthenium, Ru-00424

$C_{16}H_{14}O_4Ru_2$
Di-μ-carbonyldicarbonylbis[(1,2,3,4,5-η)-1-methyl-2,4-cyclopentadien-1-yl]diruthenium, Ru-00425

$C_{16}H_{14}Ru$
Phenylruthenocene, Ru-00426

$C_{16}H_{15}NO_{12}Os_4$
Dodecacarbonyltri-μ-hydrotetraosmate(1−); Tetramethylammonium salt, *in* Os-00163

$C_{16}H_{16}ClF_6N_2PRu$
($η^6$-Benzene)chlorobis(pyridine)ruthenium(1+); Hexafluorophosphate, *in* Ru-00427

$C_{16}H_{16}ClN_2Ru^{\oplus}$
($η^6$-Benzene)chlorobis(pyridine)ruthenium(1+), Ru-00427

$C_{16}H_{16}Cl_4O_2Ru_2$
Bis(bicyclo[2.2.1]hepta-2,5-diene)-dicarbonyltetrachlorodiruthenium, Ru-00428

$C_{16}H_{18}B_2F_8O_2Ru$
[(1,2,3,4,5,6-η)-Benzoato][(1,2,3,4,5,6-η)-1,3,5-trimethylbenzene]ruthenium(1+) hydrogen tetrafluoroborate(1:1:2), *in* Ru-00429

$C_{16}H_{18}O_2Ru^{\oplus\oplus}$
($η^6$-Benzoic acid)($η^6$-1,3,5-trimethylbenzene)-ruthenium(2+), Ru-00429

$C_{16}H_{18}O_{16}P_2Ru_3$
Decacarbonylbis(trimethyl phosphite-*P*)triruthenium, Ru-00430

$C_{16}H_{20}N_2O_6Os_2$
Hexacarbonyl[μ-[$η^2$-*N*,*N*′-1,2-ethanediylidenebis[2-methyl-2-propanamine]-*N*,*N*′]]diosmium, Os-00280

$C_{16}H_{20}N_2O_6Ru_2$
Hexacarbonyl[μ-[$η^2$-*N*,*N*′-1,2-ethanediylidenebis[2-methyl-2-propanamine]-*N*,*N*′]]diruthenium, Ru-00431

$C_{16}H_{20}Ru$
[(2,3,4,5-η)Bicyclo[4.2.0]octa-2,4-diene][(1,2,3,4,5,6-η)-1,3,5-cyclooctatriene]ruthenium, Ru-00432

$C_{16}H_{22}O_{16}P_2Ru_4$
Decacarbonyltetra-μ-hydrobis(trimethyl phosphite-*P*)tetraruthenium, Ru-00433

$C_{16}H_{22}Os$
[(1,2,5,6-η)-1,5-Cyclooctadiene][(1,2,3,4,5,6-η)-1,3,5-cyclooctatriene]osmium, Os-00281

$C_{16}H_{22}Ru$
Bis[(1,2,3,4,5-η)-2,4-cyclooctadien-1-yl]ruthenium, Ru-00434
[(1,2,5,6-η)-1,5-Cyclooctadiene][(1,2,3,4,5,6-η)-1,3,5-cyclooctatriene]ruthenium, Ru-00435

$C_{16}H_{23}F_6O_3OsP$
Dicarbonyl[(1,2,3,4,5-η)-1,2,3,4,5-pentamethyl-2,4-cyclopentadien-1-yl](tetrahydrofuran)osmium(1+); Hexafluorophosphate, *in* Os-00282

$C_{16}H_{23}O_3Os^{\oplus}$
Dicarbonyl[(1,2,3,4,5-η)-1,2,3,4,5-pentamethyl-2,4-cyclopentadien-1-yl](tetrahydrofuran)osmium(1+), Os-00282

$C_{16}H_{24}F_{12}N_4P_2Ru$
Tetrakis(acetonitrile)[(1,2,5,6-η)-1,5-cyclooctadiene]-ruthenium(2+); Bis(hexafluorophosphate), *in* Ru-00436

$C_{16}H_{24}N_4Ru^{\oplus\oplus}$
Tetrakis(acetonitrile)[(1,2,5,6-η)-1,5-cyclooctadiene]-ruthenium(2+), Ru-00436

$C_{16}H_{26}O_2RuSi_2$
Dicarbonyl(trimethylsilyl)[(1,2,3,6,7-η)-8-(trimethylsilyl)-2,4,6-cyclooctatrien-1-yl]-ruthenium, Ru-00437

$C_{16}H_{26}Ru$
Bis($η^2$-ethene)[(1,2,3,4,5,6-η)-hexamethylbenzene]-ruthenium, Ru-00438
[(1,2,5,6-η)-1,5-Cyclooctadiene]bis[(1,2,3-η)-2-methyl-2-propenyl]ruthenium, Ru-00439

$C_{16}H_{27}ClF_6OP_2Ru$
Carbonylchloro[(1,2,3,4,5,6-η)-hexamethylbenzene]-(trimethylphosphine)ruthenium(1+); Hexafluorophosphate, *in* Ru-00440

$C_{16}H_{27}ClOPRu^{\oplus}$
Carbonylchloro[(1,2,3,4,5,6-η)-hexamethylbenzene]-(trimethylphosphine)ruthenium(1+), Ru-00440

$C_{16}H_{27}OPRu$
Carbonyl[(1,2,3,4,5,6-η)-hexamethylbenzene]-(trimethylphosphine)ruthenium, Ru-00441

$C_{16}H_{28}F_6OP_2Ru$
Carbonyl[(1,2,3,4,5,6-η)-hexamethylbenzene]-hydro(trimethylphosphine)ruthenium(1+); Hexafluorophosphate, *in* Ru-00443

$C_{16}H_{28}NO_{17}P_3Ru_3$
Heptacarbonyl-μ-hydro-μ-nitrosyltris(trimethyl phosphite-*P*)triruthenium, Ru-00442

$C_{16}H_{28}OPRu^{\oplus}$
Carbonyl[(1,2,3,4,5,6-η)-hexamethylbenzene]-hydro(trimethylphosphine)ruthenium(1+), Ru-00443

$C_{16}H_{30}Ge_4O_6Os_2$
Hexacarbonylbis[μ-(dimethylgermylene)]-bis(trimethylgermyl)diosmium, Os-00283

$C_{16}H_{30}Ge_4O_6Ru_2$
Hexacarbonylbis[μ-(dimethylgermylene)]-bis(trimethylgermyl)diruthenium, Ru-00444

$C_{16}H_{30}O_6Os_2Si_4$
Hexacarbonylbis[μ-(dimethylsilylene)]-bis(trimethylsilyl)diosmium, Os-00284

$C_{16}H_{30}O_6Ru_2Si_4$
Hexacarbonylbis[μ-(dimethylsilylene)]-bis(trimethylsilyl)diruthenium, Ru-00445

$C_{16}H_{30}O_6Ru_2Sn_4$
Hexacarbonylbis[μ-(dimethylstannylene)]-bis(trimethylstannyl)diruthenium, Ru-00446

$C_{16}IO_{15}Os_5^{\ominus}$
[$μ_5$-(Carbon)]pentadecacarbonyliodopentaosmate(1−), Os-00285

$C_{16}O_{15}Os_5$
(Carbon)pentadecacarbonylpentaosmium, Os-00286

$C_{16}O_{15}Ru_5$
(Carbon)pentadecacarbonylpentaruthenium, Ru-00447

$C_{16}O_{16}Os_5$
Hexadecacarbonylpentaosmium, Os-00287

$C_{16}O_{16}Os_6S_2$
Hexadecacarbonyl-μ_3-thioxo-μ_4-thioxohexaosmium, Os-00288

$C_{16}O_{22}Os_6$
Hexadecacarbonylhexaoxohexaosmium, Os-00289

$C_{17}H_2O_{16}Ru_6$
[μ_6-(Carbon)]hexadecacarbonyldihydrohexaruthenium, Ru-00448

$C_{17}H_4NO_{15}Os_3Re$
[(Acetonitrile)decacarbonyl-μ-hydrotriosmium]-pentacarbonylrhenium, Os-00290

$C_{17}H_4O_{15}Ru_5$
Pentadecacarbonyl-μ-ethenylidenedi-μ-hydropentaruthenium, Ru-00449

$C_{17}H_5O_{15}PRu_5$
Pentadecacarbonyl[μ_4-(ethylphosphoranetetrayl)]-pentaruthenium, Ru-00450

$C_{17}H_7O_{17}Os_5P$
[μ_5-(Carbon)]tetradecacarbonyl[μ-(dimethyl phosphito-*O''*:*P*)]-μ-hydropentaosmium, Os-00291

$C_{17}H_8N_2O_{10}Os_3$
Decacarbonyl-μ-hydro[μ-[(4-methylphenyl)azo]]triosmium, Os-00292

$C_{17}H_8O_7Ru_3$
[μ-[(1,2,3,3*a*,8*a*-η:4,5,6-η:6,7,8-η)-Azulene]]-heptacarbonyltriruthenium, Ru-00451

$C_{17}H_8O_9Os_3$
Nonacarbonyldi-μ-hydro[μ_3-[(1-η:1-η:1,2-η)-2-phenylethenylidene]]triosmium, Os-00293

$C_{17}H_9F_{18}O_5PRu$
Dicarbonyl[(1,2,3,4-η)-hexakis(trifluoromethyl)benzene]-(trimethyl phosphite-*P*)ruthenium, Ru-00452

$C_{17}H_{12}ORu$
(η^5-2,4-Cyclopentadien-1-yl)[(1,2,3,3*a*,8*a*-η)-1,8-dihydro-8-oxocyclopenta[*a*]inden-1-yl]ruthenium, Ru-00453

$C_{17}H_{12}O_{11}Os_4$
Undecacarbonyl[μ-[(1-η:1,2-η)-1-cyclohexen-1-yl]]tri-μ-hydrotetraosmium, Os-00294

$C_{17}H_{14}OOs$
Benzoylosmocene, Os-00295

$C_{17}H_{14}ORu$
Benzoylruthenocene, Ru-00454

$C_{17}H_{14}O_9Os_3$
Nonacarbonyl[μ_3-[(1-η:1,2-η:2-η)cyclooctyne]]di-μ-hydrotriosmium, Os-00296

$C_{17}H_{15}F_6PRu$
[(1,2,3,4,5,6-η)-1,1'-Biphenyl](η^5-2,4-cyclopentadien-1-yl)ruthenium(1+); Hexafluorophosphate, *in* Ru-00455

$C_{17}H_{15}NORu$
Ruthenocenecarboxylic acid; Anilide, *in* Ru-00185

$C_{17}H_{15}Ru^{\oplus}$
[(1,2,3,4,5,6-η)-1,1'-Biphenyl](η^5-2,4-cyclopentadien-1-yl)ruthenium(1+), Ru-00455

$C_{17}H_{16}BF_6PRu$
(η^6-Benzene)[(1,2,3,4,5,6-η)-1-phenylboratabenzene]-ruthenium(1+); Hexafluorophosphate, *in* Ru-00456

$C_{17}H_{16}BRu^{\oplus}$
(η^6-Benzene)[(1,2,3,4,5,6-η)-1-phenylboratabenzene]-ruthenium(1+), Ru-00456

$C_{17}H_{16}OOs$
(1-Hydroxy-1-phenylmethyl)osmocene, Os-00297

$C_{17}H_{16}ORu$
α-Phenylruthenocenemethanol, Ru-00457

$C_{17}H_{16}O_6Ru_2Si$
Hexacarbonyl[μ-[[(2,3,6,7-η:3,4,5,6-η)-1,3,5,7-cyclooctatetraen-1-yl]trimethylsilane]]-diruthenium, Ru-00458

$C_{17}H_{16}O_6Ru_3S$
Hexacarbonyl[μ_3-[(1,2,3-η:4,5-η:6,7-η)-2,4,6-cycloheptatrien-1-yl]][μ_3-(2-methyl-2-propanethiolato)]triruthenium, Ru-00459

$C_{17}H_{16}O_7Os_3P_2$
[μ_3-[(1-η:1,2-η:2-η)-Benzene]]heptacarbonylbis[μ-(dimethylphosphino)]triosmium, Os-00298

$C_{17}H_{16}Os$
Benzylosmocene, Os-00299

$C_{17}H_{16}Ru$
Benzylruthenocene, Ru-00460

$C_{17}H_{20}O_2Ru_2$
Dicarbonylbis(η^5-2,4-cyclopentadien-1-yl)-μ-ethylidene[μ-(1-methylethylidene)]diruthenium, Ru-00461

$C_{17}H_{20}Os$
[(1,2,3,4-η)-1,3,5,7-cyclooctatetraene][(1,2,3,4,5,6-η)-1,3,5-trimethylbenzene]osmium, Os-00300

$C_{17}H_{22}ORu$
Carbonyl[(1,2,5,6-η)-1,5-cyclooctadiene][(1,2,5,6-η)-1,3,5-cyclooctatriene]ruthenium, Ru-00462

$C_{17}H_{23}B_2F_8NRu$
[(1,2,3,4,5,6-η)-*N*,*N*-Dimethylaniline][(1,2,3,4,5,6-η)-1,3,5-trimethylbenzene]ruthenium(2+); Bis(tetrafluoroborate), *in* Ru-00463

$C_{17}H_{23}F_6PRu$
(η^5-2,4-Cyclopentadien-1-yl)[(1,2,3,4,5,6-η)-hexamethylbenzene]ruthenium(1+); Hexafluorophosphate, *in* Ru-00464

$C_{17}H_{23}NRu^{\oplus\oplus}$
[(1,2,3,4,5,6-η)-*N*,*N*-Dimethylaniline][(1,2,3,4,5,6-η)-1,3,5-trimethylbenzene]ruthenium(2+), Ru-00463

$C_{17}H_{23}Ru^{\oplus}$
(η^5-2,4-Cyclopentadien-1-yl)[(1,2,3,4,5,6-η)-hexamethylbenzene]ruthenium(1+), Ru-00464

$C_{17}H_{30}F_6OP_2Ru$
Carbonyl[(1,2,3,4,5,6-η)-hexamethylbenzene]-methyl(trimethylphosphine)ruthenium(1+); Hexafluorophosphate, *in* Ru-00465

$C_{17}H_{30}OPRu^{\oplus}$
Carbonyl[(1,2,3,4,5,6-η)-hexamethylbenzene]-methyl(trimethylphosphine)ruthenium(1+), Ru-00465

$C_{17}H_{40}F_6O_9P_4Ru$
[(1,2,5,6-η)-1,5-Cyclooctadiene]hydrotris(trimethyl phosphite-*P*)ruthenium(1+); Hexafluorophosphate, *in* Ru-00466

$C_{17}H_{40}O_9P_3Ru^{\oplus}$
[(1,2,5,6-η)-1,5-Cyclooctadiene]hydrotris(trimethyl phosphite-*P*)ruthenium(1+), Ru-00466

$C_{17}O_{16}Ru_6^{\ominus\ominus}$
[μ_6-(Carbon)]tri-μ-carbonyltridecacarbonylhexaruthenate(2−), Ru-00467

$C_{18}HO_{18}Os_6^{\ominus}$
Octadecacarbonyl-μ_3-hydrohexaosmate(1−), Os-00301

$C_{18}HO_{18}Ru_6^{\ominus}$
Octadecacarbonyl-μ_6-hydrohexaruthenate(1−), Ru-00468

$C_{18}H_2O_{18}Os_6$
Octadecacarbonyldihydrohexaosmium, Os-00302

$C_{18}H_2O_{18}Ru_6$
Octadecacarbonyldi-μ_3-hydrohexaruthenium, Ru-00469

$C_{18}H_4O_{16}Ru_6$
Hexadecacarbonyl-μ-ethenylidenedi-μ-hydrohexaruthenium, Ru-00470

$C_{18}H_8O_{12}Ru_4$
Dodecacarbonyl[μ_4-(1-η:1,2-η:1,2-η:2-η)-1-cyclohexene-1,2-diyl]tetraruthenium, Ru-00471

$C_{18}H_9NO_{10}Os_3$
Decacarbonyl-μ-hydro[μ-[(methylimino)phenylmethyl-*C*:*N*]]-triosmium, Os-00303

$C_{18}H_{10}N_2O_6Ru_2$
Hexacarbonyl[μ-[*N*-phenyl-1,2-benzenediaminato(2−)]]-diruthenium, Ru-00472

$C_{18}H_{10}O_6Ru_2Se_2$
Bis(μ-benzeneselenolato)hexacarbonyldiruthenium, Ru-00473

$C_{18}H_{10}O_6Ru_2Te_2$
Bis(μ-benzenetellurolato)hexacarbonyldiruthenium, Ru-00474

$C_{18}H_{10}O_8Ru_3$
Octacarbonylbis(η^5-2,4-cyclopentadien-1-yl)-triruthenium, Ru-00475

$C_{18}H_{11}O_{18}Os_5P$
Pentadecacarbonyldi-μ-hydro(trimethyl phosphite-*P*)pentaosmium, Os-00304

$C_{18}H_{12}O_6Ru_3$
Di-μ-carbonyltetracarbonyl[μ-[(1,2,3,4-η:1,5,6,7-η)-2,4,6-cycloheptatrien-1-yl]](η^5-2,4-cyclopentadien-1-yl)triruthenium, Ru-00476

$C_{18}H_{12}O_{10}Ru_3$
Decacarbonyl(1,3-cyclooctadiene)triruthenium, Ru-00477

$C_{18}H_{14}CrO_6Ru$
Pentacarbonyl(ethoxyruthenocenylmethylene)chromium, Ru-00478

$C_{18}H_{14}MoO_6Ru$
Pentacarbonyl(ethoxyruthenocenylmethylene)molybdenum, Ru-00479

$C_{18}H_{14}N_2ORu$
Carbonyl(η^5-2,4-cyclopentadien-1-yl)[(η^3-*C*,*N*,*N*′)2-(phenylazo)phenyl]ruthenium, Ru-00480

$C_{18}H_{14}O_6RuW$
Pentacarbonyl(ethoxyruthenocenylmethylene)tungsten, Ru-00481

$C_{18}H_{14}Ru$
Bis[(1,2,3,3*a*,7*a*-η)-1*H*-inden-1-yl]ruthenium, Ru-00482

$C_{18}H_{15}Cl_5OsP^{\ominus}$
Pentachloro(triphenylphosphine)osmate(1−), Os-00305

$C_{18}H_{21}Cl_2N_3O_2Ru$
Dicarbonyldichloro[2-(phenylazo)phenyl]ruthenate(1−); Tetramethylammonium salt, *in* Ru-00331

$C_{18}H_{22}Cl_2N_2Ru$
Dichloro[(1,2,5,6-η)-1,5-cyclooctadiene]bis(pyridine)-ruthenium, Ru-00483

$C_{18}H_{22}Ru$
Bis[(1,2,3,3*a*,7*a*-η)-1*H*-4,5,6,7-tetrahydroinden-1-yl]-ruthenium, Ru-00484

$C_{18}H_{23}ClN_2Ru$
Chloro[(1,2,5,6-η)-1,5-cyclooctadiene]-hydrobis(pyridine)ruthenium, Ru-00485

$C_{18}H_{24}B_2F_8Ru$
(η^6-Benzene)[(1,2,3,4,5,6-η)-hexamethylbenzene]-ruthenium(2+); Bis(tetrafluoroborate), *in* Ru-00488

$C_{18}H_{24}Cl_4O_2Ru_2$
Dicarbonyldi-μ-chlorodichlorobis[(1,2,5,6-η)-1,5-cyclooctadiene]diruthenium, Ru-00486

$C_{18}H_{24}Ge_2O_4Ru_2$
Tetracarbonyl[μ-[(1,2,3,3*a*,6*a*-η:3*a*,4,5,6,6*a*-η)-pentalene]]bis(trimethylgermyl)diruthenium, Ru-00487

$C_{18}H_{24}Ru^{\oplus\oplus}$
(η^6-Benzene)[(1,2,3,4,5,6-η)-hexamethylbenzene]-ruthenium(2+), Ru-00488
Bis[(1,2,3,4,5,6-η)-1,3,5-trimethylbenzene]-ruthenium(2+), Ru-00489

$C_{18}H_{26}F_{12}P_2RuS$
[η^6-(1-Isopropyl-4-methylbenzene)](η^5-tetramethylthiophene)ruthenium(2+); Bis(hexafluorophosphate), *in* Ru-00492

$C_{18}H_{26}O_4Ru$
[(1,2,5,6-η)-1,5-Cyclooctadiene]bis(2,4-pentanedionato-*O*,*O*′)ruthenium, Ru-00490

$C_{18}H_{26}Ru$
[(1,2,3,4-η)-1,3-Cyclohexadiene][(1,2,3,4,5,6-η)-hexamethylbenzene]ruthenium, Ru-00491

$C_{18}H_{26}RuS^{\oplus\oplus}$
[η^6-(1-Isopropyl-4-methylbenzene)](η^5-tetramethylthiophene)ruthenium(2+), Ru-00492

$C_{18}H_{27}F_6PRu$
[(1,2,3-η)-2-Cyclohexen-1-yl][(1,2,3,4,5,6-η)-hexamethylbenzene]ruthenium(1+); Hexafluorophosphate, *in* Ru-00494

$C_{18}H_{27}O_{18}P_3Ru_3$
Nonacarbonyltris(trimethyl phosphite-*P*)triruthenium, Ru-00493

$C_{18}H_{27}Ru^{\oplus}$
[(1,2,3-η)-2-Cyclohexen-1-yl][(1,2,3,4,5,6-η)-hexamethylbenzene]ruthenium(1+), Ru-00494

$C_{18}H_{31}O_{18}P_3Ru_4$
Nonacarbonyltetra-μ-hydrotris(trimethyl phosphite-*P*)tetraruthenium, Ru-00495

$C_{18}H_{33}ClPRu$
(η^6-Benzene)chloro(tributylphosphine)ruthenium(1+), *in* Ru-00704

$C_{18}H_{34}Cl_2N_2Ru_2$
μ-Chlorochlorobis[(1,2,5,6-η)-1,5-cyclooctadiene][μ-(1,1-dimethylhydrazine-*N*:*N*′)]-μ-hydrohydrodiruthenium, Ru-00496

$C_{18}H_{36}Br_4NO_2Os$
Tetrabromodicarbonylosmate(1−); Tetrabutylammonium salt, *in* Os-00005

$C_{18}H_{36}I_4NO_2S$
Dicarbonyltetraiodoosmate(1−); Tetrabutylammonium salt, *in* Os-00009

$C_{18}H_{40}Br_4N_2O_2Os$
Tetrabromodicarbonylosmate(2−); Bis(tetraethylammonium) salt, *in* Os-00006

$C_{18}H_{40}Cl_4N_2O_2Os$
Dicarbonyltetrachloroosmate(2−); Bis(tetraethylammonium) salt, *in* Os-00007

$C_{18}H_{42}F_{12}P_6Ru$
Tris(trimethylphosphine)[(1,2,3,4,5-η)-6-(trimethylphosphonio)-2,4-cyclohexadien-1-yl]-ruthenium(2+); Bis(hexafluorophosphate), *in* Ru-00497

$C_{18}H_{42}P_4Ru^{\oplus\oplus}$
Tris(trimethylphosphine)[(1,2,3,4,5-η)-6-(trimethylphosphonio)-2,4-cyclohexadien-1-yl]-ruthenium(2+), Ru-00497

$C_{18}H_{45}Cl_2N_2OsP_3$
Dichloro(dinitrogen)tris(triethylphosphine)osmium, Os-00306

$C_{18}H_{54}O_{18}P_6Ru^{\oplus\oplus}$
Hexakis(trimethyl phosphite-*P*)ruthenium(2+), Ru-00498

$C_{18}H_{58}P_6Ru_2$
Di-μ-hydrodihydrohexakis(trimethylphosphine)-diruthenium, Ru-00499

$C_{18}O_{17}Ru_6$
μ_6-Carbon-μ-carbonylhexadecacarbonylhexaruthenium, Ru-00500

$C_{18}O_{18}Os_6^{\ominus\ominus}$
Octadecacarbonylhexaosmate(2−), Os-00307
Octadecacarbonylhexaosmium, Os-00308

$C_{18}O_{18}Ru_6^{\ominus\ominus}$
Di-μ-carbonyldi-μ_3-carbonyltetradecacarbonylhexaruthenate(2−), Ru-00501

$C_{19}H_5O_{13}PRu_4$
Tridecacarbonyl(μ_3-phenylphosphinidene)tetraruthenium, Ru-00502

$C_{19}H_{10}O_{11}Ru_4$
Undecacarbonyl[(1,2-η:5-η:5,6-η:5,6-η:6-η)-1-cycloocten-5-yne]tetraruthenium, Ru-00503

$C_{19}H_{14}N_2O_2Ru$
Dicarbonyl(η^5-2,4-cyclopentadien-1-yl)[2-(phenylazo)-phenyl]ruthenium, Ru-00504

$C_{19}H_{14}O_3Ru$
Tricarbonyl(η^4-1,4-diphenyl-1,3-butadiene)ruthenium, Ru-00505

$C_{19}H_{14}O_6Ru_3$
Di-μ-carbonyltetracarbonyl[μ-(2,4,6-cyclooctatrien-1-yl)](η^5-2,4-cyclopentadien-1-yl)triruthenium, Ru-00506

$C_{19}H_{15}O_2PRu$
Dicarbonyl(η^5-2,4-cyclopentadien-1-yl)-(diphenylphosphino)ruthenium, Ru-00507

$C_{19}H_{16}ORu$
1-Phenyl-3-ruthenocenyl-2-propen-1-one, Ru-00508
3-Phenyl-1-ruthenocenyl-2-propen-1-one, Ru-00509

$C_{19}H_{21}NO_{11}Ru_3$
μ-Carbonyldecacarbonyl-μ-hydrotriruthenate(1–); Tetraethylammonium salt, *in* Ru-00168

$C_{19}H_{26}As_2Cl_2ORu$
Carbonyldichlorobis(dimethylphenylarsine)(η^2-ethene)-ruthenium, Ru-00510

$C_{19}H_{26}Cl_2OP_2Ru$
Carbonyldichlorobis(dimethylphenylphosphine)(η^2-ethene)-ruthenium, Ru-00511

$C_{19}H_{26}Ru$
[(2,3,5,6-η)-Bicyclo[2.2.1]hepta-2,5-diene]-[(1,2,3,4,5,6-η)-hexamethylbenzene]ruthenium, Ru-00512

$C_{19}H_{27}F_6PRu$
[(1,2,3,4,5-η)-2,4-Cycloheptadien-1-yl][(1,2,3,4,5,6-η)-hexamethylbenzene]ruthenium(1+); Hexafluorophosphate, *in* Ru-00514

$C_{19}H_{27}ORu^{\oplus}$
Carbonyl[(1,2,3-η)-2-cyclohexen-1-yl][(1,2,3,4,5,6-η)-hexamethylbenzene]ruthenium(1+), Ru-00513

$C_{19}H_{27}Ru^{\oplus}$
[(1,2,3,4,5-η)-2,4-Cycloheptadien-1-yl][(1,2,3,4,5,6-η)-hexamethylbenzene]ruthenium(1+), Ru-00514

$C_{19}H_{34}N_2O_3Ru$
Tricarbonyl[*N*,*N'*-1,2-ethanediylbis[2,4-dimethyl-3-pentanamine]-*N*,*N'*]ruthenium, Ru-00515

$C_{19}H_{36}Br_3NO_3Os$
Tribromotricarbonylosmate(1–); Tetrabutylammonium salt, *in* Os-00012

$C_{19}H_{36}Cl_3NO_3Os$
Tricarbonyltrichloroosmate(1–); Tetrabutylammonium salt, *in* Os-00013

$C_{19}H_{36}I_3NO_3Os$
Tricarbonyltriiodoosmate(1–); Tetrabutylammonium salt, *in* Os-00014

$C_{19}O_{19}Os_5$
Nonadecacarbonylpentaosmium, Os-00309

$C_{20}H_2O_{19}Os_7$
(Carbon)nonadecacarbonyldihydroheptaosmium, Os-00310

$C_{20}H_2O_{20}Os_3Re_2$
Decacarbonyl(decacarbonyldi-μ-hydrotriosmium)-diruthenium, Os-00311

$C_{20}H_6O_{16}Os_6$
Hexadecacarbonyl-μ_3-ethylidyne-μ_4-ethylidynehexaosmium, Os-00312

$C_{20}H_{10}O_6Ru_2$
Hexacarbonyl[μ-[η^2:η^4-1,2-phenylene(2-phenyl-1,2-ethenediyl)]]diruthenium, Ru-00516

$C_{20}H_{10}O_{12}Os_4$
Dodecacarbonyldi-μ-hydro[μ-(2-phenylethylidene)]-tetraosmium, Os-00313

$C_{20}H_{13}FeO_{12}PRu_3$
Di-μ-carbonyloctacarbonyl(dicarbonyliron)di-μ-hydro(dimethylphenylphosphine)triruthenium, Ru-00517

$C_{20}H_{15}Cl_2O_2PRu$
Dicarbonyldichloro(triphenylphosphine)ruthenium, *in* Ru-00752

$C_{20}H_{15}I_3O_2PRu^{\ominus}$
Dicarbonyltriiodo(triphenylphosphine)ruthenate(1–), Ru-00518

$C_{20}H_{16}F_{12}P_2Ru$
Bis[(1,2,3,4,4*a*,8*a*-η)naphthalene]ruthenium(2+); Bis(hexafluorophosphate), *in* Ru-00521

$C_{20}H_{16}O_4Ru_3$
Tetracarbonylbis[μ-(1,3,5,7-cyclooctatetraene)]-triruthenium, Ru-00519

$C_{20}H_{16}O_6Ru_3$
Di-μ-carbonyltetracarbonyl[(1,2,3,4,5-η)-2,4-cycloheptadien-1-yl][μ-[(1,2,3,4-η:1,5,6,7-η)-2,4,6-cycloheptatrien-1-yl]]triruthenium, Ru-00520

$C_{20}H_{16}Ru^{\oplus\oplus}$
Bis[(1,2,3,4,4*a*,8*a*-η)naphthalene]ruthenium(2+), Ru-00521

$C_{20}H_{18}HgRu_2$
Diruthenocenylmercury, Ru-00522

$C_{20}H_{18}Ru_2$
1,1''-Biruthenocene, Ru-00523

$C_{20}H_{20}O_{10}Ru$
1,2,3,4,5-Pentakis(methoxycarbonyl)ruthenocene, *in* Ru-00389

$C_{20}H_{24}O_2Ru_3S_3$
μ-Carbonylcarbonyltris(η^5-2,4-cyclopentadien-1-yl)-tris[μ-(methanethiolato)]triruthenium, Ru-00524

$C_{20}H_{26}Ru$
[(1,2,3,4-η)-1,3,5,7-Cyclooctatetraene][(1,2,3,4,5,6-η)-hexamethylbenzene]ruthenium, Ru-00525

$C_{20}H_{29}B_2F_8NRu$
(η^6-Dimethylaniline)(η^6-hexamethylbenzene)ruthenium(2+); Bis(tetrafluoroborate), *in* Ru-00526

$C_{20}H_{29}F_6PRu$
[(1,2,3,4,5-η)-2,4,6-Cyclooctatrien-1-yl][(1,2,3,4,5,6-η)-hexamethylbenzene]ruthenium(1+); Hexafluorophosphate, *in* Ru-00527
[(1,2,3,6,7-η)-2,4,6-Cyclooctatrien-1-yl][(1,2,3,4,5,6-η)-hexamethylbenzene]ruthenium(1+); Hexafluorophosphate, *in* Ru-00528

$C_{20}H_{29}NRu^{\oplus\oplus}$
(η^6-Dimethylaniline)(η^6-hexamethylbenzene)ruthenium(2+), Ru-00526

$C_{20}H_{29}Ru^{\oplus}$
[(1,2,3,4,5-η)-2,4,6-Cyclooctatrien-1-yl][(1,2,3,4,5,6-η)-hexamethylbenzene]ruthenium(1+), Ru-00527
[(1,2,3,6,7-η)-2,4,6-Cyclooctatrien-1-yl][(1,2,3,4,5,6-η)-hexamethylbenzene]ruthenium(1+), Ru-00528

$C_{20}H_{32}As_4ClF_6N_2PRu$
Chloro(dinitrogen)bis[1,2-phenylenebis[dimethylarsine]-*As*,*As'*]ruthenium(1+); Hexafluorophosphate, *in* Ru-00529

$C_{20}H_{32}As_4ClF_6N_2RuSb$
Chloro(dinitrogen)bis[1,2-phenylenebis[dimethylarsine]-*As*,*As'*]ruthenium(1+); Hexafluoroantimonate, *in* Ru-00529

$C_{20}H_{32}As_4ClN_2Ru^{\oplus}$
Chloro(dinitrogen)bis[1,2-phenylenebis[dimethylarsine]-*As*,*As'*]ruthenium(1+), Ru-00529

$C_{20}H_{32}Cl_4Ru_2$
Di-μ-chlorodichlorobis[(1,2,3,6,7,8-η)-2,7-dimethyl-2,6-octadiene-1,8-diyl]diruthenium, Ru-00530

$C_{20}H_{32}N_4O_4Ru_2$
[μ-[*N*,*N*′-Bis(1-methylethyl)-1,4-bis[(1-methylethyl)-imino]-2,3-butanediaminato(2−)-N^1,N^2,N^3:N^2,N^3,N^4]]-tetracarbonyldiruthenium, Ru-00531
Tetracarbonylbis[μ-(η[2]-*N*,*N*′-1,2-ethanediylidenebis[2-propanamine]-*N*,*N*′]]diruthenium, Ru-00532

$C_{20}H_{40}O_{20}P_4Ru_4$
Octacarbonyltetra-μ-hydrotetrakis(trimethyl phosphite-*P*)tetraruthenium, Ru-00533

$C_{20}H_{58}B_2F_8P_6Ru_2$
Di-μ-methylenehexakis(trimethylphosphine)-diruthenium(2+); Bis(tetrafluoroborate), *in* Ru-00534

$C_{20}H_{58}P_6Ru_2^{\oplus\oplus}$
Di-μ-methylenehexakis(trimethylphosphine)-diruthenium(2+), Ru-00534

$C_{20}O_{20}Os_6$
Icosacarbonylhexaosmium, Os-00314

$C_{20}O_{20}Os_7^{\ominus\ominus}$
Eicosacarbonylheptaosmate(2−), Os-00315

$C_{21}H_5O_{15}PRu_5$
Pentadecacarbonyl[μ_4-(phenylphosphoranetetrayl)]-pentaruthenium, Ru-00535

$C_{21}H_6O_{14}Ru_6$
(Benzene)[μ_6-(carbon)]tetradecacarbonylhexaruthenium, Ru-00536

$C_{21}H_6O_{16}Os_6$
Hexadecacarbonyl[μ_3-[(1-η:1,2-η:2-η)-1,2-dimethyl-1,2-ethenediyl]](μ_5-methanetetrayl)hexaosmium, Os-00316

$C_{21}H_{10}N_2O_9Ru_3$
Bis[μ_3-[benzeneaminato(2−)]]nonacarbonyltriruthenium, Ru-00537
Nonacarbonyl[μ_3-[*N*-phenyl-*o*-phenylenediaminato(2−)]-triruthenium, Ru-00538

$C_{21}H_{11}O_9PRu_3$
Nonacarbonyl[μ_3-(diphenylphosphino)]-μ-hydrotriruthenium, Ru-00539

$C_{21}H_{14}O_9Ru_4$
(η[6]-Benzene)nonacarbonyl[μ_4-(1-η:1,2-η:1,2-η:2-η)-1-cyclohexene-1,2-diyl]tetraruthenium, Ru-00540

$C_{21}H_{15}Br_2O_3OsP$
Dibromotricarbonyl(triphenylphosphine)osmium, Os-00317

$C_{21}H_{15}Br_2O_3PRu$
Dibromotricarbonyl(triphenylphosphine)ruthenium, Ru-00541

$C_{21}H_{15}Cl_2O_3OsP$
Tricarbonyldichloro(triphenylphosphine)osmium, Os-00318

$C_{21}H_{15}Cl_2O_3PRu$
Tricarbonyldichloro(triphenylphosphine)ruthenium, Ru-00542

$C_{21}H_{15}I_2O_3OsP$
Tricarbonyldiiodo(triphenylphosphine)osmium, Os-00319

$C_{21}H_{15}I_2O_3PRu$
Tricarbonyldiiodo(triphenylphosphine)ruthenium, Ru-00543

$C_{21}H_{16}ClO_3OsP$
Tricarbonylchlorohydro(triphenylphosphine)osmium, Os-00320

$C_{21}H_{16}O_9Ru_3$
Nonacarbonyl[μ_3-[(1-η:1,2,3-η:3-η)-1,6,9-cyclododecatrien-1-yl-3-ylidene]]-μ-hydrotriruthenium, Ru-00544

$C_{21}H_{17}O_3OsP$
Tricarbonyldihydro(triphenylphosphine)osmium, Os-00321

$C_{21}H_{17}O_3PRu$
Tricarbonyldihydro(triphenylphosphine)ruthenium, Ru-00545

$C_{21}H_{18}FeOOs$
(Osmocenylcarbonyl)ferrocene, Os-00322

$C_{21}H_{18}FeORu$
Ferrocenylruthenocenylmethanone, Ru-00546

$C_{21}H_{18}O_2Ru$
3-Phenyl[5]ruthenocenophan-1,5-dione, Ru-00547

$C_{21}H_{18}O_3RuSi$
Tricarbonyl[(2,3,4,5-η)-1,1-dimethyl-2,5-diphenylsilacyclopenta-2,4-diene]ruthenium, Ru-00548

$C_{21}H_{20}FeOs$
Ferrocenylmethylosmocene, Os-00323

$C_{21}H_{20}FeRu$
(Ferrocenylmethyl)ruthenocene, Ru-00549

$C_{21}H_{22}Ru$
3-Phenyl[5]ruthenocenophane, Ru-00550

$C_{21}H_{32}N_4O_5Ru_2$
[μ-[*N*,*N*′-Bis(1-methylethyl)-1,4-bis[(1-methylethyl)-imino]-2,3-butanediaminato(2−)-N^1,N^2,N^3:N^2,N^3,N^4]]-μ-carbonyltetracarbonyldiruthenium, Ru-00551

$C_{21}H_{32}OOs_2$
μ-Carbonyldi-μ-hydrobis[(1,2,3,4,5-η)-1,2,3,4,5-pentamethyl-2,4-cyclopentadien-1-yl]diosmium, Os-00324

$C_{21}H_{36}F_{12}O_3P_2Ru$
[(1,2,3,4,5,6-η)-Hexamethylbenzene]tris(2-propanone)-ruthenium(2+); Bis(hexafluorophosphate), *in* Ru-00552

$C_{21}H_{36}O_3Ru^{\oplus\oplus}$
[(1,2,3,4,5,6-η)-Hexamethylbenzene]tris(2-propanone)-ruthenium(2+), Ru-00552

$C_{21}H_{43}O_9P_3Ru$
[η[4]-1,2,3,4-Tetramethyl-5,6-bis(methylene)-1,3-cyclohexadiene]tris(trimethyl phosphite-*P*)-ruthenium, Ru-00553

$C_{21}H_{60}P_6Ru_2$
Tri-μ-methylenehexakis(trimethylphosphine)diruthenium, Ru-00554

$C_{21}H_{61}BF_4P_6Ru_2$
μ-Methyldi-μ-methylenehexakis(trimethylphosphine)-diruthenium(1+); Tetrafluoroborate, *in* Ru-00555

$C_{21}H_{61}P_6Ru_2^{\oplus}$
μ-Methyldi-μ-methylenehexakis(trimethylphosphine)-diruthenium(1+), Ru-00555

$C_{21}O_{21}Os_7$
Heneicosacarbonylheptaosmium, Os-00325

$C_{22}H_{10}O_9Ru_3$
Nonacarbonyl[μ[3]-[η:η:η[3]-3,5-cyclohexadiene-1,2-diylidene(phenylmethylidyne)]]-μ-hydrotriruthenium, Ru-00556

$C_{22}H_{10}O_{15}Ru_6$
[μ_6-(Carbon)]-μ-carbonyltetradecacarbonyl[μ-[(2,3-η:4,5-η)-2,4-hexadiene]]hexaruthenium, Ru-00557

$C_{22}H_{14}O_4Ru_2$
Di-μ-carbonyldicarbonylbis[(1,2,3,3*a*,7*a*-η)-1*H*-inden-1-yl]diruthenium, Ru-00558

$C_{22}H_{14}O_9Ru_4$
Nonacarbonyl[μ_3-(4,6,8-trimethylazulene)]-tetraruthenium, Ru-00559

$C_{22}H_{15}O_4OsP$
Tetracarbonyl(triphenylphosphine)osmium, Os-00326

$C_{22}H_{15}O_4PRu$
Tetracarbonyl(triphenylphosphine)ruthenium, Ru-00560

$C_{22}H_{15}O_4RuSb$
Tetracarbonyl(triphenylstibine)ruthenium, Ru-00561

$C_{22}H_{16}O_{10}Ru_4$
Decacarbonyl-2,8-cyclododecadiene-3,7-diyl-1-ylidenetetraruthenium, Ru-00562

$C_{22}H_{16}O_{15}Ru_6S_3$
[μ_6-(Carbon)]pentadecacarbonyltris[μ-(ethanethiolato)]-μ-hydrohexaruthenium, Ru-00563

$C_{22}H_{18}FeRu$
(Ruthenocenylethynyl)ferrocene, Ru-00564

$C_{22}H_{18}O_6Ru_3$
Di-μ-carbonyltetracarbonyl[μ-[(1,2,6,7-η:2,3,4,5,6-η)-2,4,6-cyclooctatrien-1-yl]][(1,2,3,3*a*,6*a*-η)-1,4,5,6-tetrahydro-1-pentalenyl]triruthenium, Ru-00565

$C_{22}H_{20}B_2Os$
Bis[(1,2,3,4,5,6-η)-1-phenylboratabenzene]osmium, Os-00327

$C_{22}H_{20}B_2Ru$
Bis[(1,2,3,4,5,6-η)-1-phenylboratabenzene]ruthenium, Ru-00566

$C_{22}H_{20}NO_2PRu$
Carbonylnitrosyl(η^3-2-propenyl)(triphenylphosphine)-ruthenium, Ru-00567

$C_{22}H_{22}B_2F_8Ru$
(η^6-Benzene)[[2.2]paracyclophane]ruthenium(2+); Bis(tetrafluoroborate), *in* Ru-00569

$C_{22}H_{22}O_4Ru_2$
Di-μ-carbonyldicarbonylbis[(1,2,3,3*a*,7*a*-η)-4,5,6,7-tetrahydro-1*H*-inden-1-yl]diruthenium, Ru-00568

$C_{22}H_{22}Ru^{\oplus\oplus}$
(η^6-Benzene)[[2.2]paracyclophane]ruthenium(2+), Ru-00569

$C_{22}H_{24}B_7O_2PRu$
Dicarbonyl[η^5-nonahydrodicarbanonaborato(4−)]-(triphenylphosphine)ruthenium, Ru-00570

$C_{22}H_{26}B_2F_8Ru$
[(1,2,3,4,5,6-η)-Hexamethylbenzene][(1,2,3,4,4*a*,8*a*-η)-naphthalene]ruthenium(2+); Bis(tetrafluoroborate), *in* Ru-00573

$C_{22}H_{26}O_8Ru_2Si_4$
[μ-[1,2,3,4-Benzenetetrayltetrakis[dimethylsilylene]]]-octacarbonyldiruthenium, Ru-00571
[μ-[1,2,4,5-Benzenetetrayltetrakis[dimethylsilylene]]]-octacarbonyldiruthenium, Ru-00572

$C_{22}H_{26}Ru^{\oplus\oplus}$
[(1,2,3,4,5,6-η)-Hexamethylbenzene][(1,2,3,4,4*a*,8*a*-η)-naphthalene]ruthenium(2+), Ru-00573

$C_{22}H_{32}O_2Os_2$
Dicarbonyldi-μ-hydrobis[(1,2,3,4,5-η)-1,2,3,4,5-pentamethyl-2,4-cyclopentadien-1-yl]diosmium, Os-00328

$C_{22}H_{34}Cl_2Ru_2$
Di-μ-chlorobis[(1,2,5,6-η)-1,5-cyclooctadiene]bis(η^3-propenyl)diruthenium, Ru-00574

$C_{22}H_{58}F_6O_6P_6Ru_2S_2$
Di-μ-methylenehexakis(trimethylphosphine)-diruthenium(2+); Bis(trifluoromethylsulfonate), *in* Ru-00534

$C_{22}O_{21}Os_8$
(Carbon)heneicosacarbonyloctaosmium, Os-00329

$C_{22}O_{22}Os_8^{\ominus\ominus}$
Di-μ-carbonylicosacarbonyloctaosmate(2−), Os-00330

$C_{23}H_{10}F_{10}ORu_2S_2$
μ-Carbonylbis(η^5-2,4-cyclopentadien-1-yl)bis[μ-(pentafluorobenzenethiolato)]diruthenium, Ru-00575

$C_{23}H_{12}N_2O_{10}Os_3$
Decacarbonyl[μ-(diphenylmethanone hydrazonato)]-μ-hydrotriosmium, Os-00331

$C_{23}H_{14}O_5Ru_2$
Pentacarbonyl[μ-[(η^3:η^5)-3,5-cyclohexadien-1-yl-2-ylidene(phenylmethylidyne)-2,4-cyclopentadien-1-ylidene]]diruthenium, Ru-00576

$C_{23}H_{20}BrO_2PRu$
Bromodicarbonyl(η^3-2-propenyl)(triphenylphosphine)-ruthenium, Ru-00577

$C_{23}H_{20}ClO_2PRu$
Dicarbonylchloro(η^3-2-propenyl)(triphenylphosphine)-ruthenium, Ru-00578

$C_{23}H_{20}FeORu$
1-Ferrocenyl-3-ruthenocenyl-2-propen-1-one, Ru-00579
3-Ferrocenyl-1-ruthenocenyl-2-propen-1-one, Ru-00580

$C_{23}H_{20}ORu_2$
1,3-Diruthenocenyl-2-propen-1-one, Ru-00581

$C_{23}H_{21}O_3OsP$
Tricarbonyldimethyl(triphenylphosphine)osmium, Os-00332

$C_{23}H_{22}B_2F_8Ru$
[(1,2,3,4,4*a*,9*a*-η)-Anthracene][(1,2,3,4,5,6-η)-1,3,5-trimethylbenzene]ruthenium(2+); Bis(tetrafluoroborate), *in* Ru-00582

$C_{23}H_{22}Ru^{\oplus\oplus}$
[(1,2,3,4,4*a*,9*a*-η)-Anthracene][(1,2,3,4,5,6-η)-1,3,5-trimethylbenzene]ruthenium(2+), Ru-00582

$C_{23}H_{39}N_3Os$
[(1,2,5,6-η)-1,5-Cyclooctadiene]tris(2-isocyano-2-methylpropane)osmium, Os-00333

$C_{23}O_{23}Os_8$
Tricosacarbonyloctaosmium, Os-00334

$C_{24}H_{10}F_{10}O_2Ru_2S_2$
μ-Carbonylcarbonylbis[η^5-2,4-cyclopentadien-1-yl][μ-(pentafluorobenzenethiolato)]-(pentafluorobenzenethiolato)diruthenium, Ru-00583
Dicarbonylbis(η^5-2,4-cyclopentadien-1-yl)bis[μ-(pentafluorobenzenethiolato)]diruthenium, Ru-00584

$C_{24}H_{10}O_{10}Os_3$
Decacarbonyl[μ_3-[1,1′-[(1-η:1,2-η:2-η)-1,2-ethynediyl]-bis[benzene]]]triosmium, Os-00335

$C_{24}H_{12}O_{14}Ru_6$
μ_6-Carbon-μ-carbonyltridecacarbonyl(1,3,5-trimethylbenzene)hexaruthenium, Ru-00585

$C_{24}H_{18}O_2Ru$
1,1′-Dibenzoylruthenocene, Ru-00586
1,2-Dibenzoylruthenocene, Ru-00587

$C_{24}H_{18}Ru_2$
1,4-Diruthenocenyl-1,3-butadiyne, Ru-00588

$C_{24}H_{20}BrOOsP$
Bromocarbonyl(η^5-2,4-cyclopentadien-1-yl)-(triphenylphosphine)osmium, Os-00336

$C_{24}H_{20}BrOPRu$
Bromocarbonyl(η^5-2,4-cyclopentadien-1-yl)-(triphenylphosphine)ruthenium, Ru-00589

$C_{24}H_{20}BrO_4PRu$
Bromocarbonyl(η^5-2,4-cyclopentadien-1-yl)(triphenyl phosphite-*P*)ruthenium, Ru-00590

$C_{24}H_{20}ClOPRu$
Carbonylchloro(η^5-2,4-cyclopentadien-1-yl)-(triphenylphosphine)ruthenium, Ru-00591

$C_{24}H_{20}ClO_4PRu$
Carbonylchloro(η^5-2,4-cyclopentadien-1-yl)(triphenyl phosphite-*P*)ruthenium, Ru-00592

$C_{24}H_{20}IOPRu$
Carbonyl(η^5-2,4-cyclopentadien-1-yl)-iodo(triphenylphosphine)ruthenium, Ru-00593

$C_{24}H_{20}IO_4PRu$
Carbonyl(η^5-2,4-cyclopentadien-1-yl)iodo(triphenyl phosphite-*P*)ruthenium, Ru-00594

$C_{24}H_{20}O_2Ru_2S_2$
Bis[μ-(benzenethiolato)]dicarbonylbis(η^5-2,4-cyclopentadien-1-yl)diruthenium, Ru-00595

$C_{24}H_{20}O_2Ru_2Se_2$
Bis[μ-(benzeneselenolato)]dicarbonylbis(η^5-2,4-cyclopentadien-1-yl)diruthenium, Ru-00596

$C_{24}H_{20}O_4Ru_4$
Tetra-μ_3-carbonyltetrakis(η^5-2,4-cyclopentadien-1-yl)-tetraruthenium, Ru-00597

$C_{24}H_{21}Br_2PRu$
(η^6-Benzene)dibromo(triphenylphosphine)ruthenium, Ru-00598

$C_{24}H_{21}Cl_2PRu$
(η^6-Benzene)dichloro(triphenylphosphine)ruthenium, Ru-00599

$C_{24}H_{21}I_2OsP$
(η^6-Benzene)diiodo(triphenylphosphine)osmium, Os-00337

$C_{24}H_{21}I_2PRu$
(η^6-Benzene)diiodo(triphenylphosphine)ruthenium, Ru-00600

$C_{24}H_{21}OPRu$
Carbonyl(η^5-2,4-cyclopentadien-1-yl)-hydro(triphenylphosphine)ruthenium, Ru-00601

$C_{24}H_{22}Ru$
Bis[1-phenyl-[(2,3,4,5,6-η)-cyclohexa-2,4-dien-6-yl]]-ruthenium, Ru-00602

$C_{24}H_{24}GeIO_3PRu$
Tricarbonyliodo(trimethylgermyl)(triphenylphosphine)-ruthenium, Ru-00603

$C_{24}H_{24}IO_3PRuSi$
Tricarbonyliodo(trimethylsilyl)(triphenylphosphine)-ruthenium, Ru-00604

$C_{24}H_{24}IO_3PRuSn$
Tricarbonyliodo(trimethylstannyl)(triphenylphosphine)-ruthenium, Ru-00605

$C_{24}H_{25}O_3PRuSi$
Tricarbonylhydro(trimethylsilyl)(triphenylphosphine)-ruthenium, Ru-00606

$C_{24}H_{25}O_3PRuSn$
Tricarbonylhydro(trimethylstannyl)(triphenylphosphine)-ruthenium, Ru-00607

$C_{24}H_{27}O_{23}P_3Ru_6$
[μ_6-(Carbon)]-μ-carbonyltridecacarbonyltris(trimethyl phosphite-*P*)hexaruthenium, Ru-00608

$C_{24}H_{30}O_4Ru_2$
Di-μ-carbonyldicarbonylbis[(1,2,3,4,5-η)-1,2,3,4,5-pentamethyl-2,4-cyclopentadien-1-yl]diruthenium, Ru-00609

$C_{24}H_{32}N_4O_8Ru_4$
Octacarbonylbis[μ-(η^4-*N*,*N'*-1,2-ethanediylidenebis[2-propanamine]-*N*,*N'*]]tetraruthenium, Ru-00610

$C_{24}H_{34}Ru$
[(1,2,3,4,5,6-η)-Hexamethylbenzene][1,2,3,4-η)-1,2,3,4-tetramethyl-5,6-bis(methylene)-1,3-cyclohexadiene]ruthenium, Ru-00611

$C_{24}H_{36}B_2F_8Ru$
Bis[(1,2,3,4,5,6-η)-hexamethylbenzene]ruthenium(2+); Bis-(tetrafluoroborate), *in* Ru-00613

$C_{24}H_{36}Cl_4Ru_2$
Di-μ-chlorodichlorobis[(1,2,3,4,5,6-η)-hexamethylbenzene]diruthenium, Ru-00612

$C_{24}H_{36}Ru^{\oplus\oplus}$
Bis[(1,2,3,4,5,6-η)-hexamethylbenzene]ruthenium(2+), Ru-00613
[(1,2,3,4-η)-Hexamethylbenzene][(1,2,3,4,5,6-η)-hexamethylbenzene]ruthenium, Ru-00614

$C_{24}H_{38}ClRu_2^{\oplus}$
μ-Chlorobis[(1,2,3,4,5,6-η)-hexamethylbenzene]di-μ-hydrodiruthenium(1+), Ru-00615

$C_{24}H_{38}Cl_2Ru_2$
μ-Chlorobis[(1,2,3,4,5,6-η)-hexamethylbenzene]di-μ-hydrodiruthenium(1+); Chloride, *in* Ru-00615

$C_{24}H_{62}O_{12}P_4Ru$
Dihydrotetrakis(triethyl phosphite-*P*)ruthenium, Ru-00616

$C_{25}H_{10}N_2O_{13}Os_5$
Tridecacarbonyl-μ-hydro[μ_3-[*N*-phenyl-1,2-benzenediaminato(3−)-*N*,*N'*:*N'*,*N'*]]pentaosmium, Os-00338

$C_{25}H_{17}O_7Os_2P$
Heptacarbonyldihydro(triphenylphosphine)diosmium, Os-00339

$C_{25}H_{20}BF_4O_2PRu$
Dicarbonyl(η^5-2,4-cyclopentadien-1-yl)-(triphenylphosphine)ruthenium(1+); Tetrafluoroborate, *in* Ru-00617

$C_{25}H_{20}O_2PRu^{\oplus}$
Dicarbonyl(η^5-2,4-cyclopentadien-1-yl)-(triphenylphosphine)ruthenium(1+), Ru-00617

$C_{25}H_{22}NO_2PRu$
(Aminocarbonyl)carbonyl(η^5-2,4-cyclopentadien-1-yl)-(triphenylphosphine)ruthenium, Ru-00618

$C_{25}H_{22}ORu_2$
1,5-Diruthenocenyl-1,4-pentadien-3-one, Ru-00619

$C_{25}H_{23}Cl_2PRu$
Dichloro[(1,2,3,4,5,6-η)-1,3,5-cycloheptatriene]-(triphenylphosphine)ruthenium, Ru-00620

$C_{25}H_{23}OPRu$
Carbonyl(η^5-2,4-cyclopentadien-1-yl)-methyl(triphenylphosphine)ruthenium, Ru-00621

$C_{25}H_{24}AuGeO_4PRu$
Tetracarbonyl(trimethylgermyl)[(triphenylphosphine)-aurio]ruthenium, Ru-00622

$C_{25}H_{24}AuO_4PRuSi$
Tetracarbonyl(trimethylsilyl)[(triphenylphosphine)-aurio]ruthenium, Ru-00623

$C_{25}H_{24}ClPRu$
(η^6-Benzene)chloromethyl(triphenylphosphine)ruthenium, Ru-00624

$C_{25}H_{27}N_2OPRuS_4$
Carbonylbis(dimethylcarbamodithioato-*S*,*S'*)-(triphenylphosphine)ruthenium, Ru-00625

$C_{25}H_{30}O_8Ru_3Si_3$
Octacarbonyl[μ_3-[η^2:η^2:η^5-1,3,5-tris(trimethylsilyl)-pentalene]]triruthenium, Ru-00626
Octacarbonyl[(1,2,3,3*a*,6*a*-η:3*a*,4,5,6,6*a*-η)-1,3,5-tris(trimethylsilyl)]pentalenetriruthenium, Ru-00627

$C_{25}IO_{24}Os_{10}^{\ominus}$
[μ_6-(Carbon)]tetracosacarbonyl-μ-iododecaosmate(1−), Os-00340

$C_{25}I_2O_{24}Os_{10}$
(Carbon)tetracosacarbonylbis-μ_2-iododecaosmium, Os-00341

$C_{25}O_{24}Os_{10}^{\ominus\ominus}$
[μ_6-(Carbon)]tetracosacarbonyldecaosmate(2−), Os-00342

$C_{26}H_{10}FeO_{12}Ru_3$
Nonacarbonyl[μ_4-[(1-η:1,2-η:1,2-η:2-η)-1,2-diphenyl-1,2-ethenediyl]](tricarbonyliron)triruthenium, Ru-00628

$C_{26}H_{10}O_{12}Ru_4$
Dodecacarbonyl[μ_4-[(1-η:1,2-η:1,2-η:2-η)-1,2-diphenyl-1,2-ethenediyl]]tetraruthenium, Ru-00629

$C_{26}H_{20}O_2Ru_2$
1,1'':1',1'''-Bis(3-oxo-1-propene-1,3-diyl)-bisruthenocene, Ru-00630

$C_{26}H_{23}O_2PRu$
Acetylcarbonyl(η^5-2,4-cyclopentadien-1-yl)-(triphenylphosphine)ruthenium, Ru-00631

$C_{26}H_{25}OPRu$
Carbonyl(η^5-2,4-cyclopentadien-1-yl)-ethyl(triphenylphosphine)ruthenium, Ru-00632

$C_{26}H_{25}OPRuS_2$
(η^5-2,4-Cyclopentadien-1-yl)(*O*-ethylcarbonodithioato-*S*,*S'*)(triphenylphosphine)ruthenium, Ru-00633

$C_{26}H_{25}O_2PRu$
Dicarbonyl[(1,2,3,4-η)-2,3-dimethyl-1,3-butadiene]-(triphenylphosphine)ruthenium, Ru-00634

$C_{26}H_{25}PRu$
(η5-2,4-Cyclopentadien-1-yl)(η3-2-propenyl)-(triphenylphosphine)ruthenium, Ru-00635

$C_{26}H_{27}PRu$
(η6-Benzene)dimethyl(triphenylphosphine)ruthenium, Ru-00636
Bis(η4-1,3-butadiene)(triphenylphosphine)ruthenium, Ru-00637

$C_{26}H_{30}O_8Ru_3$
Di-μ-carbonylhexacarbonyl[μ3-[(1,2,3,4-η:1,2,3,4-η:1,4-η)-1-(3,3-dimethyl-1-butenyl)-2,4-bis(1,1-dimethylethyl)-1,3-butadiene-1,4-diyl]]-triruthenium, Ru-00638

$C_{26}H_{36}O_{25}P_4Ru_6$
[μ6-(Carbon)]-μ-carbonyldodecacarbonyltetrakis(trimethylphosphite-*P*)hexaruthenium, Ru-00639

$C_{27}H_{17}O_9Os_3P$
Nonacarbonyldihydro(triphenylphosphine)triosmium, Os-00343

$C_{27}H_{24}BF_4O_2PRu$
Dicarbonyl[(1,2,3,4,5-η)-2,4-cycloheptadien-1-yl]-(triphenylphosphine)ruthenium(1+); Tetrafluoroborate, *in* Ru-00640

$C_{27}H_{24}O_2PRu^{\oplus}$
Dicarbonyl[(1,2,3,4,5-η)-2,4-cycloheptadien-1-yl]-(triphenylphosphine)ruthenium(1+), Ru-00640

$C_{27}H_{25}O_2PRu$
Dicarbonyl[(1,2,3,4-η)-1,3-cycloheptadiene]-(triphenylphosphine)ruthenium, Ru-00641
Dicarbonyl[(1,3,4,5-η)-4-cycloheptene-1,3-diyl]-(triphenylphosphine)ruthenium, Ru-00642

$C_{27}H_{27}PRu$
(η5-2,4-Cyclopentadien-1-yl)[(1,2,3-η)-2-methyl-2-propenyl)](triphenylphosphine)ruthenium, Ru-00643

$C_{28}HO_{29}Os_9^{\ominus}$
Heptacosacarbonyl[μ3-(carboxylato-*C*:*O*:*O*′)]-μ-hydrononaosmate(1−), Os-00344

$C_{28}H_{17}O_{10}Os_3P$
Decacarbonyl-μ-hydrohydro(triphenylphosphine)triosmium, Os-00345

$C_{28}H_{18}Cl_2N_4O_4Ru_2$
Tetracarbonyldi-μ-chlorobis[2-(phenylazo)phenyl-*C*,*N*′]-diruthenium, *in* Ru-00330

$C_{28}H_{22}O_2Ru$
1,1′-Bis(1-oxo-3-phenyl-2-propenyl)ruthenocene, Ru-00644

$C_{28}H_{24}Br_2O_2P_2Ru$
Dibromodicarbonyl[1,2-ethanediylbis(diphenylphosphine)-*P*,*P*′]ruthenium, Ru-00645

$C_{28}H_{24}I_2O_2OsP_2$
Dicarbonyl[1,2-ethanediylbis[diphenylphosphine]-*P*,*P*′]-diiodoosmium, Os-00346

$C_{28}H_{27}O_2PRu$
Dicarbonyl[(1,4,5,6-η)-5-cyclooctene-1,4-diyl]-(triphenylphosphine)ruthenium, Ru-00646

$C_{28}H_{30}NPRuS_2$
(η5-2,4-Cyclopentadien-1-yl)(*N*,*N*-diethylcarbamodithioato-*S*,*S*′)(triphenylphosphine)-ruthenium, Ru-00647

$C_{28}H_{32}N_4Ru^{\oplus\oplus}$
[(1,2,5,6-η)-1,5-Cyclooctadiene]tetrakis(pyridine)-ruthenium(2+), Ru-00648

$C_{28}H_{40}As_4Cl_2Fe_2Ru$
Dichlorobis[1,1′-ferrocenediylbis[dimethylarsine]]-ruthenium, Ru-00649

$C_{28}H_{40}F_6P_4Ru$
(η4-1,3-Butadiene)tris(dimethylphenylphosphine)-hydroruthenium(1+); Hexafluorophosphate, *in* Ru-00650

$C_{28}H_{40}P_3Ru^{\oplus}$
(η4-1,3-Butadiene)tris(dimethylphenylphosphine)-hydroruthenium(1+), Ru-00650

$C_{28}H_{78}B_{14}P_4Ru_2$
Bis[μ-[η6-nonahydrodicarbanonaborato(2−)-*B*,*H*]]-tetrakis(triethylphosphine)diruthenium, *in* Ru-00379

$C_{28}H_{80}B_2F_8P_8Ru_3$
Tetra-μ-methyleneoctakis(trimethylphosphine)-triruthenium(2+); Bis(tetrafluoroborate), *in* Ru-00651

$C_{28}H_{80}P_8Ru_3^{\oplus\oplus}$
Tetra-μ-methyleneoctakis(trimethylphosphine)-triruthenium(2+), Ru-00651

$C_{29}H_{15}AsO_{11}Os_3$
Undecacarbonyl(triphenylarsine)triosmium, Os-00347

$C_{29}H_{15}Fe_2O_{11}PRu$
Tricarbonyl(di-μ-carbonylhexacarbonyldiiron)-(triphenylphosphine)ruthenium, Ru-00652

$C_{29}H_{15}O_{11}Os_3P$
Undecacarbonyl(triphenylphosphine)triosmium, Os-00348

$C_{29}H_{15}O_{11}PRu_3$
Undecacarbonyl(triphenylphosphine)triruthenium, Ru-00653

$C_{29}H_{19}O_{11}PRu_4$
Undecacarbonyltetra-μ-hydro(triphenylphosphine)-tetraruthenium, Ru-00654

$C_{29}H_{25}BRu$
(η5-2,4-Cyclopentadien-1-yl)[(η6-phenyl)-triphenylborato(1−)]ruthenium, Ru-00655

$C_{29}H_{36}O_{29}Os_6P_4$
Heptadecacarbonyltetrakis(trimethylphosphite-*P*)-hexaosmium, Os-00349

$C_{30}As_2F_{20}O_6Ru_2$
Bis[μ-[bis(pentafluorophenyl)arsino]]-hexacarbonyldiruthenium, Ru-00656

$C_{30}F_{20}O_6P_2Ru_2$
Bis[μ-[bis(pentafluorophenyl)phosphino]]-hexacarbonyldiruthenium, Ru-00657

$C_{30}H_{20}O_6P_2Ru_2$
Hexacarbonylbis[μ-(diphenylphosphino)]diruthenium, Ru-00658

$C_{30}H_{24}O_2Ru_2$
1,1″:1′,1‴-Bis(3-oxo-1,4-pentadiene-1,5-diyl)-bisruthenocene, Ru-00659

$C_{30}H_{26}ClPRu$
(η6-Benzene)chlorophenyl(triphenylphosphine)ruthenium, Ru-00660

$C_{30}H_{26}Ru_3$
1,1″:1‴,1⁗-Terruthenocene, Ru-00661

$C_{30}H_{27}BRu$
[(1,2,3,4,5-η)-1-Methyl-2,4-cyclopentadien-1-yl][(η6-phenyl)triphenylborato(1−)]ruthenium, Ru-00662

$C_{30}H_{27}F_{12}P_3Ru$
(η6-Benzene)[(1,2,3,4,5-η)-6-(triphenylphosphino)-2,4-cyclohexadien-1-yl]ruthenium(2+); Bis(hexafluorophosphate), *in* Ru-00663

$C_{30}H_{27}PRu^{\oplus\oplus}$
(η6-Benzene)[(1,2,3,4,5-η)-6-(triphenylphosphino)-2,4-cyclohexadien-1-yl]ruthenium(2+), Ru-00663

$C_{30}H_{34}ClPRu$
Chloro[(1,2,3,4,5,6-η)-hexamethylbenzene]-hydro(triphenylphosphine)ruthenium, Ru-00664

$C_{30}H_{75}BrO_{15}OsP_5^{\oplus}$
Bromopentakis(triethyl phosphite-*P*)osmium(1+), Os-00350

$C_{30}H_{75}BrO_{15}P_5Ru^{\oplus}$
Bromopentakis(triethyl phosphite-*P*)ruthenium(1+), Ru-00665

$C_{30}H_{75}ClO_{15}OsP_5^{\oplus}$
Chloropentakis(triethyl phosphite-*P*)osmium(1+), Os-00351

$C_{30}H_{75}ClO_{15}P_5Ru^{\oplus}$
Chloropentakis(triethyl phosphite-*P*)ruthenium(1+), Ru-00666

$C_{30}H_{76}O_{15}P_5Ru^{\oplus}$
Hydropentakis(triethyl phosphite-*P*)ruthenium(1+), Ru-00667

$C_{31}HO_{32}Os_{10}^{\ominus}$
Triacontacarbonyl[μ_3-(carboxylato-*C,O,O'*)]-μ-hydrodecaosmate(1−), Os-00352

$C_{31}H_{20}O_3Os$
Tricarbonyl[(1,2,3,4-η)-1,2,3,4-tetraphenyl-1,3-cyclobutadiene]osmium, Os-00353

$C_{31}H_{20}O_3Ru$
Tricarbonyl(1,2,3,4-tetraphenyl-1,3-cyclobutadiene)ruthenium, Ru-00668

$C_{31}H_{21}F_{12}PRu$
(η^5-2,4-Cyclopentadien-1-yl)[(1,3,4-η)-5,5,5-trifluoro-1,2,3-tris(trifluoromethyl)-1,3-pentadienyl](triphenylphosphine)ruthenium, Ru-00669

$C_{31}H_{25}N_2O_{13}Os_5P$
Tridecacarbonyl-μ-hydro[μ_3-[*N*-phenyl-1,2-benzenediaminato(3−)-*N,N':N',N''*]](triethylphosphine)pentaosmium, Os-00354

$C_{31}H_{27}OPRu$
Benzylcarbonyl(η^5-2,4-cyclopentadien-1-yl)(triphenylphosphine)ruthenium, Ru-00670

$C_{31}H_{28}BNO_2Ru$
Amminedicarbonyl(η^5-2,4-cyclopentadien-1-yl)ruthenium(1+); Tetraphenylborate, *in* Ru-00060

$C_{31}H_{29}BrOsP_2$
Bromo(η^5-2,4-cyclopentadien-1-yl)[1,2-ethanediylbis[diphenylphosphine]-*P,P'*]osmium, Os-00355

$C_{31}H_{29}ClP_2Ru$
Chloro(η^5-2,4-cyclopentadien-1-yl)[1,2-ethanediylbis[diphenylphosphine]-*P,P'*]ruthenium, Ru-00671

$C_{31}H_{38}INO_{15}Os_5$
Pentadecacarbonyldi-μ-hydroiodopentaosmate(1−); Tetrabutylammonium salt, *in* Os-00249

$C_{32}H_{20}O_4Ru$
Tricarbonyl[(2,3,4,5-η)-2,3,4,5-tetraphenyl-2,4-cyclopentadien-1-one]ruthenium, Ru-00672

$C_{32}H_{24}ClN_2O_2PRu$
Dicarbonylchloro[2-(phenylazo)phenyl](triphenylphosphine)ruthenium, Ru-00673

$C_{32}H_{24}O_4RuSi$
Tricarbonyl[(2,3,4,5-η)-1-methyl-2,3,4,5-tetraphenylsilacyclopenta-2,4-dien-1-ol]ruthenium, Ru-00674

$C_{32}H_{25}BO_3Ru$
Tricarbonyl(η^5-2,4-cyclopentadien-1-yl)ruthenium(1+); Tetraphenylborate, *in* Ru-00072

$C_{32}H_{25}OPRu$
Carbonyl(η^5-2,4-cyclopentadien-1-yl)(phenylethynyl)(triphenylphosphine)ruthenium, Ru-00675

$C_{32}H_{29}ClOP_2Ru$
Carbonyl(η^5-2,4-cyclopentadien-1-yl)[1,2-ethanediylbis[diphenylphosphine]-*P,P'*]ruthenium(1+); Chloride, *in* Ru-00676

$C_{32}H_{29}OP_2Ru^{\oplus}$
Carbonyl(η^5-2,4-cyclopentadien-1-yl)[1,2-ethanediylbis[diphenylphosphine]-*P,P'*]ruthenium(1+), Ru-00676

$C_{32}H_{31}BRu$
[(1,2,3,5,6-η)-2,5-Cyclooctadien-1-yl][(η^6-phenyl)triphenylborato(1−)]ruthenium, Ru-00677

$C_{32}H_{32}P_2Ru$
(η^5-2,4-Cyclopentadien-1-yl)[1,2-ethanediylbis[diphenylphosphine]-*P,P'*]methylruthenium, Ru-00678

$C_{33}H_{15}O_{13}PRu_5$
Tridecacarbonyl[μ-(diphenylphosphino)][μ_4-[(1-η:1-η:1,2-η:1,2-η)phenylethynyl]]pentaruthenium, Ru-00679

$C_{33}H_{15}O_{14}PRu_5$
[μ_5-(Carbon)]tetradecacarbonyl(triphenylphosphine)pentaruthenium, Ru-00680

$C_{33}H_{30}Cl_3OPRu$
[(2,3,5,6-η)-Bicyclo[2.2.1]hepta-2,5-diene]carbonyltrichlororuthenate(1−); Benzyltriphenylphosphonium salt, *in* Ru-00078

$C_{33}H_{40}N_2O_{16}Ru_6$
[μ_6-(Carbon)]tri-μ-carbonyltridecacarbonylhexaruthenate(2−); Bis(tetraethylammonium) salt, *in* Ru-00467

$C_{34}H_{20}O_6Os_2$
Hexacarbonyl[μ-[(1,2,3,4-η:1,4-η)-1,2,3,4-tetraphenyl-1,3-butadiene-1,4-diyl]]diosmium, Os-00356

$C_{34}H_{20}O_6Ru_2$
Hexacarbonyl[μ-[(1,2,3,4-η:1,4-η)-1,2,3,4-tetraphenyl-1,3-butadiene-1,4-diyl]]diruthenium, Ru-00681

$C_{34}H_{40}N_2O_{18}Os_6$
Octadecacarbonylhexaosmate(2−); Bis(tetraethylammonium) salt, *in* Os-00307

$C_{34}H_{57}O_9P_3Ru_3Si$
Tri-μ-carbonylhexacarbonyl[μ_3-[(methylsilylidyne)tris[dibutylphosphine]-*P:P':P''*]]triruthenium, Ru-00682

$C_{34}H_{68}N_6RuTi_2$
Hexakis(diethylaminato)-μ-1,1'-ruthenocenediyldititanium, Ru-00683

$C_{34}H_{72}I_4N_2O_2Os$
Dicarbonyltetraiodoosmate(2−); Bis(tetrabutylammonium)salt, *in* Os-00010

$C_{35}H_{15}O_{16}PRu_6$
[μ_6-(Carbon)]-μ-carbonylpentadecacarbonyl(triphenylphosphine)hexaruthenium, Ru-00684

$C_{35}H_{20}O_7Ru_2$
Hexacarbonyl[μ-[(1,2,5-η:1,4,5-η)-3-oxo-1,2,4,5-tetraphenyl-1,4-pentadiene-1,5-diyl]]diruthenium, Ru-00685

$C_{35}H_{29}N_2PRu$
(η^5-2,4-Cyclopentadien-1-yl)[2-(phenylazo)phenyl](triphenylphosphine)ruthenium, Ru-00686

$C_{35}H_{31}BOs$
(η^6-Benzene)(η^5-2,4-cyclopentadien-1-yl)osmium(1+); Tetraphenylborate, *in* Os-00140

$C_{35}H_{31}BRu$
(η^6-Benzene)(η^5-2,4-cyclopentadien-1-yl)ruthenium(1+); Tetraphenylborate, *in* Ru-00189

$C_{35}H_{38}ClO_7P_2Ru^{\oplus}$
Carbonylchloro(η^4-tetraphenylcyclobutadiene)bis(trimethyl phosphite-*P*)ruthenium(1+), Ru-00687

$C_{35}H_{38}Cl_4O_7P_2Ru$
Carbonylchloro(η^4-tetraphenylcyclobutadiene)bis(trimethyl phosphite-*P*)ruthenium(1+); Trichloride, *in* Ru-00687

$C_{35}H_{38}O_7P_2Ru$
Carbonyl(η^4-tetraphenylcyclobutadiene)bis(trimethylphosphite-*P*)ruthenium, Ru-00688

$C_{36}H_{20}O_8Os_3$
Octacarbonyl[μ-[(1,2,3,4-η:1,4-η)-1,2,3,4-tetraphenyl-2-butene-1,4-diylidene]]triosmium, Os-00357

$C_{36}H_{20}O_8Ru_3$
Di-μ-carbonylhexacarbonyl(1,2,3,4-tetraphenyl-1,3-butadiene-1,4-diyl)triruthenium, Ru-00689
Octacarbonylbis[1,1'-(1,2-ethynediyl)bis[benzene]]triruthenium, Ru-00690

$C_{36}H_{23}AsO_{12}Ru_4$
Dodecacarbonyltrihydrotetraruthenate(1−); Tetraphenylarsonium salt, *in* Ru-00220

$C_{36}H_{26}F_{24}FeP_8Ru_2$
Hydrotetrakis(phosphorous trifluoride)ruthenate(1−); Tris(phenanthroline)iron(2+) salt, *in* Ru-00868

$C_{36}H_{28}O_{10}P_2Ru_4$
Decacarbonyl[μ-[1,2-ethanediylbis[diphenylphosphine]-*P*:*P*′]]tetra-μ-hydrotetraruthenium, Ru-00691

$C_{36}H_{30}As_2Br_3NOs$
Tribromonitridobis(triphenylarsine)osmium, Os-00358

$C_{36}H_{30}As_2Cl_3NOsS$
Trichlorobis(triphenylarsine)thionitrosylosmium, Os-00359

$C_{36}H_{30}As_2Cl_3NRu$
Trichloronitridobis(triphenylarsine)ruthenium, Ru-00692

$C_{36}H_{30}BClF_4N_2O_2P_2Ru$
Chlorodinitrosylbis(triphenylphosphine)ruthenium(1+); Tetrafluoroborate, *in* Ru-00697

$C_{36}H_{30}Br_3NOOsP_2$
Tribromonitrosylbis(triphenylphosphine)osmium, Os-00360

$C_{36}H_{30}Br_3NOP_2Ru$
Tribromonitrosylbis(triphenylphosphine)ruthenium, Ru-00693

$C_{36}H_{30}ClF_6N_2O_2P_3Ru$
Chlorodinitrosylbis(triphenylphosphine)ruthenium(1+); Hexafluorophosphate, *in* Ru-00697

$C_{36}H_{30}ClI_2NOP_2Ru$
Chlorodiiodonitrosylbis(triphenylphosphine)ruthenium, Ru-00694

$C_{36}H_{30}ClNOP_2Ru$
Chloronitrosylbis(triphenylphosphine)ruthenium, Ru-00695

$C_{36}H_{30}ClNO_3P_2Ru$
Chloro(dioxygen)nitrosylbis(triphenylphosphine)-ruthenium, Ru-00696

$C_{36}H_{30}ClN_2O_2P_2Ru^{\oplus}$
Chlorodinitrosylbis(triphenylphosphine)ruthenium(1+), Ru-00697

$C_{36}H_{30}Cl_2P_2Ru$
Dichlorobis(triphenylphosphine)ruthenium, Ru-00698

$C_{36}H_{30}Cl_3HgNOOsP_2$
Dichloro(chloromercurio)nitrosylbis(triphenylphosphine)-osmium, Os-00361

$C_{36}H_{30}Cl_3NOOsP_2$
Trichloronitrosylbis(triphenylphosphine)osmium, Os-00362

$C_{36}H_{30}Cl_3NOP_2Ru$
Trichloronitrosylbis(triphenylphosphine)ruthenium, Ru-00699

$C_{36}H_{30}Cl_3NOsP_2$
Trichloronitridobis(triphenylphosphine)osmium, Os-00363

$C_{36}H_{30}Cl_3NP_2Ru$
Trichloronitridobis(triphenylphosphine)ruthenium, Ru-00700

$C_{36}H_{30}Cl_3NRuSb_2$
Trichloronitridobis(triphenylstibine)ruthenium, Ru-00701

$C_{36}H_{30}Cl_4OsP_2^{\ominus}$
Tetrachlorobis(triphenylphosphine)osmate(1−), Os-00364
Tetrachlorobis(triphenylphosphine)osmium, Os-00365

$C_{36}H_{30}Cl_4P_2Ru^{\ominus}$
Tetrachlorobis(triphenylphosphine)ruthenate(1−), Ru-00702

$C_{36}H_{30}N_2O_2OsP_2$
Dinitrosylbis(triphenylphosphine)osmium, Os-00366

$C_{36}H_{30}N_2O_2P_2Ru$
Dinitrosylbis(triphenylphosphine)ruthenium, Ru-00703

$C_{36}H_{33}BORu$
(η^5-2,4-Cyclopentadien-1-yl)[(1,2,3,4,5,6-η)-methoxybenzene]ruthenium(1+); Tetraphenylborate, *in* Ru-00253

$C_{36}H_{33}BRu$
(η^5-2,4-Cyclopentadien-1-yl)[(1,2,3,4,5,6-η)-methylbenzene]ruthenium(1+); Tetraphenylborate, *in* Ru-00254

$C_{36}H_{34}BClRu$
(η^6-Benzene)chloro[(1,2,3,4-η)-1,3-cyclohexadiene]-ruthenium(1+); Tetraphenylborate, *in* Ru-00255

$C_{36}H_{66}Cl_2P_2Ru_2^{\oplus\oplus}$
Bis(η^6-benzene)di-μ-chlorobis(tributylphosphine)-diruthenium(2+), Ru-00704

$C_{36}H_{66}Cl_4P_2Ru_2$
Bis(η^6-benzene)di-μ-chlorobis(tributylphosphine)-diruthenium(2+); Dichloride, *in* Ru-00704

$C_{36}H_{90}O_{18}OsP_6^{\oplus\oplus}$
Hexakis(triethyl phosphite-*P*)osmium(2+), Os-00367

$C_{36}H_{90}O_{18}P_6Ru^{\oplus\oplus}$
Hexakis(triethyl phosphite-*P*)ruthenium(2+), Ru-00705

$C_{37}H_{20}FeO_{13}P_2Ru_3$
Nonacarbonylbis[μ-(diphenylphosphino)]-(tetracarbonyliron)triruthenium, Ru-00706

$C_{37}H_{20}O_9Os_3$
Nonacarbonyl[μ-[(1,2,3,4-η:1,4-η)-1,2,3,4-tetraphenyl-1,3-butadiene-1,4-diyl]]triosmium, Os-00368

$C_{37}H_{24}O_7Os_3P_2$
Heptacarbonylbis[μ-(diphenylphosphino)][μ_3-[(1-η:1,2-η:2-η)-1,3-cyclohexadien-5-yne]]triosmium, Os-00369

$C_{37}H_{30}As_2Cl_3ORu^{\ominus}$
Carbonyltrichlorobis(triphenylarsine)ruthenate(1−), Ru-00707

$C_{37}H_{30}BrNO_2P_2Ru$
Bromocarbonylnitrosylbis(triphenylphosphine)ruthenium, Ru-00708

$C_{37}H_{30}Br_3RuS$
Tribromocarbonothioylbis(triphenylphosphine)ruthenium, Ru-00709

$C_{37}H_{30}ClNO_2OsP_2$
Carbonylchloronitrosylbis(triphenylphosphine)osmium, Os-00370

$C_{37}H_{30}ClNO_2P_2Ru$
Carbonylchloronitrosylbis(triphenylphosphine)ruthenium, Ru-00710

$C_{37}H_{30}Cl_2OP_2Ru$
Carbonyldichlorobis(triphenylphosphine)ruthenium, *in* Ru-00863

$C_{37}H_{30}Cl_2P_2RuS$
(Carbonothioyl)dichlorobis(triphenylphosphine)-ruthenium, *in* Ru-00864

$C_{37}H_{30}Cl_3P_2RuS^{\ominus}$
(Carbonothioyl)trichlorobis(triphenylphosphine)-ruthenate(1−), Ru-00711

$C_{37}H_{30}INO_2P_2Ru$
Carbonyliodonitrosylbis(triphenylphosphine)ruthenium, Ru-00712

$C_{37}H_{31}NO_2OsP_2$
Carbonylhydronitrosylbis(triphenylphosphine)osmium, Os-00371

$C_{37}H_{32}F_6NO_2OsP_3$
Carbonyldihydronitrosylbis(triphenylphosphine)-osmium(1+); Hexafluorophosphate, *in* Os-00372

$C_{37}H_{32}NO_2OsP_2^{\oplus}$
Carbonyldihydronitrosylbis(triphenylphosphine)-osmium(1+), Os-00372

$C_{37}H_{47}IO_9P_3Ru^{\oplus}$
Iodo(η^4-tetraphenylcyclobutadiene)tris(trimethyl phosphite-*P*)ruthenium(1+), Ru-00713

$C_{37}H_{47}I_4O_9P_3Ru$
Iodo(η^4-tetraphenylcyclobutadiene)tris(trimethyl phosphite-*P*)ruthenium(1+); Triiodide, *in* Ru-00713

$C_{37}H_{47}O_9P_3Ru$
(η^4-Tetraphenylcyclobutadiene)tris(trimethyl phosphite-*P*)ruthenium, Ru-00714

$C_{37}H_{67}BrOOsP_2$
Bromocarbonylhydrobis(tricyclohexylphosphine)osmium, Os-00373

$C_{37}H_{67}ClOOsP_2$
Carbonylchlorohydrobis(tricyclohexylphosphine)osmium, Os-00374

$C_{38}H_{30}As_2Br_2O_2Os$
Dibromodicarbonylbis(triphenylarsine)osmium, Os-00375

$C_{38}H_{30}As_2Cl_2O_2Os$
Dicarbonyldichlorobis(triphenylarsine)osmium, Os-00376

$C_{38}H_{30}Br_2OP_2RuSe$
Dibromo(carbonoselenoyl)carbonylbis(triphenylphosphine)-ruthenium, Ru-00715

$C_{38}H_{30}Br_2O_2OsP_2$
Dibromodicarbonylbis(triphenylphosphine)osmium, Os-00377

$C_{38}H_{30}Br_2O_2OsSb_2$
Dibromodicarbonylbis(triphenylstibine)osmium, Os-00378

$C_{38}H_{30}Br_2O_2P_2Ru$
Dibromodicarbonylbis(triphenylphosphine)ruthenium, Ru-00716

$C_{38}H_{30}ClF_4NOP_2Ru$
Chloronitrosyl(η^2-tetrafluoroethene)-bis(triphenylphosphine)ruthenium, Ru-00717

$C_{38}H_{30}Cl_2OOsP_2S$
(Carbonothioyl)carbonyldichlorobis(triphenylphosphine)-osmium, Os-00379

$C_{38}H_{30}Cl_2OOsP_2Se$
(Carbonoselenoyl)-carbonyldichlorobis(triphenylphosphine)osmium, Os-00380

$C_{38}H_{30}Cl_2OOsP_2Te$
(Carbonotelluroyl)-carbonyldichlorobis(triphenylphosphine)osmium, Os-00381

$C_{38}H_{30}Cl_2OP_2RuS$
(Carbonothioyl)carbonyldichlorobis(triphenylphosphine)-ruthenium, Ru-00718

$C_{38}H_{30}Cl_2OP_2RuSe$
(Carbonylselenoyl)-carbonyldichlorobis(triphenylphosphine)ruthenium, Ru-00719

$C_{38}H_{30}Cl_2O_2OsP_2$
Dicarbonyldichlorobis(triphenylphosphine)osmium, Os-00382

$C_{38}H_{30}Cl_2O_2OsSb_2$
Dicarbonyldichlorobis(triphenylstibine)osmium, Os-00383

$C_{38}H_{30}Cl_2O_2P_2Ru$
Dicarbonyldichlorobis(triphenylphosphine)ruthenium, Ru-00720

$C_{38}H_{30}Cl_2O_8P_2Ru$
Dicarbonyldichlorobis(triphenyl phosphite-*P*)ruthenium, Ru-00721

$C_{38}H_{30}Cl_2P_2RuS_2$
Bis(carbonothioyl)dichlorobis(triphenylphosphine)-ruthenium, Ru-00722

$C_{38}H_{30}Cl_4OOsP_2$
Carbonyldichloro(dichloromethylene)-bis(triphenylphosphine)osmium, Os-00384

$C_{38}H_{30}F_6NO_3OsP_3$
Dicarbonylnitrosylbis(triphenylphosphine)osmium(1+); Hexafluorophosphate, *in* Os-00387

$C_{38}H_{30}F_6NO_3P_3Ru$
Dicarbonylnitrosylbis(triphenylphosphine)ruthenium(1+); Hexafluorophosphate, *in* Ru-00724

$C_{38}H_{30}I_2O_2OsP_2$
Dicarbonyldiiodobis(triphenylphosphine)osmium, Os-00385

$C_{38}H_{30}I_2O_2OsSb_2$
Dicarbonyldiiodobis(triphenylstibine)osmium, Os-00386

$C_{38}H_{30}I_2O_2P_2Ru$
Dicarbonyldiodobis(triphenylphosphine)ruthenium, Ru-00723

$C_{38}H_{30}NO_3OsP_2^{\oplus}$
Dicarbonylnitrosylbis(triphenylphosphine)osmium(1+), Os-00387

$C_{38}H_{30}NO_3P_2Ru^{\oplus}$
Dicarbonylnitrosylbis(triphenylphosphine)ruthenium(1+), Ru-00724

$C_{38}H_{30}O_2OsP_2S_2$
Dicarbonyl(dithioperoxy)bis(triphenylphosphine)osmium, Os-00388

$C_{38}H_{30}O_2OsP_2Se_2$
Dicarbonyl(diseleno)bis(triphenylphosphine)osmium, Os-00389

$C_{38}H_{30}O_2P_2RuS_2$
Dicarbonyl(dithioperoxy)bis(triphenylphosphine)-ruthenium, Ru-00725

$C_{38}H_{30}O_4OsP_2$
Dicarbonyl(dioxygen)bis(triphenylphosphine)osmium, Os-00390

$C_{38}H_{30}O_4P_2Ru$
Dicarbonyl(dioxygen)bis(triphenylphosphine)ruthenium, Ru-00726

$C_{38}H_{30}O_4P_2RuS$
Dicarbonyl[(*O*,*S*-η)-sulfur dioxide]-bis(triphenylphosphine)ruthenium, Ru-00727

$C_{38}H_{30}O_6P_2RuS$
Dicarbonyl[sulfata(2−)-*O*,*O′*]bis(triphenylphosphine)-ruthenium, Ru-00728

$C_{38}H_{30}O_6P_2RuS_2$
Dicarbonyl[(*O*,*S*-η)-sulfur dioxide]-bis(triphenylphosphine)ruthenium; SO_2 complex, *in* Ru-00727

$C_{38}H_{31}BrO_2P_2Ru$
Bromodicarbonylhydrobis(triphenylphosphine)ruthenium, Ru-00729

$C_{38}H_{31}ClO_2OsP_2$
Dicarbonylchlorohydrobis(triphenylphosphine)osmium, Os-00391

$C_{38}H_{31}ClO_2P_2Ru$
Dicarbonylchlorohydrobis(triphenylphosphine)ruthenium, Ru-00730

$C_{38}H_{32}O_2OsP_2$
Dicarbonyldihydrobis(triphenylphosphine)osmium, Os-00392

$C_{38}H_{32}O_2P_2Ru$
Dicarbonyldihydrobis(triphenylphosphine)ruthenium, Ru-00731

$C_{38}H_{33}BF_4O_3P_2Ru$
Aquadicarbonylhydrobis(triphenylphosphine)ruthenium(1+) ; Tetrafluoroborate, *in* Ru-00733

$C_{38}H_{33}Cl_2NO_2P_2Ru$
Acetyldichloronitrosylbis(triphenylphosphine)ruthenium, Ru-00732

$C_{38}H_{33}O_3P_2Ru^{\oplus}$
Aquadicarbonylhydrobis(triphenylphosphine)ruthenium(1+) , Ru-00733

$C_{38}H_{39}B_7P_2Ru$
[(4,5,6,7,8,9-η)-Nonahydro-6,8-dicarbanonaborato(4−)]-bis(triphenylphosphine)ruthenium, Ru-00734

$C_{38}H_{42}B_9ClP_2Ru$
Chlorohydrobis(triphenylphosphine)[(7,8,9,10,11-η)-undecahydro-7,8-dicarbaundecaborato(2−)]-ruthenium, Ru-00735
Chlorohydrobis(triphenylphosphine)[(7,8,9,10,11-η)-undecahydro-7,9-dicarbaundecaborato(2−)]-ruthenium, Ru-00736

$C_{38}H_{43}B_9P_2Ru$
Dihydrobis(triphenylphosphine)[(7,8,9,10,11-η)-undecahydro-7,8-dicarbaundecaborato(2−)]-ruthenium, Ru-00737
Dihydrobis(triphenylphosphine)[(7,8,9,10,11-η)-undecahydro-7,9-dicarbaundecaborato(2−)]-ruthenium, Ru-00738

$C_{38}H_{67}BN_6Ru$
[(1,2,5,6-η)-1,5-Cyclooctadiene]tris(1,1-dimethylhydrazine-N^2)hydroruthenium(1+); Tetraphenylborate, *in* Ru-00378

$C_{38}H_{67}BrO_2OsP_2$
Bromodicarbonylhydrobis(tricyclohexylphosphine)osmium, Os-00393

$C_{39}H_{29}AsO_9Ru_3$
Nonacarbonyl[μ_3-[(1-η:1,2-η:1,2-η)-3,3-dimethyl-1-butynyl]]triruthenate(1−); Tetraphenylarsonium salt, *in* Ru-00387

$C_{39}H_{30}AlCl_3I_2O_3P_2Ru$
Tricarbonyliodobis(triphenylphosphine)ruthenium(1+); Trichloroiodoaluminate, *in* Ru-00741

$C_{39}H_{30}AlCl_5O_3P_2Ru$
Tricarbonylchlorobis(triphenylphosphine)ruthenium(1+); Tetrachloroaluminate, *in* Ru-00740

$C_{39}H_{30}As_2O_3Ru$
Tricarbonylbis(triphenylarsine)ruthenium, Ru-00739

$C_{39}H_{30}ClO_3OsP_2^{\oplus}$
Tricarbonylchlorobis(triphenylphosphine)osmium(1+), Os-00394

$C_{39}H_{30}ClO_3P_2Ru^{\oplus}$
Tricarbonylchlorobis(triphenylphosphine)ruthenium(1+), Ru-00740

$C_{39}H_{30}Cl_2O_3OsP_2$
Tricarbonylchlorobis(triphenylphosphine)osmium(1+); Chloride, *in* Os-00394

$C_{39}H_{30}Cl_2O_7OsP_2$
Tricarbonylchlorobis(triphenylphosphine)osmium(1+); Perchlorate, *in* Os-00394

$C_{39}H_{30}Cl_2O_7P_2Ru$
Tricarbonylchlorobis(triphenylphosphine)ruthenium(1+); Perchlorate, *in* Ru-00740

$C_{39}H_{30}IO_3P_2Ru^{\oplus}$
Tricarbonyliodobis(triphenylphosphine)ruthenium(1+), Ru-00741

$C_{39}H_{30}I_2O_3P_2Ru$
Tricarbonyliodobis(triphenylphosphine)ruthenium(1+); Iodide, *in* Ru-00741

$C_{39}H_{30}O_3OsP_2$
Tricarbonylbis(triphenylphosphine)osmium, Os-00395

$C_{39}H_{30}O_3P_2Ru$
Tricarbonylbis(triphenylphosphine)ruthenium, Ru-00742

$C_{39}H_{30}O_9P_2Ru$
Tricarbonylbis(triphenyl phosphite-*P*)ruthenium, Ru-00743

$C_{39}H_{31}BrO_3OsP_2$
Tricarbonylhydrobis(triphenylphosphine)osmium(1+); Bromide, *in* Os-00397

$C_{39}H_{31}ClO_2OsP_2S$
Dicarbonylchloro(thioxomethyl)bis(triphenylphosphine)-osmium, Os-00396

$C_{39}H_{31}F_6O_3OsP_3$
Tricarbonylhydrobis(triphenylphosphine)osmium(1+); Hexafluorophosphate, *in* Os-00397

$C_{39}H_{31}F_6O_3P_3Ru$
Tricarbonylhydrobis(triphenylphosphine)ruthenium(1+); Hexafluorophosphate, *in* Ru-00744

$C_{39}H_{31}O_3OsP_2^{\oplus}$
Tricarbonylhydrobis(triphenylphosphine)osmium(1+), Os-00397

$C_{39}H_{31}O_3P_2Ru^{\oplus}$
Tricarbonylhydrobis(triphenylphosphine)ruthenium(1+), Ru-00744

$C_{39}H_{32}F_6NO_2OsP_3$
Carbonyl(η^2-ethyne)nitrosylbis(triphenylphosphine)-osmium(1+); Hexafluorophosphate, *in* Os-00398

$C_{39}H_{32}NO_2OsP_2^{\oplus}$
Carbonyl(η^2-ethyne)nitrosylbis(triphenylphosphine)-osmium(1+), Os-00398

$C_{39}H_{32}O_2OsP_2S$
Dicarbonyl(η^2-methanethial)bis(triphenylphosphine)-osmium, Os-00399

$C_{39}H_{32}O_3OsP_2$
Dicarbonyl(η^2-formaldehyde)bis(triphenylphosphine)-osmium, Os-00400

$C_{39}H_{33}ClOP_2RuSe_2$
Carbonylchloro[η^2-(methylseleno)selenoxomethyl]-bis(triphenylphosphine)ruthenium, Ru-00745

$C_{39}H_{33}ClO_2P_2Ru$
Acetylcarbonylchlorobis(triphenylphosphine)ruthenium, Ru-00746

$C_{39}H_{33}ClP_2RuS_2$
Carbonylchloro[η^2-(methylthio)thioxomethyl]-bis(triphenylphosphine)ruthenium, Ru-00747

$C_{39}H_{33}I_3O_2P_2Ru$
Dicarbonyltriiodo(triphenylphosphine)ruthenate(1−); Methyltriphenylphosphonium salt, *in* Ru-00518

$C_{39}H_{33}NO_4OsP_2$
Carbonyl(methoxycarbonyl)-nitrosylbis(triphenylphosphine)osmium, Os-00401

$C_{39}H_{35}NOP_2Ru$
Nitrosyl(η^3-2-propenyl)bis(triphenylphosphine)-ruthenium, Ru-00748

$C_{39}H_{39}BRu$
(η^5-2,4-Cyclohexadien-1-yl)[(1,2,3,4,5,6-η)-1,3,5-trimethylbenzene]ruthenium(1+); Tetraphenylborate, *in* Ru-00402

$C_{39}H_{41}B_9OP_2Ru$
Carbonylbis(triphenylphosphine)[(7,8,9,10,11-η)-undecahydro-7,8-dicarbaundecaborato(2−)]-ruthenium, Ru-00749
Carbonylbis(triphenylphosphine)[(7,8,9,10,11-η)-undecahydro-7,9-dicarbaundecaborato(2−)]-ruthenium, Ru-00750

$C_{39}H_{65}BBrO_{15}OsP_5$
Bromopentakis(trimethyl phosphite-*P*)osmium(1+); Tetraphenylborate, *in* Os-00266

$C_{39}H_{65}BBrO_{15}P_5Ru$
Bromopentakis(trimethyl phosphite-*P*)ruthenium(1+); Tetraphenylborate, *in* Ru-00409

$C_{39}H_{65}BClO_{15}OsP_5$
Chloropentakis(trimethyl phosphite-*P*)osmium(1+); Tetraphenylborate, *in* Os-00267

$C_{39}H_{65}BClO_{15}P_5Ru$
Chloropentakis(trimethyl phosphite-*P*)ruthenium(1+); Tetraphenylborate, *in* Ru-00410

$C_{39}H_{66}BO_{15}P_5Ru$
Hydropentakis(trimethyl phosphite-*P*)ruthenium(1+); Tetraphenylborate, *in* Ru-00412

$C_{40}H_{30}Al_2Cl_8O_4OsP_2$
Tetracarbonylbis(triphenylphosphine)osmium(2+); Bis(tetrachloroaluminate), *in* Os-00403

$C_{40}H_{30}Au_2O_4OsP_2$
Tetracarbonylbis[(triphenylphosphine)gold]osmium, Os-00402

$C_{40}H_{30}ClF_6NOP_2Ru$
Chloro[(2,3-η)-1,1,1,4,4,4-hexafluoro-2-butyne]-nitrosylbis(triphenylphosphine)ruthenium, Ru-00751

$C_{40}H_{30}Cl_4O_4P_2Ru_2$
Tetracarbonyldi-μ-chlorodichlorobis(triphenylphosphine)-diruthenium, Ru-00752

$C_{40}H_{30}O_4OsP_2^{\oplus\oplus}$
Tetracarbonylbis(triphenylphosphine)osmium(2+), Os-00403

$C_{40}H_{30}O_4OsSn_2$
Tetracarbonylbis(triphenylstannyl)osmium, Os-00404

$C_{40}H_{30}O_4RuSn_2$
Tetracarbonylbis(triphenylstannyl)ruthenium, Ru-00753

$C_{40}H_{33}ClO_4OsP_2$
Dicarbonylchloro(methoxycarbonyl)-bis(triphenylphosphine)osmium, Os-00405

$C_{40}H_{33}IO_4P_2Ru$
Dicarbonyliodo(methoxycarbonyl)bis(triphenylphosphine)-ruthenium, Ru-00754

$C_{40}H_{34}OOsP_2S$
(Carbonothioyl)carbonyl(η^2-ethene)-bis(triphenylphosphine)osmium, Os-00406

$C_{40}H_{34}O_2OsP_2$
Dicarbonyl(η^2-ethene)bis(triphenylphosphine)osmium, Os-00407

$C_{40}H_{34}O_2OsP_2S_2$
Dicarbonylhydro[(methylthio)thioxomethyl]-bis(triphenylphosphine)osmium, Os-00408

$C_{40}H_{34}O_2P_2Ru$
Dicarbonyl(η^2-ethene)bis(triphenylphosphine)ruthenium, Ru-00755

$C_{40}H_{34}Ru_4$
1,1″:1‴,1⁗:1‴″,1‴‴-Quaterruthenocene, Ru-00756

$C_{40}H_{35}ClO_2P_2Ru$
Carbonylchloro(1-oxopropyl)bis(triphenylphosphine)-ruthenium, Ru-00757

$C_{40}H_{36}Cl_2N_2P_2Ru$
Bis(acetonitrile)dichlorobis(triphenylphosphine)-ruthenium, Ru-00758

$C_{40}H_{36}O_2P_2RuS_4$
Bis(*O*-methylcarbonodithioato-*S,S′*)-bis(triphenylphosphine)ruthenium, Ru-00759

$C_{40}H_{42}Cl_4NP_2Ru$
Tetrachlorobis(triphenylphosphine)ruthenate(1−); Tetramethyl-ammonium salt, *in* Ru-00702

$C_{41}H_{34}O_6P_2Ru$
(η^5-2,4-Cyclopentadien-1-yl)[2-[(diphenoxyphosphino)-oxy]phenyl-*C,P*](triphenyl phosphite-*P*)ruthenium, Ru-00760

$C_{41}H_{34}P_2Ru$
(η^5-2,4-Cyclopentadien-1-yl)[2-(diphenylphosphino)-phenyl-*C,P*](triphenylphosphine)ruthenium, Ru-00761

$C_{41}H_{35}As_2ClRu$
Chloro(η^5-2,4-cyclopentadien-1-yl)bis(triphenylarsine)-ruthenium, Ru-00762

$C_{41}H_{35}BrO_6OsP_2$
Bromo(η^5-2,4-cyclopentadien-1-yl)bis(triphenyl phosphite-*P*)osmium, Os-00409

$C_{41}H_{35}BrOsP_2$
Bromo(η^5-2,4-cyclopentadien-1-yl)-bis(triphenylphosphine)osmium, Os-00410

$C_{41}H_{35}BrP_2Ru$
Bromo(η^5-2,4-cyclopentadien-1-yl)-bis(triphenylphosphine)ruthenium, Ru-00763

$C_{41}H_{35}ClO_6P_2Ru$
Chloro(η^5-2,4-cyclopentadien-1-yl)bis(triphenyl phosphite-*P*)ruthenium, Ru-00764

$C_{41}H_{35}ClP_2Ru$
Chloro(η^5-2,4-cyclopentadien-1-yl)-bis(triphenylphosphine)ruthenium, Ru-00765

$C_{41}H_{35}FP_2Ru$
(η^5-2,4-Cyclopentadien-1-yl)-fluorobis(triphenylphosphine)ruthenium, Ru-00766

$C_{41}H_{35}IP_2Ru$
(η^5-2,4-Cyclopentadien-1-yl)iodobis(triphenylphosphine)-ruthenium, Ru-00767

$C_{41}H_{36}P_2Ru$
(η^5-2,4-Cyclopentadien-1-yl)-hydrobis(triphenylphosphine)ruthenium, Ru-00768

$C_{41}H_{38}NP_2Ru$
(Acetonitrile)(η^3-2-propenyl)bis(triphenylphosphine)-ruthenium, Ru-00769

$C_{41}H_{39}BP_2Ru$
(η^5-2,4-Cyclopentadien-1-yl)[tetrahydroborato(1−)-*H*]-bis(triphenylphosphine)ruthenium, Ru-00770

$C_{41}H_{40}N_2O_{24}Os_{10}$
[μ_6-(Carbon)]tetracosacarbonyldecaosmate(2−); Bis(tetraethylammonium) salt, *in* Os-00342

$C_{41}H_{43}BRu$
(η^5-2,4-Cyclopentadien-1-yl)[(1,2,3,4,5,6-η)-hexamethylbenzene]ruthenium(1+); Tetraphenylborate, *in* Ru-00464

$C_{42}H_{21}AsO_{18}Ru_6$
Octadecacarbonyl-μ_6-hydrohexaruthenate(1−); Tetraphenylarsonium salt, *in* Ru-00468

$C_{42}H_{30}F_8O_2P_2Ru$
Dicarbonyl(1,1,2,2,3,3,4,4-octafluoro-1,4-butanediyl)-bis[triphenylphosphine]ruthenium, Ru-00771

$C_{42}H_{32}O_6Os_2P_2$
Hexacarbonyldihydrobis(triphenylphosphine)diosmium, Os-00411

$C_{42}H_{35}AsCl_5OsP$
Pentachloro(triphenylphosphine)osmate(1−); Tetraphenylarsonium salt, *in* Os-00305

$C_{42}H_{35}NOP_2Ru$
(Cyanato-*N*)(η^5-2,4-cyclopentadien-1-yl)-bis(triphenylphosphine)ruthenium, Ru-00772

$C_{42}H_{35}NP_2Ru$
(Cyano-*C*)(η^5-2,4-cyclopentadien-1-yl)-bis(triphenylphosphine)ruthenium, Ru-00773

$C_{42}H_{35}NP_2RuS$
(η^5-2,4-Cyclopentadien-1-yl)(thiocyanato-*N*)-bis(triphenylphosphine)ruthenium, Ru-00774

$C_{42}H_{35}OP_2Ru^{\oplus}$
Carbonyl(η^5-2,4-cyclopentadien-1-yl)-bis(triphenylphosphine)ruthenium(1+), Ru-00775

$C_{42}H_{36}ClF_6P_3Ru$
(η^6-Benzene)chlorobis(triphenylphosphine)ruthenium(1+); Hexafluorophosphate, *in* Ru-00776

$C_{42}H_{36}ClP_2Ru^{\oplus}$
(η^6-Benzene)chlorobis(triphenylphosphine)ruthenium(1+), Ru-00776

$C_{42}H_{36}F_6IOsP_3$
(η^6-Benzene)iodobis(triphenylphosphine)osmium(1+); Hexafluorophosphate, *in* Os-00412

$C_{42}H_{36}IOsP_2^{\oplus}$
(η^6-Benzene)iodobis(triphenylphosphine)osmium(1+), Os-00412

$C_{42}H_{36}OP_2Ru$
Hydro[(1,2,3,4,5-η)-6-oxo-2,4-cyclohexadien-1-yl]-bis(triphenylphosphine)ruthenium, Ru-00777

$C_{42}H_{36}OsP_2$
(η^6-Benzene)bis(triphenylphosphine)osmium, Os-00413

$C_{42}H_{37}BF_4P_2Ru$
(η^6-Benzene)hydrobis(triphenylphosphine)ruthenium(1+); Tetrafluoroborate, *in* Ru-00779

$C_{42}H_{37}ClP_2Ru$
Chloro[(1,2,3,4,5-η)-1-methyl-2,4-cyclopentadien-1-yl]-bis(triphenylphosphine)ruthenium, Ru-00778

$C_{42}H_{37}F_6OsP_3$
(η^6-Benzene)hydrobis(triphenylphosphine)osmium(1+); Hexafluorophosphate, *in* Os-00414

$C_{42}H_{37}OsP_2^{\oplus}$
(η^6-Benzene)hydrobis(triphenylphosphine)osmium(1+), Os-00414

$C_{42}H_{37}P_2Ru^{\oplus}$
(η^6-Benzene)hydrobis(triphenylphosphine)ruthenium(1+), Ru-00779

$C_{42}H_{38}P_2Ru$
[(1,2,3,4,5-η)-2,4-Cyclohexadien-1-yl]-hydrobis(triphenylphosphine)ruthenium, Ru-00780
(η^5-2,4-Cyclopentadien-1-yl)-methylbis(triphenylphosphine)ruthenium, Ru-00781

$C_{42}H_{40}O_2OsP_2$
Dicarbonyldiethylbis(triphenylphosphine)osmium, Os-00415

$C_{42}H_{40}P_2Ru$
Bis(η^3-2-propenyl)bis(triphenylphosphine)ruthenium, Ru-00782

$C_{42}H_{42}N_2P_2RuS_4$
Bis(dimethylcarbamodithioato-*S,S'*)-bis(triphenylphosphine)ruthenium, Ru-00783

$C_{42}H_{45}P_3Ru$
Hydrotris(methyldiphenylphosphine)(η^3-2-propenyl)-ruthenium, Ru-00784

$C_{42}H_{47}N_2OsP_3$
(Dinitrogen)tris(ethyldiphenylphosphine)dihydroosmium, Os-00416

$C_{43}H_{29}O_7Os_3P_2$
Heptacarbonyl[μ_3-[(5-η:5,6-η:6-η)-1,3-cyclohexadien-5-yne]][μ-(diphenylphosphino)]-hydro(triphenylphosphine)triosmium, Os-00417

$C_{43}H_{30}Cl_4N_2O_2P_2Ru$
Dicarbonyl[1,2,3,4-tetrachloro-5-(η^2-diazo)-1,3-cyclopentadiene]bis(triphenylphosphine)ruthenium, Ru-00785

$C_{43}H_{35}Cl_2NO_2P_2Ru$
Benzoyldichloronitrosylbis(triphenylphosphine)-ruthenium, Ru-00786

$C_{43}H_{36}N_2OOsP_2$
Carbonylhydro(phenylazo)bis(triphenylphosphine)osmium, Os-00418

$C_{43}H_{36}N_2OP_2Ru$
Carbonylhydro(phenylazo)bis(triphenylphosphine)-ruthenium, Ru-00787

$C_{43}H_{37}Cl_2NOP_2Ru$
Benzyldichloronitrosylbis(triphenylphosphine)ruthenium, Ru-00788

$C_{43}H_{37}Cl_3N_2P_2Ru$
Trichloro(4-methylphenylazo)bis(triphenylphosphine)-ruthenium, Ru-00789

$C_{43}H_{38}AsF_6NP_2Ru$
Acetonitrile(η^5-2,4-cyclopentadien-1-yl)-bis(triphenylphosphine)ruthenium(1+); Hexafluoroarsenate, *in* Ru-00791

$C_{43}H_{38}Cl_2P_2Ru$
[(2,3,5,6-η)-Bicyclo[2.2.1]hepta-2,5-diene]-dichlorobis(triphenylphosphine)ruthenium, Ru-00790

$C_{43}H_{38}F_6NP_3Ru$
Acetonitrile(η^5-2,4-cyclopentadien-1-yl)-bis(triphenylphosphine)ruthenium(1+); Hexafluorophosphate, *in* Ru-00791

$C_{43}H_{38}NOsP_2^{\oplus}$
Acetonitrile(η^5-2,4-cyclopentadien-1-yl)-bis(triphenylphosphine)osmium(1+), Os-00419

$C_{43}H_{38}NP_2Ru^{\oplus}$
Acetonitrile(η^5-2,4-cyclopentadien-1-yl)-bis(triphenylphosphine)ruthenium(1+), Ru-00791

$C_{43}H_{38}OP_2RuS_2$
(η^5-2,4-Cyclopentadien-1-yl)(*O*-methylcarbonodithioato-*S*)bis(triphenylphosphine)ruthenium, Ru-00792

$C_{43}H_{39}F_6OsP_3$
(η^6-Benzene)methylbis(triphenylphosphine)osmium(1+); Hexafluorophosphate, *in* Os-00420

$C_{43}H_{39}OsP_2^{\oplus}$
(η^6-Benzene)methylbis(triphenylphosphine)osmium(1+), Os-00420

$C_{43}H_{40}P_2Ru$
(η^5-2,4-Cyclopentadien-1-yl)-ethylbis(triphenylphosphine)ruthenium, Ru-00793

$C_{44}H_{30}Cl_4O_4P_2Ru$
Dicarbonyl(3,4,5,6-tetrachloro-1,2-benzenediolato-*O,O'*)-bis(triphenylphosphine)ruthenium, Ru-00794

$C_{44}H_{30}F_{12}P_2RuS_4$
Bis(1,1,1,4,4,4-hexafluoro-2,3-butanedithionato-*S,S'*)-bis(triphenylphosphine)ruthenium, Ru-00795

$C_{44}H_{35}BF_4N_2O_2P_2Ru$
Dicarbonyl(phenylazo)bis(triphenylphosphine)-ruthenium(1+); Tetrafluoroborate, *in* Ru-00798

$C_{44}H_{35}BrN_2O_2OsP_2$
Bromodicarbonyl(phenylazo)bis(triphenylphosphine)-osmium, Os-00421

$C_{44}H_{35}BrN_2O_2P_2Ru$
Bromodicarbonyl(phenylazo)bis(triphenylphosphine)-ruthenium, Ru-00796

$C_{44}H_{35}ClN_2O_2OsP_2$
Dicarbonylchloro(phenylazo)bis(triphenylphosphine)-osmium, Os-00422

$C_{44}H_{35}F_6N_2O_2OsP_3$
Dicarbonyl(phenylazo)bis(triphenylphosphine)osmium(1+); Hexafluorophosphate, *in* Os-00424

$C_{44}H_{35}F_6N_2O_2P_3Ru$
Dicarbonyl(phenylazo)bis(triphenylphosphine)-ruthenium(1+); Hexafluorophosphate, *in* Ru-00798

$C_{44}H_{35}IN_2O_2OsP_2$
Dicarbonyliodo(phenylazo)bis(triphenylphosphine)osmium, Os-00423

$C_{44}H_{35}IN_2O_2P_2Ru$
Dicarbonyliodo(phenylazo)bis(triphenylphosphine)-ruthenium, Ru-00797

$C_{44}H_{35}N_2O_2OsP_2^{\oplus}$
Dicarbonyl(phenylazo)bis(triphenylphosphine)osmium(1+), Os-00424

$C_{44}H_{35}N_2O_2P_2Ru^{\oplus}$
Dicarbonyl(phenylazo)bis(triphenylphosphine)-ruthenium(1+), Ru-00798

$C_{44}H_{36}B_2F_8N_2O_2P_2Ru$
Dicarbonyl(phenyldiazene-*N*^{2})bis(triphenylphosphine)-ruthenium(2+); Bis(tetrafluoroborate), *in* Ru-00801

$C_{44}H_{36}ClN_2O_2P_2Ru^{\oplus}$
Dicarbonylchloro(phenyldiazene-*N*^{2})-bis(triphenylphosphine)ruthenium(1+), Ru-00799

$C_{44}H_{36}Cl_2N_2O_6P_2Ru$
Dicarbonylchloro(phenyldiazene-*N*^{2})-bis(triphenylphosphine)ruthenium(1+); Perchlorate, *in* Ru-00799

$C_{44}H_{36}N_2O_2OsP_2$
Dicarbonylhydro(phenylazo)bis(triphenylphosphine)-osmium, Os-00425

$C_{44}H_{36}N_2O_2P_2Ru$
Dicarbonylhydro(phenylazo)bis(triphenylphosphine)-ruthenium, Ru-00800
Dicarbonyl(phenyldiazene-*N*^{2})bis(triphenylphosphine)-ruthenium(2+), Ru-00801

$C_{44}H_{38}Cl_2N_2OOsP_2$
(Acetylphenyldiazene-*N*^{2},*O*^{1})-dichlorobis(triphenylphosphine)osmium, Os-00426

$C_{44}H_{38}O_6P_2Ru$
Dicarbonyl[(2,3-η)-dimethyl-2-butenedioate]-bis(triphenylphosphine)ruthenium, Ru-00802
$C_{44}H_{38}P_2Ru$
[(1,2,3,4,5,6-η)-1,3,5,7-Cyclooctatetraene]-bis(triphenylphosphine)ruthenium, Ru-00803
(η^5-2,4-Cyclopentadien-1-yl)-1-propynylbis(triphenylphosphine)ruthenium, Ru-00804
$C_{44}H_{40}P_2Ru$
(η^5-2,4-Cyclopentadien-1-yl)-2-propenylbis(triphenylphosphine)ruthenium, Ru-00805
$C_{44}H_{42}P_2Ru$
(η^5-2,4-Cyclopentadien-1-yl)-propylbis(triphenylphosphine)ruthenium, Ru-00806
$C_{44}H_{44}P_2Ru$
Bis[(1,2,3-η)-2-methyl-2-propenyl]-bis(triphenylphosphine)ruthenium, Ru-00807
$C_{44}H_{48}Ge_2O_2P_2Ru$
Dicarbonylbis(trimethylgermyl)bis(triphenylphosphine)-ruthenium, Ru-00808
$C_{44}H_{48}O_2P_2RuSi_2$
Dicarbonylbis(trimethylsilyl)bis(triphenylphosphine)-ruthenium, Ru-00809
$C_{45}H_{31}NO_9Os_3P_2S$
Nonacarbonyl-μ-hydro-μ_3-thioxotriosmate(1−); Bis(triphenylphosphine)iminium salt, *in* Os-00069
$C_{45}H_{36}F_6P_2Ru$
(η^5-2,4-Cyclopentadien-1-yl)[3,3,3-trifluoro-1-(trifluoromethyl)-1-propenyl]-bis(triphenylphosphine)ruthenium, Ru-00810
$C_{45}H_{37}ClOOsP_2$
Carbonylchloro[(4-methylphenyl)methylidyne]-bis(triphenylphosphine)osmium, Os-00427
$C_{45}H_{44}P_2Ru$
Butyl(η^5-2,4-cyclopentadien-1-yl)-bis(triphenylphosphine)ruthenium, Ru-00811
$C_{46}H_{30}N_2O_{11}P_2Ru_3$
Decacarbonyl-μ-nitrosyltriruthenate(1−); Bis(triphenylphosphine)iminium salt, *in* Ru-00166
$C_{46}H_{30}O_{10}Os_3P_2$
Decacarbonylbis(triphenylphosphine)triosmium, Os-00428
$C_{46}H_{30}O_{10}P_2Ru_3$
Decacarbonylbis(triphenylphosphine)triruthenium, Ru-00812
$C_{46}H_{34}O_{10}P_2Ru_4$
Decacarbonyltetra-μ-hydrobis(triphenylphosphine)-tetraruthenium, Ru-00813
$C_{46}H_{36}BF_4NO_4P_2Ru$
[(Benzoylamino)carbonyl]-dicarbonylbis(triphenylphosphine)ruthenium(1+); Tetrafluoroborate, *in* Ru-00814
$C_{46}H_{36}NO_4P_2Ru^{\oplus}$
[(Benzoylamino)carbonyl]-dicarbonylbis(triphenylphosphine)ruthenium(1+), Ru-00814
$C_{46}H_{37}ClNO_2OsP_2^{\oplus}$
Dicarbonylchloro[4-methyl(isocyanobenzene)]-bis(triphenylphosphine)osmium(1+), Os-00429
$C_{46}H_{38}O_2OsP_2$
Dicarbonyl[(4-methyl-2,5-cyclohexadien-1-ylidene)-methylene]bis(triphenylphosphine)osmium, Os-00430
$C_{47}H_{31}NO_{11}Os_3P_2$
μ-Carbonyldecacarbonyl-μ-hydrotriosmate(1−); Bis(triphenylphosphine)iminium salt, *in* Os-00121
$C_{47}H_{31}NO_{11}P_2Ru_3$
μ-Carbonyldecacarbonyl-μ-hydrotriruthenate(1−); Bis(triphenylphosphine)iminium salt, *in* Ru-00168
$C_{47}H_{35}F_5P_2Ru$
(η^5-2,4-Cyclopentadien-1-yl)(pentafluorophenyl)-bis(triphenylphosphine)ruthenium, Ru-00815
$C_{48}H_{30}FeN_2O_{13}P_2Ru_3$
Tri-μ-carbonylnonacarbonyl(nitrosylferrate)-triruthenate(1−); Bis(triphenylphosphine)iminium salt, *in* Ru-00213
$C_{48}H_{33}NO_{12}P_2Ru_4$
Dodecacarbonyltrihydrotetraruthenate(1−); Bis(triphenylphosphine)iminium salt, *in* Ru-00220
$C_{48}H_{40}Cl_4N_4O_4P_4Ru_4$
Tetra-μ-chlorotetrakis[μ-(diphenylphosphino)]-tetranitrosyltetraruthenium, Ru-00816
$C_{48}H_{42}I_4Os_2P_2$
η^6-Benzenediiodo(triphenylphosphine)osmium, Os-00431
$C_{48}H_{42}N_2P_2S_2Ru$
Bis(2-aminobenzenethiolato-*N,S*)bis(triphenylphosphine)-ruthenium, Ru-00817
$C_{48}H_{42}P_2Ru$
Benzyl(η^5-2,4-cyclopentadien-1-yl)-bis(triphenylphosphine)ruthenium, Ru-00818
$C_{48}H_{44}P_2Ru$
(η^4-1,3-Butadiene)[(η^2-ethenyl)benzene]-bis(triphenylphosphine)ruthenium, Ru-00819
$C_{48}H_{48}P_2Ru$
[(1,2,3-η)-2,4-Cyclohexadien-1-yl][(1,2-η)-cyclohexene]-hydrobis(triphenylphosphine)ruthenium, Ru-00820
$C_{49}H_{30}ClNO_{13}P_2Ru_4$
μ-Carbonyldodecacarbonyl-μ-chlorotetraruthenate(1−); Bis(triphenylphosphine)iminium salt, *in* Ru-00281
$C_{49}H_{30}O_7Os_3$
Heptacarbonyl[μ_3-[1,1′-(η^1:η^1:η^2-1,2-ethynediyl)-bis[benzene]]][μ-(1,2,3,4-η:1,4-η)-1,2,3,4-tetraphenyl-1,3-butadiene-1,4-diyl]triosmium, Os-00432
$C_{49}H_{31}FeNO_{13}P_2Ru_3$
Di-μ-carbonylnonacarbonyl(dicarbonylferrate)-μ-hydrotriruthenate(1−); Bis(triphenylphosphine)iminum salt, *in* Ru-00282
$C_{49}H_{31}Fe_2NO_{13}P_2Ru_2$
Hexacarbonyl(μ-carbonylhexacarbonyldiferrate)-μ-hydrodiruthenate(1−); Bis(triphenylphosphine)iminium salt, *in* Ru-00283
$C_{49}H_{31}NO_{13}Os_4P_2$
μ-Carbonyldodecacarbonyl-μ-hydrotetraosmate(1−); Bis(triphenylphosphine)iminium salt, *in* Os-00205
$C_{49}H_{31}NO_{13}P_2Ru_4$
μ_3-Carbonyldodecacarbonylhydrotetraruthenate(1−); Bis(triphenylphosphine)iminium salt, *in* Ru-00284
$C_{49}H_{37}NO_{13}Os_4P_2$
Undecacarbonyltetra-μ-hydro(methoxycarbonyl)-tetraosmate(1−); Bis(triphenylphosphine)iminium salt, *in* Os-00219
$C_{49}H_{40}ClCuP_2Ru$
(η^5-2,4-Cyclopentadien-1-yl)(phenylethynyl)-bis(triphenylphosphine)ruthenium; CuCl adduct, *in* Ru-00821
$C_{49}H_{40}OsP_2$
(η^5-2,4-Cyclopentadien-1-yl)(phenylethynyl)-bis(triphenylphosphine)osmium, Os-00433
$C_{49}H_{40}P_2Ru$
(η^5-2,4-Cyclopentadien-1-yl)(phenylethynyl)-bis(triphenylphosphine)ruthenium, Ru-00821
$C_{50}H_{32}BrF_6NO_{10}Os_3P_2$
Bromodecacarbonyl[(2,3-η)-1,1,1,4,4,4-hexafluoro-2-butene]triosmate(1−); Bis(triphenylphosphine)iminium salt, *in* Os-00224
$C_{50}H_{40}As_2Cl_4O_2Ru$
Dicarbonyltetrachlororuthenate(2−); Bis(tetraphenylarsonium) salt, *in* Ru-00011
$C_{50}H_{40}Cl_2N_2P_2Ru$
Bis(benzonitrile)dichlorobis(triphenylphosphine)-ruthenium, Ru-00822

$C_{50}H_{46}N_2O_2P_4Ru_2$
Bis[μ-(diphenylphosphino)]-dinitrosylbis(methyldiphenylphosphine)-diruthenium, Ru-00823

$C_{50}H_{48}P_2Ru$
[(η^2-Ethenyl)benzene][(1,2,3,4-η)-1,3-hexadiene]-bis(triphenylphosphine)ruthenium, Ru-00824

$C_{50}H_{50}P_2Ru$
[η^2-(Ethenylbenzene)][(1,2,3-η)-2-hexenyl]-hydrobis(triphenylphosphine)ruthenium, Ru-00825

$C_{51}H_{30}INO_{15}OsP_2$
Pentadecacarbonyliodopentaosmate(1−); Bis(triphenylphosphine)iminium salt, *in* Os-00269

$C_{51}H_{31}NO_{15}Os_5P_2$
Pentadecacarbonyl-μ-hydropentaosmate(1−); Bis(triphenylphosphine)iminium salt, *in* Os-00248

$C_{51}H_{45}N_3OP_2Ru$
[1,3-Bis(4-methylphenyl)-1-triazenato-N^1,N^3]-carbonylhydrobis(triphenylphosphine)ruthenium, Ru-00826

$C_{52}H_{30}INO_{15}Os_5P_2$
[μ_5-(Carbon)]pentadecacarbonyliodopentaosmate(1−); Bis(triphenylphosphine)iminium salt, *in* Os-00285

$C_{52}H_{30}O_{15}P_2Ru_6$
[μ_6-(Carbon)]-μ-carbonyltetradecacarbonylbis(triphenylphosphine)-hexaruthenium, Ru-00827

$C_{52}H_{44}ClNOP_2Ru$
Carbonylchloro[η^2-(4-methylphenyl)[(4-methylphenyl)-imino]methyl]bis(triphenylphosphine)ruthenium, Ru-00828

$C_{52}H_{46}P_2Ru$
Bis[(η^2-ethenyl)benzene]bis(triphenylphosphine)-ruthenium, Ru-00829

$C_{52}H_{50}P_4Ru$
Bis[1,2-ethanediylbis[diphenylphosphine]-*P*,*P'*]-dihydroruthenium, Ru-00830

$C_{53}H_{44}N_2O_3P_2Ru$
[*N*,*N'*-Bis(4-methylphenyl)ureato(2−)-*N*,*N'*]-dicarbonylbis(triphenylphosphine)ruthenium, Ru-00831

$C_{54}H_{31}NO_{18}Os_6P_2$
Octadecacarbonyl-μ_3-hydrohexaosmate(1−); Bis(triphenylphosphine)iminium salt, *in* Os-00301

$C_{54}H_{31}NO_{18}P_2Ru_6$
Octadecacarbonyl-μ_6-hydrohexaruthenate(1−); Bis(triphenylphosphine)iminium salt, *in* Ru-00468

$C_{54}H_{45}Br_2OsP_3$
Dibromotris(triphenylphosphine)osmium, Os-00434

$C_{54}H_{45}Br_2P_3Ru$
Dibromotris(triphenylphosphine)ruthenium, Ru-00832

$C_{54}H_{45}Cl_2OsP_3$
Dichlorotris(triphenylphosphine)osmium, Os-00435

$C_{54}H_{45}Cl_2P_3Ru$
Dichlorotris(triphenylphosphine)ruthenium, Ru-00833

$C_{54}H_{46}BF_4P_3Ru$
Hydro[(η^6-phenyl)diphenylphosphine]-bis(triphenylphosphine)ruthenium(1+); Tetrafluoroborate, *in* Ru-00837

$C_{54}H_{46}BrP_3Ru$
Bromohydrotris(triphenylphosphine)ruthenium, Ru-00834

$C_{54}H_{46}ClP_3Ru$
Chlorohydrotris(triphenylphosphine)ruthenium, Ru-00835

$C_{54}H_{46}NOOsP_3$
Hydronitrosyltris(triphenylphosphine)osmium, Os-00436

$C_{54}H_{46}NOP_3Ru$
Hydronitrosyltris(triphenylphosphine)ruthenium, Ru-00836

$C_{54}H_{46}P_3Ru^{\oplus}$
Hydro[(η^6-phenyl)diphenylphosphine]-bis(triphenylphosphine)ruthenium(1+), Ru-00837

$C_{54}H_{47}N_2P_3Ru$
(Dinitrogen)dihydrotris(triphenylphosphine)ruthenium, Ru-00838

$C_{54}H_{49}OsP_3$
Tetrahydrotris(triphenylphosphine)osmium, Os-00437

$C_{54}H_{49}P_3Ru$
Tetrahydrotris(triphenylphosphine)ruthenium, Ru-00839

$C_{54}H_{95}BBrO_{15}OsP_5$
Bromopentakis(triethyl phosphite-*P*)osmium(1+); Tetraphenylborate, *in* Os-00350

$C_{54}H_{95}BBrO_{15}P_5Ru$
Bromopentakis(triethyl phosphite-*P*)ruthenium(1+); Tetraphenylborate, *in* Ru-00665

$C_{54}H_{95}BClO_{15}OsP_5$
Chloropentakis(triethyl phosphite-*P*)osmium(1+); Tetraphenylborate, *in* Os-00351

$C_{54}H_{95}BClO_{15}P_5Ru$
Chloropentakis(triethyl phosphite-*P*)ruthenium(1+); Tetraphenylborate, *in* Ru-00666

$C_{54}H_{96}BO_{15}P_5Ru$
Hydropentakis(triethyl phosphite-*P*)ruthenium(1+); Tetraphenylborate, *in* Ru-00667

$C_{55}H_{45}Br_2O_{10}P_3Ru$
Dibromocarbonyltris(triphenyl phosphite-*P*)ruthenium, Ru-00840

$C_{55}H_{45}Cl_2OP_3Ru$
Carbonyldichlorotris(triphenylphosphine)ruthenium, Ru-00841

$C_{55}H_{45}Cl_2O_{10}P_3Ru$
Carbonyldichlorotris(triphenyl phosphite-*P*)ruthenium, Ru-00842

$C_{55}H_{45}F_6NO_2OsP_4$
Carbonylnitrosyltris(triphenylphosphine)osmium(1+); Hexafluorophosphate, *in* Os-00438

$C_{55}H_{45}F_6NO_2P_4Ru$
Carbonylnitrosyltris(triphenylphosphine)ruthenium(1+); Hexafluorophosphate, *in* Ru-00844

$C_{55}H_{45}I_2O_{10}P_3Ru$
Carbonyldiiodotris(triphenyl phosphite-*P*)ruthenium, Ru-00843

$C_{55}H_{45}NO_2OsP_3^{\oplus}$
Carbonylnitrosyltris(triphenylphosphine)osmium(1+), Os-00438

$C_{55}H_{45}NO_2P_3Ru^{\oplus}$
Carbonylnitrosyltris(triphenylphosphine)ruthenium(1+), Ru-00844

$C_{55}H_{46}As_3ClOOs$
Carbonylchlorohydrotris(triphenylarsine)osmium, Os-00439

$C_{55}H_{46}BrOOsP_3$
Bromocarbonylhydrotris(triphenylphosphine)osmium, Os-00440

$C_{55}H_{46}BrOP_3Ru$
Bromocarbonylhydrotris(triphenylphosphine)ruthenium, Ru-00845

$C_{55}H_{46}ClOOsP_3$
Carbonylchlorohydrotris(triphenylphosphine)osmium, Os-00441

$C_{55}H_{46}ClOP_3Ru$
Carbonylchlorohydrotris(triphenylphosphine)ruthenium, Ru-00846

$C_{55}H_{47}OOsP_3$
Carbonyldihydrotris(triphenylphosphine)osmium, Os-00442

$C_{55}H_{47}OP_3Ru$
Carbonyldihydrotris(triphenylphosphine)ruthenium, Ru-00847

$C_{56}H_{36}O_{18}P_2Ru_6$
Di-μ-carbonyldi-μ_3-carbonyltetradecacarbonylhexaruthenate(2−); Bis(methyltriphenylphosphonium) salt, *in* Ru-00501

$C_{56}H_{45}O_2OsP_3$
Dicarbonyltris(triphenylphosphine)osmium, Os-00443
$C_{56}H_{45}O_2P_3Ru$
Dicarbonyltris(triphenylphosphine)ruthenium, Ru-00848
$C_{56}H_{46}O_2P_3Ru^{\oplus}$
Dicarbonylhydrotris(triphenylphosphine)ruthenium(1+), Ru-00849
$C_{56}H_{49}P_3Ru$
(η^2-Ethene)tris(triphenylphosphine)ruthenium, Ru-00850
$C_{57}H_{50}ClP_3Ru$
Chloro(η^3-2-propenyl)tris(triphenylphosphine)ruthenium, Ru-00851
$C_{58}H_{48}ClO_3P_3Ru$
Acetyldicarbonyltris(triphenylphosphine)ruthenium(1+); Chloride, *in* Ru-00852
$C_{58}H_{48}O_3P_3Ru^{\oplus}$
Acetyldicarbonyltris(triphenylphosphine)ruthenium(1+), Ru-00852
$C_{58}H_{51}P_3Ru$
(η^4-1,3-Butadiene)tris(triphenylphosphine)ruthenium, Ru-00853
$C_{60}H_{50}AsCl_4OsP_2$
Tetrachlorobis(triphenylphosphine)osmate(1−); Tetraphenylarsonium salt, *in* Os-00364
$C_{60}H_{50}N_2O_2P_4Ru_2$
Bis[μ-(diphenylphosphino)]-dinitrosylbis(triphenylphosphine)diruthenium, Ru-00854
$C_{61}H_{30}INO_{24}Os_{10}P_2$
[μ_6-(Carbon)]tetracosacarbonyl-μ-iododecaosmate(1−); Bis(triphenylphosphine)iminium salt, *in* Os-00340
$C_{61}H_{50}AsCl_3P_2RuS$
(Carbonothioyl)trichlorobis(triphenylphosphine)-ruthenate(1−); Tetraphenylarsonium salt, *in* Ru-00711
$C_{62}H_{50}BNO_3OsP_2$
Dicarbonylnitrosylbis(triphenylphosphine)osmium(1+); Tetraphenylborate, *in* Os-00387
$C_{62}H_{50}BNO_3P_2Ru$
Dicarbonylnitrosylbis(triphenylphosphine)ruthenium(1+); Tetraphenylborate, *in* Ru-00724
$C_{62}H_{52}As_2Cl_3OPRu$
Carbonyltrichlorobis(triphenylarsine)ruthenate(1−); Benzyltriphenylphosphonium salt, *in* Ru-00707
$C_{63}H_{45}O_9Os_3P_3$
Nonacarbonyltris(triphenylphosphine)triosmium, Os-00444
$C_{63}H_{45}O_9P_3Ru_3$
Nonacarbonyltris(triphenylphosphine)triruthenium, Ru-00855
$C_{63}H_{49}O_9P_3Ru_4$
Nonacarbonyltetra-μ-hydrotris(triphenylphosphine)-tetraruthenium, Ru-00856
$C_{63}H_{58}B_2Ru$
(η^6-Benzene)[(1,2,3,4,5,6-η)-1,3,5-trimethylbenzene]-ruthenium(2+); Bis(tetraphenylborate), *in* Ru-00400
$C_{64}H_{31}NO_{29}Os_9P_2$
Heptacosacarbonyl[μ_3-(carboxylato-*C*:*O*:*O'*)]-μ-hydrononaosmate(1−); Bis(triphenylphosphine)iminium salt, *in* Os-00344
$C_{66}H_{55}BOP_2Ru$
Carbonyl(η^5-2,4-cyclopentadien-1-yl)-bis(triphenylphosphine)ruthenium(1+); Tetraphenylborate, *in* Ru-00775
$C_{66}H_{64}B_2Ru$
Bis[(1,2,3,4,5,6-η)-1,3,5-trimethylbenzene]-ruthenium(2+); Bis(tetraphenylborate), *in* Ru-00489
$C_{66}H_{94}B_2O_{18}P_6Ru$
Hexakis(trimethyl phosphite-*P*)ruthenium(2+); Bis(tetraphenylborate), *in* Ru-00498
$C_{67}H_{58}BNOsP_2$
Acetonitrile(η^5-2,4-cyclopentadien-1-yl)-bis(triphenylphosphine)osmium(1+); Tetraphenylborate, *in* Os-00419
$C_{67}H_{58}BNP_2Ru$
Acetonitrile(η^5-2,4-cyclopentadien-1-yl)-bis(triphenylphosphine)ruthenium(1+); Tetraphenylborate, *in* Ru-00791
$C_{72}H_{60}Br_2O_{12}P_4Ru$
Dibromotetrakis(triphenyl phosphite-*P*)ruthenium, Ru-00857
$C_{72}H_{60}Br_2OsP_4$
Dibromotetrakis(triphenylphosphine)osmium, Os-00445
$C_{72}H_{60}Br_2P_4Ru$
Dibromotetrakis(triphenylphosphine)ruthenium, Ru-00858
$C_{72}H_{60}Cl_2P_4Ru$
Dichlorotetrakis(triphenylphosphine)ruthenium, Ru-00859
$C_{72}H_{60}Cl_4P_4Ru_2$
Di-μ-chlorodichlorobis(triphenylphosphine)diruthenium, *in* Ru-00698
$C_{72}H_{61}ClO_{12}P_4Ru$
Chlorohydrotetrakis(triphenyl phosphite-*P*)ruthenium, Ru-00860
$C_{72}H_{62}P_4Ru$
Dihydrotetrakis(triphenylphosphine)ruthenium, Ru-00861
$C_{73}H_{60}Cl_4P_4Ru_2S$
(Carbonothioyl)tri-μ-chlorochlorotetrakis(triphenylphosphine)-diruthenium, Ru-00862
$C_{74}H_{60}Cl_4O_2P_4Ru_2$
Dicarbonyldi-μ-chlorodichlorotetrakis(triphenylphosphine)-diruthenium, Ru-00863
$C_{74}H_{60}Cl_4P_4Ru_2S_2$
Bis(carbonothioyl)di-μ-chlorodichlorotetrakis(triphenylphosphine)-diruthenium, Ru-00864
$C_{76}H_{66}Cl_4OP_4Ru_2S$
(Carbonothioyl)tri-μ-chlorochlorotetrakis(triphenylphosphine)-diruthenium; Me_2CO complex, *in* Ru-00862
$C_{76}H_{72}B_2N_4Ru$
[(1,2,5,6-η)-1,5-Cyclooctadiene]tetrakis(pyridine)-ruthenium(2+); Bis(tetraphenylborate), *in* Ru-00648
$C_{78}H_{66}N_4O_4P_4Ru_2$
Dinitrosylbis(triphenylphosphine)ruthenium; Benzene adduct (2:1), *in* Ru-00703
$C_{79}H_{65}BNO_2OsP_3$
Carbonylnitrosyltris(triphenylphosphine)osmium(1+); Tetraphenylborate, *in* Os-00438
$C_{80}H_{64}O_8P_4Ru_4$
Octacarbonyltetra-μ-hydrotetrakis(triphenylphosphine)-tetraruthenium, Ru-00865
$C_{84}H_{62}N_2O_{12}Os_4P_4$
Dodecacarbonyldi-μ-hydrotetraosmate(2−); Bis[bis(triphenylphosphine)iminium] salt, *in* Os-00159
$C_{84}H_{130}B_2O_{18}OsP_6$
Hexakis(triethyl phosphite-*P*)osmium(2+); Bis(tetraphenylborate), *in* Os-00367
$C_{84}H_{130}B_2O_{18}P_6Ru$
Hexakis(triethyl phosphite-*P*)ruthenium(2+); Bis(tetraphenylborate), *in* Ru-00705
$C_{85}H_{60}N_2O_{13}P_4Ru_4$
μ_3-Carbonyldodecacarbonyltetraruthenate(2−); Bis[bis(triphenylphosphine)iminium]salt, *in* Ru-00324
$C_{87}H_{60}N_2O_{15}Os_5P_4$
Pentadecacarbonylpentaosmate(2−); Bis[bis(triphenylphosphine)iminium]salt, *in* Os-00270
$C_{89}H_{60}N_2O_{16}Ru_6P_4$
[μ_6-(Carbon)]tri-μ-carbonyltridecacarbonylhexaruthenate(2−); Bis[bis(triphenylphosphine)iminium] salt, *in* Ru-00467

$C_{90}H_{60}N_2O_{18}P_4Ru_6$
Di-μ-carbonyldi-μ3-carbonyltetradecacarbonylhexaruthenate(2–); Bis[bis(triphenylphosphine)]iminium salt, *in* Ru-00501
$C_{94}H_{60}N_2O_{22}Os_8P_4$
Di-μ-carbonylicosacarbonyloctaosmate(2–); Bis[bis(triphenylphosphine)iminium] salt, *in* Os-00330
$C_{97}H_{60}N_2O_{24}Os_{10}P_4$
[μ6-(Carbon)]tetracosacarbonyldecaosmate(2–); Bis[bis(triphenylphosphine)iminium] salt, *in* Os-00342
$Cl_5H_8N_3ORu$
Pentachloronitrosylruthenate(2–); Di-NH_4 salt, *in* Ru-00866
Cl_5K_2NORu
Pentachloronitrosylruthenate(2–); Di-K salt, *in* Ru-00866
$Cl_5NORu^{\ominus\ominus}$
Pentachloronitrosylruthenate(2–), Ru-00866
F_5P_5Ru
▷Pentakis(phosphorous trifluoride)ruthenium, Ru-00867
$F_{12}HOsP_4^{\ominus}$
Hydrotetrakis(phosphorous trifluoride)osmate(1–), Os-00446
$F_{12}HP_4Ru^{\ominus}$
Hydrotetrakis(phosphorous trifluoride)ruthenate(1–), Ru-00868
$F_{12}H_2OsP_4$
Dihydrotetrakis(phosphorous trifluoride)osmium, Os-00447
$F_{12}H_2P_4Ru$
Dihydrotetrakis(phosphorous trifluoride)ruthenium, Ru-00869
$F_{12}K_2OsP_4$
Tetrakis(phosphorous trifluoride)osmate(2–); Di-K salt, *in* Os-00448
$F_{12}K_2P_4Ru$
Tetrakis(phosphorous trifluoride)ruthenate(2–); Di-K salt, *in* Ru-00870
$F_{12}OsP_4^{\ominus\ominus}$
Tetrakis(phosphorous trifluoride)osmate(2–), Os-00448
$F_{12}P_4Ru^{\ominus\ominus}$
Tetrakis(phosphorous trifluoride)ruthenate(2–), Ru-00870
$F_{15}OsP_5$
Pentakis(phosphorous trifluoride)osmium, Os-00449

CAS Registry Number Index

1273-81-0 Osmocene, Os-00109

1287-13-4 Ruthenocene, Ru-00149

1316-92-3 Acetylruthenocene, Ru-00240

11058-58-5 Dichlorobis[1,1′-ferrocenediylbis[dimethylarsine]]ruthenium, Ru-00649

11064-20-3 Dodecacarbonyltetrahydro(iron)triruthenium, Ru-00221

11067-28-0 Hexadecacarbonylhexaoxohexaosmium, Os-00289

11067-30-4 [μ-[(1,2,3,3*a*,8*a*-η:4,5,6-η:6,7,8-η)-Azulene]]heptacarbonyltriruthenium, Ru-00451

11105-98-9 Ferrocenylmethylosmocene, Os-00323

11105-99-0 (Ferrocenylmethyl)ruthenocene, Ru-00549

11127-38-1 Nonacarbonyldi-μ-hydro-μ_3-thioxotriruthenium, Ru-00089

12089-68-8 (1-Hydroxyethyl)ruthenocene, Ru-00256

12104-84-6 Carbonyldihydrotris(triphenylphosphine)osmium, Os-00442

12108-25-7 Tricarbonyl[(1,2,3,4-η)-1,3-cyclohexadiene]ruthenium, Ru-00101

12109-37-4 Dodecacarbonylhexachlorotriruthenium, Ru-00211

12112-96-8 Decacarbonyl(1,3-cyclooctadiene)triruthenium, Ru-00477

12116-59-5 Tricarbonyl[(2,3,4,5,6-η)-1-methoxy-2,4-cyclohexadien-1-yl]ruthenium(1+); Tetrafluoroborate, *in* Ru-00140

12116-81-3 Dicarbonyl[(1,2,3,4,5-η)-2,4-cyclohexadien-1-yl](methoxycarbonyl)ruthenium, Ru-00147

12118-06-8 Bis[(1,2,3,4,4*a*,8*a*-η)naphthalene]ruthenium(2+); Bis(hexafluorophosphate), *in* Ru-00521

12119-52-7 Bis(η^6-benzene)di-μ-chlorobis(tributylphosphine)diruthenium(2+); Dichloride, *in* Ru-00704

12128-15-3 Dicarbonyl(η^5-2,4-cyclopentadien-1-yl)ruthenate(1−), Ru-00052

12131-08-7 Tetracarbonylbis(η^5-2,4-cyclopentadien-1-yl)-(mercury)diruthenium, Ru-00332

12131-10-1 Tetracarbonylbis(η^5-2,4-cyclopentadien-1-yl)diosmium, Os-00233

12132-87-5 Di-μ-carbonyldicarbonylbis(η^5-2,4-cyclopentadien-1-yl)diruthenium, Ru-00335

12170-97-7 Dichloro[(1,2,3,6,7,10,11,12-η)-2,6,10-dodecatriene-1,12-diyl]ruthenium, Ru-00267

12176-87-3 Dicarbonyldi-μ-chlorodichlorobis[(1,2,5,6-η)-1,5-cyclooctadiene]diruthenium, Ru-00486

12212-72-5 [(2,3,4,5-η)Bicyclo[4.2.0]octa-2,4-diene]-[(1,2,3,4,5,6-η)-1,3,5-cyclooctatriene]ruthenium, Ru-00432

12215-07-5 (η^6-Benzene)[(1,2,3,4-η)-1,3-cyclohexadiene]ruthenium, Ru-00259

12216-20-5 [(1,2,3,4,5,6-η)-1,3,5-Cycloheptatriene]-[(1,2,5,6-η)-1,5-cyclooctadiene]ruthenium, Ru-00404

12216-80-7 [(2,3,5,6-η)-Bicyclo[2.2.1]hepta-2,5-diene]-dichlorobis(triphenylphosphine)ruthenium, Ru-00790

12245-07-7 (Cyano-π-cyclopentadienyl)-π-cyclopentadienylruthenium, *in* Ru-00185

12277-76-8 μ_6-Carbon-μ-carbonylhexadecacarbonylhexaruthenium, Ru-00500

12278-41-0 μ_6-Carbon-μ-carbonyltridecacarbonyl(1,3,5-trimethylbenzene)hexaruthenium, Ru-00585

12282-26-7 Tricarbonyl[(1,2,3,4-η)-1,3,5,7-cyclooctatetraene]ruthenium, Ru-00176

12289-05-3 μ-Carbonyltetracarbonyl[μ-[(1,2,3,4,5,6,7,8-η)-1,3,5,7-cyclooctatetraene]]diruthenium, Ru-00296

12289-15-5 [(2,3,5,6-η)-Bicyclo[2.2.1]hepta-2,5-diene]-bis(η^3-2-propenyl)ruthenium, Ru-00322

12289-52-0 [(1,2,5,6-η)-1,5-Cyclooctadiene]bis(η^3-2-propenyl)ruthenium, Ru-00369

12289-94-0 [(1,2,5,6-η)-1,5-Cyclooctadiene]bis[(1,2,3-η)-2-methyl-2-propenyl]ruthenium, Ru-00439

12290-41-4 Tetracarbonylbis[μ-(1,3,5,7-cyclooctatetraene)]triruthenium, Ru-00519

12321-05-0 Tricarbonyl(1,2,3,4-tetraphenyl-1,3-cyclobutadiene)ruthenium, Ru-00668

12321-08-3 Tricarbonyl[(2,3,4,5-η)-2,3,4,5-tetraphenyl-2,4-cyclopentadien-1-one]ruthenium, Ru-00672

12337-22-3 Ethynylruthenocene, Ru-00233

12337-23-4 3-Chloro-3-ruthenocenyl-2-propen-1-al, Ru-00302

12337-26-7 (Ruthenocenylethynyl)ferrocene, Ru-00564

12375-04-1 Dodecacarbonyltetra-μ-hydrotetraosmium, Os-00167

12375-24-5 Tridecacarbonyldihydroirontriruthenium, Ru-00288

12375-24-5 Di-μ-carbonyldicarbonyldi-μ-hydrotris[μ_3-(tricarbonylruthenio)]iron, Ru-00287

12388-68-0 Octacarbonyl(tetracarbonyliron)diruthenium, Ru-00214

12389-47-8 Tetracarbonyl(octacarbonyldiruthenium)osmium, Os-00181

12389-50-3 Octacarbonyl(tetracarbonylruthenium)diosmium, Os-00200

12512-44-6 Propanoylruthenocene, Ru-00311

12557-93-6 Decacarbonyl-μ-chloro-μ-hydrotriosmium, Os-00093

12559-97-6 Tricarbonyl[(1,2,3,6-η)-1,3,5,7-cyclooctatetraene]osmium, Os-00132

12560-38-2 Dodecacarbonylirondiosmium, Os-00151

12560-43-9 Dodecacarbonylhydromanganesediosmium, Os-00153

12560-44-0 Dodecacarbonylhydrorheniumdiosmium, Os-00154

12563-65-4 Dodecacarbonylmanganatediosmate(1−); Tetramethylammonium salt, *in* Os-00198

12563-67-6 Dodecacarbonylrhenatediosmate(1−); Tetramethylammonium salt, *in* Os-00199

12563-74-5 Di-μ-carbonylnonacarbonyl(dicarbonyliron)di-μ-hydrotriosmium, Os-00208

12563-75-6 Tridecacarbonyltrihydromanganesetriosmium, Os-00212

12563-76-7 Tridecacarbonyltrihydrorheniumtriosmium, Os-00215

12566-57-3 Hexadecacarbonylhydromanganesetriosmium, Os-00271

12566-58-4 Hexadecacarbonylhydrorheniumtriosmium, Os-00272

12568-51-3 Octadecacarbonyldi-μ_3-hydrohexaruthenium, Ru-00469

12577-43-4 [μ_6-(Carbon)]-μ-carbonylpentadecacarbonyl(triphenylphosphine)-hexaruthenium, Ru-00684

12577-61-6 Octacarbonylbis[1,1′-(1,2-ethynediyl)-bis[benzene]]triruthenium, Ru-00690

12628-24-9 (η^6-Benzene)dichlororuthenium; Polymer, *in* Ru-00038

12679-33-3 Decachlororuthenocene, Ru-00120

13820-58-1 Pentachloronitrosylruthenate(2−); Di-NH_4 salt, *in* Ru-00866

14564-32-0 Dibromodicarbonylbis(triphenylphosphine)ruthenium, Ru-00716

14564-35-3 Dicarbonyldichlorobis(triphenylphosphine)ruthenium, Ru-00720

14564-39-7 Dicarbonyldiodobis(triphenylphosphine)ruthenium, Ru-00723

14655-23-3 Tricarbonylbis(triphenylarsine)ruthenium, Ru-00739

14741-36-7 Tricarbonylbis(triphenylphosphine)ruthenium, Ru-00742

14854-54-7 Pentachloronitrosylruthenate(2−); Di-K salt, *in* Ru-00866

14878-21-8 Dibromotetracarbonylosmium, Os-00015

14911-96-7 Bis(aniline)dicarbonyldichlororuthenium, Ru-00341

14911-97-8 Bis(aniline)dibromodicarbonylruthenium, Ru-00340

14912-60-8 Aquacarbonyltetrachlororuthenate(2−); Di-NH_4 salt, *in* Ru-00006

14978-16-6 Bis(2-aminothiophenol)dicarbonylruthenium, Ru-00336
15091-95-9 Tricarbonyldiiodo(tripheny!phosphine)ruthenium, Ru-00543
15243-33-1 Dodecacarbonyltriruthenium, Ru-00280
15349-78-7 Trichloronitrosylbis(triphenylphosphine)ruthenium, Ru-00699
15390-63-3 Carbonyldichlorotris(triphenylphosphine)ruthenium, Ru-00841
15393-10-9 Dicarbonylbis[μ-(dimethylcarbamodithioato-*S*:*S*,*S*′)]bis(dimethylcarbamodithioato-*S*,*S*′)-diruthenium, Ru-00370
15393-11-0 Dicarbonylbis(dimethylcarbamodithioato-*S*,*S*′)ruthenium, Ru-00085
15453-53-9 Nonacarbonyltris(triphenylphosphine)triruthenium, Ru-00855
15529-49-4 Dichlorotris(triphenylphosphine)ruthenium, Ru-00833
15555-76-7 Dibromotetrakis(triphenylphosphine)ruthenium, Ru-00858
15555-77-8 Dichlorotetrakis(triphenylphosphine)ruthenium, Ru-00859
15663-29-3 Dibromotetracarbonylruthenium, Ru-00022
15663-30-6 Tetracarbonyldiiodoruthenium, Ru-00028
15696-40-9 Dodecacarbonyltriosmium, Os-00201
15709-57-6 Dicarbonyldichlorobis(triphenylphosphine)ruthenium; *ab*-dicarbonyl-*df*-dichloro-*ce*-diphosphine-*form*, *in* Ru-00720
15709-73-6 Dicarbonyldichlorobis(triphenylphosphine)ruthenium; *af*-dicarbonyl-*bd*-dichloro-*ce*-diphosphine-*form*, *in* Ru-00720
15709-75-8 Dibromotris(triphenylphosphine)ruthenium, Ru-00832
15928-45-7 Tricarbonylchlorobis(triphenylphosphine)osmium(1+); *trans-form*, Chloride, *in* Os-00394
16369-40-7 Dicarbonyldichlororuthenium; Monomer, *in* Ru-00010
16369-41-8 Dibromodicarbonylruthenium, Ru-00008
16369-42-9 Dicarbonyldiiodoruthenium, Ru-00015
16369-43-0 Tetrabromodicarbonylruthenate(2−); *cis-form*, Di-Cs salt, *in* Ru-00009
16369-44-1 Dicarbonyltetraiodoruthenate(2−); *cis-form*, Di-Cs salt, *in* Ru-00016
16406-48-7 Pentacarbonylruthenium, Ru-00031
16406-49-8 Pentacarbonylosmium, Os-00027
16481-98-4 Dibromotricarbonyl(triphenylphosphine)osmium, Os-00317
16481-99-5 Tricarbonyldiiodo(triphenylphosphine)osmium, Os-00319
16520-58-4 Nonacarbonyltris(triphenylphosphine)triosmium, Os-00444
16520-59-5 Tricarbonyldichloro(triphenylphosphine)osmium, Os-00318
16591-90-5 Dicarbonyldichlorobis(triphenylphosphine)osmium, Os-00382
16592-04-4 Tricarbonyliodobis(triphenylphosphine)ruthenium(1+); Trichloroiodoaluminate, *in* Ru-00741
16594-36-8 Tricarbonylchlorobis(triphenylphosphine)ruthenium(1+); Tetrachloroaluminate, *in* Ru-00740
16594-57-3 Dicarbonylchloro(methoxycarbonyl)-bis(triphenylphosphine)osmium, Os-00405
16594-61-9 Dicarbonyliodo(methoxycarbonyl)-bis(triphenylphosphine)ruthenium, Ru-00754
16674-09-2 Tetracarbonylbis(triphenylphosphine)osmium(2+); Bis(tetrachloroaluminate), *in* Os-00403
16971-31-6 Carbonylchlorohydrotris(triphenylphosphine)osmium, Os-00441
16971-32-7 Bromocarbonylhydrotris(triphenylphosphine)osmium, Os-00440
16971-33-8 Carbonylchlorohydrotris(triphenylphosphine)ruthenium, Ru-00846
16971-34-9 Bromocarbonylhydrotris(triphenylphosphine)ruthenium, Ru-00845
17439-05-3 Dicarbonyltetrachlororuthenate(2−); *cis-form*, Di-Cs salt, *in* Ru-00011
17456-71-2 Dicarbonyldiiodobis(triphenylphosphine)osmium, Os-00385
17456-72-3 Dibromodicarbonylbis(triphenylphosphine)osmium, Os-00377
17456-75-6 Dibromodicarbonylbis(triphenylarsine)osmium, Os-00375
17456-76-7 Dicarbonyldichlorobis(triphenylarsine)osmium, Os-00376
17456-77-8 Dicarbonyldiiodobis(triphenylstibine)osmium, Os-00386
17456-78-9 Dibromodicarbonylbis(triphenylstibine)osmium, Os-00378
17568-67-1 Carbonylpentachloroosmate(3−); Tri-Cs salt, *in* Os-00004
17568-72-8 Aquacarbonyltetrachlororuthenate(2−); Di-Cs salt, *in* Ru-00006
17632-05-2 Tetracarbonyldiiodoosmium; *cis-form*, *in* Os-00021
17632-08-5 Hexacarbonyldi-μ-iododiiododiosmium, Os-00040
17633-40-8 Hexacarbonyldi-μ-chlorodichlorodiosmium, Os-00031
17633-49-7 Dicarbonyl[1,2-ethanediylbis[diphenylphosphine]-*P*,*P*′]diiodoosmium, Os-00346
17653-14-4 Tetracarbonylbis(triphenylstannyl)ruthenium; *trans-form*, *in* Ru-00753
17685-65-3 Tetracarbonyldiiodoosmium, Os-00021
17731-88-3 Di-μ-bromodibromohexacarbonyldiosmium, Os-00029
18195-25-0 Dicarbonyldichlorobis(triphenylstibine)osmium, Os-00383
18400-02-7 Dicarbonyltetrachloroosmate(2−); *cis-form*, Di-Cs salt, *in* Os-00007
18400-28-7 Tribromotricarbonylosmate(1−); Cs salt, *in* Os-00012
18475-75-7 Tetracarbonyldiiodoruthenium; *cis-form*, *in* Ru-00028
18953-16-7 Dibromotetracarbonylosmium; *cis-form*, *in* Os-00015
18972-43-5 Tricarbonyldihydro(triphenylphosphine)osmium, Os-00321
18974-23-7 Dicarbonyldihydrobis(triphenylphosphine)osmium, Os-00392
19529-00-1 Dihydrotetrakis(triphenylphosphine)ruthenium, Ru-00861
19592-09-7 Pentachloronitrosylruthenate(2−), Ru-00866
19702-30-8 ▷Pentakis(phosphorous trifluoride)ruthenium, Ru-00867
19702-31-9 Pentakis(phosphorous trifluoride)osmium, Os-00449
20332-49-4 Tricarbonylbis(triphenylphosphine)ruthenium; *trans-form*, *in* Ru-00742
20332-51-8 Tricarbonylbis(triphenylphosphine)osmium; *trans-form*, *in* Os-00395
20468-34-2 Tetracarbonyl(di-μ-carbonylhexacarbonyldiiron-)ruthenium, Ru-00215
20492-47-1 Bis[μ-[bis(pentafluorophenyl)arsino]]hexacarbonyldiruthenium, Ru-00656
20909-98-2 Bis[μ-[bis(pentafluorophenyl)phosphino]]hexacarbonyldiruthenium, Ru-00657
21029-23-2 Tetracarbonyldihydroruthenium, Ru-00027
21029-29-8 Dicarbonyldihydrobis(triphenylphosphine)ruthenium, Ru-00731
21077-76-9 Di-μ-carbonylundecacarbonyldi-μ-hydrotetraruthenium, Ru-00291
21192-23-4 Tetracarbonyl(triphenylphosphine)ruthenium, Ru-00560
21192-24-5 Tetracarbonyl(triphenylphosphine)osmium, Os-00326
21216-91-1 Tetracarbonyldi-μ-chlorodichlorobis(triphenylphosphine)diruthenium, Ru-00752
21349-95-1 μ_3-Carbonyldodecacarbonyldi-μ-hydrotetraosmium, Os-00210
21439-09-8 Tetracarbonylbis(trimethylstannyl)ruthenium, Ru-00165
21439-16-7 Dibromotetracarbonylruthenium; *cis-form*, *in* Ru-00022
21475-91-2 Tetracarbonyldichlororuthenium; *cis-form*, *in* Ru-00024
21710-53-2 Tetracarbonylbis(chloromercurio)ruthenium; *cis-form*, *in* Ru-00023
21710-55-4 Tetracarbonylbis(chloromercurio)osmium, Os-00016
21773-70-6 Dodecacarbonydichlorotriosmium, Os-00150
21868-05-3 Hexacarbonylbis[μ-(dimethylstannylene)]-bis(trimethylstannyl)diruthenium, Ru-00446

21868-07-5 Tetracarbonylbis(trimethylplumbyl)ruthenium, Ru-00163
22180-41-2 Trichloronitrosylbis(triphenylphosphine)osmium, Os-00362
22186-54-5 Carbonylchloronitrosylbis(triphenylphosphine)osmium, Os-00370
22337-78-6 Carbonyldihydrotris(triphenylphosphine)ruthenium; *Orange-form*, *in* Ru-00847
22337-84-4 (Dinitrogen)dihydrotris(triphenylphosphine)ruthenium, Ru-00838
22372-70-9 Tetracarbonyldihydroosmium, Os-00019
22377-52-2 Tricarbonylbis(trimethyl phosphite-*P*)osmium; *trans-form*, *in* Os-00083
22391-69-1 Bromohydrotris(triphenylphosphine)ruthenium, Ru-00834
22391-74-8 Bis(dimethylcarbamodithioato-*S*,*S*′)-bis(triphenylphosphine)ruthenium, Ru-00783
22391-77-1 Hexacarbonyldi-μ-iododiosmium, Os-00039
22533-60-4 Tetracarbonylhydroosmate(1−); Na salt, *in* Os-00018
22587-71-9 Octacarbonyldiiododiosmium, Os-00068
22588-62-1 Dicarbonyl(tetrafluoroethene)bis(trimethyl phosphite-*P*)ruthenium, Ru-00161
22594-69-0 Hexacarbonyldi-μ-chlorodichlorodiruthenium, Ru-00033
22594-70-3 Di-μ-bromodibromohexacarbonyldiruthenium, Ru-00032
22594-81-6 Tricarbonyltrichlororuthenate(1−); Cs salt, *in* Ru-00018
22594-82-7 Tribromotricarbonylruthenate(1−); Cs salt, *in* Ru-00017
22594-83-8 Tricarbonyltriiodoruthenate(1−); Cs salt, *in* Ru-00021
22605-82-9 Tetracarbonyliodomethylosmium; *cis-form*, *in* Os-00025
22639-01-6 Tetracarbonyldimethylosmium; *cis-form*, *in* Os-00038
22639-02-7 Tricarbonyldimethyl(triphenylphosphine)osmium, Os-00332
22639-03-8 Tetracarbonylhydromethylosmium; *cis-form*, *in* Os-00026
22804-20-2 Bromocarbonylnitrosylbis(triphenylphosphine)ruthenium, Ru-00708
22804-23-5 Chloro(dioxygen)nitrosylbis(triphenylphosphine)ruthenium, Ru-00696
22863-29-2 Tricarbonyldichloro(triphenylphosphine)ruthenium, Ru-00542
22863-33-8 Tetrachlorobis(triphenylphosphine)ruthenate(1−); *trans-form*, Tetramethylammonium salt, *in* Ru-00702
22863-37-2 Carbonylchloronitrosylbis(triphenylphosphine)ruthenium, Ru-00710
22873-63-8 Tetracarbonyldichloroosmium, Os-00017
22949-66-2 Dibromotricarbonyl(triphenylphosphine)ruthenium, Ru-00541
23254-68-4 Diacetyldibromodicarbonylosmate(2−); Bis(bromomagnesium) salt, *in* Os-00035
23733-18-8 Decacarbonyl[μ-(ethanethiolato)]-μ-hydrotriosmium, Os-00169
23733-19-9 [(μ-Benzenethiolato)]decacarbonyl-μ-hydrotriosmium, Os-00275
23733-20-2 Decacarbonyl[μ-(ethanethiolato)]-μ-hydrotriruthenium, Ru-00226
24228-59-9 Tetrahydrotris(triphenylphosphine)osmium, Os-00437
24651-66-9 Chlorohydrotetrakis(triphenyl phosphite-*P*)ruthenium, Ru-00860
24899-53-4 Dihydrotetrakis(phosphorous trifluoride)ruthenium, Ru-00869
24899-54-5 Dihydrotetrakis(phosphorous trifluoride)osmium, Os-00447
25158-72-9 Octacarbonylbis[μ-(dimethylstannylene)]diruthenium, Ru-00248
25340-26-5 Tetracarbonyldiiodoosmium; *trans-form*, *in* Os-00021
25360-32-1 Carbonyldihydrotris(triphenylphosphine)ruthenium; *White-form*, *in* Ru-00847
25685-05-6 Octacarbonyldihydrodiosmium, Os-00056
25685-06-7 Octacarbonyldichlorodiosmium, Os-00054
25685-07-8 Dibromooctacarbonyldiosmium, Os-00053
25702-65-2 Tricarbonylhydrobis(triphenylphosphine)osmium(1+); Bromide, *in* Os-00397
25702-66-3 Tricarbonylhydrobis(triphenylphosphine)osmium(1+); Hexafluorophosphate, *in* Os-00397
25702-68-5 Dicarbonyldihydrobis(triphenylphosphine)osmium; *cis,cis,trans-form*, *in* Os-00392
25929-58-2 Tetracarbonylbis(trimethylsilyl)ruthenium, Ru-00164
25930-76-1 Tetracarbonylhydro(trimethylsilyl)osmium; *cis-form*, *in* Os-00052
25931-79-7 Tetracarbonylbis(trimethylstannyl)osmium, Os-00117
26024-45-3 Tetracarbonyl(trimethylsilyl)-[(triphenylphosphine)aurio]ruthenium; *cis-form*, *in* Ru-00623
26024-47-5 Dicarbonylbis(trimethylsilyl)-bis(triphenylphosphine)ruthenium, Ru-00809
26220-98-4 Dodecacarbonyl-μ-hydrotriruthenium(1+); Hexafluorophosphate, *in* Ru-00217
26220-99-5 Dodecacarbonyl-μ-hydrotriosmium(1+); Hexafluorophosphate, *in* Os-00155
26407-33-0 Bis(acetonitrile)-dichlorobis(triphenylphosphine)ruthenium; *cis,cis,trans-form*, *in* Ru-00758
26407-38-5 Bis(benzonitrile)-dichlorobis(triphenylphosphine)ruthenium, Ru-00822
26455-77-6 Bis(2-aminobenzenethiolato-*N*,*S*)-bis(triphenylphosphine)ruthenium, Ru-00817
26553-82-2 Dichloro(dinitrogen)tris(triethylphosphine)osmium, Os-00306
26876-73-3 Tetrakis(phosphorous trifluoride)ruthenate(2−); Di-K salt, *in* Ru-00870
26876-74-4 Tetrakis(phosphorous trifluoride)osmate(2−); Di-K salt, *in* Os-00448
27497-62-7 (Dinitrogen)tris(ethyldiphenylphosphine)dihydroosmium, Os-00416
27599-26-4 Tetracarbonyldipropionylosmium; *cis-form*, *in* Os-00108
27659-99-0 Bromotetracarbonylmethylosmium; *cis-form*, *in* Os-00024
27857-55-2 Tetracarbonylosmate(2−); Di-Na salt, *in* Os-00022
27882-33-3 Tetracarbonyldiethylosmium; *cis-form*, *in* Os-00062
27882-34-4 Dicarbonyldiethylbis(triphenylphosphine)osmium, Os-00415
27882-35-5 Bromotetracarbonylethylosmium, Os-00033
28109-19-5 Di-μ-bromodecacarbonyltriosmium, Os-00090
28109-20-8 Decacarbonyldi-μ-iodotriosmium, Os-00118
28411-13-4 μ-Carbonyloctacarbonyldiosmium, Os-00086
28577-70-0 Dibromotetrakis(triphenyl phosphite-*P*)ruthenium, Ru-00857
28577-76-6 Carbonyldichlorotris(triphenyl phosphite-*P*)ruthenium, Ru-00842
28577-77-7 Dibromocarbonyltris(triphenyl phosphite-*P*)ruthenium, Ru-00840
28577-78-8 Carbonyldiiodotris(triphenyl phosphite-*P*)ruthenium, Ru-00843
28577-82-4 Dicarbonyldichlorobis(triphenyl phosphite-*P*)ruthenium, Ru-00721
28909-62-8 Chloro(dinitrogen)bis[1,2-phenylenebis[dimethylarsine]-*As*,*As*′]ruthenium(1+); *trans-form*, Hexafluorophosphate, *in* Ru-00529
28986-71-2 Dibromodicarbonylruthenium; Polymer, *in* Ru-00008
28986-72-3 Dicarbonyldiiodoruthenium; Monomer, *in* Ru-00015
29079-55-8 [1,2-Bis(benzeneseleno)ethane-*Se*,*Se*′]dibromodicarbonylruthenium, Ru-00424
29088-68-4 Dicarbonyldiiodoruthenium; Polymer, *in* Ru-00015
29091-98-3 Dicarbonyldiodobis(triphenylphosphine)ruthenium; *af*-dicarbonyl-*bd*-diiodo-*ce*-diphosphine-*form*, *in* Ru-00723
29099-76-1 [1,2-Bis(ethylthio)ethane-*S*,*S*′]dicarbonyldiiodoruthenium, Ru-00086
29099-84-1 Dibromodicarbonylbis(triphenylphosphine)ruthenium; *ab*-dibromo-*cd*-dicarbonyl-*ef*-diphosphine-*form*, *in* Ru-00716
29136-57-0 Dibromodicarbonyl[1,2-ethanediylbis(diphenylphosphine)-*P*,*P*′]ruthenium, Ru-00645
29292-11-3 Tribromonitrosylbis(triphenylphosphine)osmium, Os-00360

29826-66-2 Carbonyliodonitrosylbis(triphenylphosphine)ruthenium, Ru-00712

29826-79-7 Trichloronitrosylbis(triphenylphosphine)ruthenium; *trans-form, in* Ru-00699

29826-81-1 Chlorodiiodonitrosylbis(triphenylphosphine)ruthenium, Ru-00694

29826-82-2 Tribromonitrosylbis(triphenylphosphine)ruthenium; *trans-form, in* Ru-00693

29994-24-9 Tetracarbonylbis(trimethylsilyl)osmium, Os-00116

29994-26-1 Octacarbonylbis(trimethylsilyl)diosmium, Os-00243

29994-93-2 Dodecacarbonyltetra-μ_3-oxotetraosmium, Os-00204

30066-39-8 Bis(acetonitrile)dicarbonyldichlororuthenium, Ru-00037

30066-41-2 Bis(benzonitrile)dicarbonyldichlororuthenium, Ru-00421

30173-88-7 Undecacarbonyl(triphenylphosphine)triosmium, Os-00348

30173-89-8 Undecacarbonyl(triphenylarsine)triosmium, Os-00347

30178-79-1 Dicarbonylchloro[2-(phenylazo)phenyl](triphenylphosphine)ruthenium, Ru-00673

30191-89-0 Pentabromocarbonylruthenate(2−); Di-Cs salt, *in* Ru-00002

30258-60-7 Dichloro(chloromercurio)nitrosylbis(triphenylphosphine)osmium, Os-00361

30306-42-4 Pentacarbonylhydroosmium(1+); Hexafluorophosphate, *in* Os-00023

30347-85-4 Nonacarbonyltris[μ-(dimethylgermylene)]triruthenium, Ru-00399

30347-86-5 Nonacarbonyltris[μ-(dimethylgermylene)]triosmium, Os-00263

30347-88-7 Dinitrosylbis(triphenylphosphine)osmium, Os-00366

30352-63-7 Dinitrosylbis(triphenylphosphine)ruthenium, Ru-00703

30754-01-9 Dicarbonyldichloro[2-(phenylazo)phenyl]ruthenate(1−); Tetramethylammonium salt, *in* Ru-00331

30759-74-1 Tricarbonylbis(trimethyl phosphite-*P*)ruthenium; *trans-form, in* Ru-00115

30846-57-2 Tetracarbonyliodo(trimethylgermyl)ruthenium; *trans-form, in* Ru-00062

30846-58-3 Tetracarbonyliodo(trimethylstannyl)ruthenium; *trans-form, in* Ru-00064

30846-59-4 Bromotetracarbonyl(trimethylgermyl)osmium; *trans-form, in* Os-00047

30846-60-7 Tetracarbonyliodo(trimethylgermyl)osmium; *trans-form, in* Os-00049

30846-64-1 Tricarbonyliodo(trimethylgermyl)(triphenylphosphine)ruthenium, Ru-00603

30846-65-2 Tricarbonyliodo(trimethylstannyl)(triphenylphosphine)ruthenium, Ru-00605

30846-67-4 Tricarbonylhydro(trimethylstannyl)(triphenylphosphine)ruthenium, Ru-00607

30848-49-8 Bromotetracarbonyl(trimethylsilyl)ruthenium; *trans-form, in* Ru-00061

30848-50-1 Tetracarbonyliodo(trimethylsilyl)ruthenium, Ru-00063

30848-52-3 Bromotetracarbonyl(trimethylsilyl)osmium; *trans-form, in* Os-00048

30848-53-4 Tetracarbonyliodo(trimethylsilyl)osmium, Os-00050

30848-58-9 Tricarbonyliodo(trimethylsilyl)(triphenylphosphine)ruthenium, Ru-00604

30848-61-4 Tricarbonylhydro(trimethylsilyl)(triphenylphosphine)ruthenium, Ru-00606

30854-78-5 Tetracarbonylbis(trimethylgermyl)ruthenium, Ru-00162

30854-80-9 Octacarbonylbis(trimethylgermyl)diruthenium, Ru-00360

30854-81-0 Hexacarbonylbis[μ-(dimethylgermylene)]bis(trimethylgermyl)diosmium, Os-00283

31005-45-5 Hexacarbonyltris[μ-(dimethylgermylene)]diosmium, Os-00191

31075-99-7 Dicarbonyl[sulfata(2−)-*O,O′*]bis(triphenylphosphine)ruthenium, Ru-00728

31177-20-5 Carbonylchlorohydrobis(tricyclohexylphosphine)osmium, Os-00374

31200-91-6 Dicarbonyl[(*O,S*-η)-sulfur dioxide]bis(triphenylphosphine)ruthenium, Ru-00727

31252-76-3 Dodecacarbonyltetra-μ-fluorotetrafluorotetraruthenium, Ru-00212

31275-06-6 Tetrahydrotris(triphenylphosphine)ruthenium, Ru-00839

31447-11-7 Tricarbonylbis(triphenyl phosphite-*P*)ruthenium, Ru-00743

31532-12-4 Bis(2-isocyano-2-methylpropane)tricarbonylruthenium, Ru-00320

31541-94-3 Tricarbonylbis(trimethyl phosphite-*P*)ruthenium, Ru-00115

31725-25-4 Chloro(η^3-2-propenyl)tris(triphenylphosphine)ruthenium, Ru-00851

31725-27-6 Di-μ-carbonyldicarbonylbis(η^5-2,4-cyclopentadien-1-yl)diruthenium; *cis-form, in* Ru-00335

31725-28-7 Di-μ-carbonyldicarbonylbis(η^5-2,4-cyclopentadien-1-yl)diruthenium; *trans-form, in* Ru-00335

31741-51-2 Carbonyl(η^5-2,4-cyclopentadien-1-yl)(η^3-2-propenyl)ruthenium, Ru-00105

31741-71-6 Tricarbonyl(η^5-2,4-cyclopentadien-1-yl)ruthenium(1+); Hexafluorophosphate, *in* Ru-00072

31741-73-8 Dicarbonyl(cyanato-*N*)(η^5-2,4-cyclopentadien-1-yl)ruthenium, Ru-00070

31741-74-9 Dicarbonyl(η^5-2,4-cyclopentadien-1-yl)(thiocyanato-*S*)ruthenium, Ru-00069

31741-75-0 Dicarbonyl(η^5-2,4-cyclopentadien-1-yl)(thiocyanato-*N*)ruthenium, Ru-00068

31742-01-5 [(1,2,5,6-η)-1,5-Cyclooctadiene]bis(2,4-pentanedionato-*O,O′*)ruthenium, Ru-00490

31781-73-4 Tricarbonylchloro(η^3-2-propenyl)ruthenium, Ru-00035

31781-74-5 Bromotricarbonyl(η^3-2-propenyl)ruthenium, Ru-00034

31781-75-6 Tricarbonyliodo(η^3-2-propenyl)ruthenium, Ru-00036

31781-76-7 Tricarbonylchloro[(1,2,3-η)-2-methyl-2-propenyl]ruthenium, Ru-00057

31781-83-6 Dicarbonyl(η^5-2,4-cyclopentadien-1-yl)iodoruthenium, Ru-00051

31811-43-5 Dicarbonyl(η^5-2,4-cyclopentadien-1-yl)-2-propenylruthenium, Ru-00143

31854-77-0 Carbonyl(η^5-2,4-cyclopentadien-1-yl)iodo(triphenyl phosphite-*P*)ruthenium, Ru-00594

31870-71-0 Dicarbonyl-2,4-cyclopentadien-1-yl(η^5-2,4-cyclopentadien-1-yl)ruthenium, Ru-00229

31960-26-6 Carbonyl(η^5-2,4-cyclopentadien-1-yl)iodo(trimethyl phosphite-*P*)ruthenium, Ru-00112

32073-80-6 Carbonyltrichlororuthenium, Ru-00003

32073-81-7 Tribromocarbonylruthenium, Ru-00001

32073-82-8 Carbonyltriiodoruthenium, Ru-00007

32216-10-7 Nonacarbonyl[μ^3-[η:η:η^3-3,5-cyclohexadiene-1,2-diylidene(phenylmethylidyne)]]-μ-hydrotriruthenium, Ru-00556

32236-82-1 Hydrotetrakis(phosphorous trifluoride)osmate(1−), Os-00446

32236-95-6 Hydrotetrakis(phosphorous trifluoride)ruthenate(1−), Ru-00868

32423-80-6 Hexacarbonyldi-μ-iododiiododiruthenium, Ru-00043

32423-81-7 Tricarbonyltrichlororuthenate(1−), Ru-00018

32423-82-8 Tribromotricarbonylruthenate(1−), Ru-00017

32423-83-9 Tricarbonyltriiodoruthenate(1−), Ru-00021

32457-60-6 Tetracarbonyl(di-μ-carbonylhexacarbonyldiiron)osmium, Os-00152

32574-34-8 Decacarbonyl(μ-ethanethiolato)di-μ-hydrotriruthenium(1+), Ru-00228

32591-91-6 Nonacarbonyltri-μ-hydro-μ_3-thioxotriruthenium(1+), Ru-00092

32610-56-3 Acetyldicarbonyl(η^5-2,4-cyclopentadien-1-yl)ruthenium, Ru-00100

32610-57-4 Acetylcarbonyl(η^5-2,4-cyclopentadien-1-yl)(triphenylphosphine)ruthenium, Ru-00631

32611-12-4 Dicarbonylchloro(η^5-2,4-cyclopentadien-1-yl)ruthenium, Ru-00047

32611-25-9 Carbonyl(η^5-2,4-cyclopentadien-1-yl)iodo(triphenylphosphine)ruthenium, Ru-00593

32611-26-0 Bromocarbonyl(η^5-2,4-cyclopentadien-1-yl)(trimethylphosphite-*P*)ruthenium, Ru-00110

32611-27-1 Bromocarbonyl(η^5-2,4-cyclopentadien-1-yl)(triphenylphosphine)ruthenium, Ru-00589

32611-29-3 Carbonylchloro(η^5-2,4-cyclopentadien-1-yl)(triphenyl phosphite-*P*)ruthenium, Ru-00592

32611-30-6 Carbonylchloro(η^5-2,4-cyclopentadien-1-yl)(trimethylphosphite-*P*)ruthenium, Ru-00111

32613-25-5 Carbonylchloro(η^5-2,4-cyclopentadien-1-yl)(triphenylphosphine)ruthenium, Ru-00591

32613-42-6 Bromocarbonyl(η^5-2,4-cyclopentadien-1-yl)(triphenyl phosphite-*P*)ruthenium, Ru-00590

32613-75-5 Bromodicarbonyl(η^5-2,4-cyclopentadien-1-yl)ruthenium, Ru-00046

32629-41-7 Dicarbonyl(η^5-2,4-cyclopentadien-1-yl)[2-(phenylazo)phenyl]ruthenium, Ru-00504

32648-27-4 Bis[1,2-bis(methylthio)ethane-*S*,*S'*]dibromoruthenium; *trans-form*, *in* Ru-00088

32648-41-2 [1,2-Bis(benzenethiolato)ethane-*S*,*S'*]dibromodicarbonylruthenium, Ru-00423

32677-88-6 Dicarbonylbis[(1,2,3-η)-2-methyl-2-propenyl]ruthenium, Ru-00158

32677-89-7 Bis[(1,2,3-η)-2-methyl-2-propenyl]bis(trimethyl phosphite-*P*)ruthenium; *cis-form*, *in* Ru-00374

32698-53-6 Dichlorobis(triphenylphosphine)ruthenium; Polymer, *in* Ru-00698

32732-05-1 [(1,2,3,4-η)-Hexamethylbenzene][(1,2,3,4,5,6-η)-hexamethylbenzene]ruthenium, Ru-00614

32823-70-4 Dicarbonyl[(1,2,3,4-η)-hexakis(trifluoromethyl)benzene](trimethyl phosphite-*P*)ruthenium, Ru-00452

32825-30-2 Carbonyl(η^5-2,4-cyclopentadien-1-yl)[(η^3-*C*,*N*,*N'*)2-(phenylazo)phenyl]ruthenium, Ru-00480

32874-17-2 Tricarbonyl[(1,2,5,6-η)-1,5-cyclooctadiene]ruthenium, Ru-00192

32993-05-8 Chloro(η^5-2,4-cyclopentadien-1-yl)bis(triphenylphosphine)ruthenium, Ru-00765

32993-06-9 Bromo(η^5-2,4-cyclopentadien-1-yl)bis(triphenylphosphine)ruthenium, Ru-00763

33039-43-9 Acetylosmocene, Os-00177

33113-17-6 Tricarbonyl(η^5-2,4-cyclopentadien-1-yl)ruthenium(1+); Tetraphenylborate, *in* Ru-00072

33134-73-5 Dicarbonyl(η^2-tetrafluoroethene)bis(trimethyl phosphite-*P*)osmium, Os-00113

33152-40-8 Tetracarbonylbis(pentacarbonylmanganio)ruthenium, Ru-00380

33153-18-3 Hydronitrosyltris(triphenylphosphine)osmium, Os-00436

33153-70-7 Tetracarbonylbis(pentacarbonylrhenio)osmium, Os-00247

33182-85-3 Bis(μ-benzenetellurolato)hexacarbonyldiruthenium, Ru-00474

33182-86-4 Bis(μ-benzeneselenolato)hexacarbonyldiruthenium, Ru-00473

33292-90-9 Tetracarbonylbis(pentacarbonylmanganio)osmium, Os-00246

33306-40-0 Tricarbonyl[(1,2,3,6,7-η)-2,4,6-cyclooctatrien-1-yl]ruthenium(1+); Hexafluorophosphate, *in* Ru-00179

33306-42-2 Tricarbonyl[(1,2,3,6,7-η)-2,4,6-cyclooctatrien-1-yl]ruthenium(1+); Tetrafluoroborate, *in* Ru-00179

33307-38-9 Dicarbonylbis(λ^3-2-propenyl)ruthenium, Ru-00083

33309-01-2 Tricarbonyl[(1,2,3,4-η)-1,3,5,7-cyclooctatetraene]osmium, Os-00131

33309-02-3 Tricarbonyl[(1,2,5,6,7-η)-1,3,5-cyclooctatrien-1-yl]osmium(1+), Os-00134

33309-03-4 [(2,3,4,5,6-η)-Bicyclo[5.1.0]octa-3,5-dien-2-yl]tricarbonylosmium(1+), Os-00135

33310-08-6 Hexacarbonyl[μ-[(1,2,3,4-η:1,4-η)-1,2,3,4-tetraphenyl-1,3-butadiene-1,4-diyl]]diruthenium, Ru-00681

33310-09-7 Hexacarbonyl[μ-[(1,2,3,4-η:1,4-η)-1,2,3,4-tetramethyl-1,3-butadiene-1,4-diyl]]diruthenium, Ru-00338

33310-10-0 Hexacarbonyl[μ-[(1,2,3,4-η:1,4-η)-1,2,3,4-tetraphenyl-1,3-butadiene-1,4-diyl]]diosmium, Os-00356

33310-11-1 Hexacarbonyl[μ-[(1,2,3,4-η:1,4-η)-1,2,3,4-tetramethyl-1,3-butadiene-1,4-diyl]]diosmium, Os-00236

33573-19-2 Bis(1,1,1,4,4,4-hexafluoro-2,3-butanedithionato-*S*,*S'*)bis(triphenylphosphine)ruthenium; *trans-form*, *in* Ru-00795

33635-55-1 Tricarbonylbis(triphenylphosphine)osmium, Os-00395

33678-05-6 Tricarbonyl[(1,2,3,4,5-η)-2,4-cyclohexadien-1-yl]osmium(1+); Tetrafluoroborate, *in* Os-00073

33680-33-0 (η^5-2,4-Cyclopentadien-1-yl)[(η^6-phenyl)triphenylborato(1−)]ruthenium, Ru-00655

33792-22-2 Dicarbonylnitrosylbis(triphenylphosphine)ruthenium(1+); Tetraphenylborate, *in* Ru-00724

33846-31-0 Nonacarbonyl[μ_3-[*N*-phenyl-*o*-phenylenediaminato(2−)]triruthenium, Ru-00538

33846-32-1 Hexacarbonyl[μ-[*N*-phenyl-1,2-benzenediaminato(2−)]]diruthenium, Ru-00472

33849-20-6 Dicarbonyl[hydrotris(pyrazolato)borato(1−)]iodoruthenium, Ru-00182

33849-21-7 Bromodicarbonyl[hydrotris(pyrazolato)borato(1−)]-ruthenium, Ru-00180

33849-22-8 Dicarbonylchloro[hydrotris(pyrazolato)borato(1−)]ruthenium, Ru-00181

33881-64-0 Dicarbonylnitrosylbis(triphenylphosphine)ruthenium(1+); Hexafluorophosphate, *in* Ru-00724

33991-11-6 Hydronitrosyltris(triphenylphosphine)ruthenium, Ru-00836

34076-51-2 Dichlorobis(triphenylphosphine)ruthenium, Ru-00698

34214-62-5 (Carbonothioyl)trichlorobis(triphenylphosphine)ruthenate(1−); Tetraphenylarsonium salt, *in* Ru-00711

34269-35-7 Tetracarbonyldi-μ-chlorobis[2-(phenylazo)phenyl-*C*,*N'*]diruthenium, *in* Ru-00330

34275-37-1 Acetyldicarbonyltris(triphenylphosphine)ruthenium(1+); Chloride, *in* Ru-00852

34307-28-3 Tetracarbonylhydro(trimethylgermyl)osmium; *cis-form*, *in* Os-00051

34307-40-9 Hexacarbonyltris[μ-(dimethylgermylene)]diruthenium, Ru-00269

34307-41-0 Hexacarbonylbis[μ-(dimethylgermylene)]bis(trimethylgermyl)diruthenium, Ru-00444

34397-62-1 Dicarbonylbis(trimethylgermyl)bis(triphenylphosphine)ruthenium, Ru-00808

34397-63-2 Tetracarbonylbis(trichlorogermyl)ruthenium; *cis-form*, *in* Ru-00025

34397-64-3 Tetracarbonylbis(trimethylgermyl)osmium, Os-00114

34397-77-8 Tetracarbonylbis(trimethylgermyl)ruthenium; *cis-form*, *in* Ru-00162

34438-91-0 Dodecacarbonyltetra-μ-hydrotetraruthenium, Ru-00224

34438-92-1 Undecacarbonyltetra-μ-hydro(trimethyl phosphite-*P*)tetraruthenium, Ru-00339

34438-93-2 Decacarbonyltetra-μ-hydrobis(trimethyl phosphite-*P*)tetraruthenium, Ru-00433

34438-94-3 Nonacarbonyltetra-μ-hydrotris(trimethyl phosphite-*P*)tetraruthenium, Ru-00495

34439-98-0 Tetracarbonyl(trimethylgermyl)[(triphenylphosphine)aurio]ruthenium; *cis-form*, *in* Ru-00622

34482-43-4 Octacarbonyltetra-μ-hydrotetrakis(trimethyl phosphite-*P*)tetraruthenium, Ru-00533

34676-01-2 Bromocarbonyl(η^5-2,4-cyclopentadien-1-yl)(triphenylphosphine)osmium, Os-00336

34676-02-3 Carbonyl(η^5-2,4-cyclopentadien-1-yl)bis(triphenylphosphine)ruthenium(1+); Tetraphenylborate, *in* Ru-00775

34676-03-4 Acetonitrile(η^5-2,4-cyclopentadien-1-yl)bis(triphenylphosphine)ruthenium(1+); Hexafluorophosphate, *in* Ru-00791

34676-04-5 Acetonitrile(η^5-2,4-cyclopentadien-1-yl)bis(triphenylphosphine)ruthenium(1+); Hexafluoroarsenate, *in* Ru-00791

34676-08-9 (η^5-2,4-Cyclopentadien-1-yl)hydrobis(triphenylphosphine)ruthenium, Ru-00768

34692-08-5 Chloro(η^5-2,4-cyclopentadien-1-yl)bis(triphenylarsine)ruthenium, Ru-00762

34692-09-6 (η^5-2,4-Cyclopentadien-1-yl)methylbis(triphenylphosphine)ruthenium, Ru-00781

34692-10-9 (η^5-2,4-Cyclopentadien-1-yl)iodobis(triphenylphosphine)ruthenium, Ru-00767

34692-11-0 (η^5-2,4-Cyclopentadien-1-yl)(pentafluorophenyl)bis(triphenylphosphine)ruthenium, Ru-00815

34738-27-7 Acetonitrile(η^5-2,4-cyclopentadien-1-yl)bis(triphenylphosphine)ruthenium(1+); Tetraphenylborate, *in* Ru-00791

34738-50-6 Tricarbonyl[(1,4,5,6-η)-2-cyano-5-cyclooctene-1,4-diyl]ruthenium; exo-*form*, *in* Ru-00236

34742-23-9 Bromo(η^5-2,4-cyclopentadien-1-yl)bis(triphenylphosphine)osmium, Os-00410

34742-79-5 Undecacarbonyltetra-μ-hydro(triphenylphosphine)tetraruthenium, Ru-00654

34742-80-8 Decacarbonyltetra-μ-hydrobis(triphenylphosphine)tetraruthenium, Ru-00813

34742-81-9 Nonacarbonyltetra-μ-hydrotris(triphenylphosphine)tetraruthenium, Ru-00856

34766-74-0 Nonacarbonyl[μ-[(1,2,3,4-η:1,4-η)-1,2,3,4-tetraphenyl-1,3-butadiene-1,4-diyl]]triosmium, Os-00368

34779-38-9 [(2,3,5,6-η)-Bicyclo[2.2.1]hepta-2,5-diene]carbonyltrichlororuthenate(1−); Cs salt, *in* Ru-00078

34801-97-3 Di-μ-chlorodichlorobis[(1,2,3,6,7,8-η)-2,7-dimethyl-2,6-octadiene-1,8-diyl]diruthenium, Ru-00530

34802-04-5 (η^5-2,4-Cyclopentadien-1-yl)[(1,3,4-η)-5,5,5-trifluoro-1,2,3-tris(trifluoromethyl)-1,3-pentadienyl](triphenylphosphine)ruthenium, Ru-00669

34802-07-8 Tricarbonyl[(1,2,3,4-η)-1,3-cyclooctadiene]osmium, Os-00141

34802-75-0 Octacarbonyltetra-μ-hydrotetrakis(triphenylphosphine)tetraruthenium, Ru-00865

34807-86-8 (η^5-2,4-Cyclopentadien-1-yl)[2-(phenylazo)phenyl](triphenylphosphine)ruthenium, Ru-00686

34808-04-3 Carbonylchloro(1-oxopropyl)bis(triphenylphosphine)ruthenium, Ru-00757

34808-05-4 Acetylcarbonylchlorobis(triphenylphosphine)ruthenium, Ru-00746

34822-84-9 (1-Acetyl-1*H*-1,2-diazepine)tricarbonylruthenium, *in* Ru-00073

34822-87-2 Chloro(η^5-2,4-cyclopentadien-1-yl)bis(trimethyl phosphite-*P*)ruthenium, Ru-00206

34852-76-1 [(2,3,5,6-η)-Bicyclo[2.2.1]hepta-2,5-diene]carbonyltrichlororuthenate(1−); Benzyltriphenylphosphonium salt, *in* Ru-00078

35083-08-0 Hexakis(trimethyl phosphite-*P*)ruthenium(2+); Bis(tetraphenylborate), *in* Ru-00498

35083-09-1 Hexakis(triethyl phosphite-*P*)ruthenium(2+); Bis(tetraphenylborate), *in* Ru-00705

35083-10-4 Chloropentakis(trimethyl phosphite-*P*)ruthenium(1+); Tetraphenylborate, *in* Ru-00410

35083-11-5 Bromopentakis(trimethyl phosphite-*P*)ruthenium(1+); Tetraphenylborate, *in* Ru-00409

35083-12-6 Hydropentakis(trimethyl phosphite-*P*)ruthenium(1+); Tetraphenylborate, *in* Ru-00412

35083-13-7 Hydropentakis(triethyl phosphite-*P*)ruthenium(1+); Tetraphenylborate, *in* Ru-00667

35083-14-8 Chloropentakis(trimethyl phosphite-*P*)osmium(1+); Tetraphenylborate, *in* Os-00267

35083-15-9 Bromopentakis(trimethyl phosphite-*P*)osmium(1+); Tetraphenylborate, *in* Os-00266

35083-16-0 Chloropentakis(triethyl phosphite-*P*)osmium(1+); Tetraphenylborate, *in* Os-00351

35083-17-1 Bromopentakis(triethyl phosphite-*P*)osmium(1+); Tetraphenylborate, *in* Os-00350

35084-36-7 Chloropentakis(triethyl phosphite-*P*)ruthenium(1+); Tetraphenylborate, *in* Ru-00666

35084-37-8 Bromopentakis(triethyl phosphite-*P*)ruthenium(1+); Tetraphenylborate, *in* Ru-00665

35163-44-1 Bis[μ-(acetato-*O*:*O*′)]hexacarbonyldiosmium, Os-00102

35795-47-2 Dicarbonyl(η^2-ethene)bis(triphenylphosphine)ruthenium, Ru-00755

35796-48-6 Bis(η^6-benzene)di-μ-chlorodichlorodiruthenium, *in* Ru-00038

35796-51-1 Bis(acetonitrile)(η^6-benzene)chlororuthenium(1+); Trichloromercurate, *in* Ru-00154

35798-10-8 Dicarbonylhydrotris(triphenylphosphine)ruthenium(1+), Ru-00849

35880-55-8 Dicarbonyltris(triphenylphosphine)osmium, Os-00443

35880-56-9 Dicarbonyl(dioxygen)bis(triphenylphosphine)ruthenium, Ru-00726

35880-57-0 Dicarbonyl(dioxygen)bis(triphenylphosphine)osmium, Os-00390

36179-61-0 Carbonyldihydrotris(triphenylphosphine)osmium; *cis-dihydro-form*, *in* Os-00442

36180-35-5 Dicarbonyl(cyano-*C*)(η^5-2,4-cyclopentadien-1-yl)ruthenium, Ru-00067

36180-37-7 Dicarbonyl(η^5-2,4-cyclopentadien-1-yl)phenylruthenium, Ru-00299

36237-03-3 Chlorodinitrosylbis(triphenylphosphine)ruthenium(1+); Hexafluorophosphate, *in* Ru-00697

36429-98-8 Tetracarbonyl(phosphorous trifluoride)ruthenium, Ru-00026

36429-99-9 Tricarbonylbis(phosphorous trifluoride)ruthenium, Ru-00020

36430-00-9 Dicarbonyltris(phosphorous trifluoride)ruthenium, Ru-00012

36444-68-5 Carbonylnitrosyltris(triphenylphosphine)osmium(1+), Os-00438

36502-87-1 (Benzonitrile)tricarbonyldichlororuthenium, Ru-00125

36542-23-1 Carbonyltetrakis(phosphorous trifluoride)ruthenium, Ru-00005

36543-21-2 Dibromotris(triphenylphosphine)osmium, Os-00434

36568-57-7 Hexacarbonylbis[μ-(dimethylphosphino)]diruthenium, Ru-00156

36571-25-2 Heptacarbonylbis[μ-(diphenylphosphino)][μ_3-[(1-η:1,2-η:2-η)-1,3-cyclohexadien-5-yne]]triosmium, Os-00369

36583-24-1 Decacarbonyldi-μ-nitrosyltriruthenium, Ru-00167

36583-25-2 Decacarbonyldi-μ-nitrosyltriosmium, Os-00120

36632-57-2 Hexacarbonyltris[μ-(dimethylsilyl)]diruthenium, Ru-00272

36632-59-4 Hexacarbonylbis[μ-(dimethylsilylene)]bis(trimethylsilyl)diosmium, Os-00284

36632-60-7 Nonacarbonyltris[μ-(dimethylsilylene)]triosmium, Os-00264

36655-50-2 Nonacarbonyl[μ_3-[(1-η:1,2,3-η:3-η)-1,6,9-cyclododecatrien-1-yl-3-ylidene]]-μ-hydrotriruthenium, Ru-00544

36669-90-6 Carbonylhydronitrosylbis(triphenylphosphine)osmium, Os-00371

36835-56-0 [μ_3-[(1-η:1,2-η:2-η)-Benzene]]-heptacarbonylbis[μ-(dimethylphosphino)]triosmium, Os-00298

37190-51-5 (Benzene)[μ_6-(carbon)]tetradecacarbonylhexaruthenium, Ru-00536

37190-66-2 Tricosacarbonyloctaosmium, Os-00334

37216-50-5 Octadecacarbonylhexaosmium, Os-00308

37279-20-2 Decacarbonyl-2,8-cyclododecadiene-3,7-diyl-1-ylidenetetraruthenium, Ru-00562

37298-24-1 1,1′,2,2′,3,3′,4,4′-Octachloro-5,5′-diiodoruthenocene, Ru-00119

37298-86-5 Diruthenocenylmercury, Ru-00522

37362-03-1 Bis(η^6-benzene)di-μ-bromodibromodiruthenium, Ru-00237

37366-21-5 Undecacarbonyl[(1,2-η:5-η:5,6-η:5,6-η:6-η)-1-cycloocten-5-yne]tetraruthenium, Ru-00503

37375-79-4 Bis(η^6-benzene)di-μ-iododiiododiruthenium, Ru-00239

37668-62-5 (η^5-2,4-Cyclopentadien-1-yl)[2-(diphenylphosphino)phenyl-*C*,*P*]-(triphenylphosphine)ruthenium, Ru-00761

37668-63-6 (η^5-2,4-Cyclopentadien-1-yl)[2-[(diphenoxyphosphino)oxy]phenyl-*C,P*]-(triphenyl phosphite-*P*)ruthenium, Ru-00760

37684-75-6 [(1,2,5,6-η)-1,5-Cyclooctadiene]-tetrakis(pyridine)ruthenium(2+); Bis(tetraphenylborate), *in* Ru-00648

37820-98-7 Dicarbonyltriiodo(triphenylphosphine)ruthenate(1−); Methyltriphenylphosphonium salt, *in* Ru-00518

37834-28-9 Tetracarbonylbis(trichlorogermyl)ruthenium, Ru-00025

38386-79-7 Tetra-μ-chlorotetrakis[μ-(diphenylphosphino)]tetranitrosyltetraruthenium, Ru-00816

38685-25-5 Hexacarbonylbis[μ-(diphenylphosphino)]diruthenium, Ru-00658

38686-18-9 Nonacarbonyltris(trimethyl phosphite-*P*)triruthenium, Ru-00493

38686-52-1 Undecacarbonyl(triphenylphosphine)triruthenium, Ru-00653

38814-44-7 (η^2-Ethene)tris(triphenylphosphine)ruthenium, Ru-00850

38816-61-4 Ruthenocenesulfonyl chloride, *in* Ru-00146

38816-64-7 Ruthenoceneboronic acid, Ru-00151

38816-65-8 Bromoruthenocene, Ru-00134

38816-66-9 Phenylruthenocene, Ru-00426

38816-67-0 Acetoxyruthenocene, *in* Ru-00142

38833-39-5 Methoxyruthenocene, *in* Ru-00142

38856-98-3 Chloronitrosylbis(triphenylphosphine)ruthenium, Ru-00695

38903-39-8 Tricarbonyl[(2,3,4,5-η)-2,4-cyclohexadien-1-one]ruthenium, Ru-00094

38959-07-8 Chlororuthenocene, Ru-00135

38960-71-3 Heptacarbonyl[μ_3-[(5-η:5,6-η:6-η)-1,3-cyclohexadien-5-yne]][μ-(diphenylphosphino)]-hydro(triphenylphosphine)triosmium, Os-00417

38979-81-6 Nonacarbonyl[μ_3-[(5-η:5,6-η:6-η)-1,3-cyclohexadien-5-yne]]di-μ-hydrotriosmium, Os-00252

38979-82-7 Nonacarbonyldi-μ-hydro-μ_3-thioxotriosmium, Os-00070

38979-93-0 Octacarbonyl[μ-[(1,2,3,4-η:1,4-η)-1,2,3,4-tetraphenyl-2-butene-1,4-diylidene]]triosmium, Os-00357

39045-39-1 Bis[μ-(diphenylphosphino)]-dinitrosylbis(methyldiphenylphosphine)diruthenium, Ru-00823

39358-58-2 (η^6-Benzene)(η^5-2,4-cyclopentadien-1-yl)ruthenium(1+); Chloride, *in* Ru-00189

39395-16-9 Dichloro[(1,2,5,6-η)-1,5-cyclooctadiene]osmium, Os-00064

39427-31-1 Bis(η^5-2,4-cyclopentadien-1-yl)iodoruthenium(1+); Triiodide, *in* Ru-00141

39556-32-6 Trichloro(4-methylphenylazo)-bis(triphenylphosphine)ruthenium, Ru-00789

39732-36-0 (η^6-Benzene)diiodo(triphenylphosphine)ruthenium, Ru-00600

39732-37-1 (η^6-Benzene)dibromo(triphenylphosphine)ruthenium, Ru-00598

39732-39-3 (η^6-Benzene)chloro[(1,2,3,4-η)-1,3-cyclohexadiene]ruthenium(1+); Tetraphenylborate, *in* Ru-00255

39732-57-5 (η^6-Benzene)dichloro(triphenylphosphine)ruthenium, Ru-00599

39945-24-9 Hexacarbonylbis[μ-(dimethylsilylene)]-bis(trimethylsilyl)diruthenium, Ru-00445

40424-93-9 Dicarbonylnitrosylbis(triphenylphosphine)osmium(1+); Tetraphenylborate, *in* Os-00387

40424-95-1 Carbonylnitrosyltris(triphenylphosphine)osmium(1+); Tetraphenylborate, *in* Os-00438

40436-33-7 Carbonyl(methoxycarbonyl)-nitrosylbis(triphenylphosphine)osmium, Os-00401

40436-34-8 Carbonylhydronitrosylbis(triphenylphosphine)osmium; *trans-form*, *in* Os-00371

40464-70-8 Carbonylnitrosyltris(triphenylphosphine)ruthenium(1+); Hexafluorophosphate, *in* Ru-00844

40464-73-1 Dicarbonylnitrosylbis(triphenylphosphine)osmium(1+); Hexafluorophosphate, *in* Os-00387

40464-74-2 Carbonylnitrosyltris(triphenylphosphine)osmium(1+); Hexafluorophosphate, *in* Os-00438

40530-71-0 Tricarbonylhydrobis(triphenylphosphine)ruthenium(1+); Hexafluorophosphate, *in* Ru-00744

40672-07-9 1,1′-Dimethylosmocene, Os-00184

40695-22-5 (η^6-Benzene)chloromethyl(triphenylphosphine)ruthenium, Ru-00624

40802-32-2 Dichlorotris(triphenylphosphine)osmium, Os-00435

40999-71-1 Chloro[(2,3-η)-1,1,1,4,4,4-hexafluoro-2-butyne]-nitrosylbis(triphenylphosphine)ruthenium, Ru-00751

40999-72-2 Chloronitrosyl(η^2-tetrafluoroethene)-bis(triphenylphosphine)ruthenium, Ru-00717

41060-95-1 Dicarbonyldi-μ-chlorodichlorotetrakis(triphenylphosphine)diruthenium, Ru-00863

41326-30-1 Dicarbonyl(3,4,5,6-tetrachloro-1,2-benzenediolato-*O,O′*)-bis(triphenylphosphine)ruthenium, Ru-00794

41348-18-9 [(2,3,4,5-η)-Bicyclo[4.2.0]octa-2,4-diene]tricarbonylruthenium, Ru-00186

41357-40-8 Dicarbonyl[(1,2,3,3*a*,6*a*-η)-1,4,5,6-tetrahydro-1-pentalenyl](trimethylgermyl)ruthenium, Ru-00319

41392-83-0 Hydro[(η^6-phenyl)diphenylphosphine]-bis(triphenylphosphine)ruthenium(1+); Tetrafluoroborate, *in* Ru-00837

41424-31-1 (η^6-Benzene)chlorophenyl(triphenylphosphine)ruthenium, Ru-00660

41500-34-9 (η^6-Benzene)chloro(η^3-2-propenyl)ruthenium, Ru-00109

41509-54-0 Tetracarbonylbis[(triphenylphosphine)gold]osmium; *cis-form*, *in* Os-00402

41517-70-8 Nonacarbonyl-μ_3-ethylidynetri-μ-hydrotriruthenium, Ru-00173

41529-78-6 Carbonyl(η^2-ethyne)-nitrosylbis(triphenylphosphine)osmium(1+); Hexafluorophosphate, *in* Os-00398

41550-66-7 Tricarbonyl[(1,2,3,4-η)-1,3,5-cycloheptatriene]ruthenium, Ru-00133

41550-67-8 Tricarbonyl[(1,2,3,4-η)-1,3-cycloheptadiene]ruthenium, Ru-00144

41550-69-0 Tricarbonyl[(2,3,4,5-η)-2,4,6-cycloheptatrien-1-one]ruthenium, Ru-00127

41553-07-5 Tetracarbonylbis(triphenylstannyl)osmium; *trans-form*, *in* Os-00404

41553-08-6 Tetracarbonylbis(trimethylplumbyl)osmium, Os-00115

41654-97-1 Bis[(1,2,3-η)-2-methyl-2-propenyl]-bis(trimethyl phosphite-*P*)ruthenium, Ru-00374

41666-87-9 Bis[μ-(diphenylphosphino)]-dinitrosylbis(triphenylphosphine)diruthenium, Ru-00854

41685-13-6 [(2,3,4,6-η)Bicyclo[3.2.1]oct-3-ene-2,6-diyl]tricarbonylruthenium, Ru-00187

41753-60-0 Bis[1,2-ethanediylbis[diphenylphosphine]-*P,P′*]dihydroruthenium, Ru-00830

41753-97-3 Decacarbonylbis(triphenylphosphine)triosmium, Os-00428

41766-80-7 Decacarbonyldi-μ-hydrotriosmium, Os-00095

42230-39-7 Tetracarbonyldichloroosmium; *trans-form*, *in* Os-00017

42442-15-9 Tricarbonyl[(1,2,3,4-η)-1,3-cyclopentadiene]ruthenium, Ru-00074

42442-16-0 Dicarbonyl(η^5-2,4-cyclopentadien-1-yl)hydroosmium, Os-00045

42516-70-1 [(2,3,5,6-η)-Bicyclo[2.2.1]hepta-2,5-diene]-[(1,2,3,4,5,6-η)-1,3,5,7-cyclooctatetraene]ruthenium, Ru-00397

42516-72-3 [(1,2,5,6-η)-1,5-Cyclooctadiene][(1,2,3,4,5,6-η)-1,3,5-cyclooctatriene]ruthenium, Ru-00435

42516-74-5 [(1,2,5,6-η)-1,5-Cyclooctadiene][(1,2,3,4,5,6-η)-1,3,5-cyclooctatriene]osmium, Os-00281

42516-75-6 [(1,2,3,4,5,6-η)-1,3,5,7-Cyclooctatetraene]-bis(triphenylphosphine)ruthenium, Ru-00803

42531-48-6 Trichloronitridobis(triphenylphosphine)osmium, Os-00363

42531-50-0 Tribromonitridobis(triphenylarsine)osmium, Os-00358

42536-05-0 Hexakis(isocyanomethane)osmium(2+); Diperchlorate, *in* Os-00192

42572-84-9 Hexakis(isocyanomethane)ruthenium(2+), Ru-00270

42594-06-9 Tetrachlorobis(triphenylphosphine)osmium, Os-00365

42594-07-0 Pentachloro(triphenylphosphine)osmate(1−); Tetraphenylarsonium salt, *in* Os-00305

42608-08-2 Dibromotetrakis(isocyanomethane)osmium; *trans-form*, *in* Os-00063

42718-59-2 Dicarbonyldinitrosylosmium, Os-00011

42765-74-2 Nonacarbonyl[μ_3-[(1-η:1-η:1,2-η)ethenylidene]]-di-μ-hydrotriosmium, Os-00126

42781-57-7 Tricarbonyldihydro(triphenylphosphine)ruthenium, Ru-00545

42781-58-8 Tetracarbonyldihydroruthenium; *cis-form*, *in* Ru-00027

42790-11-4 Heptacarbonyl[μ_3-[1,1′-(η^1:η^1:η^2-1,2-ethynediyl)bis[benzene]]][μ-(1,2,3,4-η:1,4-η)-1,2,3,4-tetraphenyl-1,3-butadiene-1,4-diyl]triosmium, Os-00432

42802-19-7 Dicarbonyl(η^5-2,4-cyclopentadien-1-yl)-(diphenylphosphino)ruthenium, Ru-00507

42802-20-0 Dicarbonyl(η^5-2,4-cyclopentadien-1-yl)ruthenate(1−); Na salt, *in* Ru-00052

42868-19-9 Tricarbonyl[(1,2,3,4-η)-1,2,3,4-tetraphenyl-1,3-cyclobutadiene]osmium, Os-00353

43089-98-1 Tricarbonyl[(1,2,3,4,5-η)-2,4-cyclohexadien-1-yl]ruthenium(1+); Tetrafluoroborate, *in* Ru-00095

43184-26-5 Dicarbonyl(η^5-2,4-cyclopentadien-1-yl)-(trimethylsilyl)ruthenium, Ru-00159

43184-28-7 Carbonyl[(1,2,3,3*a*,7*a*-η)-1*H*-inden-1-yl](η^3-2-propenyl)ruthenium, Ru-00304

43198-14-7 Tetracarbonyl[1,2-ethanediylbis[dimethylsilylene]]ruthenium, Ru-00160

43198-15-8 Tetracarbonyl[1,2-ethanediylbis[dimethylsilylene]]osmium, Os-00112

46134-89-8 Tricarbonyl[(1,2,3,4,5-η)-2,4-cyclohexadien-1-yl]osmium(1+), Os-00073

47841-10-1 Tetrachlorobis(triphenylphosphine)ruthenate(1−), Ru-00702

47859-32-5 Carbonyl(η^5-2,4-cyclopentadien-1-yl)-bis(triphenylphosphine)ruthenium(1+), Ru-00775

48107-17-1 [(2,3,5,6-η)-Bicyclo[2.2.1]hepta-2,5-diene]dichlororuthenium, Ru-00058

49564-81-0 Tetracarbonyliodo(trimethylsilyl)ruthenium; *trans-form*, *in* Ru-00063

49631-67-6 Hexakis(isocyanomethane)ruthenium(2+); Bis(hexafluorophosphate), *in* Ru-00270

49631-68-7 Hexakis(isocyanomethane)ruthenium(2+); Bis(tetrafluoroborate), *in* Ru-00270

50381-48-1 Trioxo(*tert*-butylnitrido)osmium, Os-00020

50643-53-3 (Carbon)nonadecacarbonyldihydroheptaosmium, Os-00310

50648-14-1 Pentadecacarbonyldi-μ-hydropentaosmium, Os-00250

50648-17-4 Hexadecacarbonyldi-μ-hydropentaosmium, Os-00273

50648-30-1 Octadecacarbonyldihydrohexaosmium, Os-00302

50648-32-3 Tetracarbonyl[μ-[(1,2,3,3*a*,6*a*-η:3*a*,4,5,6,6*a*-η)-pentalene]]bis(trimethylgermyl)diruthenium, Ru-00487

50938-34-6 Tribromocarbonothioylbis(triphenylphosphine)ruthenium, Ru-00709

50982-13-3 Dichloro[(1,2,5,6-η)-1,5-cyclooctadiene]ruthenium, Ru-00084

51158-87-3 Nonacarbonyl-μ_3-ethylidynetri-μ-hydrotriosmium, Os-00129

51185-97-8 [μ_3-[Benzenaminato(2−)]]nonacarbonyldi-μ-hydrotriruthenium, Ru-00381

51185-99-0 [μ_3-[Benzenaminato(2−)]]-μ_3-carbonylnonacarbonyltriruthenium, Ru-00417

51205-06-2 (Carbon)pentadecacarbonylpentaosmium, Os-00286

51205-07-3 (Carbon)pentadecacarbonylpentaruthenium, Ru-00447

51205-37-9 Nonacarbonyl[μ_3-[(1-η:1,2-η:2-η)cyclooctyne]]-di-μ-hydrotriosmium, Os-00296

51233-18-2 Bis[μ_3-[benzeneaminato(2−)]]nonacarbonyltriruthenium, Ru-00537

51267-41-5 Octacarbonylbis(trimethylsilyl)diruthenium, Ru-00363

51404-01-4 Di-μ-carbonyltetracarbonyl[(1,2,3,4,5-η)-2,4-cycloheptadien-1-yl][μ-[(1,2,3,4-η:1,5,6,7-η)-2,4,6-cycloheptatrien-1-yl]]triruthenium, Ru-00520

51405-78-8 Bis(carbonothioyl)-dichlorobis(triphenylphosphine)ruthenium, Ru-00722

51418-48-5 Carbonylpentachlororuthenate(2−), Ru-00004

51508-72-6 Dicarbonyl[(1,2,3,4,5,6-η)-1,3,5-cycloheptatriene]ruthenium, Ru-00098

51508-73-7 Dicarbonyl[(1,2,3,4,5-η)-2,4-cycloheptadien-1-yl]iodoruthenium, Ru-00103

51508-74-8 Bromodicarbonyl[(1,2,3,4,5-η)-2,4-cycloheptadien-1-yl]ruthenium, Ru-00102

51831-98-2 (η^6-Benzene)dichlororuthenium, Ru-00038

51982-86-6 Dicarbonyltetrachlororuthenate(2−); *cis-form*, Bis-(tetraphenylarsonium) salt, *in* Ru-00011

52015-87-9 (η^6-Benzene)[hydrotris(1*H*-pyrazolato-*N*′)-borato(1−)-*N*^{2},*N*$^{2'}$,*N*$^{2''}$]ruthenium(1+); Hexafluorophosphate, *in* Ru-00395

52247-30-0 Tricarbonyl[(1,4,5,6-η)-5-cyclooctene-1,4-diyl]ruthenium, Ru-00193

52326-31-5 Dicarbonyl[(1,2,3,3*a*,7*a*-η)-1-*H*-inden-1-yl]ruthenate(1−); Na salt, *in* Ru-00175

52359-05-4 Bromocarbonylhydrobis(tricyclohexylphosphine)-osmium, Os-00373

52462-32-5 (η^5-2,4-Cyclopentadien-1-yl)[3,3,3-trifluoro-1-(trifluoromethyl)-1-propenyl]-bis(triphenylphosphine)ruthenium, Ru-00810

52462-47-2 Bis(η^6-benzene)tri-μ-chlorodiruthenium(1+); Hexafluorophosphate, *in* Ru-00238

52462-53-0 Tris(acetonitrile)(η^6-benzene)ruthenium(2+), Ru-00265

52462-54-1 Tris(acetonitrile)(η^6-benzene)ruthenium(2+); Bis-(tetrafluoroborate), *in* Ru-00265

52462-67-6 (η^6-Benzene)dimethyl(triphenylphosphine)ruthenium, Ru-00636

52504-75-3 Tricarbonyl(2,4-hexadiene)ruthenium, Ru-00108

52524-96-6 Bromodicarbonylhydrobis(tricyclohexylphosphine)-osmium, Os-00393

52537-43-6 (1-Hydroxy-1-methylethyl)ruthenocene, Ru-00316

52538-83-7 Bromodicarbonylhydrobis(triphenylphosphine)ruthenium, Ru-00729

52646-85-2 Bis(dimethylcarbamodithioato-*S*,*S*′)-bis(triphenylphosphine)ruthenium; *cis-form*, *in* Ru-00783

52649-43-1 Tricarbonyl(η^4-1,4-diphenyl-1,3-butadiene)ruthenium, Ru-00505

52649-53-3 Tricarbonyl[(1,2,3,4-η)-2,3-dimethyl-1,3-butadiene]ruthenium, Ru-00107

52675-73-7 Hexacarbonyl-μ-hydro[μ-[(*O*,1-η:1,2-η)-3-oxo-1-phenyl-1-butenyl]]diruthenium, Ru-00422

53062-03-6 Hexakis(triethyl phosphite-*P*)osmium(2+); Bis(tetraphenylborate), *in* Os-00367

53079-97-3 [(1,2,5,6-η)-1,5-Cyclooctadiene]bis(η^3-2-propenyl)osmium, Os-00244

53080-00-5 Bis(acetonitrile)[(1,2,5,6-η)-1,5-cyclooctadiene](η^3-2-propenyl)osmium(1+); Tetrafluoroborate, *in* Os-00265

53080-27-6 Tetrakis(acetonitrile)[(2,3,5,6-η)-bicyclo[2.2.1]hepta-2,5-diene]ruthenium(2+); Diperchlorate, *in* Ru-00403

53093-70-2 Chlorodinitrosylbis(triphenylphosphine)ruthenium(1+), Ru-00697

53128-63-5 Carbonyldihydronitrosylbis(triphenylphosphine)-osmium(1+); Hexafluorophosphate, *in* Os-00372

53228-31-2 Tetracarbonylbis(trimethylsilyl)ruthenium; *cis-form*, *in* Ru-00164

53265-34-2 Dicarbonyl[(1,2,3,3*a*,6*a*-η)-1,4,5,6-tetrahydro-1-pentalenyl](trimethylsilyl)ruthenium, Ru-00321

53449-90-4 Dicarbonyl(η^5-2,4-cyclopentadien-1-yl)methylruthenium, Ru-00079

53449-91-5 Benzyldicarbonyl(η^5-2,4-cyclopentadien-1-yl)ruthenium, Ru-00337

53449-92-6 Dicarbonyl(η^5-2,4-cyclopentadien-1-yl)(methylsulfonyl)ruthenium, Ru-00082

53495-34-4 Dihydrotetrakis(triethyl phosphite-*P*)ruthenium, Ru-00616

53537-18-1 Dicarbonylbis(η^5-2,4-cyclopentadien-1-yl)bis[μ-(methanethiolato)]diruthenium; *cis-form*, *in* Ru-00353

53537-19-2 μ-Carbonylcarbonyltris(η^5-2,4-cyclopentadien-1-yl)tris[μ-(methanethiolato)]triruthenium, Ru-00524

53540-05-9 (2-Methyl-1-oxopropyl)osmocene, Os-00241

53556-38-0 Dicarbonyl(η^5-2,4-cyclopentadien-1-yl)(pentafluorobenzenethiolato)ruthenium, Ru-00293

53556-39-1 Dicarbonyl(η^5-2,4-cyclopentadien-1-yl)(methanethiolato)ruthenium, Ru-00080

53556-40-4 Dicarbonylbis(η^5-2,4-cyclopentadien-1-yl)bis[μ-(pentafluorobenzenethiolato)]diruthenium, Ru-00584

53585-80-1 Dicarbonylbis(η^5-2,4-cyclopentadien-1-yl)bis[μ-(methanethiolato)]diruthenium; *trans-form*, *in* Ru-00353

53595-03-2 [μ_2-[Benzeneaminato(1−)]]decacarbonyl-μ-hydrotriosmium, Os-00276

53608-71-2 Carbonylhydro(phenylazo)bis(triphenylphosphine)osmium, Os-00418

53612-32-1 μ-Carbonylbis(η^5-2,4-cyclopentadien-1-yl)bis[μ-(pentafluorobenzenethiolato)]diruthenium, Ru-00575

53652-60-1 μ_3-Carbonylnonacarbonyl[μ_3-[1,1,1-trimethylsilanaminato(2−)]]triruthenium, Ru-00298

53837-24-4 Chloro[(1,2,5,6-η)-1,5-cyclooctadiene]hydrobis(pyridine)ruthenium, Ru-00485

53886-39-8 (η^6-Benzene)chloro(η^3-2-propenyl)osmium, Os-00081

53886-69-4 (1-Hydroxy-2-methylpropyl)osmocene, Os-00242

54071-76-0 Tetrakis(acetonitrile)[(1,2,5,6-η)-1,5-cyclooctadiene]ruthenium(2+); Bis(hexafluorophosphate), *in* Ru-00436

54163-49-4 Carbonyltrichlorobis(triphenylarsine)ruthenate(1−); Benzyltriphenylphosphonium salt, *in* Ru-00707

54206-37-0 [1,3-Bis(4-methylphenyl)-1-triazenato-N^1,N^3]carbonylhydrobis(triphenylphosphine)ruthenium, Ru-00826

54217-59-3 Bis(carbonothioyl)di-μ-chlorodichlorotetrakis(triphenylphosphine)diruthenium, Ru-00864

54275-51-3 (Carbonothioyl)tri-μ-chlorochlorotetrakis(triphenylphosphine)diruthenium, Ru-00862

54327-52-5 Octacarbonyl[(1,2,3,3*a*,6*a*-η;3*a*,4,5,6,6*a*-η)-pentalene]triruthenium, Ru-00418

54438-62-9 Osmocenium; Tetrafluoroborate, *in* Os-00110

54469-35-1 Di-μ-carbonyltetracarbonyl[μ-[(1,2,6,7-η:2,3,4,5,6-η)-2,4,6-cyclooctatrien-1-yl]][(1,2,3,3*a*,6*a*-η)-1,4,5,6-tetrahydro-1-pentalenyl]triruthenium, Ru-00565

54538-51-1 Ruthenocenium, Ru-00150

54628-71-6 1,1′-Diacetylruthenocene, Ru-00346

54733-59-4 Dicarbonyl(phenylazo)bis(triphenylphosphine)ruthenium(1+); Tetrafluoroborate, *in* Ru-00798

54734-25-7 Dicarbonyl(phenyldiazene-N^2)bis(triphenylphosphine)ruthenium(2+); Bis(tetrafluoroborate), *in* Ru-00801

54866-27-2 Dichloro[(1,2,3,4-η)-1,3-cycloheptadiene]ruthenium, Ru-00065

54866-29-4 Dichloro[(1,2,3,4-η)-1,3,5-cycloheptatriene]ruthenium, Ru-00059

54873-26-6 Bis[(1,2,3,4,5-η)-2,4-cycloheptadien-1-yl]ruthenium, Ru-00366

54873-27-7 [(1,2,3,4,5-η)-2,4-Cycloheptadien-1-yl][(1,2,3,4,5-η)-2,4,6-cycloheptatrien-1-yl]ruthenium, Ru-00356

54890-53-8 Chlorodinitrosylbis(triphenylphosphine)ruthenium(1+); Tetrafluoroborate, *in* Ru-00697

54913-13-2 (η^6-Benzene)[(1,2,5,6-η)-1,5-cyclooctadiene]ruthenium, Ru-00365

54932-36-4 (η^6-Benzene)carbonylbis(trichlorogermyl)ruthenium, Ru-00053

54933-53-8 Dicarbonylchlorohydrobis(triphenylphosphine)osmium, Os-00391

55073-02-4 Nonacarbonyl[μ_3-[(1-η:1,2-η:1,2-η)ethynyl]]-μ-hydrotriosmium, Os-00122

55102-19-7 Chlorohydrotris(triphenylphosphine)ruthenium, Ru-00835

55272-32-7 (Cyano-*C*)(η^5-2,4-cyclopentadien-1-yl)bis(triphenylphosphine)ruthenium, Ru-00773

55272-36-1 Chloro[(1,2,3,4,5-η)-1-methyl-2,4-cyclopentadien-1-yl]bis(triphenylphosphine)ruthenium, Ru-00778

55326-91-5 [(1,2,3,4,5-η)-1-Methyl-2,4-cyclopentadien-1-yl][(η^6-phenyl)triphenylborato(1−)]ruthenium, Ru-00662

55397-90-5 Methoxycarbonylosmocene, *in* Os-00136

55399-52-5 Tetrabromodicarbonylosmate(1−); *trans-form*, Tetrabutylammonium salt, *in* Os-00005

55399-54-7 Dicarbonyltetraiodoosmate(1−); *trans-form*, Tetrabutylammonium salt, *in* Os-00009

55399-57-0 Dicarbonyltetraiodoosmate(2−); *trans-form*, Di-Cs salt, *in* Os-00010

55399-59-2 Tricarbonyltriiodoosmate(1−); *fac-form*, Tetrabutylammonium salt, *in* Os-00014

55411-27-3 Trichloronitridobis(triphenylstibine)ruthenium, Ru-00701

55411-35-3 (Aminocarbonyl)ruthenocene, *in* Ru-00185

55449-95-1 Trichloronitridobis(triphenylarsine)ruthenium, Ru-00692

55449-96-2 Trichloronitridobis(triphenylphosphine)ruthenium, Ru-00700

55450-03-8 Tetrachlorobis(triphenylphosphine)osmate(1−); Tetraphenylarsonium salt, *in* Os-00364

55729-62-9 Tricarbonyl[(1,2,3,5,6-η)-2,5-cyclooctadien-1-yl]ruthenium(1+); Tetrafluoroborate, *in* Ru-00188

55729-67-4 Tricarbonyl[(1,2,3,4,5-η)-2,4-cyclooctadien-1-yl]osmium(1+); Tetrafluoroborate, *in* Os-00138

55763-66-1 Dicarbonyl[(1,4,5,6-η)-5-cyclooctene-1,4-diyl](triphenylphosphine)ruthenium, Ru-00646

55763-67-2 Tricarbonyl[(1,2,3,5,6-η)-2,5-cyclooctadien-1-yl]osmium(1+), Os-00139

55802-01-2 Carbonylchlorohydrotris(triphenylarsine)osmium, Os-00439

55802-24-9 Nonacarbonyl[μ-[(1-η:1-η:1,2-η)-ethenylidene]]-di-μ-hydrotriruthenium, Ru-00170

55938-92-6 Tricarbonyl[(1,2,3,5-η)-4-cyano-2-cyclooctene-1,4-diyl]osmium, Os-00175

55938-93-7 Tricarbonyl[(1,3,4,5-η)-4-cyclooctene-1,3-diyl]osmium, Os-00142

56027-95-3 Carbonylpentachloroosmate(2−); Di-Cs salt, *in* Os-00003

56093-36-8 Nonacarbonyldihydro(triphenylphosphine)triosmium, Os-00343

56116-49-5 Octacarbonyl[μ_3-[η^2:η^2:η^5-1,3,5-tris(trimethylsilyl)pentalene]]triruthenium, Ru-00626

56144-64-0 Dicarbonylchlorohydrobis(triphenylphosphine)ruthenium, Ru-00730

56214-65-4 Nonacarbonyldi-μ-hydro[μ_3-[(1-η:1-η:1,2-η)-2-phenylethenylidene]]triosmium, Os-00293

56259-36-0 [(Benzoylamino)carbonyl]dicarbonylbis(triphenylphosphine)ruthenium(1+); Tetrafluoroborate, *in* Ru-00814

56260-08-3 Carbonylchloro[η^2-(methylseleno)selenoxomethyl]bis(triphenylphosphine)ruthenium, Ru-00745

56260-11-8 Dibromo(carbonoselenoyl)carbonylbis(triphenylphosphine)ruthenium, Ru-00715

56260-16-3 Carbonylchloro[η^2-(methylthio)thioxomethyl]bis(triphenylphosphine)ruthenium, Ru-00747

56260-17-4 (Carbonothioyl)carbonyldichlorobis(triphenylphosphine)osmium, Os-00379

56260-19-6 (Carbonylselenoyl)carbonyldichlorobis(triphenylphosphine)ruthenium, Ru-00719

56285-09-7 Dicarbonyl(η^5-2,4-cyclopentadien-1-yl)(η^2-ethene)ruthenium(1+); Tetrafluoroborate, *in* Ru-00104

56298-19-2 (Carbonothioyl)carbonyldichlorobis(triphenylphosphine)ruthenium, Ru-00718

56335-36-5 Dicarbonyl(phenylazo)bis(triphenylphosphine)osmium(1+); Hexafluorophosphate, *in* Os-00424

56335-38-7 Dicarbonylchloro(phenylazo)bis(triphenylphosphine)osmium, Os-00422

56335-39-8 Bromodicarbonyl(phenylazo)bis(triphenylphosphine)osmium, Os-00421

56335-44-5 Dicarbonylhydro(phenylazo)bis(triphenylphosphine)osmium, Os-00425

56335-47-8 Bromodicarbonyl(phenylazo)bis(triphenylphosphine)ruthenium, Ru-00796

56335-49-0 Dicarbonyliodo(phenylazo)bis(triphenylphosphine)ruthenium, Ru-00797

56335-50-3 Dicarbonylhydro(phenylazo)bis(triphenylphosphine)ruthenium, Ru-00800

56335-53-6 Carbonylhydro(phenylazo)bis(triphenylphosphine)ruthenium, Ru-00787

56363-47-4 Dicarbonyliodo(phenylazo)bis(triphenylphosphine)osmium, Os-00423

56389-71-0 Dicarbonyl(phenylazo)bis(triphenylphosphine)ruthenium(1+); Hexafluorophosphate, *in* Ru-00798

56398-24-4 Undecacarbonyl-μ-hydrohydrotriosmium, Os-00125

56398-26-6 Decacarbonyl-μ-hydrohydro(triphenylphosphine)triosmium, Os-00345

56465-05-5 Carbonylbis(triphenylphosphine)[(7,8,9,10,11-η)-undecahydro-7,8-dicarbaundecaborato(2−)]ruthenium, Ru-00749

56465-06-6 Tricarbonyl[(7,8,9,10,11-η)undecahydro-7,8-dicarbaundecaborato(2−)]ruthenium, Ru-00030

56669-40-0 Dodecacarbonyltrihydrotetraruthenate(1−); Tetraphenylarsonium salt, *in* Ru-00220

56781-02-3 Bromodicarbonyl(η^3-2-propenyl)(triphenylphosphine)ruthenium, Ru-00577

56802-15-4 Dodecacarbonyl[μ_4-[(1-η:1,2-η:1,2-η:2-η)-1,2-diphenyl-1,2-ethenediyl]]tetraruthenium, Ru-00629

56846-19-6 Bis(*O*-methylcarbonodithioato-*S*,*S′*)bis(triphenylphosphine)ruthenium; *cis-form*, *in* Ru-00759

56870-60-1 Carbonylbis(dimethylcarbamodithioato-*S*,*S′*)(triphenylphosphine)ruthenium, Ru-00625

57177-16-9 Pentacarbonyl[μ-[(η^3:η^5)-3,5-cyclohexadien-1-yl-2-ylidene(phenylmethylidyne)-2,4-cyclopentadien-1-ylidene]]diruthenium, Ru-00576

57307-34-3 Dicarbonylchloro(phenyldiazene-N^2)bis(triphenylphosphine)ruthenium(1+); Perchlorate, *in* Ru-00799

57327-06-7 Carbonyl(η^5-2,4-cyclopentadien-1-yl)hydro(triphenylphosphine)ruthenium, Ru-00601

57327-07-8 Tricarbonyl(η^5-2,4-cyclopentadien-1-yl)ruthenium(1+); Tetrafluoroborate, *in* Ru-00072

57327-08-9 Dicarbonyl(η^5-2,4-cyclopentadien-1-yl)(triphenylphosphine)ruthenium(1+), Ru-00617

57327-09-0 Dicarbonyl(η^5-2,4-cyclopentadien-1-yl)(triphenylphosphine)ruthenium(1+); Tetrafluoroborate, *in* Ru-00617

57349-59-4 Dicarbonyl(η^5-2,4-cyclopentadien-1-yl)hydroruthenium, Ru-00055

57373-35-0 μ-Carbonylnonacarbonyl[μ_3-[(1-η:1,2-η:2-η)-1,2-ethenediyl]]triosmium, Os-00156

57373-36-1 [μ-[(1,2,3,4-η:1,4-η)-1,3-Butadiene-1,4-diyl]]nonacarbonyltriosmium, Os-00216

57383-59-2 Hexacarbonyl[μ-[(1,2,3-η:4,5,6-η)-1,3,5-cycloheptatriene]]diruthenium, Ru-00297

57398-60-4 Tetracarbonylruthenate(2−); Di-Na salt, *in* Ru-00029

57405-60-4 Hexacarbonyl[μ-[(1,2,4,5-η:1,5-η)-1,2,4,5-tetramethyl-3-oxo-1,4-pentadiene-1,5-diyl]]diosmium, Os-00260

57512-16-0 Tricarbonyl[(1,2,3,4-η)-1,3-cyclopentadiene]osmium, Os-00059

57512-74-0 Tetracarbonyl[3,4-dimethyl-2(3*H*)-thiazolylidene]osmium, Os-00072

57556-81-7 Dicarbonyl[(1,2,3,4-η)-1,3-cycloheptadiene](triphenylphosphine)ruthenium, Ru-00641

57583-96-7 Tricarbonyl[(2,3,4,5-η)-1,1-dimethyl-2,5-diphenylsilacyclopenta-2,4-diene]ruthenium, Ru-00548

57603-32-4 Tricarbonyl[(1,3,4,5-η)-4-cycloheptene-1,3-diyl]ruthenium, Ru-00145

57603-33-5 Dicarbonyl[(1,3,4,5-η)-4-cycloheptene-1,3-diyl](triphenylphosphine)ruthenium, Ru-00642

57603-47-1 (η^7-Cycloheptatrienylium)[(1,2,5,6-η)-1,5-cyclooctadiene]ruthenium(1+); Tetrafluoroborate, *in* Ru-00401

57603-49-3 [(1,2,3,4,5-η)-2,4-Cycloheptadien-1-yl][(1,2,3,4,5,6-η)-1,3,5-cycloheptatriene]ruthenium(1+); Tetrafluoroborate, *in* Ru-00359

57603-50-6 [(1,2,3,4,5-η)-2,4-Cycloheptadien-1-yl][(1,2,3,4,5,6-η)-1,3,5-cycloheptatriene]ruthenium(1+); Hexafluorophosphate, *in* Ru-00359

57603-52-8 [(1,2,3,4,5-η)-6-Cyano-2,4-cycloheptadien-1-yl][(1,2,3,4,5-η)-2,4-cycloheptadien-1-yl]ruthenium, Ru-00398

57650-19-8 Dicarbonyl[(1,2,3,4,5-η)-2,4-cycloheptadien-1-yl](trimethylgermyl)ruthenium, Ru-00268

57650-20-1 Dicarbonyl[(1,2,3,4,5-η)-2,4-cyclohexadien-1-yl](trimethylgermyl)ruthenium, Ru-00200

57659-27-5 Tricarbonyl[(1,2,3,4,5-η)-2,4-cycloheptadien-1-yl]ruthenium(1+); Tetrafluoroborate, *in* Ru-00139

57659-28-6 Dicarbonyl[(1,2,3,4,5-η)-2,4-cycloheptadien-1-yl](trimethylsilyl)ruthenium, Ru-00271

57659-30-0 Dicarbonyl[(1,2,3,4,5-η)-2,4-cycloheptadien-1-yl](triphenylphosphine)ruthenium(1+); Tetrafluoroborate, *in* Ru-00640

57659-31-1 Dicarbonyl[(1,2,3,4,5-η)-2,4-cyclohexadien-1-yl](trimethylsilyl)ruthenium, Ru-00201

57673-31-1 Nonacarbonyl[μ_3-[(1-η:1,2-η:1,2-η)-3,3-dimethyl-1-butynyl]]-μ-hydrotriruthenium, Ru-00388

57902-93-9 Trichlorobis(triphenylarsine)thionitrosylosmium, Os-00359

57903-36-3 [μ-[(1,2,3,3*a*,8*a*-η:4,5,6-η)-azulene]]pentacarbonyldiruthenium, Ru-00383

58023-41-9 Decacarbonyl[μ_3-[1,1′-[(1-η:1,2-η:2-η)-1,2-ethynediyl]bis[benzene]]]triosmium, Os-00335

58073-96-4 Tetracarbonylbis(trimethylstannyl)ruthenium; *cis-form*, *in* Ru-00165

58280-32-3 Octacarbonylbis(trimethylstannyl)diruthenium, Ru-00364

58320-36-8 Tricarbonyl[(1,2,3,4,5-η)-2,4-cycloheptadien-1-yl]ruthenium(1+); Hexafluorophosphate, *in* Ru-00139

58355-23-0 (η^5-2,4-Cyclopentadien-1-yl)(phenylethynyl)bis(triphenylphosphine)ruthenium, Ru-00821

58355-41-2 Dicarbonyl(trimethylsilyl)[(1,2,3,6,7-η)-8-(trimethylsilyl)-2,4,6-cyclooctatrien-1-yl]ruthenium, Ru-00437

58369-31-6 (η^6-Benzene)carbonylbis(trichlorosilyl)ruthenium, Ru-00054

58410-32-5 Benzoyldichloronitrosylbis(triphenylphosphine)ruthenium, Ru-00786

58410-34-7 Benzyldichloronitrosylbis(triphenylphosphine)ruthenium, Ru-00788

58411-69-1 Decacarbonyl-μ-hydro[μ-(2-pyridinyl-*C*:*N*)]triosmium, Os-00251

58428-52-7 Dicarbonyltetrachloroosmate(2−); *cis-form*, Bis(tetraethylammonium) salt, *in* Os-00007

58428-54-9 Tetrabromodicarbonylosmate(2−); Bis(tetraethylammonium) salt, *in* Os-00006

58441-24-0 (Acetylphenyldiazene-N^2,O^1)dichlorobis(triphenylphosphine)osmium, Os-00426

58448-30-9 Acetyldichloronitrosylbis(triphenylphosphine)ruthenium, Ru-00732

58755-32-1 Tricarbonyltrichloroosmate(1−); *mer-form*, Tetrabutylammonium salt, *in* Os-00013

58776-23-1 Tribromotricarbonylosmate(1−); *mer-form*, Tetrabutylammonium salt, *in* Os-00012

58776-25-3 Tricarbonyltriiodoosmate(1−); *mer-form*, Tetrabutylammonium salt, *in* Os-00014

58800-92-3 Tricarbonyltrichloroosmate(1−); *fac-form*, Tetrabutylammonium salt, *in* Os-00013
58800-94-5 Tribromotricarbonylosmate(1−); *fac-form*, Tetrabutylammonium salt, *in* Os-00012
59113-94-9 (Carbon)heneicosacarbonyloctaosmium, Os-00329
59140-13-5 Carbonyl(η^5-2,4-cyclopentadien-1-yl)ethyl(triphenylphosphine)ruthenium, Ru-00632
59158-84-8 Heneicosacarbonylheptaosmium, Os-00325
59169-04-9 Tricarbonyliodobis(triphenylphosphine)ruthenium(1+); Iodide, *in* Ru-00741
59327-83-2 Tetracarbonyl[1,2-phenylenebis[dimethylsilylene]]ruthenium, Ru-00354
59369-25-4 [μ-[1,2,4,5-Benzenetetrayltetrakis[dimethylsilylene]]]octacarbonyldiruthenium, Ru-00572
59421-52-2 Pentacarbonyl-μ-2,4,6-cycloheptatrien-1-yl(trimethylgermyl)diosmium, Os-00261
59421-53-3 Pentacarbonyl-μ-2,4,6-cycloheptatrien-1-yl(trimethylsilyl)diosmium, Os-00262
59423-89-1 Dicarbonyl[(1,2,3,4,5-η)-2,4-cyclopentadien-1-yl](trimethylsilyl)osmium, Os-00193
59423-90-4 Dicarbonyl[(1,2,3,4,5-η)-2,4-cycloheptadien-1-yl]iodoosmium, Os-00078
59423-94-8 Dicarbonyl[(1,2,3,4,5-η)-2,4-cycloheptadien-1-yl](trimethylgermyl)osmium, Os-00190
59423-95-9 Bromodicarbonyl[(1,2,3,4,5-η)-2,4-cycloheptadien-1-yl]osmium, Os-00077
59423-97-1 Carbonyl[(1,2,3,4,5-η)-2,4-cycloheptadien-1-yl]-bis(trimethyl phosphite-*P*)osmium(1+); Iodide, *in* Os-00245
59432-90-5 [μ-[1,2,3,4-Benzenetetrayltetrakis[dimethylsilylene]]]octacarbonyldiruthenium, Ru-00571
59453-42-8 Decacarbonyl[μ-[(1-η:1,2-η)ethenyl]]-μ-hydrotriosmium, Os-00164
59464-98-1 [μ-[(1,2,3,4-η:1,4-η)-1,3-Butadiene-1,4-diyl]]hexacarbonyldiosmium, Os-00098
59526-98-6 Tricarbonylchlorohydro(triphenylphosphine)osmium, Os-00320
59532-85-3 Decacarbonyl[μ-[(dimethyliminio)methylene]]-μ-hydrotriruthenium, Ru-00294
59557-52-7 Chloro(η^5-2,4-cyclopentadien-1-yl)-bis(triphenyl phosphite-*P*)ruthenium, Ru-00764
59557-58-3 (η^5-2,4-Cyclopentadien-1-yl)-fluorobis(triphenylphosphine)ruthenium, Ru-00766
59559-91-0 Tricarbonyltrifluororuthenium, Ru-00019
59568-94-4 Benzyl(η^5-2,4-cyclopentadien-1-yl)-bis(triphenylphosphine)ruthenium, Ru-00818
59584-57-5 Dicarbonyl[(1,2,3,3*a*,6*a*-η)-1,4,5,6-tetrahydro-1-pentalenyl](trimethylsilyl)osmium, Os-00222
59584-58-6 Dicarbonyl[(1,2,3,3*a*,6*a*-η)-1,4,5,6-tetrahydro-1-pentalenyl](trimethylgermyl)osmium, Os-00221
59599-35-8 Tetracarbonyldimethylosmium, Os-00038
59621-15-7 Hexacarbonyl[μ-[(1,2,3,4,5,8-η)-1,3,5,7-cyclooctatetraene]]diosmium, Os-00231
59676-07-2 Tricarbonyl[(1,2,3,4-η)-1,3-cycloheptadiene]osmium, Os-00105
59687-85-3 Hexacarbonyl[μ-(η^2:η^4-1,2-ethenediyl-1,2-phenylene)]diruthenium, Ru-00327
59707-01-6 Hexacarbonyl[μ-(η^2:η^4-1,2-ethenediyl-1,2-phenylene)]diosmium, Os-00227
59992-85-7 Dichloro[(1,2,5,6-η)-1,5-cyclooctadiene]-bis(pyridine)ruthenium, Ru-00483
60020-25-9 Aquadicarbonylhydrobis(triphenylphosphine)ruthenium(1+); Tetrafluoroborate, *in* Ru-00733
60105-91-1 Dicarbonyltetrachlororuthenate(2−), Ru-00011
60109-68-4 Hexadecacarbonylpentaosmium, Os-00287
60120-83-4 Octadecacarbonyl-μ_3-hydrohexaosmate(1−), Os-00301
60120-85-6 Octadecacarbonylhexaosmate(2−), Os-00307
60146-16-9 [μ-[(1,2-η:3,4-η)-1,3-Butadiene]]decacarbonyltriosmium; *s-cis-form*, *in* Os-00228
60172-17-0 Dihydrobis(triphenylphosphine)[(7,8,9,10,11-η)-undecahydro-7,9-dicarbaundecaborato(2−)]-ruthenium, Ru-00738
60172-19-2 Carbonylbis(triphenylphosphine)[(7,8,9,10,11-η)-undecahydro-7,9-dicarbaundecaborato(2−)]-ruthenium, Ru-00750
60172-21-6 Dihydrobis(triphenylphosphine)[(7,8,9,10,11-η)-undecahydro-7,8-dicarbaundecaborato(2−)]-ruthenium, Ru-00737
60175-44-2 [μ-[(1,2-η:3,4-η)-1,3-Butadiene]]decacarbonyltriosmium, Os-00229
60268-95-3 Decacarbonyl[(1,2,3,4-η)-1,3-cyclohexadiene]triosmium, Os-00278
60269-72-9 [μ-[(1,2-η:3,4-η)-1,3-Butadiene]]decacarbonyltriosmium, Os-00228
60317-82-0 Nonacarbonyl[μ_3-[(1,2-η:3-η:4,5-η)-2,4-cyclohexadien-1-yl]]-μ-hydrotriosmium, Os-00256
60321-20-2 Decacarbonyl[(1,2,3,4,5-η)-2,4-cyclohexadien-1-yl]triosmium(1+); Tetrafluoroborate, *in* Os-00277
60373-24-2 (Benzene)nonacarbonyl-μ-hydrotriosmium(1+); Tetrafluoroborate, *in* Os-00254
60429-09-6 Undecacarbonyl[μ_3-[(1-η:1,2-η:2-η)-1,2-ethenediyl]]di-μ-hydrotetraosmium, Os-00217
60442-72-0 Dodecacarbonyldimethyltriosmium, Os-00230
60447-29-2 Dibromotetrakis(triphenylphosphine)osmium; *trans-form*, *in* Os-00445
60479-38-1 Tetrachlorobis(triphenylphosphine)osmium; *trans-form*, *in* Os-00365
60496-59-5 μ-Carbonyldecacarbonyl-μ-hydrotriruthenate(1−), Ru-00168
60552-36-5 Bis[(1,2,3,4,5,6-η)-1-phenylboratabenzene]ruthenium, Ru-00566
60552-37-6 Bis[(1,2,3,4,5,6-η)-1-methylboratabenzene]ruthenium, Ru-00266
60595-73-5 Carbonyldichlorobis(dimethylphenylphosphine)-(η^2-ethene)ruthenium, Ru-00511
60674-61-5 Di-μ-carbonyldicarbonylbis[(1,2,3,3*a*,7*a*-η)-4,5,6,7-tetrahydro-1*H*-inden-1-yl]diruthenium, Ru-00568
60687-58-3 Di-μ-carbonyldicarbonylbis[(1,2,3,4,5-η)-1-methyl-2,4-cyclopentadien-1-yl]diruthenium, Ru-00425
60687-59-4 Di-μ-carbonyldicarbonylbis[(1,2,3,3*a*,7*a*-η)-1*H*-inden-1-yl]diruthenium, Ru-00558
60764-69-4 Bis[(1,2,3,4,5,6-η)-1-phenylboratabenzene]osmium, Os-00327
60767-64-8 Tri-μ-carbonylheptacarbonyl[μ-(pyridazine-N^1:N^2)]triruthenium, Ru-00326
60829-94-9 (η^6-Benzene)trichlororuthenate(1−); Cs salt, *in* Ru-00039
60843-19-8 Bis(bicyclo[2.2.1]hepta-2,5-diene)dicarbonyltetrachlorodiruthenium, Ru-00428
60879-64-3 (η^6-Benzene)chlorobis(pyridine)ruthenium(1+); Hexafluorophosphate, *in* Ru-00427
60898-12-6 Hexakis(diethylaminato)-μ-1,1′-ruthenocenediyldititanium, Ru-00683
60898-13-7 1,1′-Dilithioruthenocene, Ru-00132
61003-55-2 (η^4-1,3-Butadiene)tris(dimethylphenylphosphine)-hydroruthenium(1+); Hexafluorophosphate, *in* Ru-00650
61024-44-0 [(1,2,5,6-η)-1,5-Cyclooctadiene]-hydrotris(trimethyl phosphite-*P*)ruthenium(1+); Hexafluorophosphate, *in* Ru-00466
61081-79-6 [(4,5,6,7,8,9-η)-Nonahydro-6,8-dicarbanonaborato(4−)]-bis(triphenylphosphine)ruthenium, Ru-00734
61127-38-6 Tricarbonylchloro[(1,2,3-η)-2,4,7-cycloctatrien-1-yl]ruthenium, Ru-00197
61127-38-6 Tricarbonylchloro[(1,2,3-η)-2,4,7-cyclooctatrien-1-yl]ruthenium, Ru-00177
61182-07-8 μ-Carbonyldecacarbonyl-μ-hydrotriosmate(1−), Os-00121
61182-08-9 μ-Carbonyldecacarbonyl-μ-hydrotriosmate(1−); Bis(triphenylphosphine)iminium salt, *in* Os-00121
61182-40-9 1,1′-Bis(1-oxo-3-phenyl-2-propenyl)ruthenocene, Ru-00644
61182-41-0 3-Phenyl[5]ruthenocenophan-1,5-dione, Ru-00547
61182-42-1 3-Phenyl[5]ruthenocenophane, Ru-00550
61199-98-2 μ-Bromodecacarbonyl-μ-hydrotriosmium, Os-00092
61226-84-4 (η^4-1,3-Butadiene)tris(triphenylphosphine)ruthenium, Ru-00853

61245-30-5 μ-Carbonyldecacarbonyl-μ-hydrotriosmate(1−); Tetramethylammonium salt, *in* Os-00121
61250-78-0 Heptacosacarbonyl[μ_3-(carboxylato-*C*:*O*:*O'*)]-μ-hydrononaosmate(1−), Os-00344
61312-93-4 Heptacosacarbonyl[μ_3-(carboxylato-*C*:*O*:*O'*)]-μ-hydrononaosmate(1−); Bis(triphenylphosphine)-iminium salt, *in* Os-00344
61647-76-5 Dicarbonyltris(triphenylphosphine)ruthenium, Ru-00848
61816-94-2 Di-μ-carbonylhexacarbonyl(dicarbonyliron)di-μ-hydro(tricarbonylruthenium)diosmium, Os-00207
61817-37-6 Hydro[(1,2,3,4,5-η)-6-oxo-2,4-cyclohexadien-1-yl]bis(triphenylphosphine)ruthenium, Ru-00777
61817-93-4 Bis(acetonitrile)decacarbonyltriosmium, Os-00226
61839-26-7 Pentakis(trimethyl phosphite-*P*)ruthenium, Ru-00411
61863-84-1 Di-μ-carbonyltricarbonyl(dicarbonyliron)-(hexacarbonyl-μ-hydrodiruthenium)-μ-hydroosmium, Os-00206
61880-52-2 Pentakis(trimethyl phosphite-*P*)osmium, Os-00268
61930-76-5 [μ-(Carbonothioyl)]-μ-carbonyldicarbonylbis(η^5-2,4-cyclopentadien-1-yl)diruthenium; *trans-form*, *in* Ru-00334
61930-77-6 [μ-(Carbonothioyl)]-μ-carbonyldicarbonylbis(η^5-2,4-cyclopentadien-1-yl)diruthenium; *cis-form*, *in* Ru-00334
61930-78-7 Bis[μ-(carbonothioyl)]dicarbonylbis(η^5-2,4-cyclopentadien-1-yl)diruthenium; *cis-form*, *in* Ru-00333
61930-80-1 Dicarbonyl(η^5-2,4-cyclopentadien-1-yl)-[(methylthio)thioxomethyl]ruthenium, Ru-00099
61930-82-3 (Carbonothioyl)carbonyl(η^5-2,4-cyclopentadien-1-yl)iodoruthenium, Ru-00050
61930-84-5 (Carbonothioyl)dicarbonyl(η^5-2,4-cyclopentadien-1-yl)ruthenium(1+); Trifluoromethanesulfonate, *in* Ru-00071
61989-65-9 Bis[μ-(carbonothioyl)]dicarbonylbis(η^5-2,4-cyclopentadien-1-yl)diruthenium; *trans-form*, *in* Ru-00333
62086-86-6 Octadecacarbonyl-μ_3-hydrohexaosmate(1−); Bis-(triphenylphosphine)iminium salt, *in* Os-00301
62252-84-0 Pentadecacarbonylpentaosmate(2−), Os-00270
62414-47-5 Pentadecacarbonylpentaosmate(2−); Bis[bis(triphenylphosphine)iminium]salt, *in* Os-00270
62422-34-8 (η^6-Benzene)[(1,2,3,4-η)-1,3,5,7-cyclooctatetraene]ruthenium, Ru-00348
62422-36-0 [(1,2,3,4-η)-1,3,5,7-Cyclooctatetraene]-[(1,2,3,4,5,6-η)-hexamethylbenzene]ruthenium, Ru-00525
62434-66-6 Octadecacarbonyl-μ_6-hydrohexaruthenate(1−), Ru-00468
62434-67-7 Octadecacarbonyl-μ_6-hydrohexaruthenate(1−); Bis-(triphenylphosphine)iminium salt, *in* Ru-00468
62449-05-2 Di-μ-carbonyldi-μ_3-carbonyltetradecacarbonylhexaruthenate(2−), Ru-00501
62501-17-1 Di-μ-carbonyldi-μ_3-carbonyltetradecacarbonylhexaruthenate(2−); Bis[bis(triphenylphosphine)]iminium salt, *in* Ru-00501
62662-80-0 (η^6-Benzene)hydrobis(triphenylphosphine)ruthenium(1+); Tetrafluoroborate, *in* Ru-00779
62745-23-7 Tricarbonyl[(2,3,4,5-η)methyl-1*H*-azepine-1-carboxylate]ruthenium, *in* Ru-00128
62777-06-4 Dicarbonyldiiodobis(triphenylphosphine)ruthenium; *ab*-dicarbonyl-*cd*-diiodo-*ef*-diphosphine-*form*, *in* Ru-00723
62793-22-0 Dicarbonylchloro(thioxomethyl)-bis(triphenylphosphine)osmium, Os-00396
62793-33-3 Hexacarbonyldihydrobis(triphenylphosphine)diosmium, Os-00411
62793-34-4 Di-μ-bromohexacarbonyldiosmium, Os-00028
62795-57-7 Tricarbonyl[(2,3,4,5-η)-1-methyl-2,3,4,5-tetraphenylsilacyclopenta-2,4-dien-1-ol]ruthenium, Ru-00674
62795-62-4 (Acetonitrile)carbonyldichloro[(1,2,5,6-η)-1,5-cyclooctadiene]ruthenium, Ru-00199
62863-43-8 Dodecacarbonyldihydrotriosmium, Os-00158
62883-45-8 (η^4-Butadiene)tricarbonylruthenium, Ru-00056
62927-94-0 Bis(η^4-1,3-butadiene)(triphenylphosphine)ruthenium, Ru-00637
62961-03-9 Tribromonitrosylbis(triphenylphosphine)ruthenium, Ru-00693
62976-01-6 Heptacarbonyldihydro(triphenylphosphine)diosmium, Os-00339
62976-02-7 Hexacarbonyldi-μ-chlorodiosmium, Os-00030
63022-50-4 Dicarbonyl[(1,2,3,4,5-η)-2,4-cyclohexadien-1-yl](methoxycarbonyl)osmium, Os-00106
63022-51-5 Tricarbonyl[(1,2,3,4-η)-5-methoxy-1,3-cyclohexadiene]osmium; *exo-form*, *in* Os-00107
63128-11-0 μ-Carbonyloctacarbonyldiruthenium, Ru-00116
63166-12-1 Tricarbonyl[(1,2,3,4-η)-1,3-cyclohexadiene]osmium, Os-00075
63174-13-0 Bis[2-methyl-2-propanaminoato(2−)]dioxoosmium, Os-00067
63180-66-5 Tricarbonyl[(1,2,3,4-η)-5-methoxy-1,3-cyclohexadiene]osmium; *endo-form*, *in* Os-00107
63280-32-0 Hexadecacarbonyl-μ-ethenylidenedi-μ-hydrohexaruthenium, Ru-00470
63280-43-3 Nonacarbonyltri-μ-hydro-μ_3-methylidynetriruthenium, Ru-00124
63322-98-5 Pentadecacarbonyl-μ-ethenylidenedi-μ-hydropentaruthenium, Ru-00449
63322-99-6 Decacarbonyl-μ-ethenylidenedi-μ-hydrotriruthenium, Ru-00222
63363-66-6 Decacarbonyl-μ-hydro[μ-[(methylimino)-phenylmethyl-*C*:*N*]]triosmium, Os-00303
63373-89-7 μ-Amidodecacarbonyl-μ-hydrotriosmium, Os-00097
63374-18-5 Dicarbonyl[(2,3-η)-dimethyl-2-butenedioate]-bis(triphenylphosphine)ruthenium, Ru-00802
63374-19-6 Dicarbonyl(1,1,2,2,3,3,4,4-octafluoro-1,4-butanediyl)bis[triphenylphosphine]ruthenium, Ru-00771
63374-84-5 [μ_4-(Boryl-*B*:*B*:*H*:*H'*)]dodecacarbonyl-μ-hydrotetraruthenium, Ru-00219
63374-85-6 Dodecacarbonyl[μ_4-(1-η:η^2:η^2:2-η-1,2-ethenediyl)]tetraruthenium, Ru-00325
63395-36-8 Bis[(1,2,3,4,5-η)-2,4-cyclooctadien-1-yl]ruthenium, Ru-00434
63625-18-3 Bis(acetonitrile)chloro[(1,2,3,4,5,6-η)-1,3,5-cycloheptatriene]ruthenium(1+); Tetrafluoroborate, *in* Ru-00198
63630-81-9 Di-μ-chlorodichlorobis(1,3,5-cycloheptatriene)diruthenium, *in* Ru-00059
63630-82-0 Dichloro[(1,2,3,4,5,6-η)-1,3,5-cycloheptatriene](triphenylphosphine)ruthenium, Ru-00620
63866-74-0 Dicarbonylhydro[(methylthio)thioxomethyl]-bis(triphenylphosphine)osmium, Os-00408
64012-09-5 Nonacarbonyltri-μ-hydro-μ_3-methylidynetriosmium, Os-00099
64052-01-3 Decacarbonyl-μ-hydro-μ-methyltriosmium, Os-00127
64279-31-8 Pentabromocarbonylosmate(2−); Di-K salt, *in* Os-00001
64279-35-2 Pentabromocarbonylosmate(3−), Os-00002
64279-36-3 Pentabromocarbonylosmate(2−); Bis(tetramethylammonium) salt, *in* Os-00001
64331-16-4 Carbonyl[(1,2,5,6-η)-1,5-cyclooctadiene]-[(1,2,5,6-η)-1,3,5-cyclooctatriene]ruthenium, Ru-00462
64387-08-2 Decacarbonyldi-μ-chlorotriosmium, Os-00091
64479-10-3 Dicarbonyl(dithioperoxy)bis(triphenylphosphine)-osmium, Os-00388
64479-11-4 Dicarbonyl(dithioperoxy)bis(triphenylphosphine)-ruthenium, Ru-00725
64722-25-4 [(1,2,5,6-η)-1,5-Cyclooctadiene]tris(1,1-dimethylhydrazine-N^2)hydroruthenium(1+); Hexafluorophosphate, *in* Ru-00378
64799-85-5 Pentaammine(η^2-ethyne)ruthenium(2+); Dithionate, *in* Ru-00013
64883-53-0 (Carbonothioyl)carbonyl(η^2-ethene)-bis(triphenylphosphine)osmium, Os-00406
64884-14-6 3-Phenyl-1-ruthenocenyl-2-propen-1-one; (*E*)-*form*, *in* Ru-00509

64885-82-1 Decacarbonyl(decacarbonyldi-μ-hydrotriosmium)diruthenium, Os-00311
64972-45-8 Octacarbonyldinitrosyl(trimethyl phosphite)triosmium, Os-00133
64972-46-9 Nonacarbonyldi-μ-nitrosyl(trimethyl phosphite-*P*)triosmium, Os-00172
64973-41-7 Nonacarbonyldinitrosyltriosmium, Os-00084
65230-38-8 Nitrosyl(η^3-2-propenyl)bis(triphenylphosphine)ruthenium, Ru-00748
65230-39-9 Carbonylnitrosyl(η^3-2-propenyl)-(triphenylphosphine)ruthenium, Ru-00567
65297-47-4 [(1,2,3,5,6-η)-2,5-Cyclooctadien-1-yl][(η^6-phenyl)triphenylborato(1−)]ruthenium, Ru-00677
65630-15-1 [(1,2,3,4,5-η)-2,4-Cyclohexadien-1-yl](η^5-2,4-cyclopentadien-1-yl)ruthenium, Ru-00195
65702-94-5 (Acetonitrile)undecacarbonyltriosmium, Os-00213
65724-47-2 μ-Carbonylhexacarbonyl(μ-carbonylpentacarbonyldiiron)di-μ-hydrodiruthenium, Ru-00289
65908-54-5 Decacarbonyl-μ-hydro[μ-(acetyl-*C*:*O*)]triosmium, Os-00165
66098-55-3 Undecacarbonyl(trimethyl phosphite-*P*)triosmium, Os-00232
66115-12-6 Carbonylpentachlororuthenate(2−); Di-K salt, *in* Ru-00004
66172-60-9 Decacarbonylbis(triphenylphosphine)triruthenium, Ru-00812
66320-89-6 μ-Chlorochlorobis[(1,2,5,6-η)-1,5-cyclooctadiene][μ-(1,1-dimethylhydrazine-*N*:*N'*)]-μ-hydrohydrodiruthenium, Ru-00496
66321-02-6 Tri-μ-methylenehexakis(trimethylphosphine)diruthenium, Ru-00554
66322-95-0 Decacarbonyl[μ-[1,2-ethanediylbis[diphenylphosphine]-*P*:*P'*]]-tetra-μ-hydrotetraruthenium, Ru-00691
66429-64-9 Dodecacarbonyldiiodotriosmium, Os-00197
66507-20-8 [3]Ruthenocenophan-1-one, Ru-00305
66507-21-9 [3]Ruthenocenophane, Ru-00313
66507-22-0 [4]Ruthenocenophan-2-one, Ru-00345
66507-23-1 [4]-Ruthenocenophan-1-one, Ru-00344
66507-24-2 [4]Ruthenocenophane, Ru-00357
66507-26-4 3-Ruthenocenyl-2-propenoic acid, Ru-00307
66515-97-7 [μ_3-(2(3*H*)-Benzothiazolethionato-N^3:S^2:S^2)]-nonacarbonyl-μ-hydrotriruthenium, Ru-00416
66560-23-4 3-Ruthenocenylpropanoic acid, Ru-00312
66705-59-7 Bromooctacarbonylmethyldiosmium, Os-00071
66705-63-3 Bromododecacarbonylmethyltriosmium, Os-00211
66749-30-2 Dibromododecacarbonyltriosmium, Os-00149
66926-04-3 Dicarbonyl(η^2-methanethial)-bis(triphenylphosphine)osmium, Os-00399
67126-04-9 1,1''-Biruthenocene, Ru-00523
67269-69-6 Tri-μ-carbonylnonacarbonyldi-μ-hydrotetraruthenate(2−); Di-K salt, *in* Ru-00218
67328-91-0 Chlorohydrobis(triphenylphosphine)-[(7,8,9,10,11-η)undecahydro-7,8-dicarbaundecaborato(2−)]ruthenium, Ru-00735
67350-80-5 Chlorohydrobis(triphenylphosphine)-[(7,8,9,10,11-η)-undecahydro-7,9-dicarbaundecaborato(2−)]ruthenium, Ru-00736
67420-77-3 Bis(η^2-ethene)[(1,2,3,4,5,6-η)-hexamethylbenzene]ruthenium, Ru-00438
67420-82-0 [(1,2,3-η)-2-Cyclohexen-1-yl][(1,2,3,4,5,6-η)-hexamethylbenzene]ruthenium(1+); Hexafluorophosphate, *in* Ru-00494
67420-83-1 Carbonyl[(1,2,3-η)-2-cyclohexen-1-yl]-[(1,2,3,4,5,6-η)-hexamethylbenzene]ruthenium(1+), Ru-00513
67421-01-6 [(1,2,3,4-η)-1,3-Cyclohexadiene][(1,2,3,4,5,6-η)-hexamethylbenzene]ruthenium, Ru-00491
67421-02-7 Di-μ-chlorodichlorobis[(1,2,3,4,5,6-η)-hexamethylbenzene]diruthenium, Ru-00612
67421-25-4 Bis[1,2-ethanediylbis[dimethylphosphine]-*P*,*P'*]-(η^2-ethene)ruthenium, Ru-00377
67422-61-1 [μ_5-(Carbon)]tetradecacarbonyl[μ-(dimethyl phosphito-*O''*:*P*)]-μ-hydropentaosmium, Os-00291
67430-20-0 [(1,2,3,4,5-η)-2,4-Cycloheptadien-1-yl]-[(1,2,3,4,5,6-η)-hexamethylbenzene]ruthenium(1+); Hexafluorophosphate, *in* Ru-00514
67507-27-1 (Acetonitrile)(η^3-2-propenyl)-bis(triphenylphosphine)ruthenium, Ru-00769
67507-29-3 Hydrotris(methyldiphenylphosphine)(η^3-2-propenyl)ruthenium, Ru-00784
67612-85-5 (η^6-Benzene)dichloro(trimethylphosphine)ruthenium, Ru-00113
67612-92-4 (η^6-Benzene)bis(trimethylphosphine)ruthenium, Ru-00276
67612-94-6 (η^6-Benzene)bis(trimethyl phosphite-*P*)ruthenium, Ru-00275
67966-16-9 Dodecacarbonyltri-μ-hydro-μ-iodotetraosmium, Os-00162
67993-17-3 Undecacarbonyltetra-μ-hydro(methoxycarbonyl)tetraosmate(1−); Bis(triphenylphosphine)iminium salt, *in* Os-00219
68088-86-8 Hexadecacarbonyl-μ_3-ethylidyne-μ_4-ethylidynehexaosmium, Os-00312
68088-93-7 Bis[(η^2-ethenyl)benzene]bis(triphenylphosphine)ruthenium, Ru-00829
68091-54-3 Hexadecacarbonyl[μ_3-[(1-η:1,2-η:2-η)-1,2-dimethyl-1,2-ethenediyl]](μ_5-methanetetrayl)hexaosmium, Os-00316
68124-76-5 Carbonylchloro[η^2-(4-methylphenyl)[(4-methylphenyl)imino]methyl]-bis(triphenylphosphine)ruthenium, Ru-00828
68193-71-5 Triammine(η^6-benzene)ruthenium(2+); Bis(hexafluorophosphate), *in* Ru-00042
68246-45-7 [η^6-(1-Isopropyl-4-methylbenzene)](η^5-tetramethylthiophene)ruthenium(2+); Bis(hexafluorophosphate), *in* Ru-00492
68569-22-2 Tetracarbonyl(η^2-ethene)osmium, Os-00032
68581-83-9 Pentadecacarbonyliodopentaosmate(1−); Bis(triphenylphosphine)iminium salt, *in* Os-00269
68643-61-8 Dicarbonyl(η^5-2,4-cyclopentadien-1-yl)(η^2-ethene)ruthenium(1+); Hexafluorophosphate, *in* Ru-00104
68643-64-1 Dicarbonyl(η^5-2,4-cyclopentadien-1-yl)-(isocyanomethane)ruthenium(1+); Hexafluorophosphate, *in* Ru-00097
68643-65-2 Dicarbonyl(η^5-2,4-cyclopentadien-1-yl)[(1,2-η)-1-propene]ruthenium(1+); Hexafluorophosphate, *in* Ru-00152
68643-67-4 Dicarbonyl[(1,2-η)-cyclohexene](η^5-2,4-cyclopentadien-1-yl)ruthenium(1+); Hexafluorophosphate, *in* Ru-00314
68663-43-4 Amminedicarbonyl(η^5-2,4-cyclopentadien-1-yl)ruthenium(1+); Tetraphenylborate, *in* Ru-00060
68663-45-6 Carbonyl(η^5-2,4-cyclopentadien-1-yl)[1,2-ethanediylbis[diphenylphosphine]-*P*,*P'*]ruthenium(1+); Chloride, *in* Ru-00676
68663-47-8 (Acetonitrile)dicarbonyl(η^5-2,4-cyclopentadien-1-yl)ruthenium(1+); Hexafluorophosphate, *in* Ru-00096
68796-11-2 Nonacarbonyl(η^2-ethene)-μ-hydro[μ-(methanethiolato)]triosmium, Os-00171
68796-12-3 Nonacarbonyl-μ-hydro[μ_3-(methanethiolato)]triosmium, Os-00100
68796-19-0 Nonacarbonyl-μ-hydro-μ_3-thioxotriosmate(1−), Os-00069
69047-78-5 Bis[μ-(benzenethiolato)]dicarbonylbis(η^5-2,4-cyclopentadien-1-yl)diruthenium; *cis-form*, *in* Ru-00595
69047-79-6 Bis[μ-(benzeneselenolato)]dicarbonylbis(η^5-2,4-cyclopentadien-1-yl)diruthenium; *cis-form*, *in* Ru-00596
69047-82-1 (Benzenethiolato)dicarbonyl(η^5-2,4-cyclopentadien-1-yl)ruthenium, Ru-00300
69047-83-2 (Benzeneselenolato)dicarbonyl(η^5-2,4-cyclopentadien-1-yl)ruthenium, Ru-00301
69048-01-7 Decacarbonyl-μ-hydro[μ-(methoxymethylidyne-)]triosmium, Os-00166
69058-50-0 (η^6-Benzene)dichloro[sulfinylbis[methane]-*O*]ruthenium, *in* Ru-00038
69058-55-5 (η^6-Benzene)chlorobis(triphenylphosphine)ruthenium(1+); Hexafluorophosphate, *in* Ru-00776
69074-64-2 Dodecacarbonyltri-μ-hydrotetraosmate(1−); Tetramethylammonium salt, *in* Os-00163
69085-99-0 Nonacarbonyl-μ-hydro-μ_3-thioxotriosmate(1−); Bis(triphenylphosphine)iminium salt, *in* Os-00069
69088-23-9 Bis[μ-(benzenethiolato)]dicarbonylbis(η^5-2,4-cyclopentadien-1-yl)diruthenium; *trans-form*, *in* Ru-00595

69088-24-0 Bis[μ-(benzeneselenolato)]dicarbonylbis(η^5-2,4-cyclopentadien-1-yl)diruthenium; *trans-form*, *in* Ru-00596

69175-65-1 (η^5-2,4-Cyclopentadien-1-yl)-1-propynylbis(triphenylphosphine)ruthenium, Ru-00804

69191-10-2 [(1,2,3,4-η)-Benzene](η^6-benzene)ruthenium, Ru-00250

69210-23-7 Di-μ-carbonyltetracarbonyl[μ-[(1,2,3,4-η:1,5,6,7-η)-2,4,6-cycloheptatrien-1-yl]]-(η^5-2,4-cyclopentadien-1-yl)triruthenium, Ru-00476

69215-27-6 Dodecacarbonyldi-μ-hydrotetraosmate(2−); Bis[bis-(triphenylphosphine)iminium] salt, *in* Os-00159

69251-43-0 Dicarbonyliodo[(1,2,3,3*a*,6*a*-η)-1,4,5,6-tetrahydro-1-pentalenyl]ruthenium, Ru-00136

69403-53-8 [*N,N'*-Bis(4-methylphenyl)ureato(2−)-*N,N'*]-dicarbonylbis(triphenylphosphine)ruthenium, Ru-00831

69450-46-0 [(1,2,3-η)-2,4-Cyclohexadien-1-yl][(1,2-η)-cyclohexene]hydrobis(triphenylphosphine)ruthenium, Ru-00820

69450-47-1 [(1,2,3,4,5-η)-2,4-Cyclohexadien-1-yl]-hydrobis(triphenylphosphine)ruthenium, Ru-00780

69450-61-9 (η^4-1,3-Butadiene)[(η^2-ethenyl)benzene]-bis(triphenylphosphine)ruthenium, Ru-00819

69450-63-1 [(η^2-Ethenyl)benzene][(1,2,3,4-η)-1,3-hexadiene]bis(triphenylphosphine)ruthenium, Ru-00824

69462-16-4 Bis(η^6-benzene)di-μ-chlorodichlorodiosmium, *in* Os-00036

69500-43-2 [η^2-(Ethenylbenzene)][(1,2,3-η)-2-hexenyl]-hydrobis(triphenylphosphine)ruthenium, Ru-00825

69553-71-5 Di-μ-carbonyltetracarbonyl[μ-(2,4,6-cyclooctatrien-1-yl)](η^5-2,4-cyclopentadien-1-yl)triruthenium, Ru-00506

69558-33-4 Dicarbonyl(η^2-formaldehyde)-bis(triphenylphosphine)osmium, Os-00400

69700-33-0 Tri-μ-carbonylhexacarbonyl[μ_3-[(methylsilylidyne)tris[dibutylphosphine]-*P*:*P'*:*P''*]]triruthenium, Ru-00682

69891-00-5 (η^6-Benzene)[(1,2,3,4,5,6-η)-1-phenylboratabenzene]ruthenium(1+); Hexafluorophosphate, *in* Ru-00456

69942-19-4 Decacarbonyl-μ-hydrohydro(2-isocyano-2-methylpropane)triosmium, Os-00259

70073-17-5 Dodecacarbonyltrihydrotetraruthenate(1−); Bis(triphenylphosphine)iminium salt, *in* Ru-00220

70084-45-6 Nonacarbonyldi-μ-hydro(2-isocyano-2-methylpropane)triosmium, Os-00234

70210-62-7 Iodoruthenocene, Ru-00137

70210-63-8 1,1'-Diiodoruthenocene, Ru-00131

70317-01-0 Bis(η^6-benzene)tri-μ-chlorodiosmium(1+); Tetrafluoroborate, *in* Os-00176

70377-88-7 [μ-(*tert*-Butylcarbonimidoyl)]decacarbonyl-μ-hydrotriosmate(1−); Hydrogen compd., *in* Os-00257

70379-01-0 Eicosacarbonylheptaosmate(2−), Os-00315

70379-03-2 Triacontacarbonyl[μ_3-(carboxylato-*C,O,O'*)]-μ-hydrodecaosmate(1−), Os-00352

70586-96-8 Benzylcarbonyl(η^5-2,4-cyclopentadien-1-yl)-(triphenylphosphine)ruthenium, Ru-00670

70669-56-6 Di-μ-carbonyldicarbonylbis[(1,2,3,4,5-η)-1,2,3,4,5-pentamethyl-2,4-cyclopentadien-1-yl]diruthenium, Ru-00609

70755-71-4 (η^6-Benzene)nonacarbonyl[μ_4-(1-η:1,2-η:1,2-η:2-η)-1-cyclohexene-1,2-diyl]tetraruthenium, Ru-00540

70755-72-5 Dodecacarbonyl[μ_4-(1-η:1,2-η:1,2-η:2-η)-1-cyclohexene-1,2-diyl]tetraruthenium, Ru-00471

70772-42-8 Nonacarbonyl[μ_3-[(1-η:1,2-η:2-η)-1,3-cyclohexadiene-1,2-diyl]]di-μ-hydrotriruthenium, Ru-00384

70804-60-3 Pentadecacarbonyl-μ-hydropentaosmate(1−), Os-00248

70804-61-4 Pentadecacarbonyl-μ-hydropentaosmate(1−); Bis-(triphenylphosphine)iminium salt, *in* Os-00248

71191-30-5 Pentadecacarbonyl[μ_4-[methyl phosphinito(2−)-*P*:*P*:*P*:*P*]]pentaosmium, Os-00274

71296-09-8 Di-μ-methylenehexakis(trimethylphosphine)diruthenium(2+); Bis(tetrafluoroborate), *in* Ru-00534

71303-84-9 [(1,2,3,4,5,6-η)-Hexamethylbenzene]tris(2-propanone)ruthenium(2+); Bis(hexafluorophosphate), *in* Ru-00552

71324-44-2 μ-Methyldi-μ-methylenehexakis(trimethylphosphine)diruthenium(1+); Tetrafluoroborate, *in* Ru-00555

71343-96-9 Undecacarbonyl[μ-(η:η^2-ethenyl)]tri-μ-hydrotetraosmium, Os-00218

71356-99-5 Tetracarbonyl(triphenylstibine)ruthenium, Ru-00561

71397-10-9 Chloro(η^5-2,4-cyclopentadien-1-yl)[1,2-ethanediylbis[diphenylphosphine]-*P,P'*]ruthenium, Ru-00671

71397-13-2 Bromo(η^5-2,4-cyclopentadien-1-yl)[1,2-ethanediylbis[diphenylphosphine]-*P,P'*]osmium, Os-00355

71397-23-4 Bromo(η^5-2,4-cyclopentadien-1-yl)bis(triphenyl phosphite-*P*)osmium, Os-00409

71397-25-6 Acetonitrile(η^5-2,4-cyclopentadien-1-yl)-bis(triphenylphosphine)osmium(1+); Tetraphenylborate, *in* Os-00419

71397-30-3 (η^5-2,4-Cyclopentadien-1-yl)tris(trimethyl phosphite-*P*)ruthenium(1+); Hexafluorophosphate, *in* Ru-00375

71438-17-0 Carbonyl(η^5-2,4-cyclopentadien-1-yl)-methyl(triphenylphosphine)ruthenium, Ru-00621

71562-47-5 Nonacarbonyltri-μ-hydro[μ_3-(methoxymethylidyne)]-triruthenium, Ru-00174

71562-48-6 Nonacarbonyltri-μ-hydro[μ_3-(methoxymethylidyne)]-triosmium, Os-00130

71633-32-4 Bromo(η^5-2,4-cyclopentadien-1-yl)bis(trimethyl phosphite-*P*)osmium, Os-00147

71652-17-0 Carbonyldichloro[(1,2,3,4,5,6-η)-hexamethylbenzene]ruthenium, Ru-00318

71652-19-2 Carbonylchloro[(1,2,3,4,5,6-η)-hexamethylbenzene](trimethylphosphine)ruthenium(1+); Hexafluorophosphate, *in* Ru-00440

71652-20-5 Carbonyl[(1,2,3,4,5,6-η)-hexamethylbenzene]-(trimethylphosphine)ruthenium, Ru-00441

71652-22-7 Carbonyl[(1,2,3,4,5,6-η)-hexamethylbenzene]-hydro(trimethylphosphine)ruthenium(1+); Hexafluorophosphate, *in* Ru-00443

71685-08-0 Carbonyl[(1,2,3,4,5,6-η)-hexamethylbenzene]-methyl(trimethylphosphine)ruthenium(1+); Hexafluorophosphate, *in* Ru-00465

71685-17-1 (η^6-Benzene)(η^2-ethene)(trimethylphosphine)ruthenium, Ru-00203

71685-19-3 (η^6-Benzene)(η^2-ethene)-hydro(trimethylphosphine)ruthenium(1+); Hexafluorophosphate, *in* Ru-00204

71685-21-7 (η^6-Benzene)(η^2-ethene)-methyl(trimethylphosphine)ruthenium(1+); Hexafluorophosphate, *in* Ru-00273

71685-25-1 (η^6-Benzene)chloro(η^2-ethene)-(trimethylphosphine)ruthenium(1+); Hexafluorophosphate, *in* Ru-00202

71737-42-3 Decacarbonyl-μ-hydro[μ-(methoxymethylidyne-)]triruthenium, Ru-00223

71744-55-3 [(1,2,5,6-η)-1,5-Cyclooctadiene]tris(2-isocyano-2-methylpropane)osmium, Os-00333

71817-46-4 Octacarbonylbis(η^5-2,4-cyclopentadien-1-yl)triruthenium, Ru-00475

71825-64-4 [(1,2,3,4,5,6-η)-Benzoato][(1,2,3,4,5,6-η)-1,3,5-trimethylbenzene]ruthenium(1+) hydrogen tetrafluoroborate(1:1:2), *in* Ru-00429

71825-72-4 (η^6-Benzene)[(1,2,3,4,5,6-η)-hexamethylbenzene]ruthenium(2+); Bis(tetrafluoroborate), *in* Ru-00488

71825-80-4 [(1,2,3,4,5,6-η)-Hexamethylbenzene]-[(1,2,3,4,4*a*,8*a*-η)-naphthalene]ruthenium(2+); Bis(tetrafluoroborate), *in* Ru-00573

71825-82-6 [(1,2,3,4,4*a*,9*a*-η)-Anthracene][(1,2,3,4,5,6-η)-1,3,5-trimethylbenzene]ruthenium(2+); Bis-(tetrafluoroborate), *in* Ru-00582

71825-88-2 (η^6-Benzene)[(1,2,3,4,5,6-η)-methoxybenzene]ruthenium(2+); Bis(tetrafluoroborate), *in* Ru-00310
71825-94-0 (η^6-Benzene)(η^6-chlorobenzene)ruthenium(2+); Bis(tetrafluoroborate), *in* Ru-00234
71825-96-2 (η^6-Benzene)(η^6-phenol)ruthenium(2+); Bis(tetrafluoroborate), *in* Ru-00241
71826-04-5 (η^6-Benzene)[(1,2,3,4,5,6-η)-methylbenzoate]-ruthenium(2+) bis(tetrafluoroborate), *in* Ru-00306
71836-19-6 [(1,2,3,4,5,6-η)-*N*,*N*-Dimethylaniline]-[(1,2,3,4,5,6-η)-1,3,5-trimethylbenzene]ruthenium(2+); Bis(tetrafluoroborate), *in* Ru-00463
71851-93-9 (η^6-Dimethylaniline)(η^6-hexamethylbenzene)ruthenium(2+); Bis(tetrafluoroborate), *in* Ru-00526
71861-31-9 Bis[(1,2,3,4,5,6-η)-hexamethylbenzene]ruthenium(2+); Bis(tetrafluoroborate), *in* Ru-00613
71861-33-1 (η^6-Benzene)[1-(η^6-phenyl)ethanone]ruthenium(2+); Bis(tetrafluoroborate), *in* Ru-00342
71933-81-8 μ-Chlorobis[(1,2,3,4,5,6-η)-hexamethylbenzene]-di-μ-hydrodiruthenium(1+); Chloride, *in* Ru-00615
71933-82-9 [(2,3,5,6-η)-Bicyclo[2.2.1]hepta-2,5-diene]-[(1,2,3,4,5,6-η)-hexamethylbenzene]ruthenium, Ru-00512
71935-53-0 Pentadecacarbonyldi-μ-hydro(trimethyl phosphite-*P*)pentaosmium, Os-00304
71935-55-2 Pentadecacarbonyldi-μ-hydroiodopentaosmate(1−), Os-00249
71936-70-4 μ-Carbonyldecacarbonyl-μ-hydrotriruthenate(1−); Bis(triphenylphosphine)iminium salt, *in* Ru-00168
71936-71-5 μ-Carbonyldecacarbonyl-μ-hydrotriruthenate(1−); Tetraethylammonium salt, *in* Ru-00168
71957-41-0 (η^5-2,4-Cyclopentadien-1-yl)(phenylethynyl)-bis(triphenylphosphine)osmium, Os-00433
72042-46-7 Dicarbonyl(η^5-2,4-cyclopentadien-1-yl)ethylruthenium, Ru-00106
72098-73-8 Pentaamine(η^2-ethene)ruthenium(2+); Dithionate, *in* Ru-00014
72155-83-0 Dicarbonyl(diseleno)bis(triphenylphosphine)osmium, Os-00389
72250-50-1 Octadecacarbonyl-μ_6-hydrohexaruthenate(1−); Tetraphenylarsonium salt, *in* Ru-00468
72268-06-5 Dodecacarbonyldi-μ-hydrodi-μ_3-selenoxotetraosmium, Os-00161
72268-07-6 Nonacarbonyldi-μ-hydro-μ_3-selenoxotriruthenium, Ru-00090
72268-08-7 Nonacarbonyldi-μ-hydro-μ_3-telluroxotriruthenium, Ru-00091
72277-78-2 1-Phenyl-3-ruthenocenyl-2-propen-1-one; (*E*)-*form*, *in* Ru-00508
72282-38-3 Nonacarbonyldi-μ_3-thioxotriruthenium, Ru-00117
72282-39-4 Nonacarbonyldi-μ_3-selenoxotriruthenium, Ru-00118
72282-40-7 Nonacarbonyldi-μ_3-thioxotriosmium, Os-00087
72282-41-8 Nonacarbonyldi-μ_3-selenoxotriosmium, Os-00088
72282-42-9 Nonacarbonyldi-μ_3-telluroxotriosmium, Os-00089
72314-28-4 Dodecacarbonyldi-μ-hydrodi-μ_3-thioxotetraosmium, Os-00160
72477-51-1 (Aminocarbonyl)dicarbonyl(η^5-2,4-cyclopentadien-1-yl)ruthenium, Ru-00077
72643-21-1 Octacarbonyl[μ_3-[(1,2-η:2,3-η:3*a*,4,5,6,6*a*-η)-pentalene]]triruthenium, Ru-00419
72693-30-2 Octacarbonyl[(1,2,3,3*a*,6*a*-η:3*a*,4,5,6,6*a*-η)-1,3,5-tris(trimethylsilyl)]pentalenetriruthenium, Ru-00627
72707-18-7 Tricarbonyl[(2,3,4,5-η)-2,4-hexadiene]osmium, Os-00080
72707-23-4 Tricarbonyl[(1,2,3,4-η)-2,3-dimethyl-1,3-butadiene]osmium, Os-00079
72763-42-9 Amminedicarbonyl(η^5-2,4-cyclopentadien-1-yl)ruthenium(1+); Hexafluorophosphate, *in* Ru-00060
72763-49-6 (Aminocarbonyl)dicarbonyl(η^5-2,4-cyclopentadien-1-yl)osmium, Os-00060
72763-52-1 (Aminocarbonyl)carbonyl(η^5-2,4-cyclopentadien-1-yl)(triphenylphosphine)ruthenium, Ru-00618
72765-28-7 (Acetonitrile)(η^6-benzene)dichloroosmium, *in* Os-00036
72765-30-1 [(1,2,3,4-η)-1,3,5,7-cyclooctatetraene]-[(1,2,3,4,5,6-η)-1,3,5-trimethylbenzene]osmium, Os-00300
72765-31-2 (η^6-Benzene)[(1,2,3,4-η)-1,3,5,7-cyclooctatetraene]osmium, Os-00240
72768-82-2 Dicarbonylchloro(η^5-2,4-cyclopentadien-1-yl)osmium, Os-00042
72804-54-7 (η^6-Benzene)(η^5-2,4-cyclopentadien-1-yl)ruthenium(1+); Tetraphenylborate, *in* Ru-00189
72804-56-9 (η^5-2,4-Cyclopentadien-1-yl)[(1,2,3,4,5,6-η)-methylbenzene]ruthenium(1+); Tetraphenylborate, *in* Ru-00254
72804-59-2 [(1,2,3,4,5,6-η)-1,1′-Biphenyl](η^5-2,4-cyclopentadien-1-yl)ruthenium(1+); Hexafluorophosphate, *in* Ru-00455
72804-61-6 (η^6-Chlorobenzene)(η^5-2,4-cyclopentadien-1-yl)ruthenium(1+); Tetrafluoroborate, *in* Ru-00183
72804-78-5 (η^5-2,4-Cyclopentadien-1-yl)[(1,2,3,4,5,6-η)-hexamethylbenzene]ruthenium(1+); Hexafluorophosphate, *in* Ru-00464
72910-30-6 Nonacarbonyldi-μ-nitrosyl(trimethylamine)triosmium, Os-00173
72918-60-6 [μ-[*N*,*N*′-Bis(1-methylethyl)-1,4-bis[(1-methylethyl)imino]-2,3-butanediaminato(2−)-N^1,N^2,N^3:N^2,N^3,N^4]]-μ-carbonyltetracarbonyldiruthenium, Ru-00551
72918-62-8 [μ-[*N*,*N*′-Bis(1-methylethyl)-1,4-bis[(1-methylethyl)imino]-2,3-butanediaminato(2−)-N^1,N^2,N^3:N^2,N^3,N^4]]tetracarbonyldiruthenium, Ru-00531
72952-08-0 Di-μ-carbonyldi-μ_3-carbonyltetradecacarbonylhexaruthenate(2−); Bis(methyltriphenylphosphonium) salt, *in* Ru-00501
73046-96-5 (η^6-Benzene)(η^2-ethene)bis(trimethylphosphine)ruthenium(2+); Bis(hexafluorophosphate), *in* Ru-00372
73085-12-8 Hexacarbonyl[μ_3-[(1,2,3,4-η:5,6-η:7,8-η)-1,3,5,7-cyclooctatetraene]]-μ_3-thioxotriruthenium, Ru-00329
73107-21-8 Hexacarbonyl[μ_3-[(1,2,3-η:4,5-η:6,7-η)-2,4,6-cycloheptatrien-1-yl]][μ_3-(2-methyl-2-propanethiolato)]triruthenium, Ru-00459
73107-22-9 Tetracarbonyl[μ-(1,2,3,4-η:1,5,6,7-η)-2,4,6-cycloheptatrien-1-yl][μ-(2-methyl-2-propanethiolato)]diruthenium, Ru-00396
73178-90-2 1,1″:1‴,1⁗-Terruthenocene, Ru-00661
73178-91-3 1,1″:1‴,1⁗:1‴‴,1‴‴″-Quaterruthenocene, Ru-00756
73230-18-9 Heptacarbonyl-μ-hydro-μ-nitrosyltris(trimethyl phosphite-*P*)triruthenium, Ru-00442
73230-19-0 Decacarbonyl-μ-hydro-μ-nitrosyltriruthenium, Ru-00121
73238-03-6 Decacarbonyl-μ-hydro-μ-nitrosyltriosmium, Os-00094
73249-64-6 Carbonyldichloro(dichloromethylene)-bis(triphenylphosphine)osmium, Os-00384
73249-66-8 (Carbonoselenoyl)-carbonyldichlorobis(triphenylphosphine)osmium, Os-00380
73267-55-7 (Carbonotelluroyl)-carbonyldichlorobis(triphenylphosphine)osmium, Os-00381
73413-27-1 [μ_6-(Carbon)]tri-μ-carbonyltridecacarbonylhexaruthenate(2−); Bis[bis(triphenylphosphine)iminium] salt, *in* Ru-00467
73413-28-2 [μ_6-(Carbon)]tri-μ-carbonyltridecacarbonylhexaruthenate(2−); Bis(tetraethylammonium) salt, *in* Ru-00467
73531-62-1 Dicarbonyl[1,2,3,4-tetrachloro-5-(η^2-diazo)-1,3-cyclopentadiene]-bis(triphenylphosphine)ruthenium, Ru-00785
73555-60-9 (Carbonothioyl)heptacarbonyldi-μ-hydrodi-μ_3-thioxotriosmium, Os-00055
73746-95-9 [μ_3-(Bromomethylidyne)]nonacarbonyltri-μ-hydrotriruthenium, Ru-00122
73746-96-0 [μ_3-(Bromomethylidyne)]nonacarbonyltri-μ-hydrotriosmium, Os-00096
73746-97-1 Nonacarbonyl[μ_3-(chloromethylidyne)]tri-μ-hydrotriruthenium, Ru-00123
73746-98-2 Nonacarbonyltri-μ-hydro[μ_3-(2-methoxy-2-oxoethylidyne)]triruthenium, Ru-00227
74029-08-6 Di-μ-carbonylicosacarbonyloctaosmate(2−), Os-00330
74091-69-3 Di-μ-carbonylicosacarbonyloctaosmate(2−); Bis-[bis(triphenylphosphine)iminium] salt, *in* Os-00330
74095-85-5 Hexacarbonylbis[μ-(dimethylarsino)]diruthenium, Ru-00153

74128-10-2 Di-μ-carbonylnonacarbonyl(dicarbonylferrate)-μ-hydrotriruthenate(1−), Ru-00282
74128-11-3 Di-μ-carbonyloctacarbonyl(dicarbonyliron)di-μ-hydro(dimethylphenylphosphine)triruthenium, Ru-00517
74219-48-0 Dihydrotetrakis(trimethyl phosphite-*P*)ruthenium; *cis-form*, *in* Ru-00277
74231-12-2 μ-Carbonyldodecacarbonyl-μ-chlorotetraruthenate(1−); Bis(triphenylphosphine)iminium salt, *in* Ru-00281
74238-32-7 Hexacarbonyl[μ-[[(2,3,6,7-η:3,4,5,6-η)-1,3,5,7-cyclooctatetraen-1-yl]trimethylsilane]]diruthenium, Ru-00458
74325-95-4 Decacarbonyl[μ-[(dimethylamino)carbonyl-*C*:*O*]]-μ-hydrotriruthenium, Ru-00295
74325-98-7 Octacarbonyldi-μ-hydrohydro-μ_3-thioxotriosmate(1−), Os-00057
74333-23-6 Hexacarbonyl[μ-[(1,2,3,4-η:1,4,5,8-η)-1,3,5,7-cyclooctatetraene]]diruthenium, Ru-00328
74481-94-0 (η^5-2,4-Cyclopentadien-1-yl)[1,2-ethanediylbis[diphenylphosphine]-*P*,*P′*]methylruthenium, Ru-00678
74504-98-6 [μ_6-(Carbon)]pentadecacarbonyltris[μ-(ethanethiolato)]-μ-hydrohexaruthenium, Ru-00563
74521-00-9 [μ_6-(Carbon)]-μ-carbonyldodecacarbonyltetrakis(trimethylphosphite-*P*)hexaruthenium, Ru-00639
74535-09-4 [μ_6-(Carbon)]-μ-carbonyltridecacarbonyltris(trimethyl phosphite-*P*)hexaruthenium, Ru-00608
74552-68-4 Hexacarbonyl[μ-[η^2-*N*,*N′*-1,2-ethanediylidenebis[2-methyl-2-propanamine]-*N*,*N′*]]diruthenium, Ru-00431
74552-72-0 Tetracarbonylbis[μ-(η^2-*N*,*N′*-1,2-ethanediylidenebis[2-propanamine]-*N*,*N′*]]diruthenium, Ru-00532
74558-74-0 Chloro(η^5-2,4-cyclopentadien-1-yl)-bis(trimethylphosphine)ruthenium, Ru-00207
74573-16-3 (η^5-2,4-Cyclopentadien-1-yl)-2-propenylbis(triphenylphosphine)ruthenium, Ru-00805
74573-22-1 (η^5-2,4-Cyclopentadien-1-yl)(η^3-2-propenyl)-(triphenylphosphine)ruthenium, Ru-00635
74573-23-2 (η^5-2,4-Cyclopentadien-1-yl)[(1,2,3-η)-2-methyl-2-propenyl)](triphenylphosphine)ruthenium, Ru-00643
74790-26-4 Undecacarbonyl-μ-hydrohydrotriruthenium, Ru-00169
74904-04-4 [μ_6-(Carbon)]-μ-carbonyltetradecacarbonylbis(triphenylphosphine)-hexaruthenium, Ru-00827
75026-12-9 [μ_6-(Carbon)]tetracosacarbonyldecaosmate(2−), Os-00342
75117-74-1 [μ_6-(Carbon)]tetracosacarbonyldecaosmate(2−); Bis[bis(triphenylphosphine)iminium] salt, *in* Os-00342
75182-14-8 Chloro[(1,2,3,4,5,6-η)-hexamethylbenzene]-hydro(triphenylphosphine)ruthenium, Ru-00664
75182-16-0 Bromo(η^5-2,4-cyclopentadien-1-yl)-bis(trimethylphosphine)ruthenium, Ru-00205
75182-17-1 (η^5-2,4-Cyclopentadien-1-yl)-iodobis(trimethylphosphine)ruthenium, Ru-00208
75182-21-7 (η^5-2,4-Cyclopentadien-1-yl)-nitrosylbis(trimethylphosphine)ruthenium(2+); Bis(hexafluorophosphate), *in* Ru-00209
75182-23-9 Dicarbonyl(η^5-2,4-cyclopentadien-1-yl)-(trimethylphosphine)ruthenium(1+); Hexafluorophosphate, *in* Ru-00157
75182-25-1 Carbonyl(η^5-2,4-cyclopentadien-1-yl)-bis(trimethylphosphine)ruthenium(1+); Hexafluorophosphate, *in* Ru-00274
75182-27-3 (η^5-2,4-Cyclopentadien-1-yl)-tris(trimethylphosphine)ruthenium(1+); Hexafluorophosphate, *in* Ru-00376
75215-44-0 μ_3-Carbonyldodecacarbonyltetraruthenate(2−); Di-K salt, *in* Ru-00324
75215-45-1 μ_3-Carbonyldodecacarbonyltetraruthenate(2−), Ru-00324
75215-46-2 μ_3-Carbonyldodecacarbonyltetraruthenate(2−); Bis-[bis(triphenylphosphine)iminium]salt, *in* Ru-00324
75280-67-0 Pentacarbonyl(ethoxyruthenocenylmethylene)tungsten, Ru-00481
75280-68-1 Pentacarbonyl(ethoxyruthenocenylmethylene)molybdenum, Ru-00479
75280-69-2 Pentacarbonyl(ethoxyruthenocenylmethylene)chromium, Ru-00478
75293-20-8 Bromotetracarbonyl(ruthenocenylmethylidyne)tungsten; *trans-form*, *in* Ru-00385
75330-60-8 Dodecacarbonyl[μ_4-[(1-η:1,2-η:1,2-η:2-η)-1,2-ethenediyl]]tetraosmium, Os-00225
75345-68-5 (η^6-Benzene)hydrobis(triphenylphosphine)osmium(1+); Hexafluorophosphate, *in* Os-00414
75346-82-6 Carbonylchloro[(4-methylphenyl)methylidyne]-bis(triphenylphosphine)osmium, Os-00427
75353-17-2 (η^6-Benzene)iodobis(triphenylphosphine)osmium(1+); Hexafluorophosphate, *in* Os-00412
75353-18-3 (η^6-Benzene)bis(triphenylphosphine)osmium, Os-00413
75353-19-4 (η^6-Benzene)bis(trimethyl phosphite-*P*)osmium, Os-00195
75353-21-8 (η^6-Benzene)methylbis(triphenylphosphine)osmium(1+); Hexafluorophosphate, *in* Os-00420
75353-25-2 (η^6-Benzene)(η^2-ethene)iodo(trimethylphosphine)-osmium(1+); Hexafluorophosphate, *in* Os-00144
75353-28-5 (η^6-Benzene)(η^2-ethene)(trimethylphosphine)osmium, Os-00145
75353-31-0 (η^6-Benzene)(η^2-ethene)-hydro(trimethylphosphine)osmium(1+); Hexafluorophosphate, *in* Os-00146
75353-35-4 (η^6-Benzene)(η^2-ethene)-methyl(trimethylphosphine)osmium(1+); Hexafluorophosphate, *in* Os-00194
75353-42-3 (η^6-Benzene)iodomethyl(trimethylphosphine)osmium, Os-00111
75368-02-4 (η^6-Benzene)diiodo(triphenylphosphine)osmium, Os-00337
75535-78-3 Tricarbonyl[(1,3,4,5-η)-4-cyclohexene-1,3-diyl]osmium, Os-00076
75535-79-4 Tricarbonyl[(2,3,4,5-η)-2,4-cyclohexadiene-1-carbonitrile]osmium, Os-00104
75535-80-7 Tricarbonyl[(1,3,4,5-η)-2-cyano-4-cyclohexene-1,3-diyl]osmium, Os-00103
75592-69-7 Carbonyl(η^5-2,4-cyclopentadien-1-yl)-(phenylethynyl)(triphenylphosphine)ruthenium, Ru-00675
75687-45-5 Tricarbonylbis(trimethylphosphine)ruthenium; *trans-form*, *in* Ru-00114
75699-40-0 Dodecacarbonyldi-μ-hydro[μ-(2-phenylethylidene)]-tetraosmium, Os-00313
75700-75-3 (η^6-Benzene)chloro(1,2-ethanediamine-*N*,*N′*)osmium(1+); Hexafluorophosphate, *in* Os-00065
75700-79-7 Ammine(η^6-benzene)(1,2-ethanediamine-*N*,*N′*)osmium(2+); Bis(hexafluorophosphate), *in* Os-00066
75700-95-7 (η^6-Benzene)dichloro(pyridine)osmium, Os-00137
75700-97-9 (η^6-Benzene)carbonyldichloroosmium, Os-00044
75701-03-0 η^6-Benzenedibromoosmium, Os-00034
75701-04-1 (η^6-Benzene)diiodoosmium, Os-00037
75701-06-3 Triaqua(η^6-benzene)ruthenium(2+), Ru-00041
75756-64-8 (η^4-1,3-Butadiene)tricarbonylosmium, Os-00046
75756-65-9 Tricarbonyl[(1,2,3,4-η)-1,3-pentadiene]ruthenium, Ru-00081
75756-66-0 Tricarbonyl[(1,2,3,4-η)-1,3-pentadiene]osmium; *syn-form*, *in* Os-00061
75764-07-7 Bromodecacarbonyl[(2,3-η)-1,1,1,4,4,4-hexafluoro-2-butene]triosmate(1−); Bis(triphenylphosphine)iminium salt, *in* Os-00224
75826-70-9 Dicarbonyl[(1,2,3,4-η)-2,3-dimethyl-1,3-butadiene](triphenylphosphine)ruthenium, Ru-00634
75864-71-0 Di-μ-carbonylhexacarbonyl[μ_3-[(1,2,3,4-η:1,2,3,4-η:1,4-η)-1-(3,3-dimethyl-1-butenyl)-2,4-bis(1,1-dimethylethyl)-1,3-butadiene-1,4-diyl]]triruthenium, Ru-00638
75901-26-7 Di-μ-carbonylnonacarbonyl(dicarbonylruthenium)-di-μ-hydrotriosmium, Os-00209
75952-48-6 μ-Carbonyldicarbonylbis(η^5-2,4-cyclopentadien-1-yl)(μ-ethylidyne)diruthenium(1+); *cis-form*, *in* Ru-00393

75952-49-7 μ-Carbonyldicarbonylbis(η^5-2,4-cyclopentadien-1-yl)-μ-ethenylidenediruthenium, Ru-00390

75963-10-9 Tetracarbonyl[μ-[η^4-*N*,*N'*-1,2-ethanediylidenebis[2-propanamine]-*N*,*N'*]]-μ-1,2-ethenediyldiruthenium, Ru-00361

76000-21-0 (Carbonothioyl)octacarbonyldi-μ_3-thioxotriosmium, Os-00085

76171-47-6 Di-μ-hydrodihydrohexakis(trimethylphosphine)diruthenium, Ru-00499

76171-49-8 Dihydrotetrakis(trimethylphosphine)ruthenium, Ru-00278

76190-51-7 Tridecacarbonyl-μ-hydro[μ_3-[*N*-phenyl-1,2-benzenediaminato(3−)-*N*,*N'*:*N'*,*N'*]]pentaosmium, Os-00338

76210-26-9 Tetra-μ-methyleneoctakis(trimethylphosphine)triruthenium(2+); Bis(tetrafluoroborate), *in* Ru-00651

76297-42-2 (η^5-2,4-Cyclopentadien-1-yl)[(1,2-η)-1-propene]-bis(trimethylphosphine)ruthenium(1+); Hexafluorophosphate, *in* Ru-00373

76451-72-4 Dicarbonyl[(4-methyl-2,5-cyclohexadien-1-ylidene)methylene]bis(triphenylphosphine)-osmium, Os-00430

76468-65-0 μ-Carbonyldodecacarbonyl-μ-hydrotetraosmate(1−); Bis(triphenylphosphine)iminium salt, *in* Os-00205

76550-98-6 [μ_5-(Carbon)]pentadecacarbonyliodopentaosmate(1−); Bis(triphenylphosphine)iminium salt, *in* Os-00285

76581-49-2 Nonacarbonyldi-μ-hydro[μ_3-[methanaminato-(2−)]]triosmium, Os-00101

76581-50-5 Decacarbonyl-μ-hydro[μ-(methanaminato)]triosmium, Os-00128

76601-70-2 Dicarbonyl[η^5-nonahydrodicarbanonaborato(4−)]-(triphenylphosphine)ruthenium, Ru-00570

76612-05-0 Dicarbonyl[(*O*,*S*-η)-sulfur dioxide]bis(triphenylphosphine)ruthenium; SO_2 complex, *in* Ru-00727

76680-98-3 Bis(η^6-benzene)osmium(2+), Os-00179

76687-31-5 Bis(η^6-Benzene)ruthenium(2+); Bis(hexafluorophosphate), *in* Ru-00251

76721-83-0 Nonacarbonyl[μ_3-[(1-η:1,2-η:2-η)-1-(1,1-dimethylethyl)-1,2-ethenediyl]]di-μ-hydrotriruthenium(2+), Ru-00392

76722-36-6 (η^6-Benzene)[(1,2,3,4,5,6-η)-6-(dimethoxyphosphinyl)-2,4-cyclohexadien-1-yl]ruthenium(1+); Hexafluorophosphate, *in* Ru-00362

76736-33-9 Dicarbonylchloro(η^3-2-propenyl)-(triphenylphosphine)ruthenium, Ru-00578

76741-75-8 Nonacarbonyl[μ_3-[(1-η:1,2-η:1,2-η)-3,3-dimethyl-1-butynyl]]triruthenate(1−); Tetraphenylarsonium salt, *in* Ru-00387

76756-17-7 Decacarbonyl-μ-hydro[μ-[(4-methylphenyl)azo-]]triosmium, Os-00292

76786-22-6 [(1,2,3,4-η)-1,3-Cyclohexadiene]tris(trimethyl phosphite-*P*)ruthenium, Ru-00408

76786-24-8 [(1,2,3,4-η)-1,3-Cyclohexadiene]-tris(phosphorous trifluoride)ruthenium, Ru-00040

76791-95-2 Nonacarbonyl-μ-hydro-μ_4-nitrido(tricarbonyliron)triruthenium, Ru-00216

76798-39-5 Bis[μ-[η^6-nonahydrodicarbanonaborato(2−)-*B*,*H*]]-tetrakis(triethylphosphine)diruthenium, *in* Ru-00379

76830-28-9 Bis(η^6-benzene)osmium(2+); Bis(hexafluorophosphate), *in* Os-00179

76917-52-7 μ_3-Carbonyldodecacarbonylhydrotetraruthenate(1−), Ru-00284

76917-53-8 μ_3-Carbonyldodecacarbonylhydrotetraruthenate(1−); Bis(triphenylphosphine)iminium salt, *in* Ru-00284

76917-54-9 μ_3-Carbonyldodecacarbonylhydrotetraruthenate(1−); K salt, *in* Ru-00284

76932-90-6 (η^6-Benzene)[(1,2,3,4,5-η)-6-(triphenylphosphino)-2,4-cyclohexadien-1-yl]ruthenium(2+); Bis(hexafluorophosphate), *in* Ru-00663

76934-04-8 [μ_6-(Carbon)]-μ-carbonyltetradecacarbonyl[μ-[(2,3-η:4,5-η)-2,4-hexadiene]]hexaruthenium, Ru-00557

77089-76-0 (η^6-Benzene)[[2.2]paracyclophane]ruthenium(2+); Bis(tetrafluoroborate), *in* Ru-00569

77170-74-2 (η^5-2,4-Cyclopentadien-1-yl)[(1,2,3,4,5,6-η)-methoxybenzene]ruthenium(1+); Tetrafluoroborate, *in* Ru-00253

77170-78-6 (η^6-Aniline)(η^5-cyclopentadienyl)ruthenium(1+); Tetrafluoroborate, *in* Ru-00190

77208-32-3 μ-Carbonyldecacarbonyl-μ-methylenetriosmium, Os-00157

77308-83-9 Tricarbonyl(di-μ-carbonylhexacarbonyldiiron)-(triphenylphosphine)ruthenium, Ru-00652

77347-77-4 Carbonylchloro(η^4-tetraphenylcyclobutadiene)-bis(trimethyl phosphite-*P*)ruthenium(1+); Trichloride, *in* Ru-00687

77347-82-1 (η^4-Tetraphenylcyclobutadiene)tris(trimethyl phosphite-*P*)ruthenium, Ru-00714

77347-86-5 Iodo(η^4-tetraphenylcyclobutadiene)-tris(trimethyl phosphite-*P*)ruthenium(1+); Triiodide, *in* Ru-00713

77357-74-5 [(1,2,3,6,7-η)-2,4,6-Cyclooctatrien-1-yl]-[(1,2,3,4,5,6-η)-hexamethylbenzene]ruthenium(1+); Hexafluorophosphate, *in* Ru-00528

77357-79-0 (η^6-Benzene)[(1,2,3,4,5-η)-2,4,6-cyclooctatrien-1-yl]ruthenium(1+); Hexafluorophosphate, *in* Ru-00358

77370-83-3 Carbonyl(η^4-tetraphenylcyclobutadiene)-bis(trimethylphosphite-*P*)ruthenium, Ru-00688

77400-18-1 [(1,2,3,4,5-η)-2,4,6-Cyclooctatrien-1-yl]-[(1,2,3,4,5,6-η)-hexamethylbenzene]ruthenium(1+); Hexafluorophosphate, *in* Ru-00527

77424-05-6 Tricarbonyltri-μ-hydro(nonacarbonyldi-μ-hydrotriosmium)rhenium, Os-00168

77495-49-9 Tricarbonylbis[(2,3-η)-2-methyl-2-propenoate]ruthenium, Ru-00194

77924-84-6 Decacarbonyl[μ-(diphenylmethanone hydrazonato)]-μ-hydrotriosmium, Os-00331

77933-78-9 Carbonyl(η^5-2,4-cyclopentadien-1-yl)[(1,2,3-η)-2-methyl-2-propenyl]ruthenium; *endo-form*, *in* Ru-00155

77983-17-6 Carbonyl(η^5-2,4-cyclopentadien-1-yl)(η^3-2-propenyl)ruthenium; *endo-form*, *in* Ru-00105

77983-18-7 Carbonyl(η^5-2,4-cyclopentadien-1-yl)(η^3-2-propenyl)ruthenium; *exo-form*, *in* Ru-00105

78037-33-9 Carbonyl(η^5-2,4-cyclopentadien-1-yl)[(1,2,3-η)-2-methyl-2-propenyl]ruthenium; *exo-form*, *in* Ru-00155

78166-65-1 Nonadecacarbonylpentaosmium, Os-00309

78168-08-8 Decacarbonylbis(trimethyl phosphite-*P*)triruthenium, Ru-00430

78179-75-6 Tridecacarbonyl[μ-(diphenylphosphino)][μ_4-[(1-η:1-η:1,2-η:1,2-η)phenylethynyl]]pentaruthenium, Ru-00679

78192-37-7 Hexacarbonyl[μ-[η^2-*N*,*N'*-1,2-ethanediylidenebis[2-methyl-2-propanamine]-*N*,*N'*]]diosmium, Os-00280

78199-25-4 Tricarbonyl[*N*,*N'*-1,2-ethanediylbis[2,4-dimethyl-3-pentanamine]-*N*,*N'*]ruthenium, Ru-00515

78199-36-7 Octacarbonylbis[μ-(η^4-*N*,*N'*-1,2-ethanediylidenebis[2-propanamine]-*N*,*N'*]]tetraruthenium, Ru-00610

78248-93-8 Nonacarbonyl[μ_4-[(1-η:1,2-η:1,2-η:2-η)-1,2-diphenyl-1,2-ethenediyl]](tricarbonyliron)triruthenium; *axial-form*, *in* Ru-00628

78248-94-9 Nonacarbonyl[μ_4-[(1-η:1,2-η:1,2-η:2-η)-1,2-diphenyl-1,2-ethenediyl]](tricarbonyliron)triruthenium; *equatorial-form*, *in* Ru-00628

78319-35-4 Tetracarbonyl[(2,3-η)-methyl 2-propenoate]ruthenium, Ru-00076

78319-36-5 (Dimethyl maleate)tetracarbonylruthenium, *in* Ru-00066

78319-38-7 [(3,4-η)-3-Buten-2-one]tetracarbonylruthenium, Ru-00075

78319-39-8 Tetracarbonyl[(2,3-η)-2-propenenitrile]ruthenium, Ru-00044

78319-40-1 Tetracarbonyl[(3,4-η)-2,5-furandione]ruthenium, *in* Ru-00066

78330-36-6 Pentadecacarbonyl[μ_4-(phenylphosphoranetetrayl)]-pentaruthenium, Ru-00535

78330-37-7 Pentadecacarbonyl[μ_4-(ethylphosphoranetetrayl)]-pentaruthenium, Ru-00450

78343-01-8 Tris(trimethylphosphine)[(1,2,3,4,5-η)-6-(trimethylphosphonio)-2,4-cyclohexadien-1-yl]ruthenium(2+); Bis(hexafluorophosphate), *in* Ru-00497

78343-09-6 (η^6-Benzene)tris(trimethylphosphine)ruthenium(2+); Bis(hexafluorophosphate), *in* Ru-00407

78343-13-2 (Acetonitrile)(η^6-benzene)-bis(trimethylphosphine)ruthenium(2+); Bis(hexafluorophosphate), *in* Ru-00371

78557-65-0 Undecacarbonyltetra-μ-hydro(trimethyl phosphite-*P*)tetraosmium, Os-00238

78571-90-1 Di-μ-carbonylnonacarbonyl(dicarbonylferrate)-μ-hydrotriruthenate(1−); Bis(triphenylphosphine)iminum salt, *in* Ru-00282

78571-92-3 Hexacarbonyl(μ-carbonylhexacarbonyldiferrate)-μ-hydrodiruthenate(1−); Bis(triphenylphosphine)iminium salt, *in* Ru-00283

78618-87-8 (η^6-Benzene)[(1,2,3,4,5-η)-1-ethyl-2,4-cyclopentadien-1-yl]ruthenium(1+); Hexafluorophosphate, *in* Ru-00315

78618-91-4 [(1,2,3,4,5-η)-1-Acetyl-2,4-cyclopentadien-1-yl](η^6-benzene)ruthenium(1+); Hexafluorophosphate, *in* Ru-00308

78618-95-8 (η^5-2,4-Cyclopentadien-1-yl)[(1,2,3,4,5,6-η)-methylbenzene]ruthenium(1+); Hexafluorophosphate, *in* Ru-00254

78618-96-9 (η^6-Chlorobenzene)(η^5-2,4-cyclopentadien-1-yl)ruthenium(1+); Hexafluorophosphate, *in* Ru-00183

78643-30-8 [(1,2,3,4,5-η)-1-Acetyl-2,4-cyclopentadien-1-yl](η^6-chlorobenzene)ruthenium(1+); Hexafluorophosphate, *in* Ru-00303

78697-98-0 Nonacarbonyltri-μ-hydro[μ_3-(2-methoxy-2-oxoethylidyne)]triosmium, Os-00170

78770-95-3 Dodecacarbonyl[μ_3-[methanaminato(2−)]]tetraosmium, Os-00214

78919-48-9 Bromopentadecacarbonyl-μ-hydro-μ_5-methanetetraylpentaruthenium, Ru-00414

79043-00-8 Nonacarbonylbis[μ-(diphenylphosphino)]-(tetracarbonyliron)triruthenium, Ru-00706

79056-45-4 Hexacarbonyldi-μ-chlorodichlorodiruthenium; Trichlorostibine adduct, *in* Ru-00033

79061-97-5 Nonacarbonyl[μ_3-[(1-η:1-η:1,2-η)-3,3-dimethyl-1-butenylidene]]di-μ-hydrotriruthenium, Ru-00391

79075-59-5 Nonacarbonyl[μ_3-(3,3-dimethylbutylidyne)]tri-μ-hydrotriruthenium, Ru-00394

79085-63-5 Decacarbonyl-μ-nitrosyltriruthenate(1−); Bis(triphenylphosphine)iminium salt, *in* Ru-00166

79151-49-8 1,2,3,4,5-Pentakis(methoxycarbonyl)ruthenocene, *in* Ru-00389

79230-48-1 Nonacarbonyldi-μ-hydro[μ_3-(phenylphosphinidene)]-triosmium, Os-00255

79230-49-2 Nonacarbonyldi-μ-hydro[μ_3-(phenylphosphinidene)]-triruthenium, Ru-00382

79234-38-1 Chloro(η^5-2,4-cyclopentadien-1-yl)bis(2-isocyano-2-methylpropane)ruthenium, Ru-00406

79235-61-3 Decacarbonyl-μ-hydro[μ-(phenylphosphino)]triruthenium, Ru-00420

79255-11-1 [(Acetonitrile)decacarbonyl-μ-hydrotriosmium]pentacarbonylrhenium, Os-00290

79354-72-6 1,1″:1′,1‴-Bis(3-oxo-1,4-pentadiene-1,5-diyl)bisruthenocene; *(all-E)-form*, *in* Ru-00659

79354-75-9 1,3-Diruthenocenyl-2-propen-1-one; (*E*)*-form*, *in* Ru-00581

79354-76-0 3-Ferrocenyl-1-ruthenocenyl-2-propen-1-one; (*E*)-*form*, *in* Ru-00580

79354-77-1 1,5-Diruthenocenyl-1,4-pentadien-3-one; (*E*,*E*)-*form*, *in* Ru-00619

79354-80-6 1,1′-Ruthenocenedicarboxaldehyde, Ru-00230

79354-83-9 1,1′-Bis(hydroxymethyl)ruthenocene, Ru-00258

79361-63-0 1,1″:1′,1‴-Bis(3-oxo-1-propene-1,3-diyl)bisruthenocene, Ru-00630

79408-42-7 (η^5-2,4-Cyclopentadien-1-yl)-[tetrahydroborato(1−)-*H*]-bis(triphenylphosphine)ruthenium, Ru-00770

79408-44-9 (η^5-2,4-Cyclopentadien-1-yl)(thiocyanato-*N*)-bis(triphenylphosphine)ruthenium, Ru-00774

79408-45-0 (Cyanato-*N*)(η^5-2,4-cyclopentadien-1-yl)-bis(triphenylphosphine)ruthenium, Ru-00772

79408-47-2 (η^5-2,4-Cyclopentadien-1-yl)(*O*-ethylcarbonodithioato-*S*,*S′*)-(triphenylphosphine)ruthenium, Ru-00633

79408-50-7 (η^5-2,4-Cyclopentadien-1-yl)(*N*,*N*-diethylcarbamodithioato-*S*,*S′*)-(triphenylphosphine)ruthenium, Ru-00647

79594-92-6 (η^5-2,4-Cyclopentadien-1-yl)-ethylbis(triphenylphosphine)ruthenium, Ru-00793

79594-93-7 Butyl(η^5-2,4-cyclopentadien-1-yl)-bis(triphenylphosphine)ruthenium, Ru-00811

79619-57-1 (η^5-2,4-Cyclopentadien-1-yl)-propylbis(triphenylphosphine)ruthenium, Ru-00806

79643-37-1 1-Ferrocenyl-3-ruthenocenyl-2-propen-1-one; (*E*)-*form*, *in* Ru-00579

79725-76-1 Dicarbonyl(η^2-ethene)bis(triphenylphosphine)osmium, Os-00407

79845-73-1 [μ_5-(Carbon)]pentadecacarbonylchloropentaruthenate(1−), Ru-00413

79846-24-5 Tridecacarbonyl-μ-hydro[μ_3-[*N*-phenyl-1,2-benzenediaminato(3−)-*N*,*N′*:*N′*,*N′*]]-(triethylphosphine)pentaosmium, Os-00354

79949-33-0 μ-Carbonyldecacarbonyltriruthenate(2−); Ca salt, *in* Ru-00210

79949-34-1 μ-Carbonyldecacarbonyltriosmate(2−); Di-K salt, *in* Os-00148

80030-38-2 μ-Carbonyldecacarbonyl[μ-(trimethylsilyl)-methylene]triosmium, Os-00258

80050-10-8 [μ_5-(Carbon)]-tetradecacarbonyl(triphenylphosphine)pentaruthenium, Ru-00680

80303-44-2 Dicarbonylbis(η^5-2,4-cyclopentadien-1-yl)-μ-ethylidene[μ-(1-methylethylidene)]diruthenium, Ru-00461

81408-64-2 [η^4-1,2,3,4-Tetramethyl-5,6-bis(methylene)-1,3-cyclohexadiene]tris(trimethyl phosphite-*P*)-ruthenium, Ru-00553

81554-80-5 Dicarbonyl[(1,2,3,4,5-η)-1,2,3,4,5-pentamethyl-2,4-cyclopentadien-1-yl](tetrahydrofuran)osmium(1+); Hexafluorophosphate, *in* Os-00282

81554-85-0 Tricarbonyl(η^5-2,4-cyclopentadien-1-yl)osmium(1+); Tetrafluoroborate, *in* Os-00058

81554-87-2 Tricarbonyl[(1,2,3,4,5-η)-1,2,3,4,5-pentamethyl-2,4-cyclopentadien-1-yl]osmium(1+); Tetrafluoroborate, *in* Os-00220

81554-88-3 Bromodicarbonyl(η^5-2,4-cyclopentadien-1-yl)osmium, Os-00041

81554-89-4 Bromodicarbonyl[(1,2,3,4,5-η)-1,2,3,4,5-pentamethyl-2,4-cyclopentadien-1-yl]osmium, Os-00186

81554-96-3 Dicarbonylhydro[(1,2,3,4,5-η)-1,2,3,4,5-pentamethyl-2,4-cyclopentadien-1-yl]osmium, Os-00188

81554-97-4 Dicarbonyl(η^5-2,4-cyclopentadien-1-yl)iodoosmium, Os-00043

81554-98-5 Dicarbonyl[(1,2,3,4,5-η)-1,2,3,4,5-pentamethyl-2,4-cyclopentadien-1-yl]iodoosmium, Os-00187

81555-01-3 Dicarbonyl(1,3,5-cycloheptatrien-1-yl)(η^5-2,4-cyclopentadien-1-yl)osmium, Os-00235

81555-04-6 (Acetonitrile)dicarbonyl(η^5-2,4-cyclopentadien-1-yl)osmium(1+); Tetrafluoroborate, *in* Os-00074

81571-98-4 Dicarbonyl[(1,2-η)-1,3,5-cycloheptatriene](η^5-2,4-cyclopentadien-1-yl)osmium(1+); Tetrafluoroborate, *in* Os-00237

81740-97-8 Dicarbonyldi-μ-hydrobis[(1,2,3,4,5-η)-1,2,3,4,5-pentamethyl-2,4-cyclopentadien-1-yl]diosmium, Os-00328

81740-98-9 μ-Carbonyldi-μ-hydrobis[(1,2,3,4,5-η)-1,2,3,4,5-pentamethyl-2,4-cyclopentadien-1-yl]diosmium, Os-00324

81753-10-8 Carbonyltrihydro[(1,2,3,4,5-η)-1,2,3,4,5-pentamethyl-2,4-cyclopentadien-1-yl]osmium, Os-00143

81799-11-3 (μ_3-Anilinato)nonacarbonyldi-μ-hydrotriosmium, Os-00253

82080-90-8 Dodecacarbonyl-μ_3-thioxotetraosmium, Os-00202
82093-50-3 Dodecacarbonyldi-μ_3-thioxotetraosmium, Os-00203
82190-35-0 Hexadecacarbonyl-μ_3-thioxo-μ_4-thioxohexaosmium, Os-00288
82466-20-4 (η^5-2,4-Cyclopentadien-1-yl)[(1,2,3,4,5,6-η)-methoxybenzene]ruthenium(1+); Tetraphenylborate, *in* Ru-00253
82466-21-5 (η^5-2,4-Cyclopentadien-1-yl)[(1,2,3,4,5,6-η)-hexamethylbenzene]ruthenium(1+); Tetraphenylborate, *in* Ru-00464
82639-15-4 Octadecacarbonylhexaosmate(2−); Bis(tetraethylammonium) salt, *in* Os-00307
82639-16-5 [μ_6-(Carbon)]tetracosacarbonyldecaosmate(2−); Bis(tetraethylammonium) salt, *in* Os-00342
82647-79-8 Nonacarbonyl[μ_3-(diphenylphosphino)]-μ-hydrotriruthenium, Ru-00539
82704-20-9 (η^6-Benzene)[(7,8,9,10,11-η)-undecahydro-7,8-dicarbaundecaborato(2−)]ruthenium, Ru-00087
83005-40-7 [(1,2,3,4,5,6-η)-Hexamethylbenzene][1,2,3,4-η)-1,2,3,4-tetramethyl-5,6-bis(methylene)-1,3-cyclohexadiene]ruthenium, Ru-00611
83111-40-4 Tricarbonylchlorobis(triphenylphosphine)ruthenium(1+); Perchlorate, *in* Ru-00740
83114-82-3 Tricarbonylchlorobis(triphenylphosphine)osmium(1+); *trans-form*, Perchlorate, *in* Os-00394
83242-24-4 (Carbonothioyl)tri-μ-chlorochlorotetrakis(triphenylphosphine)diruthenium; Me_2CO complex, *in* Ru-00862
83573-03-9 Decacarbonyl-μ-hydro-μ_3-methylidynetriosmium, Os-00123
83585-34-6 Nonacarbonyldi-μ-hydro[μ_3-(oxoethenylidene-)]triosmium, Os-00124
83721-25-9 Tricarbonyltriiodoruthenate(1−); Tetraethylammonium salt, *in* Ru-00021
84108-59-8 Dihydrotetrakis(trimethylphosphine)osmium, Os-00196